Experimental Methods in Wastewater Treatment

Experimental Methods in Wastewater Treatment

Mark C. M. van Loosdrecht
Per H. Nielsen
Carlos M. Lopez-Vazquez
Damir Brdjanovic

Published by: **IWA Publishing**
Republic - Export Building
1 Clove Crescent
London E14 2BA, UK
T: +44 (0) 20 7654 5500
F: +44 (0) 20 7654 5555
E: publications@iwap.co.uk
I: www.iwaponline.com

First published 2016
© 2016 IWA Publishing

Apart from any fair dealing for the purposes of research or private study, or criticism or review, as permitted under the UK Copyright, Designs and Patents Act (1998), no part of this publication may be reproduced, stored or transmitted in any form or by any means, without the prior permission in writing of the publisher, or, in the case of photographic reproduction, in accordance with the terms of licences issued by the Copyright Licensing Agency in the UK, or in accordance with the terms of licenses issued by the appropriate reproduction rights organization outside the UK. Enquiries concerning reproduction outside the terms stated here should be sent to IWA Publishing at the address printed above.

The publisher makes no representation, express or implied, with regard to the accuracy of the information contained in this book and cannot accept any legal responsibility or liability for errors or omissions that may be made.

Disclaimer
The information provided and the opinions given in this publication are not necessarily those of IWA and IWA Publishing and should not be acted upon without independent consideration and professional advice. IWA and IWA Publishing will not accept responsibility for any loss or damage suffered by any person acting or refraining from acting upon any material contained in this publication.

British Library Cataloguing in Publication Data
A CIP catalogue record for this book is available from the British Library

Library of Congress Cataloguing in Publication Data
A catalogue record for this book is available from the Library of Congress

Cover design: Peter Stroo
Graphic design: Hans Emeis

ISBN 9781780404745 (Hardback)
ISBN 9781789065275 (Paperback)
ISBN 9781780404752 (eBook)

Preface

Wastewater treatment is a core technology for water resources protection and reuse, as is clearly demonstrated by the great success of its consequent implementation in many countries worldwide. During the last decennia scientific research has made vast progress in understanding the complex and interdisciplinary aspects of the biological, biochemical, chemical and mechanical processes involved. It can be concluded that the global application of existing knowledge and experience in wastewater treatment technology will represent a cornerstone in future water management, as expressed in the Strategic Development Goals accepted by the UN in September 2015.

Only about one fifth of the wastewater produced globally is currently being adequately treated. To achieve the goal for sustainable water management by 2030 would require extra wastewater treatment facilities for about 600,000 people each day. I am convinced that this book will make its own significant contribution to meeting this ambitious goal.

In the near future, most of the global population will live in cities and in low and middle-income countries, where most wastewater is not adequately treated. Probably the most limiting factor in achieving the goals for sustainable water management is the lack of qualified, well-trained professionals, able to comprehend the scientific research results and transfer them into practice. It is therefore of prime importance to make currently available scientific advances and proven experiences in wastewater treatment technology applications easily accessible worldwide. This was one of the drivers for the development of this book, which represents an innovative contribution to help overcome such a capacity development challenge. The book is most definitely expected to contribute to bridging the gaps between the science and technology, and their practical applications.

The great collection of authors and reviewers represents an interdisciplinary team of globally acknowledged experts. The book will therefore make a major contribution to establishing a common professional language, enhancing global communication between wastewater professionals. In addition, the authors have linked the description of the scientific basis for wastewater treatment processes with a video-based online course for the training of students, researchers, engineers, laboratory technicians and treatment plant operators, demonstrating commonly accepted experimentation procedures and their application for lab-, pilot-, and full-scale treatment plant operation.

From the perspective of the IWA this book also has the great potential to enhance the development of a new generation of researchers and enable them to communicate on a global scale and beyond their specific field of expertise. Both aspects are urgently needed to develop adapted solutions for specific local conditions and to make them globally available for implementation.

There has been a trend for some time that scientific research and practice have been growing apart from each other. Part of the reason for this is the global implementation of an academic assessment method that primarily focuses on the impact of publications on the progress in scientific research. Applied research results with an impact on practice in water quality management are not yet being sufficiently rewarded as their impact is not always reflected by citations in scientific journals. This book attempts to overcome this problem as it aims to enhance the dialogue and co-operation between scientists and practitioners. Scientists are encouraged to deal with the practical problems with scientific methods, while the practitioners are encouraged to understand the scientific background of all the processes relevant for treatment plant optimization.

While conventional wastewater treatment plant operation was driven by effluent quality and cost minimization, this book fully incorporates the paradigm shift towards material and energy recovery from wastewater. In this respect the book is also very relevant for developed countries, as the new paradigm will heavily influence the future development of wastewater management worldwide.

As IWA president I want to congratulate the authors of this book on their great achievement and also thank the Bill & Melinda Gates Foundation and the Dutch government for their financial support.

Prof. Dr. Helmut Kroiss
President International Water Association

Contributors

Carlos M. Lopez-Vazquez	UNESCO-IHE Institute for Water Education, The Netherlands	1. 2.
Damir Brdjanovic	UNESCO-IHE Institute for Water Education, The Netherlands	1. 2.
Eldon R. Rene	UNESCO-IHE Institute for Water Education, The Netherlands	2.
Elena Ficara	Milan University of Technology, Italy	2.
Elena Torfs	Université Laval, Canada	6.
Eveline I.P. Volcke	Ghent University, Belgium	4.
George A. Ekama	University of Cape Town, South Africa	3.
Glen T. Daigger	University of Michigan, United States of America	6.
Gürkan Sin	Technical University of Denmark, Denmark	5.
Henri Spanjers	Delft University of Technology, The Netherlands	3.
Holger Daims	University of Vienna, Austria	8.
Ilse Y. Smets	Catholic University of Leuven, Belgium	6.
Imre Takács	Dynamita, France	6.
Ingmar Nopens	Ghent University, Belgium	6.
Jeppe L. Nielsen	Aalborg University, Denmark	7.
Jiři Wanner	University of Chemistry and Technology Prague, Czech Republic	7.
Juan A. Baeza	Universitat Autònoma de Barcelona, Spain	5.
Kartik Chandran	Columbia University, United States of America	4.
Krist V. Gernaey	Technical University of Denmark, Denmark	5.
Laurens Welles	UNESCO-IHE Institute for Water Education, The Netherlands	2.
Mads Albertsen	Aalborg University, Denmark	8.
Mari K.H. Winkler	University of Washington, United States of America	6.
Mark C.M. van Loosdrecht	Delft University of Technology, The Netherlands	1. 2. 4.
Mathieu Spérandio	Institut National des Sciences Appliquées de Toulouse, France	3.
Morten S. Dueholm	Aalborg University, Denmark	8.
Nancy G. Love	University of Michigan, United States of America	2.
Per H. Nielsen	Aalborg University, Denmark	1. 7. 8.
Peter A. Vanrolleghem	Université Laval, Canada	3. 4. 6.
Piet N.L. Lens	UNESCO-IHE Institute for Water Education, The Netherlands	2.
Rasmus H. Kirkegaard	Aalborg University, Denmark	8.
Robert J. Seviour	La Trobe University, Australia	7.
Sebastiaan C.F. Meijer	Yuniko BV, The Netherlands	5.
Sophie Balemans	Ghent University, Belgium	6.
Søren M. Karst	Aalborg University, Denmark	8.
Sylvie Gillot	IRSTEA, France	4.
Tessa P.H. van den Brand	KWR Watercycle Research Institute, The Netherlands	2.
Tommaso Lotti	Milan University of Technology, Italy	2.
Yves Comeau	École Polytechnique de Montréal, Canada	2.

Chapter author
Chapter reviewer

About the editors

Mark C.M. van Loosdrecht is a well-renown scientist recognised for his significant contributions to the study of reducing energy consumption and the footprint of wastewater treatment plants through his patented and award-winning technologies Sharon®, Anammox® and Nereda®. His main work focuses on the use of microbial cultures within the environmental process-engineering field, with a special emphasis on nutrient removal, biofilm and biofouling. Currently he is a full professor and Group Leader of Environmental Biotechnology at TU Delft. A fellow of the Royal Dutch Academy of Arts and Sciences (KNAW), the Netherlands Academy of Technology and Innovation (AcTI) and the International Water Association (IWA), Professor van Loosdrecht has won numerous prestigious awards. His research interests include granular sludge systems, microbial storage polymers, wastewater treatment, gas treatment, soil treatment, microbial conversion of inorganic compounds, production of chemicals from waste, and modelling. Apart from his other achievements, he has published over 500 papers, supervised 65 PhD students so far and is an honorary professor at the University of Queensland. He is also currently the Editor-in-Chief for Water Research and Advisor to IWA Publishing.

Carlos M. Lopez-Vazquez is Associate Professor in Wastewater Treatment Technology at UNESCO-IHE Institute for Water Education. In 2009 he received his doctoral degree on Environmental Biotechnology (cum laude) from Delft University of Technology and UNESCO-IHE Institute for Water Education. During his professional career, he has taken part in different advisory and consultancy projects for both public and private sectors concerning municipal and industrial wastewater treatment systems. After working for a couple of years in the Water R&D Department of Nalco Europe on industrial water and wastewater treatment applications, he re-joined UNESCO-IHE's Sanitary Engineering Chair Group in 2009. Since then, he has been involved in education, capacity building and research projects guiding dozens of MSc and several PhD students. By applying mathematical modelling as an essential tool, he has a special focus on the development and transfer of innovative and cost-effective wastewater treatment technologies to developing countries, countries in transition and industrial applications.

Per H. Nielsen is a full professor at the Department of Chemistry and Bioscience at Aalborg University, Denmark where he heads the multidisciplinary Centre for Microbial Communities. He is also a visiting scientist at the Singapore Centre on Environmental Life Sciences Engineering, Nanyang Technological University, Singapore. Prof. Nielsen's research group has been active in environmental biotechnology for over 25 years, focusing on the microbial ecology of biological wastewater treatment, bioenergy production, bioremediation, biofilms, infection of implants and the development of system microbiology approaches based on new sequencing technologies. He chaired the IWA specialist group Microbial Ecology and Water Engineering for eight years (2005-2013) and is Chair of the IWA BioCluster. He is a Fellow of the Danish Academy of Technical Sciences (ATV) and the International Water Association (IWA) and has received several prestigious awards. He has published more than 230 peer-reviewed publications and supervised 25 PhD students. His main research interest is microbial ecology in water engineering, particularly related to wastewater treatment where he has developed and applied several novel methods to study uncultured microorganisms, e.g. by using next-generation sequencing technologies. He is the initiator and responsible for the MiDAS field guide open resource for wastewater microbiology.

Damir Brdjanovic is Professor of Sanitary Engineering at UNESCO-IHE and Endowed Professor at Delft University of Technology in the Environmental Biotechnology Group. Areas of his expertise include pro-poor and emergency sanitation, faecal sludge management, urban drainage, and wastewater treatment. He is a pioneer in the practical application of models in wastewater treatment practice in developing countries. He invented the Shit Killer® device for excreta management in emergencies, the award-winning eSOS® Smart Toilet and associated software eSOS View®, with funding by the Bill & Melinda Gates Foundation (BMGF). He has initiated the development and implementation of innovative didactic approaches and novel educational products (including e-learning) at UNESCO-IHE. In 2015, together with the BMGF, he founded the Global Faecal Sludge Management e-learning Alliance. Currently his chair group consists of ten staff members, three post-doctoral fellows and 22 PhD students. In addition, in excess of 100 MSc students have graduated under his supervision so far. Prof. Brdjanovic has a sound publication record, is co-initiator of the IWA Journal of Water, Sanitation and Hygiene for Development, and is the initiator, author and editor of five books in the wastewater treatment and sanitation field. In 2015 he became an International Water Association Fellow.

About the book and online course

Over the past twenty years, the knowledge and understanding of wastewater treatment has advanced extensively and moved away from empirically-based approaches to a fundamentally-based first-principles approach embracing chemistry, microbiology, and physical and bioprocess engineering, often involving experimental laboratory work and techniques. Many of these experimental methods and techniques have matured to the degree that they have been accepted as reliable tools in wastewater treatment research and practice. For sector professionals, especially the new generation of young scientists and engineers entering the wastewater treatment profession, the quantity, complexity and diversity of these new developments can be overwhelming, particularly in developing countries where access to advanced level laboratory courses in wastewater treatment is not readily available. In addition, information on innovative experimental methods is scattered across scientific literature and only partially available in the form of textbooks or guidelines. This book seeks to address these deficiencies. It assembles and integrates the innovative experimental methods developed by research groups and practitioners around the world and broadly applied in wastewater treatment research and practice.

Experimental Methods in Wastewater Treatment book forms part of the internet-based curriculum in sanitary engineering at UNESCO-IHE and, as such, may also be used together with video recordings of methods and approaches performed and narrated by the authors, including guidelines on best experimental practices. The book is written for undergraduate and postgraduate students, researchers, laboratory staff, plant operators, consultants, and other sector professionals.

The idea of making this book and the online learning course was conceived in 2009 when UNESCO-IHE agreed to utilize some of the programmatic funds provided by the Dutch Ministry of Foreign Affairs to develop innovative learning methods and products. However it took until 2011 to acquire the additional funds from the Bill & Melinda Gates Foundation (BMGF) that enabled the original idea to be fully executed. The conceptual framework for the book, and the online course that it is part of, was agreed upon in Montreal during the IWA World Water Congress and Exhibition in September 2010 and further detailed during the IWA event in Essen, Activated Sludge – 100 Years and Counting. The latter was the occasion when the concept was introduced of also having established reviewers in the field to provide critical feedback on the manuscripts and improve the quality of the final product, in addition to the esteemed groups of experts writing the chapters of the book. Besides providing chapters in the book, authors were requested to prepare presentation slides, tutorial exercises and to deliver scenarios and narration for video-recorded lectures and execution of experimental procedures at UNESCO-IHE and partner laboratories. These materials have been compiled into a digital package available to those registered for the online course. IWA Publishing has agreed to publish the book and market both the book and online learning course. It has also been agreed that the book and online course digital materials are available free of charge. The online course is delivered once or twice a year depending on the demand (please consult the UNESCO-IHE website for further information on how to embark on the course or download the course materials). The book is also used for teaching as part of a lecture series in the Sanitary Engineering specialization of the UNESCO-IHE's Master's Program in Urban Water and Sanitation. It is conceptualized in such a way that it can be used as a self-contained textbook or as an integral part of the online learning course.

A number of individuals deserve to be singled out as their support was crucial in this development and is highly appreciated: Dr. Roshan Shrestha, Dr. Doulaye Koné, Dr. Frank Rijsberman and Dr. Brian Arbogast (BMGF), and Dr. Wim Duven and Jetze Heun (UNESCO-IHE). The book was edited by Peter Stroo, Hans Emeis, Claire Taylor, Michelle Jones, and Maggie Smith. The credit for the content goes to all the authors, reviewers and enthusiastic group of editors. Further, I acknowledge the contributors who allowed their data, images and photographs to be used in this book and the course.

Finally, I hope that this book and the training materials will be useful in your research or practical work, be it at a laboratory-, pilot- or full-scale wastewater treatment plant.

Prof. Dr. Damir Brdjanovic
Professor of Sanitary Engineering

Table of contents

1. INTRODUCTION — 1
Mark C.M. van Loosdrecht, Per H. Nielsen, Carlos M. Lopez-Vazquez and Damir Brdjanovic (aut.)

2. ACTIVATED SLUDGE ACTIVITY TESTS — 7
Carlos M. Lopez-Vazquez, Laurens Welles, Tommaso Lotti, Elena Ficara, Eldon R. Rene, Tessa P.H. van den Brand, Damir Brdjanovic and Mark C.M. van Loosdrecht (aut.)
Yves Comeau, Piet N.L. Lens and Nancy G. Love (rev.)

2.1 INTRODUCTION — 7
2.2 ENHANCED BIOLOGICAL PHOSPHORUS REMOVAL — 9
- 2.2.1 Process description — 9
- 2.2.2 Experimental set-up — 11
  - 2.2.2.1 Reactors — 11
  - 2.2.2.2 Activated sludge sample collection — 16
  - 2.2.2.3 Activated sludge sample preparation — 16
  - 2.2.2.4 Substrate — 17
  - 2.2.2.5 Analytical procedures — 19
  - 2.2.2.6 Parameters of interest — 22
- 2.2.3 EBPR batch activity tests: Preparation — 22
  - 2.2.3.1 Apparatus — 22
  - 2.2.3.2 Materials — 24
  - 2.2.3.3 Media preparation — 24
  - 2.2.3.4 Material preparation — 25
  - 2.2.3.5 Activated sludge preparation — 28
- 2.2.4 Batch activity tests: Execution — 29
  - 2.2.4.1 Anaerobic EBPR batch activity tests — 30
  - 2.2.4.2 Anoxic EBPR batch tests — 33
  - 2.2.4.3 Aerobic EBPR batch tests — 34
- 2.2.5 Data analysis — 36
  - 2.2.5.1 Estimation of stoichiometric parameters — 36
  - 2.2.5.2 Estimation of kinetic parameters — 41
- 2.2.6 Data discussion and interpretation — 42
  - 2.2.6.1 Anaerobic batch activity tests — 42
  - 2.2.6.2 Aerobic batch activity tests — 45
  - 2.2.6.3 Anoxic batch activity tests — 46
- 2.2.7 Example — 47
  - 2.2.7.1 Description — 47
  - 2.2.7.2 Data analysis — 47
- 2.2.8 Additional considerations — 51
  - 2.2.8.1 GAO occurrence in EBPR systems — 51
  - 2.2.8.2 The effect of carbon source — 51
  - 2.2.8.3 The effect of temperature — 51
  - 2.2.8.4 The effect of pH — 52
  - 2.2.8.5 Denitrification by EBPR cultures — 52
  - 2.2.8.6 Excess and shortage of intracellular compounds — 52
  - 2.2.8.7 Excessive aeration — 53
  - 2.2.8.8 Shortage of essential ions — 53
  - 2.2.8.9 Toxicity/inhibition — 53

2.3 BIOLOGICAL SULPHATE REDUCTION — 54
- 2.3.1 Process description — 54
- 2.3.2 Sulphide speciation — 56
- 2.3.3 Effects of environmental and operating conditions on SRB — 57
  - 2.3.3.1 Carbon source — 57
  - 2.3.3.2 COD to SO_4^{2-} ratio — 58
  - 2.3.3.3 Temperature — 58
  - 2.3.3.4 pH — 59
  - 2.3.3.5 Oxygen — 59
- 2.3.4 Experimental set-up — 60
  - 2.3.4.1 Estimation of volumetric and specific rates — 60
  - 2.3.4.2 The reactor — 60
  - 2.3.4.3 Mixing — 61
  - 2.3.4.4 pH control — 61
  - 2.3.4.5 Temperature control — 61
  - 2.3.4.6 Sampling and dosing ports — 62
  - 2.3.4.7 Sample collection — 62
  - 2.3.4.8 Media — 62
- 2.3.5 Analytical procedures — 63
  - 2.3.5.1 $COD_{organics}$ and COD_{total} — 63
  - 2.3.5.2 Sulphate — 64
  - 2.3.5.3 Sulphide — 64
- 2.3.6 SRB batch activity tests: preparation — 65
  - 2.3.6.1 Apparatus — 65
  - 2.3.6.2 Materials — 65
  - 2.3.6.3 Media — 65
  - 2.3.6.4 Material preparation — 66
  - 2.3.6.5 Mixed liquor preparation — 67
  - 2.3.6.6 Sample collection and treatment — 68
- 2.3.7 Batch activity tests: execution — 68
- 2.3.8 Data analysis — 69
  - 2.3.8.1 Mass balances and calculations — 69
  - 2.3.8.2 Data discussion and interpretation — 70
- 2.3.9 Example — 70
- 2.3.10 Practical recommendations — 72

2.4 BIOLOGICAL NITROGEN REMOVAL — 73
- 2.4.1 Process description — 73
  - 2.4.1.1 Nitrification — 74
  - 2.4.1.2 Denitrification — 75
  - 2.4.1.3 Anaerobic ammonium oxidation (Anammox) — 76
- 2.4.2 Process-tracking alternatives — 76
  - 2.4.2.1 Chemical tracking — 77
  - 2.4.2.2 Titrimetric tracking — 77
  - 2.4.2.3 Manometric tracking — 78
- 2.4.3 Experimental set-up — 79
  - 2.4.3.1 Reactors — 79
  - 2.4.3.2 Instrumentation for titrimetric tests — 79
  - 2.4.3.3 Instrumentation for manometric tests — 80
  - 2.4.3.4. Activated sludge sample collection — 81
  - 2.4.3.5 Activated sludge sample preparation — 82
  - 2.4.3.6 Substrate — 82
  - 2.4.3.7 Analytical procedures — 83
  - 2.4.3.8 Parameters of interest — 83
  - 2.4.3.9 Type of batch tests — 86
- 2.4.4 Nitrification batch activity tests: Preparation — 86
  - 2.4.4.1 Apparatus — 86
  - 2.4.4.2 Materials — 86
  - 2.4.4.3 Media preparation — 86
- 2.4.5 Nitrification batch activity tests: Execution — 87
- 2.4.6 Denitrification batch activity tests: Preparation — 92
  - 2.4.6.1 Apparatus — 92
  - 2.4.6.2 Materials — 93
  - 2.4.6.3 Working solutions — 93
  - 2.4.6.4 Materials preparation — 93
- 2.4.7 Denitrification batch activity tests: Execution — 93
- 2.4.8 Anammox batch activity tests: Preparation — 99

2.4.8.1 Apparatus	99
2.4.8.2 Materials	99
2.4.8.3 Working solutions	99
2.4.8.4 Materials preparation	100
2.4.9 Anammox batch activity tests: Execution	100
2.4.10 Examples	103
2.4.10.1 Nitrification batch activity test	103
2.4.10.2 Denitrification batch activity test	105
2.4.10.3 Anammox batch activity test	107
2.4.11 Additional considerations	109
2.4.11.1 Presence of other organisms	109
2.4.11.2 Shortage of essential micro- and macro-nutrients	109
2.4.11.3 Toxicity or inhibition effects	110
2.4.11.4 Effects of carbon source on denitrification	110
2.5 AEROBIC ORGANIC MATTER REMOVAL	**111**
2.5.1 Process description	111
2.5.2 Experimental set-up	112
2.5.2.1 Reactors	112
2.5.2.2 Activated sludge sample collection	112
2.5.2.3 Activated sludge sample preparation	113
2.5.2.4 Media	113
2.5.2.5 Analytical tests	114
2.5.2.6 Parameters of interest	114
2.5.3 Aerobic organic matter batch activity tests: Preparation	115
2.5.3.1 Apparatus	115
2.5.3.2 Materials	115
2.5.3.3 Working solutions	115
2.5.3.4 Material preparation	116
2.5.3.5 Activated sludge preparation	117
2.5.4 Aerobic organic matter batch activity tests: Execution	117
2.5.5 Data analysis	118
2.5.6 Example	119
2.5.6.1 Description	119
2.5.6.2 Data analysis	119
2.5.7 Additional considerations and recommendations	121
2.5.7.1 Simultaneous storage and microbial growth	121
2.5.7.2 Lack of nutrients	121
2.5.7.3 Toxicity or inhibition	121

3. RESPIROMETRY 133

Henry Spanjers and Peter A. Vanrolleghem (aut.)
George A. Ekama and M. Spérandio (rev.)

3.1 INTRODUCTION	**133**
3.1.1 Basics of respiration	134
3.1.2 Basics of respirometry	135
3.2 GENERAL METHODOLOGY OF RESPIROMETRY	**136**
3.2.1 Basics of respirometric methodology	136
3.2.2 Generalized principles: beyond oxygen	136
3.2.2.1 Principles based on measuring in the liquid phase	136
3.2.2.2 Principles based on measuring during the gas phase	138
3.3 EQUIPMENT	**141**
3.3.1 Equipment for anaerobic respirometry	141
3.3.1.1 Biogas composition	141
3.3.1.2 Measuring the gas flow	142
3.3.2 Equipment for aerobic and anoxic respirometry	143
3.3.2.1 Reactor	143
3.3.2.2 Measuring arrangement	143
3.3.2.3 Practical implementation	144
3.4 WASTEWATER CHARACTERIZATION	**150**
3.4.1 Biomethane potential (BMP)	150
3.4.1.1 Purpose	150
3.4.1.2 General	150

3.4.1.4 Data processing	151
3.4.1.5 Recommendations	151
3.4.2 Biochemical oxygen demand (BOD)	152
3.4.2.1 Purpose	152
3.4.2.2 General	152
3.4.2.3 Test execution	153
3.4.3 Short-term biochemical oxygen demand (BOD_{st})	157
3.4.3.1 Test execution	158
3.4.3.2 Calculations	160
3.4.4 Toxicity and inhibition	160
3.4.4.1 Purpose	160
3.4.4.2 Test execution	160
3.4.4.3 Calculations	161
3.4.4.4 Biodegradable toxicants	162
3.4.5 Wastewater fractionation	163
3.4.5.1 Readily biodegradable substrate (S_B)	166
3.4.5.2 Slowly biodegradable substrate (XC_B)	167
3.4.5.3 Heterotrophic biomass (X_{OHO})	168
3.4.5.4 Autotrophic (nitrifying) biomass (X_{ANO})	168
3.4.5.5 Ammonium (S_{NHx})	168
3.4.5.6 Organic nitrogen fractions ($XC_{B,N}$ and $S_{B,N}$)	168
3.5 BIOMASS CHARACTERIZATION	**169**
3.5.1 Volatile suspended solids	169
3.5.2 Specific methanogenic activity (SMA)	169
3.5.2.1 Purpose	169
3.5.2.2 General	169
3.5.2.3 Test execution	169
3.5.2.4 Data processing	170
3.5.3 Specific aerobic and anoxic biomass activity	171
3.5.3.1 Maximum specific nitrification rate (AUR)	171
3.5.3.2 Maximum specific aerobic heterotrophic respiration rate (OUR)	173
3.5.3.3 Maximum specific denitrification rate (NUR)	173

4. OFF-GAS EMISSION TESTS 177

Kartik Chandran, Eveline I.P. Volcke, Mark C.M. van Loosdrecht (aut.)
Peter A. Vanrollegem and Sylvie Guillot (rev.)

4.1 INTRODUCTION	**177**
4.2 SELECTING THE SAMPLING STRATEGY	**178**
4.2.1 Plant performance	178
4.2.2 Seasonal variations in emissions	178
4.2.3 Sampling objective	179
4.3 PLANT ASSESSMENT AND DATA COLLECTION	**179**
4.3.1 Preparation of a sampling campaign	179
4.3.2 Sample identification and data sheet	180
4.3.3 Factors that can limit the validity of the results	181
4.3.4 Practical advice for analytical measurements	181
4.3.5 General methodology for sampling	182
4.3.6 Sampling in the framework of the off-gas measurements	183
4.3.7 Testing and measurements protocol	185
4.4 EMISSION MEASUREMENTS	**185**
4.5 N_2O MEASUREMENT IN OPEN TANKS	**186**
4.5.1 Protocol for measuring the surface flux of N_2O	188
4.5.1.1 Equipment, materials and supplies	188
4.5.1.2 Experimental procedure	188
4.5.1.3 Sampling methods for nitrogen GHG emissions	189
4.5.1.4 Direct measurement of the liquid-phase N_2O content	191
4.6 MEASUREMENT OF OFF-GAS FLOW IN OPEN TANKS	**191**
4.6.1 Protocol for aerated or aerobic zone	192
4.6.2 Protocol for non-aerated zones	192
4.7 AQUEOUS N_2O and CH_4 CONCENTRATION DETERMINATION	**192**

4.7.1 Measurement protocol for dissolved N$_2$O measurement using polarographic electrodes	193
4.7.1.1 Equipment	193
4.7.1.2 Experimental procedure	193
4.7.2 Measurement protocol for dissolved gasses using gas chromatography	194
4.7.3 Measurement protocol for dissolved gas measurement by the salting-out method	194
4.7.3.1 Equipment	195
4.7.3.2 Sampling procedure	195
4.7.3.3 Measurement procedure	195
4.7.3.4 Calculations	196
4.7.4 Measurement protocol for dissolved gas measurement by the stripping method	196
4.7.4.1 Operational principle	196
4.7.4.2 Equipment	197
4.7.4.3 Calibration batch test	198
4.7.4.4 Measurement accuracy	198
4.7.4.5 Calculation of the N$_2$O formation rate in the stripping device	198
4.8 DATA ANALYSIS AND PROCESSING	**199**
4.8.1 Determination of fluxes	199
4.8.2 Determination of aggregated emission fractions	199
4.8.3 Calculation of the emission factors	200

5. DATA HANDLING AND PARAMETER ESTIMATION   201
Gürkan Sin and Krist V. Gernaey (aut.)
Sebastiaan C.F. Meijer and Juan A. Baeza (rev.)

5.1 INTRODUCTION	**201**
5.2 THEORY AND METHODS	**202**
5.2.1 Data handling and validation	202
5.2.1.1 Systematic data analysis for biological processes	202
5.2.1.2 Degree of reduction analysis	203
5.2.1.3 Consistency check of experimental data	204
5.2.2 Parameter estimation	205
5.2.2.1 Manual trial and error method	205
5.2.2.2 Formal statistics methods	205
5.2.3 Uncertainty analysis	209
5.2.3.1 Linear error propagation	209
5.2.3.2 The Monte Carlo method	209
5.2.4 Local sensitivity analysis and identifiability analysis	210
5.2.4.1 Local sensitivity analysis	210
5.2.4.2 Identifiability analysis using the collinearity index	210
5.3 METHODOLOGY AND WORKFLOW	**211**
5.3.1 Data consistency check using elemental balance and a degree of reduction analysis	211
5.3.2 Parameter estimation workflow for non-linear least squares method	212
5.3.3 Parameter estimation workflow for the bootstrap method	212
5.3.4 Local sensitivity and identifiability analysis workflow	213
5.3.5 Uncertainty analysis using the Monte Carlo method and linear error propagation	213
5.4 ADDITIONAL EXAMPLES	**214**
5.5 ADDITIONAL CONSIDERATIONS	**232**

6. SETTLING TESTS   235
Elena Torfs, Ingmar Nopens, Mari K.H. Winkler, Peter A. Vanrolleghem, Sophie Balemans and Ilse Y. Smets (aut.)
Glenn T. Daigger and Imre Takács (rev.)

6.1 INTRODUCTION	**235**
6.2 MEASURING SLUDGE SETTLEABILITY IN SSTs	**236**
6.2.1 Sludge settleability parameters	237
6.2.1.1 Goal and application	237
6.2.1.2 Equipment	237
6.2.1.3 The sludge volume index (SVI)	237
6.2.1.4 The diluted sludge volume index (DSVI)	237
6.2.1.5 The stirred specific volume index (SSVI$_{3.5}$)	238
6.2.2 The batch settling curve and hindered settling velocity	238
6.2.2.1 Goal and application	238
6.2.2.2 Equipment	239
6.2.2.3 Experimental procedure	239
6.2.2.4 Interpreting a batch settling curve	240
6.2.2.5 Measuring the hindered settling velocity	241
6.2.3 v_{hs}-X relation	241
6.2.3.1 Goal and application	241
6.2.3.2 Equipment	242
6.2.3.3 Experimental procedure	242
6.2.3.4 Determination of the zone settling parameters	243
6.2.3.5 Calibration by empirical relations based on SSPs	244
6.2.4 Recommendations for performing batch settling tests	245
6.2.4.1 Shape and size of the batch reservoir	245
6.2.4.2 Sample handling and transport	245
6.2.4.3 Concentration range	245
6.2.4.4 Measurement frequency	245
6.2.5 Recent advances in batch settling tests	245
6.3 MEASURING FLOCCULATION STATE OF ACTIVATED SLUDGE	**246**
6.3.1 DSS/FSS test	246
6.3.1.1 Goal and application	246
6.3.1.2 Equipment	246
6.3.1.3 DSS test	246
6.3.1.4 FSS test	247
6.3.1.5 Interpretation of a DSS/FSS test	248
6.3.2 Recommendations	249
6.3.2.1 Flocculation conditions	249
6.3.2.2 Temperature influence	249
6.3.2.3 Supernatant sampling	249
6.3.3 Advances in the measurement of the flocculation state	250
6.4 MEASURING THE SETTLING BEHAVIOUR OF GRANULAR SLUDGE	**250**
6.4.1 Goal and application	250
6.4.2 Equipment	251
6.4.3 Density measurements	251
6.4.4 Granular biomass size determination	252
6.4.4.1 Sieving	252
6.4.4.2 Image analyser	253
6.4.5 Calculating the settling velocity of granules	253
6.4.6 Recommendations	254
6.4.6.1 Validation of results	254
6.4.6.2 Application for flocculent sludge	255
6.5 MEASURING SETTLING VELOCITY DISTRIBUTION IN PSTs	**255**
6.5.1 Introduction	255
6.5.2 General principle	255
6.5.3 Sampling and sample preservation	256
6.5.4 Equipment	256
6.5.5 Analytical protocol	257
6.5.6 Calculations and result presentation	258
6.5.6.1 Mass balance check	258
6.5.6.2 Calculation of the settling velocity distribution	258
6.5.6.3 Recommendations	259

7. MICROSCOPY   263
Jeppe L. Nielsen, Robert J. Seviour and Per H. Nielsen (aut.)
Jiří Wanner (rev.)

7.1 INTRODUCTION	**263**
7.2 THE LIGHT MICROSCOPE	**263**
7.2.1 Standard applications of light microscopy	265

7.2.2 Low power objective	265
7.2.3 High power objective	265
7.2.4 Immersion objective	265
7.2.5 Important considerations	266
7.2.6 Bright-field and dark-field illumination	266
7.2.7 Fluorescence microscopy	267
7.2.8. Confocal laser scanning microscopy	269
7.3 MORPHOLOGICAL INVESTIGATIONS	**269**
7.3.1 Microscopic 'identification' of filamentous microorganisms	270
7.3.2 'Identification' of protozoa and metazoa	271
7.4 EXAMINING ACTIVATED SLUDGE SAMPLES MICROSCOPICALLY	**272**
7.4.1 Mounting the activated sludge sample	272
7.4.2 Gram staining	273
7.4.2.1 Reagents and solutions for Gram staining	273
7.4.2.2 Procedure	274
7.4.3 Neisser staining	274
7.4.3.1 Reagents and solutions for Neisser staining	274
7.4.3.2 Procedure	275
7.4.4 DAPI staining	275
7.4.4.1 Reagents and solutions for DAPI staining	275
7.4.4.2 Procedure	275
7.4.5 CTC staining	276
7.4.5.1 Reagents and solutions for CTC staining	276
7.4.5.2 Procedure	276
7.5 FLUORESCENCE *in situ* HYBRIDIZATION	**276**
7.5.1 Reagents and solutions for FISH	277
7.5.2 Procedure	278
7.6 COMBINED STAINING TECHNIQUES	**280**
7.6.1 FISH-DAPI staining	281
7.6.1.1 Reagents and solutions for DAPI staining	281
7.6.1.2 Procedure	281
7.6.2 FISH-PHA staining	282
7.6.2.1 Reagents and solutions for PHA staining	282
7.6.2.2 Procedure	282

8. MOLECULAR METHODS 285

Søren M. Karst, Mads Albertsen, Rasmus H. Kirkegaard, Morten S. Dueholm and Per H. Nielsen (aut.)
Holger Daims (rev.)

8.1 INTRODUCTION	**285**
8.2 EXTRACTION OF DNA	**286**
8.2.1 General considerations	286
8.2.2 Sampling	286
8.2.3 DNA extraction	286
8.2.3.1 Cell lysis	286
8.2.3.2 Nuclease activity inhibition and protein removal	287
8.2.3.3 Purification	287
8.2.3.4 Elution and storage	287
8.2.4 Quantification and integrity	287
8.2.5 Optimised DNA extraction from wastewater activated sludge	288
8.2.5.1 Materials	288
8.2.5.2 DNA extraction	288
8.3 REAL-TIME QUANTITATIVE PCR (qPCR)	**289**
8.3.1 General considerations	289
8.3.2 Materials	291
8.3.3 Methods	292
8.3.4 Data handling	294
8.3.5 Data output and interpretation	294
8.3.6 Troubleshooting	295
8.3.7 Example	295
8.3.7.1 Samples	295
8.3.7.2 qPCR reaction setup	296
8.3.7.3 Results	296
8.4 AMPLICON SEQUENCING	**297**
8.4.1 General considerations	297
8.4.2 The 16S rRNA gene as a phylogenetic marker gene	297
8.4.3 PCR amplification	299
8.4.3.1 PCR reaction	299
8.4.3.2 PCR biases	300
8.4.3.3 Primer choice	300
8.4.4 DNA sequencing	301
8.4.4.1 Sequencing platform	301
8.4.4.2 Sequencing depth	301
8.4.5 Bioinformatic processing	301
8.4.5.1 Available software	301
8.4.5.2 Raw data	302
8.4.5.3 Quality scores and filtering	303
8.4.5.4 Merging paired end-reads	303
8.4.5.5 OTU clustering	303
8.4.5.6 Chimera detection and removal	304
8.4.5.7 Taxonomic classification	304
8.4.5.8 The OTU table	304
8.4.6 Data analysis	304
8.4.6.1 Defining the goal of the data analysis	304
8.4.6.2 Data validation and sanity check	305
8.4.6.3 Communities or individual species?	305
8.4.6.4 Identifying core and transient species	306
8.4.6.5 Explorative analysis using multivariate statistics	306
8.4.6.6 Correlation analysis	307
8.4.6.7 Effect of treatments on individual species	307
8.4.7 General observations	307
8.4.7.1 A relative analysis	307
8.4.7.2 Copy number bias	307
8.4.7.3 Primer bias	307
8.4.7.4 Standardization	308
8.4.7.5 Impact of the method	308
8.4.8 Protocol: Illumina V1-3 16S rRNA amplicon libraries	308
8.4.8.1 Apparatus	308
8.4.8.2 Materials	308
8.4.8.3 Protocol	309
8.4.9 Interpretation and troubleshooting	311
8.4.9.1 Sample DNA quality control and dilution	311
8.4.9.2 Library PCR	312
8.4.9.3 Library cleanup	312
8.4.9.4 Library quality control	313
8.4.9.5 Library pooling	314
8.4.9.6 Pool quality control and dilution	314
8.4.9.7 Storage	314
8.4.10 Protocol: Illumina V1-3 16S amplicon sequencing	314
8.4.10.1 Apparatus	314
8.4.10.2 Reagents	314
8.4.10.3 Protocol	314
8.4.10.4 Interpretation and troubleshooting	315
8.4.11 Design of Illumina 16S amplicon sequencing adaptors	317
8.5 OTHER METHODS	**319**

LIST OF SYMBOLS AND ABBREVIATIONS

1

INTRODUCTION

Authors:
Mark C.M. van Loosdrecht
Per H. Nielsen
Carlos M. Lopez-Vazquez
Damir Brdjanovic

Wastewater treatment forms a crucial link in the services that the sanitation sector delivers to society. For centuries, sanitation largely consisted of transporting fresh, clean water to the cities, and using this water to transport the waste out of the city and discharge it into the natural environment. However, with the increase in human populations in cities as a result of the industrial revolution in the 19th century, this could no longer be maintained. The occurrence of epidemic diseases facilitated the development of wastewater treatment facilities and their implementation since the early 20th century. This development has been largely an empirical activity with theoretical approaches following experimental observations (Figure 1.1).

Figure 1.1 Noyes Laboratory on the campus of the University of Illinois in Urbana was arguably the most important in promoting research in wastewater in the early 20th century (photo: University of Illinois, 1902).

The discovery and development of activated sludge technology (described in detail in Jenkins and Wanner, 2014) was crucial as it triggered the rapid development and application of various analytical and experimental methods. Experimental work in the Lawrence Experimental Station in Massachusetts, USA, which at that time (1912) was a unique facility aimed at the experimental verification of different possible wastewater treatment procedures, inspired Gilbert Fowler to request Edward Ardern and William Lockett to repeat the experiments with wastewater aeration in the UK that he had seen in the USA. In 1913 and 1914 Lockett and Ardern carried out lab-scale experiments at the Manchester - Davyhulme wastewater treatment plant (Figure 1.2). Glass bottles were used to represent lab-scale aeration basins 'fed' by sewage from different districts of Manchester. Contrary to the experiments that Fowler saw in Massachusetts, in the Manchester aeration tests the sediment that remained after decantation was left in the bottle and a new dose of sewage was added to the sediment for the next batch. Lockett and Ardern soon found that the amount of the sediment increased with the increasing number of batches. At the same time the aeration time necessary for 'full oxidation' of sewage (full oxidation was a term used to describe the removal of degradable organics and for complete nitrification) was reduced. By using this technique of repeated batch aeration with the sediment remaining in the bottle, Lockett and Ardern were able to shorten the required aeration time for 'full oxidation' from a few weeks to less than one day, which made the process technically feasible. The sediment formed during the aeration of sewage was called activated sludge due to its appearance and activity. Lockett and Ardern published their results in a famous series of three papers (Ardern and Lockett 1914a, 1914b, 1915). This was the 'birth' of activated sludge, which is today the workhorse of wastewater treatment and the most widely applied sewage treatment technology in the world.

Examples are the commonly used chemical or biological oxygen demand tests. The iconic 'Standard Methods for the Examination of Water and Wastewater' (APHA *et al.*, 2012, Figure 1.3) has for generations of sanitary engineers been the resource for analysing their experimental systems and full-scale operations. These methods focus heavily on the chemical characterization and measurement of specific microorganisms.

Figure 1.2 The Davyhulme Sewage Works Laboratory, where the activated sludge process was developed in the early 20th century (photo: United Utilities).

Figure 1.3 The Standard Methods for the Examination of Water and Wastewater. The first edition appeared in 1905 (image: APHA *et al.*, 2012).

Wastewater engineering is a profession that is extremely experiment-based, and therefore it has always had the need to develop and standardise methods. This seemingly simple activity is strongly hampered by two factors, namely: (*i*) wastewater engineering is a typical interdisciplinary activity where chemical engineers, civil engineers, microbiologists and chemists interact to develop and understand the processes; the challenge here is to integrate methods and approaches from these disciplines, and, (*ii*) in addition, wastewater and its treatment processes are by their nature difficult to define with exactitude. It is for instance virtually impossible to measure all the individual compounds in the wastewater itself. Identifying all the relevant microorganisms in the processes has long been impossible and is still a complicated challenge. Defining all the potentially occurring chemical conversions is, due to the myriad of chemicals present, again an almost impossible task.

Due to the undefined nature of the experimental system, research has tended to progress slowly and it heavily depends on standardised methods that may not be exact but, when used in a standardised way, are very helpful and useful to compare experimental results.

Societal demands on the efficiency of wastewater treatment plants have advanced, moving from public health protection to water resources and environmental protection and nowadays to integrated resource and energy recovery. Therefore the need to accurately characterize the microbial processes in the wastewater treatment processes has increased over recent decades. Certainly, it is a challenge to develop standardized methods for experimental work that can be easily repeated in different laboratories. In many cases, the exact handling is important, but it is not easy to be written down in a practical protocol.

Therefore, to avoid these problems, it was decided to develop not only a book describing all the experimental methods but also a video catalogue with the methods described in this book actually being demonstrated in the laboratory. This book and its associated video-based material are designed to support the research and development field with a manual for characterizing the biological processes in wastewater treatment. The editors have decided in this first edition of the 'Experimental methods in wastewater treatment' book to focus on the activated sludge process since this is worldwide by far the most applied technology. Nevertheless, most of the methods presented in this book can also be applicable to biofilm-based technologies or anaerobic digestion processes.

The decision to focus on experimental methods related to the activated sludge process has resulted in seven chapters describing the key experimental methods. The content and focus of these chapters are summarised in Table 1.1. Activated sludge consists of a myriad of microorganisms, converting a range of important compounds (organic matter, oxygen, nitrogen and phosphate compounds). The first three chapters focus on characterizing the conversion capacities of the microbial communities for the major microbial processes. A distinction has been made between full liquid-phase-based methods and methods where the conversion are characterized by measuring the respiration of the organisms, usually gas-phase measurements. Since there is an increasing focus on and interest in assessing the environmental impact of wastewater treatment plants, a separate chapter has been added for measuring greenhouse gas emissions from wastewater treatment plants. These chapters are followed by a chapter describing data handling techniques. Measurements often, certainly from full-scale or pilot plants, have relatively large uncertainties. With adequate data handling techniques the measurements can be used to derive associated (difficult to directly measure) process data or to minimise their uncertainty.

Activated sludge processes mainly depend on settling of the flocculent sludge to separate the biomass from the cleaned wastewater. This is often the Achilles heel of the treatment process and a key factor in the process design. One chapter is therefore devoted to characterization of the sludge settling properties.

As said earlier, microorganisms are the workhorses in the activated sludge process. Therefore the microscope is unavoidably the main technique to observe them directly, not only for individual organisms but also for the floc morphology related to settling characteristics. For a long time the microscope has been the main method of choice when observing which bacteria are present in activated sludge. However, although very helpful, it cannot show the full complexity of the microbial community. The last decade's advance in molecular DNA-based techniques has revolutionized the way one can observe microorganisms. These generic novel methods are described in the final chapter of this book.

Within the chapters the authors have tried to describe especially those methods that are experimentally complex and not standard analytical procedures. Therefore, standard analytical methods for e.g. organic matter, ammonium, phosphate etc. are not described in detail. On the other hand, it was also decided to include some analytical techniques recently developed and/or improved that are becoming frequently used but are scattered across scientific literature (e.g. glycogen and poly-hydroxy-alkanoates determination). In addition, methods that could be of academic interest but currently have limited practical application have not been included in detail in the text.

In terms of symbols and notation, an attempt has been made to standardize them as much as possible. While this was achieved at the chapter level, full standardization was not possible across all the chapters due to their diverse nature and heterogeneity of items as well as lack of global agreement on the use of symbols and notations, although the most common guidelines were quite closely followed (e.g. Corominas *et al.*, 2010).

The book is conceptualized so as to satisfy users with high demands who are able to handle complex analytical and experimental equipment. However, the content is equally suited to the requirements of less advanced laboratories and less experienced experimenters; in particular, the complementary, freely available video materials address the execution of experiments in more challenging environments, such as those usually prevailing in most less developed countries.

"To measure is to know."
Lord Kelvin

Table 1.1 A simplified overview of the experimental methods presented in the book per process of interest.

Process	Chapter							
	Introduction	Activated sludge activity tests	Respirometry	Off-gas emission tests	Data handling and parameter estimation	Settling tests	Microscopy	Molecular methods
Organic matter removal	Overview and rationale to experimental methods	Kinetics	Biochemical oxygen demand (BOD) Short-term biochemical oxygen demand Wastewater characterization and fractionation Biomass characterization Toxicity and inhibition					
Nitrification		AOO and NOO activity Kinetics Stoichiometry	Wastewater characterization and fractionation Biomass characterization AOO and NOO activity Toxicity and inhibition Kinetics Stoichiometry	Sampling methods for nitrogen GHG emissions Methods for off-gas measurements Aqueous N_2O and CH_4 concentration determination methods Gas measurement methods in open tanks	Data handling and validation Parameter estimation Uncertainty analysis Local sensitivity analysis and identifiability analysis	Settling velocity distributions in primary settling tanks Sludge settleability in secondary settling tanks Flocculation properties Settling behaviour of granular sludge	Light microscopy Confocal microscopy Morphological investigations Staining techniques Fluorescence in situ Hybridization (FISH) Combined staining techniques	DNA extraction Real-time quantitative PCR Amplicon sequencing
Denitrification		Denitrification over NO_2 and NO_3 Denitrification on RBCOD and SBCOD Stoichiometry Kinetics	Denitrification over NO_2 and NO_3 Toxicity and inhibition Stoichiometry Kinetics					
Anammox		AMX activity Kinetics Stoichiometry						
EBPR		PAO, GAO, and DPAO activity Kinetics Stoichiometry	Aerobic kinetics and stoichiometry Toxicity and inhibition					
Anaerobic treatment		SRB activity Kinetics Stoichiometry	Specific methanogenic activity Biomethane potential Toxicity and inhibition Kinetics Stoichiometry					
Settling								

AMX Anammox organisms
AOO Ammonium oxidizing organisms
CH_4 Methane
DNA Deoxyribonucleic acid
DPAO Denitrifying poly-phosphate accumulating organisms
EBPR Enhanced biological phosphorus removal
FISH Fluorescence in situ hybridization
GAO Glycogen accumulating organisms
GHG Greenhouse gas emissions

N_2O Nitrous oxide
NO_2 Nitrite
NO_3 Nitrate
NOO Nitrite oxidizing organisms
PAO Poly-phosphate accumulating organisms
PCR Polymerase chain reaction
RBCOD Readily biodegradable COD also known as readily biodegradable organics
SBCOD Slowly biodegradable COD also known as slowly biodegradable organics
SRB Sulphate reducing bacteria or SRO Sulphate reducing organism

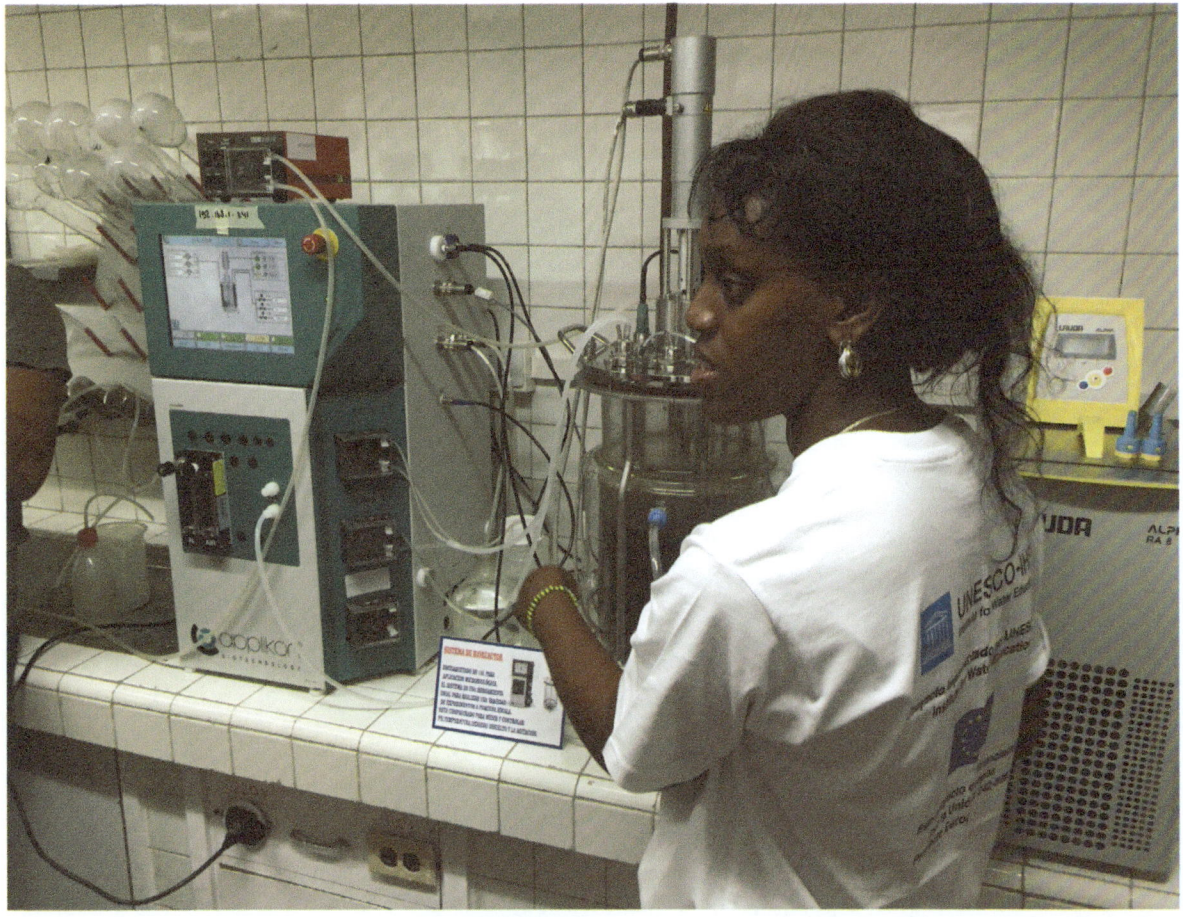

Figure 1.4 The mission of UNESCO-IHE is to contribute to the education and training of professionals, to expand the knowledge base through research and to build the capacity of sector organizations, knowledge centres and other institutions active in the fields of water, the environment and infrastructure in developing countries and countries in transition. The photos depict the illustrative example of the Institute's latest project in Cuba where the laboratory of the Instituto de Investigaciones para la Industria Alimenticia (IIIA) in Havana has been equipped with new state-of-the-art technology and where the local staff has been trained on how to operate the equipment and prepare and carry out experimental work (photo: Brdjanovic, 2015).

References

American Public Health Association (APHA), American Water Works Association (AWWA), and Water Environment Federation (WEF) (2012). Standard Methods for the Examination of Water and Wastewater, 22nd Edition. New York. ISBN 9780875530130.

Ardern, E., Lockett, W.T. (1914a) Experiments on the Oxidation of Sewage without the Aid of Filters. *J. Soc. Chem. Ind.*, 33: 523.

Ardern, E., Lockett, W.T. (1914b) Experiments on the Oxidation of Sewage without the Aid of Filters, Part II. *J. Soc. Chem. Ind.*, 33: 1122.

Ardern, E., Lockett, W.T. (1915) Experiments on the Oxidation of Sewage without the Aid of Filters, Part III. *J. Soc. Chem. Ind.*, 34: 937.

Corominas, L.L., Rieger, L., Takács, I., Ekama, A.G., Hauduc, H., Vanrolleghem, P.A., Oehmen, A., Gernaey, K.V., van Loosdrecht, M.C.M., Comeau Y. (2010). New framework for standardized notation in wastewater treatment modelling. *Water Sci Technol.* 61(4): 841-57.

Jenkins, D. and Wanner, J. Eds. (2014) 100 years of activated sludge and counting. IWA Publishing, London, ISBN 9781780404936, pg. 464.

The section on activated sludge historical development presented in this chapter is adapted from Jenkins and Wanner (2014).

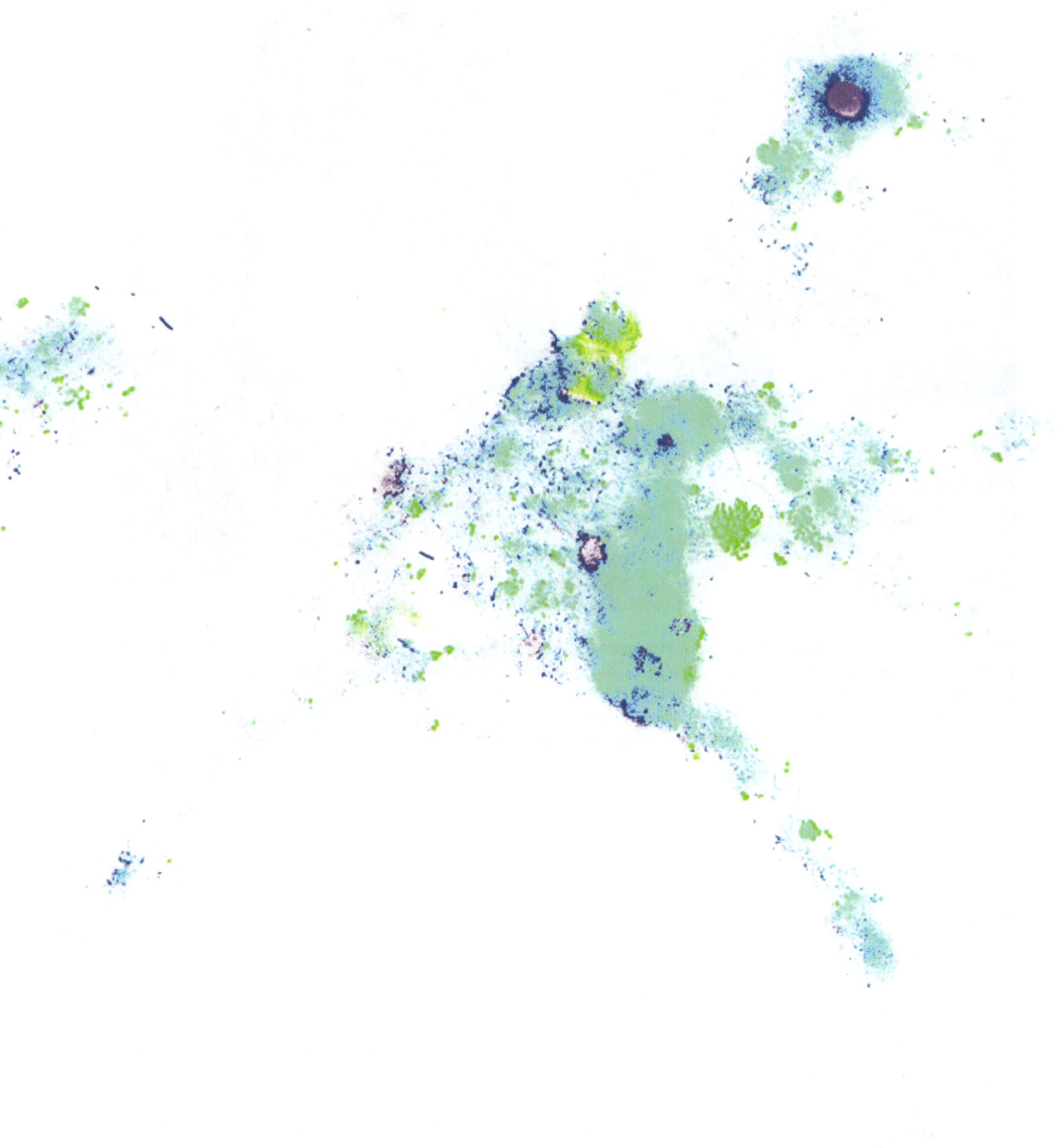

2

ACTIVATED SLUDGE ACTIVITY TESTS

Authors:
Carlos M. Lopez-Vazquez
Laurens Welles
Tommaso Lotti
Elena Ficara
Eldon R. Rene
Tessa P.H. van den Brand
Damir Brdjanovic
Mark C.M. van Loosdrecht

Reviewers:
Yves Comeau
Piet N.L. Lens
Nancy G. Love

2.1 INTRODUCTION

Different conditions and factors affect the degree and rate (speed) at which the compounds and contaminants of concern are removed by microbial populations in biological wastewater treatment systems. Certainly, the plant configuration and operational conditions play a major role in the prevalence of specific microbial populations and their activities, but factors as diverse and broad as wastewater characteristics and environmental and climate conditions have a strong influence as well. Eventually, in any biological wastewater treatment system, there will be a need to assess, define and understand the plant performance with regard to the removal of certain contaminants and the response of the sludge to inhibitory or toxic compounds of interest. Moreover, from a modelling perspective it is also of interest to assess and determine the stoichiometry and kinetic rates of the conversion processes performed by specific microbial populations (e.g. ordinary heterotrophic organisms: OHOs; denitrifying ordinary heterotrophic organisms: dOHOs; ammonium-oxidizing organisms: AOOs; nitrite-oxidizing organisms: NOOs; phosphate-accumulating organisms: PAOs; sulphate-reducing bacteria: SRB, also identified as sulphate-reducing organisms, SRO (Corominas et al., 2010); or, anaerobic ammonium-oxidizing organisms: anammox. Thereby, the execution of batch activity tests can be rather useful to: (i) study the biodegradability of a given wastewater stream (municipal or industrial), (ii) determine the stoichiometric and kinetic parameters involved in the conversion of a specific compound, (iii) study the potential interactions (e.g. symbiosis and competition) between microbial populations and (iv) assess the potential inhibitory or toxic effects of certain wastewaters, compounds or substances.

The nature and type of the batch activity tests can differ depending upon the compounds of interest and the metabolism and physiology of the microbial populations involved in the removal or conversion processes. For instance, they can range from relatively simple aerobic tests where organic matter removal by OHOs is measured to more complex alternating anaerobic-anoxic-aerobic batch tests to assess the activity of PAOs under the presence of different electron acceptors (such as nitrate, nitrite and oxygen) from activated sludge systems performing enhanced biological phosphorus removal (EBPR).

This chapter presents an overview of the most common batch activity tests and protocols and their execution with the aim of assessing the conversion processes involved in: (*i*) enhanced biological phosphorus removal by PAOs under alternating anaerobic-aerobic conditions, (*ii*) denitrification via nitrate or nitrite by PAOs, (*iii*) reduction of sulphate by SRBs, (*iv*) removal of organics under aerobic conditions by OHOs, (*v*) denitrification by dOHOs using nitrate or nitrite as final electron acceptor, (*vi*) oxidation of ammonia and nitrite by AOOs and NOOs under aerobic conditions and (*vii*) nitrogen removal by anammox bacteria. These experimental protocols aim to serve as a useful guide that establishes a basis for standardizing batch activity tests for use on existing, emerging and innovative treatment processes. It was decided to start the order of presentation with EBPR systems involving PAOs as the processes are complex and include all three biochemical activated sludge environments: anaerobic, anoxic and aerobic.

Figure 2.1.1 Experimental facilities for activated sludge activity tests at UNESCO-IHE Institute for Water Education in the Netherlands (photo: UNESCO-IHE).

2.2 ENHANCED BIOLOGICAL PHOSPHORUS REMOVAL

2.2.1 Process description

Enhanced biological phosphorus removal (EBPR) can be implemented in activated sludge wastewater treatment systems by introducing an anaerobic stage at the start of the wastewater treatment lines. High P-removal efficiency, lower operational costs, lower sludge production and the potential recovery of phosphorus have contributed to its application and popularity (Mino *et al.*, 1998; Henze *et al.*, 2008; Oehmen *et al.*, 2007). EBPR is performed by phosphorus (polyphosphate)-accumulating organisms (PAOs) (Comeau *et al.*, 1987; Mino *et al.*, 1998) that, by intracellular accumulation of polyphosphate (poly-P), can remove higher quantities of phosphorus (0.35-0.38 g P g VSS^{-1} of PAOs) than OHOs (0.03 g P g VSS^{-1} of OHOs) (Wentzel *et al.*, 2008). The scientific, microbiological and engineering characteristics of the EBPR process have been the main focus of research carried out during the last few decades by different research groups (Wentzel *et al.*, 1986, 1987; Comeau *et al.*, 1986, 1987; Smolders *et al.*, 1994a,b; Mino *et al.*, 1987, 1998; Oehmen *et al.*, 2005a, 2005c, 20i306, 2007; Nielsen *et al.*, 2010). In particular, efforts have focused on developing a better understanding of the actual EBPR metabolic mechanisms, to unravel the microbial identity of the organisms involved, and to optimize the required process configurations, all with the aim of improving and increasing the EBPR process efficiency and reliability.

PAOs are heterotrophic organisms. However, unlike OHOs, PAOs have the unique capability of using intracellularly stored poly-P to produce the required energy (adenosine tri-phosphate, ATP) under anaerobic conditions to store readily biodegradable organic matter (RBCOD), such as volatile fatty acids (VFA) like acetate (Ac) and propionate (Pr), as intracellular poly-β-hydroxy-alkanoates (PHAs). Stored PHAs are later utilized under anoxic or aerobic conditions for enhanced phosphorus uptake, glycogen synthesis, biomass growth and maintenance. This feature gives PAOs a competitive advantage over other microbial populations of relevance. Thus, PAOs can be enriched to achieve EBPR by recycling activated sludge through alternating the anaerobic and anoxic or aerobic stages, while directing the influent which is usually rich in VFA to the anaerobic stage. A schematic representation of the PAOs' metabolism is shown in Figure 2.2.1.

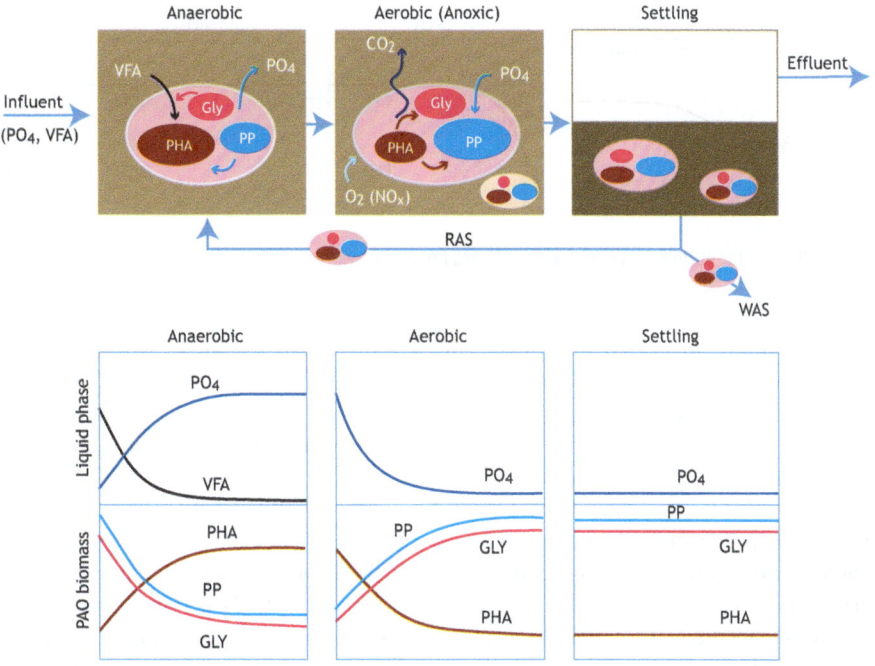

Figure 2.2.1 Conceptual scheme of an activated sludge wastewater treatment plant performing EBPR, illustrating the activity of PAOs (Lopez-Vazquez, 2009; adapted from Meijer, 2004).

In the anaerobic stage, PAOs store intracellularly the readily biodegradable organics present in the raw influent or settled sewage (mostly VFA) as PHAs using two other intracellularly stored polymers that take part in the aforementioned metabolism: poly-P and glycogen (a polymer of glucose). Poly-P is hydrolysed and utilized by PAOs to provide the required energy (as ATP) for the transport and storage of VFA as PHAs (Wentzel et al., 1986), while glycogen is used to supply the required reducing power for the conversion of VFA into PHAs as well as to provide the additional required energy (as ATP) (Comeau et al., 1986, 1987; Smolders et al., 1994a; Mino et al., 1998). Thus, the anaerobic uptake of VFA by PAOs results in the storage of PHAs and simultaneous hydrolysis of poly-P and glycogen. The most common PHA polymers stored by PAOs are poly-β-hydroxybutyrate (PHB), poly- -hydroxyvalerate (PHV) and poly-β-hydroxy-2-methylvalerate (PH_2MV). Their presence and amount depends on the VFA composition (Ac or Pr). When Ac is the most abundant VFA in the media, PAOs store mostly PHB (up to 90 % of the stored PHAs) (Smolders et al., 1994a), but when Pr is the dominant VFA, then PHAs exist mostly as PHV and PH_2MV (Oehmen et al., 2007).

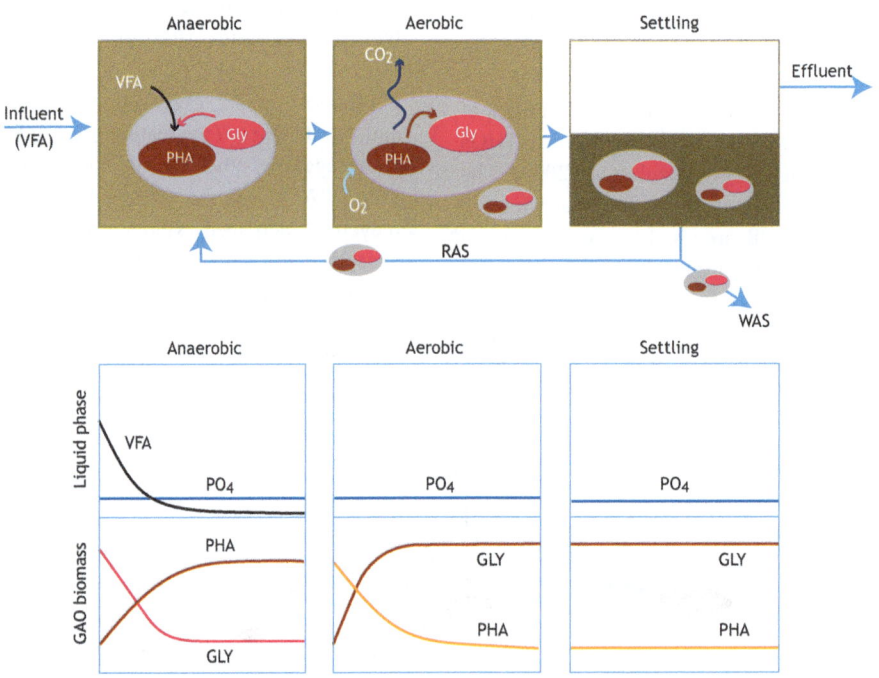

Figure 2.2.2 Conceptual scheme of the microbial activity of GAOs (adapted from Lopez-Vazquez, 2009).

In addition to VFA uptake, the anaerobic hydrolysis of poly-P and glycogen also provides the energy required by PAOs to cover their anaerobic maintenance requirements without carbon uptake. Consequently, the hydrolysis of poly-P leads to the release of orthophosphate (PO_4) into the bulk liquid, which is reflected in an increase in orthophosphate concentration in the liquid phase during the anaerobic stage (Figure 2.2.1). In addition to the uptake of VFA present in the influent of the activated sludge system, PAOs can also store VFA generated by fermentative organisms in the anaerobic stage from fermentable organics present in the influent. Once PAOs reach the aerobic stage, they utilize the PHAs stored in the anaerobic phase as a carbon and energy source using oxygen as the electron acceptor; the energy from this reaction is used to take up and store a higher amount of PO_4 than the amount previously released in the anaerobic stage (Figure 2.2.1). This results in the aerobic uptake and removal of phosphorus from the liquid phase. In the aerobic stage, PHAs are also used to: (*i*) replenish the intracellular glycogen pool, (*ii*) support biomass growth, and (*iii*) cover the aerobic maintenance energy needs of PAOs (Smolders et al., 1994b). Net P-removal from wastewater is achieved through the wastage of activated sludge (WAS) at the end of the aerobic phase, when the sludge contains a high poly-P content (Figure 2.2.1). Alternatively, denitrifying phosphorus-accumulating organisms (DPAOs) exist which can also take up PO$_4$ under anoxic conditions using nitrate or nitrite as electron acceptors (Vlekke *et al.*, 1988; Kuba *et al.*, 1993; Hu *et al.*, 2002; Kerr-Jespersen *et al.*, 1993; Guisasola *et al.*, 2009). Also, PAOs, being heterotrophic organisms, are able to take up carbon sources under aerobic conditions, releasing orthophosphate while the carbon source is available and removing PO$_4$ afterwards (Guisasola *et al.*, 2004; Ahn *et al.*, 2007). However, eventually PAO can lose the competition against OHOs due to their metabolic adaptation to permanent aerobic conditions (Pijuan *et al.*, 2006). For a deeper understanding of the metabolism and factors affecting the EBPR process, the reader is referred to materials published elsewhere (Comeau *et al.*, 1986; Mino *et al.*, 1998; Oehmen *et al.*, 2007).

The proliferation of glycogen-accumulating organisms (GAOs) has been observed in EBPR systems under certain conditions (e.g. when acetate or propionate are present as the sole carbon source, when temperatures exceed 20 °C, at pHs below 7.0, and/or at dissolved oxygen (DO) concentrations higher than 2 mg L^{-1}) (Oehmen *et al.*, 2007; Lopez-Vazquez *et al.*, 2009a,b; Carvalheira *et al.*, 2014). GAOs have an apparently similar metabolism to that of PAOs, but they rely solely on their intracellularly-stored glycogen pools as the source of energy and reducing equivalents that drive the anaerobic storage of VFA as PHAs without any contribution from poly-P (Figure 2.2.2). Their presence is often associated with suboptimal EBPR performance because they do not contribute to phosphorus removal, but compete with PAOs for substrate under anaerobic conditions leading to the deterioration of EBPR systems (Saunders *et al.*, 2003; Thomas *et al.*, 2003). Therefore, GAOs are assumed to be an undesirable population in EBPR systems.

2.2.2 Experimental setup

2.2.2.1 Reactors

To assess the EBPR process performance, batch activity tests can be carried out under anaerobic, aerobic and anoxic conditions depending upon the parameters of interest and nature of the study. In any case, the bioreactor(s) used for the execution of tests must: (*i*) avoid oxygen intrusion under anaerobic and anoxic conditions, (*ii*) provide enough aeration capacity to produce DO concentrations higher than 2 mg L^{-1} under aerobic conditions, (*iii*) provide complete mixing conditions, (*iv*) allow temperature control; (*v*) allow pH control, and (*vi*) have ports for sample collection and the addition of influent, solutions, gases and any other liquid media or substrate used in the test (Figures 2.2.4 and 2.2.5).

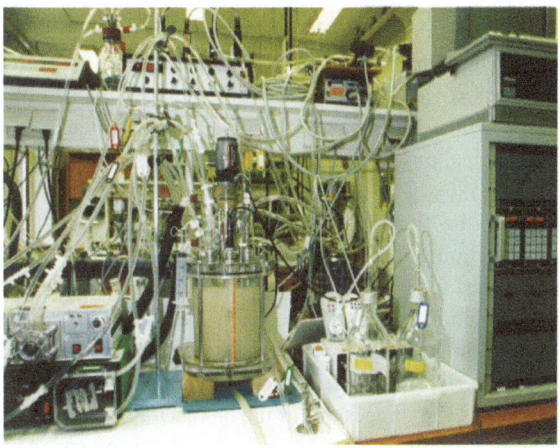

Figure 2.2.3 Vintage EBPR experimental setup with the characteristic yellowish colour of highly enriched biomass with PAOs used at Delft University of Technology in the early 1990s for the development of the TUDelft bio-P metabolic model (Smolders *et al.*, 1994a, 1994b; Murnleitner *et al.*, 1997) and pioneering research on the impact of temperature on EBPR (Brdjanovic *et al.*, 1998a) (photo: Brdjanovic, 1994).

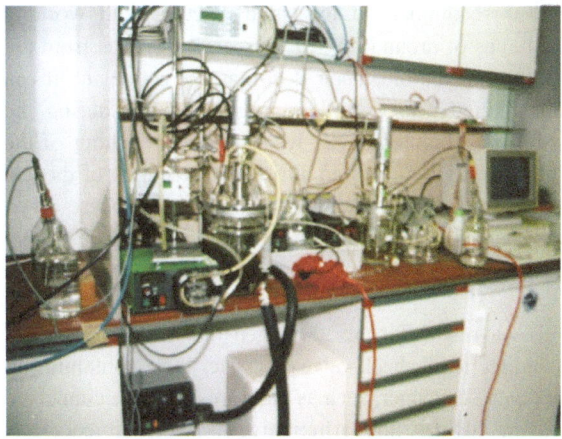

Figure 2.2.4 Temporary experimental setup used to carry out batch tests with activated sludge at the WWTP Haarlem Waarderpolder in the Netherlands. This was the first study (Brdjanovic *et al.*, 2000) in which a separate validation of the TUDelft bio-P metabolic model was carried out using various batch tests with mixed culture biomass from a full-scale plant (photo: Brdjanovic, 1997).

Anaerobic conditions

The experimental setup used for EBPR batch activity tests must be able to create and maintain strict anaerobic conditions. This means that no electron acceptors (namely oxygen, nitrate or nitrite) should be available to the biomass during the anaerobic phase.

A redox probe can be used to monitor the creation of anaerobic conditions when the redox values are lower than -300 mV. The lab setup should be airtight and equipped with an off-gas exit connected to the lid of the bioreactor. Usually, there are three undesirable sources of oxygen: (*i*) the oxygen dissolved in the influent, (*ii*) the residual oxygen present in the activated sludge itself, and (*iii*) oxygen intrusion from the head space. To remove the first two listed sources, N_2 gas should be sparged under mixing conditions from the bottom of the bioreactor for 5 to 10 min prior to the beginning of the test and during influent addition. Sparging time will depend on the mass transfer properties of the gas-liquid interface, which depends on a number of factors including: the dimensions of the bioreactor, the presence and location of baffles, dimensions and stirring speed of mixing blades, gas diffuser configuration and flow rate, and medium composition. To avoid oxygen intrusion, the headspace can be flushed either by the N_2 gas already sparged at the bottom of the bioreactor to the activated sludge or by flushing the head space for 5 to 10 min, depending on the volume of the head space and the gas flow rate. A N_2-gas flow rate of around 30 L h^{-1} is commonly used in lab-scale fermenters with an operating volume of up to 3 L, while a lower flow rate of about 6 L h^{-1} is recommended for batch reactors with working volumes of around 0.5-1.0 L. Sparging N_2 gas from the bottom of the bioreactor is common practice and it can be applied prior to, at the beginning, and during the execution of the activity tests, whereas flushing of the head space is often used during the execution of the test to avoid oxygen intrusion from the atmosphere when mixing the activated sludge. Combining these two approaches is both unusual and unnecessary.

To avoid diffusion of oxygen into the bioreactor, a unidirectional check-valve or a water-lock (containing an oxygen scavenger, such as $NaSO_2$) should be connected to the off-gas line. Alternatively, if the bioreactor is continuously sparged with N_2 gas, resulting in positive pressure inside the bioreactor and a continuous off-gas flow, a check valve or water-lock are not essential for ensuring anaerobic conditions. If N_2 gas is unavailable or cannot be continuously supplied due to limitations of the equipment, the activated sludge should be gently yet completely mixed at slower speeds (much lower than 300 rpm) under airtight conditions until the dissolved oxygen (DO) concentration drops below the detection limit (practically zero) and the redox values are lower than -300 mV. In addition, the volume of headspace should be reduced to minimize the risk of oxygen intrusion by filling the fermenter to the maximum working volume and/or by reducing the surface area of the gas-liquid interface by adding non-reactive, floating polyurethane foam or sponge beads. Silicon rubber stoppers and seals, plastic and aluminium foils, among other materials, are usually used to create airtight conditions.

In addition to avoiding oxygen intrusion, the presence and availability of other electron acceptors (such as nitrate or nitrite) must be prevented to keep strict anaerobic conditions throughout the test. Their prevention is not as straightforward as the removal of oxygen. It often requires an adequate handling of the activated sludge sample prior to the execution of the test. This may involve the controlled addition of nitrifying inhibitors under non-aerated and aerated conditions, or sludge 'washing', as explained later in Section 2.2.3.5).

Anoxic conditions

The creation of strict anaerobic conditions means that no electron acceptors should be present inside the bioreactor, the creation of anoxic conditions indicates that although DO must not be present, other electron acceptors, such as nitrite or nitrate, must be available. This also means that the creation of anoxic conditions must reduce or eliminate oxygen intrusion as is done for anaerobic phases. The experimental setup should allow the (controlled) availability (presence) of the electron acceptors of interest. The desired electron acceptors can be generated either in the system itself, via a preceding nitrification step, or be externally added as nitrate - or nitrite - solutions at defined concentrations at the start of the batch test or during the test. If available, DO concentrations below the detection limit together with redox values in between -200 to 0 mV can indicate that the required anoxic conditions have been reached. Usually, the latter is the most common practice because the experimental configuration can be simplified by doing so, and there is a better control of the required dosing time and concentration. Nevertheless, a combination of several reactors and experimental stages/phases can be made to incorporate a nitrification step between the anaerobic and anoxic phases to provide the required electron acceptors to drive the anoxic metabolism of EBPR cultures (Kuba *et al.*, 1993). When external electron donors are added, the system must have an adequate dosing port and a way to release the resulting extra pressure created by injecting the liquid volume.

Aerobic conditions

Most commercially available fermenters have gas spargers usually located at the bottom of the fermenter just below the stirring blades of the mixer. When supplying compressed air (e.g. either from a central or a local/portable air compressor), these arrangements can provide a satisfactory oxygen supply leading to DO concentrations reaching far above the limiting conditions of the microbial processes. As a general rule of thumb, DO concentrations of at least 2 mg L^{-1} are considered adequate for most applications. For batch reactors with a working volume of about 3 L, a compressed air flow rate of around 60 L h^{-1} (1 L min^{-1}) can usually provide the required aeration. However, the biomass composition and concentration, wastewater characteristics, organic matter and intracellular PHA content (in the case of EBPR) may increase the DO requirements. Under these conditions, the air flow rate should be increased so as to maintain the DO concentration above 2 mg L^{-1} throughout the test. Alternatively, a pure oxygen supply can be used instead of compressed air to increase the DO availability under specific conditions (e.g. in industrial applications).

In more advanced applications, it may be necessary or desirable to carry out aerobic batch activity tests at a constant (set) DO concentration. For these applications, a two-way DO control can be used to define a DO set point and keep it stable throughout the aerobic batch test. Most advanced fermenters are equipped with such a two-way control operated by at least two solenoids with an on/off function that alternatively supplies air or N_2 gas, depending upon the actual measured DO in the liquid phase.

Less advanced fermenters used for the execution of aerobic batch activity tests can be equipped with a portable air compressor that provides an adequate air flow rate. Aquarium stones can be placed at the bottom of the fermenters in line with the mixing/stirring system to distribute bubbles for good oxygen transfer. As previously discussed, the air supply should be able to produce and sustain a bulk liquid DO concentration of at least 2 mg L^{-1} within the first 10 min.

The two most common commercially available DO probes are the membrane-type and the optical-type. Prior to use (and preferably also after use), they should be calibrated according to the manufacturer's or supplier's instructions. In addition, all connections should be checked. In the case of the membrane-type DO probe, the membrane should be clean, should not have any damage, and the probe should be properly filled with fresh electrolytes. Moreover, no bubbles should accumulate or be trapped on the membrane surface. The surface of optical probes also needs to be cleaned periodically and the head cap replaced annually.

Mixing

Mixing of the bioreactor's content must be generous to favour a homogenous distribution of the activated sludge (mixture of liquid phase and biomass) and wastewater as well as other substances (e.g. orthophosphate, nitrate or nitrite solutions). Commonly, in most 3 L fermenters, a mixing speed of up to 500 rpm can be applied, while slower mixing speeds of around 100 rpm are used in larger fermenters (of 10 L and larger). Excessive mixing can lead to floc breakage, reducing the mass transfer resistance through the flocs and the settling properties. On the other hand, insufficient mixing can result in dead zones, large flocs, sludge stratification, limited diffusion of substrates and oxygen, and, in extreme cases, in settling. Slower mixing speeds can be used as long as the bulk liquid is well mixed and neither stratification nor accumulation of solids is observed. Advanced fermenters can have an automatic control to regulate the mixing speed in time using a vertical axis with blade propellers. Furthermore, the use and installation of vertical blades or baffles connected to the inner side of the fermenter's lid or the choice of a more efficient impeller can further improve the mixing conditions. In less sophisticated systems, mixing can be provided by positioning the fermenters on stirring plates and using magnetic stirrers. As previously mentioned, to reduce the potential oxygen intrusion in anaerobic batch tests, the stirring speed can be reduced as long as it does not compromise the good mixing conditions.

Temperature control

Temperature has a strong effect on the metabolism of PAOs and their competitors (e.g. GAOs) (Brdjanovic et al., 1997; Lopez-Vazquez et al., 2009a,b). Therefore, adequate and stable temperature-controlled conditions are advisable for the execution of the batch activity tests. Advanced fermenters are usually equipped with a double glass wall (double-jacketed reactors) and usually water (of a temperature similar to the target temperature in the bioreactor) is recirculated through the double wall. The water temperature is adjusted in the controlling and operating console of the fermenter (through internal heaters, heat exchangers and condensers) or by using external heating jackets or a water bath and recirculation devices. Depending upon the desired working temperature, other fluids rather than water can be recommended (e.g. anti-freeze solutions for temperatures lower than 5 °C or oils for temperatures higher than 30 °C). Advanced systems can automatically measure the bulk water temperature inside the bioreactor and adjust accordingly to keep a stable temperature. However, it is important to keep in mind that the temperature of the cooling/heating fluid is often different (by a couple of degrees Celsius) from the actual temperature measured in the activated sludge. These differences occurr due to thermal exchange efficiency and thermal exchange between the recirculated fluid and the air during its transport from the controller to the fermenter, in particular when the operating temperature is significantly different from the ambient (room) temperature. Under such conditions, it is recommended to adjust the fluid temperature in the controller unit until the target temperature in the liquid phase is reached and remains stable.

Besides the individual temperature control that a fermenter may have, the entire experimental setup can be located inside a temperature-controlled room set at the target temperature. Nevertheless, if temperature is not an issue and the batch tests can be performed at room or ambient temperature in a defined location, there should be the certainty that the temperature will not fluctuate considerably (less than ± 1-2 °C) from the preparation until the end of the test. In any case, the temperature of execution must be always recorded and reported.

Last but not least, it is important to mention that both the activated sludge and the wastewater or synthetic medium of the study (whenever applicable) must have the same temperature prior to the execution of the activity tests to avoid temperature shocks and fluctuations that may compromise the outcomes of the tests. Under these circumstances, all the activated sludge, wastewater and solutions need to be exposed to the working temperature and their actual temperature must be monitored until they reach the target temperature. The temperature adjustment of the activated sludge must be carried out with no electron donor present (i.e. no external carbon source). Usually, only a short exposure time of the biomass to the desired temperature of maximum 1 to 2 h is necessary. Whenever needed, the samples could also be acclimatized for longer periods (up to 3-4 h) until the target temperature is reached, but special attention must be paid to avoid compromising the metabolic activity of PAOs (e.g. leading to the consumption of the intracellular compounds). If the temperature difference between the activated sludge and substrate media or wastewater is high or if the temperature effects are of interest (e.g.

higher than 5 °C to assess a potential temperature shock), the tests must be conducted as soon as the target temperature is reached.

Usually, most of the tests are executed around 20 °C, but it can be as low as 5 °C (to assess the biomass activity under winter/cold climate conditions) (Brdjanovic *et al.*, 1997) or as high as 30 - 35 °C for tropical conditions or industrial applications (Cao *et al.*, 2009; Ong *et al.*, 2014), and even up to 55 °C for thermophilic conditions (Lopez-Vazquez *et al.*, 2014). Tests are rarely performed below 5 °C because in practice the temperature of municipal wastewater is seldom colder and is usually around 7 - 12 °C.

pH control

pH is an important operating parameter for EBPR (and many other) processes. This is particularly because the metabolism of PAOs during the anaerobic uptake of carbon (VFA) will result in higher P-release levels at higher pH and lower P release at lower pH (Smolders *et al.*, 1994a). Also, mixing and the vigorous sparging of N_2 gas or compressed air can strip the dissolved CO_2 out of the solution and raise the pH above 7.0. (e.g. in the range of 7.8-8.5), affecting a number of biological and physical processes.

On the other hand, the biological removal of constituents, such as phosphate by PAOs, tends to decrease the buffering capacity of the liquid and change the pH during an experiment. The alkalinity of the wastewater or other solutions added can increase the buffering capacity of the bulk liquid and reduce the fluctuations. CO_2 sparging can compensate for the CO_2 that strips out when mixing or sparging compressed air or N_2 gas. Thus, similar to temperature, pH must be stable prior to, throughout and until the end of the EBPR batch test.

Under certain circumstances, different pH set points can be applied during different experimental phases (e.g. a pH of 7.5 under anaerobic conditions followed by a pH of 7.0 in the aerobic stage). An acceptable fluctuation pH range is assumed to be ± 0.1. In this regard, the use of a two-way pH controller (for acid and base addition, such as HCl and NaOH, respectively) is recommended.

Advanced bioreactor systems usually have pH control settings, but simpler yet reliable external pH controllers can also be used. In less advanced systems, pH levels can be controlled through the manual addition of acid and base solution. The usual molarity of acid and base solutions for pH control is around 0.2 to 0.4 M. If manual pH control is applied, the concentrations can be lower (e.g. 0.1 M). Depending on the activity of the sludge, different molarities can be used. If the molarity of the solutions is too high, it may lead to sudden pH changes, where the pH values may drop or increase drastically around the set point, crossing the lower or upper limit of the pH control settings and even oscillating below and above the pH set point. Lower molarities may lead to a slow response to adjust the pH to the desired pH set point, which in extreme cases might not be reached and create a considerable dilution of the activated sludge in the bioreactor.

Due to the fast initial speed of microbial conversions, the potential acid or base consumption will be higher at the beginning of the tests or when switching from one phase to another (e.g. from anaerobic to aerobic), but it will usually stabilize by the end of the test. In any case, a considerable deviation from the pH set point (e.g. of more than ± 0.10) must be corrected, preferably within 5-10 sec. The actual pH measured in the liquid phase during the experiment should always be reported.

Certain pH controllers have specific settings that should be adjusted to maintain a stable pH, such as the volume (stroke) of the pulses of acid and base addition and the response time in between acid and base addition pulses. The time in between the acid or base addition pulses should be adjusted to the time that is needed for the system to obtain homogenously mixed conditions after the addition of acid or base.

Similarly to temperature, if pH shocks are to be studied, the EBPR batch activity tests must start as fast as the pH of the activated sludge reaches the target pH of the study. Any required pH adjustment must be performed preferably in less than 5 min prior to the start of the test to avoid any premature or side effects on the metabolism of PAOs (e.g. leading to certain P release or consumption of polymers stored intracellularly). The use of sulphuric acid and alkaline, phosphate buffers and Tris(hydroxymethyl)aminomethane (Tris) solutions must be avoided. This is mostly since they can lead to interferences such as benefiting sulphate-reducing bacteria (SRB) over PAOs (Saad et al., 2013, Rubio-Rincon et al., 2016, submitted), enhancing P precipitation with carbonate species (chemical P-precipitation) (Barat et al., 2008) or increasing the salinity levels beyond those that PAOs can withstand (Welles et al., 2014). It is needless to say that proper pH control is essential for the success of experiments as even very short exposure of biomass to extreme pH (low or high) will quite certainly affect the biomass irreversibly.

All pH meters and sensors should be calibrated immediately (and preferably checked afterwards) according to the manufacturer's or supplier's instructions and all the connections should be checked. Special attention must be paid to the selection and use of pH sensors that can stand the particular characteristics of the wastewater and EBPR sludge subject to study. For instance, high salinity, high chlorides or high H_2S concentrations can lead to interferences if the pH sensors cannot tolerate the higher concentrations. The reader should always verify in manuals, booklets and/or with providers and suppliers if the pH sensors and meters to be used are suitable for the particular wastewater characteristics to be tested.

Sampling and dosing ports

The reactors/fermenters used to carry out the batch activity tests should also have conveniently located sampling and dosing ports to ensure the collection of representative samples from the liquid phase as well as favour a fast dispersion or mixture of any substance or solution added to the bulk of the liquid. The sampling ports can be composed of flexible (rubber or plastic) tubing with an inner diameter that makes it possible to connect different syringes of 5, 10 or 20 mL volume, but also of smaller or bigger volumes (e.g. of 1 or even 50 mL). The sampling port inside the fermenter needs to reach a favourable depth and location to allow a satisfactory sampling before, during and at the end of the test. Usually, the sampling port can be located at the middle level of the lowest third or quarter of the bioreactor's working volume subject to the provision of well-mixing conditions. The most important requirement is to obtain a representative sample from a well-mixed bioreactor.

Regarding the dosing ports, they need to be located and positioned in such a way that they allow a fast dispersion of the solutions or substances added into the bioreactor. They can be well defined injection ports located on the lid of the bioreactor or flexible openings (e.g. through the septum). A formal and structured location and integration of the sampling ports into the fermenter configuration is mostly required when working with airtight reactors to avoid oxygen intrusion. Moreover, the use of lab clips (or similar devices) is recommended to close the tubing of the sampling ports or temporarily unused dosing ports and avoid spillages and splashes caused by a possible increase in the internal pressure in the bioreactor. To counteract any potential under or overpressure, a needle can be inserted into a septum located on the lid of the bioreactor and a tedlar bag (or a similar flexible container filled with an inter gas

- e.g. nitrogen) can be connected to the needle to connect the gas phase of the headspace with the inert gas.

2.2.2.2 Activated sludge sample collection

Contrary to some wastewater treatment processes, the sampling time and location of an EBPR activated sludge sample is highly dependent on the type of batch activity test to be conducted. The latter is based on the alternating anaerobic-(anoxic)-aerobic conditions required by the physiology of PAOs. Thus, a fresh sample should preferably be collected at the end of the preceding reaction stage. Thereby, for an anaerobic batch test, the activated sludge sample should be collected at the end of the aerobic phase at the full- or pilot-scale wastewater treatment plant or 'parent' laboratory bioreactor, whereas, for an aerobic batch test, the sample can be collected at the end of the anaerobic or anoxic phase depending on the system configuration. For the execution of anoxic tests, the sludge can be collected at the end of the anaerobic stage. Alternatively, samples collected in the aerobic phase can be used to execute sequential anaerobic-aerobic, anaerobic-anoxic or anaerobic-anoxic-aerobic batch tests.

Certainly, the sampling location will depend on the system configuration. In full- and pilot-scale wastewater treatment plants, the physical borders between stages must be identified prior to sampling. In extreme cases, where the phases are not (physically) well defined, the redox limits or boundaries need to be determined with the use of a DO meter, redox meter and/or by determination of the nitrate and nitrite concentrations. In lab-scale systems (usually operated on a time-base mode), the sample collection can be relatively easier, since the reaction time defines the length of the stages. To obtain homogenous and representative samples, the sludge samples must be collected in sampling spots where well-mixed conditions take place. Ideally, batch activity tests must be performed as soon as possible after collection (in less than 1-2 h for tests to be conducted with sludge collected at the end of an aeration tank/phase or in a few minutes (2-3 min) for sludge collected at the end of an anaerobic or anoxic phase). In lab-scale systems, in principle, this should not be a problem if the batch activity tests are performed in the same laboratory and their execution is coordinated and synchronized with the operation cycle of the lab bioreactor. Also, at full- and pilot-scale treatment plants, batch activity tests can be performed *in situ* shortly after the collection of activated sludge if the sewage plant laboratory is conditioned and equipped with the required experimental and analytical equipment (Figure 2.2.4).

If the batch activity tests cannot be performed *in situ* on the same day, an activated sludge sample can be collected at the end of the aerobic stage. Afterwards, the sampling bucket can be properly stored and transported in a fridge or in ice (below or close to 4 °C) under non-aerated conditions and the activity tests should be performed not later than 24 h after sampling. The sampling collection at the end of the aerobic stage and storage under non-aerated conditions at the lower temperature can help to preserve the original biomass condition by slowing down the bacterial metabolism. Therefore, in principle it is advised not to aerate the activated sludge samples since this could lead to P release and oxidation of intracellular compounds (such as PHAs, glycogen and even poly-P). This also implies that the biomass present in the activated sludge sample needs to be 're-activated' and acclimatized to the target pH and temperature of interest prior to the execution of the batch activity tests. In any case, the *in situ* execution of the batch activity tests is preferable. The total volume of activated sludge to be collected depends on the number of tests, bioreactor volume and total volume of samples to be collected to assess the biomass activity. Often 10-20 L of activated sludge from full-scale wastewater treatment plants can be collected. On the other hand, samples collected from lab-scale reactors rarely contain more than 1 L because lab-scale systems are usually smaller (from 0.5 to 2.2 L and under certain cases up to 8-10 L) and the maximum volume that can be withdrawn from lab-scale reactors is often set by the daily withdrawal of the excess of sludge from the system (which is directly related to the applied sludge retention time (SRT) and, consequently, defined by the growth rate of the organisms).

2.2.2.3 Activated sludge sample preparation

For batch activity tests performed *in situ*, in principle, the sludge will be merely transferred from the parent bioreactor (in the case of a lab-enriched sludge) or reaction tank (in the case of pilot-scale or full-scale plants) to the fermenter or bioreactor where the activity tests will take place. Usually, the sludge transfer must take place before the end of the reaction phase that precedes the reaction phase of interest. Then, the batch activity test can start as soon as the desired pH, redox conditions and temperature are adjusted. During the adjustment until the test starts, the same or similar conditions to those prevailing when the sludge samples were collected should be kept. This means that sludge samples collected at the end of the anaerobic stage should be kept under anaerobic conditions and therefore must not be aerated or exposed to the presence of any electron acceptor. Similarly, samples collected in the aerobic stage must be aerated and sludge samples collected in the anoxic stage should not be aerated. If desirable, a few milligrams of nitrate can be added to activated sludge samples collected in the anoxic tanks (to a final concentration of ~5 mg NO_3-N L^{-1}) to maintain anoxic conditions as long as it is necessary.

If only EBPR tests will be executed and nitrification tests are not of interest, then a nitrification inhibitor can be added to the sludge sample immediately after the sludge is transferred to the fermenter (e.g. Allyl-N-thiourea: ATU to a final concentration of 20 mg L^{-1}). This will restrain nitrification and consequently (*i*) avoid higher oxygen consumption in aerobic EBPR batch tests and, (*ii*) limit the accumulation of nitrate (or nitrite) if samples are aerated prior to the execution of anaerobic tests. Should the real and actual conditions be assessed, then the corresponding batch activity tests must be conducted right away after sludge collection with the minimum adjustments and stable conditions required (e.g. for pH and temperature). A comprehensive sampling procedure must be carried out before, during, and after the tests to document the results obtained. However, in addition, the execution of batch activity tests under favourable conditions to PAOs is always recommended. This can help to (*i*) assess the EBPR potential that the system can have, (*ii*) benchmark the EBPR plant activity, (*ii*) detect interferences, and (*iv*) contribute to the definition of improvement strategies.

As described elsewhere, interferences to PAOs can be (but are not limited to) the presence of nitrate or nitrite in the aerobic sludge samples collected to execute anaerobic batch tests, the existence of RBCOD in anaerobic samples for the performance of aerobic or anoxic tests, or the detection of nitrite in anoxic samples intended to carry out aerobic batch tests. Thus, if tests are designed to be conducted under favourable conditions to PAOs then such interferences should be avoided. After this, sludge samples can be exposed shortly (for 1 or maximum 2 h) to a pre-treatment or preparation step as a troubleshooting strategy. For instance, to remove the nitrate present in an aerobic sample intended for an anaerobic batch test (~5-10 mg NO_3-N L^{-1}), after collection the sludge can be transferred to the airtight batch bioreactor and gently mixed under non-aerated conditions. The nitrate (and nitrite) concentration can be monitored until it drops below the detection limits. Rapid detection techniques, such as nitrate and/or nitrite detection paper strips (e.g. Sigma-Aldrich), can be rather useful here. Once nitrate is no longer observed, the corresponding anaerobic batch test can start. If RBCOD is detected in a sample taken from an anaerobic tank, the anaerobic conditions can be extended after the sludge is transferred to the airtight bioreactor until no more RBCOD is observed, before the anoxic or aerobic test starts. Anoxic samples to carry out aerobic tests where nitrate is observed do not need pre-treatment since nitrate is innocuous to PAOs under aerobic conditions. However, if nitrite is detected, it must be removed because it has been proven to be rather inhibitory and even toxic to PAOs under certain aerobic conditions (Pijuan *et al.*, 2010; Zhou *et al.*, 2012; Yoshida *et al.*, 2006; Saito *et al.*, 2004; Zeng *et al.*, 2014). To avoid nitrite, a similar approach like the one previously described for the removal of nitrate can be applied.

When the batch tests cannot be performed *in situ* and sludge samples are stored under cold conditions (at around 4 °C), sludge samples need to be 're-activated' because the cold temperature considerably slows down the bacterial metabolism. Due to the particular physiology of PAOs, to reactivate the sludge it must be aerated for 1-2 h at the pH and, particularly, at the temperature of execution of the batch activity tests. This procedure will help to remove residual biodegradable organics. If only EBPR activity tests will be carried out, a nitrification inhibitor should be added (e.g. ATU at a final concentration in the bioreactor of 20 mg L^{-1}) prior to aeration. If a potential interference is detected (e.g. nitrate, nitrite or RBCOD) prior to the 1-2 h aeration period, the sludge reactivation can start with a pre-treatment step at the temperature and pH of interest. Afterwards, even if the objective is only to assess the anoxic or aerobic activity of PAOs, the EBPR activity tests should start with an anaerobic incubation stage using synthetic or real wastewater as a feed. This practice will ensure that the PAOs will have PHAs intracellularly stored to carry out their aerobic or anoxic metabolisms.

Nevertheless, in general, the objective of the experimental plan should be to minimize as much as possible the needs for transport, cooling, storage and reactivation of the sludge (among other potential steps). Whenever possible, it is best to always use 'fresh' sludge (and substrate/media).

2.2.2.4 Substrate

When real wastewater (either raw or settled) is used for the execution of activity tests, it can be fed in a relatively straightforward manner to the bioreactor/fermenter. For normal (regular) conditions, the feeding step takes place at the beginning of the anaerobic stage to favour the VFA uptake by PAOs and the intracellular availability of PHAs. If judged necessary, a rough filtration step (using 10 µm pore size filters) can be used to remove the remaining debris and large particles present in the raw wastewater. If the activity tests need to be performed at different biomass concentrations, the treated effluent from the plant can be collected and used for dilution (assuming that solids effluent concentrations are relatively low, e.g. 20-30 mg TSS L^{-1}). If different carbon or phosphorus sources and concentrations are to be studied, the plant effluent can also be used to prepare a semi-synthetic media containing a RBCOD concentration of between 50 and 100 mg COD L^{-1}.

Batch activity tests are frequently performed with synthetic wastewater: (*i*) to ensure a better control of the experimental conditions, (*ii*) to create the desired redox conditions, (*iii*) to study and assess the effects of different wastewater composition, or (*iv*) to evaluate the inhibitory or toxic effects of certain solutions or compounds. However, such practice can be expensive due to the potentially large amount of chemicals needed.

Depending upon the nature, purpose and sequence of the activity tests (anaerobic, anoxic or aerobic), the carbon and phosphorus concentrations present in a synthetic wastewater can vary since they are usually the subject of removal and study. Moreover, the concentrations may be adjusted proportionally to the length or duration of the test. Usually, concentrations of up to 50 and 100 mg RBCOD L^{-1} are used when activated sludge samples are obtained from full-scale plants and of up to 400 mg L^{-1} in the case of lab-enriched cultures (though higher concentrations are sometimes applied). Regarding the phosphorus concentrations, synthetic solutions can contain none or low P concentrations when assessing the anaerobic P release or up to 100-120 mg PO_4-P L^{-1} when testing the maximum aerobic P-uptake activities. However, regardless of the nature of the activity test, synthetic wastewater must contain the required macro- and micro-elements (especially potassium and magnesium but also calcium, iron, zinc, cobalt, among others) in sufficient amounts and suitable species that PAOs require in order to avoid any metabolic limitation that may jeopardize the outcomes of the batch activity tests (Brdjanovic *et al.*, 1996). A suggested synthetic wastewater recipe for an initial orthophosphate concentration of 20 mg P L^{-1} can contain per litre (Smolders *et al.*, 1994a): 107 mg NH_4Cl, 90 mg $MgSO_4·7H_2O$, 14 mg $CaCl_2·2H_2O$, 36 mg KCl, 1 mg yeast extract and 0.3 mL of a trace element solution (that includes per litre 10 g EDTA, 1.5 g $FeCl_3·6H_2O$, 0.15 g H_3BO_3, 0.03 g $CuSO_4·5H_2O$, 0.12 g $MnCl_2·4H_2O$, 0.06 g $Na_2MoO_4·2H_2O$, 0.12 g $ZnSO_4·7H_2O$, 0.18 g KI and 0.15 g $CoCl·6H_2O$). Overall, it is important to underline that the minimal concentrations of K and Mg need to be proportional to the phosphorus concentration following a molar 1:1:3 Mg:K:P ratio. This is mostly because Mg and K are essential for poly-P formation since they serve as counter-ions in poly-P. If desired, the synthetic wastewater can be concentrated, sterilized in an autoclave (for 1 h at 110 ºC) and used as a stock solution if several tests will be performed in a defined period of time. However, the solution must be discarded if any precipitation or loss of transparency is observed.

For experiments performed with lab-enriched cultures, it is best to execute the tests with the same (synthetic) wastewater used for the cultivation according to the carbon or phosphorus concentrations of the study. Alternatively and similar to full-scale samples, the effluent from the bioreactor can be collected, filtered through rough pore size filters to remove any particles, and used to prepare the required media with the desired carbon and phosphorus concentrations for the execution of the activity tests.

For the execution of (conventional) anoxic tests, nitrate and nitrite solutions can be prepared to create the required anoxic redox conditions. For this purpose, different stock solutions can be prepared using nitrate salts and nitrite salts. However, for practical applications, it is recommended to follow a step-wise approach and carefully monitor their addition so their concentrations in the bulk water do not exceed more than 10 mg L^{-1} in the case of nitrite and 20 mg L^{-1} for nitrate. This will avoid a potential inhibitory effect due to nitrate or nitrite accumulation as described elsewhere (Saito *et al.*, 2004; Yoshida *et al.*, 2006; Pijuan *et al.*, 2010, Zhou *et al.*, 2007, 2012). Moreover, pH levels lower than 7.0, in combination with higher nitrite concentrations, can be comparatively more inhibiting to PAOs because nitrite can be present as free nitrous acid (FNA) - the 'protonated' species of nitrite. At pH 7.0, Zhou *et al.* (2007) and Pijuan *et al.* (2010) observed 50 % inhibition of the anoxic and aerobic metabolism of PAOs at FNA concentrations of 0.01 and 0.0005 mg HNO_2-N L^{-1}, respectively (equivalent to 45 and 2 mg NO_2-N L^{-1} at pH 7.0). Therefore to ensure their availability during the execution of the anoxic tests, the nitrite or nitrate concentrations need to be monitored during the tests and depending on their concentrations, nitrite or nitrate solutions will need to be added in different steps.

When tests are conducted to assess the potential inhibitory or toxic effects of given compounds at different concentrations, concentrated stock solutions can be prepared and added during the test at the concentrations of interest. Tests performed to assess whether the inhibitory or toxic effects are reversible must be carried out after 'washing' the biomass to remove the inhibiting or toxic compound(s). The washing step is often performed by consecutive settling and re-suspension of the sludge sample in a carbon-free media (either fully synthetic or using a treated effluent after filtration) under the redox conditions of interest. Similarly, when maintenance tests need to be executed, a carbon- and phosphorus-free media can be used at the redox conditions and operating conditions of the study.

2.2.2.5 Analytical procedures

Most of the analytical procedures required (for the determination of total P, PO_4, NH_4, NO_2, NO_3, MLSS, MLVSS, among others) should be performed following standardized and commonly applied analytical protocols detailed in Standard Methods (APHA et al., 2012). VFA determination (for acetate, propionate and even other volatile fatty acids) can be conducted by gas chromatography (GC), high pressure liquid chromatography (HPLC) or by applying regular analytical determination protocols. However, contrary to most of the analytical parameters of interest, the determination of PHAs and glycogen requires a more demanding sample preparation and sophisticated equipment and procedures, and in addition, their determination procedures are only to be found in specialized scientific literature. Therefore, the analytical procedures for the determination of PHAs and glycogen are described in more detail in the following paragraphs in this chapter.

- **PHA**

As mentioned earlier, the most common PHA polymers stored by PAOs are poly-β-hydroxy-butyrate (PHB), poly-β-hydroxy-valerate (PHV) and poly-β-hydroxy-2-methyl-valerate (PH_2MV). Their relative presence and stored amount depends on the VFA composition (Ac or Pr) and the type of metabolism involved in the storage (PAO or GAO metabolism). For enriched lab cultures performing EBPR (where PAOs are the dominant organisms composed of more than 90 % of the total biomass), and when Ac is the most abundant VFA, PAOs store VFA mostly as PHB (up to 90 % of PHAs) (Smolders et al., 1994a). However, when Pr is the dominant VFA, then PHV and PH_2MV can be composed of up to 45 % and 53 % of the total PHAs stored, respectively (Oehmen et al., 2005c). When GAOs are present in EBPR systems, the sludge stores higher amounts of PHV too. For instance, lab-enriched GAO cultures (comprising more than 90 % of the total biomass) cultivated with Ac as VFA leads to a PHB and PHV accumulation of around 73 % and 26 %, respectively (Zeng et al., 2003a, Lopez-Vazquez et al., 2007, 2009a), whereas an enriched PAO culture cultivated under similar conditions contains mostly PHB and less than 10 % PHV (Smolders et al., 1994a). Meanwhile, GAO lab-systems fed with Pr result in practically no PHB accumulation, but up to 43 % PHV and 54 % PH_2MV (Oehmen et al., 2006). It is important to underline that the PHA analytical determination technique provides the PHA contents of the MLSS quite accurately. This implies that a precise determination of MLSS is equally important to obtain correct PHA concentrations and to accurately determine net conversions during a biochemical stage.

Compared with full-scale EBPR systems, the determination of PHAs in lab-scale enriched EBPR systems is usually easier since lab-scale systems are smaller and, more importantly, EBPR cultures are enriched with PAOs (> 90 %) (Oehmen et al., 2004, 2006; Lopez-Vazquez et al., 2007) and consequently, the intracellular PHA contents can reach up to 10 % of the total MLSS concentration depending upon the VFA type available (Lopez-Vazquez et al., 2009a). On the other hand, in the best case, the PHA contents accumulated in the mixed biomass from full-scale systems reach between 1 and 2 % of the total MLSS concentration because PAOs (and GAOs) hardly comprise more than 15 % of the total bacterial population (Lopez-Vazquez et al., 2008a). This implies that the analytical determination of PHAs from full-scale samples may not always be suitable, reliable or therefore representative of the direct collection of grab or composite samples. In extreme cases, PHA contents may fall below the detection limit. From an economic perspective and in view of the required resources (in terms of analytical equipment, costs of chemical consumables and highly qualified lab staff), the PHAs determination will probably not be (cost) effective when performed on samples from a full-scale plant. Alternatively, to assess the potential accumulation of PHAs in full-scale systems, real full-scale sludge samples can be used to execute batch activity tests under more favourable and controlled conditions for EBPR that can maximize accumulation of PHAs and facilitate its analytical determination (Lanham et al., 2014). Nevertheless, this latter approach still requires the analytical determination of PHAs to be performed with high precision.

Regarding the analytical determination of the different PHA polymers, it has been a matter of discussion and improvement since the late 1990s (Baetens *et al.*, 2002). So far, the most reliable method involves two slightly different procedures (Oehmen *et al.*, 2005b): (*i*) one for the determination of PHB and PHV polymers, and (*ii*) another for the determination of PHV and PH$_2$MV.

For both determination procedures, activated sludge samples must be collected *in situ* in (15 mL) centrifugation tubes. The sample volume should be sufficient to obtain around 20 mg of TSS. To preserve the sample, 4-5 drops of paraformaldehyde (37 % concentration) have to be added to the plastic centrifugation tube (in a fume hood) prior to collection of the sample, and once the sample is taken, it should be stored temporarily at 0-4 °C for around 2 h. To remove the remaining paraformaldehyde and dissolved solids in the liquid phase, samples need to be washed twice with tap water. The washing steps include: (*i*) centrifugation (for 10 min at 4,500 rpm), (*ii*) careful withdrawal of the supernatant by decanting (if a solid pellet is formed) or otherwise with a pipette, avoiding the removal of any particle or solid, (*iii*) tap water addition (10 mL), and (*iv*) re-suspension with a vortex. After the second washing step, the sample must be centrifuged one more time, and supernatant must be discarded. Afterwards, the sample must be stored at -20 °C and subsequently freeze-dried in a lyophilizer at -80 °C and 0.1 mbar for 48 h (or longer), until the sample is fully dried. Once the sample has been freeze-dried, the digestion, esterification and extraction procedures can start.

As described by Oehmen *et al.* (2005b), for PHB and PHV determination, 20 mg of the freeze-dried sample can be transferred to a digestion tube and added to 2 mL of an acidified methanol solution containing a 3 % sulphuric acid (H$_2$SO$_4$) concentration and approximately 100 mg L^{-1} of sodium benzoate. Afterwards, samples are digested and esterified for 2 h at 100 °C. After digestion and esterification, samples are cooled down to room temperature; distilled water is added and mixed vigorously. 1 h of settling time must be provided to achieve a phase separation. The chloroform phase can be transferred to a vial, dried with 0.5-1.0 g of granular sodium sulphate pellets and separated from the solid phase. Standard solutions can be prepared in parallel at defined concentrations using commercial co-polymers of R-3-hydroxybutyric acid (3HB) and R-3-hydroxyvaleric acid (3HV) copolymer (7:3). After extraction and esterification, 3 µL of the liquid phase can be injected into a chromatograph. Certain recommended characteristics and operating conditions of the chromatograph are: (*i*) to be equipped with a DB-5 column (30 m length × 0.25 mm I.D. × 0.25 µm film), (*ii*) to apply a 1:15 split injection ratio, (*iii*) to use helium (He) as the carrier gas at a flow rate of 1.5 mL min^{-1}, (*iv*) to be equipped with a flame ionization detector (FID) operated at 300 °C with an injection port at 250 °C, and, (*v*) to vary the oven temperature starting at 80 °C for 1 min, increasing 10 °C min^{-1} up to 120 °C, and then to further increase it at a temperature pace of 45 °C min^{-1} up to 270 °C, and hold it at 270 °C for 3 min. When following this procedure and conditions, the PHB and PHV peaks will show up around 2 and 3 min after injection.

Alternatively, for PHB and PHV determination, another procedure followed by Smolders *et al.* (1994a) involves the addition of 20 mg of freeze-dried biomass to 1.5 mL of dichloroethane, and 1.5 mL of concentrated HCl as well as 1-propanol 1:4 (in volume). 1 mg of benzoic acid in the 1-propanol solution is added as internal standard. Samples are digested and esterified for 2 h at 100 °C and, at least every 30 min, samples are vortexed. After cooling, 3 mL of distilled water is added. The contents are mixed vigorously on a vortex and afterwards centrifuged for a few minutes to obtain a satisfactory and well-defined phase separation. About 1 mL of the lower (organic) phase is drawn off and filtered over a small column of dried water-free sodium sulphate into GC sample vials. As a recommendation, 3 standards must be run for every series of 15 samples. When using this method, 1 µL of the lower liquid phase from the solution can be injected into a gas chromatograph equipped and operated as follows: (*i*) using a HP Innowax column (30 m length × 0.32 mm I.D. × 0.25 µm film), (*ii*) applying a 1:10 split injection ratio, (*iii*) using He as the carrier gas (at a flow rate of 6.3 mL min^{-1}), (*iv*) operating a FID at 250 °C, applying an injection temperature of 200 °C, (*v*) with an initial oven temperature of 80 °C kept for 1 min, that increases to 130 °C at temperature pace of 25 °C min^{-1}, and then to 210 °C at 15 °C min^{-1} and finally held at 210 °C for 12 min. This long final time is recommended to elute propylesters of no interest (e.g. from cell wall constituents). Last but not least, the PHA contents of the biomass is reported as a percentage of the MLSS concentrations, which is used to calculate the PHA concentrations.

If the activated sludge samples contain high concentrations of salts, a saline washing solution with a similar osmotic strength like that of the original sample should be used instead of tap water. This will avoid the cytolysis of the cells and preserve the intracellular compounds (such as PHAs and glycogen) avoiding their dissolution and potential loss through the supernatant. However, when a saline washing solution is used, the high concentration of total dissolved solids (TDS) in the remaining liquid (after centrifugation) may precipitate and lead to apparent deviations in the MLSS concentrations of the original sample, from which the PHA contents will be determined. To compensate, a correction factor will be needed to take into account the potential effect of the TDS on the final solids sample when the PHA concentrations are determined.

Although the two previous analytical procedures can be rather accurate for the determination of PHB and PHV, none of them can, without any further modification, be satisfactorily used for the determination of PH$_2$MV (of particular importance when propionate is present as a carbon source in EBPR cultures). Thus, to improve the PH$_2$MV extraction, Oehmen et al. (2005b) recommend applying the same procedure described for PHB and PHV determination, but using an acidified methanol solution containing 10 % H$_2$SO$_4$ (instead of 3 % H$_2$SO$_4$) and extending the digestion phase at 100 °C to 20 h. Since a commercial product to be used as a direct standard for PH$_2$MV determination is not available, Oehmen et al. (2005b) recommended the use of 2-hydroxycaproic acid which is assumed to have a similar relative response to that of PH$_2$MV (based on the fact that these two molecules are isomers of each other). This procedure has proven useful for the simultaneous determination of PHV and PH$_2$MV, but not for PHB. As a consequence, if the three polymers (PHB, PHV and PH$_2$MV) must be determined, the two different determination procedures must be performed. Further details about the analytical PHA determination techniques can be found in the original sources (Baetens et al., 2002; Oehmen et al., 2005b). From a microscopic visualization perspective, Nile blue A stain can be used to qualitatively visualize PHAs and Neisser stain for poly-P (Mino et al., 1998; Mesquita et al., 2013). Further details about the microscopic observation of these and other intracellular polymers and the use of different stains can be found in Chapter 7 on Microscopy.

- **Glycogen**

EBPR cultures utilize glycogen as a source of energy and reducing power for the storage of PHAs. Glycogen ($C_6H_{10}O_5$) is a multi-branched polysaccharide of glucose ($C_6H_{12}O_6$) similar to starch and cellulose but with a different glycosidic bond and geometry between molecules (Dircks et al., 2001; Wentzel et al., 2008). Its relative presence and intracellular storage by EBPR cultures depends on the VFA composition (Ac or Pr), influent P/C ratio, and dominant organisms (either PAOs or GAOs) (Schuler and Jenkins, 2003; Oehmen et al., 2007). In enriched lab PAO cultures cultivated with Ac as carbon source (where PAOs compose of more than 90 % of total biomass), at influent P/C ratio lower than 0.04 mol mol^{-1}, the glycogen fractions can reach up to 20 % of the total MLVSS concentration, whereas at influent P/C ratios higher than 0.04, the glycogen fractions are usually lower than 15 % (Smolders et al., 1995; Schuler and Jenkins, 2003; Welles et al., 2016, submitted). Similarly, lab-enriched PAOs cultures cultivated with Pr as carbon source tend to store less intracellular glycogen that often does not reach more than 15 % MLVSS since PAOs' anaerobic metabolism on Pr requires less glycogen hydrolysis for anaerobic P release and intracellular PHA storage (Oehmen et al., 2005c). Conversely, GAO cultures enriched in the laboratory using Ac or Pr as carbon source can have glycogen fractions as high as 30 % MLVSS regardless of the carbon source fed (Filipe et al., 2001b; Zeng et al., 2003a; Oehmen et al., 2005a,c; Dai et al., 2007; Lopez-Vazquez et al., 2009a). Similar to PHA determination, the determination of the intracellular glycogen content may be easier to estimate in lab-scale systems (where EBPR cultures can comprise more than 90 % of the total microbial population) but it will not be so straightforward in full-scale systems since PAOs (and GAOs) hardly comprise more than 15 % of the total bacterial population (Lopez-Vazquez et al., 2008a). Consequently, the glycogen fractions present in full-scale EBPR systems may hardly reach more than 5 % of the total MLVSS concentrations, which makes its determination more difficult and challenging when compared to lab-scale systems. Nevertheless, it may be still feasible but it requires analytical determination of high precision (Lanham et al., 2014). Glycogen ($C_6H_{10}O_5$) is a multi-branched polysaccharide; it should be hydrolysed and extracted prior to its determination. Thus, different methods have been proposed for the analytical determination of glycogen, ranging from enzymatic hydrolysis tests (Parrou and Francois, 1997) to biochemically-based (Brdjanovic et al., 1997) and through its indirect determination by high-performance liquid chromatography (HPLC) as glucose after an acid hydrolysis and extraction (Smolders et al., 1994a; Lanham et al., 2012). Unfortunately, a direct method is not yet available. For practical reasons and after several improvements throughout the years, the HPLC method after acid hydrolysis and extraction is one of the most frequently applied procedures. The HPLC method, after acid hydrolysis and extraction for the determination of glycogen, consists of the digestion of an activated sludge sample diluted with 6 M HCl, leading to a final HCl concentration of 0.6 M HCl, and digested at 100 °C for 5 h. After digestion, the sample is allowed to cool down to room temperature under quiescent conditions and the supernatant is filtered through 0.2 or 0.45 μm pore size filters. The filtered supernatant is poured into a vial and glycogen can be quantified by HPLC as glucose (Smolders *et al.*, 1994a). The latter is because glycogen ($C_6H_{10}O_5$) shares the same carbon content as glucose ($C_6H_{12}O_6$) (on a carbon mole basis). However, the determination of glycogen as glucose content extracted from the biomass is not entirely accurate as the non-glycogen glucose-containing content of the biomass (cells) will also make a part of the extracted material, and the glycogen of other glycogen-containing populations beside EBPR (e.g. GAOs) will do the same too. Recently Lanham *et al.* (2012) improved the glycogen extraction technique. Freeze-dried samples prepared like those for PHA determination can be used: activated sludge can be collected *in situ*, added to a 15 mL centrifugation tube containing 4-5 drops of formaldehyde (37 % concentrated), stored at 0-4 °C for around 2 h, washed with tap water and freeze-dried. They recommend using a ratio of 1 mg freeze-dried sludge to 1 mL 0.9 M HCl solution to improve the acid hydrolysis and extraction of glycogen for its further determination as glucose. Then, depending upon the sludge aggregation, the sludge samples can be digested for 2 h in the case of flocculant sludge, 5 h for granular sludge and 3 h if the aggregation state is not known or if it varies. Later on, 5 mg of the freeze-dried sample can be added to 5 mL of a 0.9 M HCl solution, digested for 5 h at 100 °C, supernatant filtered through 0.2 μm pore size filters and measured as glucose by HPLC. If the latter procedure is applied, then the 5 mg of the freeze-dried sample should be carefully and precisely weighed and the results will be reported as a percentage of MLSS. The previous HPLC determination technique has proven to be sufficiently accurate and reliable in lab-enriched cultures where EBPR populations are dominant. However, as previously discussed, their determination in sludge samples from full-scale systems may not be accurate enough. Tentatively, Periodic Acid-Schiff (PAS) stain can be used to get a rough microscopic qualitative estimation of glycogen and other carbohydrate granules present in the cells (Mesquita *et al.*, 2013).

2.2.2.6 Parameters of interest

To determine and assess the metabolic activities of PAOs, different stoichiometric ratios and kinetic rates for the anaerobic, anoxic and aerobic stages can be estimated based on the data collected from the execution of the batch activity tests. Table 2.2.1 shows a description of the expected parameters of interest.

2.2.3 EBPR batch activity tests: preparation

This section describes not only the different steps but also the apparatus characteristics and materials needed for the execution of the batch activity tests.

2.2.3.1 Apparatus

1. An (airtight) batch bioreactor or fermenter equipped with a mixing system and adequate sampling ports (as described in Section 2.2.2.1).
2. A nitrogen gas supply (recommended).
3. An oxygen supply (compressed air or pure oxygen sources).
4. A pH electrode (if not included/incorporated in the batch bioreactor setup).
5. A 2-way pH controller via HCl and NaOH addition (alternatively a one-way control - generally for HCl addition - or manual pH control can be applied through the manual addition of HCl and NaOH).
6. A thermometer (recommended temperature working range of 0 to 40 °C).
7. A temperature control system (if not included in the batch bioreactor setup).
8. A DO meter with an electrode (if not included/incorporated in the batch bioreactor setup).
9. An automatic 2-way dissolved oxygen controller via nitrogen and oxygen gas supplies (if not included in the batch bioreactor setup and if tests must be performed at a defined dissolved oxygen concentration).
10. Confirm that all electrodes and meters (pH, temperature and DO) are calibrated less than 24 h before execution of the batch activity tests in accordance with the guidelines and recommendations from the manufacturers and/or suppliers.
11. A centrifuge with a working volume capacity of at least 250 mL to carry out the sludge washing procedure (if required).
12. A stop watch.

ACTIVATED SLUDGE ACTIVITY TESTS

Table 2.2.1 Stoichiometric and kinetic parameters of interest for activated sludge samples performing EBPR.

Parameter	Symbol	Typical unit on a mole basis	Typical unit on a mg or g basis
ANAEROBIC PARAMETERS			
Stoichiometric			
Anaerobic orthophosphate release to VFA uptake ratio	$Y_{VFA\_PO4,An}$	P-mol C-mol^{-1}	mg P mg VFA^{-1}
Anaerobic glycogen utilization to VFA uptake ratio	$Y_{Gly/VFA,An}$	C-mol C-mol^{-1}	mg C mg VFA^{-1}
Anaerobic PHA production to VFA uptake ratio	$Y_{VFA\_PHA,An}$	C-mol C-mol^{-1}	mg C mg VFA^{-1}
Anaerobic PHB formation to VFA uptake ratio	$Y_{VFA\_PHB,An}$	C-mol C-mol^{-1}	mg C mg VFA^{-1}
Anaerobic PHV formation to VFA uptake ratio	$Y_{VFA\_PHV,An}$	C-mol C-mol^{-1}	mg C mg VFA^{-1}
Anaerobic PH$_2$MV formation to VFA uptake ratio	$Y_{VFA\_PH2MV,An}$	C-mol C-mol^{-1}	mg C mg VFA^{-1}
Anaerobic PHV formation to PHB formation ratio	$Y_{PHV/PHB,An}$	C-mol C-mol^{-1}	mg C mg C^{-1}
Kinetic			
Maximum specific anaerobic VFA uptake rate	$q_{VFA,An}$	C-mol C-mol^{-1} h^{-1}	mg VFA mg active biomass^{-1} h^{-1}
Maximum specific anaerobic PO$_4$ release rate	$q_{PP\_PO4,An}$	P-mol C-mol^{-1} h^{-1}	mg P mg active biomass^{-1} h^{-1}
Maximum specific anaerobic PHA production rate	$q_{VFA\_PHA,An}$	C-mol C-mol^{-1} h^{-1}	mg PHA mg active biomass^{-1} h^{-1}
Anaerobic PO$_4$ release maintenance rate	$m_{PP\_PO4,An}$	P-mol C-mol^{-1} h^{-1}	mg P mg active biomass^{-1} h^{-1}
Anaerobic ATP maintenance coefficient	$m_{ATP,An}$	mol ATP C-mol^{-1} h^{-1}	mg ATP mg active biomass^{-1} h^{-1}
Anaerobic secondary PO$_4$ release rate	$m_{PP\_PO4,Sec,An}$	P-mol C-mol^{-1} h^{-1}	mg P mg active biomass^{-1} h^{-1}
ANOXIC PARAMETERS			
Stoichiometric			
Anoxic PHA degradation to NO$_X$ consumption ratio	$Y_{NOx\_PHA,Ax}$	C-mol N-mol^{-1}	mg C mg NO$_X^{-1}$
Anoxic glycogen formation to NO$_X$ consumption ratio	$Y_{NOx\_Gly,Ax}$	C-mol N-mol^{-1}	mg C mg NO$_X^{-1}$
Anoxic poly-P formation to NO$_X$ consumption ratio	$Y_{NOx\_PP,Ax}$	P-mol N-mol^{-1}	mg P mg NO$_X^{-1}$
Anoxic biomass growth to NO$_X$ consumption ratio	$Y_{NOx,Bio,Ax}$	C-mol N-mol^{-1}	mg C mg NO$_X^{-1}$
Anoxic glycogen formation to PHA consumption ratio	$Y_{PHA\_Gly,Ax}$	C-mol C-mol^{-1}	mg C mg C^{-1}
Anoxic poly-P formation to PHA consumption ratio	$Y_{PHA\_PP,Ax}$	P-mol C-mol^{-1}	mg P mg C^{-1}
Anoxic biomass growth to PHA consumption ratio	$Y_{PHA\_Bio,Ax}$	C-mol C-mol^{-1}	mg C mg C^{-1}
Kinetic			
Maximum specific anoxic PHA degradation rate	$q_{PHA,Ax}$	C-mol C-mol^{-1} h^{-1}	mg PHA mg active biomass^{-1} h^{-1}
Maximum specific anoxic glycogen formation rate	$q_{PHA\_Gly,Ax}$	C-mol C-mol^{-1} h^{-1}	mg Gly mg active biomass^{-1} h^{-1}
Maximum specific anoxic poly-P formation rate	$q_{PO4\_PP,Ax}$	P-mol C-mol^{-1} h^{-1}	mg PP mg active biomass^{-1} h^{-1}
Maximum specific anoxic biomass growth rate	$q_{Bio,Ax}$	C-mol C-mol^{-1} h^{-1}	mg active biomass mg active biomass^{-1} h^{-1}
Anoxic ATP maintenance coefficient	$m_{ATP,Ax}$	mol ATP C-mol^{-1} h^{-1}	mg ATP mg active biomass^{-1} h^{-1}
Anoxic endogenous respiration rate	m_{NOx}	N-mol C-mol^{-1} h^{-1}	mg NOx mg active biomass^{-1} h^{-1}
AEROBIC PARAMETERS			
Stoichiometric			
Aerobic PHA degradation to O$_2$ consumption ratio	Y_{PHA}	C-mol mol O$_2^{-1}$	mg C mg O$_2^{-1}$
Aerobic Glycogen formation to O$_2$ consumption ratio	Y_{Gly}	C-mol mol O$_2^{-1}$	mg C mg O$_2^{-1}$
Aerobic Poly-P formation to O$_2$ consumption ratio	Y_{PP}	P-mol mol O$_2^{-1}$	mg P mg O$_2^{-1}$
Aerobic PAO biomass growth to O$_2$ consumption ratio	Y_{PAO}	C-mol mol O$_2^{-1}$	mg C mg O$_2^{-1}$
Aerobic glycogen formation to PHA consumption ratio	$Y_{PHA\_Gly,Ox}$	C-mol C-mol^{-1}	mg C mg C^{-1}
Aerobic Poly-P formation to PHA consumption ratio	$Y_{PHA\_PP,Ox}$	P-mol C-mol^{-1}	mg P mg C^{-1}
Aerobic biomass growth to PHA consumption ratio	$Y_{PHA\_Bio,Ox}$	C-mol C-mol^{-1}	mg C mg C^{-1}
Kinetic			
Maximum specific aerobic PHA degradation rate	$q_{PHA,Ox}$	C-mol C-mol^{-1} h^{-1}	mg PHA mg active biomass^{-1} h^{-1}
Maximum specific aerobic glycogen formation rate	$q_{PHA\_Gly,Ox}$	C-mol C-mol^{-1} h^{-1}	mg Gly mg active biomass^{-1} h^{-1}
Maximum specific aerobic poly-P formation rate	$q_{PO4\_PP,Ox}$	P-mol C-mol^{-1} h^{-1}	mg PP mg active biomass^{-1} h^{-1}
Maximum specific aerobic biomass growth rate	$q_{Bio,Ox}$	C-mol C-mol^{-1} h^{-1}	mg active biomass mg active biomass^{-1} h^{-1}
Aerobic ATP maintenance coefficient	$m_{ATP,Ox}$	mol ATP C-mol^{-1} h^{-1}	mg ATP mg active biomass^{-1} h^{-1}
Aerobic endogenous respiration rate of a culture	m_{O2}	mol O$_2$ C-mol^{-1} h^{-1}	mg O$_2$ mg active biomass^{-1} h^{-1}

2.2.3.2 Materials

1. Two graduated cylinders of 1 or 2 L (depending upon the sludge volumes used) to hold the activated sludge and wash the sludge if required.
2. At least 2 plastic syringes (preferably of 20 mL or at least of 10 mL volume) for the collection and determination of soluble compounds (after filtration).
3. At least 3 plastic syringes (preferably of 20 mL) for the collection of solids, particulate or intracellular compounds (without filtration).
4. 0.45 μm pore size filters. Preferably not of cellulose-acetate because they may release traces of cellulose or acetate into the collected water samples. Consider having at least twice as many filters as the number of samples that need to be filtered for the determination of soluble compounds.
5. 10 or 20 mL transparent plastic cups to collect the samples for the determination of soluble compounds (e.g. soluble COD, acetate, propionate, orthophosphate, nitrate, nitrite).
6. 10 or 20 mL transparent plastic cups to collect the samples for the determination of mixed liquor suspended solids and volatile suspended solids (MLSS and MLVSS, respectively). Consider the collection of these samples by triplicate due to the variability of the analytical technique.
7. 15 mL plastic tubes for centrifugation for the determination of PHAs and/or glycogen.
8. A plastic box or dry ice box filled with ice with the required volume to temporarily store (for up to 1-2 h after the conclusion of the batch activity test) the plastic cups and plastic tubes for centrifugation after the collection of the samples.
9. Plastic gloves and safety glasses.
10. Pasteur or plastic pipettes for HCl and/or NaOH addition (when pH control is carried out manually).
11. Metallic lab clips or clamps to close the tubing used as a sampling port when samples are not collected from the bioreactor/fermenter.

2.2.3.3 Media preparation

- **Real wastewater**

    If real wastewater will be used to carry out the batch activity test, the sample needs to be collected at the influent of the corresponding wastewater treatment plant and the batch activity test performed as soon as possible after collection. Depending on the nature of the test, the researcher should decide whether to take a sample of raw sewage or settled sewage (if the plant employs primary settling). If due to location, transportation issues or other logistics, tests cannot be performed in less than 1 or 2 h immediately after collection, then one should keep the wastewater sample cold until the test is conducted (e.g. by placing the bucket or jerry can in a fridge at 4 °C). Nevertheless, prior to the execution of the test, the temperature of the wastewater needs to be adjusted to the target temperature at which the batch activity test will be executed (preferably reached in less than 1 h). A water bath or a temperature-controlled room can be used for this purpose, as described in Section 2.2.2.1.

- **Synthetic influent media or substrate**

    If tests can be or are desired to be performed with synthetic wastewater, depending on the type of tests (anaerobic, anoxic or aerobic), the synthetic influent media can contain a mixture of carbon and orthophosphate sources plus necessary (macro and micro) nutrients. Generally, they can be mixed all together in the same media (for anaerobic-(anoxic)-aerobic tests); split in two solutions (*i*) C source and (*ii*) P source (plus nutrient solution); or prepared separately if they need to be added in different phases or time. The usual compositions and concentrations are:

    a. Carbon source solution: This is usually composed of a RBCOD source, preferably volatile fatty acids such as acetate or propionate, depending on the nature or goal of the test and the corresponding research questions. Sometimes, more complex substrates are used, containing a mixture of RBCOD and slowly biodegradable COD (SBCOD); however, these are not applied in the tests described in this chapter, and thus are omitted. For anaerobic batch activity tests, the COD concentration in the feed needs to be set to a level that ensures that all the COD is consumed within the anaerobic stage. For batch activity tests performed with activated sludge from a full-scale plant, usually COD concentrations not higher than 100 mg L^{-1} are recommended. For lab-scale activated sludge samples, the COD concentrations can be as high as the influent COD concentration of the lab-scale system (and even sometimes 2 to 3 times higher) as long as the RBCOD fed is fully removed in the anaerobic stage and is not toxic or inhibitory to PAOs.

    b. Orthophosphate source solution: The orthophosphate concentrations can be adjusted as desired depending on the purpose of the experiment. For single anaerobic batch test experiments only, orthophosphate concentrations can be as low as 2-3 mg PO$_4$-P L^{-1} or even be excluded, whereas concentrations as high as 75 mg PO$_4$-P L^{-1} for full-scale samples or more than 120 mg PO$_4$-P L^{-1} for lab-enriched cultures (Wentzel et al., 1987) can be added to assess the maximum P-uptake capacity of sludge under anoxic or aerobic conditions (when an anaerobic phase precedes the anoxic or aerobic test).

c. The nutrient solution: This should contain all the required macro (ammonium, magnesium, sulphate, calcium, potassium) and micronutrients (iron, boron, copper, manganese, molybdate, zinc, iodine, cobalt) to ensure that cells are not limited by their absence and avoid obtaining wrong results and, in extreme cases, the failure of the test. Thus, despite the fact that their concentrations may seem very low, it is necessary to make sure that all of the constituents are added to the solution in the required amounts. The following composition (amounts per litre of nutrient solution) is recommended (based on Smolders et al., 1994a): 107 mg NH$_4$Cl, 90 mg MgSO$_4$·7H$_2$O, 14 mg CaCl$_2$·2H$_2$O, 36 mg KCl, 1 mg yeast extract and 0.3 mL of a trace element solution (that includes per litre 10 g EDTA, 1.5 g FeCl$_3$·6H$_2$O, 0.15 g H$_3$BO$_3$, 0.03 g CuSO$_4$·5H$_2$O, 0.12 g MnCl$_2$·4H$_2$O, 0.06 g Na$_2$MoO$_4$·2H$_2$O, 0.12 g ZnSO$_4$·7H$_2$O, 0.18 g KI and 0.15 g CoCl·6H$_2$O). Similar nutrient solutions can be used as long as they contain all the previously reported required nutrients.

- **Nitrate or nitrite solution**
  When the batch tests comprise an anoxic stage, nitrate or nitrite solutions (as required) can be used to create the anoxic conditions in the batch activity tests (Section 2.2.2.4). Nitrate and nitrite salts can be used for this purpose (e.g. KNO$_3$ or NaNO$_2$, respectively). Nevertheless, their addition must be carefully monitored to ensure their presence and availability without creating any inhibitory or even toxic effect on the biomass. Thus, it is recommended to keep their concentrations below 20 mg NO$_3$-N L^{-1} and 10 mg NO$_2$-N L^{-1} (at pH 7.0).

- **Washing media**
  If the sludge sample must be 'washed' to remove an undesirable compound (which may be even inhibitory or toxic), it is necessary to prepare a nutrient solution to wash the sludge that contains per litre (Smolders et al., 1994a): 107 mg NH$_4$Cl, 90 mg MgSO$_4$·7H$_2$O, 14 mg CaCl$_2$·2H$_2$O, 36 mg KCl, 1 mg yeast extract and 0.3 mL of a trace element solution (that includes per litre 10 g EDTA, 1.5 g FeCl$_3$·6H$_2$O, 0.15 g H$_3$BO$_3$, 0.03 g CuSO$_4$·5H$_2$O, 0.12 g MnCl$_2$·4H$_2$O, 0.06 g Na$_2$MoO$_4$·2H$_2$O, 0.12 g ZnSO$_4$·7H$_2$O, 0.18 g KI and 0.15 g CoCl·6H$_2$O). The washing process can be repeated twice or three times. Afterwards, the following preparation steps of the batch activity tests can be performed. In special cases when sludge from a full-scale plant is used, plant effluent may be used for washing purposes (given that its composition allows for this).

- **Formaldehyde solution**
  To prepare and preserve the samples for the analytical determination of PHAs and glycogen, a commercial formaldehyde solution (37 % concentration) is needed.

- **ATU (Allyl-N-thiourea) solution**
  To inhibit nitrification, an ATU solution can be prepared to reach an initial concentration of around 20 mg L^{-1} (after addition to the sludge). The ATU solution must be added before the sludge is exposed to any aerobic conditions (including the sludge sample preparation or acclimatization).

- **Acid and base solutions**
  These should be 100-250 mL of 0.2 M HCl and 100-250 mL 0.2 M NaOH solutions for automatic or manual pH control, and 10-50 mL of 1 M HCl and 10-50 mL 1 M NaOH solutions for initial pH adjustment if the desired operational pH is very different from the pKa value of the buffering agent.

- The working and stock solutions required to carry out the determination of the analytical parameters of interest must be also prepared in accordance with Standard Methods (APHA et al., 2012) and the corresponding protocols.

- It is recommended to take a sample of the media prior to execution of the experiment to confirm/check the initial (desired) concentration of the parameter(s) of interest (e.g. COD, orthophosphate, etc.).

2.2.3.4 Material preparation

- The number of samples (and their volume) to be collected should be defined in accordance with the type of analysis, the volumes required for the analytical determination and the number of replicates of the dissolved parameters of interest:

a. E.g. a 3 mL sample needs to be collected if this is the minimum volume required to determine the acetate concentration in that sample by duplicate.
  b. When two or more parameters will be determined using the same collected sample, then the required volumes must be summed up: e.g. if 3 mL is needed for Ac, 5 mL for ammonia, and 6 mL for orthophosphate determination, then at least a 14 mL sample must be collected.

  In a well-mixed system, liquid samples are only taken once (although the analysis may be conducted in replicate) because the sampling itself does not usually affect the quality of the sample (unless the samples are not filtered by omission).

- Also, the number of samples (and their volume) to be collected should be defined in accordance with the type of analysis, the volumes or masses required for the analytical determination of the particular parameters of interest and the number of replicates per analysis (e.g. 3 samples of 10 mL each for MLSS/MLVSS determination by triplicate, 20 mg TSS per every PHA analysis and 5 mg TSS for each glycogen analysis). In the case of samples focused on solids concentrations, there is deviation in the solids concentrations from sample to sample due to the nature of sampling procedures. Therefore, the sampling and analysis preferably need to be conducted in triplicate.
- Frequency of sample collection:
  a. If the maximum specific (initial) kinetic rates are to be determined (e.g. maximum specific acetate uptake rate or maximum orthophosphate uptake rate) then a higher number of samples must be taken at the beginning of the corresponding stage or phase (anaerobic, anoxic or aerobic). In particular, samples may need to be collected every 5 min during the first 30-40 min of duration of the batch activity test. The 5 min period is the minimum practical period needed for taking and handling a single (set of) sample(s).
  b. If only the stoichiometric ratios need to be assessed and not the kinetics (e.g. the anaerobic P-released/Ac-uptake ratio or anaerobic PHB formation/Ac uptake ratio), then the samples can be collected only at the beginning and end of each stage or phase to determine the total conversions of interest.
- To increase the data reliability and know the initial conditions of the sludge, it is strongly recommended to collect a series of sludge samples before any media is added. In particular for PHAs and glycogen, which require a long sampling time considering the fast conversion rates of the biomass, similar to MLSS and MLVSS analyses, samples are often collected in triplicate together with MLSS and MLVSS samples.
- Carefully define the maximum and minimum working volumes of the bioreactor:
  a. The estimation of the minimum final volume at the end of the test (after the collection of all the samples) will avoid problems with sampling (e.g. when the height of the final volume is lower than the height of the sampling port/tubing) and controlling the operational conditions (e.g. uncontrolled acid and base addition will occur if the tip of the pH electrode is not submerged leading to a pH shock, without noticing any problems in the reading of the pH meter). It can also prevent inadequate or insufficient aeration and mixing (extremely low volumes can lead to high oxygen intrusion, dead volumes, and biomass losses if splashed on the bioreactor walls).
  b. To estimate the minimum initial volume at the beginning of the test based on the minimum final volume and the volume required for sampling (taking into account the initial activated sludge volume and the addition of media and other solutions) and to verify that this volume does not exceed the maximum working volume of the bioreactor which will allow to avoid spillages and flooding. The potential increase in volume due to gas sparging (e.g. nitrogen or compressed air) must also be considered to define the maximum working volume.
- Once the number and frequency of the sample collections have been defined, then label all the plastic cups. Preferably, define a nomenclature and/or abbreviation that will allow you to easily identify and recognize the batch test, the sampling time and the parameter(s) of interest to be determined with that sample. Labelling both the plastic cups and the cover will help to easily identify the sample.
- A simple working sheet created in a spreadsheet can be rather useful to execute and keep track of the sampling collection and batch test execution. Furthermore, it can be used to keep a database of the different batch tests carried out. Table 2.2.10 contains an example of a working sheet.
- Organize all the required material within a relatively close radius of action around the batch setup so any delay in handling and preparing the samples can be avoided. Otherwise, it will be difficult to respect the initial 5 min frequency of sampling.

- If samples for PHA and glycogen determination are to be collected, carefully add 4-5 drops of the 37 % formaldehyde solution using a plastic Pasteur pipette. Add the formaldehyde solution inside a fume hood or at least in a well ventilated place. After addition, close the tubes immediately and keep them closed until the samples are added. Always wear plastic gloves and treat the used materials contaminated with formaldehyde as chemical residue in accordance with your local lab regulations.
- Calibrate all the meters (pH, DO and thermometer) less than 24 h prior to the execution of the tests and store the electrode/sensors in appropriate solutions until the execution of the tests, following the particular recommendations of the corresponding manufacturer or supplier and confirm that the readings are reliable.
- One should be aware that each sample taken needs immediate attention, handling and proper storage, before the next sample is taken (Table 2.2.2).

Table 2.2.2 Suggestions for the storage and preservation of samples as a function of the analytical determination of the parameter of interest.

EBPR-related parameter of interest	Material of sample container	Method of preservation	Maximum recommended time between sampling, preservation procedure and analysis
Total BOD	Plastic or glass	Cool to 1-5 °C; or, freeze to -20 °C and store in the dark.	24 h for samples stored at 1-5 °C; up to 1 month for frozen samples.
Soluble/dissolved BOD	Plastic or glass	Filter immediately after collection through 0.45 μm pore size filters and cool to 1-5 °C; or, freeze to -20 °C and store in the dark.	24 h for samples stored at 1-5 °C; up to 1 month for frozen samples.
Total COD	Plastic or glass	Cool to 1-5 °C; or, add concentrated H_2SO_4 to lower pH to 1-2, freeze to -20 °C and store in the dark.	24 h for samples stored at 1-5 °C; up to 6 months for acidified samples frozen and stored in the dark.
Soluble/dissolved COD	Plastic or glass	Filter immediately after collection through 0.45 μm pore size filters, cool to 1-5 °C.	24 h for samples stored at 1-5 °C; up to 6 months for acidified samples frozen and stored in the dark.
VFA	Plastic or glass	Filter immediately after collection through 0.45 μm pore size filters, cool to 1-5 °C; or, add concentrated H_2SO_4 to lower pH to 1-2, freeze to -20 °C and store in the dark.	24 h for samples stored at 1-5 °C; up to 6 months for acidified samples frozen and stored in the dark.
Total P	Plastic or glass acid washed (0.1 M HCl)	Add concentrated H_2SO_4 to lower pH to 1-2 and freeze to -20 °C.	6 months.
PO_4	Plastic or glass acid washed (0.1 M HCl)	Filter immediately after collection through 0.45 μm pore size filters and cool to 1-5 °C.	24 h.
NH_4	Plastic or glass	Filter immediately after collection through 0.45 μm pore size filters, cool to 1-5 °C; or, add concentrated H_2SO_4 to lower pH to 1-2, freeze to -20 °C and store in the dark.	24 h for samples stored at 1-5 °C; up to 21 days for acidified samples frozen and stored in the dark.
NO_2	Plastic or glass	Filter immediately after collection through 0.45 μm pore size filters, cool to 1-5 °C.	24 h.
NO_3	Plastic or glass	Filter immediately after collection through 0.45 μm pore size filters, cool to 1-5 °C; or, add concentrated HCl to lower pH to 1-2, freeze to -20 °C and store in the dark.	24 h for samples stored at 1-5 °C; 7 days for acidified samples frozen and stored in the dark.
PHAs	Plastic	After corresponding sampling and preparation procedure (see Section 2.2.2.5) store at -20 °C or -80 °C; freeze-dried samples can also be stored at -20 or -80 °C.	Up to 6 months.
Glycogen	Plastic or glass	After corresponding sampling and preparation procedure (see Section 2.2.2.5) store at -20 °C; after digestion also store at -20 °C.	Up to 6 months.
MLSS	Plastic or glass	Cool to 1-5 °C.	2 days.
MLVSS	Plastic or glass	Cool to 1-5 °C.	24 h.

2.2.3.5 Activated sludge preparation

These procedures consider that batch activity tests can be performed as soon as possible after the collection of samples from full- or lab-scale systems or, in the worst-case scenario, within 24 h after collection. The execution of batch tests 24 h after the collection of activated sludge samples is not recommended due to potential changes that EBPR culture can experience during handling (unless the exposure time after collection is of particular interest for the execution of tests). Bearing in mind the previous comments, the following three procedures are recommended to prepare the activated sludge samples for the execution of batch activity tests:

- If batch activity tests can be executed in less than 1 h after collection of the sludge sample and if the sludge sample does not need to be washed:
  a. Adjust the temperature of the batch bioreactor where the tests will take place to the target temperature of the study.
  b. Collect the sludge:
     i. At the end of the aerobic tank or stage to carry out anaerobic batch tests.
     ii. At the end of the anaerobic stage or tank to perform anoxic or aerobic batch tests.
  c. Transfer the sludge sample to the bioreactor or fermenter where the batch activity tests will take place.
  d. Add the ATU solution (if applicable for the objective of the test) to a final concentration of 20 mg L^{-1} (see Section 2.2.3.3).
  e. Start a gentle mixing (50 - 100 rpm) and follow the temperature of the sludge sample by placing an external thermometer inside the bioreactor (if the setup does not have a built-in thermometer).
  f. Keep mixing until the sludge has reached the target temperature of the study.
  g. Keep the same redox conditions prevailing during collection until the batch activity tests are ready to start:
     i. Avoid the aeration of samples collected under anaerobic conditions. If available, use an airtight bioreactor and sparge nitrogen gas to avoid/reduce oxygen intrusion.
     ii. Avoid the aeration of samples collected under anoxic conditions and preferably add a nitrate solution to a final concentration of around 10 mg NO_3-N L^{-1} to preserve the anoxic environment while mixing.
     iii. Aerate the sludge samples collected in the aerobic tank or under aerobic conditions, keeping a dissolved oxygen concentration higher than 2 mg L^{-1}.
  h. Sludge for the execution of anaerobic batch tests: if nitrate is detected in the aerobic samples collected to perform these tests (see Section 2.2.2.3), the aeration must stop and the sludge sample should be gently mixed until nitrate is no longer observed. As soon as nitrate is no longer observed, the sludge can be immediately used to execute the anaerobic batch activity tests. Nitrate or nitrite detection strips (Sigma-Aldrich) can be used for a quick estimation of the presence of these compounds.
  i. Sludge to perform anoxic and aerobic batch activity tests can then be used to carry out the tests within 1 h of collection.

- If batch activity tests can be conducted in less than 1 h after collection of the sludge sample but the activated sludge sample needs to be washed:
  a. Adjust the temperature of the batch bioreactor where the tests will take place to the target temperature of the study. Collect the sludge at the end of the aerobic tank or stage.
  b. Wash the sludge in a mineral solution (see Section 2.2.3.3) as follows:
     i. Separate the biomass by settling or mild centrifugation (2,000-3,000 rpm for 5 min) and carefully remove the supernatant volume while avoiding losing the biomass.
     ii. Replace the supernatant volume with the same volume of nutrient solution and mix gently for 5 min.
     iii. Repeat the previous washing procedure at least one more time.
     iv. After the last washing cycle, separate the biomass by settling or mild centrifugation (2,000-3,000 rpm for 5 min) and re-suspend the sludge in the same volume of mineral solution previously added.
     v. The washed sample should have the same MLSS concentration as in the lab- or full-scale system where it was taken from. Thus, define and adjust the volume of the mineral solution added to re-suspend the washed sludge in order to reach the same MLSS concentration like in the original source.
  c. Transfer the washed sludge sample to the bioreactor or fermenter where the batch activity tests will take place.
  d. Add the ATU solution (if applicable for the objective of the tests) to a final concentration of 20 mg L^{-1} (see Section 2.2.3.3).

e. Start a gentle mixing (50-100 rpm) and follow the temperature of the sludge sample by placing an external thermometer inside the bioreactor (if the setup does not have a built-in thermometer).
f. Keep mixing until the sludge is exposed to the target temperature of the study for at least 30 min.
g. Start to aerate the sludge sample, keeping a dissolved oxygen concentration not lower than 2 mg L^{-1} while providing a gentle mixing until the batch activity test starts.
h. Due to the exposure of the sludge to aerated and non-aerated conditions, it is only recommended to use this sludge to execute batch activity tests that start with an anaerobic phase (anaerobic-(anoxic)-aerobic). The execution of batch activity tests that start with an anoxic or aerobic phase is not recommended because the washing steps may decrease the intracellular PHA contents and poly-P contents due to handling during the washing steps.

- If due to location and distance issues, the tests cannot be performed within less than 1 or 2 h after collection (but within 24 h):
  a. Adjust the temperature of the batch bioreactor to the target temperature of the study.
  b. Keep the sludge sample cold until the test is executed (e.g. by placing the bucket or jerry can in a fridge at 4 °C), avoiding aerating the sludge sample.
  c. Prior to the execution of the test, take the sludge sample out of the fridge, cool box or cold room.
  d. Mix the content gently in order to obtain a homogenous and representative sample with a similar MLSS concentration as in the original lab- or scale system where it was collected from.
  e. If the sample needs to be washed, wash the sludge in a mineral solution (see Section 2.2.3.3) as follows (otherwise, skip the washing step):
     i. Separate the biomass by settling or mild centrifugation (2,000-3,000 rpm for 5 min) and carefully remove the supernatant volume while avoiding losing the biomass.
     ii. Replace the supernatant volume with the same volume of nutrient solution and mix gently for 5 min.
     iii. Repeat the previous washing procedure at least one more time.
     iv. After the last washing cycle, separate the biomass by settling or mild centrifugation (2,000-3,000 rpm for 5 min) and re-suspend the sludge in the same volume of nutrient solution previously added.
     v. Since the washed sample should have the same MLSS concentration as in the lab- or full-scale system where it was taken from, define and adjust the volume of the mineral solution to obtain the same MLSS concentration as in the original source.
  f. Transfer the washed sludge to the bioreactor or fermenter where the batch activity test will take place.
  g. Add the ATU solution to a final concentration of 20 mg L^{-1} (see Section 2.2.3.3).
  h. Start to aerate the sludge sample, keeping the dissolved oxygen concentration higher than 2 mg L^{-1} while mixing gently.
  i. Follow the temperature of the sludge sample by placing an external thermometer inside the bioreactor (if the setup does not have a built-in thermometer).
  j. Keep aerating and mixing for at least 1 h (maximum 2 h) but ensure that the sludge is exposed to the target temperature of the study for at least 30 min.
  k. If nitrate is detected after the procedure to adjust the temperature, (see Section 2.2.2.3 on activated sludge sample preparation), the aeration must stop. Mix gently until nitrate is no longer observed. Afterwards, the sludge can be immediately used to start and execute an anaerobic phase test.
  l. If the sample was 'washed', it is only recommended to use this sludge to execute batch activity tests that start with an anaerobic phase (e.g. anaerobic-(anoxic)-aerobic), because the washing steps may decrease the intracellular PHA contents and poly-P contents due to handling during the washing steps. Thus, an anaerobic stage is always needed to replenish the PHA contents of the biomass.

2.2.4 Batch activity tests: execution

Once the experimental setup, materials, solutions, and activated sludge are ready, the corresponding batch activity test can be conducted. To facilitate the execution and for data track record and archiving purposes, an experimental implementation plan should be prepared in advance. Table 2.2.10 presents a template for an experimental implementation plan that can be used with necessary modifications for the execution of each of the batch activity tests described in the following sections. Due to the particular metabolism of EBPR cultures, EBPR batch activity tests can range from anaerobic to anoxic and aerobic tests, including different combinations among them, depending on the purpose or goal of the test.

Thus, the following EBPR batch activity tests are presented in this chapter:

Test code no.	Redox conditions	Short description and purpose
EBPR.ANA.1	Anaerobic	Executed under the absence of an external carbon source to assess the endogenous anaerobic maintenance activity of EBPR cultures.
EBPR.ANA.2	Anaerobic	Performed after the addition of a defined concentration of carbon to determine the maximum anaerobic activity of EBPR cultures.
EBPR.ANA.3	Anaerobic	Carried out after the addition of a carbon source in excess to estimate the maximum activity of EBPR cultures under non-limiting carbon conditions.
EBPR.ANOX.1	Anoxic	Performed with activated sludge samples collected at the end of the anaerobic stage/phase.
EBPR.ANOX.2	Combined anaerobic-anoxic	Anaerobic-anoxic test conducted with sludge collected at the end of an aerobic stage.
EBPR.AER.1	Aerobic	Performed with sludge collected at the end of an anaerobic or anoxic stage to assess the aerobic EBPR activity.
EBPR.AER.2	Combined anaerobic-aerobic	Anaerobic-aerobic test conducted with sludge collected at the end of an aerobic stage.
EBPR.AER.3	Combined anaerobic-anoxic-aerobic in series	Anaerobic-anoxic-aerobic test carried out with sludge collected at the end of an aerobic stage to assess the sequential anaerobic, anoxic and aerobic EBPR activities.
EBPR.AER.4	Combined anaerobic-anoxic-aerobic in parallel	After the conduction of a common anaerobic phase, one anoxic and one aerobic test are performed in parallel with the same sludge to assess the anoxic and aerobic EBPR activities for comparison purposes.

2.2.4.1 Anaerobic EBPR batch activity tests

The length of an anaerobic batch test can last from 1 h to more than 8 h. Biomass is sensitive to pH and temperature, so the tests should be conducted at the temperature and pH of interest and fluctuations should be avoided. Tests should be executed under the absence of any electron acceptor (molecular oxygen, nitrate or nitrite) (e.g. truly anaerobic conditions). To avoid or minimize oxygen intrusion, it is recommended to use airtight reactors/fermenters and, if available, sparge N_2 gas continuously throughout the execution of the test. To remove nitrate present in the sample, ATU must be added during the sample preparation (before the sample is aerated) and the sludge can be gently mixed for a few minutes to remove any residual nitrate. Depending upon the purpose of the anaerobic batch activity test, the availability and presence of electron donors can vary. The following anaerobic tests are widely performed:

1. **Test EBPR.ANA.1** Performed under the absence of external carbon source to assess the endogenous anaerobic maintenance activity of EBPR cultures.
2. **Test EBPR.ANA.2** Carried out after the addition of a defined concentration of carbon (which should be fully consumed within the duration of the anaerobic test): to determine the maximum carbon uptake rate, maximum P-release rate, half-saturation constant for carbon uptake, and associated anaerobic stoichiometry such as P-released to carbon consumed ratio.
3. **Test EBPR.ANA.3** Executed after the addition of a carbon source in excess of the concentration that an EBPR culture could consume within the duration of the test: to estimate the maximum concentration of phosphorus that can be released and the maximum concentration of carbon that the EBPR cultures can consume under non-limiting carbon conditions.

Since the presence and type of carbon source play a major role in EBPR processes, sludge collection and preparation are of major importance. It is preferable to execute these tests with activated sludge collected at the end of the aerobic stage (to minimize the presence of an originally present carbon source) and/or after following a washing procedure. Thus, the following protocols for the execution of anaerobic batch tests are suggested depending upon the presence and availability of an external carbon source:

Test EBPR.ANA.1 Anaerobic batch EBPR tests performed under the absence of an electron donor

a. After sludge has been collected, prepared and transferred to the batch bioreactor (see Section 2.2.3.5), keep the sample aerated for at least 30 min while confirming that the pH and temperature are at the target value of interest. Otherwise, set up the corresponding set points (if automatic pH and temperature controllers are applied) or adjust manually. Wait until stable conditions are reached.
b. Once stable operating conditions are reached, around 20 min before the start of the test take the first samples of the water phase and biomass to determine the initial concentrations of the parameters of interest: C-source, total P, PO_4 and MLSS and MLVSS concentrations. Samples for the determination of PHAs and glycogen can also be collected to assess the anaerobic stoichiometric conversions. It is recommended to also take samples of the media to check and verify the initial concentrations.
c. For sampling, connect the syringe, open or release the lab clip or clamp that closes the sampling port, and pull and push the syringe several times until a homogenous sample is collected (usually around 5 times are required). Next, when the syringe is full, close the clip and remove the syringe.
d. Samples for the determination of soluble components must be immediately filtered (through 0.45 µm pore size filters). Other samples (e.g. PHAs, glycogen) need to be prepared in accordance with the corresponding protocols explained earlier in the chapter.
e. During the test execution, temporarily store the samples at 4 °C in the fridge or preferably in a cool box with ice.
f. 10 min before the start of the test stop the aeration and close the bioreactor.
g. If available, start to sparge N_2 gas and continue sparging until the end of the batch test. Provide an adequate outlet to allow the exit of N_2 gas and avoid building up overpressure (which can lead to flooding and biomass loss). If a continuous N_2 gas sparging throughout the batch test is not feasible, sparge N_2 gas for 10 min (alternatively another suitable gas could be used to remove the oxygen present and avoid its intrusion) and afterwards keep the bioreactor under airtight conditions.
h. Start the execution of the anaerobic test at 'time zero'. Keep track of the execution time with the stopwatch.
i. Duration and sampling:

    i. If only the anaerobic endogenous maintenance P-release rate must be determined, the test can last for 1 h or maximum 2 h with continuous sampling every 15 min for the determination of PO_4 released by biomass (PAOs) under the absence of external carbon.
    ii. If the anaerobic endogenous maintenance glycogen conversion rate and stoichiometric conversions are also of interest, the anaerobic tests must be extended for up to 6 and 8 h (recommended). Samples for PO_4 determination can be collected every 30 min together with samples for PHA and glycogen determination.
    iii. Conclude the anaerobic test with the collection of samples for the determination of COD, Total P, PO_4 and MLSS and MLVSS concentrations, as well as for PHAs and glycogen (if applicable).

j. Ensure that considerable temperature and pH variations (higher than 1 °C or ± 0.1, for temperature and pH, respectively) do not take place during the execution of the batch test, and that DO readings remain below the detection limits. Make sure that all the electrodes used are recently calibrated before the execution of the test.
k. Organize the samples and ensure that all the samples are complete and properly labelled to avoid mixing of samples and other trivial mistakes.
l. Until the collected samples are analysed, preserve and store them as recommended by the corresponding analytical procedures.
m. Clean up the apparatus and take appropriate measures to keep and preserve the different sensors, equipment and materials.
n. Keep (part of) the sludge used in the test for possible further use (e.g. for microbial identification, see chapters 7 and 8).

Test EBPR.ANA.2 Anaerobic batch EBPR tests performed under a defined addition of an electron donor

a. Repeat steps 'a' to 'g' from Test EBPR.ANA.1.
b. For anaerobic tests performed with the presence of an external carbon source (electron donor), the tests start at 'time zero' with the addition of the real or synthetic wastewater (as a carbon source solution).
c. To execute the tests, the following MLVSS and RBCOD concentrations are suggested depending on the origin of the sludge samples:
    i. Sludge samples from full-scale activated sludge systems: add the RBCOD source to reach an initial RBCOD-to-MLVSS ratio in the bioreactor of between 0.025 and 0.050 mg COD mg VSS^{-1}.

For instance, mix the RBCOD source with the fresh activated sludge in such a way that the initial RBCOD concentration in the bioreactor is between 50 and 100 mg COD L^{-1} and the initial MLVSS concentration is around 2,000 mg VSS L^{-1}. A low RBCOD-to-MLVSS ratio is preferable to ensure the RBCOD consumption for the duration of the anaerobic test.

  ii. Sludge samples from lab-scale activated sludge systems: the RBCOD source can be added to reach an initial RBCOD-to-MLVSS ratio in the bioreactor of between 0.05 and 0.10 mg COD mg VSS^{-1}. For example, the initial RBCOD and MLVSS concentrations after mixing can range between 100 and 300 mg COD L^{-1} and 2,000 and 3,000 mg VSS L^{-1}, respectively. Higher concentrations may also be acceptable as long as the COD is fully consumed within the length of the anaerobic stage. However, avoid COD concentrations higher than 800 mg COD L^{-1} because this may be inhibitory to biomass (author's personal observations).

d. After the addition of the wastewater, keep track of the execution and sampling times with a stopwatch.
e. Duration and sampling:
  i. Tests can last between 2 and 4 h.
  ii. To determine the anaerobic kinetic parameters, samples for the determination of soluble COD and PO_4 should be collected every 5 min in the first 30-40 min of execution of the test. After this period, the sampling frequency can be reduced to 10 or 15 min during the first 1 h, and later on to every 15 or 30 min until the test is finished.
  iii. If the anaerobic kinetic conversion of PHAs and glycogen is of interest, samples should be collected at the same time as the samples for COD and PO_4 determination.
  iv. For stochiometric conversions, samples for PHA and glycogen determination must be taken at the beginning and end of the test.
  v. Conclude the anaerobic test with the collection of samples for the determination of C-source and/or COD (as desired depending upon the parameters of interest), total P, PO_4 and MLSS and MLVSS concentrations, as well as for PHAs and glycogen (if applicable).
f. Repeat steps 'j' to 'n' from Test EBPR.ANA.1.

Test EBPR.ANA.3 Anaerobic batch EBPR tests performed after the addition of an electron donor in excess

a. Repeat steps 'a' to 'g' from Test EBPR.ANA.1.
b. For anaerobic tests performed with the presence of an external carbon source (electron donor), the tests start at 'time zero' with the addition of the real or synthetic wastewater (as a carbon source solution).
c. To execute the tests, the following MLVSS and RBCOD concentrations are suggested depending on the origin of the sludge samples:
  i. Sludge samples from full-scale activated sludge systems: Perform the anaerobic tests with an initial RBCOD-to-MLVSS ratio in the bioreactor higher than 0.15 mg COD mg VSS^{-1} after mixing the RBCOD source and the sludge. For instance, for sludge samples with MLVSS concentrations of around 2,000, add 300 mg COD L^{-1} of the RBCOD source of interest. Higher concentrations can be added but avoid adding more than 800 mg COD L^{-1} since this has been shown to be inhibitory for EBPR cultures (author's personal observation). If all RBCOD is consumed, more RBCOD can be added until it is not totally consumed. Monitoring the PO_4 concentration's profile during the execution of the test can be used as an indirect method to assess whether the sludge has reached its maximum RBCOD removal/uptake capacity. This can be applied if after an additional dose of RBCOD source there is not a considerable increase in PO_4 concentrations (e.g. less than 2-3 mg PO_4-P L^{-1} during 30-60 min, which corresponds to anaerobic endogenous P release).
  ii. Sludge samples from lab-scale activated sludge systems: similarly to full-scale sludge samples, apply an initial RBCOD-to-MLVSS ratio in the bioreactor higher than 0.2 mg COD mg VSS^{-1} (after mixing the COD source and the sludge). Lab-scale cultures, particularly from EBPR systems, have a considerably high RBCOD removal capacity, which may require repeating the COD addition more than twice until no further COD uptake is observed. For instance, for sludge samples with a MLVSS concentration of between 2,000 and 3,000 mg L^{-1}, add at least 400 mg COD L^{-1} of the RBCOD source of interest, but avoid adding more than 800 mg COD L^{-1} (due to the potentially inhibitory effects on EBPR cultures). If all RBCOD is consumed, more RBCOD can be added until a residual COD is observed. Monitoring the PO_4 concentrations profile during

the execution of the test can be used as an indirect method to assess whether the sludge has reached its maximum RBCOD removal capacity. This approach can be applied if after an additional dose of RBCOD source there is not a considerable increase in PO$_4$ concentrations (e.g. less than 2-3 mg PO$_4$-P L^{-1} after 30-60 min).
d. Start to keep track of the execution time with a stopwatch just after the addition of the wastewater.
e. Duration and sampling:
  i. Tests can last more than 2-4 h for samples from full-scale systems and even longer for lab-scale systems depending on the poly-P and glycogen content of the biomass.
  ii. To determine the anaerobic kinetic parameters, samples for the determination of C-source and/or COD (depending upon the parameters of interest) and PO$_4$ must be collected every 5 min in the first 30 min of execution of the test. After this period, the sampling frequency can be reduced to every 10 or 15 min during the first 1 h, and later on to every 15 or 30 min until the test is finished.
  iii. If the anaerobic kinetic conversions of PHAs and glycogen are of interest, samples can be collected at the same time as the samples for C-source and/or COD and PO$_4$ determination.
  iv. For stochiometric conversions, samples for the determination of PHAs and glycogen must be taken at the beginning and end of the test.
  v. Prior to the end of the tests, add an additional concentration of COD. Wait for 10-15 min and take the last sample for COD and PO$_4$ determination. If no additional COD uptake and P release are observed, this will help to confirm whether the test was indeed performed under non-limiting COD conditions.
  vi. Conclude the anaerobic test with the collection of samples for the determination of C-source and/or COD, total P, PO$_4$ and MLSS and MLVSS concentrations, as well as for PHAs and glycogen (if applicable).
f. Repeat steps 'j' to 'n' from Test EBPR.ANA.1.

2.2.4.2 Anoxic EBPR batch tests

Anoxic EBPR batch tests are conducted to assess the simultaneous removal of orthophosphate and nitrate (or nitrite) by PAOs. They can be performed with activated sludge samples from full- or lab-scale systems, using real or synthetic wastewater/solutions. It is important that the EBPR biomass has enough intracellularly stored PHAs available when exposed to anoxic conditions as a carbon and energy source for P-uptake, glycogen formation, biomass growth and maintenance (Figure 2.2.1). In any case, no external carbon source should be added. Sludge samples can be collected at the end of the anaerobic stage or phase, as long as RBCOD has been fully consumed (and therefore absent from the activated sludge). To ensure the availability of PHAs, it is strongly advised to avoid washing the activated sludge or biomass in between the sludge collection and handling at the end of the anaerobic stage and the start of the anoxic batch test. Under exceptional circumstances (e.g. the suspected presence of toxic compounds), activated sludge samples may be washed as long as strictly anaerobic conditions can be created during the washing procedure (a complicated procedure in practice). Instead, the sludge can be collected at the end of the aerobic stage and the anoxic EBPR test can start with a preceding anaerobic batch test under the defined addition of an electron donor (similar to Test EBPR.ANA.2, described in Section 2.2.4.1). Thus, the following anoxic activity batch tests are suggested:

1. **Test EBPR.ANOX.1** Single anoxic EBPR test performed with activated sludge samples collected at the end of the anaerobic stage/phase.
2. **Test EBPR.ANOX.2** Combined anaerobic-anoxic EBPR batch test. The anoxic EBPR batch test is conducted after a preceding anaerobic phase with sludge collected at the end of the aerobic stage.

The description of each test is presented below:

Test EBPR.ANOX.1 Single anoxic EBPR batch tests

a. Collect the activated sludge sample in the full- or lab-scale plant at the end of the anaerobic stage. Prepare and transfer it to the batch bioreactor as described in Section 2.2.3.5. Keep the sample under anaerobic conditions for at least 30 min while confirming that the pH, DO and temperature are at the target values of interest (otherwise adjust and wait until stable conditions are reached).
b. Repeat steps 'b' to 'e' from Test EBPR.ANA.1.
c. If available, sparge N$_2$ gas continuously until the end of the batch test. Provide an adequate outlet to allow the exit of N$_2$ gas and avoid any overpressure. If continuous N$_2$ gas sparging throughout the batch test cannot be applied, sparge N$_2$ gas for 10 min (alternatively another

suitable gas can be used to remove the oxygen present and avoid its intrusion) and afterwards keep the bioreactor airtight.

d. Start the execution of the anoxic EBPR test at 'time zero' with the addition of nitrate or nitrite (depending upon the final electron acceptor of interest). The same operating conditions like those applied for the anaerobic tests must be applied to avoid the intrusion of oxygen (see Test EBPR.ANA.1).

e. Duration and sampling of the anoxic EBPR batch test:

   i. The anoxic EBPR test can last between 2 and 4 h.

   ii. Depending upon the final electron acceptor of interest:

   - Anoxic EBPR tests performed with nitrate (NO_3^-) as the final electron acceptor:

   For activated sludge samples exposed to the presence of nitrate, up to 20 mg NO_3-N L^{-1} can be added at the beginning of the anoxic stage. On the other hand, it is not recommended to add more than 20 mg NO_3-N L^{-1} to activated sludge samples not regularly exposed to high nitrate concentrations. Nevertheless, if all the nitrate is consumed, an additional 20 mg NO_3-N L^{-1} can be added to extend the length of the anoxic stage until no further anoxic P-uptake is observed.

   - Anoxic EBPR tests performed with nitrite (NO_2^-) as the final electron acceptor:

   Usually, activated sludge samples are not exposed to high nitrite concentrations, unless they are acclimatized to the presence of this electron acceptor (in exceptional/particular cases). Thus, for activated sludge samples exposed to the presence of nitrite, up to 20 mg NO_2-N L^{-1} can be added at the beginning of the anoxic stage. On the other hand, it is not recommended to add more than 10 mg NO_2-N L^{-1} to activated sludge samples not regularly exposed to high nitrite concentrations. In the latter case, if all the nitrite is consumed, an additional 10 mg NO_2-N L^{-1} can be added to ensure the availability of electron acceptor and the anoxic stage can be extended until no further anoxic P-uptake is observed.

   iii. Since the analytical determination of nitrate or nitrite can not be determined as quickly as needed to monitor their presence during the test, nitrate and/or nitrite detection strips (Sigma-Aldrich) can be used to rapidly estimate their presence and to a certain degree of accuracy their concentration, and to assess whether the anoxic conditions are still present or if additional nitrate or nitrite must be dosed.

   iv. For the estimation of the anoxic kinetic parameters, samples for the determination of C-source (or soluble COD depending upon the analytical parameter of interest) and PO_4 must be collected every 5 min in the first 30 min of execution of the test. After this period, the sampling frequency can be reduced to every 10 or 15 min during the first 1 h, and later on to every 15 or 30 min until the test is finished.

   v. If the anoxic kinetic conversions of PHAs and glycogen are of interest, samples can be collected at the same time as the samples for PO_4 determination.

   vi. To estimate the anoxic stochiometric conversions, samples for the analytical determination of total P, PO_4, NO_3 (or NO_2, and MLSS and MLVSS concentrations, as well as for PHAs and glycogen (if applicable), must be collected both at the start and at the end of the anoxic phase.

f. Conclude the anoxic test with the collection of the last samples needed for the estimation of the anoxic stoichiometric conversions.

g. Repeat steps 'j' to 'n' from Test EBPR.ANA.1.

Test EBPR.ANOX.2 Combined anaerobic-anoxic EBPR batch tests

a. Repeat steps 'a' to 'g' from Test EBPR.ANA.1 and steps 'b' to 'f' from Test EBPR.ANA.2. Afterwards, continue with the execution of the anoxic stage.

b. Immediately after the anaerobic stage is completed, the anoxic EBPR batch test can start immediately with the addition of nitrate or nitrite (Test EBPR.ANOX.1, step 'd'). The same operating conditions like those applied for anaerobic tests (e.g. Test EBPR.ANA.1) must be applied to avoid the intrusion of oxygen.

c. Duration and sampling of the anoxic EBPR batch test: repeat step 'e' from Test EBPR.ANOX.1.

d. Repeat steps 'j' to 'n' from Test EBPR.ANA.1.

2.2.4.3 Aerobic EBPR batch tests

Aerobic EBPR batch tests can be executed to assess the orthophosphate uptake by PAOs in activated sludge systems. They can be performed with activated sludge samples from full- or lab-scale systems, using real or synthetic wastewater/solutions. Similar to anoxic BPR tests, it is important that the EBPR biomass has enough intracellularly stored PHAs available when exposed to aerobic conditions as a carbon and energy source for P-uptake, glycogen formation, biomass growth and maintenance (Figure 2.2.1). To ensure the availability of intracellularly stored PHAs, aerobic batch tests must be (*i*) performed with activated sludge samples collected at the end of the anaerobic or anoxic stage/phase (depending upon the wastewater treatment plant configuration), or, (*ii*) carried out after an anaerobic batch test and before the aerobic test is conducted. In any case, no external carbon source should be added during the aerobic phase. Thus, samples can be collected at the end of the anaerobic stage or phase, as long as RBCOD is not present. Alternatively, the aerobic test can be performed in a sequential mode after an anaerobic-anoxic test (Test EBPR.ANOX.1) resulting in an anaerobic-anoxic-aerobic test or, in parallel to an anoxic test after the execution of an anaerobic test (Test EBPR.ANA.2) (Wachtmeister *et al.*, 1997). It is strongly recommended to avoid washing the activated sludge or biomass in between the sludge collection and handling from the anaerobic or anoxic stage and the start of the aerobic batch test to reduce the potential oxidation of the PHAs. If the biomass needs to be washed (e.g. due to the suspected presence of toxic compounds), then an anaerobic test should always be executed prior to the aerobic batch test under the defined presence of an electron donor (Test EBPR.ANA.2). Thus, the following aerobic EBPR batch tests (with and without preceding anaerobic or anoxic stages) can be proposed:

1. **Test EBPR.AER.1** A single aerobic EBPR test performed with sludge collected at the end of an anaerobic or anoxic stage to assess the aerobic EBPR activity on the intracellular polymers stored in the original source of sludge.
2. **Test EBPR.AER.2** Combined anaerobic-aerobic EBPR batch tests conducted to ensure a defined concentration of intracellular PHAs to cover the aerobic metabolic requirements using sludge collected at the end of an aerobic stage.
3. **Test EBPR.AER.3** Combined anaerobic-anoxic-aerobic EBPR batch tests in series carried out with sludge collected at the end of an aerobic stage to assess the sequential anaerobic, anoxic and aerobic EBPR activities after securing a defined concentration of intracellularly stored PHAs.
4. **Test EBPR.AER.4** Combined anaerobic-anoxic-aerobic EBPR batch tests in parallel performed with sludge collected at the end of an aerobic stage to assess the anoxic and aerobic EBPR activities in parallel after an anaerobic phase. Since the anaerobic test is common, the sludge has the same content of intracellularly stored polymers at the beginning of the anoxic and aerobic tests, so both anoxic and aerobic EBPR activities can be compared to each other and in some cases even conducted in parallel (usually the privilege of more experienced experimenters).

The description of the different steps involved in the execution of aerobic EBPR tests is as follows:

Test EBPR.AER.1 Single aerobic EBPR test

a. Collect the sludge in the anaerobic stage or anoxic stage following the recommendations for sampling and activated sludge preparation described in Section 2.2.3.5. It is important to notice that single aerobic tests can only be performed when the tests can be executed immediately after collection (and preferably avoiding a washing procedure).
b. After the activated sludge has been transferred, keep the same redox conditions as those prevailing in the collection tank (e.g. anaerobic or anoxic) as described in Section 2.2.3.5 for at least 30 min while confirming that the pH and temperature are at the target values of interest (otherwise adjust and wait until stable conditions are reached). Do not start to aerate the sample and, if available, sparge N_2 gas (or another gas available) to avoid oxygen intrusion.
c. Repeat steps 'b' to 'e' from Test EBPR.ANA.1.
d. The test starts at 'time zero' with the supply of air (or pure oxygen), ensuring that the DO concentration with regard to the DO saturation concentration at local conditions reaches at least 2.0 mg DO L^{-1} within the first 10 min of execution of the aerobic test and around 4-5 mg DO L^{-1} onwards.
e. After the air supply starts, keep track of the execution and sampling time with a stopwatch.
f. Duration and sampling of the aerobic stage:
  i. The aerobic EBPR test can last between 2 - 4 h.
  ii. The DO concentration in the bulk liquid can be monitored throughout the test with the use of a DO probe.
  iii. For the estimation of the aerobic kinetic parameters, samples for the determination of PO_4 must be collected every 5 min in the first 30-40 min of execution of the test. After this period, the sampling frequency can be reduced to every 10 or

15 min during the first 1 h, and later on to every 15 or 30 min until the test is finished.

iv. If the aerobic kinetic conversions of PHAs and glycogen are of interest, samples can be collected at the same time as the samples for PO_4 determination.

v. To estimate the aerobic stochiometric conversions, samples for the analytical determination of total P, PO_4, and MLSS and MLVSS concentrations, as well as for PHAs and glycogen (if applicable), must be collected both at the start and at the end of the aerobic phase. However, the total oxygen consumption must be determined by respirometry (as presented in Chapter 3).

g. Conclude the aerobic EBPR batch test with the collection of the last samples needed for the estimation of the aerobic stoichiometric conversions.

h. Repeat steps 'j' to 'n' from Test EBPR.ANA.1.

Test EBPR.AER.2 Combined anaerobic-aerobic EBPR batch tests

a. Execute the anaerobic test as follows:
   i. Repeat steps 'a' to 'g' from Test EBPR.ANA.1.
   ii. Afterwards, execute steps 'b' to 'e' from Test EBPR.ANA.2.
b. Immediately after the execution of the anaerobic test, the aerobic EBPR batch test can start with the sparging of compressed air or pure oxygen ensuring that the DO concentration with regard to the DO saturation concentration at local conditions reaches at least 2.0 mg DO L^{-1} within the first 10 min of execution of the aerobic test and around 4-5 mg DO L^{-1} afterwards.
c. Repeat step 'f' from Test EBPR.AER.1.
d. Repeat steps 'j' to 'n' from Test EBPR.ANA.1.

Test EBPR.AER.3 Combined anaerobic-anoxic-aerobic EBPR batch tests in series

a. Execute the anaerobic test as follows:
   i. Repeat steps 'a' to 'g' from Test EBPR.ANA.1.
   ii. Afterwards, execute steps 'b' to 'e' from Test EBPR.ANA.2.
b. Immediately continue with the execution of the anoxic test by repeating steps 'd' and 'e' from Test EBPR.ANOX.1.
c. After the anoxic test, continue with the execution of the aerobic stage by repeating steps 'd' to 'g' from Test EBPR.AER.1.
d. Repeat steps 'j' to 'n' from Test EBPR.ANA.1.

Test EBPR.AER.4 Combined anaerobic-anoxic-aerobic EBPR batch tests in parallel

a. This test requires two (preferably identical) reactors: one for the execution of an anaerobic-anoxic stage and another for the execution of a single aerobic test. This is achieved by transferring, at the end of the anaerobic test, a defined volume (usually 50 %) from the bioreactor where the anaerobic-anoxic test will take place to the second bioreactor where the aerobic test will be conducted. For this purpose, the steps described here below are recommended.
b. Execute the anaerobic test as follows:
   i. Repeat steps 'a' to 'g' from Test EBPR.ANA.1.
   ii. Afterwards, execute steps 'b' to 'e' from Test EBPR.ANA.2.
c. Once the anaerobic stage is completed, transfer 50 % of the activated sludge present in the anaerobic bioreactor to the empty bioreactor (aerobic bioreactor).
d. Continue with the execution of the anoxic test in the same bioreactor where the anaerobic test took place by repeating steps 'd' and 'e' from Test EBPR.ANOX.1.
e. In parallel, carry out the execution of the aerobic test in the second bioreactor (where sludge was transferred) by repeating steps 'd' to 'g' from Test EBPR.AER.1. Note that the execution of two (anoxic and aerobic) tests in parallel requires more advanced experimentation skills. Two alternative approaches are that either two persons execute the tests, or to carry out the tests one after another.
f. Repeat steps 'j' to 'n' from Test EBPR.ANA.1.

2.2.5 Data analysis

2.2.5.1 Estimation of stoichiometric parameters

Prior to estimation of the stoichiometric and kinetic parameters, a COD balance should be conducted to validate the results and confirm their quality and reliability (Barker and Dold, 1995). A rather important tool to assess the reliability of data is the COD balance. Theoretically, the COD must be conserved in a truly anaerobic system so that the total COD entering an anaerobic stage must be equal to the total COD that leaves the anaerobic stage (Wentzel et al., 2008). Thus, assuming that soluble COD, intracellular PHAs and glycogen are the only carbon components involved in the chemical transformations, a COD balance can be performed as follows:

$$COD_{B,cons} + COD_{GLY,cons} = COD_{PHA,prod} \quad \text{Eq. 2.2.1}$$

Where:
- $COD_{B,cons}$ is the concentration of biodegradable substrate, as COD, consumed during the duration of the anaerobic batch activity test, in mg COD L^{-1}.
- $COD_{GLY,cons}$ is the concentration of intracellular glycogen consumed during the duration of the anaerobic batch activity test, in mg COD L^{-1}.
- $COD_{PHA,prod}$ is the concentration of intracellular PHAs formed or stored during the duration of the anaerobic batch activity test, in mg COD L^{-1}.

Also, the percentage of error to close the COD balance (ΔCOD (%)) can be estimated as:

$$\% \text{ COD balance} = \left[1 - \frac{COD_{B,cons} + COD_{GLY,cons} - COD_{PHA,prod}}{COD_{B,cons} + COD_{GLY,cons} + COD_{PHA,prod}}\right] \cdot 100 \quad \text{Eq. 2.2.2}$$

Ideally, ΔCOD (%) should be lower than 1-5 % but values as high as 15 % are often reported and considered acceptable in view of, and depending on, data quality and the uncertainty that the determination of certain parameters creates (in particular glycogen). The reader is invited to consult Chapter 5 for more information on the assessment of data quality.

Similarly to the anaerobic transformations, the COD balance can be an important tool to assess the reliability of the data obtained in anoxic and aerobic batch activity tests (Ekama and Wentzel, 2008a,b). Thus, the total amount of final electron acceptors consumed should equal the total amount of electron donors oxidized. For the aerobic transformations that take place during an aerobic EBPR test, a COD balance can be performed as follows:

$$\Delta COD_{cons} = COD_{input} - COD_{output} = \Delta O_{2,cons} \quad \text{Eq. 2.2.3}$$

Assuming that PHAs, glycogen (GLY) and biomass (Bio) are the only COD components that change during the anoxic or aerobic phase, the net COD consumption can be calculated as follows:

$$COD_{PHA,cons} - COD_{GLY,prod} - COD_{Bio,prod} = \Delta O_{2,cons} \quad \text{Eq. 2.2.4}$$

Where:
- $COD_{PHA,cons}$ is the total concentration of PHA consumption during the aerobic batch activity test, in mg COD L^{-1}.
- $COD_{GLY,prod}$ is the total concentration of glycogen produced during the aerobic batch activity test, in mg COD L^{-1}.
- $COD_{Bio,prod}$ is the total concentration of biomass produced during the duration of the aerobic batch activity test, in mg COD L^{-1}.
- $\Delta O_{2,cons}$ is the total concentration of oxygen consumed in the aerobic batch activity test estimated based on respirometry and oxygen uptake rates (see Chapter 3 on Respirometry) in mg COD L^{-1}.

The percentage of error to close the COD balance (ΔCOD (%)) can be estimated as:

$$\% \text{ COD balance} = \left[1 - \frac{COD_{PHA,cons} - COD_{GLY,prod} - COD_{Bio,prod} - \Delta O_{2,cons}}{COD_{PHA,cons} + COD_{GLY,prod} + COD_{Bio,prod} + \Delta O_{2,cons}}\right] \cdot 100$$
$$\text{Eq. 2.2.5}$$

Ideally, similar to the determination of COD balances for the anaerobic stage of EBPR batch activity tests, ΔCOD(%) should be lower than 1-5 % but values as high as 10 % can be considered acceptable. Chapter 5 provides different tools and approaches to assess the quality of the data obtained and execute the estimation of the COD balances with a smaller uncertainty about their reliability.

COD balances can also be determined and applied to anoxic EBPR activity tests where, for instance, nitrate or nitrite act as the final electron acceptor. When applying a COD balance on EBPR batch activity tests executed with a different final electron acceptor rather than oxygen, the approach would be similar, but: (*i*) the equivalent COD concentrations of the final electron acceptor should be determined and expressed in COD units based on its capacity to accept electrons, and (*ii*) the corresponding maximum stoichiometric yields (Y) of the metabolic conversions on the final electron acceptor of interest (for instance, nitrate or nitrite) should be known. For tests conducted with nitrate as the final electron acceptor, the maximum stoichiometric yields estimated by Kuba *et al.* (1996) can be used. Last but not least, the total consumption of the final electron acceptor during the execution of the anoxic batch EBPR test should be estimated based on the experimental methods described in Chapter 3.

Once the quality and reliability of the data have been confirmed and validated, the stoichiometric and kinetic parameters can be computed. Often, Net P-released, glycogen conversion, and PHAs produced per organic carbon or COD consumed ($Y_{VFA\_PO4,An}$, $Y_{Gly/VFA,An}$, $Y_{VFA\_PHA,An}$, $Y_{VFA\_PHB,An}$ and $Y_{VFA\_PHV,An}$ ratios, respectively) are the stoichiometric parameters of interest to assess the anaerobic stoichiometry of the EBPR processes (Table 2.2.1).

Regarding the anaerobic stoichiometric parameters, often the most common carbon source used for the execution of EBPR batch activity tests is acetate (Ac). Figure 2.2.5 shows a graphic representation of the determination of both the stoichiometry and kinetic parameters for Ac consumed and orthophosphate released in an anaerobic batch activity test. It should be noted that the Net $P_{released}$ due to Ac uptake should be estimated after excluding the secondary PO_4 release ($r_{PP\_PO4,Sec,An}$). However, because the secondary PO_4 release occurs continuously throughout the anaerobic test as a consequence of the anaerobic maintenance requirements of the cells (though usually it can only be observed after the carbon source is depleted), the accumulated PO^4 released due to anaerobic maintenance should be excluded from the total PO_4 release observed.

Thus, the Net $P_{released}$ can be estimated as follows:

Net $P_{released}$ = [(Total PO_4-$P_{released}$)] −

[($r_{PP\_PO4,Sec,An}$) · (test duration)] Eq. 2.2.6

Where:

Net $P_{released}$ is the PO_4 released due to Ac uptake only, mg PO_4-P L^{-1}.

$r_{PP\_PO4,Sec,An}$ corresponds to the PO_4 release rate due to anaerobic maintenance requirements of the biomass, mg PO_4-P L^{-1} h^{-1}.

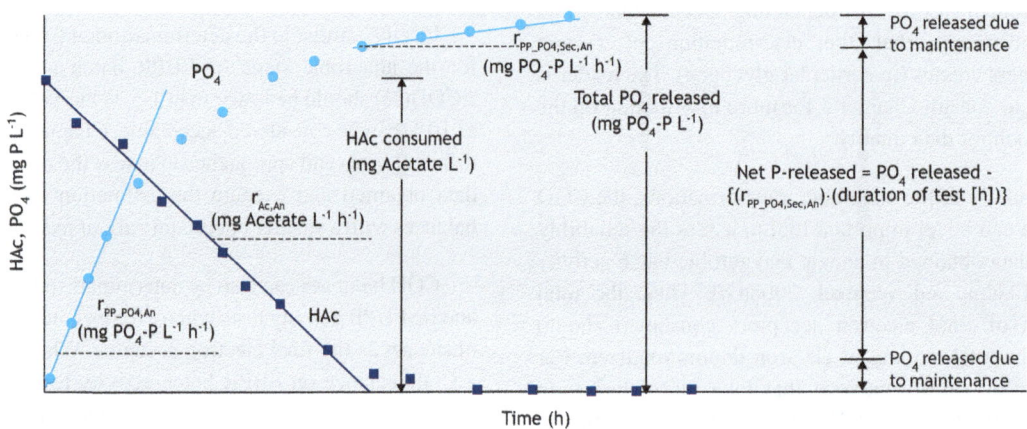

Figure 2.2.5 Example of the determination of the maximum anaerobic volumetric kinetic rates for acetate consumption (Ac) and orthophosphate released (PO_4) in an anaerobic batch activity test where all the carbon is consumed. For the estimation of the Net P released, the secondary P release ($r_{PP\_PO4,Sec,An}$) (corresponding to the anaerobic endogenous maintenance requirements which occur throughout the anaerobic test) must be excluded (Net P released = [(Total PO_4 released)] - [($r_{PP\_PO4,Sec,An}$)·(duration of the test)]). The Net P released should be used for the determination of the anaerobic PO_4-to-Ac stoichiometric ratio ($Y_{Ac\_PO4,An}$).

The net P released should be used for the determination of the anaerobic PO_4/Ac stoichiometric ratio ($Y_{Ac\_PO4,An}$). Thus, for instance, if in a batch activity test the carbon is fully consumed, the anaerobic net P-released/Ac-consumed ratio (which can also be referred to as the P/C ratio) of a given culture can be calculated as follows:

$$Y_{Ac_{PO4},An} = \frac{\text{Net } P_{released}}{S_{Ac,cons}} = \frac{S_{PO4,ini} - S_{PO4,final}}{S_{Ac,ini} - S_{Ac,final}} \quad \text{Eq. 2.2.7}$$

Where:

$S_{Ac,cons}$ is the concentration of acetate consumed in the batch activity test, mg L^{-1}.

$S_{PO4,ini}$ is the orthophosphate concentration in the bulk liquid at the beginning of the batch activity test, mg PO_4-P L^{-1}.

$S_{PO4-P,final}$ is the orthophosphate concentration in the bulk liquid at the time when the acetate concentration is consumed or at the end of the anaerobic batch activity tests if not all the acetate is depleted, mg PO_4-P L^{-1}.

$S_{Ac,ini}$ is the concentration of acetate in the bulk liquid at the beginning of the batch activity test, mg L^{-1}.

$S_{Ac,final}$ is the concentration of acetate in the bulk liquid at the end of the batch activity test, mg L^{-1}.

However, if the carbon is not depleted (e.g. when carbon is added in excess or the duration of the test is relatively too short to allow the full consumption of the carbon source), then the difference between the initial and final concentration of the compounds can be divided by the difference between the initial and final concentrations of carbon (net carbon consumed). A similar approach can be applied for the determination of other anaerobic stoichiometric ratios of interest (like the anaerobic glycogen hydrolysis and PHAs produced per C-source consumed ratios). Often, different units are used to report the stoichiometry of the anaerobic conversions, like the use of C-mol or P-mol instead of mg Ac or mg PO_4-P in scientific publications. For this purpose, Appendix I contains a series of coefficients for the conversion of units of certain compounds of interest (e.g. to convert the units of orthophosphate from g PO_4-P to P-mmol). Unlike the anaerobic stoichiometric parameters, the anoxic and aerobic stoichiometric parameters cannot be determined in a similar straightforward manner. Because EBPR cultures have different intracellular metabolic conversions occurring simultaneously under anoxic or aerobic conditions (poly-P uptake, glycogen replenishment, growth and maintenance) (Table 2.2.1), the net consumption of electron donors (PHAs) and final electron acceptors (oxygen, nitrate or nitrite) is the combined result of these four overlapping metabolic activities (Smolders *et al.*, 1994b).

Nevertheless, Smolders *et al.* (1994b) and Kuba *et al.* (1993, 1996) proved that using metabolic modelling, the four aerobic or anoxic metabolic processes are dependent on the ratio between the ATP produced per NADH consumed during aerobic or anoxic respiration, the so called 'δ-ratio' or 'δ-value' (Smolders *et al.*, 1994b; Kuba *et al.*, 1996). This means that for EBPR cultures, the aerobic or anoxic value of δ can be determined to consequently estimate the values of the different aerobic or anoxic stochiometric parameters, respectively. Furthermore, despite different δ values having been reported in literature for diverse microbial populations (ranging from 1.3 to 2.2) (Lopez-Vazquez *et al.*, 2009a), for EBPR cultures, the effect of the δ-value is rather insensitive to the aerobic and anoxic stoichiometric ratios within this range of δ values. The latter suggests that the aerobic and anoxic stoichiometric parameters are also insensitive and there may not be any need for their determination if the plant or lab-scale system operates under regular operating and environmental conditions. However, if the determination of the δ value is needed, two tests can be performed with and without the presence of orthophosphate in the bulk liquid and by measuring the oxygen uptake rate (OUR) by respirometry (Chapter 3). By computing the differences in P-uptake rates and OUR between the tests performed with and without the presence of orthophosphate in the bulk liquid, the δ value can be estimated as described by Smolders *et al.* (1994b). Alternatively, mathematical modelling can be applied to estimate the δ value based on the anoxic and/or aerobic profiles of PHAs, glycogen, orthophosphate, growth rate and maintenance requirements (Lopez-Vazquez *et al.*, 2009a). Should there be interest in the determination of δ for EBPR cultures, the reader may need to refer to cited references for further reading since this procedure falls out of the scope of this book.

As a reference and guide, Table 2.2.3 shows different parameters of interest for lab-scale EBPR cultures enriched under different operating conditions (e.g. carbon source) and dominated by either PAOs or GAOs.

Table 2.2.3 Typical stoichiometric parameters of interest for lab-scale enriched EBPR cultures cultivated under standard conditions (20 °C, pH 7, 7-8 d SRT).

	Stoichiometric parameter	Common notation	Units	Typical values	Reference
	Lab-scale PAO culture enriched with acetate				
ANAEROBIC	Anaerobic orthophosphate release to acetate uptake ratio	$Y_{Ac\_PO4,An}$	P-mol C-mol^{-1}	0.50	Smolders *et al.* (1994a)
	Anaerobic glycogen utilization to acetate uptake ratio	$Y_{Gly/Ac,An}$	C-mol C-mol^{-1}	0.50	
	Anaerobic PHA formation to acetate uptake ratio	$Y_{Ac\_PHA,An}$	C-mol C-mol^{-1}	1.22	
	Anaerobic PHB formation to acetate uptake ratio	$Y_{Ac\_PHB,An}$	C-mol C-mol^{-1}	1.10	
	Anaerobic PHV formation to acetate uptake ratio	$Y_{Ac\_PHV,An}$	C-mol C-mol^{-1}	0.12	
	Anaerobic PH$_2$MV formation to acetate uptake ratio	$Y_{Ac\_PH2MV,An}$	C-mol C-mol^{-1}	N/A	

	Parameter	Symbol	Units	Value	Reference
AEROBIC	Aerobic poly-P formation to PHA consumption ratio	$Y_{PHA\_PP,Ox}$	P-mol C-mol^{-1}	3.68	Smolders et al. (1994b)
	Aerobic glycogen formation to PHA consumption ratio	$Y_{PHA\_Gly,Ox}$	C-mol C-mol^{-1}	0.90	
	Aerobic PAO biomass growth to PHA consumption ratio	$Y_{PHA\_PAO,Ox}$	C-mol C-mol^{-1}	0.74	
	Aerobic maintenance rate of PAO	$m_{PAO,Ox}$	C-mol C-mol^{-1} h^{-1}	4x10^{-3}	
	Aerobic poly-P formation to oxygen consumption ratio	Y_{PP}	P-mol C-mol^{-1}	3.27	
	Aerobic glycogen formation to oxygen consumption ratio of PAO	$Y_{Gly,PAO}$	C-mol mol-O$_2^{-1}$	3.92	
	Aerobic PAO biomass growth to oxygen consumption ratio	Y_{PAO}	C-mol mol-O$_2^{-1}$	2.44	
	Aerobic endogenous respiration rate of PAO	$m_{PAO,O2}$	mol-O$_2$ C-mol^{-1} h^{-1}	4.5x10^{-3}	
	Lab-scale PAO culture enriched with propionate				
ANAEROBIC	Anaerobic orthophosphate release to propionate uptake ratio	$Y_{Pr\_PO4,An}$	P-mol C-mol^{-1}	0.42	Oehmen et al. (2005c)
	Anaerobic glycogen utilization to propionate uptake ratio	$Y_{Pr\_Gly,An}$	C-mol C-mol^{-1}	0.32	
	Anaerobic PHA formation to propionate uptake ratio	$Y_{Pr\_PHA,An}$	C-mol C-mol^{-1}	1.23	
	Anaerobic PHB formation to propionate uptake ratio	$Y_{Pr\_PHB,An}$	C-mol C-mol^{-1}	0.04	
	Anaerobic PHV formation to propionate uptake ratio	$Y_{Pr\_PHV,An}$	C-mol C-mol^{-1}	0.55	
	Anaerobic PH$_2$MV formation to propionate uptake ratio	$Y_{Pr\_PH2MV,An}$	C-mol C-mol^{-1}	0.65	
AEROBIC	Aerobic poly-P formation to PHA consumption ratio	$Y_{PHA\_PP,Ox}$	P-mol C-mol^{-1}	3.34	Oehmen et al. (2007)
	Aerobic glycogen formation to PHA consumption ratio	$Y_{PHA\_Gly,Ox}$	C-mol C-mol^{-1}	1.06	
	Aerobic PAO biomass growth to PHA consumption ratio	$Y_{PHA\_PAO,Ox}$	C-mol C-mol^{-1}	0.80	
	Aerobic maintenance rate of PAO	$m_{PAO,Ox}$	C-mol C-mol^{-1} h^{-1}	4x10^{-3}	
	Aerobic Poly-P formation to oxygen consumption ratio	Y_{PP}	P-mol C-mol^{-1}	3.34	
	Aerobic Glycogen formation to oxygen consumption ratio of PAO	$Y_{Gly,PAO}$	C-mol mol-O$_2^{-1}$	6.16	
	Aerobic Biomass growth to oxygen consumption ratio	Y_{PAO}	C-mol mol-O$_2^{-1}$	2.03	
	Aerobic endogenous respiration rate of PAO	$m_{PAO,O2}$	mol-O$_2$ C-mol^{-1} h^{-1}	4.5x10^{-3}	
	Lab-scale DPAO culture enriched with acetate				
ANAEROBIC	Anaerobic orthophosphate release to acetate uptake ratio	$Y_{Ac\_PO4,An}$	P-mol C-mol^{-1}	0.50	Smolders et al. (1994a), Kuba et al. (1996)
	Anaerobic glycogen utilization to acetate uptake ratio	$Y_{Gly/Ac,An}$	C-mol C-mol^{-1}	0.50	
	Anaerobic PHA formation to acetate uptake ratio	$Y_{Ac\_PHA,An}$	C-mol C-mol^{-1}	1.22	
	Anaerobic PHB formation to acetate uptake ratio	$Y_{Ac\_PHB,An}$	C-mol C-mol^{-1}	1.10	
	Anaerobic PHV formation to acetate uptake ratio	$Y_{Ac\_PHV,An}$	C-mol C-mol^{-1}	0.12	
	Anaerobic PH$_2$MV formation to acetate uptake ratio	$Y_{Ac\_PH2MV,An}$	C-mol C-mol^{-1}	N/A	
ANOXIC	Anoxic poly-P formation to PHA consumption ratio	$Y_{PHA\_PP,Ax}$	P-mol C-mol^{-1}	0.46	Kuba et al. (1996)
	Anoxic glycogen formation to PHA consumption ratio	$Y_{PHA\_Gly,Ax}$	C-mol C-mol^{-1}	1.27	
	Anoxic PAO biomass growth to PHA consumption ratio	$Y_{PHA\_PAO,Ax}$	C-mol C-mol^{-1}	1.63	
	Anoxic maintenance rate of PAO	$m_{PAO,Ax}$	C-mol C-mol^{-1} h^{-1}	3.64x10^{-3}	
	Anoxic poly-P formation to NO$_3$ consumption ratio	$Y_{NO3\_PP,Ax}$	P-mol N-mol^{-1}	0.414	
	Anoxic glycogen formation to NO$_3$ consumption ratio	$Y_{NO3\_Gly,Ax}$	C-mol N-mol^{-1}	0.35	
	Anoxic PAO biomass growth to NO$_3$ consumption ratio	$Y_{NO3\_PAO,Ax}$	C-mol N-mol^{-1}	0.57	
	Anoxic endogenous respiration rate of PAO on NO$_3$	$m_{PAO,NO3}$	N-mol C-mol^{-1} h^{-1}	3.27x10^{-3}	
	Lab-scale GAO culture enriched with acetate				
ANAEROBIC	Anaerobic glycogen utilization to acetate uptake ratio	$Y_{Gly/Ac,An}$	C-mol C-mol^{-1}	1.12	Zeng et al. (2003a)
	Anaerobic PHA formation to acetate uptake ratio	$Y_{Ac\_PHA,An}$	C-mol C-mol^{-1}	1.86	
	Anaerobic PHB formation to acetate uptake ratio	$Y_{Ac\_PHB,An}$	C-mol C-mol^{-1}	1.36	
	Anaerobic PHV formation to acetate uptake ratio	$Y_{Ac\_PHV,An}$	C-mol C-mol^{-1}	0.46	
	Anaerobic PH$_2$MV formation to acetate uptake ratio	$Y_{Ac\_PH2MV,An}$	C-mol C-mol^{-1}	0.04	
AEROBIC	Aerobic glycogen formation to PHA consumption ratio	$Y_{PHA\_Gly,Ox}$	C-mol C-mol^{-1}	0.95	Zeng et al. (2003a)
	Aerobic GAO biomass growth to PHA consumption ratio	$Y_{PHA\_GAO,Ox}$	C-mol C-mol^{-1}	0.75	
	Aerobic maintenance rate of GAO	$m_{GAO,Ox}$	C-mol C-mol^{-1} h^{-1}	3.06x10^{-3}	
	Aerobic glycogen formation to oxygen consumption ratio	Y_{Gly}	C-mol mol-O$_2^{-1}$	4.89	
	Aerobic PHA degradation to oxygen consumption ratio of GAO	$Y_{PHA,GAO}$	C-mol mol-O$_2^{-1}$	2.18	
	Aerobic endogenous respiration rate of GAO	$m_{GAO,O2}$	mol-O$_2$ C-mol^{-1} h^{-1}	3.51x10^{-3}	

2.2.5.2 Estimation of kinetic parameters

Regarding the anaerobic kinetics, the most important parameters are the maximum specific consumption rate of carbon source or VFA ($q_{VFA,An}$), the maximum specific P-release rate ($q_{PP\_PO4,An}$), PHA formation rate ($q_{VFA\_PHA,An}$) and the endogenous ATP maintenance coefficient ($m_{ATP,An}$) (Table 2.2.1). They can be computed by plotting the experimental data (y-axis) versus time (x-axis) and fitting the experimental data obtained in the anaerobic batch activity tests using linear regression. Because one is interested in the maximum rates, a linear regression approach can be applied by fitting the very first set or group of experimental data points obtained at the beginning of the batch activity test. This is the main reason why the sampling frequency within the first 30-40 min of execution of the batch activity tests is set to 5 min. Preferably the linear regression must be carried out by plotting the concentrations and fitting more than 4-5 experimental data points by linear regression while achieving a statistical determination coefficient (R^2) not lower than 0.90-0.95. With the estimation of the linear regression equation (of the form: '$y = Ax + B$'), the maximum volumetric kinetic rates of the parameters of interest can be determined with the 'A' coefficient of the linear regression expression which corresponds to the 'slope' of the set of data points. This will result in the determination of the maximum volumetric rates (usually reported in units such as mg L^{-1} h^{-1} or g m^{-3} d^{-1}). Figure 2.2.5 illustrates the estimation of the maximum anaerobic kinetic rates for an EBPR culture. The reader is referred to Chapter 5 for a more detailed description of other alternative and advanced statistical methods and tools for the determination of the rates. For activated sludge samples from full-scale systems, the rates can be expressed as maximum specific kinetic rates by dividing the volumetric rates (or values of the slopes) by the concentration of activated sludge volatile suspended solids (VSS). However, due to the particular dynamic behaviour of the intracellular compounds present in EBPR cultures, often the maximum specific kinetic rates are expressed in terms of the active biomass fraction (excluding the presence of intracellular compounds) quantified according to the following approximate expression:

Active biomass fraction = MLVSS − PHA − glycogen

Eq. 2.2.8

From a microbiological perspective, the previous equation is not an accurate expression due to the potential accumulation of non-biodegradable organics and, logically, the intrinsic metabolic activity of every single cell. Nevertheless, it is a commonly accepted estimate for experimental purposes in the wastewater treatment practice. Similarly to carbon and phosphate compounds (particularly for lab-cultivated cultures), the active biomass fraction can also be expressed in C-mol units instead of mg VSS. For this purpose the elemental composition of the biomass is applied (see Appendix I with the corresponding conversion unit factors for different compounds). Commonly, once the maximum specific kinetic rates are found they can be reported for the compound of interest as mg g VSS^{-1} h^{-1} or mg g VSS^{-1} d^{-1}. Similarly, the anaerobic endogenous ATP maintenance coefficient ($m_{ATP,An}$) can be determined based on the profile of orthophosphate concentration recorded during the execution of an anaerobic EBPR batch activity test performed under the absence of any carbon source (Section 2.2.4.1). This maximum volumetric rate resembles or corresponds to the endogenous P-release rate ($m_{PP\_PO4,Sec,An}$) observed in anaerobic tests once the carbon source is depleted (Fig. 2.2.6). Then, $m_{ATP,An}$ is equivalent to $m_{PP\_PO4,Sec,An}$ (Wentzel et al., 1989a; Smolders et al., 1994a). A similar linear regression approach can be used for the determination of the maximum specific kinetic rates for the aerobic and anoxic batch activity tests (tables 2.2.2 and 2.2.3). Usually, orthophosphate uptake, biomass growth, PHA degradation, glycogen formation and aerobic maintenance requirements are the kinetic parameters of interest dependent on the final electron acceptor available. Figure 2.2.6 displays an example to illustrate the determination of the maximum kinetic rates in an aerobic (or anoxic) batch activity test.

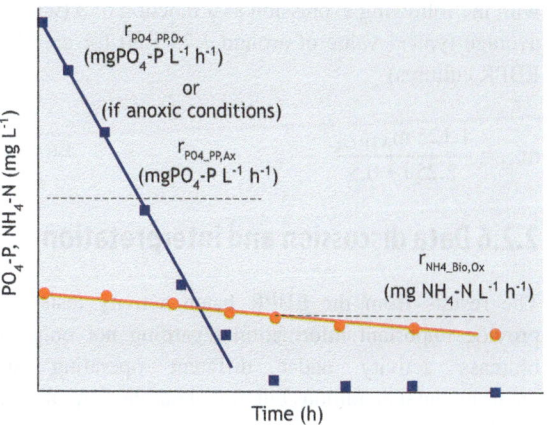

Figure 2.2.6 Example of the determination of the maximum aerobic (or anoxic) volumetric kinetic rates for orthophosphate uptake (PO_4) and ammonia consumption (NH_4) in an aerobic (or anoxic) batch test.

It is important to mention that the maximum specific biomass growth rate ($q_{PHA\_Bio,Ox}$) cannot be computed straightforwardly by following the increase in biomass concentrations during the cycle. Instead, the maximum specific ammonia (NH$_4$) consumption rate ($q_{NH4\_Bio,Ox}$) divided by the nitrogen composition of the active biomass fraction (0.20 N-mol per C-mol biomass) (Smolders et al., 1995; Zeng et al., 2003a; Lopez-Vazquez et al., 2007; Welles et al., 2014) can be used. However, the latter approach may be valid as long as nitrification is absent, no chemical precipitation or adsorption of NH$_4$ occurs and the relatively small differences allow a satisfactory determination of NH$_4$. Often, differences in NH$_4$ concentrations may be negligible or fall into the standard error of the analytical technique. This may complicate the determination process.

The experiments presented in this chapter focus on the relative conversions of intracellular and soluble compounds present in the water phase. It does not cover or tackle the consumption profiles nor the analysis of final electron acceptors (such as oxygen consumption, oxygen uptake rates, or nitrate uptake rates). These parameters are presented and discussed in Chapter 3 dealing with respirometry.

The determination of the aerobic maintenance coefficient of EBPR cultures is of major importance. For the determination of this parameter, extended aeration tests (regularly of at least 24 h) must be executed under the absence of an external carbon source. After 24 h, the aerobic ATP requirements ($m_{ATP,Ox}$) can be determined based on the oxygen consumption or uptake rate (OUR) (as O$_2$ mol consumed per mol active biomass per hour) and the aerobic maintenance requirements be computed with the following expression as a function of δ (with an average typical value of around 1.75-1.80 for enriched EBPR cultures):

$$m_{O2} = \frac{1.125\, m_{ATP,Ox}}{2.25\delta + 0.5} \qquad \text{Eq. 2.2.9}$$

2.2.6 Data discussion and interpretation

The results from the EBPR batch activity tests will provide important information regarding not only the biomass activity under different operating and environmental conditions but also about the general state of the biomass and some directions regarding the dominant microbial populations present in the sludge sample. Concerning the latter, the reader is invited to consult chapters 7 and 8. Furthermore, the identification of the dominant microbial species is an important complement to a better understanding of the EBPR process activities.

2.2.6.1 Anaerobic batch activity tests

As PAOs are the only organisms known to release phosphate during the anaerobic uptake of VFA, the anaerobic P-released/C-uptake or P-released/COD ratios, $Y_{C\_PO4,An}$ such as $Y_{VFA\_PO4,An}$, $Y_{Ac\_PO4,An}$ or $Y_{Pr\_PO4,An}$, may be considered one of the most suitable indicators to assess the PAO and GAO activity of EBPR cultures. A high P-released/C-uptake ratio has been considered as an indicator for PAO presence, whereas a low ratio has been considered to indicate the significant presence of GAOs. Contradictorily, such a consideration has been the subject of controversy (Schuler and Jenkins, 2003; Oehmen et al., 2007; Lopez-Vazquez et al., 2008b; Welles et al., 2015b). Such a controversy arises due to the broad range of ratios reported in literature, ranging from values as low as 0.025 to even 0.75 P-mol C-mol^{-1} or higher (Schuler and Jenkins, 2003). Several studies with highly enriched PAO cultures revealed that factors other than the presence of GAOs affect the anaerobic P released/C uptake as well. For instance, the carbon source, the pH and the poly-P content of the PAOs and the specific clades of PAOs enriched in the study have been shown to affect the P-released/C-uptake ratio (Smolders et al., 1994a; Filipe et al., 2001a; Zhou et al., 2008; Acevedo et al., 2012). Consequently, it is not advisable to use only this ratio as the direct and sole indicator to assess the EBPR activity of an activated sludge. Nevertheless, it may well be used to provide a rough estimation of the dominant metabolisms prevailing in the activated sludge. If supported by molecular techniques (chapters 7 and 8 from this book), it can offer an adequate and more complete overview of the biomass activity. Thus, based on observations drawn from past research, different P/C ratios performed under standard conditions (20 °C, pH 7.0) and fed with acetate may indicate (Schuler and Jenkins, 2003):

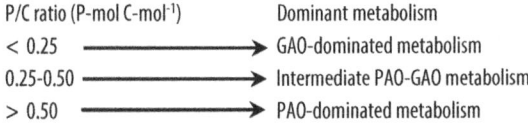

Figure 2.2.7 shows different anaerobic P-released/C-uptake ratios reported in literature as a function of the amount of phosphorus accumulated in the sludge on a TSS concentration basis expressed as the P/TSS ratio of

the sludge. Interestingly, the phosphorus content of the sludge (in terms of mg P g VSS^{-1}) has been shown to have a strong correlation with the anaerobic P-released/C-uptake ratio. As observed in Figure 2.2.7, an anaerobic P/C ratio higher than 0.50 is usually observed when the P/TSS content of the enriched PAO sludge is higher than 0.10 g P g TSS^{-1}. Thus, probably an activated sludge system will have a satisfactory EBPR activity when the anaerobic P-released/C-uptake and sludge P/TSS ratios are around or higher than 0.50 P-mol C-mol^{-1} and 0.10 mg P mg TSS^{-1}, respectively. Lower values could suggest that the EBPR activity of the system may be limited by certain operational or environmental factors such as (*i*) the relatively abundant presence of GAOs instead of PAOs, (*ii*) intrusion of electron acceptors into the anaerobic phase (like oxygen, NO$_3$, NO$_2$), (*iii*) the addition of Al of Fe salts for chemical P-removal, or occasionally (*iv*) the presence of inhibitory or toxic compounds. Under such circumstances the activated sludge system must be carefully revised to implement the corresponding corrective measures.

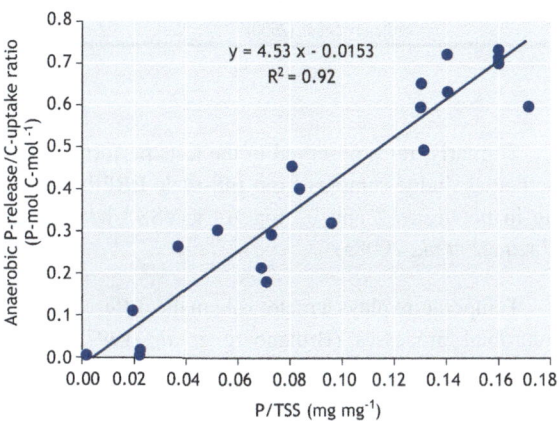

Figure 2.2.7 The anaerobic P-released/C-uptake ratio as a function of the P/TSS ratio of the biomass as reported in literature (adapted from Schuler and Jenkins, 2003; and Welles *et al.*, 2015b).

Under certain circumstances, the EBPR activity tests may be conducted under non-standard conditions (pH different to 7.0) using real wastewater or other carbon sources rather than acetate. For comparison and benchmarking purposes, the P/C ratio may be corrected using the expressions developed by Smolders *et al.* (1994a) or Filipe *et al.* (2001a) respectively, for acetate-fed cultures:

$$Y_{Ac\_PO4,An} = 0.19 \cdot pH - 0.85 \qquad \text{Eq. 2.2.10}$$

$$Y_{Ac\_PO4,An} = 0.16 \cdot pH - 0.55 \qquad \text{Eq. 2.2.11}$$

Also, the use of real wastewater or other C sources rather than acetate (like propionate, butyrate or glucose) will lead to a lower anaerobic P/C ratio than those reported for EBPR cultures where the PAO metabolism (or also PAOs) is dominant. Such lower P/C ratios can be, besides a low P/TSS content as previously discussed, a consequence of the lower energy needed for PHA storage or due to a higher involvement of GAO metabolism (possibly caused by the presence of GAOs). If the glycogen consumption/C-uptake ratio does not increase and remains within the commonly reported ratio for PAO-dominated systems of 0.35-0.50 C-mol C-mol^{-1} (Smolders *et al.*, 1994b; Schuler and Jenkins, 2003), the system can be assumed to be robust and stable, particularly if the intracellular P/TSS ratio is not considerably lower than 0.10 in enriched cultures. However, if under such circumstances the glycogen consumption/C-uptake ratio, $Y_{GLY/Ac,An}$, increases above 0.35-0.50 C-mol C-mol^{-1}, then the GAO metabolism (or the presence of GAOs themselves) may start to dominate the system and ultimately lead to the process deterioration as long as this ratio continues to increase (ultimately being able to reach a $Y_{GLY/Ac,An}$ of 1.12 C-mol C-mol^{-1}) in combination with a decrease in the P/TSS content below 0.10 mg P mg TSS^{-1}.

In parallel, PHA synthesis/C-uptake ratios higher than 1.33 C-mol C-mol^{-1} will be observed as a consequence of a higher synthesis of PHV and PH$_2$MV. PHV/C-uptake ratios higher than 0.10 and up to 0.25-0.30 may be observed together with the formation of PH$_2$MV. These values probably indicate the potential deterioration of the EBPR activity and efficiency. Regarding the anaerobic kinetic rates, in particular the initial maximum anaerobic uptake rate of carbon sources, different values have been reported for PAO and GAO lab-scale dominated cultures (mostly sequencing batch reactors, SBR) and full-scale systems such as modified UCT (University of Cape Town), Phoredox (which stands for phosphorus reduction oxidation), and PhoStrip (from phosphorus stripping) (Wentzel *et al.*, 2008) (Table 2.2.4).

Table 2.2.4 Maximum initial specific anaerobic C-source uptake rates reported in literature for lab-scale and full-scale EBPR systems.

Lab-scale EBPR systems		
Dominant microorganism/metabolism and system	$q_{VFA,An}$ in C-mol C-mol^{-1} h^{-1}	Reference
PAOs - SBR	0.27	Smolders et al. (1994a)
	0.20	Filipe et al. (2001b)
	0.20	Kuba et al. (1996)
	0.20	Brdjanovic et al. (1997)
	0.20	Lopez-Vazquez et al. (2007)
GAOs - SBR	0.24	Filipe et al. (2001a)
	0.16-0.18	Zeng et al. (2003a,b)
	0.20	Lopez-Vazquez et al. (2007)
	0.19	Lopez-Vazquez et al. (2009a)

Full-scale EBPR systems		
Dominant microorganism/metabolism	$q_{VFA,An}$ in mg Ac g VSS^{-1} h^{-1}	Reference
PAOs - Modified UCT	22	Lopez-Vazquez et al. (2008a)
	19	Lopez-Vazquez et al. (2008a)
	47	Kuba et al. (1997a,b)
	7-31	Kuba et al. (1997b)
PAOs - Phoredox	14	Lopez-Vazquez et al. (2008a)
	21	Lopez-Vazquez et al. (2008a)
	11	Lopez-Vazquez et al. (2008a)
	14	Lopez-Vazquez et al. (2008a)
PAOs - Sidestream PhoStrip	9	Lopez-Vazquez et al. (2008a)
	23	Brdjanovic et al. (2000)

Similarly to the P-released/C-uptake ratio, the kinetic rates seem to be dependent on the poly-P content of the sludge (Schuler and Jenkins 2003, Welles et al., 2016, submitted), ranging from 0.02 to 0.20 C-mol C-mol^{-1} h^{-1}. However, in most studies conducted with medium poly-P contents and P/Ac ratios around 0.5 P-mol C-mol^{-1}, the observed uptake rates converge around 0.20 C-mol C-mol^{-1} h^{-1}.

Similarly, rates observed in the tests performed with activated sludge samples from full-scale EBPR systems lie in between 17 and 22 mg Ac g VSS^{-1} h^{-1} (Lopez-Vazquez et al., 2008a).

Temperature plays a major role in the different EBPR microbial processes (Brdjanovic et al., 1997, 1998c, Table 2.2.5).

Table 2.2.5 Arrhenius temperature coefficients (θ) reported in literature to describe the maximum specific kinetic rates of the different EBPR metabolic processes occurring in lab-scale and full-scale EBPR systems (Meijer, 2004).

Parameter	θ^*	Reference
$q_{VFA,An}$	1.094 ($e^{0.090}$)	Brdjanovic et al. (1998c); Meijer (2004)
$m_{ATP,An}$	1.071 ($e^{0.069}$)	Smolders et al. (1995); Murnleitner et al. (1997)
$q_{PHA,Ox}$	1.129 ($e^{0.121}$)	Brdjanovic et al. (1998); Meijer (2004)
$q_{PHA\_Gly,Ox}$	1.125 ($e^{0.118}$)	Meijer (2004)
$q_{PO4\_PP,Ox}$	1.031 ($e^{0.031}$)	Murnleitner et al. (1997); Brdjanovic et al. (1998c)
$q_{PAO,Ox}$	1.081 ($e^{0.078}$)	Brdjanovic et al. (1997)
$m_{ATP,Ax}$	1.071 ($e^{0.090}$)	Murnleitner et al. (1997)
$m_{ATP,Ox}$	1.071 ($e^{0.090}$)	Murnleitner et al. (1997)

* The number in brackets displays the value of the Arrhenius temperature coefficient (θ) in terms of the Euler number.

2.2.6.2 Aerobic batch activity tests

For comparison purposes, the observed maximum kinetic rates must be standardized using Arrhenius coefficients. Based on observations from Brdjanovic *et al.* (1997, 1998) and other authors (Smolders *et al.*, 1995, and Murnleitner *et al.*, 1997), Meijer (2004), through the development of the TUDelft model and the execution of several modelling studies, defined suitable Arrhenius temperature coefficients to best fit the maximum kinetic rates of EBPR cultures for different operating lab- and full-scale systems.

Although the aerobic kinetic rates reported in literature for lab-scale EBPR systems are consistent (Table 2.2.6), the values observed in full-scale EBPR systems may vary widely (from modified UCT systems to BIODENIPHO -biological denitrification nitrification and phosphorus removal- plants) (Table 2.2.7).

Table 2.2.6 Maximum aerobic kinetic rates reported in literature for enriched EBPR cultures.

Culture or system	$q_{PO4\_PP,Ox}$ P-mol C-mol^{-1} h^{-1}	$q_{PHA\_Gly,Ox}$ C-mol C-mol^{-1} h^{-1}	$q_{PAO,Ox}$ C-mol C-mol^{-1} h^{-1}	$m_{ATP,Ox}$ mol ATP C-mol^{-1} h^{-1}	Reference
SBR	0.055	0.080	0.014-0.016	1.9×10^{-3}	Smolders *et al.* (1994b)
SBR	0.046	-	0.13	1.2×10^{-3}	Brdjanovic *et al.* (1997)
SBR	0.083	-	-	1.7×10^{-3}	Welles *et al.* (2014)

Table 2.2.7 Maximum initial specific aerobic P-uptake rates reported in literature for full-scale EBPR systems.

Dominant microorganism/metabolism	$q_{PO4\_PP,Ox}$ mg P g VSS^{-1} h^{-1}	Reference
PAOs - Modified UCT	19.2	Lopez-Vazquez *et al.* (2008a)
	9.0	Lopez-Vazquez *et al.* (2008a)
	13	Kuba *et al.* (1997a, b)
	4-6	Kuba *et al.* (1997b)
PAOs - Phoredox	8.0	Lopez-Vazquez *et al.* (2008a)
	9.1	Lopez-Vazquez *et al.* (2008a)
	6.2	Lopez-Vazquez *et al.* (2008a)
	6.3	Lopez-Vazquez *et al.* (2008a)
PAOs - Sidestream PhoStrip	9.8	Lopez-Vazquez *et al.* (2008a)
	2.2	Brdjanovic *et al.* (2000)
PAOs - Pilot-scale BIODENIPHO	4	Meinhold *et al.* (1999)

Such considerable differences are logical since the lab-scale systems are highly enriched with PAOs (based on similar studies PAOs can probably compose more than 80-90 % of the total populations) with little variability in the percentage of enrichment (Lopez-Vazquez *et al.*, 2009a; Welles *et al.*, 2014, 2015a), whereas in full-scale systems PAOs may comprise between 3 and 20 % of the total active biomass fraction (Lopez-Vazquez *et al.*, 2008a). Moreover, in full-scale systems performing biological nitrogen and phosphorus removal, EBPR cultures are exposed to alternating anaerobic-anoxic-aerobic (A^{2}O) stages that may reduce the availability of intracellular PHAs after the sequential exposure to anoxic and aerobic conditions (since both anoxic and aerobic metabolic activities require PHAs as the carbon and energy source). Based on the data provided in Table 2.2.7, it can be seen that values higher than 10 mg P g VSS^{-1} h^{-1} are not common. Such kinetic rates can be considered rather fast since, as previously discussed, full-scale systems often tend to have aerobic hydraulic retention times (HRT) of several hours (e.g. of at least 6-8 h and longer) which will favour the uptake of orthophosphate from the bulk liquid. Aerobic P-uptake rates ($q_{PO4\_PP,Ox}$) lower than 10 mg P g VSS^{-1} h^{-1} will not be nsatisfactory depending upon the length of the aerobic stage (though extended aeration periods must be avoided)

(Brdjanovic et al., 1998c). Another important aspect to consider is whether the plant is prone to chemical P-precipitation because its occurrence will favour the chemical P- removal to the detriment of the EBPR process.

2.2.6.3 Anoxic batch activity tests

The simultaneous removal of orthophosphate and nitrate (or nitrite) is highly desirable due to the potential savings in energy and operational costs, while at the same time keeping or maintaining a P-removal activity that can be comparable to that observed under anaerobic-aerobic conditions (Kuba et al., 1996). However, to a certain extent, it has been a matter of controversy particularly due to the rather variable and inconsistent anoxic P-removal activities observed in full-scale systems (Hu et al., 2002) and theoretically decreased P-removal potentially caused by lower PAO biomass yields. Table 2.2.8 shows an overview of different anoxic P-removal activities observed in full-scale systems (anaerobic-anoxic (A^2) and anaerobic-anoxic-oxic (A^2O) plants) that are compared to the aerobic P-removal activities of the same systems (following the protocols presented in this chapter as introduced by Murnleitner et al., 1997).

Table 2.2.8 Maximum initial anoxic kinetic rates reported in literature for lab-enriched denitrifying EBPR cultures.

Culture or system	$q_{PO4\_PP,Ax}$ P-mol C-mol^{-1} h^{-1}	$q_{PHA\_Gly,Ax}$ C-mol C-mol^{-1} h^{-1}	$q_{PAO,Ax}$ C-mol C-mol^{-1} h^{-1}	m_{Ax} C-mol C-mol^{-1} h^{-1}	Reference
SBR, PAOs, A^2 system	0.1	0.8	0.05	3.6x10^{-3}	Kuba et al. (1996)
SBR, PAOs, A^2 system	0.02-0.63^a	0.0025	-	-	Carvalho et al. (2007)
SBR, PAOs, A^2 system	0.58^a	0.9^a	-	-	Zeng et al. (2003b)
SBR, PAOs, A^2O system	0.33^a	-	-	-	Saito et al. (2004)

^{a} Units: P-mmol g VSS^{-1} h^{-1}

Table 2.2.9 Maximum initial specific anoxic P-uptake rate and its relationship with maximum aerobic P-uptake rate reported for full-scale EBPR systems.

Dominant microorganism/metabolism	$q_{PO4\_PP,Ox}$ mg P g VSS^{-1} h^{-1}	$q_{PO4\_PP,Ax}$ mg P g VSS^{-1} h^{-1}	$q_{PO4\_PP,Ax}$ / $q_{PO4\_PP,Ox}$ %	Reference
PAOs - Modified UCT	19.2	5.9	31 %	Lopez-Vazquez et al. (2008a)
	9.0	2.1	23 %	Lopez-Vazquez et al. (2008a)
	13	6	46 %	Kuba et al. (1997a,b)
	4-6	1.2-1.6	20-40 %	Kuba et al. (1997b)
PAOs - Phoredox	8.0	1.9	23 %	Lopez-Vazquez et al. (2008a)
	9.1	4.4	48 %	Lopez-Vazquez et al. (2008a)
	6.2	0.6	9 %	Lopez-Vazquez et al. (2008a)
	6.3	0.0	0 %	Lopez-Vazquez et al. (2008a)
PAOs - Sidestream PhoStrip	9.8	3.3	34 %	Lopez-Vazquez et al. (2008a)
	2.2	1.7	80 %	Brdjanovic et al. (2000)
PAOs - Pilot-scale BIODENIPHO	4	2	54 %	Meinhold et al. (1999)

As observed in Table 2.2.9, the anoxic P-uptake rate scarcely reaches more than 5 mg PO$_4$-P g VSS^{-1} h^{-1} and, even under certain circumstances, it is very low or absent. In any case, the different anoxic P activities are a combined reflection of (i) the level of enrichment of denitrifying PAOs capable of using oxygen and nitrate (and/or of other EBPR cultures and side-populations involved in the denitrification process) (Kerr-Jespersen and Henze, 1993; Meinhold et al., 1999; Saad et al., 2016, submitted), and/or (ii) a measure of the level of denitrifying capacity induced in PAOs (Kuba et al., 1996, 1997; Wachtmeister et al., 1997). In any case, the level of exposure of the activated sludge system will favour the growth of denitrifying EBPR populations and their induction. Thus, higher anoxic P-uptake activities can be expected in plant configurations operated with defined pre-denitrification stage.

2.2.7 Example

2.2.7.1 Description

To illustrate the execution of an EBPR batch activity test, data from an anaerobic-aerobic test (Test EBPR.AER.2) performed at 10 °C with a lab-enriched EBPR culture is presented in this section. Test EBPR.AER.2 was carried out to determine the anaerobic stoichiometry and the anaerobic and aerobic kinetics of the EBPR processes. Thus, the batch activity test was performed in a 2.5 L bioreactor. All the equipment, apparatus and materials were prepared as described in Section 2.2.3. pH and DO sensors were calibrated less than 24 h before the test execution. The test lasted 4.5 h and was composed of a 2.25 h anaerobic stage (created by continuously sparging N_2 gas throughout the test) followed by 2.25 h aerobic stage (created by supplying compressed air in excess, reaching a DO concentration higher than 4 mg L^{-1}). Prior to the batch test, 1.25 L of concentrated EBPR sludge collected at the end of the aerobic phase of a lab-scale bioreactor was transferred to the bioreactor and acclimatized for 30 min at 10 °C under slow mixing (100 rpm) at pH 7.0 following the recommendations described in Section 2.2.3.5. Activated sludge preparation for tests was performed in less than 1 h after sludge collection. Afterwards, 20 min before the start of the test, samples for the determination of the parameters of interest were collected (in accordance with the execution of Test EBPR.AER.2).

The test started with the addition of 1.25 L synthetic media containing 350 mg COD L^{-1} as Ac (other macro- and micro-nutrients as well as 20 mg L^{-1} ATU were included in synthetic media in accordance with Section 2.2.3.3). Because the test was executed at 10 °C, the temperature of the synthetic media was adjusted to 10 °C in a water bath operated at the same temperature before addition. Because the main objective was to determine the anaerobic stoichiometry and the anaerobic and aerobic kinetics of the EBPR processes, samples were collected more frequently (every 5 min) in the first 30 min of each anaerobic and aerobic phase. Immediately after collection, all the samples were prepared, preserved and stored prior to the analytical determination of the parameters of interest (e.g. PO_4, MLSS, MLVSS, and PHAs, among other parameters) as described in Section 2.2.3.4 "Material preparation". All the collected samples were analysed as described in Section 2.2.2.5. In particular, the intracellularly stored PHAs were determined following the protocol for the determination of PHB and PHV, and glycogen by the acid-hydrolysis and extraction methods (Smolders *et al.*, 1994a). Other PHA compounds, like PH_2MV, were not measured because acetate was the carbon source supplied and therefore it was expected that PHB and PHV would comprise most of the PHAs. Table 2.2.10 shows the experimental implementation plan of the execution of the test.

2.2.7.2 Data analysis

Following up on the results from the experiment shown in Table 2.2.10, Figure 2.2.8 shows the results from the test displayed in the implementation plan and also an estimation of the maximum volumetric kinetic rates by applying linear regression. The different anaerobic and aerobic conversions of the parameters of interest are shown in Table 2.2.11, while Table 2.2.12 displays an estimation of the different anaerobic and aerobic stoichiometric and kinetic parameters of interest.

Overall, the results of the batch activity test (Figure 2.2.8) show the typical phenotype of a PAO-dominated sludge: full Ac uptake in the anaerobic stage coupled with anaerobic P release, PHA production and glycogen consumption, while full PO_4 uptake was observed in the aerobic phase together with PHA utilization, glycogen formation and slight NH_4 consumption. Furthermore, the relatively low VSS/TSS ratio observed at the beginning and end of the test (of around 0.72-0.73) is typical for EBPR systems due to poly-P accumulation (which is reflected in a higher ash content) when compared to systems that perform organic matter removal only (with VSS/TSS ratios usually not lower than 0.80) (Wentzel *et al.*, 2008).

Table 2.2.10 Example of an experimental implementation plan for the execution of a batch activity test (Type Test No. EBPR.AER.2) performed with a lab-enriched EBPR sludge at 10 °C using synthetic influent at pH 7.0.

Combined anaerobic-aerobic EBPR batch tests — Code: EBPR.AER.2

Date:	Thursday 17.12.2015 9:00 h
Description:	Tests at 10 °C, pH 7, artificial substrate and enriched PAO culture
Test No.:	3 of 6
Duration	4,5 h (270 min)
Substrate:	Synthetic: Acetate (350 mg L^{-1}) + minerals
Sampling point:	Middle mixed liquor height in the SBR
Samples No.:	EBPR.AER.2(1-22)
Total sample volume:	305 mL (10 mL for MLVSS, 12 mL for PHA, 4.5 mL for Glycogen, 6 mL for other samples)
Reactor volume:	2.5 L

Experimental procedure in short:

Step	Time (h:min)
1. Confirm availability of sampling material and required equipment	08:00
2. Confirm calibration and functionality of the system, meters and sensors	08:10
3. Transfer 1.25 L sludge to the batch reactor	08:20
4. Keep aerobic conditions with gentle mixing and air sparging at T and pH set	08:40
5. 20 min before starting, take sample for initial conditions (EBPR.AER.2(3.1))	08:40
6. Stop aeration and start sparging by N_2 gas	08:50
7. Start cycle, add 1.25 L of sysnthetic media (0 min)	09:00
8. Continue sampling program according to schedule (5 min)	09:05
9. Stop sparging with N_2 gas, start addition of air (135 min)	11:15
10. Stop sampling and aeration (after 270 min)	13:30
11. Organize the samples and clean the experimental equipment and space	13:45
12. Ensure all equipment is switched off and samples are handled properly	14:00

Sampling schedule

Time (min)	-20	0	5	10	15	20	25	30	40	50	60	90	135
Time (h)	-0.33	0.00	0.08	0.17	0.25	0.33	0.42	0.50	0.67	0.83	1.00	1.50	2.25
Sample No.	1	2	3	4	5	6	7	8	9	10	11	12	13
Parameter						ANAEROBIC PHASE							
HAc (C-mmol L^{-1})	5.83[1]		4.85	4.57	3.98	3.48	2.87	2.21	1.35	0.43	0	0	0
PO_4-P (P-mmol L^{-1})	0[1]		0.24	0.45	1.01	1.35	1.69	2.14	2.85	3.01	3.04	3.07	3.11
NH_4-N (N-mmol L^{-1})	1.32[1]			1.34								1.26	1.39
PHA (C-mmol)	12.27											20.04	20.08
Glycogen (C-mmol)	15.09											12.68	12.71
MLSS and MLVSS (mg L^{-1})	See table												See table

[1] Average value of the concentration present in the synthetic substrate and and liquid phase of the sludge sample prior the start of the test

Sampling schedule (continued)

Time (min)	140	145	150	155	160	165	180	195	215	270
Time (h)	2.33	2.42	2.50	2.58	2.67	2.75	3.00	3.25	3.58	4.50
Sample No.	13	14	15	16	17	18	19	20	21	22
Parameter					AEROBIC PHASE					
HAc (C-mmol L^{-1})										
PO_4-P (P-mmol L^{-1})	3.00	2.73	2.45	2.20	1.90	1.64	0.84	0	0	0
NH_4-N (N-mmol L^{-1})				1.11			1.09	1.07	1.06	1.05
PHA (C-mmol)							15.55			11.72
Glycogen (C-mmol)							15.04			15.90
MLSS and MLVSS (mg L^{-1})										See table

MLSS & MLVSS measurements

Sampling point	Cup No.	W1	W2	W3	W2-W1	W2-W3	MLSS	MLVSS	Ratio
Start anaerobic phase[2]	1	0.08835	0.16525	0.10792	0.07690	0.05733	7,690	5,733	0.75
	2	0.08835	0.16553	0.10997	0.07718	0.05556	7,718	5,556	0.72
	3	0.08834	0.16435	0.10903	0.07601	0.05532	7,601	5,532	0.73
						Average	7,670	5,607	0.73
End anaerobic/Start aerobic phase	4	0.08858	0.12437	0.09606	0.03579	0.02831	3,579	2,831	0.79
	5	0.08848	0.12564	0.09646	0.03716	0.02918	3,716	2,918	0.79
	6	0.08914	0.12527	0.09648	0.03613	0.02879	3,613	2,879	0.80
						Average	3,636	2,876	0.79
End aerobic phase	7	0.08868	0.12859	0.09952	0.03991	0.02907	3,991	2,907	0.73
	8	0.08764	0.12716	0.09881	0.03952	0.02835	3,952	2,835	0.72
	9	0.08722	0.12622	0.09800	0.03900	0.02822	3,900	2,822	0.72
						Average	3,948	2,855	0.72

[2] Sample taken before substrate addition

Biomass composition

Sampling point	Start Aner.	End Aner.	End Aer.
MLSS (mg L^{-1})	3,835	3,636	3,948
MLVSS (mg L^{-1})	2,804	2,876	2,855
Ratio	0.73	0.79	0.72
Ash (mg L^{-1})	1,031	760	1,093
PHB (mg L^{-1})	241.7	392.0	232.1
PHV (mg L^{-1})	20.9	37.3	18.6
PHA (mg L^{-1})	262.6	429.3	250.8
Glycogen (mg L^{-1})	423.7	343.3	429.2
%(PHA+Gly) MLVSS^{-1}	32.0	37.0	31.0
Active biomass (mg L^{-1})	2,117	2,103	2,175
Active biomass (Cmmol L^{-1})	81.4	80.9	83.6

Note:
- Acetate (CH_2O): 30.03 mg C-mmol^{-1}
- Ortho-Phosphate (PO_4-P): 31.00 mg P-mmol^{-1}
- Ammonium (NH_4-N): 14.00 mg N-mmol^{-1}
- PHB ($CH_{1.5}O_{0.5}$): 21.52 mg C-mmol^{-1}
- PHV ($CH_{1.6}O_{0.4}$): 20.02 mg C-mmol^{-1}
- Glycogen ($CH_{10/6}O_{5/6}$): 27.00 mg C-mmol^{-1}
- Biomass ($CH_{2.09}O_{0.54}N_{0.20}$): 26.00 mg C-mmol^{-1}

Table 2.2.11 Summary of the anaerobic and aerobic conversions observed in the example of the batch activity test (Type Test No. EBPR.AER.2) performed with a lab-enriched EBPR sludge at 10 °C using synthetic influent at pH 7.0.

Parameter	Unit	Anaerobic phase			Aerobic phase		
		Start [Time: 0]	End [Time: 135 min]	Anaerobic conversion	Start [Time: 135 min]	End [Time: 270 min]	Aerobic conversion
Ac	C-mmol L^{-1}	5.20	0.00	-5.20	0.00	0.00	0.00
PO$_4$-P	P-mmol L^{-1}	0.00	3.11	3.11	3.11	0.00	-3.11
NH$_4$-N	N-mmol L^{-1}	1.32	1.39	0.07	1.39	1.05	-0.34
PHB	C-mmol L^{-1}	11.23	18.21	6.99	18.21	10.79	-7.43
PHV	C-mmol L^{-1}	1.05	1.86	0.82	1.86	0.93	-0.93
PHAs (PHB+PHV)	C-mmol L^{-1}	12.27	20.08	7.80	20.08	11.72	-8.36
Glycogen	C-mmol L^{-1}	15.69	12.71	-2.98	12.71	15.90	3.18

Table 2.2.12 Summary of the anaerobic and aerobic stoichiometric and kinetic parameters observed in the example of the batch activity test (Type Test No. EBPR.AER.2) performed with a lab-enriched EBPR sludge at 10 °C using synthetic influent at pH 7.0.

Conversion	Symbol	Unit	Estimated value
Anaerobic stoichiometry			
Net P-released to Ac uptake ratio[a]	$Y_{Ac\_PO4,An}$	P-mol C-mol^{-1}	0.57
PHA production to Ac uptake ratio	$Y_{Ac\_PHA,An}$	C-mol C-mol^{-1}	1.50
PHV production to PHB production ratio	$Y_{PHV/PHB,An}$	C-mol C-mol^{-1}	0.12
Glycogen consumption to Ac uptake ratio	$Y_{Gly/Ac,An}$	C-mol C-mol^{-1}	0.57
Anaerobic kinetic rates[b]			
Maximum volumetric Ac uptake rate	$r_{Ac,An}$	C-mmol L^{-1} h^{-1}	6.15
Maximum specific Ac uptake rate	$q_{Ac,An}$	C-mol C-mol^{-1} h^{-1}	0.075
Maximum volumetric PO$_4$ release rate	$r_{PP\_PO4,An}$	P-mmol L^{-1} h^{-1}	4.42
Maximum specific PO$_4$ release rate	$q_{PP\_PO4,An}$	P-mmol L^{-1} h^{-1}	0.054
Secondary anaerobic PO$_4$ release rate	$r_{PP\_PO4,Sec,An}$	P-mmol L^{-1} h^{-1}	0.063
Anaerobic maintenance coefficient	$m_{PP\_PO4,An}$	P-mol C-mol^{-1} h^{-1}	7.69 E-04
Aerobic kinetic rates			
Maximum volumetric PO$_4$ uptake rate	$r_{PO4\_PP,Ox}$	P-mmol L^{-1} h^{-1}	3.12
Maximum specific PO$_4$ uptake rate	$q_{PO4\_PP,Ox}$	P-mol C-mol^{-1} h^{-1}	0.038
Volumetric aerobic NH$_4$ consumption rate[c]	$r_{NH4\_Bio,Ox}$	N-mol L^{-1} h^{-1}	0.15
Maximum specific biomass growth rate[d]	$q_{PAO,Ox}$	C-mol C-mol^{-1} h^{-1}	0.009

[a] Excluding the secondary P release by multiplying $r_{PP\_PO4,Sec,An}$ by the duration of the anaerobic phase (0.063 P-mmol L^{-1} h^{-1} · 2.25 h = 0.142 P-mmol).
[b] Estimated by dividing the maximum volumetric conversion rates by the active biomass concentration at the beginning of the test of 81.4 C-mmol.
[c] Estimated: the aerobic consumption of NH$_4$ divided by the duration of the aerobic phase.
[d] Estimated: the volumetric aerobic NH$_4$ consumption rate divided by the N-content of the biomass (0.20 N-mol) and the initial active biomass concentration (81.4 C-mmol L^{-1}).

As observed in Table 2.2.12, the anaerobic stoichiometry, in particular the net P-released/Ac uptake ($Y_{Ac\_PO4,An}$, PO$_4$/Ac or P/C) ratio of 0.57 P-mol C-mol^{-1} in combination with the glycogen consumption to Ac uptake ratio of 0.57 C-mol C-mol^{-1} indicates that the observed biomass activity (physiology) corresponds to that dominated by PAO metabolism (Table 2.2.3). This can be confirmed by the relatively low PHV production/PHB production ($Y_{PHV/PHB,An}$) ratio of 0.12 since PHV/PHB ratios close to or lower than 0.10 are commonly observed in PAO-enriched systems (Smolders et al., 1994a) because of the lower glycogen consumption for Ac uptake when compared to GAO-dominated systems (Zeng et al., 2003a). Furthermore, when comparing the observed PO$_4$/Ac ratio with the values reported in Table 2.2.3 (for enriched EBPR cultures) and

Figure 2.2.5, it appears that PAOs (or their activity) are dominant in the sludge. However, this is a mere rough estimation and should be cross-checked with the use of molecular techniques (chapters 7 and 8). The P/TSS ratio of the sludge can be also used to assess the expected PO_4/Ac ratio (Figure 2.2.7). Nevertheless, based on the previous data, the sludge sample shows a satisfactory EBPR activity.

An important tool to assess the data consistency and quality is the COD balance. In this example, a COD balance performed to assess the anaerobic EBPR conversions shows that about 5.83 C-mmol Ac and 2.98 C-mmol of glycogen are consumed, while 7.80 C-mmol PHAs are produced. Using the COD conversion factor of 32 mg COD C-mmol^{-1} for both Ac and glycogen and that of 36 mg COD mg C-mmol^{-1} for PHAs, a COD balance error (ΔCOD(%)) of about 0.2 % is estimated using Eq. 2.2.2. A similar COD balance can be performed with the aerobic COD conversions if a respirometry test is performed as indicated in Chapter 3.

Regarding the anaerobic kinetic rates, the $q_{Ac,An}$ and $m_{PP\_PO4,An}$ values of 0.075 C-mol C-mol^{-1} h^{-1} and 7.69 × 10^{-4} P-mol C-mol^{-1} h^{-1}, respectively, appear to be lower than the typical values reported for PAO-enriched systems of around 0.20 C-mol C-mol^{-1} h^{-1} and 2.1 × 10^{-3} P-mol C-mol^{-1} h^{-1}, correspondingly (Table 2.2.12). However, one should realize that the batch activity test was conducted at a temperature of 10 °C, whereas previous values reported in literature have mostly been obtained from tests performed at 20 °C. Thus, for comparison purposes with tests executed at 20 °C (and also at other temperatures), the temperature Arrhenius coefficients (θ) proposed by Meijer (2004) can be applied (of 1.094 for carbon uptake and 1.071 for anaerobic maintenance) to estimate the equivalent kinetic rates at 20 °C based on the tests performed at 10 °C. Therefore, the equivalent kinetic rates at 20 °C are: 0.18 C-mol C-mol^{-1} h^{-1} for the maximum acetate uptake rate (estimated as $q_{Ac,20,An} = q_{Ac,10,An} / \theta^{(10-20)}$) and 1.53 × 10^{-3} P-mol C-mol^{-1} h^{-1} ($m_{PP\_PO4,20,An} = m_{PP\_PO4,10,An} / \theta^{(10-20)}$).

Similarly, for the maximum aerobic specific uptake rate ($q_{PO4\_PP,Ox}$) of 0.038 P-mol C-mol^{-1} h^{-1} observed at 10 °C, an equivalent maximum specific rate of 0.050 P-mol C-mol^{-1} h^{-1} is estimated at 20 °C which is within previous reported values in literature (Table 2.2.6). As previously mentioned, biomass growth cannot be directly determined from the increase in MLVSS since the potentially low biomass increase will probably fall into the standard error of the MLVSS analytical determination technique. Instead, it is estimated based on the NH$_4$ consumption observed in the test (ensuring that NH$_4$ is not removed by any other biological or chemical process). Thus, a maximum specific biomass growth rate ($q_{Bio,Ox}$) of 0.009 C-mol C-mol^{-1} h^{-1} (C-mol new biomass produced per C-mol of existing biomass) is determined. After re-calculating the approximate biomass growth rate from 10 °C at 20 °C (using the Arrhenius temperature coefficients for biomass growth of 1.081), the estimated biomass growth rate is 0.020 C-mol C-mol^{-1} h^{-1}, which is

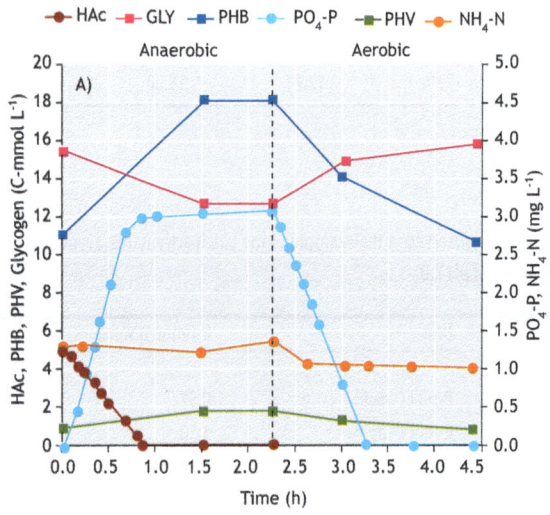

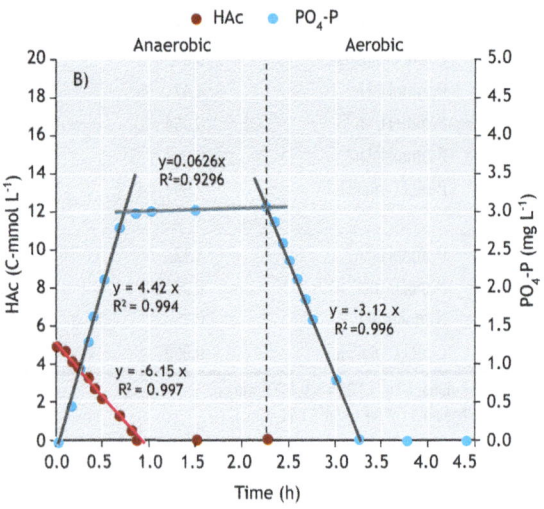

Figure 2.2.8 Graphic representation of the data obtained in the example of the experimental implementation plan for the execution of a batch activity test (Test EBPR.AER.2, Table 2.2.10) performed with a lab-enriched EBPR sludge at 10 °C using synthetic influent at pH 7.0: A) profiles of the experimental data of interest; and, B) experimental data of Ac and PO_4 showing the main trend lines of the conversion rates for the further estimation of the maximum kinetic rates.

in the range of previously reported values for EBPR sludge (0.016 C-mol C-mol^{-1} h^{-1}) (Smolders, 1995). Overall, the stoichiometric and kinetic parameters obtained in the test are comparable to similar values previously reported in literature for EBPR cultures (Table 2.2.3), strongly indicating that the EBPR activity observed in the test was typical of an enriched EBPR culture.

2.2.8 Additional considerations

2.2.8.1 GAO occurrence in EBPR systems

Due to the harmful effects that they can have on EBPR performance, the occurrence of GAOs has been the subject of extensive research in past years. So far, (*i*) the availability of a sole COD source in the influent (either acetate or propionate), (*ii*) a temperature higher than 20 °C, and (*iii*) pH values lower than 7.0 have been suggested as key factors favouring the presence of GAOs in (lab-scale) EBPR systems (Filipe *et al.*, 2001b; Oehmen *et al.*, 2004; Lopez-Vazquez *et al.*, 2009b). The most characteristic aspects that suggest the presence of GAOs are anaerobic P-released/C-uptake ratios much lower than 0.50 P-mol C-mol^{-1} and incomplete aerobic PO$_4$ uptake. Despite the apparently frequent occurrence of GAOs in lab-scale EBPR systems (possibly as a consequence of operating the lab-scale systems at the boundaries of the aforementioned parameters: with a single carbon source, usually acetate, at pH 7.0 and 20 °C), abundant GAO populations have rarely been found in full-scale municipal EBPR systems (Thomas *et al.*, 2003; Saunders *et al.*, 2003; Lopez-Vazquez *et al.*, 2008a; Lopez-Vazquez, 2009; Kong *et al.*, 2006), unless certain particular conditions (such as the discharge of industrial effluents) take place (Burow *et al.*, 2007). Although a low anaerobic P-released/C-uptake ratio could suggest a higher activity or involvement of GAOs, as described in the text below, recent observations have provided evidence that PAOs, under limiting intracellular poly-P conditions, may be able to perform a similar metabolism like GAOs under anaerobic conditions but still achieve aerobic full P-removal (Schuler and Jenkins, 2003; Zhou *et al.*, 2008; Acevedo *et al.*, 2012; Welles *et al.*, 2015b). Therefore, an anaerobic P-released/C-uptake ratio considerably lower than 0.50 P-mol C-mol^{-1} (e.g. around 0.35 P-mol C-mol^{-1}) observed in EBPR batch activity tests does not strictly imply that GAOs are more abundant than PAOs. While the actual mechanisms influencing the use of a GAO-type metabolism by PAOs will probably continue to be a matter of future research efforts, the use of microscopic and molecular techniques (chapters 7 and 8) is strongly recommended to identify the dominant EBPR microbial population.

2.2.8.2 The effect of carbon source

It is well known that RBCOD fed to the anaerobic stage (containing mainly volatile fatty acids such as acetate and propionate) enhances the growth of EBPR biomass (Comeau *et al.*, 1986; Mino *et al.*, 1998; Oehmen *et al.*, 2004). Other RBCOD sources, such as glucose, are not suitable since they appear to enhance the growth of GAOs or G-bacteria (Cech and Hartman, 1993). Also, it is assumed that more complex substrates need to be hydrolysed and fermented to VFA to become available to EBPR biomass (Wentzel *et al.*, 2008). In this regard, it is likely that more complex COD sources will not be fully consumed in the anaerobic stage and will 'leak' into the anoxic or aerobic stages, influencing the EBPR anoxic and aerobic results (for instance, when executing combined anaerobic-anoxic-aerobic tests, such as Tests EBPR.ANOX.2 or EBPR.AER.2). Ideally, for a satisfactory interpretation of the experimental data obtained, no COD source may be present in the anoxic or aerobic stage of an EBPR batch activity test (unless it is also a subject of interest). In addition, recent developments hypothesize that, besides the apparently known PAOs (*Candidatus Accumulibacter phosphatis*), other organisms, such as *Actinobacteria* or S-PAOs (Kong *et al.*, 2005; Wu *et al.*, 2014), are able to perform an excessive P-uptake like *Accumulibacter* using alternative electron donors (e.g. aminoacids or H$_2$S). For day-to-day or regular tests on well-known EBPR systems, the use of synthetic media containing VFA can be good enough to provide a satisfactory assessment of EBPR activity, as presented in this chapter. However, the use and application of more complex COD sources, which may be undoubtedly present in raw or settled municipal wastewater, can lead to either sub-optimal EBPR activity or (yet to be known) different EBPR metabolisms. The latter has been and will continue to be a matter of extensive research. Nevertheless, the reader should be aware that such conditions can lead to results that may differ from those presented in this chapter.

2.2.8.3 The effect of temperature

While extensive research has been advocated to assess the temperature dependencies of EBPR cultures, suggesting that temperatures lower than 20 °C enhance the growth of PAOs whereas higher temperatures favour the development of GAOs (Brdjanovic *et al.*, 1997, 1998b; Lopez-Vazquez *et al.*, 2009a), certain observations have indicated that stable EBPR systems can operate at temperatures higher than 25 °C (Cao et al., 2009). Though case-specific combinations between wastewater composition, operating and environmental conditions will play a major role, the long-term operation and acclimatization of EBPR cultures to those particular conditions can also lead to the development and enrichment of PAO cultures (or similar organisms sharing the PAOs' phenotype) capable of performing stable EBPR at a higher temperature. In this regard, the identification of these organisms is of major importance where the methods and techniques presented in Chapter 7 and particularly in Chapter 8 will be needed and applicable to elucidate the identity of these organisms.

2.2.8.4 The effect of pH

As discussed, pH has a direct influence on EBPR cultures (Smolders et al., 1994a, Filipe et al., 2001a). During EBPR batch activity tests, it should be carefully monitored and well-controlled to obtain reliable data (avoiding pH fluctuations higher than ± 0.1-0.2). However, higher pH levels (particularly above pH 8.0) combined with the presence of RBCOD can lead to (expectedly) higher anaerobic P release and favour chemically-induced calcium phosphate precipitation or even struvite formation depending on the wastewater composition (NH_4MgPO_4 or $KMgPO_4$) (Lin et al., 2012; Mañas et al., 2011). Also, it cannot be disregarded that the presence of aluminum salts in the wastewater produced by the discharge of drinking water sludge can play a role and lead to the precipitation of phosphorus, particularly if there is enough retention time in the sewer network. These processes will reduce the biological availability of phosphorus for PAOs, resulting in the potential deterioration of the EBPR process as PAOs will not be able to replenish their intracellular poly-P pools. Such conditions will lead to different values to those presented in this chapter. On the other hand, although the previous process may not be desirable and, thus, should be avoided in continuous conventional alternating anaerobic-anoxic-aerobic EBPR systems, it can offer interesting alternative options for P recovery that may be worth exploring in view of the essential role of phosphorus in the food chain and the potential depletion of the conventional world's phosphorus sources.

2.2.8.5 Denitrification by EBPR cultures

Denitrifying EBPR sludge has been the subject of extensive debate since the 1990s. While satisfactory simultaneous NO_3 removal and PO_4 uptake has been observed in the anoxic stage of several EBPR systems (Vlekke et al., 1988; Kuba et al., 1993, 1996, 1997a, 1997b; Wachtmeister et al., 1997; Brdjanovic et al., 2000; Zeng et al., 2003b), limited or inconsistent simultaneous denitrification and PO_4 uptake has been observed in other lab- and full-scale studies (Kerrn-Jespersen and Henze, 1993; Hu et al., 2003; Carvalho et al., 2007; Lopez-Vazquez et al., 2008a). Actually, the discovery of the existence of two different *Accumulibacter* clades (known PAOs) seemed to have been encouraged by assessing the denitrifying capabilities of EBPR biomass (Flowers et al., 2008). Flowers et al. (2008) proposed that the so-called *Accumulibacter* clade Type I has a full-denitrifying capability (able to denitrify from nitrate to di-nitrogen gas), whereas *Accumulibacter* clade Type II appears to be only able to denitrify from nitrite onwards (in line with the first observations drawn by Kerrn-Jespersen and Henze, 1993). As suggested by Kuba et al. (1996) and Lopez-Vazquez et al. (2008a), the simultaneous NO_3 removal and anoxic PO_4 uptake capability of an EBPR sludge may be a reflection of both the induction of the required denitrifying enzymes (nitrate and nitrite reductase) and the development of EBPR and side populations able to denitrify. Therefore, when executing anoxic EBPR activity tests, the relative anoxic EBPR activities observed can vary widely from practically zero to considerably high anoxic activities (Table 2.2.8, Table 2.2.9) as a function of the previously discussed exposure to anoxic conditions and the development of the denitrifying populations in the lab- or full-scale systems.

2.2.8.6 Excess/shortage of intracellular compounds

Although the depletion of the intracellular poly-P pools was initially assumed to be a limiting factor for anaerobic VFA uptake by PAOs (Brdjanovic et al., 1997), later developments showed that PAOs can utilize higher amounts of intracellularly stored glycogen as an energy source for anaerobic VFA uptake to compensate for the limiting poly-P availability, and still perform satisfactory fully aerobic P-uptake (Schuler and Jenkins, 2003; Zhou et al., 2008, Acevedo et al., 2012; Welles et al., 2014, 2015b, 2016, submitted). This is reflected in anaerobic P-released/C-uptake ratios much lower than 0.50 P-mol C-mol^{-1} (Figure 2.2.7) and higher anaerobic glycogen/C ratios (Table 2.2.3). Based on observations from Schuler and Jenkins (2003) and Welles et al. (2015b), the shift from the metabolic utilization of poly-P to glycogen appears to take place as soon as the biomass total P/TSS ratio drops below 0.08 mg P g TSS^{-1} (Acevedo et al., 2014). Furthermore, Welles et al. (2015b) have observed that the maximum kinetic rates of the two known *Accumulibacter* clades (Flowers *et al.*, 2008) appear to be affected in a different manner depending on the intracellular poly-P availability, with *Accumulibacter* Type I more affected than *Accumulibacter* Clade Type II. This can explain potential deviations from those presented in this chapter, which can be supported by the determination of the intracellular poly-P and glycogen contents or at least by the estimation of the biomass P/TSS ratio.

2.2.8.7 Excessive aeration

The exposure of EBPR biomass to extended aeration periods (e.g. longer than 12-24 h) can lead to the sequential utilization of PHAs, glycogen and intracellular poly-P under aerobic conditions (Brdjanovic *et al.*, 1998c; Lopez *et al.*, 2006), as a consequence of the biomass needs to cover their aerobic maintenance requirements. Consequently, it will lead to an eventual aerobic P release as soon as the intracellularly stored poly-P pools start to be hydrolysed. Thus, if EBPR sludge samples are aerated for extensive periods of time prior to the execution of the batch activity tests (e.g. overnight or during transportation), it is likely that a lower EBPR activity will be observed due to the potentially low(er) poly-P and glycogen contents of the sludge. Therefore, over-aeration conditions should be avoided. Furthermore, it deserves particular attention for the satisfactory operation of full-scale activated sludge EBPR systems because excessive aeration periods during low loading conditions (e.g. weekends or holidays periods) can result in undesired aerobic P release, affecting the treated effluent quality.

2.2.8.8 Shortage of essential ions

Though it may be trivial, the presence of macro- and micro-nutrients in the right concentration and (bio-) availability is essential for EBPR activated sludge systems. For instance, potassium, magnesium, iron, and calcium, among others, are rather important to regulate the microbial EBPR metabolism and support the storage of intracellular compounds (Brdjanovic *et al.*, 1997; Burow *et al.*, 2007; Barat *et al.*, 2008). Their absence, for instance of potassium, can lead to the deterioration of the EBPR process (Brdjanovic *et al.*, 1997), but their excess can influence the metabolism of PAOs inducing a GAO metabolism (Jobaggy *et al.*, 2006; Barat *et al.*, 2008), possibly due to the chemical precipitation of phosphorus with the aforementioned elements. This will reduce their bio-availability and therefore the aerobic replenishment of poly-P pools. The absence or presence in excessive concentrations of the aforementioned elements should be checked if one suspects that their concentrations differ from those regularly observed in municipal wastewater treatment systems.

2.2.8.9 Toxicity/inhibition

A limited number of compounds have been identified to be toxic or inhibiting to the EBPR process. The presence of nitrate or nitrite in the anaerobic zone is considered harmful for the EBPR process since they enhance the activity of ordinary denitrifying organisms that can consume the available RBCOD to the detriment of PAOs (Wentzel *et al.*, 2008). Also, the presence of nitrite in the aerobic zone in concentrations as low as 6-8 mg L^{-1} has been proven to inhibit PAOs (Saito *et al.*, 2004), which can worsen the negative effects at lower pH due to the increase in free nitrous acid (FNA) (Zhou *et al.*, 2008; Pijuan *et al.*, 2010). The latter favours the occurrence of GAOs over PAOs since GAOs appear to be more tolerant to the presence of FNA (Pijuan *et al.*, 2011), resulting in the deterioration of the EBPR process. Salinity is another factor that may inhibit the activity of PAOs in the short-term (hours) (Welles *et al.*, 2014, 2015a). Welles *et al.* (2014) observed that, after a short-term exposure (hours), chloride concentrations as low as 10,000 mg NaCl L^{-1} (1 % salinity) can lead to more than 50 % inhibition of the anaerobic metabolism of PAOs and practically fully inhibit the aerobic metabolism. Similar circumstances can take place due to a sudden saline intrusion into the sewage or saline intrusion to the plant (particularly in coastal regions), or caused by industrial effluent discharges. Nevertheless, the long-term exposure of EBPR sludge to high salinity concentrations (> 35,000 mg NaCl L^{-1}, 3.5 % salinity) can enhance the acclimatization of EBPR biomass which can become salt-tolerant and be able to perform satisfactory EBPR at salinity concentrations equivalent to those observed in seawater. Another compound that is potentially inhibitory or toxic to EBPR sludge is H_2S, which may be formed in the sewage (due to saline intrusion or industrial discharges) or in the anaerobic stage of the EBPR activated sludge system. Its presence can be rather inhibitory to the anaerobic metabolism of PAOs at concentrations as low as 20-25 mg H_2S L^{-1} leading to 50 % inhibition (Saad *et al.*, 2013; Rubio-Rincon *et al.*, 2016, submitted). Overall, inhibiting effects will be reflected in limiting or sub-optimal EBPR activity. If feasible, the sludge can be washed in a mineral solution (as explained in Section 2.2.3.5) to avoid the potentially inhibiting or toxic compounds.

2.3 BIOLOGICAL SULPHATE-REDUCTION

2.3.1 Process description

Sulphate (SO_4^{2-}) is naturally present in surface water and ground water and depending on the geographical location, its concentrations in drinking water supply networks can vary. Consequently, the presence of other sulphur compounds namely organic sulphides, including mercaptans, dimethyl sulphides and dimethyl disulphides is also very common in domestic wastewater. The concentration of sulphate in domestic sewage usually ranges from 20 to 60 mg L^{-1} (Moussa et al., 2006). However, the sulphate concentration in sewage can reach as high as 500 mg L^{-1}, due to the discharge of sulphate-rich industrial effluents, seawater-based toilet flushing or intrusion of saline water into the sewer (Lens et al., 1998; Chen et al., 2010; Ekama et al., 2010).

The metabolism of sulphate-reducing bacteria (SRB), beside its applications in domestic sewage treatment, can be exploited beneficially in specific industrial effluent treatment processes that generate sulphate-rich wastewater (Lens et al., 1998; Muyzer and Stams, 2008). Such industries are, among others, potato starch production, pulp and paper mills, food and fermentation industries and seafood processing facilities. In wastewater treatment applications, sulphate is usually completely reduced to sulphide, as this conversion yields the highest free Gibb's energy ($\Delta G°'$, Table 2.3.1).

Table 2.3.1 Sulphate transformation reactions performed by SRB (Jørgensen, 2006; Liamleam and Annachhatre, 2007).

Reactions	$\Delta G°'$ (KJ mol^{-1})
$SO_4^{2-} + 4H_2 + H^+ \rightarrow HS^- + 4H_2O$	-152.2
$SO_4^{2-} + H_2 + 2H^+ \rightarrow HSO_3^- + H_2O$	+19.7
$HSO_3^- + 3H_2 \rightarrow HS^- + 3H_2O$	-171.7
$3HSO_3^- + H_2 + H^+ \rightarrow S_3O_6^{2-} + 3H_2O$	-46.3
$S_3O_6^{2-} + H_2 \rightarrow S_2O_3^{2-} + HSO_3^- + H^+$	-123.0
$S_2O_3^{2-} + H_2 \rightarrow HS^- + HSO_3^-$	-2.1
$SO_4^{2-} + 2H^+ + ATP \rightarrow APS + PP_i$	+46.0
$PP_i + H_2O \rightarrow 2P_i$	-21.9
$APS + H_2 \rightarrow HSO_3^- + AMP + H^+$	-68.0

Dissimilatory sulphate-reduction is the most important anaerobic process in many different environments (Balk et al., 2008). The initial step of biological sulphate-reduction involves the transfer of exogenous sulphate through the bacterial cell membrane into the cell. The sulphate dissimilation process proceeds via the action of adenosine triphosphate (ATP) sulphurylase (Figure 2.3.1). ATP produces the highly activated molecule adenosine phosphosulphate (APS), and pyrophosphate (PPi) in the presence of sulphate, which yields inorganic phosphate. Further, APS is rapidly converted to bisulphite (HSO_3^-) by the cytoplasmic enzyme APS reductase. Pyrophosphate is hydrolyzed and the sulphate moiety of APS is reduced to bisulphite, together with adenosine monophosphate (AMP). Bisulphite in turn may be reduced via a number of intermediates to form the sulphide ion. Bisulphite is reduced to bisulfide (HS^-) via bisulphite reductase. By another mechanism, bisulphite reduction via the enzymes bisulphite reductase, trithionate reductase and thiosulfate reductase yields trithionate ($S_3O_6^{2-}$) and thiosulfate ($S_2O_3^{2-}$) as free intermediates. The physiology and growth of these bacteria has been studied in depth and is well documented (Cypionka, 1987; Gibson, 1990; Hansen, 1994; Rabus et al., 2006).

The organisms responsible for the reduction of sulphur compounds belong to both bacteria and prokaryotes (Postgate, 1965; Muyzer and Stams, 2008); nonetheless, in literature, the term SRB is used. In this chapter, the term SRB includes both bacteria and prokaryotes. The bacterial sulphate reducers are categorized into different branches, the *Deltaproteobacteria* with more than 25 genera, the Gram-positive bacteria that include *Desulfotomaculum* and *Thermodesulfobium* and Gram-negative sulphate reducers that include *Thermodesulfobacterium* and *Thermodesulfatator* (Mori et al., 2003; Moussard et al., 2004; Balk et al., 2008). In general, SRB are found present and active in sewerage systems and wastewater treatment plants. Several authors have also reported SRB activities in freshwater, marine, hypersaline and oil/hydrocarbon polluted sites (Cravo-Laureau et al., 2004; Almeida et al., 2006; Kjeldsen et al., 2007).

SRB are facultative anaerobes that live in oxygen-free or depleted environments and utilize sulphate as a terminal electron acceptor to produce hydrogen sulphide (H_2S) as one of its metabolic end products. SRB can survive under extreme environmental and operating conditions over a rather wide range of pH (4.0 to 9.5), temperature (25-75 °C) and pressures of up to 500 atm (Madigan et al., 2009; Tang et al., 2009). Sulphate is redox sensitive and the production of sulphide is an indicator of SRB activity which depends on several factors including sulphate concentrations, the concentration of organic matter/nutrients, pH and temperature, among others. In sewerage and municipal wastewater treatment plants, the presence of SRB is considered to be undesirable, due to corrosion and instability in methanogenic activity, causing an insufficient digestion process (Oude Elfrenink et al., 1994). SRB have been recognized as the major microbiologically-influenced corrosion-causing bacteria in oil/gas pipelines and sewer systems (Al Abbas et al., 2013). Microbiologically-influenced corrosion is aggravated due to the synergistic interaction of different microbes such as iron and manganese-reducing bacteria, carbon dioxide-reducing bacteria that co-exist and through cooperative metabolism with SRB (Little and Lee, 2007). According to Kjeldsen et al. (2004), the presence of SRB in activated sludge is of interest because sulphate-reduction can have negative effects on the wastewater treatment plant operation. Other negative side-effects of the activity of SRB is that the produced sulphide will inhibit other main microorganisms involved in the treatment process, such as methanogenic bacteria (MET), phosphorus-accumulating organisms (PAO), nitrifiers and others. Exceedingly high sulphide levels can be toxic for microorganisms performing methanogenesis and sulphate-reduction. High sulphide concentration can also have a deteriorating effect on the activated sludge floc structure, for instance, deflocculation, by reducing Fe(III) to Fe(II) as FeS (Caccavo et al., 1996; Nielsen and Keiding, 1998). The authors attributed this phenomenon to the better flocculating properties of Fe(III) rather than Fe(II), mainly due to its valance and lower solubility.

On the other hand, sulphide production can lead to the growth of filamentous bacteria and result in concomitant sludge bulking (Yamamoto et al., 1991; Zeitz et al., 1995; Kjeldsen et al., 2004).

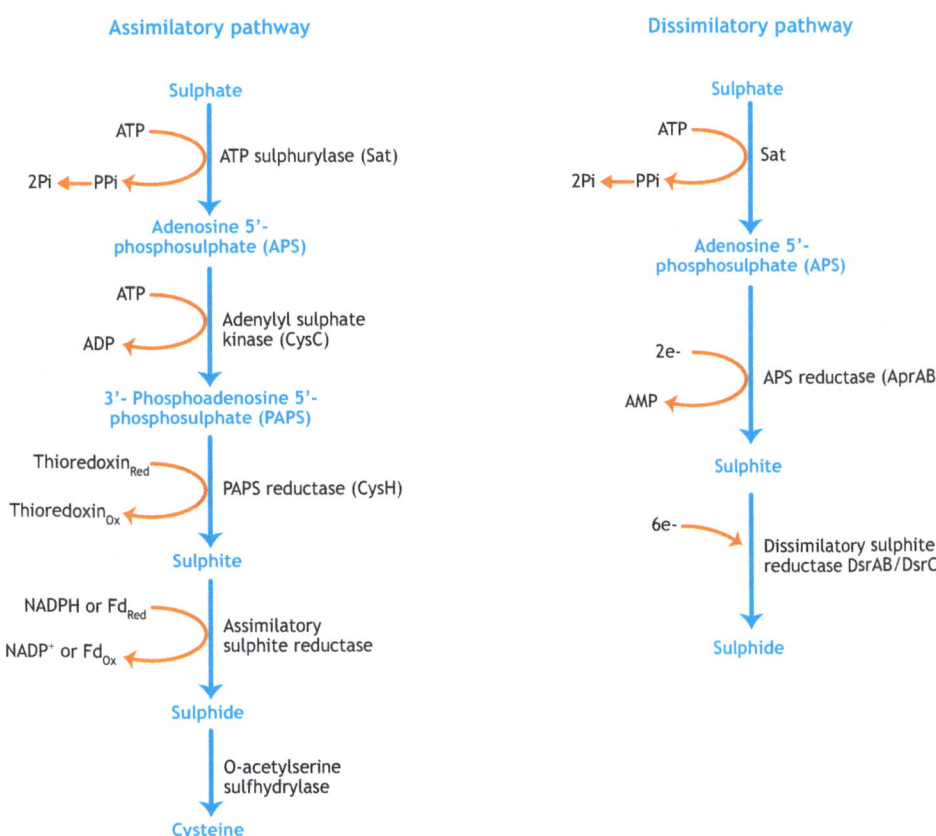

Figure 2.3.1 Prokaryotic assimilatory and dissimilatory pathways for sulphate-reduction (Grein et al., 2013).

Driven by the potable water supply stress mitigated by using seawater for toilet flushing, the research group of Hong Kong University of Science and Technology (HKUST) recognized the benefits of sulphate present in saline sewage and successfully developed a process that takes a big advantage of sulphate and SRB: the SANI® process (Sulphate-reduction, Autotrophic denitrification and Nitrification Integrated process); the only technology that applies SRB for municipal wastewater treatment purposes. Its application has led to lower sludge production, higher coliform (pathogen) and heavy metal removal while using less space and energy (Wang *et al.*, 2009; Abdeen *et al.*, 2010). In addition, SRB-based processes have been used as a pre-treatment step to enhance the digestion process. Recently, Daigger *et al.* (2015) tested a pilot-scale membrane bioreactor (MBR) to remove elemental sulphur from an anaerobically pre-treated pulp and paper effluent containing high concentrations of dissolved sulphide. Although SRB play a major role in treatment plants, they are not frequently studied in domestic wastewater treatment. To understand the positive and negative effects of SRB activity in a domestic wastewater treatment plant, activity tests should be conducted. Batch tests can be useful to estimate to what extent SRB are present and active in sewage and treatment plants and to provide an insight to develop measures that can stimulate or suppress SRB activity, depending on the chosen process design and operational conditions. Therefore, from a process performance perspective, the stability and continuation of the treatment process is highly dependent on (among others) the concentration of the sulphide (and pH) present in the liquid phase.

2.3.2 Sulphide speciation

Sulphide can be present in wastewater in various states (H_2S, HS^- and S^{2-}). The unionized H_2S is known to have the strongest inhibitory effect, due to its ability to permeate the cell membrane. pH is the main factor that determines the proportion of S^{2-} present in wastewater. Figure 2.3.2A indicates the relation between pH and sulphide speciation. As the pH of wastewater is usually around 7.6, sulphide will mostly be present as HS^-. The Eh-pH diagram depicted in Figure 2.3.2B shows the dominant aqueous species and stable solid phases on a plot defined by the Eh and pH axes. It is noteworthy that, under anaerobic conditions, the zero valent dissolved and suspended sulphur may exist in aqueous solutions as colloidal sulphur or as metal-ligated or free polysulphides and hydropolysulphides (Kamyshny *et al.*, 2008).

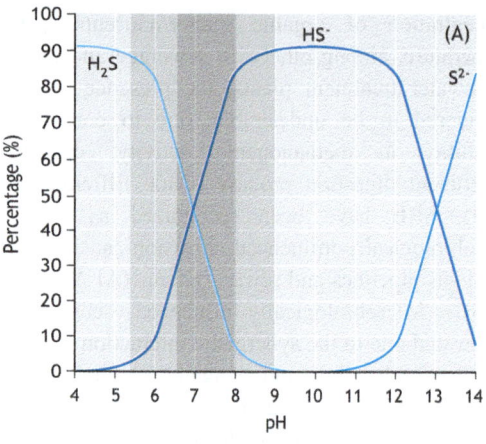

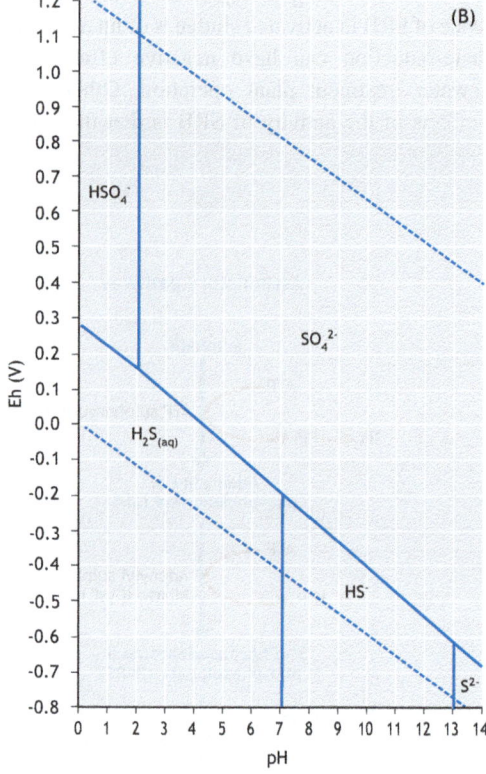

Figure 2.3.2 Sulphide speciation in wastewater: (A) pH influence at 25 °C, as adapted from Rintala and Puhakka (1994); and (B) Eh-pH diagram of various S species in aqueous solution (FACT database, adapted from Bale *et al.*, 2002).

Sulphide speciation can also be influenced by the partition coefficient value, i.e. the ratio of concentration between the gas phase and the liquid phase, which is also heavily pH and temperature-dependent. Other parameters which might influence sulphide speciation are temperature and salt concentration.

The distribution of sulphide in the gas phase (g) and liquid phase (l) can be represented by the following equation (Hulshoff Pol *et al.*, 1998):

$$S_{H2S} = \alpha \cdot C_{H2S} \qquad \text{Eq. 2.3.1}$$

Where, α (alpha) is the dimensionless distribution coefficient for H_2S liquid-gas phase equilibrium. In the liquid phase, S_{H2S}, H_2S exists as uninonized H_2S or in its ionized forms (as bisulfide, HS^- or sulphide, S^{2-}). Even small variations in the pH value will significantly affect the free H_2S concentrations. Again, the Henry's law constant for H_2S at 25 °C is ~3.4 mg L^{-1} atm^{-1}, which indicates the large volatility of this species. The dependence of α on the temperature and pKa values (pKa = - log Ka) is shown in Figure 2.3.3.

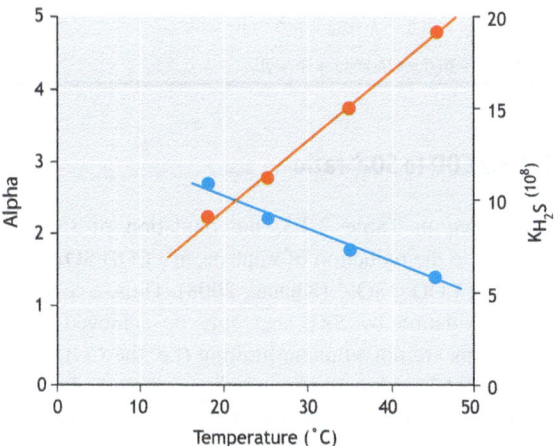

Figure 2.3.3 Temperature dependency on the distribution coefficient alpha (●) and dissociation constant pKa (●) of H_2S (adapted from Hulshoff Pol *et al.*, 1998).

2.3.3 Effects of environmental and operating conditions on SRB

2.3.3.1 Carbon source

To obtain energy for growth and maintenance, SRB can oxidize a wide range of substrates acting as the electron donor that include, among others, hydrogen, alcohols, fatty acids, aromatic and aliphatic compounds (Liamleam and Annachhatre, 2007), while sulphate is used as an external electron acceptor. In certain (industrial) plants, when the wastewater contains an insufficient electron donor/carbon source and there is interest in promoting the removal of organics by SRB, an external carbon source is usually added. Liu and Peck (1981) determined appropriate electron donors for SRB when grown as a pure culture. Several electron donors have been studied as energy and carbon sources for SRB (Table 2.3.2).

Table 2.3.2 Compounds used as energy substrates by SRB (adapted from Hansen, 1993).

Compound	Substrate
Inorganic	Hydrogen, carbon monoxide, etc.
Monocarboxylic acids	Lactate, acetate, butyrate, formate, propionate, isobutyrate, 2- and 3- methylbutyrate, higher fatty acids up to C18, pyruvate.
Dicarboxylic acids	Succinate, fumarate, malate, oxalate, maleinate, glutarate, pimelate.
Alcohols	Methanol, ethanol, propanol, butanol, ethylene glycol, 1,2- and 1,3-propanediol, glycerol.
Amino acids	Glycine, serine, cysteine, threonine, valine, leucine, isoleucine, aspartate, glutamate, phenylalanine.
Miscellaneous	Choline, furfural, oxamate, fructose, benzoate, 2-, 3- and 4-OH-benzoate, cyclohexanecarboxylate, hippurate, nicotinic acid, indole, anthranilate, quinoline, phenol, p-cresol, catechol, resorcinol, hydroquinone, protocatechuate, phloroglucinol, pyrogallol, 4-OH-phenyl-acetate, 3-phenylpropionate, 2-aminobenzoate, dihydroxyacetone

Most electron donors are products of fermentation, monomers or cell components from other sources. The three major criteria for selecting a suitable electron donor for sulphate-reduction process are: (*i*) efficiency of sulphate removal in the effluent complemented by a low COD, (*ii*) the electron donor availability, and (*iii*) cost of unit sulphate converted to sulphide (van Houten *et al.*, 1996). With greater interest in the development of sustainable SRB technologies, hydrogen has been identified as one of the most important substrates for SRB. *Desulfovibrio* species have a high affinity for hydrogen, and this is considered to be the reason why they are able to out-compete hydrogenotrophic methanogens in sulphate-rich environments (Widdel, 2006).

Recently used energy sources for biological sulphate-reduction for industrial applications include complex organic carbon sources. Van Houten *et al.* (1996) used synthetic gas (mixtures of H_2, CO, and CO_2) as the energy source in laboratory scale gas-lift reactors and showed biological sulphate-reduction by SRB. Numerous organic waste matrices (sources) have also been used as carbon sources and electron donors. These include leaf mulch, wood chips, sewage sludge, sawdust, compost, animal manure, whey, vegetable compost, and other agricultural wastes (Liamleam and Annachhatre, 2007). Lactate and molasses, though cost-effective, are not completely oxidized by SRB, generating high COD in the effluent. Hydrogen and ethanol, despite being more expensive, are still being used for sulphate loads greater than 200 kg SO_4^{2-} h^{-1}. However, due to safety reasons, ethanol is preferred to hydrogen. Interestingly, one of the first carbon sources to be considered was waste sewage sludge (Butlin *et al.*, 1956) and COD is usually present in relatively higher concentrations in municipal wastewater, suggesting that for municipal wastewater treatment applications no external electron donors may be needed. During the fermentation of COD in the sewerage system or in the anaerobic zones of wastewater treatment plants, volatile fatty acids (VFA) such as acetate and propionate may contribute to most of the readily biodegradable COD (RBCOD). Other microorganisms like methanogens or acetogens can also utilize VFA as a carbon source under anaerobic conditions, competing against SRB for VFA. The interactions between SRB, methanogens and acetogens depend on the type of VFA and other substrates present in the wastewater (Figure 2.3.4).

From the equations and Gibb's free energy values ($\Delta G°'$) shown in Table 2.3.3, it can be observed that SRB can utilize a broad range of carbon sources (lactate, hydrogen, acetate, propionate and butyrate), whereas methanogens rely mostly on hydrogen and acetate, while acetogens can utilize lactate, propionate and butyrate. Thus, SRB will compete against acetogens for lactate, propionate and butyrate, and against methanogens for hydrogen and acetate. Overall, SRB can have bigger advantages over other anaerobic organisms, as long as there is sufficient sulphate available.

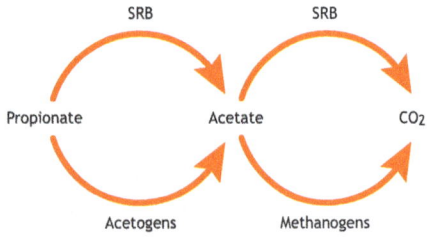

Figure 2.3.4 The competition between sulphate-reducing bacteria (SRB), methanogens and acetogens for VFA in municipal wastewater.

Table 2.3.3 Typical reactions for sulphur-reducing bacteria, methanogens and acetogens: $\Delta G°'$ (kJ mol^{-1}) values are adapted from Thauer *et al.* (2007).

Reaction	$\Delta G°'$ (kJ mol^{-1})
Sulphate-reducing bacteria	
$Lactate^- + 0.5SO_4^{2-} \rightarrow Acetate^- + HCO_3^- + 0.5HS^-$	-80.2
$4H_2 + SO_4^{2-} + H^+ \rightarrow HS^- + 4H_2O$	-36.4
$Acetate^- + SO_4^{2-} \rightarrow 2HCO_3^- + HS^-$	-47.6
$1.33Propionate^- + SO_4^{2-} \rightarrow 1.33Acetate^- + 1.33HCO_3^- + 0.75HS^- + 1.33H^+$	-50.3
$2Butyrate^- + SO_4^{2-} \rightarrow 4Acetate^- + HS^- + H^+$	-55.6
Methanogens	
$4H_2 + HCO_3^- + H^+ \rightarrow CH_4 + 3H_2O$	-33.9
$Acetate^- + H_2O \rightarrow CH_4 + HCO_3^-$	-31.0
Acetogens	
$Lactate^- + 2H_2O \rightarrow Acetate^- + HCO_3^- + H^+ + 2H_2$	-4.2
$Propionate^- + 3H_2O \rightarrow Acetate^- + HCO_3^- + H^+ + 3H_2$	+71.6
$Butyrate^- + 3H_2O \rightarrow 2Acetate^- + H^+ + 2H_2$	+9.6

2.3.3.2 COD to SO_4^{2-} ratio

As shown in Table 2.3.3, the oxidation of COD is coupled to the reduction of sulphate, in a COD/SO_4^{2-} ratio of 0.67 g COD g SO_4^{-1} (Khanal, 2008). Thus, a complete COD oxidation by SRB can only be achieved if the conditions are not sulphate-limiting (i.e. the COD/SO_4^{2-} ratio is 0.67 or lower). For medium-strength domestic wastewaters (for e.g. soluble COD concentrations of approximately 300 mg COD L^{-1}), the minimal amount of sulphate required in the influent can easily be achieved due to the intrusion of brackish water into the sewer, seawater-based toilet flushing and/or discharge of sulphate-rich (industrial) wastewaters. On the other hand, a higher COD/SO_4^{2-} ratio will favour the growth of methanogenic bacteria that will consume part of the available RBCOD.

2.3.3.3 Temperature

The sulphate-reduction rate of SRB and their doubling time are highly dependent on temperature (Hulshoff Pol *et al.*, 1998; Pikuta *et al.*, 2000). The doubling time of microorganisms capable of performing intense sulphidogenesis has been observed to drastically vary between 10 and 118 h within a temperature range of 30 to 60 °C (Pikuta *et al.*, 2000). Table 2.3.4 shows the optimum temperature for the growth of some SRB.

Table 2.3.4 Temperature range for some SRB (Tang et al., 2009).

SRB	Temperature (°C)	
	Range	Optimum
Desulfobacter	28–32	–
Desulfobulbus	28–39	–
Desulfomonas	–	30
Desulfosarcina	33–38	–
Desulfovibrio	25–35	–
Thermodesulforhabdus norvegicus	44–74	60
Desulfotomaculum luciae	50–70	–
Desulfotomaculum solfataricum	48–65	60
Desulfotomaculum thermobenzoicum	45–62	55
Desulfotomaculum thermocisternum	41–75	62
Desulfotomaculum thermosapovorans	35–60	50
Desulfacinum infernum	64	–

The effect of temperature on bacterial sulphate-reduction rates can be evaluated using the Arrhenius model (Isaksen and Jørgensen, 1996). The activation energy for sulphate-reduction can be determined by plotting the logarithm of the sulphate-reduction rate versus the inverse of temperature as follows:

$$\ln(r_{SO4,An}) = \ln(A) + \left(\frac{-E_a}{RT}\right) \quad \text{Eq. 2.3.2}$$

Where, E_a is the activation energy (J mol^{-1}), $r_{SO4,An}$ is the sulphate-reduction rate (nmol cm^{-3} d^{-1}), Where, E_a is the activation energy (J mol^{-1}), k is the sulphate-reduction rate (nmol cm^{-3} d^{-1}), A is a constant, R is the molecular gas constant (8.314 J K^{-1} mol^{-1}), and T is the absolute temperature (K).

2.3.3.4 pH

Another parameter that may influence the competition between SRB and methanogens is pH. pH affects the different metabolic processes of SRB, methanogens and acetogens. Also, pH affects the extent to which VFA and sulphide are inhibitory or toxic to SRB. A slight increase in pH (e.g. from 7.6 to 7.8) helps SRB to become dominant. Both VFA and sulphide tend to be more inhibitory or toxic at low pH values because the protonated species of VFA and total solids (TS) become abundant at these pH and, due to their neutral charge, they can freely diffuse through the membranes of the organisms and affect their intracellular constituents such as enzyme mechanisms. In any case, methanogens are more inhibited by the presence of sulphide than SRB. Therefore, batch activity tests regarding the VFA content, COD/SO$_4^{2-}$ ratio and pH effect on SRB activity are of practical interest. A typical example showing the effect of pH and temperature on the doubling time of SRB is illustrated in Figure 2.3.5.

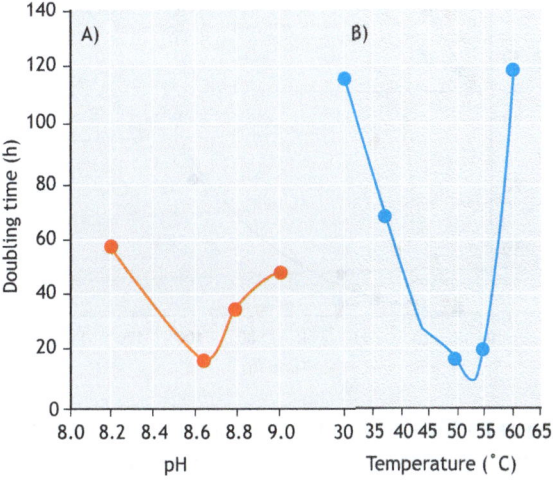

Figure 2.3.5 Influence of (A) pH and (B) temperature on the doubling time of bacterial strains capable of performing high rate sulphidogenesis (adapted from Pikuta et al., 2000).

Pikuta et al. (2000) showed the influence of pH on the doubling time of a strain (referred to as S1^T) capable of showing high rate sulphidogenesis. As illustrated in Figure 2.3.5, the optimum pH for growth was between 8.0 and 9.15, while the doubling times varied between 20 and 58 h depending on the initial pH values. The doubling time of another sulphate-reducing strain HHQ 20 isolated using enrichments with hydroxyhydroquinone and sediment from Venice was lowest (40 h) in the pH range of 6.9-7.2, while at pH 5.0 and 8.0 the doubling times were threefold higher (Reichenbecher and Schink, 1997).

2.3.3.5 Oxygen

SRB are known to be anaerobic organisms. However, sulphate-reduction activity has been reported in oxic regions of microbial mats and sea sediments but, to date, there is no known pure culture capable of performing dissimilatory sulphate-reduction in the presence of oxygen concentrations larger than 1 µM (Cypionka, 2000; Kjeldsen et al., 2004). Kjeldsen et al. (2004) performed short and long term oxygen exposure experiments and monitored sulphate-reduction rates under anoxic incubation conditions (Figure 2.3.6).

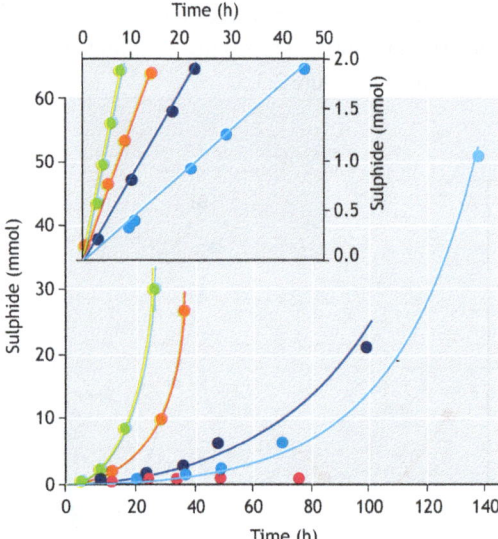

Figure 2.3.6 Sulphide production profiles in batch incubations of sludge samples exposed to 0 h (●), 33 h (●), 73 h (●), and 121 h (●) of constant aeration. The symbol (●) represents a sample that was sterilized and exposed to 121 h of aeration. The inset represents the initial linear phases of the respective curves of the same experiment (adapted from Kjeldsen et al., 2004).

The sulphate-reduction rates, calculated from the initial linear increase in sulphide concentrations, decreased from 0.24 to 0.04 µM h^{-1} when the aeration time was increased from 3 to 121 h. This was attributed to temporary inactivation of an increasing fraction of the SRB population. As shown in Figure 2.3.6, the sulphide production curves follow a biphasic profile, i.e. an initial linear increase resulting from the activity of a constant number of SRB followed by an exponential increase representing the reactivation of SRB. The authors, who monitored the oxygen concentration in the bulk liquid phase and not inside the sludge flocs, raised an important question on whether the SRB were protected from oxygen exposure in anoxic micro-niches created by high respiratory activity and diffusion resistance inside the flocs during aeration experiments. The results from their experiment suggest that SRB are capable of adapting well to the situation and short exposure times to oxygen do not necessarily affect their sulphate-reduction activities.

2.3.4 Experimental setup

To estimate the sulphate-reduction rate or sulphide production rate of a specific activated sludge reactor, one activity test should be performed if the reactor is operated under steady-state conditions for at least three SRT. However, if one would like to envisage the evolution of the rate over time, after changing the bioreactor's process conditions, the rate should be checked more frequently. For instance, depending on (*i*) the research question to be addressed, (*ii*) the operating conditions, (*iii*) substrate and (*iv*) toxic compounds (if any) present, SRB activity tests should be performed in triplicates, at least once every two weeks.

2.3.4.1 Estimation of volumetric and specific rates

Table 2.3.5 gives an overview of parameters of interest. The volumetric rates of sulphate-reduction or sulphide production can be calculated from the highest slope of the exponential phase for each experiment. The equations for the volumetric and specific activity rates are:

$$r_{SO4,An} = \ln\left(\frac{S_{SO4,final}}{S_{SO4,initial}}\right) / \text{Time} \qquad \text{Eq. 2.3.3}$$

$q_{SO4,An}$ = Volumetric activity / Initial biomass concentration

or,

$$q_{SRB,SO4,An} = \frac{r_{SO4,An}}{X_{SRB}} \qquad \text{Eq. 2.3.4}$$

Where, $S_{SO4,final}$ is the final concentration of sulphate (mg SO$_4^{2-}$ L^{-1}) and $S_{SO4,initial}$ is the initial concentration of sulphate (mg SO$_4^{2-}$ L^{-1}) in a batch activity test, respectively.

Table 2.3.5 Stoichiometric and kinetic parameters of interest for SRB-containing activated sludge (AB: active biomass).

Parameter	Symbol	Typical units	
Stoichiometric			
SO$_4^{2-}$ reduction to VFA consumption ratio	$Y_{SO4/VFA,An}$	S-mol C-mol^{-1}	mg SO$_4^{2-}$ mg VFA^{-1}
Kinetic			
Specific VFA consumption rate	$q_{SRB,VFA,An}$	C-mol C-mol^{-1} h^{-1}	mg VFA mg AB^{-1} h^{-1}
Specific SO$_4^{2-}$ uptake rate	$q_{SRB,SO4,An}$	S-mol C-mol^{-1} h^{-1}	mg SO$_4^{2-}$ mg AB^{-1} h^{-1}

2.3.4.2 The reactor

To assess the presence and activity of SRB, batch tests should be conducted under anaerobic conditions either using airtight reactors (see Sections 2.2.2.1 and 2.2.4.1) or serum bottles (Figure 2.3.7). The reactor(s) or serum bottles used for the execution of tests must have the required means to: (*i*) avoid oxygen intrusion under anaerobic conditions, (*ii*) provide satisfactory mixing conditions, (*iii*) maintain an adequate and desirable temperature, (*iv*) control pH, and (*v*) have additional ports for sample collection and addition of influent, solutions, gases and any other liquid medium or substrate used in the test.

Figure 2.3.7 Batch serum bottles containing SRB sludge placed in a shaker (photo: van den Brand, 2015).

Moreover, several serum bottles can be used in parallel, as an alternative to determine the kinetic rates in batch activity tests. This practice is known as using sacrificial bottles. They are of particular interest when pure SRB cultures are to be used to avoid contamination but also to assess the kinetic conversion rates under strict anaerobic conditions and without jeopardizing the gas-liquid equilibrium phase, which can be easily altered when sampling. In this case, when using sacrificial bottles, all the batch experiments should be started with identical initial biomass concentrations that will allow the estimation of sulphate-reduction rates and sulphide production rates. In order to perform these tests and to maintain the purity of the original SRB (when required), different glass vials should be used maintaining similar experimental conditions and sacrificing one vial for each sample. By maintaining multiple vials for the same test condition, the possibility of contamination during sampling can be completely avoided and the gas and liquid phase volumes will remain constant. As explained previously, sufficient care should be taken to provide the strictly anaerobic environments required for SRB growth. Apart from measuring sulphide in the liquid phase, the gas-phase hydrogen sulphide profiles should also be monitored in order to close the sulphur balance within the test bottles.

2.3.4.3 Mixing

When airtight reactors are used, mixing conditions can be provided as described in Section 2.2.2.1 which covers EBPR. When serum bottles are used, mixing conditions are provided by means of an orbital shaker in which several batch bottles can be placed and operated at the same time. Proper mixing should be ensured and stratification, which might result due to variations in settling capacity of activated sludge, should be avoided. The shaker speed should be adjusted in such a way that no disturbance is caused to the floc structure.

2.3.4.4 pH control

pH affects the different metabolic processes of SRB, MET and AC and therefore needs to be carefully controlled and monitored. In reactors, the pH can be controlled as described in Section 2.2.2.1 on EBPR. In serum bottles, pH may be controlled by the presence of a buffering agent in the medium. This can be done by the addition of buffers such as $CaCO_3$ or phosphate buffer during the start of the incubation. Some authors have recommended the addition of sodium bicarbonate buffers or Tris buffer made from an amine base $(HOCH_2)_3CNH_2$ to control pH involving microbiological works. In bioreactors, solutions made of KOH, $NaOH$, NH_4OH, $Ca(OH)_2$, HCl or H_2SO_4 can be automatically added to maintain the pH (Mohan *et al.*, 2005).

2.3.4.5 Temperature control

The sulphate-reduction rates of SRB and their doubling time are dependent on temperature (Hulshoff Pol *et al.*, 1998; Pikuta *et al.*, 2000). Therefore, it is recommended to operate the reactors with automatic temperature control, following the procedures and equipment described in Section 2.2.2.1 on EBPR. For experiments with serum bottles, often a water bath, incubator or a small room with temperature control is used. This makes it possible to control the temperature of many serum bottles simultaneously.

2.3.4.6 Sampling and dosing ports

When experiments are conducted in reactors, the requirements for the sampling and dosing ports will be the same as described in Section 2.2.2.1 on EBPR. When serum bottles are used, gas or liquid samples are taken by syringes with hypodermic needles through the rubber stoppers. If there is enough headspace pressure in the flask, no oxygen intrusion into the system will occur. Alternatively, to avoid the formation of micro-holes in the stopper or septa due to repeated piercing with the needle during sampling, a three-way plastic valve fitted to a needle can be permanently fixed to the septa.

2.3.4.7 Sample collection

For specific cases, if there is interest to measure SRB activity in the sewage, raw influent samples should be taken. In this case, most of the SRB activity in the sewerage will be linked to SRB present in the biofilm attached to the sewer conduits so that the activity of SRB measured in the raw influent in the lab may not be representative of the actual activity taking place in the sewerage. Specific protocols should be followed while collecting biofilm sewer samples (Flemming *et al.*, 2000).

At pilot and full-scale treatment plants, sludge samples can be collected in different places where the SRB activity needs to be addressed: primary settling tanks, anaerobic selectors, anoxic or aerobic tanks, thickeners or anaerobic digesters. In cases where experiments are conducted with laboratory-enriched SRB cultures, it is also possible to convert the continuous bioreactor or sequencing batch reactor into a batch operation mode to measure the activities. This is often considered to be beneficial as the setup is usually adequately equipped, and often more options for controlling and sampling are present, as compared to regular batch flasks. The disadvantage of performing a batch experiment in the 'parent' reactor (as is generally applicable for all microbial cultures of interest) is the possible disturbance of the system because of (a combination of) several factors such as the withdrawal of mixed liquor due to additional sampling, the application of different operating conditions required by the batch tests, changing the substrate composition and concentration, or changing the duration of phases.

2.3.4.8 Media

When activated sludge is used as a source of SRB, real wastewater can be used as substrate, following the procedures described in Sections 2.2.2.4 and 2.2.3.3 on EBPR. The medium composition depends heavily on the objective of the test. For determining the rates of SRB under the conditions applied in the parent reactor, the medium composition should be the same as the feed applied to the parent reactor. For comparison of the rate to other conditions, it could be a consideration to use an equal standard medium for each condition. However, when the research question includes an investigation of the effect of certain compounds on the SRB activity, the compound of interest should be either added to or removed from the media. If many tests of a similar nature are to be executed, it is recommended to prepare a sufficient volume of media in order to save on preparation time. As described in Section 2.3.4.4, suitable pH buffers can be added initially to the batch incubation bottles to maintain the pH.

Depending on the objectives of the activity tests, the COD composition and concentrations, as well as the sulphate concentrations present in synthetic wastewater, can vary since they are usually the subject of investigation. Moreover, the concentrations may be adjusted proportionally for the duration of the test and the sludge concentration used. Usually, concentrations of up to 50 and 100 mg RBCOD L^{-1} are used for tests with full-scale activated sludge samples and of up to 400 mg RBCOD L^{-1} with lab-enriched cultures. Regardless of the nature of the activity test, synthetic wastewater must contain the required macro- and micro-elements (such as potassium, magnesium, calcium, iron, zinc, cobalt, among others) in sufficient amounts for SRB to avoid any metabolic limitation that may jeopardize the outcomes of the batch activity.

Postgate's medium is the most commonly used medium for the growth of SRB and to carry out SRB activity tests (Postgate, 1984). It consists of (in g L^{-1}): K_2HPO_4 (0.5), NH_4Cl (1.0), Na_2SO_4 (1.2), $FeSO_4 \cdot 7H_2O$ (0.5), COD (3.0), $MgSO_4 \cdot 7H_2O$ (2.0), yeast extract (1.0), ascorbic acid (0.1), thioglycollic acid (0.1), and Na_2SO_4 (1.0). Sodium acetate (NaAcetate·$3H_2O$) can be used as the carbon source. However, several synthetic wastewater recipes for performing SRB activity tests have also been described in the literature depending on the carbon source used (Villa-Gomez *et al.*, 2011). van den Brand *et al.* (2014a) recommended the following media composition: NaAcetate·$3H_2O$ (486.2 mg L^{-1}), NaPropionate (147.1 mg L^{-1}), K_2HPO_4 (37.8 mg L^{-1}), KH_2PO_4 (14.2 mg L^{-1}), NH_4Cl (382 mg L^{-1}), $MgCl_2 \cdot 6H_2O$ (93.4 mg L^{-1}), $CaCl_2$ (58.4 mg L^{-1}), trace elements (1 mL L^{-1}), and an adequate sulphate source (for example: $MgSO_4$), resulting in a sulphate concentration of 500 mg

L^{-1}. For fermentative SRB activity tests, Villa-Gomez et al. (2011) used the following media composition in batch and continuous bioreactors: KH$_2$PO$_4$ (500 mg L^{-1}), NH$_4$Cl (200 mg L^{-1}), CaCl$_2$·2H$_2$O (2,500 mg L^{-1}), FeSO$_4$·7H$_2$O (50 mg L^{-1}), MgSO$_4$·7H$_2$O (2,500 mg L^{-1}) and lactate as the electron donor. However, it is noteworthy to remember that a combination of complex carbon sources (lactate and VFA) and sulphate can be used as long as the COD/SO$_4^{2-}$ ratio is < 0.67. However, while carrying out these tests, blackening of the media due to sulphate-reduction and sulphide production can easily be noticed. Other examples of compounds that can be used to introduce sulphate in synthetic wastewater are seawater or sodium sulphate taking into account that the COD/SO$_4^{2-}$ ratio is 0.67. Trace elements can be prepared according to Lau et al. (2006). If a test requires a particular COD/SO$_4^{2-}$ value in order to investigate the effect of limited or excess sulphate levels, the media can be simply adjusted by adding low amounts of sulphate or more COD. When the media is to be prepared in advance, follow the recommended separation of media containing COD source and N source in order to avoid biomass growth in the prepared media. The final colour of this media is transparent. If required, the synthetic wastewater can be concentrated, sterilized in an autoclave (for 1 h at 110 °C) and used as a stock solution if several tests will be performed in a defined period of time. However, the solution must be discarded if any precipitation or loss of transparency is observed.

For experiments performed with lab-enriched cultures, it is recommended to perform the tests with the same (synthetic) media that is used for the cultivation. Alternatively and similar to the case when sludge from full-scale plants is used, the effluent from the reactor can be collected, filtered through rough pore size filters to remove larger particles, and used to prepare the required media with the desired carbon and sulphate concentrations for the execution of the activity tests.

2.3.5 Analytical procedures

Analytical tests required for the SRB activity test can be performed by following standardized and commonly applied analytical protocols as described in Section 2.2.2.5 on EBPR.

However, for estimating the activity of SRB only for a few selected parameters of interest, a modification of standard techniques/protocols is required. Therefore, a detailed description of the methods for COD and sulphide determination is given below. The analysis for sulphate is also included, as this parameter is of interest in SRB activity tests. However, for each test, it is important to realize that some compounds interfere with analytical tests. For instance, COD determination is very sensitive when chloride is present in the samples. The effect of such compounds on the analytical tests should be verified by carrying out the analysis without this compound.

Concerning sulphide concentration measurements, this could be done by making a standard sulphide concentration with a known concentration and a solution with the suspected disturbing compound. Then two tests can be performed in parallel. The first test should be carried out using the standard solution and the second test using the standard solution plus the suspected compound at concentration levels that are usually expected in the mixed liquor. If the sulphide concentration is the same in both analytical tests, the effect of the potentially disturbing compound is not significant in the tests at the applied concentrations and therefore, the analytical test should be considered as appropriate for analyses.

2.3.5.1 COD$_{organics}$ and COD$_{total}$

The method to measure COD has been extensively described elsewhere (APHA et al., 2012), though some important additions are required. It is usually straightforward to determine the total COD of the sample or only the soluble COD fraction. To measure the soluble COD fraction, the sample should be filtered using a 0.45 μm filter. The sulphide also contributes to the COD measurements, and often this contribution (interference) is not desired. Therefore, two analyses should be performed, named hereafter as COD$_{organics}$ and COD$_{total}$. The COD$_{total}$ (also referred to as the total COD concentration in a soluble state), in this particular case, also includes sulphide, while in the COD$_{organics}$ analysis, interferences due to sulphide contribution can be eliminated from the sample by applying a mathematical correction and thus the COD concentration is only associated with the content of organic matter present in the sample. As recommended by Boyles (1997), prepare a separate standard for sulphide and perform a COD test on the standard, and apply a mathematical correction to the result of COD$_{total}$.

Another method to completely remove sulphide from the sample is based on the precipitation based procedure of Poinapen et al. (2009). A few drops of 10 M NaOH solution are added to the sample, as an increase in pH will induce precipitation. The H$_2$S/HS$^-$ system has a pK$_{s1}$ of 7.05 at 25 °C (Bjerrum et al., 1985) and at a pH of 10.0 most of the sulphide present in the sample is in the form of HS⁻ species with a very small proportion in the form of S²⁻ species (pK_{s2} = 12.92 at 25 °C). Neither HS⁻ nor S²⁻ species can escape from the sample since both compounds are very well dissociated. Subsequently, zinc sulphate (ZnSO₄) is added to precipitate with sulphide as zinc sulphide (ZnS), which should be separated from the sample by filtration. Finally, the measured concentrations should be corrected for the dilution with the ZnSO₄ solution.

2.3.5.2 Sulphate

Sulphate can be measured by various techniques, such as gravimetric, ion chromatography, methylthymol blue method and turbidimetric. These methods are described elsewhere (APHA *et al.*, 2012). The most commonly used method is a Hach Lange kit, a turbidimetric-based procedure or using a high performance liquid chromatography (HPLC) equipped with an AS9-SC Column and an ED 40 electrochemical detector (Dionex).

2.3.5.3 Sulphide

Depending on the pH, sulphide can be present in various states. Therefore, it is important to perform a sulphide test in both the liquid as well as the gas phase.

Liquid samples

When samples are taken for sulphide measurements, the procedures applied are crucial as sulphide is easily lost during sampling due to stripping or chemical reactions. It is therefore important to solubilize the volatile fraction in the liquid phase, for example using NaOH solution. For the analysis of sulphide, two main procedures can be used. The first procedure is referred to as the indirect COD method, while the second procedure is the direct methylene-blue method. Both these procedures will be described in this section. 1 g of sulphide corresponds to ~2 g COD. The sulphide concentration could therefore be expressed by its COD content.

This could be achieved by measuring the COD concentration of the sample in terms of COD_{total} and $COD_{organics}$. The COD concentration should be analysed using the standard method, according to the specific treatment steps described previously (Section 2.3.5.1). The sulphide concentration could then be calculated by subtracting the $COD_{organics}$ value from the COD_{total} value, and corrected for the COD/sulphide weight ratio. The expression to calculate the sulphide content (S_{H2S} in g L⁻¹) is:

$$S_{H2S} = (COD_{total} - COD_{organics}) / 2 \qquad \text{Eq. 2.3.5}$$

Where, 2 is the sulphide to COD ratio (*w/w*).

In the direct methylene-blue method, specific for sulphide analyses, zinc acetate is used to ensure a higher degree of sulphide fixation. Subsequently, the reagents containing dimethyl-parafenyl-diamine and iron are used to form methylene blue, resulting in a blue solution. The intensity of the colour can be analysed photometrically to quantify the sulphide concentration in the sample.

Gas-phase samples

Gas-phase samples should be collected from the batch incubation bottles using airtight glass syringes (for example: Hamilton, USA). The most commonly employed method to estimate gas-phase H₂S concentrations is using gas chromatography (GC). Li *et al.* (2013) used a GC (Agilent 6890 N, USA) fitted with a flame photometric detector (FPD) and DB-1701 capillary column (30 m × 0.32 mm × 0.25 μm, Hewlett Packard, USA) to estimate H₂S concentrations. The temperature of the oven, injection, and detector were set at 100 °C, 50 °C, and 200 °C, respectively, while nitrogen was used as the carrier gas. H₂S concentrations can also be determined by titration using a standard potassium iodide-iodate as the titrant and starch indicator (APHA *et al.*, 2012). Another quick and easy method to monitor the H₂S concentrations in wastewater treatment plants is the use of gas detector tubes (Kitagawa, Japan). H₂S concentrations can also be measured using Jerome 631-X Hydrogen Sulphide Analyser (Arizona Instruments, USA). However, in cases where it is difficult to perform the analysis immediately, Tedlar bags (Tedlar gas sampling bags, Sigma-Aldrich) should be used for collecting gas samples from continuously operated reactors.

Temperature and pressure are two important factors that need to be considered during the calibration of GC for gas-phase H₂S measurements. If S²⁻ in liquid phase is added to a closed system, the resulting gas-phase H₂S concentration (C_{H2S}) in the system can be calculated as follows:

$$C_{H2S} = 22.4 + 10^6 \cdot \frac{\rho \cdot V_L}{MW \cdot V} + \frac{T}{273} + \frac{760}{P} \qquad \text{Eq. 2.3.6}$$

Where, ρ is liquid density (g mL⁻¹), V_L is the volume of liquid (mL), T is the temperature (K), P is the pressure (torr), MW is the molecular weight (g mol⁻¹) and V is the volume of the closed system (L).

2.3.6 SRB batch activity tests: preparation

This section describes not only the different steps but also the apparatus characteristics and materials needed for the execution of the SRB batch activity tests.

2.3.6.1 Apparatus

If the experiments are conducted using a reactor, the description given in Section 2.2.3 on EBPR is applicable. If the tests are conducted in serum bottles, the following apparatus is required:

1. Serum vials, including a rubber stop and aluminium crimps to secure an anaerobic environment.
2. A nitrogen gas supply.
3. A pH electrode.
4. A 2-way pH controller to add either HCl and/or NaOH (alternatively, a one-way control for HCl addition or manual pH control can be applied through the manual addition of HCl and/or NaOH).
5. A thermometer (recommendable temperature working range of 0 to 40 °C). Confirm that the electrodes or meters (pH, thermometer) are calibrated less than 24 h before execution of the batch activity tests in accordance with guidelines and recommendations from suppliers.
6. A room or incubator to control the temperature at the desired temperature.
7. A shaker in which the rpm can be set up to 300 rpm.
8. A pipette and tips to measure the exact volume of the samples taken.
9. Syringes and needles to take the sample from the serum vials by making a vacuum.
10. Filters with a pore size of 0.45 μm for sample filtering and for preservation purpose.
11. Additionally, all the materials required for analysis. These are elaborated in Section 2.3.5.

2.3.6.2 Materials

1. Two graded cylinders of 1 or 2 L volume (depending upon the sludge volume used) to hold the sample and wash it, if required.
2. At least 2 plastic syringes (preferably of 20 mL or at least of 10 mL volume) for the collection and determination of soluble compounds (after filtration).
3. At least 3 plastic syringes (preferably of 20 mL volume) for the collection of solids, particulate or intracellular compounds (without filtration).
4. At least 3 glass syringes (preferably of 5 mL volume) for the collection of gas-phase samples.
5. 0.45 μm pore size filters. Preferably not of cellulose acetate because they may release some traces of cellulose or acetate into the collected water samples. Consider having at least twice as many filters as the number of samples that need to be filtered for the determination of soluble compounds.
6. Transparent plastic cups with a volume of 10 or 20 mL to collect the samples for the determination of soluble compounds (e.g. soluble COD, acetate, propionate, sulphate, and sulphide).
7. Transparent plastic cups with a volume of 10 or 20 mL to collect the samples for the determination of MLSS and MLVSS. Consider the collection of these samples in triplicates due to the variability of the analytical technique.
8. A plastic box or dry ice box filled with ice with the required volume to temporarily store (for up to 1-2 h after the conclusion of the batch activity test) the plastic cups and plastic tubes for centrifugation after the collection of the samples.
9. Plastic gloves and safety glasses.
10. Pasteur or plastic pipettes for HCl and/or NaOH addition (when the pH is controlled manually).
11. Metallic lab clips or clamps to close the tubing used as a sampling port when samples are not collected from the reactor/fermenter.

2.3.6.3 Media

- **Real wastewater**
  See Section 2.2.3.3 on EBPR.

- **Synthetic media or substrate**
  If tests require synthetic wastewater, depending on the type of the tests, the synthetic influent media could contain a mixture of carbon and sulphate sources plus relevant (macro- and micro-) nutrients. Generally, they can be mixed all together in the same media or split in two solutions (*i*) COD source, and (*ii*) SO_4^{2-} and N source plus the nutrient solution).
  a. COD source solution: It must be composed of RBCOD, preferably of VFA such as acetate or propionate, depending on the nature or goal of the test and the corresponding test objective (research questions). For anaerobic batch activity tests, the concentration needs to be adjusted to ensure that the COD is consumed within the duration of the test. For batch activity tests performed with activated sludge from a full-scale plant, usually COD concentrations not greater than 100 mg L^{-1} are recommended. For lab-scale mixed liquor sludge samples, the COD concentrations can be

as high as the influent COD concentration of the lab-scale system (and even sometimes 2 to 3 times higher). An example of a COD solution is NaAcetate·3H$_2$O (486.2 mg L^{-1}) or Na-Propionate (147.1 mg L^{-1}).

b. Sulphate source solution: Sulphate can be added as sodium sulphate. However, in some cases and especially in the case of saline wastewater treatment, sulphate is accompanied by salinity. To mimic those conditions, it is also possible to add sea salt (containing sulphate). The sulphate concentrations can be adjusted, as desired, depending on the purpose of the experiment. The nutrient solution should contain all the essential macro- (ammonium, magnesium, calcium, potassium) and micro-nutrients (iron, boron, copper, manganese, molybdate, zinc, iodine, cobalt) to ensure that SRB metabolism is not limited. This can lead to the wrong interpretation of results and in extreme cases, a near complete failure of the test. Thus, one must make sure that all the required compounds are added to the solution and in the correct amounts. The media composition is recommended in Section 2.3.4.8.

c. Washing media: If the sludge sample must be washed to remove the presence of an undesirable compound (which may be even inhibitory or toxic), prepare a fresh nutrient solution to wash the sludge containing the medium composition described above, without VFA. The washing process can be repeated two or three times following a similar procedure likewise in Section 2.2.3.5. Thereafter, the necessary preparation steps of the batch activity tests can be performed.

d. Preparing acid and base solutions: see Section 2.2.3.3 on EBPR.

If analyses are not outsourced to specialized analytical labs, the required stock and working solutions to carry out the determination of the analytical parameters of interest must be also prepared in accordance with standard methods (APHA *et al.*, 2012) and the corresponding protocols.

2.3.6.4 Material preparation

- For information on the number of samples, see Section 2.2.3.4 on EBPR batch activity tests.
- Frequency of sample collection:
    a. If the maximum specific kinetic rates must be determined (e.g. maximum specific acetate or propionate uptake rate or maximum sulphate-reduction rate), then increase the frequency of sampling during the initial few hours or days of incubation. In general, samples should be collected once every 5 min during the first 30-40 min of duration of the batch activity test, similar to the method described for EBPR tests.
    b. To ascertain the stoichiometric ratios and not the kinetic ones (e.g. anaerobic SO$_4^{2-}$ reduction/HAc uptake ratio), then the samples can be collected only during the beginning and at the end of each phase to determine the conversions of interest within the selected period (phase).
- To increase the reliability of data collected and to determine the initial conditions of the sludge, it is strongly recommended to collect a series of samples before any media is added. Carefully define the maximum and minimum working volumes of the reactor, as described in Section 2.2.2 on EBPR.
- Preparation of sample cups can be performed as described in Section 2.2.3.2 on EBPR. A simple working plan created in a spreadsheet (Section 2.3.9) can be rather useful to perform and keep track of the sample collection frequencies. Furthermore, this spreadsheet can be used to maintain a database of the different batch tests carried out. Organize all the required material within a relatively close radius of action, around the batch setup, so that delay in handling and preparing the samples can be avoided. Calibrate all the meters (pH and thermometer) less than 24 h prior to the execution of the tests and store them in proper solutions until the execution of the tests, following the particular recommendations of the corresponding manufacturer or supplier and confirm that their readings are reliable before the start of the tests. Samples must be properly stored and preserved until they are analysed (Table 2.3.6).

Table 2.3.6 Recommended sample storage and preservation procedure for the determination of SRB activity tests.

Parameter	Material of sample container	Method of preservation	Maximum recommended time between sampling, preservation procedure and analyses
Sulphate	Plastic or glass	Filter immediately after sample collection thorough 0.45 μm pore-sized filter and cool to 1-5 °C, or freeze to -20 °C.	24 h for samples stored at 1-5 °C; up to 1 month for frozen samples.
Sulphide	Plastic or glass	Immediately solubilize in a drop of 1M NaOH.	Recommended to perform these measurements as quickly as possible.

2.3.6.5 Mixed liquor preparation

These procedures consider that batch activity tests can be performed as soon as possible after collection of samples from full- or lab-scale systems or, in the worst case scenario, within 24 h after collection. Performing the batch tests 24 h after the collection of mixed liquor sludge samples is not recommended because the SRB culture can undergo potential biochemical changes during handling (unless the exposure time after collection is of particular interest for performing these tests). Bearing in mind the previous comments, the following three protocols are recommended to prepare the mixed liquor samples for conducting batch activity tests:

1. If batch activity tests can be performed in less than 1 h after collection of the sludge sample and if the sludge sample does not need to be washed:
   a. Adjust the temperature of the batch reactor where the tests will take place to the target temperature of the study.
   b. Collect the sample at a wastewater treatment plant (WWTP) from one of the following:
      - From the raw influent.
      - At the end of the aerobic tank or stage.
      - At the outlet of the primary or secondary sludge thickeners or anaerobic digesters.
      - Take sludge from a laboratory reactor.
   c. Transfer the sludge sample to the reactor or serum bottles where the batch activity tests will take place.
   d. Start a gentle mixing (50-100 rpm in the case of the reactor) and follow up the temperature of the sludge sample by placing a thermometer inside the reactor or serum bottle (if the setup does not have a built-in thermometer).
   e. Maintain mixing conditions until the sludge has reached the target temperature of the study.
   f. Maintain the anaerobic conditions by using an airtight reactor and sparge N_2 gas to avoid/reduce oxygen intrusion.
   g. If nitrate is detected in the aerobic samples collected to perform these tests then follow instructions presented in Section 2.2.2.3 on EBPR. Alternatively, for nitrate concentrations greater than 10 mg NO_3^--N L^{-1}, 9 mg COD could be added per every mg of nitrate detected (9 mg COD mg NO_3^--N^{-1}). As soon as nitrate is no longer observed, the sludge can be immediately used to conduct the anaerobic batch activity tests. Nitrate or nitrite detection strips (Sigma-Aldrich) can be used for a quick estimation of the presence of these compounds.

2. If batch activity tests can be performed in less than 1 h after collection of the sludge sample but the mixed liquor sludge sample needs to be washed, see washing procedures described in Section 2.2.3.5.
   a. Transfer the 'washed' sludge sample to the reactor or fermenter where the batch activity tests will be performed.
   b. Start a gentle mixing (50-100 rpm in the case of the reactor) and follow up the temperature of the sludge sample by placing a thermometer inside the reactor (if the set-up does not have a built-in thermometer).
   c. Maintain mixing conditions until the sludge has been exposed to the target temperature of study for at least 30 min.

3. In some cases, due to location and distance issues, the tests cannot be performed in less than 1 or 2 h after collection (but within 24 h):
   a. Adjust the temperature of the batch reactor to the target temperature of the study.
   b. Keep the sludge sample cold until the commencement of the test (e.g. by placing the bucket or jerry can in a refrigerator at 4 °C).
   c. Prior to the test, take out the sludge sample from the refrigerator, cool box or cold room.
   d. Mix the contents gently in order to obtain a homogenous and representative sample with a similar MLSS concentration as in the original lab- or scale system from where it was originally collected.
   e. If the sample needs to be washed, wash the sludge in a mineral solution (see Section 2.3.6.3 on working solutions).
   f. Transfer the washed sludge to the reactor or fermenter where the batch activity test will take place.
   g. Start to flush N_2 gas in the sludge sample while mixing gently to remove any dissolved oxygen.
   h. Follow the temperature of the sludge sample by placing a thermometer inside the reactor (if the setup does not have a built-in thermometer).
   i. Keep flushing N_2 gas and mix the contents for at least 1 h (maximum 2 h), but ensure that the sludge is exposed to the target temperature of the study for at least 30 min.
   j. After the procedure of adjusting the temperature, if nitrate is detected (see Section 2.3.6.3), add a solution containing readily biodegradable COD at a 9 mg COD mg NO_3^--N^{-1} ratio. As soon as nitrate is no longer observed, the sludge can be immediately used to start and conduct the anaerobic batch activity tests.

2.3.6.6 Sample collection and treatment

For the purpose of this activity test, three types of samples are required: *(i)* a filtered sample, *(ii)* a filtered sample with NaOH and *(iii)* a biomass sample. Table 2.3.7 provides a list of analyses corresponding to different sample collection and treatment steps.

Table 2.3.7 List of analyses performed on each sample.

Sample	Parameter
Filtered sample	COD_{total}; $COD_{organics}$; Sulphate
Filtered sample with NaOH	Sulphide
Biomass sample	MLVSS; MLSS

- **Filtered sample**

The filtered sample is used for the analyses of COD_{total}, $COD_{organics}$ and sulphate. A well-mixed sample is taken from the serum vial with a needle and a syringe. The needle is injected into the rubber stopper. Then the serum bottle is turned upside down, so that the needle is immersed in the liquid phase. The syringe is pulled in order to create a vacuum and collect the sample. Subsequently, the sample is directly filtered using a 0.45 µm filter, to avoid further conversions. Thereafter, the COD_{total}, $COD_{organics}$ and sulphate concentration are measured according the protocol described in Section 2.3.5.

- **Filtered sample with NaOH**

The filtered sample in which a few drops of NaOH are added is used for the analyses of the sulphide concentration. The sample should be transferred into a tube containing three drops of 1 M NaOH directly after sampling. This is done by using a needle and a syringe. This needle is pierced into the rubber stopper (septa). Then the serum bottle is turned upside down, such that the needle is immersed in the liquid phase. The syringe is pulled back in order to create a vacuum and collect the sample. The sooner the sample is mixed with NaOH, the less sulphide is lost during sampling. Thereafter, the sample is filtered using a 0.45 µm pore-size filter and analysed as described in Section 2.3.5. This analysis should be performed directly after collecting the sample, to avoid sulphide losses.

- **Gas sample**

Gas-phase samples should be collected from the batch incubation bottles using air-tight glass syringes. A calibration plot should be prepared in the GC using standard H_2S gas cylinders (example: 25 to 1,000 ppm). Gas-phase analysis should also be performed immediately after collecting the sample from the glass bottles in order to avoid losses.

- **Biomass sample**

In order to estimate MLVSS and MLSS concentrations, a well-mixed sample from the reactor should be collected. The easiest method is to use a syringe and needle arrangement. Ensure that you know the exact volume taken from the reactor. The MLVSS and MLSS should be analysed in triplicate as described elsewhere (APHA *et al.*, 2012). The frequency of analysing COD, sulphate and sulphide depends on the biomass activity. After data processing, a linear line (as presented in Figure 2.3.8) is required for proper analyses. The more data points present on this linear line, the better, but a balance between investments (in money and time) suggests that four points on the linear line are already sufficient. The safest option is to take several samples during the first hour, once every 10 or 15 min. The best way to deal with this is to perform extensive sampling in the first tests (as a trial), and based on these results, a design can be made to ascertain how much data points are actually required to perform subsequent tests.

2.3.7 Batch activity tests: execution

This section describes how a batch activity test of SRB should be executed. The description consists of a material and chemical list, protocol to prepare media, sampling scheme (time frame), sample collection, and analysis. This section concludes with a step-by-step approach to performing sulphate-reducing bacteria activity tests. The material and chemical list specific for the analytical tests, as well as how the tests should be conducted, is described in Section 2.3.6.

The general advice as provided in Section 2.2.4 on EBPR should be following accordingly for the SRB activity tests. Table 2.3.8 gives examples of typical tests performed to ascertain SRB activities.

Table 2.3.8 Typical tests for SRB activity analyses.

Test code	Redox	Short description and purpose
SRB.ANA.1	Anaerobic	Performed with real wastewater, for instance, the raw influent, to determine the exact rate of SRB under that particular condition.
SRB.ANA.2	Anaerobic	Conducted using a defined standard media (see Section 2.3.4.8), to compare the sulphate-reduction rate from SRB under different conditions.

Test SRB.ANA.1 Anaerobic SRB activity test

The first step of conducting a SRB activity test is to ensure the correct composition of medium and following a correctly designed experimental schedule (Section 2.3.8.2). Both steps heavily rely on the goal and scope of one's research, but also on the origin of the SRB sludge (enriched culture, mixed culture or culture collected from a full-scale or lab-scale system). The required volume of the sludge sample depends on the objective of the study and the research questions. It should be ensured that the biomass and liquid/gas phase concentrations are comparable as much as possible to the original situation, unless required otherwise. After the experimental design, the steps shown below should be followed. The same approach is also valid for SRB activity tests in reactors.

1. Collect the items on the material and chemical lists.
2. Prepare the required stock solutions for conducting analytical tests.
3. Prepare the media stock solutions.
4. Fill the serum vials with media and correct the pH, if necessary, with HCl and/or NaOH.
5. Transfer the SRB sludge into the serum bottles (preserving the anaerobic conditions as much as possible, for example, by avoiding turbulence).
6. Close the serum bottles with a rubber stopper and aluminium crimp.
7. Flush both the headspace and the liquid phase in the serum bottles with N_2 gas. This is done by injecting a needle with the gas supply and placing another needle slightly above the liquid phase from which the overpressure can escape. This latter needle should not touch the liquid phase; otherwise, the sludge will start to escape. For 80-110 mL bottles, 1 min flushing is enough to obtain the required anaerobic conditions, but this could be checked by the addition of resazurin (dying colour). The media becomes colourless after becoming anaerobic. Remove both needles from the septa at exactly the same time. The bottles should now be anaerobic.
8. Now take a sample at time t = 0 and handle this and other samples immediately. The sample is collected using a syringe. Then turn the bottle such that the needle touches the liquid (avoid taking out biomass). Pull the syringe back to collect the sample.
9. Collect the sulphide sample in 1 drop of 1 M NaOH, and the rest can be used for COD and sulphate analyses. Sulphide samples should be solubilized in NaOH and immediately analysed.
10. Follow the time scheme of the sampling.
11. After finishing the time scheme, open the serum bottles and collect the biomass sample and perform MLVSS and MLSS analyses.
12. If the experiments cannot afford SRB sludge losses and it is desired to put the sludge back into the parent reactor, wash it first three times with the media used to feed the parent reactor. If the media composition is not known, use demineralised water for washing.
13. Then the data should be analysed using the procedure described in Section 2.3.8.
14. Note: if the reactor is used for batch activity tests, a similar procedure as described for EBPR anaerobic tests should be followed (see Section 2.2.4).

Test SRB.ANA.2 Anaerobic SRB activity test

A similar approach should be followed for SRB.ANA.2 using synthetic media (Section 2.3.6.3) to estimate the SRB activity.

2.3.8 Data analysis

2.3.8.1 Mass balances and calculations

The experimental data can be used to calculate the sulphate-reducing activity of SRB (Section 2.3.8.2). Sulphide formation is a result of SRB activity and the variability of rate coefficients depends on the carbon source, initial sulphate concentrations and the presence of specific SRB genera. The additional COD, sulphate and sulphide concentration measurements taken at time t = 0 and at the end of the measurement (effluent) should be used to check the mass balances. Rate determination can only be reliable with the correct mass balances in place. The mass balance equations for COD and sulphur compounds can be represented as follows:

COD balance
$$COD_{organics,in} + S_{H2S,in} = COD_{organic,out} + S_{H2S,out} \quad \text{Eq. 2.3.6}$$

Sulphur balance
$$S_{SO4,in} + S_{H2S,in} = S_{SO4,out} + S_{H2S,out} \quad \text{Eq. 2.3.7}$$

Also it is possible to calculate an electron balance.

To calculate the biological sulphate-reduction rate (only when the mass balances are correct), the following calculations should be executed:

a. Use a standard calibration curve to convert the measured optical density values (OD_{675}) from the spectrophotometer to concentration units (mg L^{-1}).

b. Convert the sulphide concentration from mg L^{-1} into mM (if desired). The molar mass of completely dissolved sulphide (S^{2-}) is 34 g mol^{-1}.
c. Take the average value of the triplicate.
d. To obtain a profile indicating the amount of sulphide produced, correct for the sulphide concentration present at time t = 0 min (by subtracting this value from all the values).
e. Determine the slope from the first samples, in which a linear line is expected.
f. Divide the slope value by the amount of biomass present (g VSS), to determine the rate in mmol SO$_4^{2-}$ g VSS^{-1} h^{-1}). The formation of 1 mol sulphide is related to the removal of 1 mol sulphate. Also a correction from molarity to mol is required.

2.3.8.2 Data discussion and interpretation

The results of COD, sulphate and sulphide analyses of the samples taken during the start and the end of the test can be used to check the mass balances. As sulphide is easily lost during sampling, a mass balance of 100 % is not easy to obtain. Therefore, a mass balance fit of 95 % is considered to be satisfactory in order to continue data analyses. If only sulphide formation in a batch test is analysed to measure the biological sulphate-reduction rate by SRB, it is important to perform the test at least in triplicate. It is also very important to check the mass balances, which is possible by executing COD and sulphate analyses at time t = 0 and at the end of the test. The methylene-blue method is a very sensitive and reliable technique. Therefore, when the mass balance fit is around 95 %, the sulphide accumulation profile can be the basis for the calculation of the reduction rate. It is also possible from this type of analyses to determine the characteristic conversion parameters, such as the maximal conversion rate. To measure the maximal conversion rate, it is crucial that all compounds are present in excess. Then, the maximal conversion rate can be determined as described in Sections 2.3.9 and 2.2.5.2. The Lineweaver-Burkplot method can be used to determine the characteristic parameters. This method is described in detail by Nelson and Cox (2005).

There are only limited results available for an SRB activity test, specifically developed for domestic wastewater treatment. Some available literature results are presented in Table 2.3.9 and evidently a wide range of sulphate-reduction rates have been reported depending on the operating conditions that facilitated SRB activity. Although the SRB are strictly anaerobic microbes, some studies (e.g. Lens *et al.*, 1995) are based on aerobic wastewater treatment, and sometimes with low sulphate concentration in the influent, resulting in rates that were significantly lower than the ones observed by van den Brand (2014a; 2014b; 2015).

Table 2.3.9 Sulphate-reduction rates ($q_{SRB,SO4,An}$) reported in the literature from sludge taken from reactors after long-term operation.

System	Scale	$q_{SRB,SO4,An}$ (mg SO$_4^{2-}$ g VSS^{-1} h^{-1})	References
RBC	Full	1-11	Lens *et al.* (1995)
CAS	Full	0-3.1	Lens *et al.* (1995)
CAS	Laboratory	1.9	Lens *et al.* (1995)
ANS	Full	6-7.3	Lens *et al.* (1995)
ANS	Laboratory	11.6	Lens *et al.* (1995)
ANS	Laboratory	100-220	van den Brand *et al.* (2014a; 2014b; 2015)

RBC: Rotating Biological Contactor
CAS: Conventional Activated Sludge
ANS: Anaerobic Sludge

2.3.9 Example

Table 2.3.10 shows an example of worksheet that can be used for different SRB activity tests. The example concerns the tests solely based on the production of sulphide and COD, sulphate and VSS measurements taken during the beginning and end of the experiments. Figure 2.3.8 is an example of a test performed with synthetic wastewater that also includes the analysis of VFA (acetate and propionate).

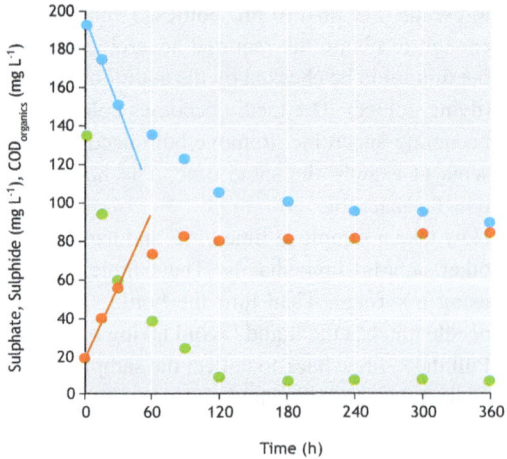

Figure 2.3.8 Example of a graph depicting sulphide (●), sulphate (●) and COD$_{organics}$ (●) concentrations during the SRB.ANA.2 activity test from which the sulphate-reduction rate can be calculated.

For this test, it is suggested to perform several (at least 10) COD and sulphate measurements, and VSS/TSS measurements from the sludge at the end of the test, as described in Section 2.3.7.

Using the protocol described in Section 2.3.5, a graph is obtained, from which the slope can be determined. A typical profile obtained from an SRB activity test is presented in Figure 2.3.8. The slope calculated in this example is 225 mg SO_4^{2-} h^{-1}, which corresponds to 2.34 mmol SO_4^{2-} h^{-1}. The amount of biomass in the reaction vessel was 4.5 g MLVSS. The working volume of the batch reactor was 2.5 L. The actual biological sulphate-reduction rate was subsequently calculated as follows:

$$q_{SRB,SO4,An} = \frac{2.34}{4.5} = 0.53 \text{ mol } SO_4 \text{ g VSS}^{-1} \text{ h}^{-1} \qquad Eq.2.3.8$$

SRB activity tests could also be used to investigate the effect of compounds on the sulphate-reduction rate. Figure 2.3.9 presents an example of a study comparing different nitrogen compounds (ammonium, nitrate and nitrite) at different N concentrations as performed by van den Brand *et al.* (2015). By calculating the sulphate-reduction rate under specific conditions as performed in Figure 2.3.9, and by presenting all the SRB rate results in one figure, the effect of nitrogen compounds on the sulphate-reduction rate can be analysed. In this case error bars are included to show the accuracy of the measurements. The sulphate-reduction rate decreased in time with an increased nitrogen concentration (Figure 2.3.9).

Table 2.3.10 A typical example of a worksheet for an SRB activity test (type SRB.ANA.2). Note that when more information regarding the kinetics is desired, sulphate and $COD_{organics}$ analysis should be performed during all the selected times. However, instead of $COD_{organics}$, the actual acetate and propionate concentrations should be measured.

Anaerobic sulphate reduction batch tests — Code: SRB.ANA.2

Date:	Wednesday, 09.10.2015 10:00 h
Description:	Test at 10 °C, pH 7, synthetic substrate and enriched SRB culture
Test No.:	1 of 3
Duration	6.0 h (360 min)
Substrate:	Synthetic: acetate and propionate (total 400 mg L^{-1})
Sampling point:	Serum vial using needle
Samples No.:	SRB.ANA.2(1.1-1.11)
Total sample volume:	80 mL (20 mL for MLVSS, 6 mL for other samples)
Reactor volume:	2.5 L source sludge, 80 mL serum vial

Experimental procedure in short:
1. Confirm availability of sampling material and required equipment.
2. Confirm calibration and functionality of systems, meters and sensors.
3. Fill serum vial with media and correct for pH.
4. Take sample for COD (acetate and propionate: $COD_{organics}$) in the feed.
5. Transfer 100 mL of sludge to serum vial.
6. Close serum vial with a rubber stop and aluminium cap.
7. Flush serum vial headspace with N_2 gas.
8. Place vial on mixing plate and take sample at t = 0.
9. Take other samples following the sampling scheme below.
10. After last sample is taken, measure biomass concentration of the remaining vial content.
11. Verify that all systems are swtiched off.

Sampling schedule

Time (min)	-20	0	15	30	60	90	120	180	240	300	360
Time (h)	-0.33	0.00	0.25	0.50	1.00	1.50	2.00	3.00	4.00	5.00	6.00
Sample No.	1	2	3	4	5	6	7	8	9	10	11
Parameter					ANAEROBIC PHASE						
COD (mg L^{-1})	396[1]	136	95	59	39	25	9	8	8	8	8
Sulphide (mg L^{-1})			18	40	58	74	83	81	83	84	85
Sulphate (mg L^{-1})			195	176	153	137	124	107	102	96	91
MLSS and MLVSS (mg L^{-1})											See table

[1] Average value of the COD concentration present in the synthetic substrate prior the start of the test

MLSS and MLVSS measurements

Sampling point	Cup No.	W1	W2	W3	W2-W1	W2-W3	MLSS	MLVSS	Ratio
End anaerobic phase	1	0.08835	0.10741	0.08849	0.01906	0.01892	1,906	1,892	0.99
	2	0.08835	0.10759	0.09018	0.01924	0.01742	1,924	1,742	0.91
	3	0.08834	0.10683	0.08940	0.01849	0.01742	1,849	1,742	0.94
						Average	1,893	1,792	0.95

Biomass composition

Sampling point	End anaerobic phase
MLSS (mg L^{-1})	1893
MLVSS (mg L^{-1})	1792
Ratio	0.95
Ash (mg L^{-1})	101

Note:
Acetate (CH_2O)	30.03 mg C-mmol^{-1}
Propionate ($CH_2O_{0.67}$)	24.69 mg C-mmol^{-1}
Sulphide (H_2S)	34.08 mg S-mmol^{-1}
Sulphate (SO_4^{2-})	96.06 mg S-mmol^{-1}

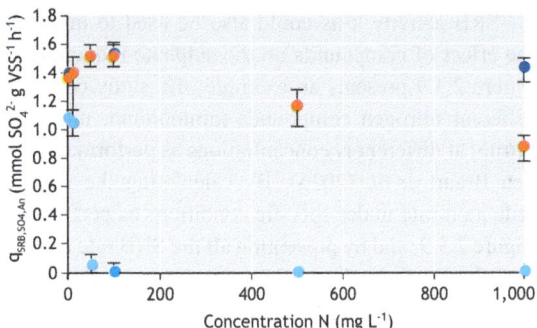

Figure 2.3.9 Example of a graph that shows the biological sulphate-reduction rate ($q_{SRB,SO4,An}$) in the presence of ammonium (●), nitrate (●) or nitrite (●) as N source in the feed, respectively.

2.3.10 Practical recommendations

Volumetric sulphate-reduction rates and SRB activities can be estimated by performing experiments under controlled laboratory conditions. Figure 2.3.11 is an example of a suspended growth bioreactor, fitted with adequate monitoring devices for pH and temperature control. Anew, such reactor configurations will facilitate the ease of collecting both liquid and gas phase samples. If an aerobic test is conducted to ascertain the influence of oxygen on SRB activity, then the same tests should be carried out, except for headspace flushing with N_2-gas. If a toxicity test is desired, then the tests using synthetic wastewater should be performed and the toxic parameter added to the experimental design. The test should be repeated by using the same sludge that was exposed to that toxic compound as the inoculum to investigate the ability of the sludge to recover from toxic stress. From a microbial perspective, if one wishes to determine the time-dependent development of the SRB community structure within the sludge, molecular biology tools like fluorescence in situ hybridization (FISH) and denaturing gradient gel electrophoresis (DGGE) of PCR-amplified 16S ribosomal DNA (rDNA) can be used (see Chapter 8).

It is also possible to study the relation between the sulphide concentration and the actual sulphate-reduction rate. Therefore, it is important to determine the moving average along the profile, in order to know the rate for the actual average sulphide concentration present in the sample. The moving average is a result of the average of three rates. These rates correspond to the sulphide concentration at which the moving average is calculated, and one rate at higher and lower concentrations. Figure 2.3.10 is an example of a typical relation between the sulphate-reduction rate and the actual sulphide concentration, based on the moving average method. This type of graph can also be prepared for the actual sulphate or COD concentration.

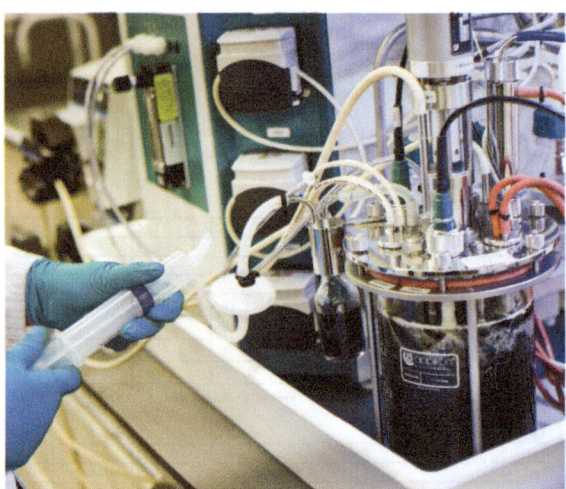

Figure 2.3.11 Collection of a liquid sample to determine the activity of sulphate-reducing bacteria (SRB). Note the characteristic black colour of the enriched SRB biomass in the reactor (photo: KWR Watercycle Research Institute, 2014).

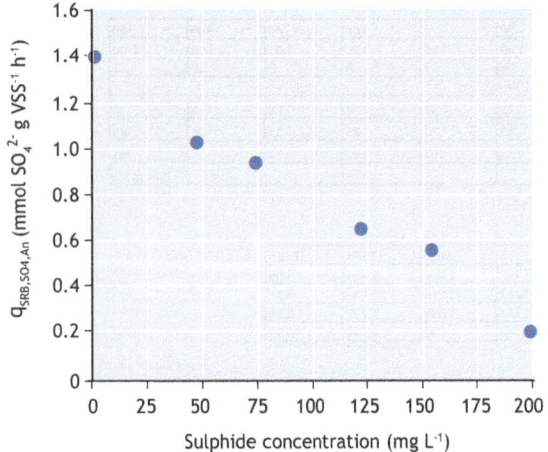

Figure 2.3.10 The relation between the sulphate-reduction rate ($q_{SRB,SO4,An}$) and the actual sulphide concentration, calculated by moving average method.

2.4 BIOLOGICAL NITROGEN REMOVAL

2.4.1 Process description

The need to remove nitrogen from wastewaters arises from its potential toxic effect on aquatic life in receiving water bodies as free ammonia (NH_3), its effect on the nitrogenous oxygen demand and on its role as a nutrient in enhancing eutrophication especially in marine environments (Metcalf and Eddy, 2003). Nitrogen is mainly present in wastewaters in its reduced form as ammonium (NH_4) and can be removed by different physicochemical and biological processes. The selection of the best alternative is commonly based on cost effectiveness. In general, physicochemical methods such as ammonium air-stripping, breakpoint chlorination and selective ion exchange are characterized by higher operational costs that are considered economically feasible only when the ammonium concentrations are higher than 5 g N L^{-1} (Mulder, 2003). In wastewater streams containing less than 5 g N L^{-1}, as in several industrial effluents and most municipal wastewaters (where typical nitrogen concentrations are usually lower than 100 mg N L^{-1}), biological nitrogen removal processes are usually preferred due to lower operational costs. Different biological processes, and combinations thereof, can be applied involving the relative metabolic pathways. Among the resulting alternatives, the wastewater characteristics in terms of the influent COD/N ratio will provide guidance in determining the most suitable biological processes for nitrogen removal. Arguably, three influent COD/N ratio ranges can be distinguished:

(i) For high influent COD/N ratios (> 20 g COD g N^{-1}), the nitrogen requirements or nitrogen assimilation of heterotrophic bacteria (ordinary heterotrophic organisms: X_{OHO}) for biomass synthesis during COD (organic matter) removal is usually sufficient to achieve the required nitrogen concentration in the effluent.

(ii) For influent COD/N ratios comprised between 5 and 20 g COD g N^{-1}, the combination of nitrogen assimilation for microbial growth and the application of conventional nitrification and heterotrophic denitrification processes can be applied.

(iii) For COD/N ratios lower than 5 g COD g N^{-1}, conventional nitrification and heterotrophic denitrification processes can hardly reach satisfactory nitrogen removal levels. In particular, the heterotrophic denitrification process will be limited by the lack of organic matter and an additional carbon source needs to be externally dosed. Thus, non-conventional nitrogen removal processes performing via the so-called 'nitrite-route' are more suitable for nitrogen removal because of their lower (or even absent) COD requirements. For this reason, besides other technical and economic considerations, processes such as partial nitritation-denitrification (also known as nitrite shunt) or partial nitrification-anammox (PNA) are nowadays state-of-the-art biological nitrogen removal processes for wastewaters with low influent COD/N ratio. Due to the relatively lower growth rate of anaerobic ammonium oxidizing bacteria, the anammox (anaerobic ammonium oxidation) process is currently mainly applied for the treatment of warm streams (> 25 °C) such as the supernatant from the anaerobic sludge digestion systems in municipal and industrial WWTPs.

Overall, depending on the wastewater characteristics and local conditions, the removal of nitrogen from wastewater can be performed by several technologies and combinations thereof. In spite of the different technologies and N-removal processes, in practically all of them ammonium is first (fully or partially) oxidized to nitrite (nitritation) or to nitrate (nitrification) and then the oxidized form of nitrogen is reduced to dinitrogen gas which is released into the atmosphere via either denitrification (from nitrate to N_2), denitritation (from nitrite to N_2) or anammox (from nitrite and ammonium to N_2) processes.

The nitrogen removal efficiency of the biological processes depends on an adequate balance between the activities of the different microbial groups. In this regard, the execution of batch activity tests to assess and determine the stoichiometry and kinetic rates of these biological conversion processes represents a useful tool for monitoring and controlling the nitrogen removal processes. Before describing in detail the batch test methodologies and procedures, the biological processes of nitrification, denitrification and anaerobic ammonium oxidation are briefly presented.

For a deeper understanding of different process configurations, operational conditions and factors affecting each process as well as the metabolism involved in the biological nitrogen removal cycle, the reader is referred to standard textbooks (e.g. Henze *et al.*, 2008; Grady *et al.*, 2011).

2.4.1.1 Nitrification

Nitrification is the name given to the production of nitrate by Schlœsing and Müntz (1877) who first recognized the biological nature of the process. One decade later, Winogradsky (1890) isolated for the first time an ammonium-oxidizing bacterium showing that specific groups of bacteria were responsible for nitrification. The immense work of bacterial cultivation performed by Winogradsky (1892) led to the isolation of *Nitrosomonas europea* and *Nitrobacter*, showing that the production of nitrate from the oxidation of ammonium was actually divided into two distinct microbiological processes performed by two phylogenetically independent groups of chemolithoautotrophic aerobic bacteria: ammonium-oxidizing organisms (X_{AOO}) producing nitrite and nitrite-oxidizing organisms (X_{NOO}) producing nitrate. Ammonium oxidation is mostly performed by chemolithoautotrophic ammonium-oxidizing bacteria that use ammonium as their energy and nitrogen source and inorganic carbon as the carbon source. Ammonium oxidation has been reported to be performed also by certain Archea (X_{AOA}, Könneke *et al.*, 2005) and by heterotrophic bacteria (van Niel *et al.*, 1993), showing that the diversity of ammonium-oxidizing organisms is larger than previously assumed. However, due to the dominance of the chemolithoautotrophic pathway performed by X_{AOO} in wastewater treatment systems, this is the ammonium oxidation process considered in this chapter.

At the microbial process level, the oxidation of ammonium to nitrite proceeds through the formation of hydroxylamine (NH_2OH) as the intermediate via the enzyme ammonia monooxygenase (AMO):

$$NH_4^+ + O_2 + H^+ + 2e^- \rightarrow NH_2OH + H_2O \quad \text{Eq. 2.4.1}$$

Hydroxylamine is then further oxidized to nitrite by the enzyme hydroxylamine oxidoreductase (HAO):

$$NH_2OH + H_2O \rightarrow NO_2^- + 5H^+ + 4e^- \quad \text{Eq. 2.4.2}$$

The combination of the two redox processes is known as nitritation:

$$NH_4^+ + 1.5O_2 \rightarrow NO_2^- + H_2O + 2H^+ \quad \text{Eq. 2.4.3}$$

This equation represents the catabolic macro-chemical reaction equation of the nitritation process. When the anabolism is also considered, then the metabolic (catabolism plus anabolism) macro-chemical reaction equation becomes:

$$NH_4^+ + 1.383O_2 + 0.09HCO_3^- \rightarrow 0.982NO_2^- + 1.036H_2O + 0.018C_5H_7O_2N + 1.892H^+ \quad \text{Eq. 2.4.4}$$

As presented in the equation above, about 2 moles of protons are produced per mole of ammonium oxidized. The carbonate system is usually the pH buffer available in the wastewater that neutralizes the production of protons through CO_2 stripping. When the carbonate buffer, usually measured in terms of alkalinity as calcium carbonate equivalents ($CaCO_3$, meq L^{-1}), is not available or is insufficient in the wastewater (e.g. in the case of municipal wastewater, alkalinity lower than about 100 mg $CaCO_3$ L^{-1} or 2 meq L^{-1}), then the pH may drop below pH 7.0 (Ekama and Wenzel, 2008).

In the second stage of the nitrification process, also known as the nitratation process, nitrite is oxidized to nitrate by X_{NOO} by means of the enzyme nitrite oxidoreductase (Nir):

$$NO_2^- + 0.5O_2 \rightarrow NO_3^- \quad \text{Eq. 2.4.5}$$

X_{NOO} are aerobic chemolithoautotrophic bacteria using, for synthesis, inorganic carbon as a carbon source (e.g. HCO_3^-) and ammonium as a nitrogen source. When including bacterial growth in the above equation, the following metabolic macro chemical reaction equation is obtained:

$$NO_2^- + 0.003NH_4^+ + 0.485O_2 + 0.015HCO_3^- + 0.012H^+ \rightarrow NO_3^- + 0.009H_2O + 0.003C_5H_7O_2N \quad \text{Eq. 2.4.6}$$

The combination of the nitritation process performed by X_{AOO} and the nitratation process performed by X_{NOO} constitutes the nitrification process. The nitrification stoichiometry can also be expressed according to the standard ASM (activated sludge model) as:

$$\left(\frac{1}{Y_{ANO}} + i_{N,ANO}\right) \cdot S_{NHx} + \frac{4.75 - Y_{ANO}}{Y_{ANO}} \cdot S_{O2} + \left(\frac{i_{N,ANO}}{14} + \frac{2}{14 \cdot Y_{ANO}}\right) \cdot S_{IC} \rightarrow X_{ANO} + \frac{1}{Y_{ANO}} \cdot S_{NO3} \quad \text{Eq. 2.4.7}$$

Where, S_{NO3} is the nitrate concentration (mg N L^{-1}); S_{IC} is the alkalinity concentration (mmol L^{-1}), S_{O2} is the DO concentration (mg O_2 L^{-1}), S_{NHx} is the ammonium concentration (mg N L^{-1}), X_{ANO} is the concentration of nitrifying organisms (mg COD L^{-1}), Y_{ANO} is the growth yield of nitrifying microorganisms (g COD g N^{-1}), and

$i_{N,ANO}$ is the nitrogen content in the nitrifying organism's cells (g N g COD^{-1}).

2.4.1.2 Denitrification

In the 1850s, first Reiset (1856) and then Pasteur (1859) reported that the reduction of nitrate was of a biological nature, marking the beginning of research into the biological nitrogen cycle. Even if Pasteur erroneously attributed nitrate reduction to 'lactic yeast', he understood the role of organics. Reiset (1856) observed that nitrogen is released into the atmosphere during the decay of plant and animal residues. Shortly after, Gayon and Dupetit (1883) named the process denitrification, organics were experimentally proven to be required in the process (Munro, 1886), and nitrite, nitric oxide (NO) and nitrous oxide (N$_2$O) were identified as intermediates (Payne, 1986). It was observed that not only bacteria but also eukaryotes and archaea can grow on the energy gained by the oxidation of organics or inorganic substrates coupled with the reduction of nitrate to nitrite, NO, N$_2$O and finally dinitrogen gas (N$_2$) (Risgaard-Petersen et al., 2006; Pina-Choa et al., 2010). Certain microorganisms capable of performing heterotrophic nitrification were also shown to be able to carry out denitrification under aerobic and anoxic conditions, the so-called *aerobic denitrification* (Robertson et al., 1995). Denitrification may also proceed without N$_2$O as an intermediate as in the recently discovered denitrification using methane as the electron donor (Ettwig et al., 2010). Nitrifier denitrification where X_{AOO} reduce nitrite to N$_2$O has also been reported (Bock et al., 1995). In contrast to other microorganisms in the nitrogen cycle (e.g. X_{AOO}, X_{NOO}, anammox), several ordinary heterotrophic organisms (X_{OHO}) are facultative denitrifiers that preferentially use oxygen as an electron acceptor due to the higher energy yield and only switch to denitrification when low oxygen levels prevail in the presence of nitrate or nitrite (Zumft, 1997). Independent of the electron donor used, the overall biochemical pathway for denitrification involves the same enzymes in each of these reduction steps from nitrate to dinitrogen gas: nitrate reductase (NAR), nitrite reductase (NIR), nitric oxide reductase (NOR) and nitrous oxide reductase (NOS):

$$NO_3^-(aq) \xrightarrow{NAR} NO_2^-(aq) \xrightarrow{NIR} NO(g) \xrightarrow{NOR} N_2O(g) \xrightarrow{NOS} N_2(g)$$
Eq. 2.4.8

If either the denitrifying microorganisms do not express all the enzymes for the complete denitrification chain or alternatively under certain environmental conditions, then the intermediates NO and N$_2$O can be emitted, both of which have a negative impact on the environment due to their toxicity and direct or indirect contribution to the greenhouse effect.

In wastewater treatment systems, biological nitrogen removal is mostly carried out by X_{OHO} using organic matter as the electron donor. When the organics naturally present in the wastewater are not sufficient to achieve complete denitrification, usually external electron donors (such as acetic acid or methanol, among others) are dosed. Due to its prevalence in wastewater applications, only denitrification catalysed by X_{OHO} (heterotrophic denitrification) is considered in this chapter, which occurs by the following generic catabolic pathway (Mateju et al., 1992):

$$NO_3^- + 1.25CH_2O \rightarrow 0.5N_2 + OH^- + 0.75H_2 + 1.25CO_2$$
Eq. 2.4.9

This equation illustrates that heterotrophic denitrification renders the environment more alkaline due to the production of hydroxide ions.

X_{OHO} can use both nitrate and nitrite as an electron acceptor. The process of nitrate reduction to dinitrogen gas is called denitrification, while the process of nitrite reduction to dinitrogen gas is called denitritation. According to standard activated sludge model (ASM) notation, when a generic soluble and biodegradable substrate (S_B) is used as the carbon source, the denitrification and denitritation stoichiometry can be described as follows, respectively:

$$\frac{1}{Y_{OHO,Ax}} \cdot S_B + \frac{1-Y_{OHO,Ax}}{2.86 \cdot Y_{OHO,Ax}} \cdot S_{NO3} + i_{N,OHO} \cdot S_{NHx} \rightarrow$$
$$X_{OHO} + \left(\frac{1-Y_{OHO,Ax}}{2.86 \cdot 14 \cdot Y_{OHO,Ax}} - \frac{i_{N,OHO}}{14}\right) \cdot S_{IC}$$
$$+ \frac{1-Y_{OHO,Ax}}{2.86 \cdot Y_{OHO,Ax}} \cdot S_{N2} \quad\quad \text{Eq. 2.4.10}$$

$$\frac{1}{Y_{OHO,Ax}} \cdot S_B + \frac{1-Y_{OHO,Ax}}{1.71 \cdot Y_{OHO,Ax}} \cdot S_{NO2} + i_{N,OHO} \cdot S_{NHx} \rightarrow$$
$$X_{OHO} + \left(\frac{1-Y_{OHO,Ax}}{1.71 \cdot 14 \cdot Y_{OHO,Ax}} - \frac{i_{N,OHO}}{14}\right) \cdot S_{IC}$$
$$+ \frac{1-Y_{OHO,Ax}}{1.71 \cdot Y_{OHO,Ax}} \cdot S_{N2} \quad\quad \text{Eq. 2.4.11}$$

Where, $Y_{OHO,Ax}$ is the heterotrophic growth yield under anoxic conditions (g COD g COD^{-1}), X_{OHO} is the concentration of heterotrophic microorganisms (mg COD L^{-1}), S_{NHx} is the ammonium concentration (mg N L^{-1}), S_{NO2} the nitrite concentration (mg N L^{-1}), S_{NO3} the nitrate concentration (mg N L^{-1}), S_{N2} is the concentration of dissolved dinitrogen gas (mg N L^{-1}), and S_{IC} is the alkalinity concentration (mmol L^{-1}).

2.4.1.3 Anaerobic ammonium oxidation (anammox)

The anammox process can be regarded as a peculiar type of denitrification, in which the oxidation of ammonium is coupled to the reduction of nitrite. Its discovery in the 1990s radically changed the understanding of the biological nitrogen cycle (Kuypers et al., 2005), refuting the conventional assumption at that time that ammonium was chemically inert, and that its oxidation required oxygen and a mixed-function oxygenase enzyme (van de Graaf et al., 1996). Anammox bacteria belong to the genus *Planctomycetes* and have been detected in several wastewater treatment plants and natural environments all around the world, showing their ubiquitous distribution. Ammonium, nitrite and bicarbonate are the main substrates in the anammox process (van de Graaf et al., 1996). The catabolic reaction catalysed by anammox bacteria couples the nitrogen atoms from ammonium and nitrite to form dinitrogen gas (N_2):

$$NH_4^+ + NO_2^- \rightarrow N_2 + 2H_2O \qquad \text{Eq. 2.4.12}$$

In the absence of oxygen, anammox bacteria activate the stable ammonium molecule through the oxidizing power of nitric oxide (NO). Briefly, anaerobic ammonium oxidation is a three-step process with NO and hydrazine as intermediates: first, nitrite is reduced to NO by the enzyme nitrite oxidoreductase (Nir), then the produced NO reacts with ammonium to form hydrazine (N_2H_4), catalysed by the unique hydrazine synthase enzyme (HZS), and finally hydrazine is oxidized to N_2 by hydrazine dehydrogenase (HDH) (Kartal et al., 2011). Anammox bacteria are autotrophs and thus make use of inorganic carbon as the carbon source for the production of biomass. In the anabolic reaction the reducing equivalents for the reduction of inorganic carbon originate from the oxidation of nitrite to nitrate as illustrated in the equation below (ammonium is considered as the N source):

$$HCO_3^- + 2.1NO_2^- + 0.2NH_4^+ + 0.8H^+ \rightarrow$$
$$CH_{1.8}O_{0.5}N_{0.2} + 2.1NO_3^- + 0.4H_2O \qquad \text{Eq. 2.4.13}$$

The metabolic macro chemical reaction equation of the anammox process is still subject to debate. Due to the difficulties in cultivating a pure culture of anammox bacteria, stoichiometric equations derived by substrates/products via mass balance are intrinsically not completely precise (Lotti et al., 2014). However, the first reaction stoichiometry reported by Strous and co-authors (1998) is widely used for reactor design and operation purposes:

$$NH_4^+ + 1.32NO_2^- + 0.066HCO_3^- + 0.13H^+ \rightarrow$$
$$1.02N_2 + 0.066CH_2O_{0.5}N_{0.15} + 0.26NO_3^- + 2.03H_2O$$
$$\text{Eq. 2.4.14}$$

Since nitrogen is usually present in wastewater as ammonium but the anammox metabolism requires both ammonium and nitrite as substrates, the anammox process has to be combined with another process to generate the required nitrite. For this purpose, the partial nitritation (PN) process is usually applied. Nitrogen removal processes based on the combination of PN and anammox process are currently part of the state of the art with about 100 full-scale implementations worldwide treating mostly the effluent from anaerobic sludge digesters as well as a variety of ammonium-rich municipal and industrial wastewaters: leather tanning, food processing, and the semiconductor, fermentation, yeast production, distilling, and winemaking industries (Lackner et al., 2014). Furthermore, encouraging results were reported for pilot-scale installations treating black water digestate (de Graaff et al., 2011), digested manure (Villegas et al., 2011), urine (Udert et al., 2008) and pharmaceutical wastewaters (Tang et al., 2011). Recently, positive results have also been obtained in the treatment of aerobically pre-treated sewage, opening up new perspectives for converting municipal WWTPs from energy-depleting to energy-generating systems (Lotti et al., 2015a).

2.4.2 Process-tracking alternatives

According to the stoichiometry of the processes involved in nitrogen removal, as presented in Section 2.4.1, various alternatives are available to assess the process kinetics and stoichiometric parameters of interest during a batch test, such as:

- Chemical tracking by assessing the nitrite, nitrate or ammonium concentrations over time; the choice of the optimal chemical species to be tracked depends on the specific process of interest.
- Titrimetric tracking by applying pH-static titration to those processes that relevantly affect the solution pH.
- Manometric tracking, applicable to processes involving soluble gaseous species of low solubility such as N_2.
- Respirometry, applicable to aerobic processes that affect the DO concentrations.

The last alternative is presented not in this chapter, but in Chapter 3 on Respirometry, which is fully dedicated to this technique.

The other tracking alternatives are briefly presented hereafter.

2.4.2.1 Chemical tracking

Tracking the concentration over time of substrates and products is the most common way to assess the kinetics of a process.

When nitrification is to be tracked, ammonium and nitrate concentrations are monitored over time. For the separate assessment of the nitritation or nitratation rates, ammonium and nitrite or nitrite and nitrate concentrations have to be measured over time. Finally, when nitritation and nitratation rates have to be assessed simultaneously, ammonium, nitrite and nitrate concentrations need to be monitored over time.

For denitrification, the most common batch test that applies the chemical tracking procedure is the so-called nitrate uptake rate (NUR) test. NUR tests make it possible to assess several parameters of practical interest, such as nitrate utilization rates, the utilization of organics for denitrification and the anoxic biomass yield coefficient (Naidoo et al., 1998; Kujawa and Klapwijk, 1999). As described in Section 2.2.4, samples are taken during the course of the batch tests and then chemical analyses of the main substrates (e.g. nitrite, nitrate and COD) are performed in order to get sufficient information for the assessment of relevant kinetic and stoichiometric parameters.

Finally, when this method is applied to the evaluation of the anammox process kinetics, ammonium, nitrite and nitrate concentrations are monitored over time. Besides the anammox process rate, this method also makes it possible to assess stoichiometric parameters of interest such as the ratio between nitrite and ammonium consumption (NO_2/NH_4 ratio) and the ratio between nitrate production and ammonium consumption (NO_3/NH_4 ratio) which can be calculated as the ratio of the corresponding conversion rates (Lotti et al., 2014).

2.4.2.2 Titrimetric tracking

The pH-static titration technique consists of the controlled addition of an appropriately diluted solution of acid or base to maintain a constant pH (therefore 'static'

pH) in a biological system where the pH is affected by different reactions. Under these conditions, the titration rate is proportional to the reaction rate via a stoichiometric factor. In principle, this technique is applicable to any biological or physical-chemical reaction affecting the proton concentration (thus, linked with pH), i.e. any reaction converting neutral substrates into acid or basic products, or acid or basic substrates into neutral products. There are several biological reactions of environmental interest that affect the pH of the suspension where they take place. Attempts to use pH-static titration have been mainly focused on nitrification (Gernaey et al., 1997, 1998; Massone et al., 1998) and denitrification (Massone et al., 1996; Rozzi et al., 1997; Bogaert et al., 1997; Foxon et al., 2002). Although these are consolidated applications of the pH-static titration technique, others can be foreseen, such as those involving the $CO_2/HCO_3^-/CO_3^{2-}$ equilibria, e.g. the heterotrophic degradation of organic substrates that produces CO_2 (Ficara and Rozzi, 2004) and acetoclastic methanogenesis that produces bicarbonate (Rozzi et al., 2002).

Nitrification monitoring is the first and most consolidated application of this technique since the relationship between the ammonium oxidised and the proton produced Y_{NH4_H+} can be calculated from the reaction stoichiometry, as assumed by ASM1 (Henze et al., 2000):

$$Y_{NH\_H+} = \frac{14}{2 + i_{N,ANO} \cdot Y_{ANO}} \approx 6.92 \text{ g N mol (protons)}^{-1}$$

Eq. 2.4.15

The N/H ratio is therefore the stoichiometric factor that allows the conversion of the titration rate into the ammonium consumption rate. The pH-static titration technique can also be applied to denitrification since this process is a 'pH-affecting reaction'. However, the assessment of the ratio between nitrite or nitrate consumption and proton production Y_{NOx_H+} based on stoichiometry is not as straightforward as for nitrification, mostly because it depends on many more factors, such as the carbon source, the sludge characteristics, and the pH set point. To theoretically calculate (Y_{NOx_H+}, a conceptual model was proposed by Petersen et al. (2002), which considers that the following four processes have a pH-dependent effect on proton production during denitrification: (*i*) uptake of weak organic acids as carbon source, (*ii*) uptake of nitrate, (*iii*) uptake of ammonia for cell synthesis, and (*iv*) production of carbon dioxide from organic carbon oxidation. Based on these assumptions, the following reaction stoichiometry was proposed to assess the ratio between the net proton production and nitrate consumption:

$$\frac{1}{Y_{OHO,Ax}} \cdot S_B + \frac{1 - Y_{OHO,Ax}}{\beta \cdot Y_{OHO,Ax}} \cdot S_{NOx} + i_{N,OHO} \cdot S_{NHx} \rightarrow$$

$$X_{OHO} + \frac{1 - Y_{OHO,Ax}}{\beta \cdot Y_{OHO,Ax}} \cdot S_{N2} + \left[-\frac{a}{C \cdot Y_{OHO,Ax}} - \frac{1 - Y_{OHO,Ax}}{\beta \cdot 14 \cdot Y_{OHO,Ax}} + \frac{b \cdot i_{N,OHO}}{14} + \frac{C \cdot (1 - Y_{OHO,Ax}) \cdot x}{C \cdot Y_{OHO,Ax}} \right] \cdot H^+$$

Eq. 2.4.16

Where, S_{NOx} is the nitrite or nitrate concentration; x is the number of carbon moles per mole of organic substrate, C is a factor (in g COD mol^{-1} organic substrate) to express the organic carbon in COD units, β is the oxygen equivalent of oxidized nitrogen; and a, b, c are pH-dependent factors which take into account the dissociation equilibria of weak acids/bases (a for organic acids - HA, b for carbonic acid, and c for ammonium):

$$a = \frac{[A^-]}{[HA] + [A^-]} = \frac{10^{-pKa}}{10^{-pH} + 10^{-pKa}} \qquad \text{Eq. 2.4.17}$$

$$b = \frac{10^{pH - pK1} \cdot (1 + 2 \cdot 10^{pH - pK2})}{1 + 10^{pH - pK1} \cdot (1 + 10^{pH - pK2})} \qquad \text{Eq. 2.4.18}$$

$$c = \frac{[NH_4^+]}{[NH_4^+] + [NH_3]} = \frac{10^{-pH}}{10^{-pH} + 10^{-pK_{NH4}}} \qquad \text{Eq. 2.4.19}$$

Where, pK_a is the dissociation constant for acetic acid (4.75 at 25 °C), pK_1 the dissociation constant for carbonic acid (6.352 at 25 °C), pK_2 the dissociation constant for bicarbonate (10.33 at 25 °C), and pK_{NH4} the dissociation constant for ammonium (9.25 at 25 °C).

By substituting the correct value of β (i.e. 2.86 g COD N^{-1} for N-NO$_3$ and 1.72 g COD N^{-1} for N-NO$_2$), it follows that Y_{NOx_H+}, in g N mol^{-1}, in the presence of nitrate, Y_{NO3_H+}, or nitrite, Y_{NO2_H+}, as an electron acceptor, can be expressed as:

$$Y_{NO3\_H+} = \frac{1 - Y_{OHO,Ax}}{2.86 \cdot Y_{OHO,Ax}} \cdot$$

$$\left[-\frac{a}{C \cdot Y_{OHO,Ax}} - \frac{1 - Y_{OHO,Ax}}{2.86 \cdot 14 \cdot Y_{OHO,Ax}} + \frac{c \cdot i_{N,OHO}}{14} + \frac{b \cdot (1 - Y_{OHO,Ax}) \cdot x}{C \cdot Y_{OHO,Ax}} \right]^{-1}$$

Eq. 2.4.20

$$Y_{NO2\_H+} = \frac{1 - Y_{OHO,Ax}}{1.72 \cdot Y_{OHO,Ax}} \cdot$$

$$\left[-\frac{a}{C \cdot Y_{OHO,Ax}} - \frac{1 - Y_{OHO,Ax}}{1.72 \cdot 14 \cdot Y_{OHO,Ax}} + \frac{c \cdot i_{N,OHO}}{14} + \frac{b \cdot (1 - Y_{OHO,Ax}) \cdot x}{C \cdot Y_{OHO,Ax}} \right]^{-1}$$

Eq. 2.4.21

These equations show that the assessment of Y_{NOx_H+} makes it necessary to know the chemical composition of the carbon source (C and x), which is seldom the case in practical applications, and the anoxic biomass growth yield coefficient, $Y_{OHO,Ax}$. As such, its evaluation is theoretically possible, but difficult in practice.

Fortunately, Y_{NOx_H+} can also be experimentally evaluated by measuring the amount of titration solution (normally acid) dosed under pH-static conditions to denitrify a known amount of nitrite or nitrate and in the presence of the carbon source of interest. Once the Y_{NOx_H+} ratio is measured, the titration rate can be easily converted into the nitrate or nitrite uptake rate.

Theoretically, even the anammox process can be monitored by pH-static titration. However, there is very little available experience of this, and therefore this alternative is not discussed in this chapter.

2.4.2.3 Manometric tracking

According to this technique, the rate of a bioprocess that produces a poorly soluble gaseous component is proportional to the rate of increase in pressure, provided that the bio-reaction takes place in a gas-tight reactor. The relationship between the generated overpressure, P(t), and the volumetric gas production, $V_G(t)$, can be obtained by assuming that the gas transfer from the liquid to the gas phase is not rate-limiting (sludge mixing allows the quick transfer of gaseous species) and that no relevant amounts of the gaseous species remain in solution. Under these conditions and at constant temperature, according to the gas law, the following relationship applies:

$$V_G(t) = \frac{P(t) - P_{atm}}{P_{atm}} \cdot V_{HS} \qquad \text{Eq. 2.4.22}$$

Where, V_{HS} is the volume of the headspace in the reactor and P_{atm} is the atmospheric pressure.

This measuring principle was proven to be applicable and advantageous in the monitoring of denitrification (Sánchez et al., 2000; Ficara et al., 2009) and of anaerobic ammonia oxidation (Dapena Mora et al., 2007;

Scaglione *et al.*, 2009; Bettazzi *et al.*, 2010; Lotti *et al.*, 2012) since both processes produce dinitrogen gas. As for denitrification, a CO_2 adsorbent should be used that is typically located in the gas headspace (e.g. NaOH pellets) so that overpressure data are only related to the release of N_2.

2.4.3 Experimental setup

2.4.3.1 Reactors

Independently of the technology applied to remove nitrogen from wastewater, to assess the performance of the biological nitrogen removal process, batch activity tests can be carried out under aerobic (nitrification) or anoxic conditions (denitrification and anammox) depending upon the parameters of interest and nature of the study. In any case, the reactor(s) used for the execution of tests must have the required means to: (*i*) avoid oxygen intrusion under anoxic conditions, (*ii*) secure satisfactory S_{O2} availability under aerobic conditions (e.g. S_{O2} higher than 2 mg L^{-1}), (*iii*) provide satisfactory mixing conditions, (*iv*) maintain an adequate and desirable temperature, (*v*) provide precise pH control, and (*vi*) have additional ports for sample collection and addition of influent, solutions, gases and any other liquid media or substrate used in the test. For the requirements needed to ensure proper anoxic conditions, aerobic conditions, mixing, temperature control, pH control and sampling and dosing ports during the execution of the tests, the reader is referred to Section 2.2.2.1. However, when titrimetric or manometric experiments are to be performed, special apparatus should be available, as described below.

2.4.3.2 Instrumentation for titrimetric tests

To perform set point titration tests, an automated titration unit is required. As for respirometers, such systems are readily available on the market. However, they can be easily implemented by using conventional laboratory equipment and basic signal acquisition and control units. Specifically, an automated titration unit should be made up of the following components (see Figure 2.4.1):
- A reaction vessel: a well-mixed reactor with a thermostat or temperature control to host the activated sludge sample with an operative volume of 0.5 to 1.5 L. The reaction vessel does not need to be gas-tight when performing nitrification tests. However, a reduced solid-liquid contact area is preferable to limit gas-liquid transfer of oxygen and carbon dioxide. Gas tightness for proper anoxic operation is required when performing denitrification tests.
- Probes to assess: temperature (with a resolution of 0.1 °C), pH (with a resolution of 0.01 pH units) and, possibly, DO (± 0.02 mg L^{-1} around the selected set point value).
- An aeration system (for aerobic processes), typically with an aeration capacity of 50-200 L L^{-1} h^{-1}, and stone diffuser for fine bubble aeration;
- An automated titration system capable of maintaining the pH within a narrow range (e.g. at a defined pH set point ± 0.02) and to record the volume of the titration solution dosed over time with a suggested resolution of 0.1 mL and a minimum logging frequency of 1 datum per minute. Manual recording of the added titration solution volume can also be obtained by storing the titration solution in a graduated cylinder and by manually reading the remaining volume at regular intervals (every few minutes for nitrification tests) or by installing the solution on a balance with its weight either read or recorded automatically. A solution of 0.05-0.02 N NaOH can be used as an alkaline titration solution, while a 0.05-0.02 N HCl solution can serve as an acidic titration solution.

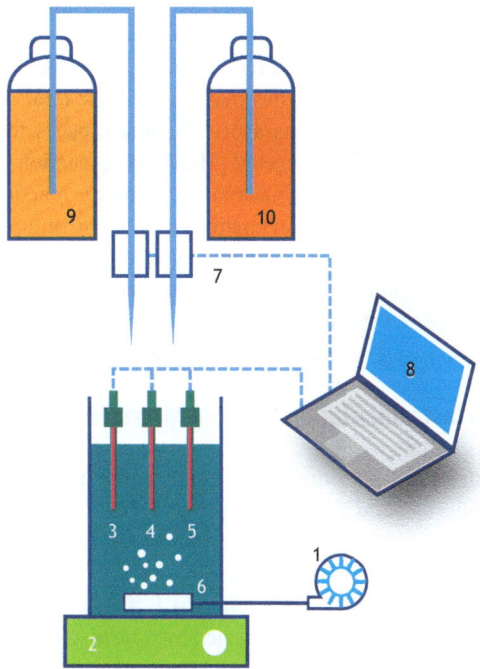

Figure 2.4.1 Scheme of a pH-static titration system: 1. aerator; 2. mixer; 3. temperature probe; 4. pH probe; 5. DO probe; 6. reaction vessel; 7. titration-solution dosing system; 8. signal acquisition and recording; 9. alkaline titration solution; 10. acidic titration solution.

When aerobic bioprocesses are involved, the pH-static system can be conveniently upgraded into a pH/DO-stat system, in which a secondary titration unit provides an H_2O_2 diluted solution, serving as an oxygenated titration solution. For this purpose, the system described in Figure 2.4.1 should be upgraded by integrating:
- A DO probe, with a minimum resolution of 0.1 mg L^{-1}.
- An additional automated dosing or titration system capable of maintaining the S_{O2} within a narrow range ($\pm$ 0.1 mg L^{-1} around the selected set point value). A 0.05-0.2 M H_2O_2 solution is appropriate when performing pH/DO-static titration tests on conventional activated sludge samples. When dosed, the H_2O_2 solution will be converted into molecular oxygen (O_2) and water by peroxydases produced by aerobic bacteria to counteract oxidative stress, and thus making oxygen available for bacterial respiration. As a matter of fact, it has been observed that diluted H_2O_2 solutions can be used for short-term respirometric tests without significant bacteria inhibition (e.g. Ficara *et al.*, 2000).

The scope of this DO-static titration unit is to maintain the DO value at a predefined set point level (DO-set point) by titrating the oxygenated titration solution, thus meeting the following objectives:
- To maintain the desired redox condition without the need for air bubbling; this makes it possible to avoid CO_2 stripping which is a pH-affecting process that overlaps with other targeted pH-affecting reactions.
- To assess the oxygen consumption rate of the reaction that, under DO-static operation, equals the titration rate of the oxygenated titration solution. This is additional information that can be used to check or complement the titration rate of the alkaline/acidic solution, as described in detail later on in this chapter.

2.4.3.3 Instrumentation for manometric tests

Tests should be performed by using gas-tight apparatus. Typically, systems applied to perform BOD tests are used. The minimum requirements are the following:
- A glass bottle with (see Figure 2.4.2):
  a. A working volume of around 1 L.
  b. Two lateral openings, sealed by rubber septa kept in place by plastic or aluminium gear, for substrate injections and gas flushing/discharge.
  c. A container for NaOH pellets located in the bottle headspace and serving as a CO_2 trap.
  d. A manometric measuring device, possibly featured with a data logger, and fixed on the top of a glass bottle with a resolution of 1-3 mbar.
- A constant temperature incubator that limits temperature oscillations to $\pm$ 0.2 °C. It is mandatory that temperature is very well controlled during the course of the test since temperature variations cause changes in the overpressure values that are not associated with gas release and this would therefore result in data noise.
- A magnetic stirrer that can operate at around 100-200 rpm. Alternatively, a thermostatic orbital shaker can be used to ensure both temperature control and mixing. In the case of anammox, a magnetic stirrer is advisable only in the case of suspended anammox biomass (100-200 rpm), while for both hybrid, biofilm on carriers and granular anammox biomass types (Hu *et al.*, 2013), a shaker is preferred in order to avoid deterioration of the anammox biofilm due to the shear forces caused by the magnetic stirrer.

Figure 2.4.2 Commercially available apparatus to perform manometric tests for denitrification purposes (photo: Lotti, 2016).

2.4.3.4 Activated sludge sample collection

The sampling time and location of an activated sludge sample performing nitrification and denitrification is highly dependent on the type of batch activity test to be conducted. The removal of nitrogen operated via nitrification and denitrification processes is based on the alternating aerobic-anoxic conditions. Thus, preferably, a fresh sample should be collected at the end of the relative reaction stage: aerobic for nitrification, anoxic for denitrification. Certainly, the sampling location will depend on the system configuration. For instance, the sampling time and location of an activated sludge sample from a PN/anammox system depends on the type of technology used; obviously when PN and anammox processes are divided into two separate stages, anammox biomass should be sampled by the anammox stage. In full- and pilot-scale wastewater treatment plants, the physical boundaries between stages must be identified prior to sampling. In extreme cases, where the phases are not (physically) well defined, the redox limits or boundaries need to be determined with the use of a DO meter, redox meter and/or by determination of the nitrate and nitrite concentrations. In lab-scale systems (usually operated on a time-base mode), the sample collection can be relatively easier, since the reaction time defines the length of the stages. To obtain homogenous and representative samples, the sludge samples must be collected in sampling spots where well-mixed conditions take place.

When anammox is the targeted process, then sampling from the outlet of the anammox tank is typically adequate. This sampling point ensures that the sludge samples collected contain limited amounts of residual ammonium/nitrite concentrations. When granular anammox biomass is considered, the outlet of the anammox tank normally contains very few granules because a granular system is usually equipped with a biomass retention system (e.g. a three-phase separator, hydro-cyclone, settling phase in a sequencing batch reactor (SBR) cycle, etc.). In the case of anammox granular systems then sampling the mixed liquor directly from the anammox tank may be adequate in continuously operated systems (CSTR, continuous stirred tank reactor), while for SBR systems, sampling should be performed before the settling phase when the reactor is completely mixed to ensure completely mixed conditions and the presence of limited amounts of residual ammonium/nitrite concentrations.

Ideally, batch activity tests must be performed as soon as possible after sample collection. In lab-scale systems, in principle, this should not be a problem if the batch activity tests are performed in the same laboratory and their execution is coordinated and synchronized with the operations of the lab reactor. Also, at full- and pilot-scale treatment plants, batch activity tests can be performed *in situ* shortly after mixed liquor collection if the sewage plant laboratory is conditioned and equipped with the required experimental and analytical equipment. If the batch activity tests cannot be performed *in situ* on the same day, a mixed liquor sample can be collected and transferred to the location where the tests will be executed. Thereafter, the sampling bucket can be properly stored and transported in a fridge or in ice box (below or close to 4 °C) under non-aerated conditions and the activity tests should be performed no later than 24 h after sampling. In order to avoid the creation of anaerobic conditions during storage and the undesired production of toxic sulphide through the reduction of sulphate, nitrate should be added to the mixed liquor at a final concentration of about 50-200 mg N L^{-1}. However, the availability of nitrate would promote endogenous biomass respiration. Therefore, it is important to stress that it is highly recommended to conduct the test as soon as possible after sample collection.

Especially after storage, the biomass present in the mixed liquor sample needs to be 'washed' to remove any added nitrate, 're-activated' and acclimatized to the target pH and temperature of interest prior to the execution of the batch activity tests. The washing step must be performed by using an appropriate 'washing medium' with a mineral composition that depends on the target microbial population, as indicated below. Tap water can also be used as a washing medium as long as its conductivity is similar to that of the cultivation medium. In any case, the *in situ* execution of the batch activity tests is preferable since this avoids the exposure of biomass to varying conditions. The total volume of activated sludge (mixed liquor) to be collected depends on the number of tests, reactor volume and total volume of samples to be collected to assess the biomass activity. Often, 10-20 L of activated sludge or mixed liquor from full-scale wastewater treatment plants is considered sufficient. On the other hand, samples collected from lab-scale reactors rarely reach more than 1 L because lab-scale systems are usually smaller (from 0.5 to 2.2 L and in certain cases up to 8-10 L) and the maximum volume that can be withdrawn from lab-scale reactors is often set by the daily withdrawal of the excess of sludge from the system. Since the maximum volume allowed to be withdrawn is directly related to the applied solids retention time (SRT), which is defined by the growth rate of the organisms, particular attention must be paid when dealing with slow-growing organisms such as nitrifiers and anammox bacteria.

Suggestions on sampling scheduling and ideal storage times have been previously described (see Section 2.2.3) and should be carefully considered.

2.4.3.5 Activated sludge sample preparation

Generally speaking, activated sludge samples can be used as such or after specific adjustments in pH (with or without the presence of a pH buffer), temperature, ammonium/nitrate/nitrite concentration, carbon source concentration, and X_{VSS}. For conventional activated sludge samples from wastewater treatment plants treating urban wastewater, a sample which has a X_{VSS} around 2-4 g VSS L^{-1} would be ideal. For anammox sludge, X_{VSS} around 5-10 g VSS L^{-1} will be preferable. For very diluted or concentrated sludge samples, a pre-concentration step (e.g. by decanting into an Imhoff cone for 30 min or by centrifuging at 4,000 rpm for a few minutes) or dilution with the secondary effluent of the same wastewater treatment plant may be helpful. This will avoid the occurrence of too slow or too fast conversion rates.

When such procedures are implemented on anoxic or anaerobic activated sludge samples, N_2 sparging should be performed immediately afterwards in order to re-establish proper anoxic/anaerobic conditions. When the anammox process is considered, a mixture of N_2/CO_2 gases (usually 95/5 % is used in practice) can be used for sparging instead of N_2 in order to avoid excessive CO_2 stripping which would cause a pH increase and may limit anammox activity during the batch test due to limiting inorganic carbon concentration. As for pH and temperature, in principle, the closer the set point pH value is to the typical operational pH of the plant, the more the resulting process rate will be representative of the operational process rate. The same concept applies to the selection of the temperature value. Since the effect of the anammox process on the pH is rather limited (0.13 mole of protons consumed per mole of ammonium converted), limited variations on the pH are expected during the execution of a batch test (e.g. from 7.5 to 7.9 according to Lotti et al., 2012). Nevertheless a pH buffer such as Hepes (N-2-hydroxyethyl-piperazine-N0-2-ethane sulfonic acid) or phosphate can be used to maintain a constant pH throughout the duration of the batch test (Dapena-Mora et al., 2007; Lotti et al., 2012).

While a Hepes buffer can be used at concentrations up to 25 mM without affecting anammox activity (Lotti et al., 2012), the concentration of the phosphate buffer should be carefully decided since it may result in process inhibition (Dapena-Mora et al., 2007; Oshiki et al., 2011). Previous reports have shown that a phosphate buffer concentration of 5.3 mM is suitable for the conduction of anammox batch tests (Dapena-Mora et al., 2007; Lotti et al., 2012).

In general, the objective of the sampling and experimental campaigns should be to minimize as much as possible the need for transportation, cooling, storage and reactivation of the sludge. Whenever possible, it is advisable to use 'fresh' sludge (and substrate/media). When actual operational conditions are to be tested, then the corresponding batch activity tests must be executed right away after sludge collection with the minimum adjustments of the operational conditions (e.g. for pH and temperature). When the batch tests cannot be performed *in situ* or shortly after collection, sludge samples must be stored at around 4 °C for preservation purposes during transportation and storage. In the case of anammox biomass, ambient temperature can be adopted for preservation purposes during transportation and storage. Storage at ambient temperature is suggested to avoid temperature shocks from the usual operative temperature of an anammox system (25-35 °C) to 4 °C, which is usually considered an adequate storage temperature for conventional activated sludge samples. Under these circumstances, the batch activity tests should preferably be executed in less than 24 h after sludge collection and after 'reactivation' by keeping the sludge at the pH and temperature of interest (after N_2 flushing in case of denitrification and anammox).

The addition of limited amounts of the substrate can favour bacterial metabolic reactivation. However, the preparation of the activated sludge is test-dependent and specific suggestions/recommendations are described in the following paragraphs.

2.4.3.6 Substrate

When real wastewater (either raw or settled) is used for the execution of activity tests, it can be fed in a relatively straightforward manner to the reactor/fermenter. A rough filtration step (using 10 µm pore filter size) can be used to remove the remaining debris and large particles present in the raw wastewater.

If different carbon sources and concentrations are to be studied, the plant effluent can also be used to prepare a semi-synthetic media containing a S_B concentration of between 50 and 100 mg COD L^{-1}.

For the execution of conventional denitrification tests, nitrate and nitrite solutions can be prepared to create the required anoxic conditions. For this purpose, different stock solutions can be prepared using nitrate- and nitrite-salts.

For the execution of anammox tests, ammonium and nitrite need to be dosed to the wastewater at the beginning of the test in order to provide the desired amount of substrate. For this purpose, different stock solutions can be prepared using ammonium- and nitrite-salts (e.g. 1-10 g N L^{-1}); the most commonly used are ammonium sulphate and sodium nitrite, respectively. Particular attention needs to be paid when considering the initial substrate concentrations. Nitrite in fact, besides being a substrate for the anammox process, is also an inhibitor (Lotti et al., 2012; Puyol et al., 2014). Initial nitrite concentration around 50-70 mg N L^{-1} is usually considered adequate. Nevertheless, lower initial nitrite concentrations (e.g. 10 mg N L^{-1}) are recommended if the anammox biomass originates from systems operated at very low (few mg N L^{-1}) nitrite concentrations (e.g. DEMON systems, Wett et al., 2007). In fact, the nitrite inhibition effect and resilience seems to depend on the 'cultivation history' of the anammox biomass, being the cultures cultivated under strict nitrite limitation more prone to nitrite inhibition (Lotti et al., 2012).

When tests are executed to assess the potential inhibitory or toxic effect of a given compound at different concentrations, concentrated stock solutions can be prepared and added during the test to obtain the concentrations of interest. Tests performed to assess whether the inhibitory or toxic effects are reversible must be carried out after washing the biomass to remove the inhibiting or toxic compound(s). Often, the washing step is performed by consecutive settling and re-suspension of the sludge sample in carbon-free and nitrite-free media (either fully synthetic or using a treated effluent after filtration) under anoxic conditions.

2.4.3.7 Analytical procedures

Analytical procedures of interest (NH$_4$, NO$_2$, NO$_3$, MLSS, MLVSS, COD, BOD) should be performed following standardized and commonly applied analytical protocols detailed in Standard Methods (APHA et al., 2012).

If a specific carbon source is used, its determination should be performed according to the relevant analytical method. However, most of the time, the determination of soluble COD can be appropriate to follow the carbon source utilization.

2.4.3.8 Parameters of interest

Nitrification

The most significant kinetic parameter of the aerobic ammonium oxidation process (nitritation) performed by AOO is the maximum biomass-specific ammonium oxidation rate ($q_{AOO,NH4}$). Similarly, for the aerobic nitrite oxidation process (nitratation) performed by NOO, the main parameter of interest is the maximum biomass-specific nitrite oxidation rate ($q_{NOO,NO2\_NO3}$). Table 2.4.1 presents typical kinetic parameter values found in literature for both the aerobic ammonium and nitrite oxidation processes. Literature values of the biomass growth yield of both AOO and NOO are also reported in Table 2.4.1.

Table 2.4.1 Expected stoichiometric and kinetic parameters of interest for activated sludge performing aerobic ammonium and nitrite oxidation to nitrite and nitrate. The kinetic parameters are reported considering a reference temperature of 20 °C.

Aerobic ammonium oxidation–nitritation process		
$q_{AOO,NH4}$	Y_{AOO}	Reference
g N g VSS^{-1} d^{-1}	g VSS g N^{-1}	
0.11	0.14	Blackburne et al. (2007)
0.09	0.11	Jones et al. (2007)
0.24	0.13	Jubany et al. (2008)
0.27	0.15	Koch et al. (2000)
0.21	0.11	Lochtman (1995)
0.22	0.15	Wiesmann (1994)
Aerobic nitrite oxidation–nitratation process		
$q_{NOO,NO2\_NO3}$	Y_{NOO}	Reference
g N g VSS^{-1} d^{-1}	g VSS g N^{-1}	
0.21	0.07	Blackburne et al. (2007)
0.13	0.07	Jones et al. (2007)
0.39	0.06	Jubany et al. (2008)
1.78	0.02	Koch et al. (2000)
0.45	0.03	Lochtman (1995)
0.78	0.04	Wiesmann (1994)
1.07	0.03	Wik and Breitholtz (1996)

The kinetics reported in Table 2.4.1 refer to an operational temperature of 20 °C and were calculated from the original values according to the Arrhenius equation reported below.

$$k_S(T) = k_S(T_{ref}) \cdot \exp\left(\frac{E_{a,S} \cdot (T_K - T_{ref})}{R \cdot T_K \cdot T_{ref}}\right) \quad \text{Eq. 2.4.23}$$

Where, $k_S(T)$ is the maximum biomass-specific consumption rate of the substrate S evaluated at the desired operative absolute temperature T_K (K), T_{ref} is the reference absolute temperature, R is the ideal gas constant (8.31 J mol^{-1} K^{-1}), $E_{a,S}$ is the activation energy of the considered bioprocess consuming the substrate S. $E_{a,NH4}$ = 68 kJ mol-NH$_4^{-1}$ K^{-1} and $E_{a,NO2}$ = 44 kJ mol-NO$_2^{-1}$ K^{-1} are typical activation energy values for the nitritation process performed by AOO and the nitratation process performed by NOO, respectively.

As it can be observed in Table 2.4.1, the aerobic consumption rates for ammonium and nitrite oxidation reported in literature can vary widely, especially for the nitratation process catalysed by NOO. The main reason can be the active biomass fraction present in the activated sludge biomass, conventionally referred to as the total MLVSS concentration. The larger the fraction of active nitrifying biomass, the higher the specific biomass conversion rate expected. In a conventional activated sludge system this can be directly related to the COD/N ratio in the influent, with the lower COD/N ratio corresponding to the higher fraction of active nitrifying biomass. As depicted in Table 2.4.1, the reported growth yield values are higher for the ammonium-oxidizing bacteria than for the nitrite-oxidizing bacteria.

Denitrification

To quantify the activity of OHOs under anoxic conditions, the relevant stoichiometric parameters and kinetic constants have to be known. As for stoichiometry, the most relevant parameter to be defined is the heterotrophic growth yield under anoxic conditions $Y_{OHO,Ax}$ (in g COD-biomass per g COD-substrate). Like the aerobic growth yield, this parameter may depend on various factors, such as the organic carbon source quantity and quality, and the environmental conditions. Typical values for this parameter are listed in Table 2.4.2. This parameter can be easily determined by setting up appropriately designed batch activity tests, as explained later on. When performing denitrification, the carbon source required per nitrate/nitrite to be removed (expressed as the COD/N ratio or as the C/N ratio) can be used instead of $Y_{OHO,Ax}$, given that a stoichiometric relationship exists between the two of them. The COD/N ratio represents the denitrification capacity of a carbon source or of a wastewater and can be a more practical parameter than the $Y_{OHO,Ax}$.

Table 2.4.2 Expected stoichiometric and kinetic parameters of interest for activated sludge wastewater treatment systems performing denitrification.

Parameter (symbol)	Remark	Value	Reference
Heterotrophic anoxic growth yield ($Y_{OHO,Ax}$)	Acetate	0.66 g COD g COD^{-1}	Ficara and Canziani (2007)
	Wastewater	0.50 g COD g COD^{-1}	Orhon et al. (1996)
	Acetate	0.66 g VSS g COD^{-1}	Kujawa and Klapwijk (1999)
	Ethanol	0.22 g VSS g COD^{-1}	Hallin et al. (1996)
	Methanol	0.18 g VSS g COD^{-1}	Tchobanoglous et al. (2003)
COD to nitrogen ratio (COD/N)	Methanol	4.6 g COD g N^{-1}	Bilanovic et al. (1999)
	Methanol	4.7 g COD g N^{-1}	Mokhayeri et al. (2006)
	Ethanol	3.5 g COD g N^{-1}	Mokhayeri et al. (2006)
	Acetate	3.4 g COD g N^{-1}	Mokhayeri et al. (2006)
Maximum biomass-specific denitrification rate (q_{NOx_N2})	Acetate,	10-19 mg N g VSS^{-1} h^{-1}	Ficara and Canziani (2007)
	Acetate, nitrite	15-28 mg N g VSS^{-1} h^{-1}	Ficara and Canziani (2007)
	Methanol, 13 °C	9.2 mg N g VSS^{-1} h^{-1}	Mokhayeri et al. (2008)
	Ethanol, 13 °C	30.4 mg N g VSS^{-1} h^{-1}	Mokhayeri et al. (2008)
	Acetate, 13 °C	31.7 mg N g VSS^{-1} h^{-1}	Mokhayeri et al. (2008)
	Acetate	1-3 mg N g VSS^{-1} h^{-1}	Kujawa and Klapwijk (1999)
	Acetate	2-10 mg N g VSS^{-1} h^{-1}	Henze (1991)

Regarding the growth kinetics, the specific substrate consumption (either nitrate/nitrite or COD) can be easily determined by batch activity tests. This value depends on the operational conditions used during the test (especially on temperature and the substrate's nature and concentrations). Therefore, these values should always be specified when reporting the results of a test. As reference conditions, the denitrification tests should be performed at 20 °C under non-limiting concentrations of carbon and nitrate/nitrite. This will allow the determination of the maximum specific denitrification rate. In practical applications, the maximum biomass specific denitrification rate q_{NOx_N2} is linked to the mixed liquor suspended solids (X_{TSS}) or, more commonly, to their volatile suspended solids content (X_{VSS}). Values reported in literature can be found in Table 2.4.2.

Anammox

The most significant kinetic and stoichiometric parameters of the anaerobic nitrogen removal process performed by anammox bacteria are the maximum specific biomass ammonium oxidation rate ($q_{AMX, NH4\_N2}$), the nitrite to ammonium ($Y_{NH4\_NO2,AMX}$) consumption ratio and the ratio between nitrate production and ammonium consumption ($Y_{NH4\_NO3,AMX}$). In Table 2.4.3 typical kinetic and stoichiometric parameters values found in literature are reported, together with the biomass growth yield.

Table 2.4.3 Stoichiometric and kinetic parameters of interest for anammox biomass performing the anaerobic ammonium oxidation process. The kinetic parameters are obtained at 30 °C. Biomass type: suspended (S), flocculent (F), granular (G). Reactor of origin: lab- (Lab) or full-scale (Full) reactor performing the anoxic stage of a 2-stage PN/anammox system (2-stage) or the 1-stage PN/anammox system (1-stage).

$q_{AMX, NH4\_N2}$ g N_2-N g VSS^{-1} d^{-1}	$Y_{AMX,NH4}$ C-mol NH_4·mol^{-1}	$Y_{NH4\_NO2,AMX}$ mol mol^{-1}	$Y_{NH4\_NO3,AMX}$ mol mol^{-1}	Biomass type	Reactor of origin	Reference
0.66	0.066	1.32	0.26	F	Lab, 2-stage	Strous et al. (1998)
0.22		1.27	0.34	G	Lab, 2-stage	Puyol et al. (2013)
0.22	0.105	1.28	0.37	F	Lab, 2-stage	Puyol et al. (2013)
2.01	0.071	1.22	0.21	S	Lab, 2-stage	Lotti et al. (2014)
3.38	0.071			S	Lab, 2-stage	Lotti et al. (2015b)
0.16				G	Full, 1-stage	Lotti et al. (2015c)
0.55				G	Full, 2-stage	Lotti et al. (2015c)

This parameter, even though it cannot be directly measured through batch tests, is useful to convert the specific biomass activity to growth rate values. Different from other bioprocesses described in this section, the anammox kinetics reported in Table 2.4.3 correspond to values measured at 30 °C which is the most common temperature at which the anammox process is operated in both lab- and full-scale systems, since it is close to the optimal temperature of these organisms (Hu et al., 2013).

As observed in Table 2.4.3, the anammox kinetic rates reported in literature can vary widely. The main reason for such variation appears to be the fraction of active anammox biomass present in the sample. In anammox reactors fed with autotrophic synthetic media, as is often the case for a lab-scale system, a lower fraction of X_{OHO} is expected compared to reactors fed with COD-containing wastewaters. The SRT applied is also expected to affect the fraction of active cells, which may be reduced by the accumulation of a significant fraction of inactive cells and non-biodegradable matter at higher SRT due to decay. Also, certain differences are observed between the kinetics of 1- or 2-stage PN/anammox systems because of the presence of X_{AOO} in the former, which contributes to a reduction in the fraction of active anammox biomass. Finally, since the anammox process is normally operated under nitrite limiting conditions in view of the inhibition potential of this substrate (Lotti et al., 2012), systems where biomass does not tend to aggregate are characterized by lower mass transfer limitations such as flocculent and (especially) suspended sludge, having higher active anammox biomass fractions compared to biofilm systems.

As observed in Table 2.4.3, also the consumption/production stoichiometric ratios may vary for different anammox systems. In literature, there is evidence that the anammox stoichiometry can be affected by the physiological state of the biomass, which can be influenced by the N-load (Dosta et al., 2008; Yang et al., 2009), temperature (Dosta et al., 2008) or pH (Carvajal-Arroyo et al., 2013).

2.4.3.9 Type of batch tests

Depending on the type of process of interest (nitrification, denitrification, anammox) and on the selected tracking technique (chemical, titrimetric or manometric), various tests can be performed to assess the nitrogen removal conversions.

A comprehensive list of tests that are described later on in this chapter is presented in Table 2.4.4.

In the following paragraphs, these tests are described in detail. First, nitrification tests are presented (Section 2.4.4), then denitrification tests (Section 2.4.5) and finally anammox tests (Section 2.4.6).

Table 2.4.4 Batch activity tests performed to assess the biological nitrogen removal conversions as a function of the process and tracking method.

Test code	Process	Tracking method	Purpose
NIT.CHE	Nitrification	Chemical	Assessing the maximum NH_4 oxidation rate
NIT.TIT.1	Nitrification	Titrimetric	Assessing the maximum NH_4 oxidation rate
NIT.TIT.2	Nitrification	Titrimetric	Assessing the maximum NH_4 and NO_2 oxidation rate and the ammonification rate
DEN.CHE.1	Denitrification	Chemical	Assessing the maximum denitrification rate and the anoxic growth yield on a specific C source
DEN.CHE.2	Denitrification	Chemical	Assessing the denitrification potential of a wastewater
DEN.MAN	Denitrification	Manometric	Assessing the maximum denitrification rate
DEN.TIT	Denitrification	Titrimetric	Assessing the maximum denitrification rate
AMX.CHE	Anammox	Chemical	Assessing the maximum anammox rate and the NO_2/NH_4 and NO_3/NH_4 ratio
AMX.MAN	Anammox	Manometric	Assessing the maximum anammox rate

2.4.4 Nitrification batch activity tests: preparation

2.4.4.1 Apparatus

Each tracking methodology (chemical, titrimetric or manometric) has specific apparatus requirements. When the chemical tracking is applied, refer to the following list of equipment:
1. A batch reactor equipped with mixing system and adequate sampling ports (Section 2.4.3.1).
2. A calibrated pH electrode (if not included/incorporated in the batch reactor setup).
3. A 2-way pH controller for HCl and NaOH addition (alternatively a one-way control - generally for HCl addition - or manual pH control can be applied through the manual addition of HCl and NaOH). For alkaline solutions that needs to be acidified, sparging with gaseous CO_2 (or a gas mixture enriched in CO_2) can be considered instead of the addition of an acidic solution since it has the advantage of avoiding the addition of the counter ions of protons (i.e. Cl⁻ when using HCl as the acidic solution).
4. A thermometer (with a recommended working temperature range of 0 °C to 40 °C).
5. A temperature control system (if not included in the batch reactor setup).
6. A DO meter with electrode (if not included/incorporated in the batch reactor setup) to verify the aerobic/anoxic conditions.
7. A stopwatch.

A list of the equipment required for titrimetric and manometric tests is described in Section 2.4.3.1.

2.4.4.2 Materials

For general instructions on material preparation, refer to Section 2.2.3.4 and Table 2.2.2. Test-specific requirements will be listed within each protocol of the test. For a complete list of the required materials refer to Section 2.2.3.2 (with the exception of points 7 and 8).

2.4.4.3 Media preparation

- **Real wastewater**

For batch tests that require the use of a real wastewater, follow the instructions reported in Section 2.2.3.3.

- **Titration solutions**

a. NaOH and HCl solutions are needed. Typically 0.05-0.1 N solutions would be suitable for most applications.
b. The H_2O_2 solution can be obtained by dilution of commercially available H_2O_2 solutions. The most common H_2O_2 solution is 3 %, corresponding to a

concentration of 0.44 mol O_2 L^{-1}. Therefore, an appropriate oxygenated titration solution can be obtained by diluting this solution 10 times (a final concentration of 44 mmol O_2 L^{-1}). To check the concentration of the H_2O_2 solution, the iodometric method can be applied (method 4,500-Cl B in APHA et al., 2012). The diluted solution should be stored in dark bottles and new solutions should be prepared every 7-10 days.

- **Ammonium and nitrite stock solutions**

These can be prepared from salts (e.g. from NH_4Cl and $NaNO_2$). An adequate concentration of the stock solutions is between 5 and 10 g N L^{-1}. The pH of the ammonium solution should be adjusted to 7.0 to reduce any potential interference during the pH-static tests.

- **Allyl-N-thiourea (ATU)**

A stock solution of approximately 5-10 g L^{-1} is generally adequate.

- **Acid and base solutions**

These should be 100-250 mL of 0.2 M HCl and 100-250 mL 0.2 M NaOH solutions for automatic or manual pH control, and 10-50 mL of 1 M HCl and 10-50 mL 1 M NaOH solutions for initial pH adjustment if the desired operational pH is very different from the pKa value of the buffering agent. Instead of NaOH, 0.2 M Na_2CO_3 can be used as the alkaline solution, which has the advantage of acting as both the base and carbon source. The use of Na_2CO_3 as the base solution is therefore recommended when the wastewater may have a deficiency of inorganic carbon. This paragraph does not apply when titrimetry is used.

- **Synthetic medium**

This should contain all the required macro- (sodium, chloride, phosphate, magnesium, sulphate, calcium, potassium) and micro-nutrients (iron, zinc, copper, manganese, boron, molybdate, cobalt iodide) to ensure that cells are not limited and in extreme cases to avoid the failure of the test. Thus, although their concentrations may seem very low, one must make sure that all of the constituents are added to the solution in the required amounts. Regarding the macro-nutrients, the following composition (amounts per litre of the nutrient solution) is recommended (based on Kampschreur et al., 2007): 72 mg NaH_2PO_4, 35 mg $MgSO_4 \cdot 7H_2O$, 5 mg $CaCl_2 \cdot 2H_2O$, 180 mg NaCl, 30 mg KCl, and 1 mg yeast extract. Micro-nutrients can be supplied by dosing 0.3 mL L^{-1} of a trace element solution containing (per litre of solution, recipe based on Kampschreur et al., 2007): 10 g EDTA, 1.5 g $FeCl_3 \cdot 6H_2O$, 0.15 g H_3BO_3, 0.03 g $CuSO_4 \cdot 5H_2O$, 0.18 g KI, 0.12 g $MnCl_2 \cdot 4H_2O$, 0.06 g $Na_2MoO_4 \cdot 2H_2O$, 0.12 g $ZnSO_4 \cdot 7H_2O$, 0.15 g $CoCl_2 \cdot 6H_2O$. Other similar nutrient solutions can be used as long as they contain all the previously reported required nutrients.

- **Washing media**

If the sludge sample must be washed to remove an undesirable compound (which may be even inhibitory or toxic), a washing media should be prepared. The same synthetic medium described above can be used as a washing medium. The washing process can be repeated twice or three times applying the procedure described in Section 2.2.3.5. Thereafter, the following preparation steps of the batch activity tests can be performed. In special cases when sludge from a full-scale plant is used, the plant effluent may be used for washing purposes (only if it does not contain toxic or inhibitory compounds).

Prior to the execution of the experiment, samples of the media and mixed liquor or activated sludge used to perform the tests should be collected to confirm/check the initial (desired) concentration of parameter(s) of interest (e.g. ammonium, nitrite, nitrate, X_{TSS}, X_{VSS}).

Finally, the required working and stock solutions to carry out the determination of the analytical parameters of interest must also be prepared in accordance with Standard Methods (APHA et al., 2012) and the corresponding protocols.

2.4.5 Nitrification batch activity tests: execution

Test NIT.CHE Nitrification chemical test: assessing the maximum ammonium oxidation rate

Activated sludge preparation
1. For sample collection please refer to Section 2.4.3.4.
2. For conventional activated sludge samples from wastewater treatment plants treating urban wastewaters, a sample with a X_{VSS} concentration of around 2-4 g VSS L^{-1} would be suitable. X_{VSS} adjustment can be performed as suggested in Section 2.4.3.5.
3. Pour a defined volume (V_{ML}) of a mixed liquor sample (typically 1 to 3 L) into the reaction vessel and start mixing, aeration and the temperature and pH control systems to maintain temperature and pH around the desired set point values. Select a desired set point pH value; typically values between 7.5 and

8.4 are adequate. If an automated pH control is not available, correct the pH to the desired value by manual addition of an acid/base solution. In principle, the closer the pH set point value to the typical operational pH at the plant, the closer and more representative the nitrification rate will be. The same principle applies to the selection of the temperature set point value. As for DO, the aeration system should provide sufficient oxygen to avoid DO-limiting conditions during the execution of the nitrification tests. This means that under endogenous conditions the observed DO value should be high (e.g. > 6 mg L^{-1}).

4. Wait for approximately 30 min to reach and ensure stable initial conditions. This pre-incubation phase will normally allow any residual nitrite remaining from the plant or source of origin of the sludge to be consumed.

Execution of the test

1. Verify that the temperature, pH, and DO readings are at the desired set point values or at least within the selected intervals. Otherwise adjust the operating conditions accordingly and wait until the system stabilizes.
2. Once stable conditions are reached, add the ammonium solution (having previously adjusted its temperature to the target temperature of the test) to achieve a neither limiting nor inhibiting ammonium concentration in the mixed liquor. Typical and adequate values are between 20 and 40 mg N L^{-1}.
3. Start the stopwatch to keep precise track of the sampling times, since formally the test starts with the addition of the ammonium solution. Collect the activated sludge samples every 20-30 min throughout the execution of the test. Note that all samples need to be filtered through 0.45 µm pore size filters (or smaller), except those used for the determination of X_{TSS} and X_{VSS} concentrations.
4. Conclude the test after 3 to 4 h or, if the sampling and analytical determination of ammonium in the collected samples allows it, when ammonium is depleted.
5. After the conclusion of the test, take a sample for the final X_{VSS} assessment.

Note that nitrite is rather unstable; therefore, nitrite concentrations have to be quickly assessed after sampling in the same day (see Table 2.2.2).

Data analysis

A typical output of this test is outlined in Figure 2.4.3. On the y-axis, the ammonium, nitrite and nitrate concentrations (in mg N L^{-1}) are reported, while time (in hours) is reported on the X-axis. The linear regression over these data allows for the assessment of the ammonium removal and nitrate production rates (in mg N L^{-1} h^{-1}). Please note that the collected data should be sufficient to reliably estimate the corresponding removal/production rate (e.g. r_{NH4} and r_{NO3}) by linear regression with a satisfactory coefficient of determination (e.g. $R^2 > 0.98$). Thus, preferably at least 4 to 5 data points are needed to carry out the linear regression, implying that a larger number of samples will need to be collected in the beginning of the test.

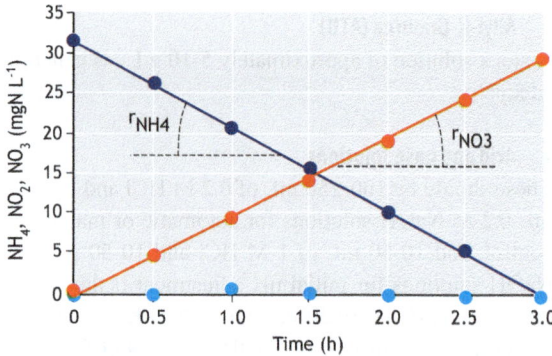

Figure 2.4.3 Typical profiles obtained in a Test NIT.CHE: ammonium (●), nitrite (●) and nitrate (●) concentrations are displayed on the y-axis. Relevant rates of interest are also displayed (e.g. ammonium removal rate r_{NH4}, and nitrate production rate r_{NO3}).

Nitrite may accumulate up to few mg N L^{-1} when conventional activated sludge is used. However, if the activated sludge is used to perform (partial) nitrification tests, then a higher nitrite accumulation will be expected and its concentration should be monitored in time, similar to ammonium and nitrate (for example see Test NIT.TIT.2). When using conventional activated sludge to perform the full oxidation of ammonium to nitrate, the ammonium removal rate should equal the nitrate production rate with a negligible accumulation of nitrite during the test. The maximum specific ammonium oxidation rate ($q_{AOO,NH4}$, as mg N g VSS^{-1} h^{-1}) can be therefore computed as follows:

$$q_{AOO,NH4} = r_{NH4} / X_{VSS} \qquad \text{Eq. 2.4.24}$$

Test NIT.TIT.1 Nitrification titration test: assessing the maximum ammonium oxidation rate

Activated sludge preparation

1. For conventional activated sludge samples from wastewater treatment plants treating urban wastewaters, a sample with a X_{VSS} of around 2-4 g VSS L^{-1} will be suitable. For very diluted or concentrated sludge samples, a concentration step (e.g. by decanting into an Imhoff cone for 30 min or by centrifuging at 4,000 rpm for a few minutes) or dilution with secondary effluent from the same wastewater treatment plant may be helpful. This will avoid having too slow or too fast nitrification rates.

2. Pour a known volume of the activated sludge sample (typically 1 L) into the reaction vessel and activate the aeration and the temperature control systems. Select a desired set point pH value. Typically, values between 7.5 and 8.4 are adequate. In principle, the closer the set point pH value to the typical operational pH of the plant or source of sludge, the closer and more representative the observed nitrification rate will be. The same principle applies to the selection of the temperature value. Concerning DO, the aeration system should provide sufficient oxygen to avoid DO-limiting conditions during the course of the nitrification tests. This means that under endogenous conditions the observed DO value should be relatively high (e.g. > 6 mg L^{-1}).

3. Activate the automated titration system for a pre-incubation period of approximately 1 h. This pre-incubation phase will ensure that: (*i*) endogenous conditions are achieved at the start of the titration test (S_B, ammonium and nitrite are oxidized during this overnight aeration phase) and (*ii*) that temperature, pH and S_{O2} are stable at the start of the test. Pre-humidified air may be used to limit significant water evaporation during this pre-incubation phase. Note that prolonged incubation periods (e.g. longer than 4 h) may reduce the nitrification rate due to fast endogenous biomass decay under aerobic conditions.

Execution of the test

1. Activate the data logging.
2. Add the ammonium chloride stock solution (having previously adjusted its temperature to the target temperature of the test) to achieve an ammonium concentration in the activated sludge that is neither limiting nor inhibiting (between 20 and 40 mg N L^{-1} are typically adequate values). NH$_4$Cl addition to an alkaline suspension has an acidifying affect (acid hydrolysis) that leads to a rapid pH drop that should be compensated by an automatic pH control system or through the manual addition of concentrated NaOH. Any addition of the titration solutions during this pH-adjustment phase should be disregarded during the data analysis. Upon the ammonium addition, nitrifying bacteria will oxidize the ammonium added and consequently an alkaline titrating solution will need to be added to compensate for the acidifying nitrification effect.

3. Record the volume of the NaOH titration solution added over time (V_{NaOH} versus time) (20-40 min are usually adequate). Check that the pH reading value remains close to the target pH set point ± 0.02 and that the S_{O2} level does not become limiting. Do not change the aeration rate or the mixing conditions since this would affect the titration rate, making the assessment of the nitrification-related titration rate cumbersome. The collected data should be sufficient to reliably estimate the titration rate (Q_{NaOH}) from a linear regression of V_{NaOH} versus time data with a satisfactory coefficient of determination ($R^2 > 0.98$).

4. Add allyl-N-thiourea in order to achieve a final ATU concentration of 10 mg L^{-1}. At this concentration, ammonium oxidation will be inhibited. Continue recording the NaOH titration rate for a further 20-30 min in order to assess the residual titration rate due to background pH affection reactions such as CO$_2$ stripping ($Q_{NaOH,final}$), if present. Note that the longer the pre-incubation period the lower the relevance of $Q_{NaOH,final}$.

5. The test can now be ended. Measure the final activated sludge volume and take a sample to assess the X_{VSS} concentration. Note that the suspension volume will change during the course of the test due to the addition of titration solutions. It is expected that the addition of titration solution does not account for more than 10 % of the final activated sludge volume.

Data analysis

A typical trend of the volume of NaOH added during the test is depicted in Figure 2.4.4. From these data (titration volume versus time data), titration rates (Q) can be computed through the slope of the titration curve using the corresponding tool in a worksheet or by applying the following formula:

$$Q = \frac{n \cdot \sum t_i \cdot V_{NaOH,i} - \sum t_i \cdot \sum V_{NaOH,i}}{n \cdot \sum t_i^2 - (\sum t_i)^2} \qquad \text{Eq. 2.4.25}$$

Where, n is the number of recorded data [t_i, $V_{NaOH,i}$] available.

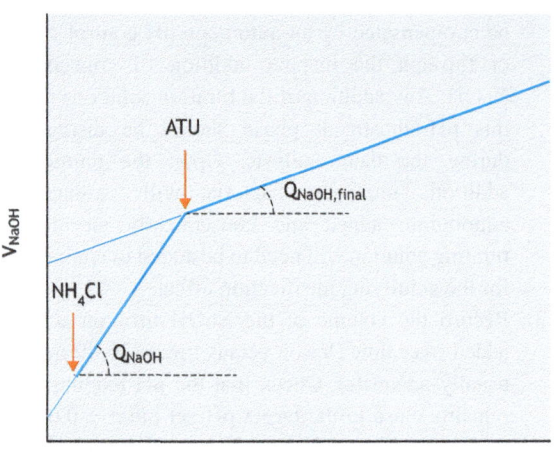

Figure 2.4.4 Example of a pH-static titration curve during a nitrification test to assess the maximum ammonium-oxidizing capacity. Arrows indicate the addition of ammonium chloride and allyl-N-thiourea. Relevant titration rates are also identified on the graph.

The NaOH titration rates (Q_{NaOH} and $Q_{NaOH,final}$ in mL min^{-1}) are used to assess the ammonium oxidation rate (F_{NHx} in mg N min^{-1}) by taking into account the concentration of the NaOH solution (N_{NaOH} in meq mL^{-1}) and the ratio between ammonium oxidation and alkalinity consumption Y_{NH4_H+} that can be assessed from Eq. 2.4.15. Therefore:

$$F_{NHx} = (Q_{NaOH} - Q_{NaOH,final}) \cdot N_{NaOH} \cdot Y_{NH4\_H+} \quad \text{Eq. 2.4.26}$$

Finally, the maximum specific ammonium oxidation rate of the sludge ($q_{AOO,NH4}$, in mg N g VSS^{-1} h^{-1}) can be computed by taking into account the X_{VSS} concentration of the sludge sample (in g VSS L^{-1}) and the suspension volume observed at the end of the test (V_{ML}):

$$q_{AOO,NH4} = 60 \cdot F_{NHx} / (V_{ML} \cdot X_{VSS}) \quad \text{Eq. 2.4.27}$$

Test NIT.TIT.2 Nitrification titration test: assessing the maximum ammonium and nitrite oxidation rates

Activated sludge preparation
Follow steps 1, 2 and 3 of the activated sludge preparation described for Test NIT.TIT.1.

Execution of the test
1. Select an appropriate set point value for S_{O2} (DO set point). Typically values between 4.0 mg N L^{-1} and 6.0 mg N L^{-1} are adequate to assess the maximum nitrification rates. Activate the data-logging system.
2. Record the volumes of the H_2O_2 and NaOH titration solutions added over time (V_{H2O2} and V_{NaOH} versus time) (20-40 min are normally adequate). Check that the pH and DO values remain within the interval pH set point $\pm$ 0.02 and S_{O2} set point $\pm$ 0.10 mg L^{-1}, respectively. The collected data should be sufficient to reliably estimate (i.e. with a satisfactory coefficient of determination, $R^2 > 0.98$) the alkaline titration rate from the linear regression of V_{NaOH} versus time data, and the oxygen titration rate (Q_{H2O2}) from the linear regression of V_{H2O2} versus time data. During this phase, titration rates are triggered by endogenous respiration which leads to DO consumption and CO_2 production; the former is compensated by H_2O_2 addition (at a rate indicated as $Q_{H2O2,ini}$), and the latter by NaOH addition (at a rate indicated as $Q_{NaOH,ini}$).
3. Add nitrite at a neither limiting nor inhibiting nitrite concentration in the activated sludge (around 10 mg N L^{-1} are typically adequate values) in order to trigger nitrite oxidation. Repeat data acquisition as described in Step 2 in order to estimate the oxygen titration rate that includes the oxygen request for nitrite oxidation ($Q_{H2O2,NO2}$). The alkaline titration rate will not change since nitrite oxidation does not significantly affect the suspension pH.
4. Add the ammonium chloride stock solution according to the instructions reported in step 2 of the test operation procedure described for Test NIT.TIT.1. This addition will trigger ammonium oxidation as well. Repeat data acquisition as described in step 2 to assess the alkaline titration rate ($Q_{NaOH,NH4}$) and the oxygen titration rate ($Q_{H2O2,NH4}$) that include the ammonium oxidation needs.
5. Add allyl-N-thiourea (ATU) to a final concentration of 10 mg L^{-1}. Ammonium oxidation will be inhibited. Continue recording the NaOH titration rate for a further 20-30 min to assess the residual titration rate due to the background pH affecting reactions such as CO_2 production ($Q_{NaOH,final}$) and oxygen-affecting reactions ($Q_{H2O2,final}$) including endogenous respiration and residual nitrite oxidation.
6. End the test according to the instructions reported in Step 4 of the test operation procedure described for Test NIT.TIT.1.

Data analysis
A typical trend of the cumulated volume of titration solutions added during the test is depicted in Figure 2.4.5. From these data (volume versus time data), the titration

rates (Q) can be computed as the slope of the titration curve using the corresponding tool in a worksheet or by applying the formula previously described in Test NIT.TIT.1.

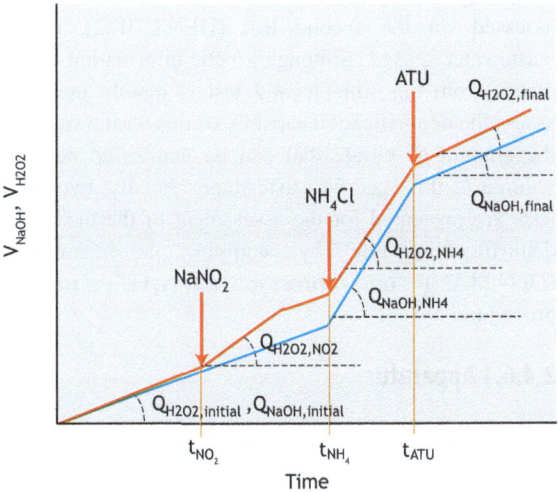

Figure 2.4.5 Example of a pH/DO-stat titration curve during a nitrification test executed to assess the maximum ammonium- and nitrite-oxidizing capacity. Arrows indicate the addition of nitrite, allyl-N-thiourea and ammonium. Relevant titration rates of interest are also displayed on the graph.

The NaOH titration rates ($Q_{NaOH,NH4}$ and $Q_{NaOH,final}$ in mL min^{-1}) can be first used to assess the ammonium oxidation rate ($F_{NHx,NaOH}$ in mg N min^{-1}) by taking into account the concentration of the NaOH titration solutions (N_{NaOH} in meq mL^{-1}) and the ratio between ammonium oxidation and alkalinity consumption, as suggested in Test NIT.TIT.1:

$$F_{NHx,NaOH} = (Q_{NaOH,NH4} - Q_{NaOH,final}) \cdot N_{NaOH} \cdot Y_{NH4\_H+}$$ Eq. 2.4.28

Similarly, oxygen titration rates ($Q_{H2O2,NH4}$ and $Q_{H2O2,final}$ in mL min^{-1}) can be first used to assess the ammonium oxidation rate ($F_{NHx,H2O2}$ in mg N min^{-1}) by taking into account the concentration of the H$_2$O$_2$ titration solutions (N_{H2O2} in mmol O$_2$ mL^{-1}) and the ratio between ammonium oxidation to nitrate and oxygen consumption, $Y_{NH4/O2\_NO3}$, that is, according to the ASM nitrification stoichiometry (Henze *et al.*, 2000):

$$F_{NHx,H2O2} = (Q_{H2O2,NH4} - Q_{H2O2,final}) \cdot N_{H2O2} \cdot 32 \cdot Y_{NH4/O2\_NO3}$$ Eq. 2.4.29

$$Y_{NH4/O2\_NO3} = \frac{1}{4.57 - Y_{AOO}} = 0.23 \text{ g N g O}_2^{-1}$$ Eq. 2.4.30

The values of $F_{NHx,NaOH}$ and $F_{NHx,H2O2}$ should be similar and their comparison can be used to validate the experimental data. Differences higher than 15 % may suggest the need for a careful verification of the experimental setup.

The oxygen titration rates collected during step 3 ($Q_{H2O2,NO2}$ in mL min^{-1}) will be used to determine the nitrite oxidation rate (F_{NO2} in mg N min^{-1}) by taking into account the ratio between nitrite oxidation to nitrate and oxygen consumption, $Y_{NO2/O2\_NO3}$, as follows (according to the two-step nitrification stoichiometry):

$$F_{NO2} = (Q_{H2O2,NO2} - Q_{H2O2,initial}) \cdot N_{H2O2} \cdot 32 \cdot Y_{NO2/O2\_NO3}$$ Eq. 2.4.31

$$Y_{NO2/O2\_NO3} = \frac{1}{1.14} = 0{,}88 \text{ g N g O}_2^{-1}$$ Eq. 2.4.32

If no difference is observed between $Q_{H2O2,NO2}$ and $Q_{H2O2,initial}$ values then either the nitrite oxidation rate is very slow or it is much slower than the ammonium oxidation rate. This may lead to nitrite accumulation during the endogenous phase since ammonium oxidation would probably take place on the ammonium released through ammonification. If so, $Q_{H2O2,initial}$ would include oxygen consumption due to the slow nitrite oxidation process and be equal to $Q_{H2O2,NO2}$. In this case, the activated sludge sample should be elutriated to remove any nitrite content by centrifugation and resuspension in a nitrite-free physiological medium to assess the nitrite oxidation rate. However, when the nitrite oxidation rate is very slow, chemical tracking over a longer testing period (a few hours) may lead to more reliable estimates and should be preferred. Note that, for very slow nitrite oxidation rates, $F_{NHx,H2O2}$ may be higher than $F_{NHx,NaOH}$. As a matter of fact, the use of $Y_{NH4/O2\_NO3}$ is no more correct since the sole ammonium oxidation to nitrite request should be taken into account, since $Y_{NH4/O2\_NO3}$ quantifies the overall ammonium plus nitrite oxidation request. Therefore, in such a case, $F_{NHx,H2O2}$ would be more correctly estimated taking into account the ratio between ammonium oxidized and oxygen consumption for ammonium oxidation to nitrite, $Y_{NH4/O2\_NO2}$, which can be expressed according to the two-step nitrification stoichiometry expression as:

$$F_{NHx,H2O2} = (Q_{H2O2,NH4} - Q_{H2O2,final}) \cdot N_{H2O2} \cdot 32 \cdot Y_{NO2/O2\_NO2}$$ Eq. 2.4.33

$$Y_{NO2/O2\_NO2} = \frac{1}{3.43 - Y_{AOO}} = 0.31 \text{ g N g } O_2^{-1}$$ Eq. 2.4.34

Moreover, the difference between $Q_{NaOH,initial}$ and $Q_{NaOH,final}$ makes it possible to estimate the ammonification rate under endogenous conditions. The first titration rate compensates for the alkaline effect of endogenous respiration and of nitrification, which is limited by the ammonification process responsible for ammonium release. The ammonification-related alkalizing effect is no longer present after ATU addition. Thus, the ammonification rate (F_{N_NHx} in mg N min^{-1}) can be estimated, by difference, according to the following equation:

$$F_{N\_NHx} = Y_{NH4/O2\_NO3} \cdot (Q_{NaOH,initial} - Q_{NaOH,final}) \cdot N_{NaOH}$$ Eq. 2.4.35

Finally, the maximum biomass-specific ammonium and nitrite oxidation rates of the sludge ($q_{AOO,NHx}$, and $q_{NOO,NO2\_NO3}$ in mg N g VSS^{-1} h^{-1}) and the specific ammonification rate (q_{N_NHx}), can be computed by taking into account the X_{VSS} concentration of the sludge sample (in g VSS L^{-1}) and the suspension volume observed at the end of the test (V_{ML}):

$$q_{AOO,NHx} = 60 \cdot F_{NHx} / (V_{ML} \cdot X_{VSS})$$ Eq. 2.4.27

$$q_{NOO,NO3\_NO2} = 60 \cdot F_{NO2} / (V_{ML} \cdot X_{VSS})$$ Eq. 2.4.36

$$q_{N\_NHx} = 60 \cdot F_{N\_NHx} / (V_{ML} \cdot X_{VSS})$$ Eq. 2.4.37

2.4.6 Denitrification batch activity tests: preparation

These tests are meant to assess the maximum denitrification rate of a sludge sample and the anoxic biomass growth yield. Various types of carbon sources can also be used such as internal carbon, external carbon sources (e.g. sugar or alcohol) or wastewater. Typically, the maximum denitrification rate is expressed when a rapidly biodegradable carbon source (to which the sludge is adapted) is used, while lower rates are observed in the presence of complex organic molecules that require a preliminary hydrolysis step or when dosing external carbon solutions that require specialized metabolic capabilities/microorganisms.

Four tests are presented. The first one (DEN.CHE.1) refers to the use of an easily biodegradable carbon source, for which both the denitrification rate and the biomass anoxic growth yield are relevant parameters to be assessed. In the second test (DEN.CHE.2), a real wastewater is used. Although kinetic information can be drawn from this, this second test is mainly meant to assess the denitrification capacity of this wastewater, i.e. the amount of nitrate that can be denitrified per unit volume of this specific wastewater. Finally, two more tests are presented for the assessment of the maximum denitrification rate by applying a manometric (DEN.MAN) or titrimetric (DEN.TIT) tracking procedure.

2.4.6.1 Apparatus

Each tracking methodology (chemical, titrimetric or manometric) has special apparatus requirements.

When the chemical tracking is applied refer to the following list of equipment:

1. An (airtight) batch reactor equipped with mixing system and adequate sampling ports (as described in Section 2.4.3.1).
2. A nitrogen gas supply (recommended).
3. A calibrated pH electrode (if not included or incorporated in the batch reactor setup).
4. A 2-way pH controller for HCl and NaOH addition (alternatively a one-way control, generally for HCl addition, or a manual pH control can be applied through the manual addition of HCl and NaOH).
5. A thermometer (with a recommended working temperature range of 0 to 40 °C).
6. A temperature control system (if not included in the batch reactor setup).
7. A DO meter with an electrode (if not included/incorporated in the batch reactor setup) to verify anoxic conditions.
8. A stopwatch.

When titrimetric tests are performed, refer to the following list of equipment:
1. The equipment for the titrimetric system described in Section 2.4.3.1.
2. A nitrogen gas supply (recommended).

When manometric tests are performed, refer to the following list of equipment:

1. The equipment for the manometric system described in Section 2.4.3.1.
2. A nitrogen gas supply (recommended).

2.4.6.2 Materials

For a complete list of required materials refer to Section 2.2.3.2 (with the exception of points 7 and 8).

2.4.6.3 Working solutions

- **Real wastewater**

For batch tests that require the use of real wastewater, follow the instructions given in Section 2.2.3.3.

- **Carbon source solution**

This is usually composed of a readily biodegradable carbon source (S_B), preferably volatile fatty acids like acetate or propionate, sugars, or alcohol solutions. The choice of the organic carbon source depends on the nature or goal of the test and the corresponding research questions. Sometimes more complex substrates are used that are more similar to real wastewaters, containing a mixture of readily and slowly biodegradable COD. For anoxic batch activity tests, the COD concentration (both total and soluble) should be known in order to select a proper dose.

- **Nitrate or nitrite solutions**

Nitrate and nitrite salts are needed to adjust the nitrate/nitrite level during denitrification tests.

- **Washing media**

If the sludge sample must be 'washed' to remove undesirable compounds, then refer to Section 2.4.3.4.

- **Acid and base solutions**

These are 100-250 mL of 0.2 M HCl and 100-250 mL 0.2 M NaOH solutions for automatic or manual pH control, and 10-50 mL of 1 M HCl and 10-50 mL 1 M NaOH solutions for an initial pH adjustment if the desired operational pH is very different from the pKa value of the buffering agent. For titrimetric tests, titration solutions should be prepared according to the instructions given in Section 2.4.4.3.

- **Nutrient solution**

This should contain all the required macro- (ammonium, magnesium, sulphate, calcium, potassium) and micro-nutrients (iron, zinc, calcium, copper, manganese, molybdate, cobalt) to ensure that cells are not short of basic nutrients for their metabolism. Thus, despite the fact that their concentrations may seem very low, one must make sure that all of the constituents are added to the solution in the required amounts. Regarding macronutrients the following composition (amounts per litre of nutrient solution) is recommended (based on Smolders et al., 1994): 107 mg NH_4Cl, 90 mg $MgSO_4 \cdot 7H_2O$, 14 mg $CaCl_2 \cdot 2H_2O$, 36 mg KCl, 1 mg yeast extract. Micronutrients can be supplied by dosing 10 mL L^{-1} of a trace element solution containing (per litre of solution) (based on Vishniac and Santer, 1957): 50 g EDTA, 22 g $ZnSO_4 \cdot 7H_2O$, 5.54 g $CaCl_2$, 5.06 g $MnCl_2 \cdot 4H_2O$, 4.99 g $FeSO_4 \cdot 7H_2O$, 1.10 g $(NH_4)_6Mo_7O_{24} \cdot 4H_2O$, 1.57 g $CuSO_4 \cdot 5H_2O$, and, 1.61 g $CoCl_2 \cdot 6H_2O$. Similar nutrient solutions can be used as long as they contain all the previously reported required nutrients.

It is recommended to take a sample of the media and sludge prior to the execution of the experiment to confirm/check the initial (desired) concentration of parameter(s) of interest (e.g. COD, nitrate/nitrate, X_{VSS}).

Finally, the required working and stock solutions to carry out the determination of the analytical parameters of interest must be also prepared in accordance to Standard Methods (APHA et al., 2012) and the corresponding protocols for their preservation and analytical determination.

2.4.6.4 Material preparation

For general instructions on how to organize the material preparation, please refer to Section 2.2.3.4 and to Table 2.2.2. Test-specific requirements will be listed within each test protocol.

2.4.7 Denitrification batch activity tests: execution

Test DEN.CHE.1 Denitrification chemical test: assessing the maximum denitrification rate and the anoxic growth yield in the presence of a specific carbon source

Activated sludge preparation

The optimal sampling point for the activated sludge would be the outlet of the post-denitrification tank.

1. For conventional activated sludge samples treating urban wastewaters, a sample with a X_{VSS} of around 2-4 g VSS L^{-1} would be suitable. The X_{VSS} can be adjusted as suggested in Section 2.4.3.5.
2. Pour a known volume (V_{ML}) of activated sludge sample (typically 1 to 3 L) into the reaction vessel and

start mixing. Also start the temperature and pH control systems to keep both parameters around the desired set point values.

3. Sparge N_2 into the reaction headspace for approximately 10 min to ensure a deoxygenated environment. Ensure a gas outlet to limit overpressure. Gas sparging can be continued until the end of the test. If this is not feasible, then a proper airtight reactor with a gas outlet preventing oxygen back-diffusion (e.g. by using a unidirectional check valve or a water lock) should be used (see Section 2.2.2.1 for more information on how to ensure anoxic conditions).

4. Wait for approximately 30 min to ensure stable initial conditions. This pre-incubation phase will usually allow the removal of any residual nitrate.

Execution of the test

Verify that your target values of temperature, pH, and S_{O2} are close to the set points or within the selected intervals. Otherwise adjust and wait until the system stabilizes.

1. Add the nitrate and S_B stock solutions. The temperature of both these solutions should have been previously adjusted to the target temperature of interest. The initial nitrate concentration should be neither limiting nor inhibitory (20 to 25 mg N L^{-1} are typically adequate values). If residual nitrate is expected to be present in the activated sludge sample, the nitrate addition should then be reduced.

2. Add a non-limiting amount of the readily biodegradable carbon source (S_B). To assess the appropriate amount of S_B, one can consider the stoichiometric relationship between the amount of S_B and nitrate consumed during denitrification:

$$Y_{NO3\_SB,Ax} = \frac{2.86}{1 - Y_{OHO,Ax}} \quad (g\ COD\ g\ N^{-1}) \quad \text{Eq. 2.4.38}$$

3. The S_B addition should guarantee that the S_B to nitrate ratio should be at least twice as much as the stoichiometric value found with the previous expression. Note that the anoxic biomass growth yield, $Y_{OHO,Ax}$, depends on the carbon source, as reported in Table 2.4.2. However, a value of 0.5 can be typically used for a rough estimation. Under this assumption, a S_B concentration in the activated sludge of 200 mg COD L^{-1} would usually be adequate. This concentration also satisfies the initial S_B to X_{VSS} ratio value (0.05-0.1 g COD g VSS^{-1}) suggested in Section 2.2.4.1. Lower values may result in too rapid carbon depletion while too high values

may lead to biomass inhibition. An ammonium stock solution can also be added to adjust the ammonium to S_B ratio to 0.05 g N g COD^{-1}.

4. The test starts with the addition of the nitrate and S_B solutions. Start the stopwatch to keep precise track of the following sampling times and start the sampling campaign.

5. Collect activated sludge samples at regular time intervals. As a general suggestion, samples for the determination of the C source (or of soluble COD depending upon the analytical parameter of interest) and of nitrite and nitrate must be collected every 10 min in the first 30 min after S_B addition, every 15 min during the following 60 min, and later on every 30 min until the end of the test.

6. Conclude the test when the nitrate and nitrite are fully depleted. Use nitrate and nitrite strip tests to quickly assess when these compounds have been consumed.

7. Take a sample for final X_{VSS} concentration assessment.

It should be noted that nitrite concentration is unstable, and therefore nitrate and nitrite concentrations have to be quickly measured, preferably on the same day.

Data analysis

The typical output of this test is outlined in Figure 2.4.6.

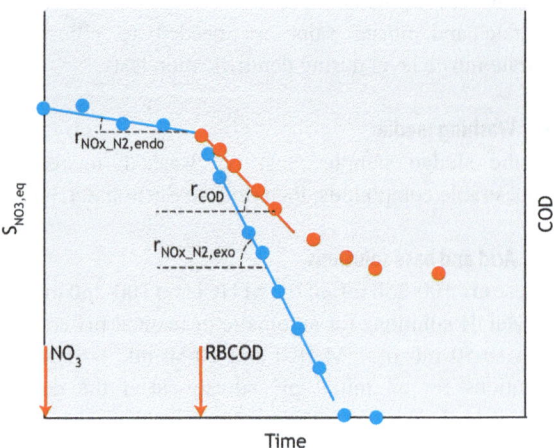

Figure 2.4.6 Typical profiles obtained in a Test DEN.CHE.1: nitrate concentrations (●) on the main y-axis, COD concentrations (●) on the secondary y-axis. Relevant rates of interest are also displayed (endogenous denitrification rate $r_{NOx\_N2,endo}$, exogenous denitrification rate, $r_{NOx\_N2,exo}$, and COD consumption rate r_{COD}). The arrow indicates the substrate addition.

On the principal y-axis, the oxidized nitrogen equivalent $S_{NO3,Eq}$ is reported, which corresponds to a weighted sum of the nitrate and nitrite concentrations:

$$S_{NO3,Eq} = S_{NO3} + 0.6 \cdot S_{NO2} \qquad \text{Eq. 2.4.39}$$

The 0.6 weight applied to the nitrite concentration corresponds to the relative electron-accepting capacity of nitrite with respect to nitrate (1.71/ 2.86 = 0.6), as suggested by Kujawa and Klapwijk (1999). On the secondary y-axis, soluble COD data are presented. During the first period (i.e. before the addition of the external carbon source), endogenous denitrification takes place and a slow-rate reduction of $S_{NO3,Eq}$ is observed.

The linear regression over these data (see Section 2.4.4.2) can be used to assess the endogenous denitrification rate ($r_{NOx\_N2,endo}$, in mg N L^{-1} min^{-1}). After the S_B addition, the availability of the exogenous carbon source speeds up the consumption of nitrate and nitrite. Collecting the nitrate/nitrite data afterwards, but before nitrate/nitrite becomes limiting, makes it possible to assess the exogenous denitrification rate ($r_{NOx\_N2,exo}$, in mg N L^{-1} min^{-1}). Similarly, the maximum COD consumption rate (r_{COD}, in mg COD L^{-1} min^{-1}) can be assessed within the same timeframe.

The maximum specific denitrification rate on S_B ($q_{NOx\_N2,SB}$ in mg N g VSS^{-1} h^{-1}) on the tested carbon source can be computed as follows:

$$q_{NOx\_N2,SB} = 60 \cdot (r_{NOx\_N2,exo} - r_{NOx\_N2,endo}) / X_{VSS} \qquad \text{Eq. 2.4.40}$$

Moreover, by combining denitrification rates and the COD consumption rate, the biomass growth yield ($Y_{OHO,Ax}$) can also be assessed according to the following formula:

$$Y_{OHO,Ax} = 1 - 2.86 \frac{(r_{NOx\_N2,exo} - r_{NOx\_N2,endo})}{r_{COD}} \qquad \text{Eq. 2.4.41}$$

Test DEN.CHE.2 Denitrification chemical test: assessing the denitrification potential of wastewater

Activated sludge preparation
Samples for this test should be collected at the outlet of the pre-denitrification tank.

1. Follow step 1 of the activated sludge preparation procedure described for Test DEN.CHE.1 Note that in this case a biomass concentration of 3-4 g VSS L^{-1} would be more adequate since a dilution effect is obtained when the wastewater is added.
2. Pour a known volume of the activated sludge sample (V_{ML}, typically 0.6-0.8 L) and keep a sample of it to assess the volatile suspended solids concentration (MLVSS in g VSS L^{-1}). Follow steps 2, 3 and 4 of the activated sludge preparation procedure described for Test DEN.CHE.1.

Test execution
1. Verify that your target values of temperature, pH, and DO are within the selected intervals. Otherwise adjust them and wait for their stabilization.
2. Select the appropriate volume of wastewater (V_{ww}) to be added. It would be ideal to add an amount of wastewater so that the final biodegradable COD concentration in the reaction vessel remains within 30 and 70 mg L^{-1}. By assuming a typical concentration of biodegradable carbon ($S_B + XC_B$) of 100-180 mg L^{-1}, a dilution factor ($V_{ML}:V_{ww}$) of 2 to 6 should be appropriate. Pour the wastewater into the reaction vessel and add the nitrate stock solution in order to achieve an initial nitrate concentration in the final mixture ($V_{ML}+V_{ww}$) of 20-25 mg N L^{-1}.
3. Start a stopwatch in order to keep precise track of the following sampling times and start the sampling campaign. As a general suggestion, samples for the determination of nitrite and nitrate concentration must be collected every 5 min during the first 30-45 min of execution of the test, every 10 or 15 for a further 30-45 min, and later on every 15 or 30 min until the end of the test.
4. Conclude the test after 3-4 h when a slow (endogenous-like) rate is observed.

Data analysis
The typical output of this test is outlined in Figure 2.4.7. On the y-axis, the oxidized nitrogen equivalent $S_{NO3,Eq}$ during the course of the test is given. Various denitrification rates can be observed. The highest rate ($r_{NOx\_N2,SB}$) is observed initially, i.e. when both S_B and XC_B are available to the heterotrophic microorganisms (interval Δt_1 in the graph). Once the S_B organic fraction is fully utilized, denitrification proceeds only on the soluble organics that are made available through the hydrolysis of XC_B, therefore on the so-called slowly biodegradable organic material (interval Δt_2 in the graph). The denitrification rate ($r_{NOx\_N2,XCB}$) is therefore limited by the hydrolysis rate of XC_B. When hydrolysable organics are fully consumed, denitrification can continue on the endogenous carbon at a slower rate ($r_{NOx\_N2,endo}$) as long as nitrate and nitrite are still available.

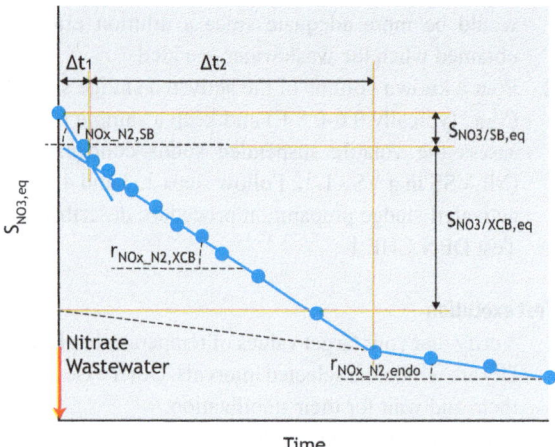

Figure 2.4.7 Typical profiles obtained in a DEN.CHE.2 test. Nitrate concentrations during the execution of the test are displayed (●). Relevant rates of interest are shown (denitrification rate on the rapidly biodegradable $r_{NOx\_N2,SB}$, and slowly biodegradable $r_{NOx\_N2,XCB}$ fractions and endogenous denitrification rate $r_{NOx\_N2,endo}$). Arrows indicate the nitrate and wastewater additions. Relevant nitrate equivalent variations are also identified ($S_{NO3/SB,eq}$ on S_B and $S_{NO3/XCB,eq}$ on XC_N).

The linear regression of $S_{NO3,Eq}$ versus time data for each interval (see Section 2.4.4.2) allows each relevant denitrification rate (in mg N L^{-1} min^{-1}) to be calculated. Specific denitrification rates (in mg N g VSS^{-1} h^{-1}) can be computed as follows:

- specific denitrification rate on S_B:

$$q_{NOx\_N2,SB} = 60 \cdot (r_{NOx\_N2,SB} - r_{NOx\_N2,endo}) / X_{VSS}$$

Eq. 2.4.42

- specific denitrification rate on XC_B:

$$q_{NOx\_N2,XCB} = 60 \cdot (r_{NOx\_N2,XCB} - r_{NOx\_N2,endo}) / X_{VSS}$$

Eq. 2.4.43

- specific endogenous denitrification rate:

$$q_{NOx\_N2,endo} = 60 \cdot (r_{NOx\_N2,endo}) / X_{VSS} \quad \text{Eq. 2.4.44}$$

The amount of nitrate equivalents that are consumed on the rapidly ($S_{NO3/SB,eq}$) and the slowly ($S_{NO3/XCB,eq}$) biodegradable carbon sources are also illustrated in the graph which can be estimated as follows:

$$S_{NO3/SB,eq} = (r_{NOx\_N2,SB} - r_{NOx\_N2,XCB}) \cdot \Delta t_1 \quad \text{Eq. 2.4.45}$$

$$S_{NO3/XCB,eq} = (r_{NOx\_N2,XCB} - r_{NOx\_N2,endo}) \cdot \Delta t_2 \quad \text{Eq. 2.4.46}$$

By considering the volume of wastewater tested, the denitrification potential of the rapidly biodegradable (DP_{SB}) and slowly biodegradable (DP_{XCB}) organic compounds can be finally assessed:

$$DP_{SB} = \frac{S_{NO3/SB,eq} \cdot (V_{ML} + V_{WW})}{V_{WW}} \quad \text{Eq. 2.4.47}$$

$$DP_{XCB} = \frac{S_{NO3/XCB,eq} \cdot (V_{ML} + V_{WW})}{V_{WW}} \quad \text{Eq. 2.4.48}$$

Test DEN.MAN Denitrification manometric test: assessing the denitrification kinetic rate

Activated sludge preparation

1. Follow step 1 of the activated sludge preparation protocol described for Test DEN.CHE.1.
2. Select the appropriate amount of activated sludge to be poured into the reaction vessel (V_{ML}). For this purpose, one should consider that the extent of the overpressure caused by gas release depends on the remaining headspace volume (V_{HS}) and on the expected N_2 generation, (amount of nitrogen to be denitrified) and the headspace volume (V_{HS}). Hence, the correct selection of this ratio is crucial to avoid extreme pressures (either too high or too low). The maximum overpressure will be achieved at the end of the test, i.e. when all the nitrate has been denitrified, described as:

$$P_{max} - P_{atm} = \frac{P_{atm}}{V_{HS}} \cdot \frac{M_{N2}}{28 \cdot 1,000} \cdot 22.4 \cdot \frac{(273 + T_C)}{273}$$

Eq. 2.4.49

Where, T_C is the temperature (°C), M_{N2} is the mass of nitrogen gas generated by denitrification during the course of the test (mg N), which depends on the nitrate concentration ($S_{NO3\_N2,Ax}$, in mg N L^{-1}) and the activated sludge volume, as follows:

$$M_{N2} = S_{NO3\_N2,Ax} \cdot V_{ML} \quad \text{Eq. 2.4.50}$$

By substituting this equation into the previous one and rearranging, the following relationship is obtained:

$$\frac{V_{ML}}{V_{HS}} = \frac{P_{max} - P_{atm}}{P_{atm}} \cdot \frac{28}{22.4} \cdot \frac{273}{(273 + T_C)} \cdot \frac{1}{\frac{S_{NO3\_N2,Ax}}{1,000}}$$

Eq. 2.4.51

The optimal value for P_{max} is normally around 0.2 atm, which means that, under typical conditions (T_C

~20 °C, $S_{NO3\_N2,Ax}$ ~20 mg N L^{-1}), a V_{ML} to V_{HS} ratio of around 11 is obtained. Therefore, for a total reactor volume of 1 L, the ideal activated sludge volume to be used during the test will be 0.92 L.

3. Pour the previously computed amount of activated sludge into the reaction vessel. Insert the magnet for the mixing. Insert NaOH pellets in the headspace for CO_2 adsorption. Flush the reactor headspace with N_2 and seal the bottle gas tightly. Place the reactor into the thermostatic chamber and under gentle mixing, wait for 30 min until the temperature stabilizes.

Test execution

1. Determine the volume of nitrate solution to be added (as a pulse or spike) to achieve a nitrate concentration in the activated sludge of 20-25 mg N L^{-1}. Using a syringe, inject the nitrate stock solution through the rubber septum. Select the amount of carbon source to be dosed (also as a pulse or spike) in order to operate under non-limiting and non-inhibiting conditions (see step 4 of the execution of Test DEN.CHE). Then inject the carbon source solution and follow up the time execution with a stopwatch.
2. Start the manometric data collection. Collect data every 15-30 min or until a bending point is observed in the overpressure curve, which indicates the exhaustion of nitrate.
3. End the test. Check the final pH and take a final sample to measure the MLVSS concentration.

Data analysis

A typical output of a manometric denitrification test using a manometer with a data logger is outlined in Figure 2.4.8.

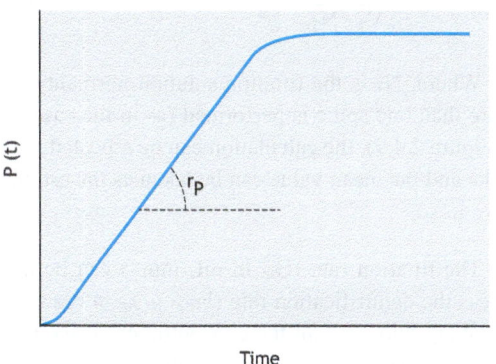

Figure 2.4.8 Typical overpressure profile obtained in a manometric denitrification test performed with a manometer equipped with a data logger. The chart displays the maximum pressure production rate r_P.

During the first 10-15 min, various phenomena may overlap caused by potential interferences such as: (*i*) intrusion into the bulk liquid of residual oxygen that may remain in the headspace, (*ii*) water vapour pressure equilibrium, (*iii*) initial N_2 accumulation in the liquid phase, and/or, (*iv*) microbiological lag phases. For this reason, the initial overpressure data should be disregarded. The following overpressure data can be used to compute the pressure production rate (r_P, atm min^{-1}) by linear regression (see Section 2.4.3.3).

From r_P, the denitrification rate (F_{NO3_N2}, mg N min^{-1}) can be assessed as follows:

$$F_{NO3\_N2} = \frac{r_p}{P_{atm}} \cdot V_{HS} \cdot \frac{28}{22.4} \cdot \frac{273}{273 + T_C} \quad \text{Eq. 2.4.52}$$

With, P_{atm} in atm and V_{HS} in mL.

Finally, the specific denitrification rate q_{NOx_N2} (mg N g VSS^{-1} h^{-1}) can be computed using the activated sludge MLVSS concentration (X_{VSS}):

$$q_{NOx\_N2} = 60 \cdot F_{NO3\_N2} / (V_{ML} \cdot X_{VSS}) \quad \text{Eq. 2.4.53}$$

Test DEN.TIT Denitrification titrimetric test: assessing the denitrification kinetic rate

Activated sludge preparation

For conventional activated sludge samples from wastewater treatment plants treating urban wastewaters, a sample with X_{VSS} of around 2-4 g VSS L^{-1} will be suitable. The sample should be collected at the outlet or end of the pre-denitrification or post-denitrification tank, depending on the target treatment section.

1. Pour a known volume of the activated sludge sample (typically 1 L) into the reaction vessel and start the temperature control systems. Select the desired pH and temperature set point values.
2. Start the automatic titration system and leave the activated sludge under these conditions for a pre-incubation period (ideally 1 h). The pre-incubation phase will encourage the consumption of any residual nitrate or nitrite remaining from the plant.

Test execution

1. Start the data logging.
2. Add the nitrate stock solution (after adjusting its temperature to the target temperature of the test) to achieve a non-limiting nitrate concentration in the activated sludge (between 10 and 20 mg N L^{-1} are typically adequate values) and the S_B solution at a

concentration that is neither limiting nor inhibiting. Note that the amount of nitrate added ($M_{NOx,ini}$ in mg N) has to be known. To assess the appropriate amount of S_B, the stoichiometric relationship between the amount of S_B and nitrate consumed can be used, which can be calculated as follows:

$$Y_{NO3\_SB,Ax} = \frac{2.86}{1 - Y_{OHO,Ax}} \quad (g\ COD\ g\ N^{-1}) \quad \text{Eq. 2.4.38}$$

Note that the same type of test can be performed using nitrite instead of nitrate. In this case, the following relationship should be applied:

$$Y_{NO2\_SB,Ax} = \frac{1.71}{1 - Y_{OHO,Ax}} \quad (g\ COD\ g\ N^{-1}) \quad \text{Eq. 2.4.54}$$

The S_B addition can therefore be calibrated in order to guarantee that the S_B to nitrate (or nitrite) ratio is at least 3-4 times the stoichiometric value reported above. Note that Y_{HD} depends on the carbon source. Nevertheless, a value of 0.5 can be typically used for this rough estimation. This means that the S_B addition should be calibrated in order to ensure a S_B concentration in the activated sludge of 350 mg COD L^{-1} when using nitrate and 200 mg COD L^{-1} when using nitrite. Upon these additions, denitrifying bacteria will become active and their activity will usually tend to increase the pH. The automatic titration system will react to decrease the pH to maintain the pH set point through acid addition.

3. Record the volume of titration solution added over time (V_{tit} versus time). Check that the pH value remains within the interval pH set point ± 0.02. DO concentration should be below the detection limit. Continue the test until a clear bending point is observed in the cumulated titration solution plot. This bending point indicates that the nitrate is fully depleted and consequently denitrification has stopped. The collected data should be sufficient to reliably estimate the titration rate (Q_{tit}) from a linear regression of V_{tit} versus time with a satisfactory coefficient of determination (normally $R^2 > 0.98$).
4. Step 3 can be repeated by adding another dose of nitrate. The addition of S_B is no longer needed since a sufficient residual concentration is still present in the activated sludge to support a second denitrification phase.
5. End the test according to the instructions reported in Step 4 of the test operation procedure described for Test NIT.TIT.1.

Data analysis

A typical trend in the cumulated volume of titration solution added during the test is depicted in Figure 2.4.9.

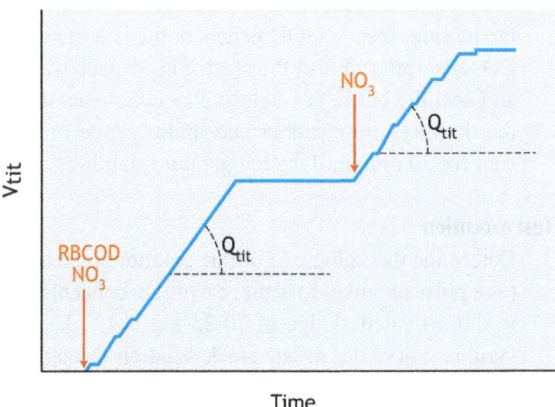

Figure 2.4.9 Example of a pH-static titration curve during a denitrification test to assess the maximum denitrification rate. Arrows indicate the addition of nitrate and S_B solutions. The relevant titration rate and is also shown in the graph.

From the data (V_{tit} versus time data), the titration rates (Q_{tit}) can be computed based on the slope of the titration curve (see Section 2.4.3).

The $Y_{NO3\_H+,Ax}$, in g N mol Protons^{-1}, can be assessed by considering the volume of titration solution added until the plateau of the titration curve is observed (V_T in Figure 2.4.9) and the mass of nitrate (or nitrite) added ($M_{NOx,ini}$):

$$Y_{NO3\_H+,Ax} = \frac{M_{NOx,ini}}{V_T \cdot N_T} \quad \text{Eq. 2.4.55}$$

Where, N_T is the titration solution normality. When more than one spike is performed (as in the case shown in Figure 2.4.9), the calculation can be repeated per each spike and the mean value can be taken as the estimate of $Y_{NO3\_H+,Ax}$.

The titration rate (Q_{tit} in mL min^{-1}) can be used to assess the denitrification rate ($F_{NO3\_H+,Ax}$ in mg N min^{-1}) by taking into account the concentration of the titration solution (N_T) and the $Y_{NO3\_H+,Ax}$ value:

$$F_{NO3\_H+,Ax} = Q_{tit} \cdot N_T \cdot Y_{NO3\_H+,Ax} \quad \text{Eq. 2.4.56}$$

Finally, the maximum specific denitrification rate of the sludge (q_{NOx_N2}, in mg N g VSS^{-1} h^{-1}), is computed by taking into account the MLVSS concentration of the sludge sample (in g VSS L^{-1}) and the suspension volume assessed at the end of the test (in L):

$$q_{NOx\_N2} = 60 \cdot F_{NO3\_H+,Ax} / (V_{ML} \cdot X_{VSS}) \qquad \text{Eq. 2.4.57}$$

2.4.8 Anammox batch activity tests: preparation

As described in Section 2.4.1, the anammox process consumes ammonium and nitrite and converts them mainly into nitrogen gas as well as a minor fraction of nitrate. When the anammox process takes place in a batch reactor then, variations in the ammonium, nitrite and nitrate concentrations are expected and can be monitored to follow the time evolution of the process. Furthermore, when a gas-tight reaction vessel is used to perform the batch test, the release of dinitrogen gas will cause a pressure increase, which can also be monitored over time to measure the reaction kinetics. Therefore, two alternatives are available to track the evolution of the anammox process:
- Chemical tracking, by assessing the ammonium/nitrite/nitrate concentration evolution in time.
- Manometric tracking by assessing the overpressure caused by dinitrogen release in a gas-tight reactor.

Each one of these alternatives is discussed hereafter in the coming two tests. The first test (AMX.CHE) refers to a batch test performed by chemical tracking to assess the maximum activity of an anammox culture fed with synthetic autotrophic medium. In this test, the stoichiometric coefficients NO$_2$/NH$_4$ and NO$_3$/NH$_4$ ratios will be also assessed. In the second test (AMX.MAN), anammox biomass is suspended in a real wastewater and a manometric tracking procedure is applied to evaluate its treatability via the anammox process. In this example, the inhibition potential of a wastewater is evaluated by comparing the maximum rate obtained in the presence of wastewater with the maximum rate observed in the presence of a synthetic medium.

2.4.8.1 Apparatus

Each tracking methodology (chemical or manometric) has special apparatus requirements.

When chemical tracking is applied, refer to the following list of equipment:

1. An (airtight) batch reactor equipped with a mixing system and adequate sampling ports (as described in Section 2.4.3.1).
2. A nitrogen gas supply (recommended).
3. A calibrated pH electrode (if not included/incorporated in the batch reactor setup).
4. A 2-way pH controller via HCl and NaOH addition (alternatively a one-way control - generally for HCl addition - or manual pH control can be applied through the manual addition of HCl and NaOH).
5. A thermometer (recommended working temperature range of 0 to 40 °C).
6. A temperature control system (if not included in the batch reactor setup).
7. A DO meter with electrode (if not included/incorporated in the batch reactor setup) to verify anoxic conditions.
8. A stopwatch.

When the manometric tests are performed, refer to the following list of equipment:
1. The equipment given for the manometric system, described in Section 2.4.3.3.
2. A nitrogen gas supply (recommended).

2.4.8.2 Materials

For a complete list of required materials refer to Section 2.2.3.2 (with the exception of points 7 and 8).

2.4.8.3 Working solutions

- **Real wastewater**

For batch tests that require the use of a real wastewater, follow the instructions given in Section 2.2.3.3.

- **Synthetic medium**

If tests can be or are desired to be performed with synthetic wastewater, the synthetic influent media could contain a mixture of ammonium, nitrite and bicarbonate plus necessary (macro- and micro-) nutrients. Generally, they can be mixed all together in the same media or prepared separately if they need to be added in different phases or times. The usual compositions and concentrations are (based on van de Graaf *et al.*, 1996): 1 g L^{-1} (NH$_4$)$_2$SO$_4$ (7.6 mM, 106 mg N L^{-1}), 0.25 g L^{-1} NaNO$_2$ (3.6 mM, 51 mg N L^{-1}), 0.6 g L^{-1} NaNO$_3$ (7.1 mM, 99 mg N L^{-1}), 1 g L^{-1} NaHCO$_3$ (11.9 mM), 0.025 g L^{-1} KH$_2$PO$_4$ (0.18 mM), 0.1 g L^{-1} MgSO$_4$·7H$_2$O (0.41 mM), 0.15 g L^{-1} CaCl$_2$·2H$_2$O (1.02 mM) and 1.25 mL L^{-1} trace elements solutions A and B (see below for trace elements A and B preparation). Similar nutrient solutions can be used as long as they contain all the previously reported required nutrients. The trace element solutions should contain all the required micro-nutrients (iron, zinc, cobalt, manganese, copper, molybdate, nickel, selenium and boron) to ensure that cells are not limited by their absence and avoid obtaining incorrect results and in extreme cases the failure of the test. Thus, despite the fact that their concentrations may seem very low, one must make sure that all of the constituents are added to the solution in the required amounts. The following composition (amounts per litre of micro-nutrient solution) is recommended (based on van de Graaf et al., 1996): Trace elements solution A: 5 g EDTA, 9.14 g $FeSO_4 \cdot 7H_2O$, trace element solution B: 15 g EDTA, 0.43 g $ZnSO_4 \cdot 7H_2O$, 0.24 g $CoCl_2 \cdot 6H_2O$, 0.99 g $MnCl_2 \cdot 4H_2O$, 0.25 g $CuSO_4 \cdot 5H_2O$, 0.22 g $Na_2MoO_4 \cdot 2H_2O$, 0.19 g $NiCl_2 \cdot 6H_2O$, 0.21 g $NaSeO_4 \cdot 10H_2O$, and 0.014 g H_3BO_3.

- **Ammonium or nitrite solutions**

For the dosage of the anammox substrates, ammonium and nitrite stock solutions can be prepared (e.g. 1-10 g N L^{-1}) using either $(NH_4)_2SO_4$ or NH_4Cl salts for ammonium and $NaNO_2$ or KNO_2 salts for nitrite. Generally, they can be mixed together in the same stock solution or prepared separately if they need to be added in different phases or at different times. When an ammonium/nitrite solution is prepared the molar ratio between the two substrates is usually set to one in order to ensure an excess of ammonium throughout the test. Bicarbonate can also be added to the batch test to avoid inorganic carbon (IC) limitation: in order to ensure IC in excess during the test, add bicarbonate to the ammonium solution up to a molar ratio equal to 0.7 mol-IC mol-NH_4^{-1} (i.e. ten times the stoichiometric requirements of 0.066 mol-IC mol-NH_4). The latter should be considered in case the wastewater tested has limiting concentrations of inorganic carbon and/or if the oxygen is removed via sparging CO_2-free gases (e.g. N_2), which will lead to CO_2 stripping.

- **Nitrate solution**

In order to ensure the correct redox conditions, nitrate can be dosed prior to the execution of the batch test using a nitrate stock solution that can be prepared (e.g. 1-10 g N L^{-1}) using either $NaNO_3$ or KNO_3 salts. This is especially needed when consecutive batch tests need to be performed using the same anammox sludge (e.g. to evaluate the long-term effect of exposure to a particular compound) in order to avoid sulphate reduction to sulphide (H_2S), which may be toxic to anammox bacteria.

- **Washing media**

If the sludge sample must be 'washed' to remove any undesirable compounds (which may even be inhibitory or toxic), a washing solution is needed. The synthetic medium described above can be used as a washing medium by simply removing the nitrite salts from the recipe. The washing process can be repeated twice or three times. Thereafter, the following preparation steps can be performed. In special cases, e.g. when sludge from a full-scale plant is used, the plant effluent may be used for washing purposes (assuming that its composition allows this).

- **Acid and base solutions**

These should be respectively 100-250 mL of 0.2 M HCl and 100-250 mL 0.2 M NaOH solutions for automatic or manual pH control, and 10-50 mL of 1 M HCl and 10-50 mL 1 M NaOH solutions for initial pH adjustment if the desired operational pH is very different from the pKa value of the buffering agent.

Finally, the required working and stock solutions to carry out the determination of the analytical parameters of interest must also be prepared in accordance with Standard Methods (APHA et al., 2012) and the corresponding protocols.

It is recommended to take samples of the media and activated sludge prior to the execution of the experiment to confirm/check the initial (desired) concentrations of parameter(s) of interest (e.g. ammonium, nitrite, nitrate, X_{TSS} and X_{VSS}).

2.4.8.4 Material preparation

For general instructions on how to organize the material preparation, refer to Section 2.2.3.4 and to Table 2.2.2. If required, test-specific requirements will be listed within each test protocol.

2.4.9 Anammox batch activity tests: execution

Test AMX.CHE Anammox chemical test: assessing the maximum anammox kinetic rate and the stoichiometric coefficient NO_2/NH_4 and NO_3/NH_4 ratios

Activated sludge preparation

For a suitable sampling point for the activated sludge, refer to Section 2.4.3.4.
1. For conventional activated sludge samples from municipal wastewater treatment plants, a sample

containing a X_{VSS} of around 2-10 g L^{-1} will be suitable. The X_{VSS} can be adjusted as suggested in Section 2.4.3.5. For anammox granular sludge, instead of using the activated sludge, granular biomass can be easily separated from the supernatant by settling (in a cylinder or in an Imhoff cone) and re-suspended in a washing solution or in the effluent of the plant. Preliminary tests can be conducted to evaluate the density of the settled granular sludge (g VSS L^{-1}) to pour a defined amount of anammox granular sludge (g VSS) into the reaction vessel.

2. Pour a defined volume of the activated sludge sample (typically 1 to 3 L) or a known amount of anammox granular sludge (re-suspended in the medium to be tested) into the reaction vessel and start mixing. Also, start the temperature and pH control systems to maintain both parameters around the desired set point values.

3. Sparge N_2 (or N_2/CO_2 gas mixture, see Section 2.4.3.5) into the headspace for approximately 10 min to ensure an oxygen-free environment. Install a gas outlet to limit the overpressure. Gas sparging can continue until the end of the test. If this is not feasible, then an airtight reactor with a device able to prevent oxygen intrusion can be used (e.g. using a unidirectional check valve or a water lock) (see Section 2.2.2.1 for further details).

4. Dose nitrate up to a final concentration in the activated sludge of 50-100 mg N L^{-1} in order to ensure an adequate redox potential and avoid sulphate reduction.

5. Wait for approximately 30 min to ensure stable initial conditions. This pre-incubation phase will normally allow the removal of any residual nitrite present.

Test execution

1. Verify that the temperature, pH and DO target values are within the desired intervals. Otherwise adjust them and wait until they stabilize.

2. Add the ammonium/nitrite stock solution (previously adjusted to the target temperature) to achieve a neither limiting nor inhibiting nitrite concentration in the activated sludge: 50 to 75 mg N L^{-1} is usually adequate for anammox biomass cultivated under non-strict nitrite limiting conditions (see Section 2.4.3.6 for further explanations). If any residual nitrite is present in the activated sludge sample, the nitrite addition should be reduced accordingly. The ammonium starting concentration is less critical considering the concentration ranges typical of anammox systems due to the lower inhibiting effect on anammox bacteria: 50 to 200 mg N L^{-1} is usually considered as adequate.

3. Start the stopwatch just after the addition of the ammonium/nitrite to track the sampling times.

4. Collect the activated sludge samples at regular time intervals. For instance, collect the samples every 20 to 30 min throughout the duration of the test (which usually lasts around 3 to 4 h).

5. Conclude the test when the nitrite is fully depleted. Use nitrite strip tests to quickly estimate the nitrite concentrations.

6. Take a sample to determine the final MLVSS concentration.

Note that the nitrite concentrations are unstable, and therefore they need to be quickly determined after sampling.

Data analysis

A typical output of this test is shown in Figure 2.4.10.

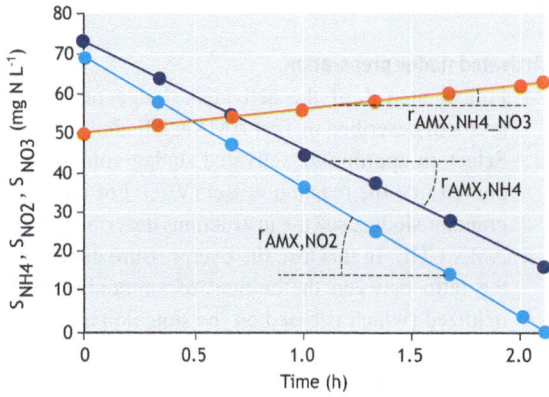

Figure 2.4.10 Typical N profiles in a Test AMX.CHE: ammonium (●), nitrite (●) and nitrate (●) concentrations are displayed on the y-axis. The relevant rates of interest are also indicated (ammonium removal rate $r_{AMX,NH4}$, nitrite removal rate $r_{AMX,NO2}$, and nitrate production rate $r_{AMX,NH4\_NO3}$).

The linear regression over these data (see Section 2.4.5) can be used to determine the ammonium ($r_{AMX,NH4}$) and nitrite removal rates ($r_{AMX,NO2}$) as well as the nitrate production ($r_{AMX,NH4\_NO3}$) rate (expressed as mg N L^{-1} h^{-1}). The anammox rate is usually expressed as the dinitrogen gas produced, which is equivalent to the nitrogen removed from the wastewater. The maximum specific anammox rate ($q_{AMX,N2}$ as mg N_2-N g VSS^{-1} h^{-1}) can therefore be computed as follows:

$$q_{AMX,N2} = \frac{r_{AMX,NH4} + r_{AMX,NO2} - r_{AMX,NH4\_NO3}}{X_{VSS}} \quad \text{Eq. 2.4.58}$$

In the previous expression, when anammox granular sludge is used to perform the test, then the biomass concentration at the denominator (the term X_{VSS}) will correspond to the g VSS added to the reactor vessel during the test preparation divided by the mixed liquor volume.

The stoichiometric coefficient ratios of interest, such as $Y_{NH4\_NO2,AMX}$ and $Y_{NH4\_NO3,AMX}$, can be easily calculated from the relative removal/production rates using the following expressions:

$$Y_{NH4\_NO2,AMX} = \frac{r_{AMX,NO2}}{r_{AMX,NH4}} \quad \text{Eq. 2.4.59}$$

$$Y_{NH4\_NO3,AMX} = \frac{r_{AMX,NH4\_NO3}}{r_{AMX,NH4}} \quad \text{Eq. 2.4.60}$$

Test AMX.MAN Anammox manometric test: assessing the maximum anammox kinetics

Activated sludge preparation
1. Follow step 1 of the activated sludge preparation protocol described in Test AMX.CHE above.
2. Select an appropriate activated sludge volume to be poured into the reaction vessel (V_{ML}). For anammox granular sludge, see the instructions described in Test AMX.CHE. In this test, the overpressure depends on the ratio between the amount of ammonium to be oxidized (which is based on the stoichiometry of the reaction; 1 mol of NH_4 consumption will lead to 1 mol of N_2 production) (see Section 2.4.1) and the headspace volume (V_{HS} in L). Hence, the correct selection of this ratio is crucial to avoid the generation of an extreme overpressure (either too low or too high). The maximum overpressure will be achieved at the end of the test, i.e. when nitrite (usually the limiting substrate) has been fully converted. Considering a $Y_{NH4\_NO2,AMX}$ stoichiometric ratio of 1.32 mol-NO_2 mol-NH_4^{-1}, the maximum overpressure (P_{max}) can be estimated as follows:

$$P_{max} - P_{atm} = \frac{P_{atm}}{V_{HS}} \cdot \frac{M_{NO2\_N2}}{14} \cdot 22.4 \cdot \frac{(273 + T_C)}{273}$$
Eq. 2.4.61

Where, M_{NO2_N2} (in mg N) is the mass of nitrite that is converted during the course of the test, dependent on the initial nitrite concentration ($S_{NO2,ini}$, in mg N L^{-1}) and activated sludge volume (V_{ML} in L):

$$M_{NO2\_N2} = S_{NO2,ini} \cdot V_{ML} \quad \text{Eq. 2.4.62}$$

By substituting this equation into the previous one and rearranging, the following relationship is obtained:

$$\frac{V_{ML}}{V_{HS}} = \frac{P_{max} - P_{atm}}{P_{atm}} \cdot \frac{14 \cdot 1.32}{22.4} \cdot \frac{273}{(273 + T_C)} \cdot \frac{1,000}{S_{NO2,ini}}$$
Eq. 2.4.63

The optimal P_{max} value usually lies around 0.2 atm, which means that, under typical test conditions ($T_C \sim 30$ °C, $S_{NO2,ini} \sim 50$ mg N L^{-1}), a V_{ML} to V_{HS} ratio of around 3 is obtained. Therefore, for a total reactor volume of 1 L ($V_{TOT} = V_{ML} + V_{HS}$), an adequate activated sludge volume will be 0.75 L.

3. Pour the previously estimated activated sludge volume into the reaction vessel. Place the reaction vessel on the shaker (see Section 2.4.3.1 for mixing requirements). Flush the reactor headspace with N_2 (or a N_2/CO_2 gas mixture, see Section 2.4.3.3) and seal the gas bottle tightly. Place the reactor into the thermostatic chamber, apply a gentle mixing and wait for 30 min until the temperature stabilizes.

Test execution
1. Select the volume of nitrite solution to be dosed at the beginning of the batch test to achieve a neither limiting nor inhibiting nitrite concentration in the activated sludge: typically, 50 to 75 mg N L^{-1} are adequate concentrations for anammox biomass cultivated under non-strict nitrite limiting conditions (see Section 2.4.3.4 for further details). Usually, the initial ammonium concentration is similar to the nitrite concentration. Higher initial ammonium concentrations can be used, as long as they remain below the reported inhibitory level for anammox bacteria (< 1 g N L^{-1}). If there is residual nitrite present in the activated sludge sample, the nitrite addition should be reduced accordingly. Using a syringe, inject the ammonium/nitrite stock solution through the rubber septum.
2. Start the manometric data collection. Collect the data every 30 to 60 min until there is a point where the overpressure curve bends, which will correspond to the exhaustion of nitrite.
3. End of the test. Check the final pH and take a final sample to determine the MLVSS concentration.

Data analysis

Using the ideal gas law, the pressure data can be converted into the gas moles (N_2 in this case) emitted to the headspace:

$$n(t) = \frac{P(t) \cdot V_{HS}}{R \cdot T_K} \quad \text{Eq. 2.4.64}$$

Where, $n(t)$ is the number of N_2 moles present in the headspace volume (V_{HS}) at time t, $P(t)$ is the pressure in the headspace at time t, R is the ideal gas constant and T_K the temperature of execution expressed in Kelvin.

Figure 2.4.11 shows a typical profile of a manometric anammox test executed with a manometer equipped with a data logger after the conversion of the recorded pressure data into moles of N_2 gas produced. Disregard the data collected in the first 10-15 min of the test since various phenomena may overlap affecting the headspace overpressure, such as: (*i*) residual oxygen intrusion from the headspace, (*ii*) water-vapour pressure equilibrium; (*iii*) initial N_2 accumulation in the liquid phase, and (*iv*) microbiological lag phases.

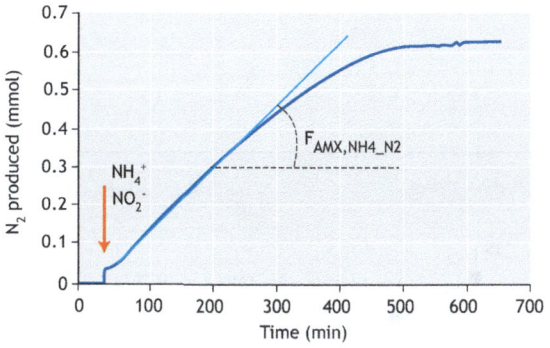

Figure 2.4.11 Typical N_2 gas profiles obtained in a Test AMX.MAN performed with a manometer equipped with a data logger. The maximum N_2 production rate $F_{AMX,NH4\_N2}$ is depicted in the figure.

The cumulative N_2 production curve can be used to compute the N_2 production rate ($F_{AMX,NH4\_N2}$, N_2-mol min^{-1}) using linear regression (see Section 2.4.5). Moreover, data can be further expressed in mg N_2-N min^{-1} using the equivalent molecular weight of dinitrogen gas (28 g N N_2-mol^{-1}). The specific anammox rate $q_{AMX,N2}$ (mg N g VSS^{-1} h^{-1}) can be computed using the MLVSS concentration:

$$q_{AMX,N2} = 60 \cdot F_{AMX,NH4\_N2} / (V_{ML} \cdot X_{VSS}) \quad \text{Eq. 2.4.65}$$

2.4.10 Examples

2.4.10.1 Nitrification batch activity test

Description

To illustrate the execution of an aerobic batch activity test for nitrification, data from a test performed at 20°C with a full-scale activated sludge sample is presented in this section. The Test NIT.CHE.1 was carried out to determine the maximum specific biomass ammonium oxidation rate. The batch activity test was performed using a 2.5 L reactor. All the equipment, apparatus and materials were prepared as described in Section 2.4.4.1. The pH and DO sensors were calibrated less than 24 h before the test execution. The batch test lasted 3 h. Including the time needed for the test preparation and the cleaning phase afterwards, the operations lasted approximately 5 h. Prior to the batch test, 2.0 L of fresh activated sludge collected at the end of the aerobic phase of a full-scale plant (according to Section 2.4.3.4) was transferred to the reactor and acclimatized for 1 h at 20 °C under slow mixing (100 rpm) following the recommendations described in Section 2.4.3.5. The pH and DO control were started up at pH and DO set points of 7.5 and 6 mg O_2 L^{-1}, respectively. Thereafter, 15 min before the start of the test, mixing was increased to 200 rpm and samples were collected for the determination of the parameters of interest (e.g. ammonium, nitrite, nitrate and MLVSS concentrations). Before the addition of the synthetic medium, its temperature was adjusted in a water bath to the target temperature of the batch test (20 °C). The first sample was collected 5 minutes before the addition of the synthetic medium to determine the initial conditions. The test started with the addition of the synthetic media (minute zero): 50 mL containing 1 g NH_4-N L^{-1} as well as other macro- and micro-nutrients as described in Section 2.4.3.6. Samples were collected every 30 min over a period of 3 h. The duration of the batch test was chosen in order to allow a complete oxidization of the ammonium present. Immediately after collection, all the samples were prepared and preserved as described in Section 2.2.3.4. All the collected samples were analysed as described in Section 2.4.3.7. Table 2.4.5 shows the experimental implementation plan and the results of the execution of the test.

Data analysis

Following on from the results from the experiment shown in Table 2.4.5, Figure 2.4.12 shows the results from the test and also an estimation of the maximum volumetric kinetic rates (by applying linear regression).

Table 2.4.5 Results from the nitrification batch activity test.

Nitrification batch activity test — Code: NIT.CHE.1

Date:	Monday 05.10.2015 10:00 h
Description:	Nitrification test at 20 °C with real activated sludge
Test No.:	1
Duration	3 h (180 min)
Substrate:	Synthetic: ammonium and nitrite (1,000 mg L^{-1}) + minerals
Sampling point:	Middle mixed liquor height in the SBR
Samples No.:	NIT.CHE 1-8
Total sample volume:	80 mL (5 mL normal sample, 20 mL for MLVSS)
Reactor volume:	2.5 L

Experimental procedure in short: — Time (h:min)
1. Confirm availability of sampling material and required equipment. — 08:00
2. Confirm calibration and functionality of systems, meters and sensors. — 08:10
3. Transfer 2.0 L of sludge to batch reactor. — 08:30
4. Start aerobic conditions with gentle mixing and air sparging at T and pH set points. — 08:40
5. 15 min before starting, take sample for initial conditions (sample NIT.CHE 1). — 09:40
6. —
7. Start batch test: add 0.05 L of synthetic influent. — 09:55
8. Take first sample for the determination of the initial conditions (minute zero) — 10:00
9. Minute 30, continue sampling program according to schedule — 10:30
10. Minute 180, stop aeration and mixing. — 13:00
11. Organize the samples and clean the system. — 13:10
12. Verify that all systems are swtiched off. — 13:20

Sampling schedule

Time (min)	-15	-5	0	30	60	90	120	150	180
Time (h)	-0.25	-0.08	0.00	0.50	1.00	1.50	2.00	2.50	3.00
Sample No.	1		2	3	4	5	6	7	8
Parameter	AEROBIC PHASE								
NH$_4$-N (mg N L^{-1})	6.8		31	27.3	20.6	17.2	12.4	4.9	0.4
NO$_2$-N (mg N L^{-1})	0[1]		0	0.4	0.9	0.6	0.3	0.4	0.1
NO$_3$-N (mg N L^{-1})	0.1		0.1	5.1	9.8	14.2	19.4	24	29.6
MLSS and MLVSS (mg L^{-1})	See table								See table

[1] Average value of the concentration present in the synthetic substrate and in the liquid phase of the sludge sample prior the start of the test

MLSS & MLVSS measurements

Sampling point	Cup No.	W1	W2	W3	W2-W1	W2-W3	MLSS	MLVSS	Ratio
Start test[2]	1	0.09630	0.16530	0.10210	0.06900	0.06320	3,450	3,160	0.92
	2	0.09580	0.16380	0.10190	0.06800	0.06190	3,400	3,095	0.91
	3	0.09640	0.16440	0.10230	0.06800	0.06210	3,400	3,105	0.91
						Average	3,417	3,120	0.91
End test	4	0.09540	0.16490	0.10200	0.06950	0.06290	3,475	3,145	0.91
	5	0.09610	0.16410	0.10180	0.06800	0.06230	3,400	3,115	0.92
	6	0.09570	0.16400	0.10220	0.06830	0.06180	3,415	3,090	0.90
						Average	3,430	3,117	0.91

[2] Concentrations corrected considering the dilution due to the synthetic medium addition.

Biomass composition

Sampling point	Start test	End test
MLSS (mg L^{-1})	3,417	3,430
MLVSS (mg L^{-1})	3,120	3,117
Ratio	0.91	0.91
Ash (mg L^{-1})	297	313

The maximum volumetric rates for ammonium removal and nitrate production are 10.3 and 9.7 mg N L^{-1} h^{-1}, respectively. Taking into account the average MLVSS concentration of 3.1 g L^{-1} between the sludge samples collected at the beginning and end of the test, a biomass-specific ammonium oxidation rate of 3.3 mg N g VSS^{-1} h^{-1} can be estimated, which corresponds to about 80 mg N g VSS^{-1} d^{-1}. It is important to note that the sum of soluble inorganic nitrogen compounds at the beginning and end of the batch test are comparable (31.1 vs 30.1 mg N L^{-1}), indicating that nitrification was the dominant process during the batch test.

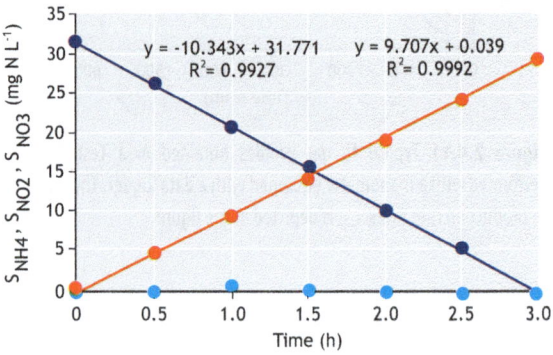

Figure 2.4.12 Graphic representation of the ammonium (●), nitrite (●) and nitrate concentrations (●) obtained in the example of the batch activity nitrification test (Test NIT.CHE.1). The test was performed with a full-scale activated sludge at 20 °C and pH 7.5 using a synthetic medium as influent.

The two trend lines show the ammonium conversion and nitrate production rates for further estimation of the corresponding maximum specific biomass kinetic rates. A decrease in the sum of ammonium, nitrite and nitrate concentrations would suggest the simultaneous occurrence of nitrogen removal processes such as denitrification or anammox, which can take place in the presence of anoxic conditions. The comparable ammonium oxidation and nitrate production rates indicate that, in the tested sludge, the activities of AOO and NOO are well-balanced and allowed to achieve full nitrification. These observations were also supported by the absence of nitrite. The measured specific ammonium oxidation rate is comparable to the kinetic rates reported in Table 2.4.1, indicating a high enrichment of nitrifying bacteria in the activated sludge sample tested.

2.4.10.2 Denitrification batch activity test

Description

In this section, a test to assess the maximum denitrification rate and the anoxic growth yield in the presence of acetate as the carbon source is described.

An airtight reactor equipped with a mechanical mixing system and automatic pH and temperature control systems was used. N_2 gas bubbling was also provided.

The day before testing, the pH and DO probes were calibrated and all the required materials were prepared as suggested in Section 2.2.3.2 (with the exception of points 7 and 8). A 10 g N L^{-1} nitrate solution was prepared with $NaNO_3$ and an acetate stock solution using sodium acetate at a concentration of 10 g COD L^{-1} (taking into account that 1 g CH_3COONa corresponds to 0.78 g COD). 1 M HCL and 1 M NaOH solutions were prepared for automatic pH correction. Moreover, 2 L of sludge were used. A preliminary working plan was defined as displayed in Table 2.4.6.

Table 2.4.6 Results from the denitrification batch activity test.

Denitrification batch activity test — Code: DEN.CHE.1

Date:	Wednesday 02.09.2015 9:00 h	
Description:	Tests at 20 °C, pH 7, artificial substrate & enriched lab culture	
Test No.:	1	
Duration	3.5 h (210 min)	
Substrate:	Synthetic: Acetate (200 mgCOD/L) + nitrate (20 mgN/L)	
Sampling point:	Middle mixed liquor height in the SBR	
Samples No.:	DEN.CHE.1 (1-18)	
Total sample volume:	240 mL (10 mL for MLVSS, and 10 mL each sample)	
Reactor volume:	2.5 L	

Experimental procedure in short:

Step	Time (h:min)
1. Day before instrumentation check, probes calibration preparation of the working plan, stock solutions, containers for sample collections, and all other materials.	
2. Sludge sampled from SBR and transferred to reaction vessel.	09:00
3. Activation of N_2 sparging.	09:05
4. Nitrate addition, fist sample taken.	09:10
5. Other 5 samples taken at 20 min intervals.	
6. Acetate addition and sampling.	10:30
7. Other 13 samples taken at 5/10 min intervals.	
8. Test stopped.	11:35
9. Sample for MLVSS measurement taken, total volume assessment.	11:40
10. Verify that all samples are correctly stored and switch off system.	11:50

Sampling schedule

Time (min)	0	20	40	60	80	85	90	95	100	105	110	120
Time (h)	0.00	0.33	0.67	1.00	1.33	1.42	1.50	1.58	1.67	1.75	1.83	2.00
Sample No.	1	2	3	4	5	6	7	8	9	10	11	12
Parameter						ANOXIC PHASE						
NO_3-N (mg N L^{-1})	23.0	23.2	22.0	21.8	20.2	18.2	15.8	14.0	12.3	11.2	10.2	7.3
NO_2-N (mg N L^{-1})	0.1	0.0	0.2	0.0	0.0	1.0	2.0	2.5	2.8	3.0	3.0	2.7
$COD_{soluble}$ (mg COD L^{-1})					418.0	407.3	397.3	385.8	378.0	366.5	353.7	336.5

Sampling schedule (continued)

Time (min)	130	140	150	160	180	185
Time (hrs)	2.17	2.33	2.50	2.67	3.00	3.08
Sample No.	13	14	15	16	17	18
Parameter			ANOXIC PHASE			
NO_3-N (mg N L^{-1})	4.9	1.8	0.3	0.0	0.0	
NO_2-N (mg N L^{-1})	2.0	2.1	1.5	1.0	0.0	
$COD_{soluble}$ (mg COD L^{-1})	315.0	303.6	285.7	275.0	271.4	
MLSS (g L^{-1})						2.51

Total biomass (gVSS)	4.58
$r_{D,end}$ (mg N L^{-1} min^{-1})	0.035
$r_{D,exog}$ (mg N L^{-1} min^{-1})	0.278
r_{COD} (mg COD L^{-1} min^{-1})	2.050
Anoxic growth yield	0.660
Denitrification rate (mg N g $MLVSS^{-1}$ h^{-1})	6.600

The initial nitrate concentration was set to 20 mg N-NO_3 L^{-1} (corresponding to the addition of 4 mL of nitrate stock solution). The initial COD concentration was set to 200 mg L^{-1} (an addition of 40 mL of acetate stock solution). According to the instructions reported in Section 2.4.7, up to 18 samples were planned to be collected. All corresponding materials and consumables (e.g. plastic cups) were prepared and labelled to avoid identification errors during the sample collection. On the day of the test, the activated sludge was sampled from a bench scale SBR at the end of its denitrification phase. Two litres of activated sludge were transferred to the reaction vessel and the temperature set point was set at 20 °C, similar to the operating temperature of the bench-scale SBR. Simultaneously, the pH-control system was started up with a set point value of 7.4. The initial pH was 7.7. The N_2 gas sparging system was turned on to remove any residual DO and ensure anoxic conditions after the nitrate addition. After 5 min, the S_{O2} concentration in the bulk liquid was below the probe detection limit and the N_2 sparging was turned off. The predefined volume of the nitrate stock solution was spiked through the sampling port. After approximately 1 min, the stopwatch was turned on and the first sample was taken (minute zero). Four more samples were taken at 20 min intervals to assess the endogenous denitrification rate. Thereafter, the COD stock solution was added and, 30 seconds afterwards, a sample was taken. Later on, 6 samples were taken at 5 min intervals and afterwards samples were taken every 10 min. The sampling campaign continued while the nitrate and nitrite profiles were continuously followed using NO_3 and NO_2 detection paper strips until the nitrate and nitrite were fully consumed. These samples were used to assess the nitrate, nitrite, and soluble COD concentration. The test finished after 180 min. One final sample was taken to assess the MLVSS concentration. Eventually, all the systems were stopped and the reactor was opened in order to measure the final volume of activated sludge. All the samples were collected and preserved as described in Section 2.2.3.4. All the collected samples were analysed as described in Section 2.4.3.7.

Data analysis

Relevant implementation data and results of the analytical determinations are reported in Table 2.4.6. In Figure 2.4.13, the COD and $S_{NO3,Eq}$ (computed as $S_{NO3,Eq}$ = S_{NO3} + 0.6 · S_{NO2}) trends are plotted.

As shown in Figure 2.4.13, the endogenous and exogenous denitrification rates were estimated by linear regression ($r_{NOx\_N2,exo}$ and $r_{NOx\_N2,endo}$ in mg N L^{-1} min^{-1}),

as well as the exogenous COD consumption rate (r_{COD}, in mg COD L^{-1} min^{-1}). Note that only data obtained in the non-limiting denitrification phase were used for the determination of the kinetic rates. For example, only those data that led to the highest linear fitting, based on the highest R^2 value, for both N-NO_{eq} and soluble COD.

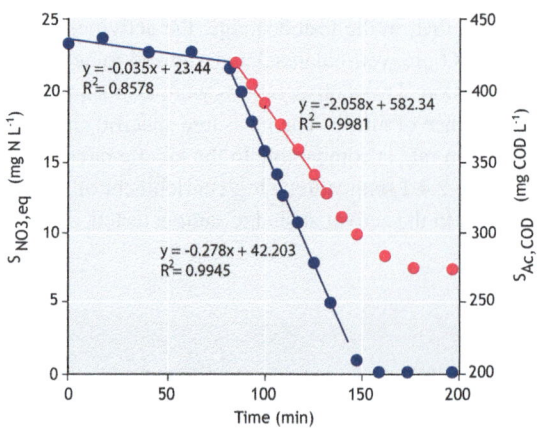

Figure 2.4.13 Graphic representation of (●) the N-NOeq and (●) soluble COD concentrations obtained in the Test DEN.CHE.1 (as presented in Table 2.4.6). The test was performed using an activated sludge sample from a SBR operated at 20 °C and pH 7.5 using acetate as the carbon source. The slopes of the trend lines were used to quantify the endogenous, exogenous denitrification and organic carbon uptake rates.

Thus, the maximum rates can be determined as:

- Maximum specific denitrification rate:

$$q_{Nox\_N2} = 60 \cdot \frac{r_{NOx\_N2,exo} - r_{NOx\_N2,endo}}{X_{VSS}}$$

$$= 60 \cdot \frac{0.28 - 0.035}{2.51} = 5.8 \text{ mg N g VSS}^{-1}\text{h}^{-1} \quad \text{Eq. 2.4.66}$$

- Anoxic growth yield on acetate:

$$Y_{OHO,Ax} = 1 - 2.86 \cdot \frac{r_{NOx\_N2,exo} - r_{NOx\_N2,endo}}{r_{COD}}$$

$$= 1 - 2.86 \cdot \frac{0.28 - 0.035}{2.05}$$

$$= 0.66 \text{ g COD}_{biomass} \text{ g COD}_{acetate}^{-1} \quad \text{Eq. 2.4.67}$$

Both obtained values are in the range of data reported in Table 2.4.2, supporting the reliability of the results.

2.4.10.3 Anammox batch activity test

Description

To illustrate the execution of an anammox batch activity test, data from a test performed at 30 °C with a full-scale anammox biomass sample is presented in this section. The Test AMX.CHE was carried out to determine the maximum specific biomass anammox rate (k_{AMX}, mg N g $MLVSS^{-1} d^{-1}$) as well as the stoichiometric parameters of interest for nitrite consumption and nitrate production with respect to ammonium consumption (NO_2/NH_4 and NO_3/NH_4 ratios described in Section 2.4.3.8). The batch activity test was performed using a 2.5 L reactor. All the equipment, apparatus and materials were prepared as described in Section 2.4.3. The pH sensor was calibrated less than 24 h before the test execution. The batch test lasted for 3 h. Including the time needed for the test preparation and the cleaning phase afterwards, the operations lasted for approximately 5 h. Prior to the batch test, 1 L of fresh anammox sludge collected at a full-scale 1-stage PN/anammox plant (according to Section 2.4.3.4) was transferred to the reactor and acclimatized for 1 h at 30 °C under slow mixing (100 rpm) following the recommendations described in Section 2.4.9. To ensure an oxygen-free environment, a N_2/CO_2 gas mixture (see Section 2.4.3.5) was sparged into the bulk liquid in the first 10 min and into the reaction headspace throughout the batch test. A gas outlet was provided to limit any overpressure and prevent oxygen back-diffusion (e.g. using a water lock). The pH control was started with a pH set point of 7.5. Thereafter, 15 min before the start of the test, mixing was increased to 200 rpm and samples were collected to characterize the anammox sludge used in the test (e.g. ammonium, nitrite, nitrate and MLVSS concentrations). The first sample was collected 5 min before the addition of a synthetic medium in order to determine the initial conditions. The test started with the addition of 70 mL of the synthetic medium containing 1 g NH_4-N L^{-1} and 1 g NO_2-N L^{-1} as well as other macro- and micro-nutrients as described in Section 2.4.9 (minute zero). In the example, nitrate was not dosed at the beginning of the test since it was already present in the anammox sludge sample. Before its addition, the temperature of the synthetic medium was adjusted in a water bath to the target temperature of the test (30 °C). Samples were collected every 30 min for a period of 3 h.

The maximum activity should be evaluated under non-limiting conditions. Considering the half-saturation constant for nitrite of 0.035 mg N L^{-1} reported in literature (in most cases the limiting substrate) (Lotti *et al.*, 2014), non-limiting nitrite concentrations are in the order of 1-2 mg N L^{-1} for flocculent or suspended anammox biomass. For biomass types characterized by higher mass transfer limitations such as biofilms attached to inert carriers and granular sludge, non-limiting conditions can be ensured at a nitrite concentration in the order of 5-10 mg N L^{-1}, depending on the density of the biofilm and on the specific anammox activity of the biofilms. When detailed information on the mass transfer characteristics of the biomass used during the batch test is not well known, one should consider for data analysis only those concentrations that can be satisfactory interpolated by linear regression ($R^2 > 0.95$). Immediately after collection, all the samples were prepared and preserved as described in Section 2.2.3.4. All the collected samples were analysed as described in Section 2.4.3.7. Table 2.4.7 shows the experimental implementation plan of the execution of the test.

Data analysis

Following the results from the experiment shown in Table 2.4.7, Figure 2.4.14 shows the results from the test displayed in the implementation plan and also an estimation of the maximum volumetric kinetic rates.

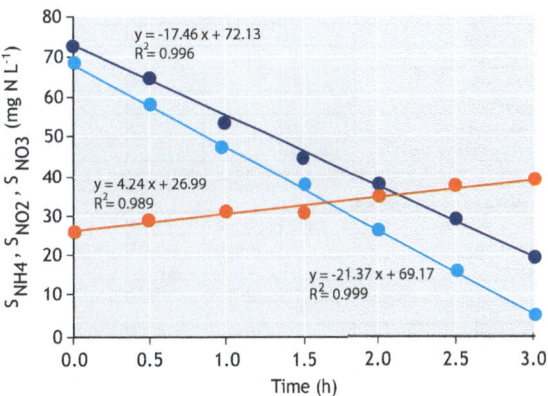

Figure 2.4.14 Typical output of an anammox batch activity test (type AMX.CHE) - ammonium (●), nitrite (●) and nitrate (●) concentrations observed during the test, resulting in depicted nitrogen conversion rates, namely: ammonium removal rate ($r_{AMX,NH4}$), nitrite removal rate ($r_{AMX,NO2}$), and nitrate production rate ($r_{AMX,NH4\_NO3}$). Note: data used in the figure were obtained from another test than presented in Table 2.4.7.

The maximum volumetric ammonium and nitrite removal, and the nitrate production rates are 17.5, 21.4 and 4.2 mg N L^{-1} h^{-1}, respectively. Taking into account the average MLVSS concentration of 2.2 g L^{-1} between the sludge samples collected at the beginning and end of the test, a specific biomass ammonium oxidation rate ($q_{AMX,NH4\_N2}$) of 7.9 mg N g VSS^{-1} h^{-1} can be estimated. This rate corresponds to 189 mg N g VSS^{-1} d^{-1}. Using the same approach, the specific biomass nitrite reduction ($q_{AMX,NO2\_N2}$) and nitrate production ($q_{AMX,NH4\_NO3}$) rates can be calculated resulting in the maximum specific rates of 9.7 and 1.9 mg N g VSS^{-1} h^{-1}, respectively, which corresponds to 232 and 46 mg N g VSS^{-1} d^{-1}, respectively. Finally, the maximum volumetric nitrogen removal rate can be calculated as the sum of the maximum volumetric ammonium and nitrite removal rate minus the maximum volumetric nitrate production, obtaining a value of 830 mg N L^{-1} d^{-1}. Similarly, the maximum specific biomass nitrogen removal rate can be computed, which results in a specific biomass kinetic rate ($q_{AMX,N2}$ or specific anammox activity - SAA) of 375 mg N g VSS^{-1} d^{-1}. Thus, the $Y_{NH4\_NO2,AMX}$ and $Y_{NH4\_NO3,AMX}$ ratios can be calculated by dividing the corresponding volumetric (or specific biomass) kinetic rates.

Table 2.4.7 Results from the anammox batch activity test.

Anammox batch activity test — Code: AMX.CHE

Date:	Tuesday 13.10.2015 10:00 h
Description:	Anammox test at 30 °C with real activated sludge
Test No.:	1
Duration:	3 h (180 min)
Substrate:	Synthetic: ammonium and nitrite (1,000 mg L^{-1} each) + minerals
Sampling point:	Middle mixed liquor height in the SBR
Samples No.:	AMX.CHE 1-8
Total sample volume:	80 mL (5 mL normal sample, 20 mL for MLVSS)
Reactor volume:	2.5 L

Experimental procedure in short: — Time (h:min)

1. Confirm availability of sampling material and required equipment. — 08:00
2. Confirm calibration and functionality of systems, meters and sensors. — 08:10
3. Transfer 1.0 L of sludge to batch reactor. — 08:30
4. Start aerobic conditions with gentle mixing and air sparging at T and pH set points. — 08:40
5. 15 min before starting, take a sample for initial conditions (AMX.CHE 1). — 09:40
6. —
7. Start batch test: add 0.07 L of synthetic influent. — 09:55
8. Take first sample for the determination of the initial conditions (minute zero) — 10:00
9. Minute 30, continue sampling program according to schedule — 10:30
10. Minute 180, stop mixing. — 13:00
11. Organize the samples and clean the system. — 13:10
12. Verify that all systems are swtiched off. — 13:20

Sampling schedule

Time (min)	-15	-5	0	30	60	90	120	150	180
Time (h)	-0.25	-0.08	0.00	0.50	1.00	1.50	2.00	2.50	3.00
Sample No.	1		2	3	4	5	6	7	8
Parameter				ANAEROBIC PHASE					
NH$_4$-N (mg N L^{-1})	8.2		73.1	63.9	52.7	44.8	38.1	29.5	19.4
NO$_2$-N (mg N L^{-1})	0[1]		69.1	58.3	47.6	38.1	25.8	16.2	4.7
NO$_3$-N (mg N L^{-1})	27.5		27.2	29.3	31.1	32.4	35.9	38	39.6
MLSS and MLVSS (mg L^{-1})	See table								See table

[1] Average value of the concentration present in the synthetic substrate and and liquid phase of te sludge sample prior the start of the test

MLSS & MLVSS measurements

Sampling point	Cup No.	W1	W2	W3	W2-W1	W2-W3	MLSS	MLVSS	Ratio
Start test[2]	1	0.09440	0.14430	0.09880	0.04990	0.04550	2,495	2,275	0.91
	2	0.09480	0.14210	0.09950	0.04730	0.04260	2,365	2,130	0.90
	3	0.09530	0.14190	0.09860	0.04660	0.04330	2,330	2,165	0.93
						Average	2,397	2,190	0.91
End test	4	0.09350	0.14360	0.09790	0.05010	0.04570	2,505	2,285	0.91
	5	0.09410	0.14330	0.09840	0.04920	0.04490	2,460	2,245	0.91
	6	0.09370	0.14200	0.09850	0.04830	0.04350	2,415	2,175	0.90
						Average	2,460	2,235	0.91

[2] Sample taken before substrate addition

Biomass composition

Sampling point	Start test	End test
MLSS (mg L^{-1})	2,397	2,460
MLVSS (mg L^{-1})	2,190	2,235
Ratio	0.91	0.91
Ash (mg L^{-1})	207	225

In this example, the observed $Y_{NH4\_NO2,AMX}$ and $Y_{NH4\_NO3,AMX}$ ratios are 1.22 and 0.24 g N g N^{-1}, respectively. The obtained specific biomass nitrogen removal rate is comparable with the kinetics reported in Table 2.4.3, indicating a high enrichment of anammox bacteria in the sludge sample tested. Also, the stoichiometric parameters of interest are in good agreement with the $Y_{NH4\_NO2,AMX}$ and $Y_{NH4\_NO3,AMX}$ reported in Table 2.4.3, indicating that the anaerobic ammonium oxidation is the dominant bioprocess occurring in the sludge sample analysed. It is important to note that when sludge samples containing COD are analysed, the $Y_{NH4\_NO2,AMX}$ and/or $Y_{NH4\_NO3,AMX}$ are expected to be different from the reference stoichiometric values due to the simultaneous occurrence of conventional heterotrophic denitrification.

2.4.11 Additional considerations

2.4.11.1 Presence of other organisms

The presence of other microorganism rather than those that carry out the biological conversions under examination may alter the results of the batch tests resulting in the under- or over-estimation of kinetics and stoichiometric parameters of interest.

In nitrification batch activity tests, the simultaneous occurrence of denitrifying and/or anammox activity may lead to an incorrect estimation of nitrite removal and/or nitrate production. Nevertheless, since both biological processes require anoxic conditions, it is sufficient to ensure the complete penetration of oxygen into the biomass. While 3-4 mg O_2 L^{-1} are considered sufficient to ensure fully aerobic conditions when testing flocculent and suspended sludge samples, higher oxygen concentrations may be required when assessing the activity of nitrifying biofilms. It is important to note that when a batch test is performed in the presence of COD, the aerobic COD removal performed by heterotrophic microorganisms would further reduce the oxygen penetration into the biofilm, thus contributing to the creation of undesirable anoxic zones (if oxygen becomes limiting).

In denitrification batch activity tests, the simultaneous occurrence of anammox activity may lead to the overestimation of nitrite and/or underestimation of the nitrate reduction kinetics. In order to avoid this inconvenience, the batch test should be carried out under ammonium-limiting conditions. It is important to note that the ammonium present during the batch activity test should be sufficient anyway to sustain the N-source requirements of denitrifying bacteria, which can be calculated in advance. When the batch test aims to assess the impact of a particular carbon source on the denitrifying kinetics, the simultaneous occurrence of the denitrification process carried out by microorganisms that can store COD intracellularly (e.g. PAO and GAO) may lead to incorrect observations. In this case, an aeration period prior to the conduction of the batch test can be used to completely remove the intracellular stored COD present in the biomass.

In anammox batch activity tests, the simultaneous occurrence of denitrifying and anammox activity may lead to an incorrect estimation of the nitrite removal and/or nitrate production kinetics, which consequently affects the $Y_{NH4\_NO2,AMX}$ and/or $Y_{NH4\_NO3,AMX}$. When OHO are the dominant denitrifying population, it is sufficient to execute the batch test in the absence of S_B to limit the presence of electron-donating compounds used for nitrite/nitrate reduction. However, when EBPR activated sludge is used instead, this would not be sufficient since intracellular storage compounds would still allow the reduction of nitrite/nitrate during the anammox batch test. A period of aeration prior to the conduction of the anammox batch test can be used to completely remove the S_B in the bulk and/or the intracellular stored COD present in the activated sludge to be tested. Nevertheless, since the anammox pathway is the only one capable of oxidizing ammonium under anoxic conditions, anammox activity can be satisfactorily assessed in a batch test even in the presence of COD by following the ammonium concentration over time.

2.4.11.2 Shortage of essential micro- and macro-nutrients

Though it may seem trivial, the presence of macro- and micro-nutrients in the right concentration and (bio-)availability is essential for the bioprocesses involved in the nitrogen removal of activated sludge systems such as the nitrification, denitrification and anammox processes described in this chapter. Commonly, macro- and micro-nutrients are present in most municipal wastewaters, but their presence should be checked and confirmed particularly if the wastewater treatment plant under examination regularly receives industrial effluents. Due to the low yield of autotrophic microorganisms such as nitrifiers and anammox bacteria, the lack of macro-nutrients is not so frequent when treating municipal sewage. Nevertheless, it should be a point of attention when these bioprocesses are applied to the treatment of industrial wastewaters and in general for high-strength wastewaters. For this purpose, a simple estimation of the nutrient requirements of the biomass as a function of the nitrogen load to be treated can be performed and compared against the influent nutrient concentrations to assess whether external addition is necessary to support the biological growth requirements and conversion rate. In some cases the external dosage of particular micro-nutrients beyond the minimal requirements can enhance the kinetics of a specific microbial population as recently reported in the case of iron dosage to anammox cultures (Chen *et al.*, 2014; Bi *et al.*, 2014).

2.4.11.3 Toxicity or inhibition effects

Several compounds were identified to be toxic or inhibitory for the bacterial communities carrying out the nitrification, denitrification and anammox processes. Since the abundance of OHOs capable of performing denitrification in activated sludge systems is broad and diverse they can acclimatise and adapt to different environmental and operating conditions and even withstand the presence of different potentially toxic or inhibitory compounds. On the other hand, autotrophic bacteria catalysing the nitrification and anammox processes are more prone to inhibition and toxicity issues. The list of compounds and concentration ranges which may be toxic or inhibitory for nitrifying and anammox bacteria is so long and complex that one should refer to specific literature. Despite the fact that microbial communities can also acclimatise and adapt to sub-optimal conditions, the inhibitory effects of certain compounds can lead to sub-optimal microbial process activity and ultimately to the failure of the bioprocess. To assess the inhibitory effect of a certain compound or of a certain particular wastewater on the nitrifying, denitrifying or anammox activity, a series of batch activity tests can be conducted as described in this chapter. The comparison between the activity measured in the presence or absence of different concentrations of potential inhibitory compounds can provide useful indications about the inhibitory potential of such compounds with regard to the particular activated sludge tested. Similarly, when the wastewater used to perform the tests is suspected to have or to generate an inhibitory effect on a particular bioprocess, two series of batch activity tests can be executed: one with the original activated sludge and another one with the same biomass but washed in a mineral solution to remove the potentially inhibiting or toxic compounds. However, it is important to note that such an approach may be successful only if the inhibiting effects of the wastewater to be tested are (rapidly) reversible.

2.4.11.4 Effects of carbon source on denitrification

It is well known that denitrification kinetics depend on the carbon sources used as electron donors (e.g. Mokhayeri *et al.*, 2006, 2008). For the regular monitoring of the denitrifying potential of activated sludge, the use of synthetic media containing S_B such as VFAs can be good enough to provide a satisfactory assessment of denitrification activity (as presented in this chapter). The use and application of more complex COD sources, which may be undoubtedly present in raw or settled municipal wastewater, can lead to sub-optimal denitrification activity. When external COD sources are needed to enhance nitrogen removal (e.g. in the post-denitrification unit), the methods described in this chapter can be used to assess the influence of different COD sources on the denitrifying kinetics of an activated sludge sample.

Figure 2.4.15 Execution of batch activity tests. Note the characteristic reddish colour of the enriched anammox culture (photo: Lotti, 2015).

2.5 AEROBIC ORGANIC MATTER REMOVAL

2.5.1 Process description

In conventional wastewater treatment systems performing aerobic organic matter removal, OHOs remove the organics present in wastewater to produce more biomass using oxygen for respiration. From a metabolic perspective, the removal process involves an anabolic (for cell synthesis) and a catabolic process (to generate the required energy for cell synthesis). In the anabolic process, OHOs obtain the required carbon for cell growth from the organic matter present in wastewater. Meanwhile, in the catabolic process, an oxidation-reduction reaction takes place involving the transfer of electrons from the organic matter (which acts as the electron donor) to oxygen (the electron acceptor), generating the required energy for cell synthesis. However, due to the rather variable mixture of biodegradable and non-biodegradable organic compounds present in wastewater, the COD is commonly used to estimate their total concentration. This is mostly because the use of COD is preferred over other analytical parameters (such as biochemical oxygen demand: BOD_5 or total organic carbon: TOC) due to several advantages. These include (Henze et al., 1997; Henze and Comeau, 2008): (*i*) the determination of the oxygen equivalence (or capacity to donate electrons) of the organic compounds, (*ii*) the more detailed and useful determination of the organic strength by being able to measure all the degradable and undegradable organics, (*iii*) the potential for the balance of organics to be closed on a COD basis (as a consequence of the previous two advantages), as well as practical implications such as (*iv*) a rapid analysis (i.e. a few hours as opposed to 5 days required for BOD_5). In general, the exact stoichiometry involved in the aerobic removal of organics is not straight forward. Nevertheless, the following equation for the aerobic consumption of glucose ($C_6H_{12}O_6$) (which neglects most of the nutrients except nitrogen) can be used to illustrate the biological aerobic removal process (Metcalf and Eddy, 2003):

$$3C_6H_{12}O_6 + 8O_2 + 2NH_3 \rightarrow 2C_5H_7NO_2 + 8CO_2 + 14H_2O$$

Eq. 2.5.1

Where, $C_5H_7NO_2$ is a simplified expression of the new cells generated from the aerobic degradation of organics (Hoover and Porges, 1952).

From a microbial growth perspective, about ⅔ of the biodegradable organics (leading to the so-called aerobic stoichiometric true yield, Y_{OHO}, of ~0.67 g COD-biomass per COD-organics consumed) are converted into new biomass via anabolism and the remaining ⅓ is oxidized using oxygen via catabolic pathways to generate the required energy for biomass growth (Marais and Ekama, 1976). In addition, for microbial growth, macronutrients (like nitrogen and phosphorus) and micronutrients (such as potassium, sodium, calcium, magnesium, zinc, manganese and iron, among others) are needed for cell synthesis (Metcalf and Eddy, 2003). The lack of any of these elements can lead to limitations of the microbial processes. It is assumed that the nitrogen and phosphorus requirements of the new biomass produced are approximately 0.10 g N g VSS^{-1} and 0.03 g P g VSS^{-1}, respectively (Ekama and Wentzel, 2008a). This means that if a wastewater contains 100 mg BCOD L^{-1} (biodegradable organics) then the nitrogen and phosphorus concentrations that need to be supplied should not be lower than 4.7 mg NH_4-N L^{-1} and 1.4 mg PO_4-P L^{-1}, respectively, to meet the nutrient requirements. This assumes a true yield of Y_{OHO} of 0.67, an observed yield of 0.40 (accounting for a sludge age of about 5 days) and a COD-to-VSS ratio of the biomass of 1.42 mg COD mg VSS^{-1}. From a practical perspective and to avoid nutrient limitations due to the different Y_{OHO} observed (see Table 2.5.1), a COD:N:P ratio of 100:5:1 is usually suggested (Metcalf and Eddy, 2003).

It is important to underline that not all the organics present in wastewater can be subject to degradation. At the most basic classification, at least four different organic fractions can be identified in a wastewater stream with different physical characteristics and degree of biodegradability that determine their removal potential in a wastewater treatment system (Ekama and Wentzel, 2008a). These are (*i*) biodegradable soluble organics rapidly converted by OHO (and thus known as readily biodegradable organics, RBCOD or S_B according to the standardized notation (Corominas et al., 2010)), (*ii*) biodegradable particulate organics that get mostly enmeshed within the activated sludge flocs and are subject to hydrolysis prior to biodegradation, and therefore commonly characterized as slowly biodegradable organics (SBCOD or XC_B in accordance with Corominas et al., 2010), (*iii*) non-biodegradable particulate organics that get mostly enmeshed in the sludge flocs and accumulate in the activated sludge system, and (*iv*) non-biodegradable soluble organics which neither get enmeshed nor degraded and remain in the soluble phase.

RBCOD can be rapidly utilized by OHOs. Although the biological breakdown of SBCOD is slow, at the SRT commonly applied in most activated sludge plants (e.g. longer than 3-4 days), SBCOD are virtually completely utilized (Ekama and Wentzel, 2008a). Thus, OHOs utilize RBCOD and SBCOD for cell synthesis, producing more biomass. The new OHO biomass generated by the removal and conversion of RBCOD and SBCOD, as well as the accumulation of the non-biodegradable particulate organics, becomes part of the organic activated sludge mass in the reactor usually measured as MLVSS. Because of the flocculation capability of the activated sludge, solids material is relatively highly settleable so it can be efficiently removed in the secondary settling tanks, providing a treated and clear effluent. The sludge mass that settles out in the secondary settling tank is returned to the biological reactor and eventually is removed via the waste of activated sludge which is a function of the plant's SRT (Arden and Lockett, 1914). However, since the non-biodegradable soluble organics cannot be efficiently removed in an activated sludge system, they leave the plant through the effluent, contributing to the effluent COD concentration.

In addition to the aerobic removal of organics, in activated sludge systems performing BNR, most of the organic matter will be removed in the anaerobic and anoxic stage that precedes the aerobic stage. For instance, in the anaerobic stage of an activated sludge system designed for EBPR, VFA are taken up by PAO, while in the anoxic stage of a BNR plant, biodegradable organics are used by denitrifying organisms for denitrification purposes using nitrate or nitrite as the electron acceptor. In addition, the potential occurrence of sulphate-reducing processes by SRBs in the anaerobic stages of an activated sludge system can also lead to the removal of organics. Furthermore, the removal of organics can also take place in anaerobic wastewater treatment systems (e.g. upflow anaerobic sludge blankets: UASB plants) under fully anaerobic conditions by strictly anaerobic organisms.

As observed, organic matter removal can occur under different environmental conditions and be performed by different groups of microorganisms. The present section focuses on the execution of batch activity tests to assess aerobic organic matter removal as the primary removal process by OHOs in conventional activated sludge systems under fully aerobic conditions. The activity assessment of other processes is presented in other sections of the present and following chapters.

The present section aims to be used as a guide for the execution of aerobic batch activity tests for the determination of the kinetic rates of RBCOD removal in activated sludge and other suspended growth systems. It does not cover the removal of SBCOD fractions because its determination requires the execution of respirometry tests described in Chapter 3. Moreover, and despite that the fractionation of the influent wastewater organic compounds into COD (as well as into N and P) is of major importance for the design, operation, modelling and evaluation of (activated sludge) wastewater treatment plants, the current chapter does not aim to elaborate on wastewater characterization and fractionation protocols. For such a purpose, the reader is referred to scientific and technical reports published elsewhere (Henze, 1992; Kappeler and Gujer, 1992; Wentzel et al., 1995; Hulsbeek et al., 2002; Roeleveld and van Loosdrecht, 2002; Vanrolleghem et al., 2003; WERF, 2003; Langergraber et al., 2004).

It is important to note that some of the wastewater characterization and fractionation protocols, as well as a thorough determination of the actual biomass yields using organics from different wastewaters, also need respirometry tests (Chapter 3).

2.5.2 Experimental setup

2.5.2.1 Reactors

To assess the aerobic organic matter removal activity by activated sludge, batch tests need to be carried out under aerobic conditions securing sufficient availability of DO (maintaining DO concentrations higher than 2 mg L^{-1}) and good mixing conditions. As for other processes, it is also important to maintain an adequate and desirable temperature, precise pH control, and have additional ports for sample collection and the addition of influent, solutions, gases and any other liquid media or substrate used in the test. In general, similar fermenters with features and characteristics of those used for the execution of EBPR batch activity tests can be used to carry out the aerobic organic matter removal tests (see Section 2.2.2.1). Similar recommendations to those described in Section 2.2.2.1 regarding aeration, mixing and pH control, location and characteristics of sampling and dosing ports can also be applied here.

2.5.2.2 Activated sludge sample collection

For the execution of aerobic organic matter removal batch activity tests, a fresh sample should be collected at the end of the aerobic tank or phase in a sampling spot where well-mixed conditions take place. Ideally, batch activity tests should be performed soon after collection (within 2 to 3 h after sampling). If the batch activity tests cannot be performed *in situ* on the same day when sampling took place, an activated sludge sample can be collected in a bucket or jerry can at the end of the aerobic stage and properly transported and stored in a fridge or using ice (to keep the temperature around 4 °C if feasible). In any case, the *in situ* execution of the batch activity tests is preferable for obvious reasons. The total volume of activated sludge to be collected depends on the number (repetition) of tests, reactor volume and total volume of samples to be collected to assess the biomass activity. Often, 10-20 L of activated sludge collected from full-scale wastewater treatment plants can be considered sufficient per batch. On the other hand, samples collected from lab-scale reactors rarely provide more than 1 L because lab-scale systems use small reactors (from 0.5 to 2.4 L and in some cases up to 8-10 L, exceptionally 15 L) and the maximum volume that can be withdrawn from lab-scale reactors is often set by the daily withdrawal of the excess of sludge from the system (which is directly related to the SRT and, consequently, defined by the growth rate of the organism(s) of interest).

2.5.2.3 Activated sludge sample preparation

For batch activity tests performed *in situ*, in principle, the sludge should be transferred from the parent reactor (in the case of a lab-enriched sludge) or reaction tank (in the case of pilot-scale or full-scale plants) to the fermenter or reactor where the activity tests will take place. Then the activated sludge should be aerated for at least 1 or 2 h to remove any residual biodegradable COD present in the system while the sample is also adjusted to the desired pH and temperature of interest (as described in Section 2.2.3.5). Alternatively, to remove any residual COD the activated sludge can be washed using the washing media and washing procedure described in sections 2.2.3.3 and 2.2.3.5, respectively.

If only an aerobic organic removal batch activity test is going to be executed (i.e. a nitrification test is not of interest), then a nitrification inhibitor can be added to the sludge sample immediately after the sludge has been transferred to the fermenter (e.g. allyl-N-thiourea: ATU to a recommended final concentration of 20 mg L^{-1}). In particular, this will restrain nitrification and consequently avoid higher oxygen consumption if respirometry tests are going to be executed in parallel (see Chapter 3). Sludge samples stored under cold conditions can also be washed with a mineral solution to remove any residual organics.

For batch tests executed with sludge samples stored under cold conditions (at around 4 °C), sludge samples need to be 're-activated' because the cold storage temperature slows down the bacterial metabolism. To re-activate the sludge, the activated sludge should be aerated for 1-2 h at the desired pH and, particularly, the temperature of study.

2.5.2.4 Media

When real wastewater (either raw or settled) is used for the execution of activity tests, it can be fed in a relatively straightforward manner to the reactor/fermenter. For normal or regular conditions, the feeding step takes place at the beginning of the test. If required, raw wastewater can be filtered (using 1 or 2 mm sieves) or settled (duration from 1 to 3 h). If the activity tests need to be performed at different solids concentration, (*i*) the treated effluent from the plant can be collected and used for dilution (assuming that solids effluent concentrations are relatively low, e.g. 20-30 mg TSS L^{-1}), or (*ii*) the activated sludge can be concentrated by decanting and discharging the supernatant in several repeated steps until reaching the MLSS of interest.

If different carbon sources and concentrations are to be studied, the plant effluent can also be used to prepare a semi-synthetic media (as long as it does not contain toxic or inhibitory compounds) containing a RBCOD concentration of interest which, for instance after a 1:1 dilution in the fermenter, can provide the target initial COD concentration. However, depending upon the nature and purpose of tests, the carbon concentrations present in a synthetic wastewater can vary and be adjusted proportionally to the duration of the test. Usually, concentrations of about 400 mg COD L^{-1} can be used with either lab- or full-scale activated sludge. Besides the carbon source, the synthetic wastewater must contain the required macro- and micro-elements. A suggested synthetic wastewater recipe for an initial COD of about 400 mg COD L^{-1} can contain per litre (Smolders *et al.*, 1994a): 862 mg NaAc·$3H_2O$ (400 mg COD), 107 mg NH_4Cl (28 mg N), 40 mg NaH_2PO_4·$2H_2O$ (8 mg P), 90 mg $MgSO_4$·$7H_2O$, 14 mg $CaCl_2$·$2H_2O$, 36 mg KCl, 1 mg yeast extract and 0.3 mL of a trace element solution (that includes per litre 10 gEDTA, 1.5 g $FeCl_3$·$6H_2O$, 0.15 gH_3BO_3, 0.03 g $CuSO_4$·$5H_2O$, 0.12 g $MnCl_2$·$4H_2O$, 0.06 g Na_2MoO_4·$2H_2O$, 0.12 g $ZnSO_4$·$7H_2O$, 0.18 g KI and 0.15 g CoCl·$6H_2O$). If desired, the synthetic wastewater can be concentrated to a higher desirable COD concentration to account for potential dilution rates, sterilized in an autoclave (for 1 h at 110 °C) and used as a stock solution if several tests are going to be performed in a defined period of time. However, the solution must be discarded if any precipitation is observed or it loses transparency.

For experiments performed with lab-enriched cultures, it is best to execute the tests with the same (synthetic) wastewater used for the cultivation with the carbon concentrations of study, unless otherwise required. Alternatively, and similar to full-scale samples, the effluent from the reactor can be collected, filtered through rough pore size filters to remove any coarse particles, and used to prepare the required media.

2.5.2.5 Analytical tests

Most of the analytical tests required (for the determination of COD, total P, PO_4, NH_4 NO_2, NO_3, TSS, VSS, etc.) can be performed following standardized and commonly applied analytical protocols detailed in Standard Methods (APHA et al., 2012). For the determination of dissolved parameters like soluble COD, PO_4, NH_4, NO_2 and NO_3, samples should be filtered immediately after collection through 0.45 µm pore size filters. Of the two most commonly applied methods for the analytical determination of COD, the dichromate method is recommended, as the permanganate method does not fully oxidize all the organic compounds (Henze and Comeau, 2008). The determination of VFA (like acetate, propionate and other volatile fatty acids) can be executed by GC. Glucose and other carbon compounds (including VFA) can be determined by HPLC.

2.5.2.6 Parameters of interest

To determine and assess the activity of OHOs, different stoichiometric ratios and kinetic rates can be estimated as displayed in Table 2.5.1.

Table 2.5.1 Expected stoichiometric and kinetic parameters of interest for activated sludge systems performing aerobic organic matter removal.

Parameter	Remark	Reference
Aerobic stoichiometric parameter Y_{OHO} (g COD-biomass g COD-substrate^{-1})		
0.67	Theoretical ratio	Ekama and Wentzel (2008a)
0.72	Acetate as organic matter source	Dircks et al. (1999)
0.37	Methanol as organic matter source	McCarty (2007)
0.65	Formate as organic matter source	McCarty (2007)
0.40 -0.80; typically 0.60	g VSS g BOD^{-1} units	Metcalf and Eddy (2003)
0.30-0.60	g VSS g RBCOD^{-1} units	Metcalf and Eddy (2003)
0.67-0.792*	Different RBCOD organic matter sources	Guisasola (2005)
0.91*	Glucose as organic matter source	Dircks et al. (1999)
0.90*	Glucose as organic matter source	Goel et al. (1999)
Aerobic kinetic parameter $q_{OHO,COD,Ox}$ (g COD substrate g COD-biomass^{-1} d^{-1})		
6	ASM2d model	Henze et al. (1999)
2-10, typically 5	g RBCOD g VSS^{-1} d^{-1}	Metcalf and Eddy (2003)
3 - 10		Kappeler and Gujer (1992)

* These values deviate considerably from the theoretical yield of 0.67 gCOD gCOD^{-1} due to the occurrence of storage processes.

It is important to mention that an Arrhenius temperature coefficient of between 1.060 and 1.123 (with a typical value of 1.070) has been suggested for the description of the aerobic organic matter removal rates (Metcalf and Eddy, 2003).

As shown in Table 2.5.1, the theoretical stoichiometric biomass yield on biodegradable organics is 0.67 g COD g COD^{-1} (Metcalf and Eddy, 2003; Ekama and Wentzel, 2008a). However, it is common to observe higher stoichiometric ratios that can apparently reach up to 0.90-0.91 g COD g COD^{-1} (Dircks et al., 1999; Goel et al., 1999). If this is the case, one should be aware that values higher than 0.67 g COD g COD^{-1} are caused by the direct storage of biodegradable organics rather than a higher biomass yield. Such processes usually occur when RBCOD (S_B) is the dominant organics in plants designed with selectors. Further details can be found in literature (Gujer et al., 1993; Henze et al., 2008). Regarding the aerobic kinetic rates for organic matter removal, those reported in literature can vary widely from 3 to 10 g COD-substrate g COD-biomass^{-1} d^{-1}. The main reason for this can be the net concentration of active biomass present in the system (with respect to the total MLVSS concentration). A short SRT (of less than 3 days) may lead to a high fraction of active biomass with respect to VSS present in the system which could be reflected in a high organic matter removal rate. However, a (very) long SRT (for instance, much higher than 20 days) will lead to a higher accumulation of non-biodegradable VSS (present in the influent) or produced by endogenous respiration by OHOs (also measurable as MLVSS) which adds to the MLVSS concentration of the sludge and leads to a lower maximum specific COD removal rate, $q_{OHO,COD,Ox}$. Thus, the highest $q_{OHO,COD,Ox}$ (up to 10 g COD-substrate g COD-biomass^{-1} d^{-1}) can be expected in tests executed with activated sludge samples containing relatively higher concentrations of active OHO biomass, typically observed in low sludge age systems.

2.5.3 Aerobic organic matter batch activity tests: preparation

2.5.3.1 Apparatus

For the execution of aerobic organic batch activity tests the following apparatus is needed:

1. A batch reactor or fermenter equipped with a mixing system and adequate sampling ports (as described in Section 2.5.2.1).
2. An oxygen supply (compressed air or pure oxygen sources).
3. A pH electrode (if not included/incorporated in the batch reactor setup).
4. A 2-way pH controller for HCl and NaOH addition (alternatively a one-way control - generally for HCl addition - or manual pH control can be applied through the manual addition of HCl and NaOH).
5. A thermometer (recommended working temperature range of 0 to 40 °C).
6. A temperature control system (if not included in the batch reactor setup).
7. A DO meter with an electrode (if not included/incorporated in the batch reactor setup).
8. An automatic 2-way controller for nitrogen and oxygen gas supply (if not included in the batch reactor setup and if tests must be performed at a defined DO concentration).
9. A centrifuge with a working volume capacity of at least 250 mL to carry out the sludge washing procedure (if required).
10. A stopwatch.

Confirm that all the electrodes and meters (pH, temperature and DO) are calibrated less than 24 h before execution of the batch activity tests in accordance with guidelines and recommendations from manufacturers and/or suppliers.

2.5.3.2 Materials

1. Two graduate cylinders of 1 or 2 L (depending upon the sludge volumes used) to hold the activated sludge and wash the sludge if required.
2. At least 2 plastic syringes (preferably of 20 mL or at least of 10 mL volume) for the collection and determination of soluble compounds (after filtration).
3. At least 3 plastic syringes (preferably of 20 mL) for the collection of solids, particulate or intracellular compounds (without filtration).
4. 0.45 μm pore size filters. Preferably not of cellulose-acetate because these may release certain traces of cellulose or acetate into the collected water samples. Consider using twice as many filters as the number of samples that need to be filtered for the determination of soluble compounds.
5. 10 or 20 mL transparent plastic cups to collect the samples for the determination of soluble compounds (e.g. soluble COD, ammonium, ortho-phosphate).
6. 10 or 20 mL transparent plastic cups to collect the samples for the determination of mixed liquor suspended solids and volatile suspended solids (MLSS and MLVSS, respectively). Consider the collection of these samples in triplicate due to the variability of the analytical technique.
7. A plastic box or dry ice box filled with ice up to the required volume to temporarily store (for up to 1-2 h after the conclusion of the batch activity test) the plastic cups and plastic tubes for centrifugation after the collection of the samples.
8. Plastic gloves and safety glasses.
9. Pasteur or plastic pipettes for HCl and/or NaOH addition (when the pH control is carried out manually).
10. Metallic lab clips or clamps to close the tubing used as a sampling port when samples are not collected from the reactor/fermenter.

2.5.3.3 Working solutions

- **Real wastewater**

  If real wastewater is used to carry out the batch activity test, there is a need to collect the sample at the influent of the wastewater treatment plant and perform the batch activity test as soon as possible after collection. If due to location and distance the tests cannot be performed in less than 1 or 2 h after collection, then one should keep the wastewater sample cold until the test is executed (e.g. by placing the bucket or jerry can in a fridge at 4 °C). Nevertheless, prior to the execution of the test, the temperature of wastewater needs to be adjusted to the target temperature at which the batch activity test will be executed (preferably reached in less than 1 h). A water bath or a temperature-controlled room can be used for this purpose.

- **Synthetic influent media or substrate**

If tests can be or are desired to be performed with synthetic wastewater, then the synthetic influent can contain a mixture of carbon and (macro- and micro-) nutrients. Generally, they can all be mixed together in the same media, as long as precipitation is not observed. The usual composition and concentration are:

a. Carbon source solution: this must be composed of a RBCOD source like VFA (such as acetate or propionate) or glucose, depending on the nature or goal of the test and the corresponding research questions. Initial COD concentration of around 400 mg L^{-1} is recommended for sludge samples from either lab- or full-scale systems.

b. The nutrient-solution: this should contain all the required macro- (ammonium, phosphorus, magnesium, sulphate, calcium, potassium) and micro-nutrients (iron, boron, copper, manganese, molybdate, zinc, iodine, cobalt) to ensure that cells are not limited by their absence and avoid obtaining erroneous results and in extreme cases the failure of the test. Thus, although their concentrations may seem trivial, one must make sure that all of the constituents are added to the solution in the required amounts. The following composition for a nutrient solution can be recommended for influents containing up to 400 mg COD L^{-1} (based on Smolders et al., 1994a) per litre: 107 mg NH_4Cl (28 mg), 90 mg $MgSO_4 \cdot 7H_2O$, 40 mg $NaH_2PO_4 \cdot 2H_2O$ (8 mg P), 14 mg $CaCl_2 \cdot 2H_2O$, 36 mg KCl, 1 mg yeast extract and 0.3 mL of a trace element solution (that includes per litre 10 g EDTA, 1.5 g $FeCl_3 \cdot 6H_2O$, 0.15 gH_3BO_3, 0.03 g $CuSO_4 \cdot 5H_2O$, 0.12 g $MnCl_2 \cdot 4H_2O$, 0.06 g $Na_2MoO_4 \cdot 2H_2O$, 0.12 g $ZnSO_4 \cdot 7H_2O$, 0.18 g KI and 0.15 g $CoCl \cdot 6H_2O$). Similar nutrient solutions can be used as long as they contain all the previously reported required nutrients.

c. Washing media: If the sludge sample must be washed to remove any residual COD or the presence of undesirable compounds (which may be even inhibitory or toxic), a nutrient solution should be prepared. The following nutrient solution to wash the sludge could be used per litre (Smolders et al., 1994a): 107 mg NH_4Cl, 40 mg $NaH_2PO_4 \cdot 2H_2O$, 90 mg $MgSO_4 \cdot 7H_2O$, 14 mg $CaCl_2 \cdot 2H_2O$, 36 mg KCl, 1 mg yeast extract and 0.3 mL of a trace element solution (that includes per litre 10 g EDTA, 1.5 g $FeCl_3 \cdot 6H_2O$, 0.15 gH_3BO_3, 0.03 g $CuSO_4 \cdot 5H_2O$, 0.12 g $MnCl_2 \cdot 4H_2O$, 0.06 g $Na_2MoO_4 \cdot 2H_2O$, 0.12 g $ZnSO_4 \cdot 7H_2O$, 0.18 g KI and 0.15 g $CoCl \cdot 6H_2O$). The washing process can be repeated two or three times. Afterwards, the following preparation steps of the batch activity tests can be performed.

d. ATU solution: To inhibit nitrification, an ATU solution can be prepared to reach an initial concentration of around 20 mg L^{-1} (after addition). The ATU solution must be added before the sludge is exposed to any aerobic conditions (including the sludge sample preparation or acclimatization).

e. Acid and base solutions: 100-250 mL of 0.2 M HCl and 100-250 mL 0.2 M NaOH solutions for automatic or manual pH control.

Finally, the required working and stock solutions to carry out the determination of the analytical parameters of interest must be also prepared in accordance with Standard Methods and the corresponding protocols.

2.5.3.4 Material preparation

1. Collect the materials to execute the batch activity tests and the definition of the number of samples to be collected can be prepared following the same steps presented in Section 2.2.3.4.
2. Determine the frequency of sample collection: for the determination of the maximum RBCOD kinetic removal rate, samples need to be collected every 5 min during the first 30-40 min of duration of the batch activity test.
3. Collect a series of samples before any media is added to increase the data reliability and to establish the initial conditions of the sludge.
4. Carefully define the maximum and minimum working volumes of the reactor according to the suggestions presented in Section 2.2.3.4.

5. Label all the plastic cups once the number and frequency of collecting the samples has been defined. Define a nomenclature and/or abbreviation to easily identify and recognize the batch test, the sampling time and the parameter(s) of interest to be determined with that sample. Labelling both the plastic cup and the cover will help to easily identify the sample.
6. Table 2.5.2 contains an example of a working sheet to execute and keep track of the sampling collection and batch test execution. Furthermore, it can be used to keep a database of the different batch tests carried out.
7. Organize all the required material within a relatively close radius of the action around the batch setup so any delay in handling and preparing the samples can be avoided.
8. Calibrate all the meters (pH, DO and thermometer) less than 24 h prior to the execution of the tests and store them in proper solutions until the execution of the tests, according to the particular recommendations of the corresponding manufacturer or supplier. Also confirm that the readings are reliable.
9. Store the samples properly and preserve them until they are analysed. Table 2.2.2 provides recommendations for sample preservation depending on the analytical determination of the parameter of interest.

2.5.3.5 Activated sludge preparation

These procedures consider that batch activity tests can be performed as soon as possible after collection of samples from full- or lab-scale systems or, in the worst case scenario, within 24 h after collection. Ideally, the execution of batch tests 24 h after the collection is not recommended due to potential changes and decay of the microorganisms. Bearing this in mind, the following procedure is recommended to prepare the activated sludge samples for the execution of batch activity tests:

1. If batch activity tests can be executed in less than 1 h after collection of the sludge sample and if the sludge sample does not need to be washed:
   a. Adjust the temperature of the batch reactor to the target temperature of the study.
   b. Collect the sludge at the end of the aerobic stage.
   c. Transfer the sludge sample to the reactor or fermenter.
   d. Add the ATU solution to inhibit nitrification (particularly if respirometry tests will be executed) to a final concentration of 20 mg L^{-1} (see Section 2.5.3.3).
   e. Start mixing (100-300 rpm) and follow the temperature and pH of the sludge sample by placing an external thermometer inside the reactor (if the setup does not have a built-in thermometer) until the sludge reaches the target temperature and pH of the study.
   f. Start to aerate the sludge samples keeping a DO concentration higher than 2 mg L^{-1}.
   g. Start the aerobic batch activity tests once the pH and DO are stable.
2. If the activated sludge sample needs to be washed then the steps presented in Section 2.2.3.5 (for tests that can be executed in less than 1 h or 24 h after collection) can be followed for the activated sludge sample preparation using the washing media indicated in Section 2.5.3.3.

2.5.4 Aerobic organic matter batch activity tests: execution

Similar to previous tests, to facilitate the execution and for data track record and archiving purposes, an experimental implementation plan should be prepared in advance similar to that presented in Table 2.5.2. Aerobic organic matter removal activity tests can be executed to assess the removal of RBCOD in activated sludge systems. They can be performed with activated sludge samples from full- or lab-scale systems, using real or synthetic wastewater. Samples should be collected at the end of the aerobic stage as long as RBCOD is not present in the sample. The following steps are proposed for the execution of the aerobic organic matter removal test:

Test OHO.AER.1. Single aerobic organic matter removal test

a. After the sludge has been collected, prepared and transferred to the batch reactor (see Section 2.5.3.5), keep the sample aerated for at least 30 min while confirming that the pH and temperature are at the target values of interest. Otherwise, set up the corresponding set points (if automatic pH and temperature controllers are applied) or adjust manually. Wait until stable conditions are reached.
b. Ensure that the DO concentration is higher than 2 mg L^{-1}.
c. Keep track of the execution and sampling time with a stopwatch.
d. Take the first samples of the water phase and biomass once stable operating conditions are reached (around 20 min before the start of the test) to determine the initial concentrations of the parameters of interest:

soluble COD, NH$_4$, PO$_4$ and MLSS and MLVSS concentrations.

e. Connect the syringe to take the samples, next open or release the lab clip or clamp that closes the sampling port, and then pull and push the syringe several times until a homogenous sample is collected (usually around 5 times are required). When the syringe is full, close the clip and remove the syringe.

f. Immediately filter the samples used for the determination of soluble COD, NH$_4$, PO$_4$ (through 0.45 µm pore size filters). Other samples (e.g. MLVSS and MLVSS) need to be prepared in accordance with the corresponding protocols.

g. Store the samples at 4 °C in the fridge or preferably in a cool box with ice before and during the test execution.

h. Start the execution of the aerobic test at 'time zero' with the addition of the real or synthetic wastewater (as the carbon source solution).

i. Add the real wastewater or synthetic influent to reach an initial RBCOD-to-MLVSS ratio in the reactor of around 0.10 mg COD mg VSS^{-1}. For example, the initial RBCOD and MLVSS concentrations after mixing can range around 400 mg COD L^{-1} for samples containing 4,000 mg VSS L^{-1}. Higher or lower ratios may also be acceptable as long as there is sufficient time for sampling.

j. Ensure that the DO readings remain above 2 mg L^{-1} after the addition of the wastewater, and that considerable temperature and pH variations (higher than 1 °C or ± 0.1 for temperature and pH, respectively) do not take place.

k. Duration and sampling:
   (i) Usually tests can last between 2 and 4 h for initial soluble COD concentrations of up to 400 mg COD L^{-1} (at pH 7.0 and 20 °C).
   (ii) To determine the aerobic kinetic parameters, samples for the determination of soluble COD should be collected every 5 min in the first 30-40 min of execution of the test. After this period, the sampling frequency can be reduced to 10 or 15 min during the first 1 h, and later on to every 15 or 30 min until the test is finished.

l. Conclude the aerobic test with the collection of samples for the determination of soluble COD, MLSS, MLVSS, NH$_4$ and PO$_4$ concentrations.

m. Organize the samples and ensure that all the samples are complete and properly labelled to avoid mixing the samples and trivial mistakes.

n. Preserve and store the samples as recommended by the corresponding analytical procedures until the collected samples are analysed.

o. Clean up the apparatus and take appropriate measures to maintain and look after the different sensors, equipment and materials.

p. Keep (part of) the sludge used in the test for possible further use (e.g. for microbial identification, see chapters 7 and 8).

2.5.5 Data analysis

The growth yield of biomass on COD is the main stoichiometric parameter of interest to assess the aerobic stoichiometry of OHO (Table 2.5.1). However, the aerobic batch activity test presented in this section cannot be used for the determination of Y$_{OHO}$. This is mostly because, for the determination of Y$_{OHO}$, respirometry tests need to be executed in parallel to the aerobic tests to assess the amount of COD being used for energy generation. This is presented elsewhere (Wentzel *et al.*, 1995; Dircks *et al.*, 1999; Goel *et al.*, 1999; Guisasola *et al.*, 2005).

The maximum specific aerobic RBCOD removal rate (q$_{OHO,COD,Ox}$) can be computed by plotting the experimental data (Y-axis) versus time (X-axis) and fitting the experimental data obtained in the aerobic batch activity tests using linear regression. Because one is interested in the maximum rates, a linear regression approach can be applied by fitting more than 4 to 5 experimental data points while achieving a statistical determination coefficient (R^2) not lower than 0.90-0.95. This is the main reason why the sampling frequency within the first 30-40 min of execution of the batch activity tests is set to 5 min. The maximum volumetric kinetic rate can be determined based on the slope of the linear regression equation. This will result in the determination of the maximum volumetric rate (usually reported in units such as mg L^{-1} h^{-1} or g m^{-3} d^{-1}). Figure 2.5.1 illustrates an estimation of the maximum aerobic kinetic removal rates for an OHO culture. The rate can be expressed as the maximum specific kinetic rate by dividing the volumetric rate (or value of the slope) by the concentration of activated sludge VSS. It is important to note that the maximum biomass-specific growth rate cannot be computed in a straightforward manner by following the increase in biomass concentrations during the cycle since it may be practically negligible and fall into the standard error of the analytical determination of MLVSS. Instead, either long-term continuous tests can be executed or the aerobic activity tests need to be combined with respirometry tests (Chapter 3) as presented elsewhere in literature (Kappeler and Gujer, 1992; Wentzel *et al.*, 1995).

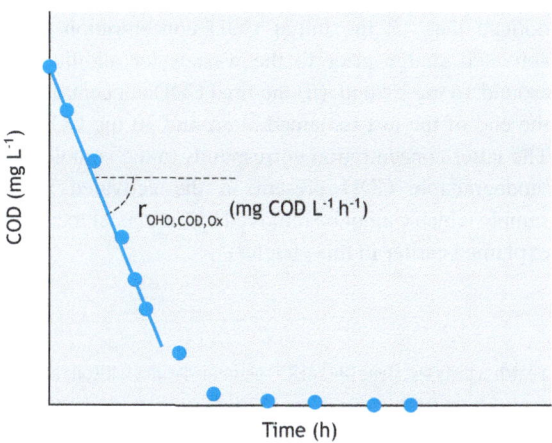

Figure 2.5.1 Example of the determination of the maximum aerobic volumetric kinetic rate for organic matter removal (expressed in mg COD L^{-1} h^{-1}) in an aerobic batch activity test.

2.5.6 Example

2.5.6.1 Description

To illustrate the execution of an aerobic batch activity test for OHO, data from a test performed at 15 °C with a full-scale activated sludge sample is presented in this section. The Test OHO.AER.1 was carried out to determine the aerobic kinetic rate of OHO for RBCOD removal. Thus, the batch activity test was performed using a 3.0 L reactor. All the equipment, apparatus and materials were prepared as described in Section 2.5.3. The pH and DO sensors were calibrated less than 24 h before the test execution. The test lasted 4 h. Prior to the batch test, 1.25 L of fresh activated sludge collected in the end of the aerobic phase of a full-scale plant was transferred to the reactor and acclimatized for 1 h at 15 °C under slow mixing (100 rpm) at pH 7.0 following the recommendations described in Section 2.5.3.5. Activated sludge preparation for tests was performed less than 1 h after sludge collection. Thereafter, 20 min before the start of the test, mixing was increased to 300 rpm and samples for the determination of the parameters of interest were collected (in accordance with the execution of Test OHO.AER.1). The test started with the addition of 1.25 L synthetic media containing 400 mg COD L^{-1} as Ac. Other macro- and micro-nutrients as well as 20 mg L^{-1} ATU were included in synthetic media in accordance with Section 2.5.3.3. Because the test was executed at 15 °C, the temperature of the synthetic media was also adjusted to 15 °C in a water bath operated at the same temperature before addition. Samples were collected every 5 min in the first 30 min of the aerobic phase. Immediately after collection, all the samples were prepared and preserved as described in Section 2.5.3.4. All the collected samples were analysed as described in Section 2.5.2.4. Table 2.5.2 shows the experimental implementation plan of the test.

2.5.6.2 Data analysis

Following the results of the batch activity test shown in Table 2.5.2, Figure 2.5.2 displays the estimation of the maximum volumetric OHO kinetic rates by applying linear regression.

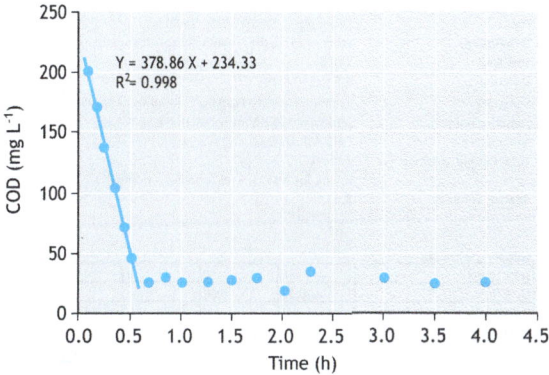

Figure 2.5.2 Graphic representation of the data obtained in the example of an experimental implementation plan for the execution of a batch activity test (Type OHO.AER.1) performed with a full-scale activated sludge at 15 °C using synthetic influent at pH 7.0. The main trend line shows the RBCOD conversion rate for the further estimation of the maximum specific kinetic rate.

Based on the maximum volumetric COD removal rate showed in Figure 2.5.2 and taking into account the MLVSS concentration of 1,890 mg L^{-1} (as an average MLVSS concentration observed between sludge samples collected at the beginning and end of the test), a specific RBCOD removal rate of 0.20 g COD g VSS^{-1} h^{-1} can be estimated, which corresponds to about 4.81 g RBCOD g VSS^{-1} d^{-1}. Using a COD-to-VSS ratio for the biomass of 1.42 g COD g VSS^{-1} (Metcalf and Eddy, 2003), a q$_{OHO,COD,Ox}$ of 3.39 g RBCOD g VSS^{-1} d^{-1} can be computed. Furthermore, since the test was not executed at 20 °C but at 15 °C, using a typical Arrhenius coefficient of 1.07 for OHO activity (Metcalf and Eddy, 2003), a q$_{OHO,COD,Ox}$ of 4.75 g RBCOD g COD^{-1} d^{-1} can be determined (3.39 g RBCOD g VSS^{-1} d^{-1} / 1.07$^{(15-20)}$). This value is in the range of other maximum specific kinetic rates reported for aerobic organic matter removal processes. It is important to note that the synthetic influent contained 400 mg COD L^{-1}, and the COD concentration measured 5 min after the start of the test was 203 mg COD L^{-1} due to the dilution of the 1.25 L of synthetic influent with the 1.25 L of activated sludge (which did not contain RBCOD) and the rapid kinetic removal rate of the biomass. Moreover, it can also be noticed that: (*i*) the initial COD concentration in the activated sludge prior to the wastewater addition was around 56 mg L^{-1} and, (*ii*) the final COD concentration at the end of the test remained at around 30 mg COD L^{-1}. The latter concentration corresponds to the soluble non-biodegradable COD present in the activated sludge sample which cannot be removed by biological means (as explained earlier in this chapter).

Table 2.5.2 Example of an experimental implementation plan for the execution of a batch activity test (Type OHO.AER.1) performed with a full-scale activated sludge at 15 °C using synthetic influent at pH 7.0.

Aerobic COD removal batch tests — Code: OHO.AER.1

Date:	Thursday 09.10.2015 10:00 h
Description:	Tests at 25 °C, pH 7, synthetic substrate and full-scale sample
Test No.:	1 of 6
Duration	4,0 h (240 min)
Substrate:	Synthetic: Acetate (350 mg L^{-1}) + mineral solution with N and P
Sampling point:	Middle mixed liquor height in the SBR
Samples No.:	OHO.AER.1(1-22)
Total sample volume:	222 mL (10 mL for MLVSS, 6 mL for other samples)
Reactor volume:	2.5 L

Experimental procedure in short:	Time (h:min)
1. Confirm the availability of sampling material and required equipment.	08:00
2. Confirm calibration and functionality of systems, meters and sensors.	08:10
3. Transfer 1.25 L of sludge to batch reactor.	08:30
4. Start with gentle mixing and air sparging at T and pH set points.	08:40
5. 20 min before starting take sample for initial conditions (OHO.AER.1.1).	09:40
7. Start the test: add 1.25 L synthetic influent (minute zero).	10:00
8. Minute 5, continue sampling program according to schedule.	10:05
9. Minute 240, stop aeration and mixing.	14:00
10. Organize the samples and clean the system.	14:15
11. Verify that all systems are swtiched off.	14:20

Sampling schedule

Time (min)	-20	0	5	10	15	20	25	30	40	50	60	75
Time (h)	-0.33	0.00	0.08	0.17	0.25	0.33	0.42	0.50	0.67	0.83	1.00	1.25
Sample No.	1		2	3	4	5	6	7	8	9	10	11
Parameter						AEROBIC PHASE						
HAc (mg L^{-1})	56[1]		203	195.5	184	173	158.5	143.5	135.5	129	119.5	83
PO$_4$-P (mg L^{-1})	1.2[1]											
NH$_4$-N (mg L^{-1})	5.3[1]											
MLSS and MLVSS (mg L^{-1})	See table											See table

[1] Average value of the concentration present in the synthetic substrate and and liquid phase of the sludge sample prior to the start of the test.

Sampling schedule (continued)

Time (min)	90	105	120	135	150	165	180	195	210	225	240
Time (h)	1.50	1.75	2.00	2.25	2.50	2.75	3.00	3.25	3.50	3.75	4.00
Sample No.	12	13	14	15	16	17	18	19	20	21	22
Parameter						AEROBIC PHASE					
HAc (mg L^{-1})	73	63	43	38	28	30.5	33	25.5	28	26.5	30
PO$_4$-P (mg L^{-1})											1.8
NH$_4$-N (mg L^{-1})											8.3
MLSS and MLVSS (mg L^{-1})											See table

MLSS & MLVSS measurements

Sampling point	Cup No.	W1	W2	W3	W2-W1	W2-W3	MLSS	MLVSS	Ratio
Start aerobic phase[2]	1	0.08835	0.10741	0.08849	0.01906	0.01892	1,906	1,892	0.99
	2	0.08835	0.10759	0.09018	0.01924	0.01742	1,924	1,742	0.91
	3	0.08834	0.10683	0.08940	0.01849	0.01742	1,849	1,742	0.94
						Average	1,893	1,792	0.95
End aerobic phase	4	0.08868	0.10758	0.08934	0.01890	0.01824	1,890	1,824	0.97
	5	0.08764	0.10617	0.08874	0.01853	0.01742	1,853	1,742	0.94
	6	0.08722	0.10648	0.08973	0.01926	0.01675	1,926	1,675	0.87
						Average	1,890	1,747	0.93

[2] Sample taken before substrate addition.

Biomass composition

Sampling point	Start Aer.	End Aer.
MLSS (mg L^{-1})	1,893	1,890
MLVSS (mg L^{-1})	1,792	1,747
Ratio	0.95	0.93
Ash (mg L^{-1})	101	142

Note:
Acetate (CH$_2$O) — 30.03 mg C-mmol^{-1}
Orthophosphate (PO$_4^{3-}$-P) — 31.00 mg P-mmol^{-1}
Ammonium (NH$_4^+$-N) — 14.00 mg N-mmol^{-1}

The presence of nitrogen (as ammonia) and phosphorus (as orthophosphate) at the beginning (of around 5.3 and 1.2 mg L^{-1} in the activated sludge, respectively) and end of the test (of 8.3 and 1.8 mg L^{-1}, accordingly) indicates that these micronutrients were not limiting. Actually, the concentrations are higher at the end of the test because the synthetic influent contained around 28 mg NH$_4$-N L^{-1} and 8 mg PO$_4$-P L^{-1}. Thus, after dilution and nutrient consumption for biomass synthesis, 8.3 mg NH$_4$-N L^{-1} and 1.8 mg PO$_4$-P L^{-1} remained at the end of the aerobic test. Should the actual nitrogen (N$_{req}$) and phosphorus (P$_{req}$) requirements be known, then they can be estimated with the following expressions:

$$N_{req} = \frac{Y_{COD} \cdot RBCOD_{removed} \cdot N_S}{f_{CV}} \qquad \text{Eq. 2.5.2}$$

$$P_{req} = \frac{Y_{COD} \cdot RBCOD_{removed} \cdot P_S}{f_{CV}} \qquad \text{Eq. 2.5.3}$$

Where, Y$_{OHO}$ is the biomass growth yield (in COD-biomass COD-substrate^{-1}), RBCOD$_{removed}$ is the concentration of RBCOD removed in the test (taking into account the initial COD concentration and the potential dilution effects), N$_S$ is the nitrogen requirement for biomass growth assumed to be around 0.10 g N g VSS^{-1}, P$_S$ is the phosphorus requirement for biomass growth assumed to be around 0.03 g P g VSS^{-1}, f$_{CV}$ is the COD-to-VSS ratio of the sludge, commonly assumed to be 1.42 or 1.48 g COD g VSS^{-1} (Metcalf and Eddy, 2003; Ekama and Wentzel, 2008a).

2.5.7 Additional considerations and recommendations

2.5.7.1 Simultaneous storage and microbial growth

Different studies performed in lab- and full-scale systems have documented the simultaneous occurrence of RBCOD storage and microbial growth under aerobic conditions (van Loosdrecht et al., 1997; Beun et al., 2000; Dircks et al., 2001; Martins et al., 2003; Sin et al., 2005). Under such conditions and when OHOs are exposed to high substrate gradients (like those observed in plug-flow reactors or in aerobic selectors in activated sludge systems), part of the RBCOD present in the bulk liquid is transported through the cell membrane and stored as intracellular polymers like PHA for their further use for microbial growth when the substrate concentrations are low or exhausted (van Loosdrecht et al., 1997; Beun et al., 2000, Dircks et al., 2001); the remaining RBCOD fraction that is not intracellularly stored is directly used for biomass growth (Sin et al., 2005). Such a combination of processes will lead to apparently higher Y$_{OHO}$ values since the aerobic RBCOD storage processes apparently require less oxygen than the direct growth on RBCOD (at least when the substrate storage process takes place). The role and importance of storage processes in activated sludge systems has been significantly recognized to such a degree that different models have been developed and discussed to provide a better description of the simultaneous occurrence of storage and growth in full-scale systems (Gujer et al., 1999; Sin et al., 2005; Guisasola et al., 2005, van Loosdrecht et al., 2015). For the execution of aerobic matter removal tests focused on the determination of the kinetic removal rates, the simultaneous occurrence of these processes should not have direct practical implications. However, the reader should be aware that they will influence the oxygen uptake rate profiles and lead to certain deviations in the suggested Y$_{OHO}$ values as thoroughly described elsewhere (Gujer et al., 1999).

2.5.7.2 Lack of nutrients

Though it may seem trivial, the presence of macro- and micro-nutrients in the right concentrations is essential for successful operation of an activated sludge system. These are usually present in sufficient amounts in most municipal wastewaters, but their presence should be checked and confirmed particularly if the sewage plant is regularly receiving industrial effluents. For this purpose, a simple estimation of the nutrient requirements of the biomass as a function of the COD removal concentration (as suggested in this chapter) can be performed and compared with the influent nutrient concentrations to assess whether there are sufficient nutrients available to cover the biological growth requirements. Lack of such nutrients can lead to severe problems including pin floc and filamentous bulking sludge (Eikelboom, 2000; Martins, 2004), affecting the effluent quality and, in extreme cases, can lead to the failure of the activated sludge system.

2.5.7.3 Toxicity or inhibition

The abundance of OHOs in activated sludge systems is so broad and diverse that they can acclimatise and adapt to different environmental and operating conditions and even stand the presence of different potentially toxic or inhibitory compounds, particularly proceeding from industrial activities. Although they can acclimatise and adapt to such diverse conditions, the inhibitory effects can lead to sub-optimal microbial process activity. To assess such potential effects, two batch activity tests can be executed: one with the original sludge and another one with the same sludge but washed in a mineral solution to remove potentially inhibiting or toxic compounds (as explained in Section 2.5.3.5). To compare the results obtained from each batch activity test, it will be very useful to assess whether the presence of certain compounds inhibits the activity of OHOs. However, one should be aware that such an approach may only be successful if the inhibiting effects are (rapidly) reversible.

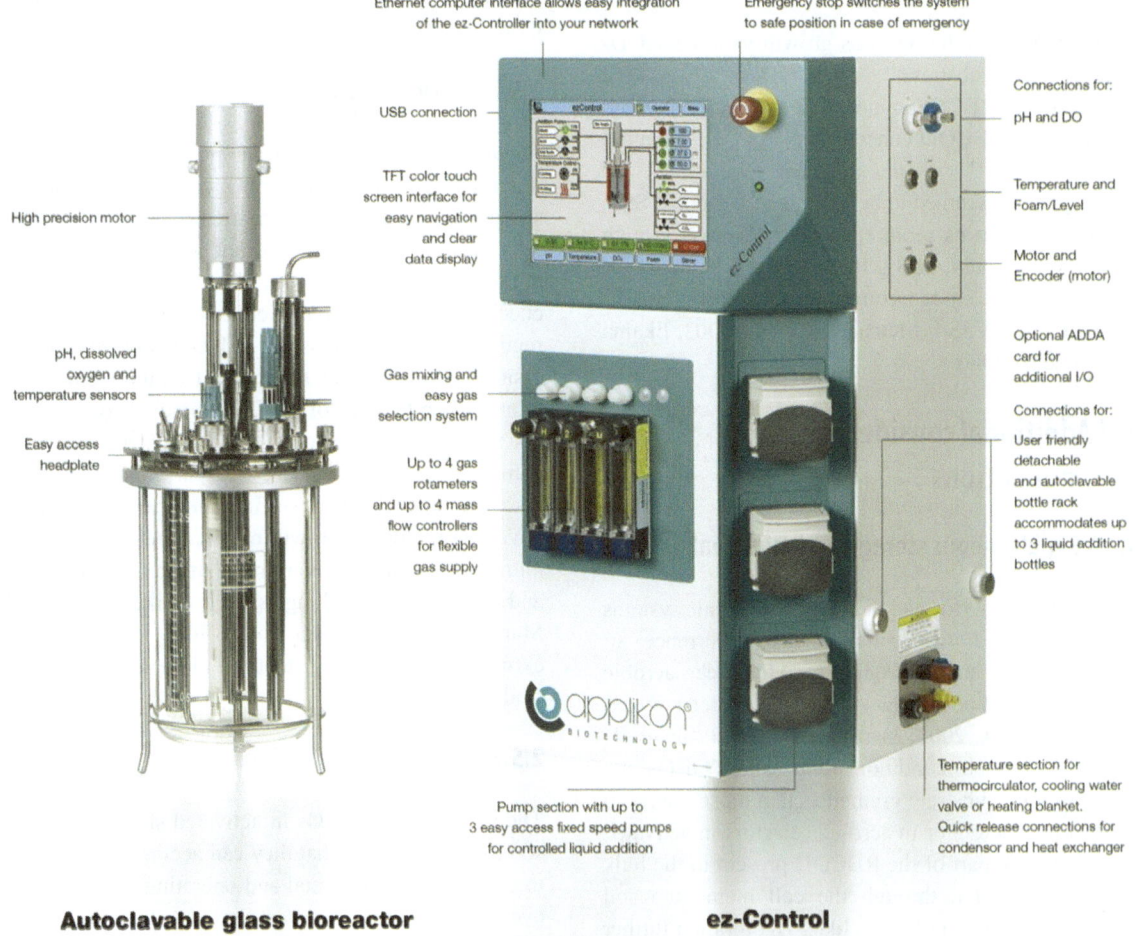

Figure 2.5.3 A high-tech bioreactor system for batch and continuous tests with microbial cultures (photo: Applikon Biotechnology B.V., 2016)

Annex I: Unit conversion coefficients

Table A1 Conversion factors (CF) for net-conversions, stoichiometric and kinetic parameters for carbon, phosphorus, nitrogen and sulphur.

Parameter	Unit (mg basis)	Unit (mole basis)	CF	Unit (COD basis)	CF
ANAEROBIC PARAMETERS					
Phosphate released/HAc uptake ratio	mg PO_4 mg $C_2H_4O_2^{-1}$	P-mol C-mol^{-1}	0.32	mg PO_4 mg COD^{-1}	0.95
Glycogen utilization/HAc uptake ratio	mg $(C_6H_{10}O_5)_n$ mg $C_2H_4O_2^{-1}$	C-mol C-mol^{-1}	1.11	mg COD mg COD^{-1}	1.12
PHB formation/HAc uptake ratio	mg $(C_4H_6O_2)_n$ mg $C_2H_4O_2^{-1}$	C-mol C-mol^{-1}	1.40	mg COD mg COD^{-1}	1.57
PHV formation/HAc uptake ratio	mg $(C_5H_8O_2)_n$ mg $C_2H_4O_2^{-1}$	C-mol C-mol^{-1}	1.50	mg COD mg COD^{-1}	1.82
PH$_2$MV formation/HAc uptake ratio	mg $(C_6H_{10}O_2)_n$ mg $C_3H_6O_2^{-1}$	C-mol C-mol^{-1}	1.58	mg COD mg COD^{-1}	2.01
Phosphate release/HPr uptake ratio	mg PO_4 mg $C_3H_6O_2^{-1}$	P-mol C-mol^{-1}	0.80	mg PO_4 mg COD^{-1}	0.65
Glycogen utilization/HPr uptake ratio	mg $(C_6H_{10}O_5)_n$ mg $C_3H_6O_2^{-1}$	C-mol C-mol^{-1}	0.92	mg COD mg COD^{-1}	0.77
PHB formation/HPr uptake ratio	mg $(C_4H_6O_2)_n$ mg $C_3H_6O_2^{-1}$	C-mol C-mol^{-1}	1.15	mg COD mg COD^{-1}	1.08
PHV formation/HPr uptake ratio	mg $(C_5H_8O_2)_n$ mg $C_3H_6O_2^{-1}$	C-mol C-mol^{-1}	1.23	mg COD mg COD^{-1}	1.25
PH$_2$MV formation/HPr uptake ratio	mg $(C_6H_{10}O_2)_n$ mg $C_3H_6O_2^{-1}$	C-mol C-mol^{-1}	1.30	mg COD mg COD^{-1}	1.39
PHV formation/PHB formation	mg $(C_5H_8O_2)_n$ mg $(C_4H_6O_2)_n^{-1}$	C-mol C-mol^{-1}	1.08	mg COD mg COD^{-1}	1.16.
Sulphate reduction/HAc uptake ratio	mg SO_4 mg $C_2H_4O_2^{-1}$	S-mol C-mol^{-1}	0.31	mg SO_4 mg COD^{-1}	0.95
Sulphate reduction/HPr uptake ratio	mg SO_4 mg $C_3H_6O_2^{-1}$	S-mol C-mol^{-1}	0.26	mg SO_4 mg COD^{-1}	0.65
Active biomass formation/Sulphate reduction	mg $CH_{2.09}O_{0.54}N_{0.20}P_{0.02}$ mg SO_4^{-1}	C-mol S-mol^{-1}	3.70	mg COD mg SO_4^{-1}	1.52
ANOXIC PARAMETERS					
PHB degradation/Nitrate removal	mg $(C_4H_6O_2)_n$ mg NO_3^{-1}	C-mol N-mol^{-1}	2.88	mg COD mg NO_3^{-1}	1.66
PHV degradation/Nitrate removal	mg $(C_5H_8O_2)_n$ mg NO_3^{-1}	C-mol N-mol^{-1}	3.10	mg COD mg NO_3^{-1}	1.92
PH$_2$MV degradation/Nitrate removal	mg $(C_6H_{10}O_2)_n$ mg NO_3^{-1}	C-mol N-mol^{-1}	3.26	mg COD mg NO_3^{-1}	2.12
Glycogen formation/Nitrate removal	mg $(C_6H_{10}O_5)_n$ mg NO_3^{-1}	C-mol N-mol^{-1}	2.30	mg COD mg NO_3^{-1}	1.18
Poly-P formation/Nitrate removal	mg $(PO_3Mg_{0.33}K_{0.33})_n$ mg NO_3^{-1}	P-mol N-mol^{-1}	0.62	-	-
Active biomass formation/Nitrate removal	mg $(CH_{2.09}O_{0.54}N_{0.20}P_{0.02})$ mg NO_3	C-mol N-mol^{-1}	2.38	mg COD mg NO_3^{-1}	1.52
Methanol degradation/Nitrate removal	mg CH_4O mg NO_3^{-1}	C-mol N-mol^{-1}	1.92	mg COD mg NO_3^{-1}	1.98
Ethanol degradation/Nitrate removal	mg C_2H_6O mg NO_3^{-1}	C-mol N-mol^{-1}	2.70	mg COD mg NO_3^{-1}	2.06
Acetate degradation/Nitrate removal	mg $C_2H_4O_2$ mg NO_3^{-1}	C-mol N-mol^{-1}	2.08	mg COD mg NO_3^{-1}	1.06
PHB degradation/Nitrite removal	mg $(C_4H_6O_2)_n$ mg NO_2^{-1}	C-mol N-mol^{-1}	2.14	mg COD mg NO_2^{-1}	1.66
PHV degradation/Nitrite removal	mg $(C_5H_8O_2)_n$ mg NO_2^{-1}	C-mol N-mol^{-1}	2.30	mg COD mg NO_2^{-1}	1.92
Anoxic PH$_2$MV degradation/Nitrite removal	mg $(C_6H_{10}O_2)_n$ mg NO_2^{-1}	C-mol N-mol^{-1}	2.42	mg COD mg NO_2^{-1}	2.12
Glycogen formation/Nitrite removal	mg $(C_6H_{10}O_5)_n$ mg NO_2^{-1}	C-mol N-mol^{-1}	1.71	mg COD mg NO_2^{-1}	1.18
Poly-P formation/Nitrite removal	mg $(PO_3Mg_{0.33}K_{0.33})_n$ mg NO_2^{-1}	P-mol N-mol^{-1}	0.46	-	-
Active biomass formation/Nitrite removal	mg $CH_{2.09}O_{0.54}N_{0.20}P_{0.02}$ mg NO_2^{-1}	C-mol N-mol^{-1}	1.77	mg COD mg NO_2^{-1}	1.52
Methanol degradation/Nitrite removal	mg CH_4O mg NO_2^{-1}	C-mol N-mol^{-1}	1.43	mg COD mg NO_2^{-1}	1.98
Ethanol degradation/Nitrite removal	mg C_2H_6O mg NO_2^{-1}	C-mol N-mol^{-1}	2.00	mg COD mg NO_2^{-1}	2.06
Acetate degradation/Nitrite removal	mg $C_2H_4O_2$ mg NO_2^{-1}	C-mol N-mol^{-1}	1.54	mg COD mg NO_2^{-1}	1.06
Nitrite removal/Ammonium removal	mg NO_2 mg NH_4^{-1}	N-mol N-mol^{-1}	0.39	mg NO_2 mg NH_4^{-1}	-
Nitrate removal/Ammonium removal	mg NO_3/mg NH_4^{-1}	N-mol N-mol^{-1}	0.29	mg NO_3 mg NH_4^{-1}	-
Act. biom. formation/Ammonium cons.	mg$(CH_{2.09}O_{0.54}N_{0.20}P_{0.02})$ mg NH_4^{-1}	C-mol N-mol^{-1}	0.69	mg COD mg NH_4^{-1}	-
AEROBIC PARAMETERS					
PHB degradation/Oxygen consumption	mg $(C_4H_6O_2)_n$ mg O_2^{-1}	C-mol mol O_2^{-1}	0.37	mg COD mg O_2^{-1}	1.66
PHV degradation/Oxygen consumption	mg $(C_5H_8O_2)_n$ mg O_2^{-1}	C-mol mol O_2^{-1}	0.40	mg COD mg O_2^{-1}	1.92
PH$_2$MV degradation/Oxygen consumption	mg $(C_6H_{10}O_2)_n$ mg O_2^{-1}	C-mol mol O_2^{-1}	0.42	mg COD mg O_2^{-1}	2.12
Glycogen formation/Oxygen consumption	mg $(C_6H_{10}O_5)_n$ mg O_2^{-1}	C-mol mol O_2^{-1}	0.29	mg COD mg O_2^{-1}	1.18
Poly-P formation/Oxygen consumption	mg $(PO_3Mg_{0.33}K_{0.33})_n$ mg O_2^{-1}	P-mol mol O_2^{-1}	0.08	mg $(PO_3Mg_{0.33}K_{0.33})_n$ mg O_2^{-1}	-
Active biom. formation/Oxygen cons.	mg $(CH_{2.09}O_{0.54}N_{0.20}P_{0.02})$ mg O_2^{-1}	C-mol mol O_2^{-1}	0.31	mg COD mg O_2^{-1}	1.52
Active biomass formation/Ammonium cons.	mg $(CH_{2.09}O_{0.54}N_{0.20}P_{0.02})$ mg NH_4^{-1}	C-mol N-mol^{-1}	0.69	mg COD mg NH_4^{-1}	1.52

GENERAL PARAMETERS					
NA	mg P mg C^{-1}	P-mol C-mol^{-1}	0.39	-	-
NA	mg C mg C^{-1}	C-mol C-mol^{-1}	1.00	-	-
NA	mg C mg N^{-1}	C-mol N-mol^{-1}	1.16	-	-
NA	mg P mg N^{-1}	P-mol N-mol^{-1}	0.45	-	-
NA	mg C mg O$_2^{-1}$	C-mol O$_2$-mol^{-1}	1.33	-	-
NA	mg P mg O$_2^{-1}$	P-mol O$_2$-mol^{-1}	1.04	-	-
MOLECULAR WEIGHTS					
HF	mg CH$_2$O$_2$	C-mol	0.022	mg COD	0.35
HAc	mg C$_2$H$_4$O$_2$	C-mol	0.033	mg COD	1.06
HPr	mg C$_3$H$_6$O$_2$	C-mol	0.041	mg COD	1.53
HBr	mg C$_4$H$_8$O$_2$	C-mol	0.045	mg COD	1.80
Lactate	mg C$_3$H$_6$O$_3$	C-mol	0.033	mg COD	1.06
Methanol	mg CH$_4$O	C-mol	0.031	mg COD	1.98
Ethanol	mg C$_2$H$_6$O	C-mol	0.043	mg COD	2.06
Glucose	mg C$_6$H$_{12}$O$_6$	C-mol	0.033	mg COD	1.06
Sulphide	mg S^{2-}	S-mol	0.031	mg COD	2.00
Carbondioxide	mg CO$_2$	C-mol	0.023	-	-
Phosphate	mg PO$_4$	P-mol	0.011	-	-
Nitrogen	mg N$_2$	N-mol	0.071	-	-
Ammonium	mg NH$_4$	N-mol	0.055	-	-
Nitrate	mg NO$_3$	N-mol	0.048	-	-
Nitrite	mg NO$_2$	N-mol	0.065	-	-
Oxygen	mg O$_2$	mol O$_2$	0.031	-	-
Sulphate	mg SO$_4$	S-mol	0.010	-	-
PHB	mg (C$_4$H$_6$O$_2$)$_n$	C-mol	0.046	mg COD	1.66
PHV	mg (C$_5$H$_8$O$_2$)$_n$	C-mol	0.050	mg COD	1.92
PH2MV	mg (C$_6$H$_{10}$O$_2$)$_n$	C-mol	0.053	mg COD	2.12
Glycogen	mg (C$_6$H$_{10}$O$_5$)$_n$	C-mol	0.037	mg COD	1.18
Poly-P	mg (PO$_3$Mg$_{0.33}$K$_{0.33}$)$_n$	P-mol	0.010	-	-
Active biomass	mg (CH$_{2.09}$O$_{0.54}$N$_{0.20}$P$_{0.02}$)	C-mol	0.038	mg COD	1.52
ADDITIONAL FACTORS					
Common active biomass fraction of VSS in EBPR enrichment cultures [a]	mg VSS	mg active biomass	0.80		-

[a] Glycogen content of the sludge varies, dependent on the poly-P content (Welles *et al.*, 2015b).

References

Abdeen, S., Di, W., Hui, L., Chen, G.-H. and van Loosdrecht, M.C.M. (2010). Fecal coliform removal in a sulfate reducing autotrophic denitrification and nitrification integrated (SANI) process for saline sewage treatment. *Water Science and Technology*, 62(11): 2564-2570.

Acevedo, B., Oehmen, A., Carvalho, G., Seco, A., Borrás, L. and Barat, R. (2012). Metabolic shift of polyphosphate-accumulating organisms with different levels of polyphosphate storage. *Water Research*, 46(6): 1889-1900.

Acevedo, B., Borrás, L., Oehmen, A. and Barat, R. (2014). Modelling the metabolic shift of polyphosphate-accumulating organisms. *Water Research*, 65: 235-244.

Al Abbas, F.M., Williamson, C., Bhola, S.M., Spear, J.R., Olson, D.L., Mishra, B. and Kakpovbia, A.E. (2013). Influence of sulfate reducing bacterial biofilm on corrosion behavior of low-alloy, high-strength steel (API-5L X80). *International Biodeterioration & Biodegradation*, 78: 34-42.

Ahn, J., Schroeder, S., Beer, M., McIlroy, S., Bayly, R.C., May, J.W., Vasiliadis, G. and Seviour, R.J. (2007). Ecology of the microbial community removing phosphate from wastewater under continuously aerobic conditions in a sequencing batch bioreactor. *Applied and Environmental Microbiology*, 73(7): 2257-2270.

Almeida, P.F., Almeida, R.C.C., Carvalho, E.B., Souza, E.R., Carvalho, A.S., Silva, C.H.T.P. and Taft, C.A. (2006). Overview of sulfate-reducing bacteria and strategies to control biosulfide generation in oil waters. In: Modern biotechnology in medicinal chemistry and industry, Taft, C.A. (Ed.). Research Signpost, Trivandrum, India.

American Public Health Association (APHA). American Water Works Association (AWWA), and Water Environment Federation (WEF). (2012). Standard methods for the examination of water and wastewater. 22nd edition, ISBN: 0875530133, Washington, D.C.

Arden, E. and Lockett, W.T. (1914). Experiments on the oxidation of sewage without the aid of filters. *Journal of the Society of Chemical Industry*, 33(10): 523-539.

Artiga, P., Gonzalez, F., Mosquera-Corral, A., Campos, J.L., Garrido, J.M., Ficara, E. and Méndez, R. (2005). Multiple analyses reprogrammable titration analyser for the kinetic characterisation of nitrifying and autotrophic denitrifying biomass. *Biochemical Engineering Journal*, 26: 176-183.

Baetens, D., Aurola, A.M., Foglia, A., Dionisi, D. and van Loosdrecht, M.C.M. (2002). Gas chromatographic analysis of polyhydroxybutyrate in activated sludge: a round-robin test. *Water Science and Technology*, 46(1-2): 357-361.

Bale, C.W., Chartrand, P., Degtrov, S.A., Eriksson, G., Hack, K., Ben Mahfoud, R., Melançon, J., Pelton, A.D. and Petersen, S. (2002). FactSage thermochemical software and databases. *Calphad*, 26: 189-228.

Balk, M., Altınbaş, M., Rijpstra, W.I.C, Sinninghe Damsté, J.S. and Stams, A.J.M. (2008). *Desulfatirhabdium butyrativorans* gen. nov., sp. nov., a butyrate-oxidizing, sulfate-reducing bacterium isolated from an anaerobic bioreactor. *International Journal of Systematic and Evolutionary Microbiology*, 58: 110-115.

Barañao, P.A. and Hall, E.R. (2004). Modelling carbon oxidation in CTMP pulp mill activated sludge systems: calibration of ASM3. *Water Science and Technology*, 50(3): 1-10.

Barat, R., Montoya, T., Borras, L., Ferrer, J. and Seco, A. (2008). Interactions between calcium precipitation and the polyphosphate-accumulating bacteria metabolism. *Water Research*, 42(13): 3415-3424.

Barker, P.S. and Dold, P.L. (1995). COD and nitrogen mass balances in activated-sludge systems. *Water Research*, 29(2): 633-643.

Bettazzi, E., Caffaz, S., Vannini, C. and Lubello, C. (2010). Nitrite inhibition and intermediates effects on Anammox bacteria: a batch-scale experimental study. *Process Biochemistry*, 45(4): 573-580.

Beun, J.J., Paletta, F., van Loosdrecht, M.C.M. and Heijnen, J.J. (2000). Stoichiometry and kinetics of Poly-B-hydroxybutyrate metabolism in aerobic, slow growing activated sludge cultures. *Biotechnology and Bioengineering*, 67: 379-389.

Bi, Z., Qiao, S., Zhou, J., Tang, X. and Zhang, J. (2014). Fast start-up of Anammox process with appropriate ferrous iron concentration. *Bioresource Technology*, 170: 506-512.

Bilanovic, D., Battistoni, P., Cecchi, F., Pavan, P. and Mata-Alvarez, J. (1999). Denitrification under high nitrate concentration and alternating anoxic conditions. *Water Research*, 33(15): 3311-3320.

Bjerrum, J., Schwarzenbach, G. and Sillén, L.G. (1958). Stability Constants. Chemical Society, London.

Blackburne, R., Vadivelu, V.M., Yuan, Z. and Keller, J. (2007). Determination of growth rate and yield of nitrifying bacteria by measuring carbon dioxide uptake rate. *Water Environment Research*, 79(12): 2437-2445.

Bock, E., Schmidt, I., Stuven, R. and Zart, D. (1995). Nitrogen loss caused by denitrifying *Nitrosomonas* cells using ammonium or hydrogen as electron donors and nitrite as electron acceptor. *Archives of Microbiology*, 163: 16-20.

Bogaert, H., Vanderhasselt, H., Gernaey, K., Yuan, Z., Thoeye, C. and Verstraete, W. (1997). A new sensor based on pH-effect of the denitrification process. *Journal of Environmental Engineering*, 123: 884-891.

Boyles, S. (1997). The Science of Chemical Oxygen Demand. Technical Information Series, Booklet No. 9, HACH Company, USA, pp: 1-23.

Brdjanovic, D., Hooijmans, C.M., van Loosdrecht, M.C.M., Alaerts, G.J. and Heijnen, J.J. (1996). The dynamic effects of potassium limitation on biological phosphorus removal. *Water Research*, 30(10): 2323-2328.

Brdjanovic, D., van Loosdrecht, M.C.M., Hooijmans, C.M., Alaerst, G. J. and Heijnen, J. J. (1997). Temperature effects on physiology of biological phosphorous removal systems. *ASCE Journal of Environmental Engineering*, 123: 144-154.

Brdjanovic, D. (1998a). *Modelling biological phosphorous removal in activated sludge systems*. PhD Thesis. Delft University of Technology, ISBN: 9054104155, Balkema Publishers, Rotterdam, the Netherlands.

Brdjanovic, D., Logemann, S., van Loosdrecht, M.C.M., Hooijmans, C.M., Alaerts, G.J. and Heijnen, J.J. (1998b). Influence of temperature on biological phosphorus removal: process and molecular ecological studies. *Water Research*, 32(4): 1035-1048.

Brdjanovic, D., Slamet, A., van Loosdrecht, M.C.M., Hooijmans, C.M., Alaerts, G.J. and Heijnen, J.J. (1998c). Impact of excessive aeration on biological phosphorus removal from wastewater. *Water Research*, 32(1): 200-208.

Brdjanovic, D., van Loosdrecht, M.C.M., Veersteeg, P., Hooijmans, C.M., Alaerts, G.J. and Heijnen, J.J. (2000). Modelling COD, N and P removal in a full-scale WWTP Haarlem Waarderpolder. *Water Research*, 34: 846-858.

Burow, L.C., Kong, Y., Nielsen, J.L., Blackall, L.L. and Nielsen, P.H. (2007). Abundance and ecophysiology of *Defluviicoccus spp.*, glycogen-accumulating organisms in full-scale wastewater treatment processes. *Microbiology*, 153(1): 178-185.

Butlin, K.R., Selwyn, S.C. and Wakerley, D.S. (1956). Sulfide production from sulphate-enriched sewage sludges. *Journal of Applied Bacteriology*, 19(1): 3-15.

Caccavo, F.Jr., Frolund, B., van Ommen Kloeke, F. and Nielsen, P.H. (1996). Deflocculation of activated sludge by the dissimilatory Fe(III)-reducing bacterium *Shewanella alga* BrY. *Applied and Environmental Microbiology*, 62: 1487-1490.

Carvajal-Arroyo, J.M., Sun, W., Sierra-Alvarez, R. and Field, J.A. (2013). Inhibition of anaerobic ammonium oxidizing (Anammox) enrichment cultures by substrates, metabolites and common wastewater constituents. *Chemosphere*, 91: 22-27.

Carvalho, G., Lemos, P.C., Oehmen, A. and Reis, M.A. (2007). Denitrifying phosphorus removal: linking the process performance with the microbial community structure. *Water Research*, 41(19): 4383-4396.

Carvalheira, M., Oehmen, A., Carvalho, G., Eusébio, M. and Reis, M.A. (2014). The impact of aeration on the competition between polyphosphate accumulating organisms and glycogen accumulating organisms. *Water Research*, 66: 296-307.

Cao, Y., Ang, C., Chua, K., Woo, F., Chi, H., Bhawna, B., Chong, C.T., Ganesan, N., Ooi, K.E. and Wah, Y. (2009). Enhanced biological phosphorus removal in the retrofitting from an anoxic selector to an anaerobic selector in a full-scale activated sludge process in Singapore. *Water Science and Technology*, 59(5): 857-865.

Cech, J.S. and Hartman, P. (1993). Competition between polyphosphate and polysaccharide accumulating bacteria in enhanced biological phosphate removal systems. *Water Research*, 27(7): 1219-1225.

Chen, G.-H., Brdjanovic, D., Ekama, G.A. and van Loosdrecht, M.C.M. (2010). Seawater as alternative water resource. In: Proceedings of the 7th IWA Leading Edge Technology Conference on Water and Wastewater Treatment, Arizona, USA, June 2-4.

Chen, H., Yu, J.-J., Jia, X.-Y., and Jin, R.-C. (2014). Enhancement of Anammox performance by Cu(II), Ni(II) and Fe(III) supplementation. *Chemosphere*, 117 (1): 610-616.

Comeau, Y., Hall, K.J., Hancock, R.E.W. and Oldham, W.K. (1986). Biochemical model for enhanced biological phosphorus removal. *Water Research*, 20(12): 1511-1521.

Comeau, Y., Rabionwitz, B., Hall, K.J and Oldham, W.K. (1987). Phosphate release and uptake in enhanced biological phosphorus removal from wastewater. *Journal (Water Pollution Control Federation)*, 59(7): 707-715.

Comeau Y. (2008). Microbial metabolism. In: Biological wastewater treatment: principles, modelling and design, Henze, M, van Loosdrecht, M.C.M., Ekama, G.A., Brdjanovic, D. (Eds.), ISBN: 9781843391883, IWA Publishing. London, UK.

Corominas, L., Rieger, L., Takács, I., Ekama, A.G., Hauduc, H., Vanrolleghem, P., Oehmen, A., Gernaey, K., van Loosdrecht M.C.M. and Comeau, Y. (2010). New framework for standardized notation in wastewater treatment modelling. *Water Science and Technolology*, 61(4): 841-857.

Cravo-Laureau, C., Matheron, R., Joulian, C., Cayol, J.-L. and Hirschler-Réa, A. (2004). *Desulfatibacillum alkenivorans* sp. nov., a novel nalkene-degrading, sulfate-reducing bacterium, and emended description of the genus *Desulfatibacillum*. *International Journal of Systematic and Evolutionary Microbiology*, 54:1639-1642.

Cypionka. (1987). Uptake of sulfate, sulfite and thiosulfate by proton-anion symport in *Desulfovibrio desulfuricans*. *Archives of Microbiology*, 148(2): 144-149.

Cypionka, H. (2000). Oxygen respiration by *Desulfovibrio* species. *Annual Reviews of Microbiology*, 54: 827-848.

Dai, Y., Yuan, Z., Wang, X., Oehmen, A. and Keller, J. (2007). Anaerobic metabolism of *Defluviicoccus vanus* related glycogen accumulating organisms (GAOs) with acetate and propionate as carbon sources. *Water Research*, 41(9): 1885-1896.

Daigger, G.T., Hodgkinson, A., Aquilina, S. and Fries, M.K. (2015). Development and implementation of a novel sulfur removal process from H_2S containing wastewaters. *Water Environment Research*, 87(7): 618-625.

Dapena-Mora, A., Fernandez, I., Campos, J.L., Mosquera-Corral, A., Mendez, R. and Jetten M.S.M. (2007). Evaluation of activity and inhibition effects on Anammox process by batch tests based on the nitrogen gas production. *Enzyme and Microbial Technology*, 40(4): 859-865.

de Graaff, M.S., Temmink, H., Zeeman, G., van Loosdrecht, M.C.M. and Buisman C.J.N. (2011). Autotrophic nitrogen removal from black water: calcium addition as a requirement for settleability. *Water Research*, 45: 63-74.

Dircks, K., Beun, J.J., van Loosdrecht, M.C.M., Heijnen, J.J. and Henze, M. (2001). Glycogen metabolism in aerobic mixed cultures. *Biotechnology and Bioengineering*, 73(2): 85-94.

Dircks, K., Henze, M., van Loosdrecht, M.C.M., Mosbaek, H. and Aspegren, H. (2001). Storage and degradation of poly-B-hydroxybutyrate in activated sludge under aerobic conditions. *Water Research*, 35: 2277-2285.

Dircks, K. Pind, P.F., Mosbaek, H. and Henze, M. (1999). Yield determination by respirometry - The possible influence of storage under aerobic conditions in activated sludge: *Water SA*, 25: 69-74.

Dosta, J., Fernandez, I., Vazquez-Padin, J. R., Mosquera-Corral, A., Campos, J.L., Mata-Alvarez, J. and Mendez, R. (2008). Short- and long-term effects of temperature on the Anammox process. *Journal of Hazardous Materials*, 154(1): 688-693.

Eckenfelder, W. (1986). Operation control and management of activated sludge plants treating industrial wastewaters. *Proceedings of a seminar sponsored by Vanderbilt University.* Tennessee.

Eikelboom, D.H. (2000). Process control of activated sludge plants by microscopic investigation. ISBN-13: 9781780406831, IWA Publishing, London, UK.

Ekama, G.A. and Wentzel, M.C. (2008a). Organic matter removal. In: Biological wastewater treatment: principles, modelling and design, Henze, M., van Loosdrecht, M.C.M., Ekama, G.A. and Brdjanovic, D. (Eds.), ISBN: 9781843391883. IWA Publishing, London, UK.

Ekama, G.A. and Wentzel, M.C. (2008b). Nitrogen removal. In: Biological wastewater treatment: principles, modelling and design, Henze, M., van Loosdrecht, M.C.M., Ekama, G.A. and Brdjanovic, D. (Eds.), ISBN: 9781843391883, IWA Publishing. London, UK.

Ekama, G.A., Wilsenach, J.A. and Chen, G.-H. (2010). Some opportunities and challenges for urban wastewater treatment. In: 7th IWA LET conference, Arizona, USA, June 2-4 (keynote presentation), Retrieved in September 2015 from http://repository.ust.hk/ir/Record/1783.1-16616

Ettwig, K.F., Butler, M.L., Le Paslier, D., Pelletier, E., Mangenot, S., Kuypers, M.M.M., Schreiber, F., Dutilh, B.E., Zedelius, J., de Beer, D., Gloerich, J., Wessels, H.J.C.T., van Alen, T., Luesken, F., Wu, M.L., van de Pas-Schoonen, K.T., Op den Camp, H.J.M., Janssen-Megens, E.M., Francoijs, K-J., Stunnenberg, H., Weissenbach, J., Jetten, M.S.M. and Strous, M. (2010). Nitrite-driven anaerobic methane oxidation by oxygenic bacteria. *Nature*, 464, 543-548.

Ficara, E. and Canziani, R. (2007). Monitoring denitrification by pH-static titration. *Biotechnology and Bioengineering*, 98(2): 368-377.

Ficara, E. and Rozzi, A. (2004). Coupling pH-static and DO-stat titration to monitor degradation of organic substrates. *Water Science and Technology*, 49(1): 69-77.

Ficara, E., Cortelezzi, P. and Rozzi, A. (2003). Theory of pH-stat titration, *Biotechnology and Bioengineering*, 82: 28-37.

Ficara, E., Musumeci, A. and Rozzi, A. (2000). Comparison and combination of titrimetric and respirometric techniques to estimate nitrification kinetics parameters. *Water SA*, 26(2): 217-224.

Ficara, E., Sambusiti, C. and Canziani, R. (2009). Manometric monitoring of biological denitrification. In: Proceeding of the 2nd IWA Specialized Conference in Nutrients Management in Wastewater Treatment Processes, Krakow, Poland, 6-9 September, 2009. Lemtech Konsulting (Ed.), Krakow. pp: 61-68.

Filipe, C.D., Daigger, G.T. and Grady, Jr, C.P. (2001a). A metabolic model for acetate uptake under anaerobic conditions by glycogen accumulating organisms: stoichiometry, kinetics, and the effect of pH. *Biotechnology and Bioengineering*, 76(1): 17-31.

Filipe, C.D., Daigger, G.T. and Grady, Jr, C.L. (2001b). pH as a key factor in the competition between glycogen-accumulating organisms and phosphorus-accumulating organisms. *Water Environment Research*, 73(2): 223-232.

Flemming, H.C., Wingender, J., Mayer, C., Korstgens, V. and Borchard, W. (2000). Cohesiveness in biofilm matrix polymers. In: Community structure and cooperation of biofilms, Allison, D., Gilbert, P., Lappin-Scott, H.M. and Wilson, M. (Eds.), SGM Symposium Series, 59, Cambridge University Press, Cambridge, UK, pp: 87-105.

Flowers, J.J., He, S., Carvalho, G., Brook, Peterson, S., López, C., Yilmaz, S., Zilles, J.L., Morgenroth, E., Lemos, P., Reis, M.A.M., Crespo, M.T.B., Noguera, D.R. and McMahon, K.D. (2008). Ecological differentiation of *Accumulibacter* in EBPR reactors. In: Proceedings of the Water Environment Federation, WEFTEC 2008, pp: 31-42.

Foxon, K.M., Brouckaert, C.J., Buckley, C.A. and Rozzi, A. (2002). Denitrifying activity measurements using an anoxic titration (pHstat) bioassay. *Water Scence and Technolology*, 46(9): 211-218.

Gayon, U. and Dupetit, G. (1883). La fermentation des nitrates. *Mem. Soc. Sci. Phys. Nat.,* Bordeaux 2(5): 35-36.

Gernaey, K., Bogaert, H., Massone, A., Vanrolleghem, P. and Verstraete, W. (1997). On-line nitrification monitoring in activated sludge with a titrimetric sensor. *Environmental Science and Technology*, 31: 2350-2355.

Gernaey, K., Bogaert, H., van Rolleghem, P., Massone, A., Rozzi, A. and Verstraete, W. (1998). A titration technique for online nitrification monitoring in activated sludge. *Water Science and Technology*, 37(12): 103-110.

Gibson, G.R. (1990). Physiology and ecology of the sulphate-reducing bacteria. *Journal of Applied Bacteriology*, 69(6): 769-797.

Goel, R., Mino, T., Satoh, H. and Matsuo, T. (1999). Modelling hydrolysis processes considering intracellular storage: *Water Science and Technology*, 39: 97-105.

Grady, Jr, L.C.P., Daigger, G.T., Love. N.G., and Filipe, C.D.M. (2011). Biological wastewater treatment. 3rd edition, ISBN 9780849396793, IWA Publishing, CRC Press, London, UK.

Grein, F., Ramos, A.R., Venceslau, S.S. and Pereira, I.A.C. (2013). Unifying concepts in anaerobic respiration: Insights from dissimilatory sulfur metabolism. *Biochimica et Biophysica Acta (BBA) - Bioenergetics*, 1827(2): 145-160.

Guisasola, A., Qurie, M., Vargas, M, Casas, C. and Baeza, J.A. (2009). Failure of an enriched nitrite-DPAO population to use nitrate as an electron acceptor. *Process Biochemistry*, 44: 689-695.

Guisasola, G., Pijuan, M., Baeza, J.A., Carrera, J., Casas, C. and Lafuente, J. (2004). Aerobic phosphorus release linked to acetate uptake in bio-P sludge: process modelling using oxygen uptake rate. *Biotechnology and Bioengineering*, 85: 721-733.

Guisasola, A., Sin, G., Baeza, J.A., Carrera, J. and Vanrolleghem, P.A. (2005). Limitations of ASM 1 and ASM 3: a comparison based on batch oxygen uptake rate profiles from different full-scale wastewater treatment plants. *Water Science and Technology*, 52(10): 69-77.

Gujer, W., Henze, M., Mino, T. and van Loosdrecht, M.C.M. (1999). Activated sludge model No. 3. *Water Science and Technology*, 39(1): 183-193.

Hallin, S., Throback, I.N., Dicksved, J. and Pell, M. (2006). Metabolic profiles and genetic diversity of denitrifying communities in activated sludge after addition of methanol or ethanol. *American Society for Microbiology*, 72(8): 5445-5452.

Hansen, T.A. (1993). Carbon metabolism of sulfate-reducing bacteria. In: The sulfate-reducing bacteria: contemporary perspectives, Odom, J.M., Singleton, Jr, R. (Eds.), ISBN 978-1-4613-9263-7, Brock/Springer Book Series in Contemporary Bioscience, Springer-Verlag, New York Inc.

Hansen, T.A. (1994). Metabolism of sulfate-reducing prokaryotes. *Antonie van Leeuwenhoek*, 66(1-3): 165-185.

Henze, M. (1991). Capabilities of biological nitrogen removal processes from wastewater. *Water Science and Techology*, 23(4-6): 669-679.

Henze, M. (1992). Characterization of wastewater for modeling of activated sludge processes. *Water Science and Technology*, 25(6): 1-15.

Henze, M., Harremoes, P., Jansen, J.L.C. and Arvin, E. (1997). Wastewater treatment: biological and chemical processes. ISBN 978-3-540-42228-0, Springer, Berlin.

Henze, M. and Comeau, Y. (2008). Wastewater characterization. In: Biological wastewater treatment: principles, modelling and design, Henze, M., van Loosdrecht, M.C.M., Ekama, G.A. and Brdjanovic, D. (Eds.), ISBN: 9781843391883, IWA Publishing, London, UK.

Henze, M., Gujer, W., Mino, T., Matsuo, T., Wentzel, M.C., Marais, G.v.R. and van Loosdrecht, M.C.M. (1999). Activated sludge model no. 2d, ASM2d. *Water Science and Technology*, 39(1): 165-182.

Henze, M., Gujer, W., Mino, T. and van Loosdrecht, M.C.M. (2000). Activated sludge models ASM1, ASM2, ASM2d and ASM3. IWA Scientific and Technical Report No. 9, IWA Publishing, London, UK.

Henze, M., van Loosdrecht, M.C.M., Ekama, G.A. and Brdjanovic, D. (2008). Biological wastewater treatment: principles, modelling and design. ISBN: 9781843391883, IWA Publishing, London, UK.

Hoover, S.R. and Porges, N. (1952). Assimilation of dairy wastes by activated sludge II: The equation of synthesis and oxygen utilization. *Sewage and Industrial wastes*, 24(3): 306-312.

Hu, Z., Lotti, T., van Loosdrecht, M.C.M. and Kartal, B. (2013). Nitrogen removal with the anaerobic ammonium oxidation process. *Biotechnology Letters*, 35(8):1145-1154.

Hu, J.Y., Ong, S.L., Ng, W.J., Lu, F. and Fan, X.J. (2003). A new method for characterizing denitrifying phosphorus removal bacteria by using three different types of electron acceptors. *Water Research*, 37(14): 3463-3471.

Hu, Z.R., Wentzel, M.C. and Ekama, G.A. (2002). Anoxic growth of phosphate-accumulating organisms (PAOs) in biological nutrient removal activated sludge systems. *Water Research*, 36(19): 4927-4937.

Hulsbeek, J.J.W., Kruit, J., Roeleveld, P.J. and van Loosdrecht, M.C.M. (2002). A practical protocol for dynamic modelling of activated sludge systems. *Water Science and Technology*, 45(6): 127-136.

Hulshoff Pol, L.W., Lens, P.N.L, Stams, A.J.M. and Lettinga, G. (1998). Anaerobic treatment of sulphate-rich wastewaters. *Biodegradation*, 9(3-4): 213-224.

Isaksen, M.F. and Jørgensen, B.B. (1996). Adaptation of psychrophilic and psychrotrophic sulfate-reducing bacteria to permanently cold marine environments. *Applied Environmental Microbiology*, 62: 408-414.

Jobbagy, A., Literathy, B., Wong, M., Tardy, G. and Liu, W. (2006). Proliferation of glycogen accumulating organisms induced by Fe (III) dosing in a domestic wastewater treatment plant. *Water Science and Technology*, 54(1): 101-109.

Jones, R., Dold, P., Takács, I., Chapman, K., Wett, B., Murthy, S. and Shaughnessy, M. (2007). Simulation for operation and control of reject water treatment processes. In: Proceedings of the Water Environment Federation WEFTEC 2007, pp: 4357-4372.

Jørgensen, B.B. (2006). Bacteria and marine biogeochemistry. In: Marine geochemistry, 2nd edition, Schulz, H.D. and Zabel, M. (Eds.), Springer-Verlag, Berlin Heidelberg, pp: 169-206.

Jubany, I., Carrera, J., Lafuente, J. and Baeza, J.A. (2008). Start-up of a nitrification system with automatic control to treat highly concentrated ammonium wastewater: Experimental results and modeling. *Chemical Engineering Journal*, 144(3): 407-419.

Kampschreur, M.J., Picioreanu, C., Tan, N.C.G., Kleerebezem, R., Jetten, M.S.M. and van Loosdrecht, M.C.M. (2007). Unraveling the source of nitric oxide emission during nitrification. *Water Environment Research*, 79: 2499-2509.

Kamyshny, Jr., A., Zilberbrand, M., Ekeltchik, I., Voitsekovski, T., Gun, J. and Lev, O. (2008). Speciation of polysulfides and zero-valent sulfur in sulfide-rich water wells in southern and central Israel. *Aquatic Geochemistry*, 14: 171-192.

Kappeler, J. and Gujer, W. (1992). Estimation of kinetic parameters of heterotrophic biomass under aerobic conditions and characterization of wastewater for activated sludge modelling. *Water Science and Technology*, 25(6): 125-139.

Kartal, B., Maalcke, W.J., de Almeida, N.M., Cirpus, I., Gloerich, J., Geerts, W., den Camp, H.J.M.O., Harhangi, H.R., Janssen-Megens, E.M., Francoijs, K.-J., Stunnenberg, H.G., Keltjens J.T., Jetten, M.S.M. and Strousm, M. (2011). Molecular mechanism of anaerobic ammonium oxidation. *Nature*, 479: 127-130.

Kerrn-Jespersen, J.P. and Henze, M. (1993). Biological phosphorus uptake under anoxic and aerobic conditions. *Water Research*, 27: 617-624

Khanal, S.K. (2008). Anaerobic Biotechnology for Bioenergy Production: Principles and Applications, Wiley-Blackwell, ISBN: 978-0-8138-2346-1, Iowa, USA.

Kjeldsen, K.U., Joulian, C. and Ingvorsen, K. (2004). Oxygen tolerance of sulfate-reducing bacteria in activated sludge. *Environmental Science & Technology*, 38(7): 2038-2043.

Kjeldsen, K.U., Loy, A., Jakobsen, T.F., Thomsen, T.R., Wagner, M. and Ingvorsen, K. (2007). Diversity of sulfate-reducing bacteria from an extreme hypersaline sediment, Great Salt Lake (Utah). *FEMS Microbiology Ecology*, 60: 287-298.

Koch, G., Egli, K., van der Meer, J.R. and Siegrist, H. (2000). Mathematical modeling of autotrophic denitrification in a nitrifying biofilm of a rotating biological contactor. *Water Science and Technology*, 41(4-5): 191-198.

Koch G., Kuhni M., Gujer W. and Siegrist H. (2000). Calibration and validation of activated sludge model No. 3 for Swiss municipal wastewater. *Water Research*, 34: 3580-3590.

Kong, Y., Nielsen, J.L. and Nielsen, P.H. (2005). Identity and ecophysiology of uncultured actinobacterial polyphosphate-accumulating organisms in full-scale enhanced biological phosphorus removal plants. *Applied and Environmental Microbiology*, 71(7): 4076-4085.

Kong, Y., Xia, Y., Nielsen, J.L. and Nielsen, P.H., (2006). Ecophysiology of a group of uncultured Gammaproteobacterial glycogen-accumulating organisms in full-scale enhanced biological phosphorus removal wastewater treatment plants. *Environmental Microbiolology*, 8(3): 479-489.

Könneke, M., Bernhard, A.E., de la Torre, J.R., Walker, C.B., Waterbury, J.B. and Stahl, D.A. (2005). Isolation of an autotrophic ammonia-oxidizing marine archaeon. *Nature*, 437(7058): 543-6.

Kuba, T., Smolders, G., van Loosdrecht, M.C.M. and Heijnen, J.J. (1993). Biological phosphorus removal from wastewater by anaerobic-anoxic sequencing batch reactor. *Water Science and Technology*, 27(5-6): 241-252.

Kuba, T., Murnleitner, E., van Loosdrecht, M.C.M. and Heijnen, J.J. (1996). A metabolic model for biological phosphorus removal by denitrifying organisms. *Biotechnology and Bioengineering*, 52: 685-695.

Kuba, T., van Loosdrecht, M.C.M., Brandse, F.A. and Heijnen, J.J. (1997a). Occurrence of denitrifying phosphorus removing bacteria in modified UCT-type wastewater treatment plants. *Water Research*, 31(4): 777-786.

Kuba, T., van Loosdrecht, M.C.M., Murnleitner, E. and Heijnen, J.J. (1997b). Kinetics and stoichiometry in the biological phosphorus removal process with short cycle times. *Water Research*, 31(4): 918-928.

Kujawa, K. and Klapwijk, B. (1999). A method to estimate denitrification potential for pre-denitrification systems using NUR batch tests. *Water Research*, 33: 2291-2300.

Kuypers, M.M.M., Lavik, G., Woebken, D., Schmid, M., Fuchs, B.M., Amann, R., Jørgensen, B.B. and Jetten, M.S.M. (2005). Massive nitrogen loss from the Benguela upwelling system through anaerobic ammonium oxidation. *Proceedings of the National Academy of Sciences of the United States of America*, 102: 6478-6483.

Lackner, S., Gilbert, E.M., Vlaeminck, S.E., Joss, A., Horn, H. and van Loosdrecht, M.C.M. (2014). Full-scale partial Nitritation/Anammox experiences - an application survey. *Water Research*, 55: 292-303.

Langergraber, G., Rieger, L., Winkler, S., Alex, J., Wiese, J., Owerdieck, C., Ahnert, M., Simon, J. and Maurer. M. (2004). A guideline for simulation studies of wastewater treatment plants. *Water Science and Technology*, 50(7): 131-138.

Lanham, A.N., Ricardo, A.R., Coma, M., Fradinho, J., Carvalheira, M., Oehmen, A., Carvalho, G. and Reis, M.A.M. (2012). Optimisation of glycogen quantification in mixed microbial cultures. *Bioresource Technology*, 118: 518-525.

Lanham, A.B., Oehmen, A., Saunders, A.M., Carvalho, G., Nielsen, P.H. and Reis, M.A. (2014). Metabolic modelling of full-scale enhanced biological phosphorus removal sludge. *Water Research*, 66: 283-295.

Lau, G.N., Sharma, K.R., Chen, G.-H. and van Loosdrecht, M.C.M. (2006). Integration of sulfate reduction autotrophic denitrification and nitrification to achieve low-cost sludge minimization for Hong Kong sewage. *Water Science and Technology*, 53(3): 227-235.

Lens, P.N., De Poorter, M.-P., Cronenberg, C.C. and Verstraete, W.H. (1995). Sulfate reducing and methane producing bacteria in aerobic wastewater treatment systems. *Water Research*, 29(3): 871-880.

Lens, P.N.L., Visser, A., Janssen, A.J.H., Hulshoff Pol, L.W. and Lettinga, G. (1998). Biotechnological treatment of sulfate-rich wastewaters. *Critical Reviews in Environmental Science and Technology*, 28(1): 41-88.

Li, L., Han, Y., Yan, X. and Liu, J. (2013). H_2S removal and bacterial structure along a full-scale biofilter bed packed with polyurethane foam in a landfill site. *Bioresource Technology*, 147: 52-58.

Liamleam, W. and Annachhatre, A.P. (2007). Electron donors for biological sulfate reduction. *Biotechnology Advances*, 25(5): 452-463.

Lin, Y.M., Bassin, J.P. and van Loosdrecht, M.C.M. (2012). The contribution of exopolysaccharides induced struvites accumulation to ammonium adsorption in aerobic granular sludge. *Water Research*, 46(4): 986-992.

Little, B.J. and Lee, J.S. (2007). Microbiologically influenced corrosion. ISBN 978-0-471-77276-7, John Wiley & Sons Inc., Hoboken, NJ, USA.

Liu, M.C and Peck, Jr., H.D. (1981). The isolation of a hexaheme cytochrome from *Desulfovibrio desulfuricans* and its identification as a new type of nitrite reductase. *Journal of Biological Chemistry*, 256(24): 13159-13164.

Lochtman, S.F.W. (1995). Proceskeuze en -optimalisatie van het SHARON proces voor slibverwerkingsbedrijf Sluisjesdijk (Process choice and optimisation of the SHARON process for the sludge treatment plant Sluisjesdijk). BODL report: TU Delft.

Lopez, C., Pons, M.N. and Morgenroth, E. (2006). Endogenous processes during long-term starvation in activated sludge performing enhanced biological phosphorus removal. *Water Research*, 40(8): 1519-1530.

Lopez-Vazquez, C.M., Song, Y.I., Hooijmans, C.M., Brdjanovic, D., Moussa, M.S., Gijzen, H.J. and van Loosdrecht, M.C.M. (2007). Short term temperature effects on the anaerobic metabolism of glycogen accumulating organisms. *Biotechnology and Bioengineering*, 97(3): 483-495.

Lopez-Vazquez, C.M., Hooijmans, C.M., Brdjanovic, D., Gijzen, H.J. and van Loosdrecht, M.C.M. (2008a). Factors affecting the microbial populations at full-scale enhanced biological phosphorus removal (EBPR) wastewater treatment plants in The Netherlands. *Water Research*, 42(10): 2349-2360.

Lopez-Vazquez, C.M., Brdjanovic, D. and van Loosdrecht, M.C.M. (2008b). Comment on "Could polyphosphate-accumulating organisms (PAOs) be glycogen accumulating organisms (GAOs)?" by Zhou, Y., Pijuan, M., Zeng, R.J, Lu, H. and Yuan. Z. Water Research (2008). doi: 10.1016/j. waterres. 2008.01.003, *Water Research*, 42(13): 3561-3562.

Lopez-Vazquez, C.M. (2009). *The competition between polyphosphate-accumulating organisms and glycogen-accumulating organisms: temperature effects and modelling*. PhD Thesis. ISBN 9780415558969, Delft University of Technology, CRC Press/Balkema, Leiden, the Netherlands.

Lopez-Vazquez, C.M., Hooijmans, C.M., Brdjanovic, D., Gijzen, H.J. and van Loosdrecht, M.C.M. (2009a). Temperature effects on glycogen accumulating organisms. *Water Research*, 43(11): 2852-2864.

Lopez-Vazquez, C.M., Oehmen, A., Hooijmans, C.M., Brdjanovic, D., Gijzen, H.J., Yuan, Z. and van Loosdrecht, M.C.M. (2009b). Modeling the PAO–GAO competition: effects of carbon source, pH and temperature. *Water Research*, 43(2): 450-462.

Lopez-Vazquez, C.M., Kubare, M., Saroj, D.P., Chikamba, C., Schwarz, J., Daims, H. and Brdjanovic, D. (2014). Thermophilic biological nitrogen removal in industrial wastewater treatment. *Applied Microbiology and Biotechnology*, 98(2): 945-956.

Lotti, T., van der Star, W.R., Kleerebezem, R., Lubello, C. and van Loosdrecht, M.C.M. (2012). The effect of nitrite inhibition on the anammox process. *Water Research*, 46(8): 2559-2269.

Lotti, T., Kleerebezem, R., Lubello, C. and van Loosdrecht, M.C.M. (2014). Physiological and kinetic characterization of a suspended cell Anammox culture. *Water Research*, 60(14): 1-14.

Lotti, T., Kleerebezem, R., Hu, Z., Kartal, B., de Kreuk, M.K., van Erp Taalman Kip, C., Kruit, J., Hendrickx, T.L.G. and van Loosdrecht, M.C.M. (2015a). Pilot-scale evaluation of anammox based main-stream nitrogen removal from municipal wastewater. *Environmental Technology*, 36(9): 1167-1177.

Lotti, T., Kleerebezem, R., Abelleira-Pereira, J.M., Abbas, B. and van Loosdrecht M.C.M. (2015b). Faster through training: the Anammox case. *Water Research*, 81: 261-268.

Lotti, T., Kleerebezem, R. and van Loosdrecht, M.C.M. (2015c). Effect of temperature change on Anammox activity. *Biotechnology and Bioengineering*, 112(1): 98-103.

Madigan, M.T., Martinko, J.M., Dunlap, P.V. and Clark, D.P. (2009). Brock biology of microorganisms, 12th edition, ISBN 0-13-232460-1, Pearson Benjamin Cummings, San Francisco, USA.

Mañas, A., Biscans, B. and Spérandio, M. (2011). Biologically induced phosphorus precipitation in aerobic granular sludge process. *Water Research*, 45(12): 3776-3786.

Marais, G.v.R. and Ekama, G.A. (1976). The activated sludge process part 1 - Steady state behaviour. *Water SA*, 2(4): 163-200.

Martins, A.M.P. (2004). *Bulking sludge control: kinetics, substrate storage, and process design aspects.* PhD Thesis. ISBN 972-9098-07-7, Delft University of Technology, Delft, The Netherlands.

Martins, A.M.P., Heijnen, J.J. and van Loosdrecht, M.C.M. (2003). Effect of feeding pattern and storage on sludge settleability under aerobic conditions. *Water Research*, 37(11): 2555-2570.

Massone, A., Antonelli, M. and Rozzi, A. (1996). The DENICON: a novel biosensor to control denitrification in biological wastewater treatment plants. Mededelingen Faculteit Landbouwkundige, University of Gent, 1709-1714.

Massone, A.G., Gernaey, K., Rozzi, A. and Verstraete, W. (1998). Measurement of ammonium concentration and nitrification rate by a new titrometric biosensor. *WEF Research Journal*, 70(3): 343-350.

Mateju, V., Cizinska, S., Krejci, J. and Janoch, T. (1992). Biological water denitrification - a review. *Enzyme and Microbial Technology*, 14: 170-183.

McCarty, P.L. (2007). Thermodynamic electron equivalents model for bacterial yield prediction: Modifications and comparative evaluations. *Biotechnology and Bioengineering*, 97: 377-388.

Meijer, S.C.F. (2004). *Theoretical and practical aspects of modelling activated sludge processes.* PhD Thesis. ISBN 90-9018027-3, Delft University of Technology, Delft, the Netherlands.

Meinhold, J., Arnold, E. and Isaacs, S. (1999). Effect of nitrite on anoxic phosphate uptake in biological phosphorus removal activated sludge. *Water Research*, 33(8): 1871-1883.

Mesquita D.P., Amaral A.L., Ferreira E.C. (2013) Activated sludge characterization through microscopy: a review on quantitative image analysis and chemometric techniques. *Analytica Chimica Acta*, 802: 14-28.

Metcalf and Eddy (2003). Wastewater engineering: treatment, disposal and reuse. 4th edition, ISBN-13:978-0070418783, McGraw-Hill, Boston, USA.

Mino, T., van Loosdrecht, M.C.M. and Heijnen, J.J. (1998). Microbiology and biochemistry of the enhanced biological phosphorus removal process. *Water Research*, 32(11): 3193-3207.

Mino, T., Arun, V., Tsuzuki, Y. and Matsuo, T. (1987). Effect of phosphorus accumulation on acetate metabolism in the biological phosphorus removal process. In: biological phosphate removal from wastewaters, Ramadori, R. (Ed.), Pergamon Press, Oxford, United Kingdom. pp: 27-38.

Mohan, S.V., Rao, N.C., Prasad, K.K. and Sarma, P.N. (2005). Bioaugmentation of an anaerobic sequencing batch biofilm reactor (AnSBBR) with immobilized sulphate reducing bacteria (SRB) for the treatment of sulphate bearing chemical wastewater. *Process Biochemistry*, 40(8): 2849-2857.

Mokhayeri, Y., Nichols, A., Murthy, S., Riffat, R., Dold, P. and Takács, I. (2006). Examining the influence of substrates and temperature on maximum specific growth rate of denitrifiers. *Water Science and Technology*, 54(8): 155-162.

Mokhayeri, Y., Riffat, R., Takács, I., Dold, P., Bott, C., Hinojosa, Bailey, W. and Murthy S. (2008). Characterizing denitrification kinetics at cold temperature using various carbon sources in lab-scale sequencing batch reactors. *Water Science and Technology*, 58(1): 233-238.

Mori, K., Kim, H., Kakegawa, T. and Hanada, S. (2003). A novel lineage of sulfate-reducing microorganisms: *Thermodesulfobiaceae* fam. nov., *Thermodesulfobium narugense*, gen. nov., sp. nov., a new thermophilic isolate from a hot spring. *Extremophiles*, 7: 283-290.

Moussa, M.S., Fuentes, O.G., Lubberding, H.J., Hooijmans, C.M., van Loosdrecht, M.C.M. and Gijzen, H.J. (2006). Nitrification activities in full-scale treatment plants with varying salt loads. *Environmental Technology*, 27(6): 635-643.

Moussa, M.S., Rojas, A.R., Hooijmans, C.M., Gijzen, H.J. and van Loosdrecht, M.C.M. (2004). Model-based evaluation of nitrogen removal in a tannery wastewater treatment plant. *Water Science and Technology*, 50(6): 251-60.

Moussard, H., L'Haridon, S., Tindall, B.J., Banta, A., Schumann, P., Stackebrandt, E., Reysenbach, A.-L., and Jeanthon, C. (2004). *Thermodesulfatator indicus* gen. nov., sp. nov., a novel thermophilic chemolithoautotrophic sulfate-reducing bacterium isolated from the Central Indian Ridge. *International Journal of Systematic and Evolutionary Microbiology*, 54: 227-233.

Mulder, A. (2003). The quest for sustainable nitrogen removal technologies. Water Science and Technology, 48(1): 67-75.

Munro, J.H.M. (1886). The formation and destruction of nitrates and nitrates in artificial solutions and in river and well waters. *Journal of the Chemical Society, Transactions*, 49: 632-681.

Murnleitner, E., Kuba, T., van Loosdrecht, M.C.M. and Heijnen, J.J. (1997). An integrated metabolic model for the aerobic and denitrifying biological phosphorus removal. *Biotechnology and Bioengineering*, 54(5): 434-450.

Muyzer, G. and Stams, A.J.M. (2008). The ecology and biotechnology of sulphate-reducing bacteria. *Nature*, 6:441-455.

Naidoo, V., Urbain, V. and Buckley, C.A. (1998). Characterisation of wastewater and activated sludge from European municipal wastewater treatment plants using the NUR test. *Water Science and Technology*, 38(1): 303-310.

Nelson, D.L. and Cox, M.M. (2005). Lehninger principles of biochemistry. ISBN-10: 1-4292-33414-8, W.H. Freeman and Company, New York, USA.

Nielsen, P.H. and Keiding K. (1998). Disintegration of activated sludge flocs in presence of sulfide. *Water Research*, 32(2): 313-320.

Nielsen, P.H., Mielczarek, A.T., Kragelund, C., Nielsen, J.L., Saunders, A.M., Kong, Y., Hansen, A.A. and Vollertsen, J.

(2010). A conceptual ecosystem model of microbial communities in enhanced biological phosphorus removal plants. *Water Research*, 44(17): 5070-5088.

Oehmen, A., Yuan, Z., Blackall, L.L. and Keller, J. (2004). Short-term effects of carbon source on the competition of polyphosphate accumulating organisms and glycogen accumulating organisms. *Water Science and Technology*, 50(10): 139-144.

Oehmen, A., Vives, M.T., Lu, H., Yuan, Z. and Keller, J. (2005a). The effect of pH on the competition between polyphosphate-accumulating organisms and glycogen-accumulating organisms. *Water Research*, 39(15): 3727-3737.

Oehmen, A., Keller-Lehmann, B., Zeng, R.J., Yuan, Z. and Keller, J. (2005b). Optimisation of poly-β-hydroxyalkanoate analysis using gas chromatography for enhanced biological phosphorus removal systems. *Journal of Chromatography A*, 1070(1-2): 131-136.

Oehmen, A., Yuan, Z., Blackall, L.L. and Keller, J. (2005c). Comparison of acetate and propionate uptake by polyphosphate accumulating organisms and glycogen accumulating organisms. *Biotechnology and Bioengineering*, 91(2): 162-168.

Oehmen, A., Saunders, A.M., Vives, M.T., Yuan, Z. and Keller, J. (2006). Competition between polyphosphate and glycogen accumulating organisms in enhanced biological phosphorus removal systems with acetate and propionate as carbon sources. *Journal of Biotechnology*, 123(1): 22-32.

Oehmen, A., Lemos, P.C., Carvalho, G., Yuan, Z., Keller, J., Blackall, L.L. and Reis, M.A. (2007). Advances in enhanced biological phosphorus removal: from micro to macro scale. *Water Research*, 41(11): 2271-2300.

Ong, Y.H., Chua, A.S.M., Fukushima, T., Ngoh, G.C., Shoji, T. and Michinaka A. (2014). High-temperature EBPR process: The performance, analysis of PAOs and GAOs and the fine-scale population study of *Candidatus* "Accumulibacter phosphatis". *Water Research*, 64: 102-112.

Orhon, D., Sözen, S. and Artan, N. (1996). The effect of heterotrophic yield on the assessment of the correction factor for anoxic growth. *Water Science and Technology*, 34(5): 67-74.

Oshiki, M., Shimokawa, M., Fujii, N., Satoh, H. and Okabe, S. (2011). Physiological characteristics of the anaerobic ammonium-oxidizing bacterium '*Candidatus Brocadia sinica*'. *Microbiology*, 157: 1706-1713.

Oude Elferink, S.J.W.H., Visser, A., Hulshoff Pol, L.W. and Stams, A.J.M. (1994). Sulfate reduction in methanogenic bioreactors. *FEMS Microbiology Reviews*, 15: 119-136.

Parrou, J.L. and François, J. (1997). A simplified procedure for a rapid and reliable assay of both glycogen and trehalose in whole yeast cells. *Analytical Biochemistry*, 248(1): 186-188.

Pasteur, L. (1859). Note sur la fermentation nitreuse. *Bulletin de la Société de chimique de Paris* (séance du 11 mars): 22-23.

Payne, W.J. (1986). 1986: Centenary of the isolation of denitrifying bacteria. *ASM News* 52(12): 627-629.

Petersen, B., Gernaey, K. and Vanrolleghem, P.A. (2002). Anoxic activated sludge monitoring with combined nitrate and titrimetric measurements. *Water Science and Technology*, 45(4-5): 181-190.

Pijuan, M., Guisasola, A., Baeza, J.A., Carrera, J., Casas, C. and Lafuente, J. (2006). Net P-removal deterioration in enriched PAO sludge subjected to permanent aerobic conditions. *Journal of Biotechnology*, 123: 117-126.

Pijuan, M., Ye, L., and Yuan, Z. (2010). Free nitrous acid inhibition on the aerobic metabolism of poly-phosphate accumulating organisms. *Water Research*, 44(20): 6063-6072.

Pijuan, M., Ye, L. and Yuan, Z. (2011). Could nitrite/free nitrous acid favour GAOs over PAOs in enhanced biological phosphorus removal systems? *Water Science and Technology*, 63(2): 345-351.

Pikuta, E., Lysenko, A., Chuvilskaya, N., Mendrock, U., Hippe, H., Suzina, N., Nikitin, D., Osipov, G. and Laurinavichius, K. (2000). *Anoxybacillus pushchinensis* gen. nov., sp. nov., a novel anaerobic, alkaliphilic, moderately thermophilic bacterium from manure, and description of *Anoxybacillus flavitherms* comb. Nov. *International Journal of Systematic and Evolutionary Microbiology*, 50(6): 2109-2117.

Piña-Ochoa, E., Hogslund, S., Geslin, E. and Risgaard-Petersen, N. (2010). Survival and life strategy of the foraminiferan *Globobulimina turgida* through nitrate storage and denitrification. *Marine Ecology Progress Series*, 417: 39-49.

Poinapen, J., Ekama, G.A. and Wentzel, M.C. (2009). Biological sulphate reduction with primary sewage sludge in an upflow anaerobic sludge bed (UASB) reactor - Part 2: Modification of simple wet chemistry analytical procedures to achieve COD and S mass balances. *Water SA*, 35(5): 535-542.

Postgate, J.R. (1965). Recent advances in the study of the sulfate-reducing bacteria. *Bacteriological Reviews*, 29(4): 425-441.

Postgate, J.R. (1984). The Sulfate-Reducing Bacteria. 2nd Edition, ISBN: 9780521257916, Cambridge University Press, UK.

Puyol, D., Carvajal-Arroyo, J.M., Garcia, B., Sierra-Alvarez, R. and Field, J.A. (2013). Kinetic characterization of *Brocadia* spp.-dominated Anammox cultures. *Bioresource Technology*, 139: 94-100.

Puyol, D., Carvajal-Arroyo, J.M., Sierra-Alvarez, R. and Field, J.A. (2014). Nitrite (not free nitrous acid) is the main inhibitor of the anammox process at common pH conditions. *Biotechnology Letters*, 36(3): 547-551.

Rabus, R., Hansen, T.A. and Widdel, F. (2006). Dissimilatory sulfate- and sulfur-reducing prokaryotes. In: The prokaryotes, Vol. 2, Dworkin, M., Falkow, S., Rosenberg, E., Schleifer, K.-H. and Stackebrandt, E. (Eds.), New York: Springer, pp: 659-768.

Ramdani, A., Dold, P., Gadbois, A., Deleris, S., Houweling D. and Comeau, Y. (2012). Characterization of the heterotrophic biomass and the endogenous residue of activated sludge. *Water Research*, 46(3): 653-668.

Rebac, S., Visser, A., Gerbens, S., van Lier, J.B., Stams, A.J.M. and Lettinga, G. (1996). The effect of sulphate on propionate and butyrate degradation in a psychrophilic anaerobic expanded granular sludge bed (EGSB) reactor. *Environmental Technology*, 17(9): 997-1005.

Reichenbecher, W. and Schink. B. (1997). *Desulfovibrio inopinatus* sp. nov., a new sulfate-reducing bacterium that degrades hydroxyhydroquinone (1,2,4-trihydroxybenzene). *Archives of Microbiology*, 168: 338-344.

Reiset, J. (1856). Experiences sur la putrefaction et sur la formation des fumiers. *Comptes rendus des séances hebdomadaires de l'Académie des Sciences*, 42: 177-180.

Rikmann, E., Zekker, I., Tomingas, M., Tenno, T., Menert, A., Loorits, L. and Tenno, T. (2012). Sulfate-reducing anaerobic ammonium oxidation as a potential treatment method for high nitrogen-content wastewater. *Biodegradation*, 23: 509-524.

Rintala, J.A. and Puhakka, J.A. (1994). Anaerobic treatment in pulp and paper-mill waste management: A review. *Bioresource Technology*, 47(1): 1-18.

Risgaard-Petersen, N., Langezaal, A.M., Ingvardsen, S., Schmid, M. C., Jetten, M.S.M., Op den Camp, H.J.M., Derksen, J.W.M., Piña-Ochoa, E., Eriksson, S.P., Nielsen, L.P., Revsbech, N.P., Cedhagen, T. and van der Zwaan, G. J. (2006). Evidence for complete denitrification in a benthic foraminifer. *Nature*, 443(7107): 93-96.

Robertson, L.A., Dalsgaard, T., Revsbech, N.P. and Kuenen, J.G. (1995). Confirmation of aerobic denitrification in batch cultures, using gas-chromatography and N-15 mass-spectrometry. *FEMS Microbiology Ecology*, 18(2): 113-119.

Roeleveld, P.J. and van Loosdrecht, M.C.M. (2002). Experience with guidelines for wastewater characterisation in The Netherlands. *Water Science and* Technology, 45(6): 77-87.

Rozzi, A., Castellazzi, L. and Speece, R.E. (2002). Acetoclastic methanogenic activity measurements by a titration biosensor. *Biotechnology and Bioengineering*, 77(1): 20-26.

Rozzi, A., Castellazzi, L. and Speece, R.E. (2002). Acetoclastic methanogenic activity measurements by a titration biosensor. *Biotechnology and Bioengineering*, 77(1): 20-26.

Rozzi, A., Ficara, E. and Rocco, A. (2003). DO-stat titration respirometry: principle of operation and validation. *ASCE-Journal of Environmental Engineering*, 129(7): 602-609.

Rozzi, A., Massone, A. and Antonelli, M. (1997). A VFA measuring biosensor based on nitrate reduction. *Water Science and Technology*, 36(6-7): 183-189.

Rubio-Rincon, F., Welles, L., López-Vázquez, C.M., van Loosdrecht, M.C.M. and Brdjanovic, D., (2016). Sulfide effects on the metabolism of *Candidatus Accumulibacter phosphatis* clade I. (*submitted*).

Saad, S.A., Welles, L., López-Vázquez, C.M., van Loosdrecht, M.C.M. and Brdjanovic. D. (2013). Sulfide effects on the anaerobic kinetics of phosphorus-accumulating organisms. In: Proceedings of 13th World Congress on Anaerobic Digestion, 25-28th June, Santiago de Compostela, Spain.

Saad, S.A., Welles, L., Abbas, B., López-Vázquez, C.M., van Loosdrecht, M.C.M. and Brdjanovic, D. (2016). Denitrification pathways of *Candidatus Accumulibacter phosphatis* clade I using different carbon sources. (*submitted*).

Saito, T., Brdjanovic, D. and van Loosdrecht, M.C.M. (2004). Effect of nitrite on phosphate uptake by phosphate accumulating organisms. *Water Research*, 38(17): 3760-3768.

Sánchez, M., Mosquera-Corral, A., Mendez, R. and Lema, J.M. (2000). Simple methods for the determination of the denitrifying activity of sludges. *Bioresource Technology*, 75(1-6): 1-6.

Saunders, A.M., Oehmen, A., Blackall, L.L., Yuan, Z. and Keller, J. (2003). The effect of GAOs on anaerobic carbon requirements in full-scale Australian EBPR plants. *Water Science and Technology*, 47(11): 37-43.

Scaglione, D., Buttiglieri, G., Ficara, E., Caffaz, S., Lubello, C. and Malpei, F. (2009). Microcalorimetric and manometric tests to assess anammox activity. *Water Science and Technology*, 60(10): 2705-2711.

Schuler, A.J. and Jenkins, D. (2003). Enhanced biological phosphorus removal from wastewater by biomass with different phosphorus contents, part I: experimental results and comparison with metabolic models. *Water Environment Research*, 75(6): 485-498.

Schloesing, T. and Müntz, A. (1877). Sur la nitrification par les ferments organics. *Comptes rendus des séances hebdomadaires de l'Académie des Sciences*, 85: 301-303.

Sin, G., Guisasola, A., De Pauw, D.J.W., Baeza, J.A., Carrera, J. and Vanrolleghem, P.A. (2003). A new approach for modelling simultaneous storage and growth processes for activated sludge systems under aerobic conditions. *Biotechnology and Bioengineering*, 92(5): 600-613.

Smolders, G.J.F., van der Meij, J., van Loosdrecht, M.C.M. and Heijnen, J.J. (1994a). Model of the anaerobic metabolism of the biological phosphorus removal process: stoichiometry and pH influence. *Biotechnology and Bioengineering*, 43(6): 461-470.

Smolders, G.J.F., van der Meij, J., van Loosdrecht, M.C.M. and Heijnen, J.J. (1994b). Stoichiometric model of the aerobic metabolism of the biological phosphorus removal process. *Biotechnology and Bioengineering*, 44(7): 837-848.

Smolders, G.J.F., van Loosdrecht, M.C.M. and Heijnen, J.J. (1995). A metabolic model for the biological phosphorus removal process. *Water Science and Technology*, 31(2): 79-93.

Strous, M., Heijnen, J.J., Kuenen, J.G. and Jetten, M.S.M. (1998). The sequencing batch reactor as a powerful tool for the study of slowly growing anaerobic ammonium-oxidizing microorganisms. *Applied Microbiology and Biotechnology*, 50: 589-596.

Tang, K., Baskaran, V. and Nemati, M. (2009). Bacteria of the sulphur cycle: An overview of microbiology, biokinetics and their role in petroleum and mining industries. *Biochemical Engineering Journal*, 44(1): 73-94.

Tang, C.-J., Zheng, P., Chen, T.-T., Zhang, J.-Q., Mahmood, Q., Ding, S., Chen, X.-G., Chen, J.-W., and Wu D.-T. (2011). Enhanced nitrogen removal from pharmaceutical wastewater using SBA-ANAMMOX process. *Water Research*, 45: 201-210.

Thauer, R.K., Jungermann, K. and Decker, K. (1977). Energy conservation in chemotrophic anaerobic bacteria. *Bacteriological Reviews*, 41: 100-180.

Thomas, M., Wright, P., Blackall, L.L., Urbain, V. and Keller, J. (2003). Optimisation of Noosa BNR plant to improve performance and reduce operating costs. *Water Science and Technology*, 47(12): 141-148.

Udert, K.M., Kind, E., Teunissen, M., Jenni, S. and Larsen, T.A. (2008). Effect of heterotrophic growth on nitritation/anammox in a single sequencing batch reactor. *Water Science and Technology*, 58: 277-284.

van de Graaf, A., De Bruijn, P., Robertson, L.A., Jetten, M.S.M. and Kuenen, J.G. (1996). Autotrophic growth of anaerobic ammonium oxidizing microorganisms in a fluidized bed reactor. *Microbiology*, 142(8): 2187-2196.

van den Brand, T.P.H., Roest, K., Brdjanovic, D., Chen, G.-H. and van Loosdrecht, M.C.M. (2014a). Influence of acetate and propionate on sulphate-reducing bacteria activity. *Journal of Applied Microbiology*, 117(6): 1839-1847.

van den Brand, T.P.H., Roest, K., Chen, G.-H., Brdjanovic, D. and van Loosdrecht, M.C.M. (2014b). Temperature effect on acetate and propionate consumption by sulphate reducing bacteria in saline wastewater. *Applied Microbiology and Biotechnology*, 98(9): 4245-4255.

van den Brand, T.P.H., Roest, K., Chen, G.-H., Brdjanovic, D. and van Loosdrecht, M.C.M. (2015). Occurence and activity of sulphate reducing bacteria in aerobic activated sludge systems. *World Journal of Microbiology and Biotechnology*, 31(3): 507-516.

van Houten, R.T., van der Spoel, H., van Aelst, A.C., Hulshoff Pol, L.W. and Lettinga, G. (1996). Biological sulfate reduction using synthesis gas as energy and carbon source. *Biotechnology and Bioengineering*, 50(2): 136-144.

van Loosdrecht, M.C.M., Lopez-Vazquez, C.M., Meijer, S.C.F., Hooijmans, C.M., and Brdjanovic, D. (2015) Twenty-five years of ASM1: past, present and future of wastewater treatment modelling. *Journal of Hydroinformatics*, 17(5):697-718.

van Loosdrecht, M.C.M., Pot, M.A. and Heijnen, J.J. (1997). Importance of bacterial storage polymers in bioprocesses. *Water Science and Technology*, 35(1): 41-47.

van Niel, E.W.J., Arts, P.A.M., Wesselink, B.J., Robertson L.A. and Kuenen, J.G. (1993). Competition between heterotrophic and autotrophic nitrifiers for ammonia in chemostat cultures. *FEMS Microbiology Ecology*, 102: 109-118.

Vanrolleghem, P.A., Insel, G., Petersen, B., Sin, G., De Pauw, D., Nopens, I., Doverman, H., Weijers, S. and Gernaey, K. (2003). A comprehensive model calibration procedure for activated sludge models. In: Proceedings of the 76th Annual WEF Conference and Exposition, October 11-15, Los Angeles.

Villa-Gomez, D., Ababneh, H., Papirio, S., Rousseau, D.P.L. and Lens, P.N.L. (2011). Effect of sulfide concentration on the location of the metal precipitates in inversed fluidized bed reactors. *Journal of Hazardous Materials*, 192(1): 200-207.

Villegas, J.D., de Laclos, H.F., Dovat, J., Membrez, Y. and Holliger, C. (2011). Nitrogen removal from digested manure in a simple one-stage process. *Water Science and Technology*, 63: 1991-1996.

Vishniac, W. and Santer, M. (1957). *Thiobacilli. Bacteriological Reviews*, 21: 195-213.

Vlekke, G.J.F.M., Comeau, Y. and Oldham W.K. (1988). Biological phosphate removal from wastewater with oxygen or nitrate in sequencing batch reactors. *Environmental Technology Letters*, 9: 791-796.

Wachtmeister, A., Kuba, T., van Loosdrecht, M.C.M. and Heijnen, J.J. (1997). A sludge characterization assay for aerobic and denitrifying phosphorus removing sludge. *Water Research*, 31(3): 471-478.

Wang, J., Lu, H., Chen, G.H., Lau, G.N., Tsang, W.L. and van Loosdrecht, M.C.M. (2009). A novel sulfate reduction, autotrophic denitrification, nitrification integrated (SANI) process for saline wastewater treatment. *Water Research*, 43(9): 2363-2372.

Welles, L., López-Vázquez, C.M., Hooijmans, C.M., van Loosdrecht, M.C.M. and Brdjanovic, D. (2014). Impact of salinity on the anaerobic metabolism of phosphate-accumulating organisms (PAO) and glycogen-accumulating organisms (GAO). *Applied Microbiology and Biotechnology*, 98(17): 7609-7622.

Welles, L., López-Vázquez, C.M., Hooijmans, C.M., van Loosdrecht, M.C.M. and Brdjanovic, D. (2015a). Impact of salinity on the aerobic metabolism of phosphate-accumulating organisms. *Applied Microbiology and Biotechnology*, 99(8): 3659-3672.

Welles, L., Tian, W.D., Saad, S., Abbas, B., López-Vázquez, C.M., Hooijmans, C.M., van Loosdrecht, M.C.M. and Brdjanovic, D. (2015b). *Accumulibacter* clades Type I and II performing kinetically different glycogen-accumulating organisms metabolisms for anaerobic substrate uptake. *Water Research*, 15(83): 354-366.

Welles, L., Abbas, B., López-Vázquez, C.M., Hooijmans, C.M., van Loosdrecht, M.C.M. and Brdjanovic, D. (2016). Metabolic response of 'Candidatus Accumulibacter phosphatis' clade II to changes in P/C ratio in their environment. (submitted).

Wentzel, M.C., Dold, P.L., Loewenthal, R.E., Ekama, G.A. and Marais, G.v.R. (1987). Experiments towards establishing the kinetics of biological excess phosphorus removal. In: Biological phosphate removal from wastewaters: Proceedings of an IAWPRC Specialized Conference, Rome, Italy, 28-30 September 28-30, 1987 (Pergamon Press, Vol. 4, p. 79)..

Wentzel, M.C., Comeau, Y., Ekama, G.A., van Loosdrecht, M.C.M. and Brdjanovic, D. (2008). Enhanced biological phosphorus removal. In: Biological wastewater treatment: principles, modelling and design, Henze, M., van Loosdrecht, M.C.M., Ekama, G.A. and Brdjanovic, D. (Eds.), ISBN: 9781843391883, IWA Publishing, London, UK.

Wentzel, M.C., Lötter, L.H., Loewenthal, R.E and Marais, G.v.R. (1986). Metabolic behaviour of *Acinetobacter* spp. in enhanced biological phosphorus removal- a biochemical model. *Water SA*, 12(4): 209-224.

Wentzel, M.C., Mbewe, A. and Ekama, G.A. (1995). Batch tests for measurement of readily biodegradable COD and active organism concentrations in municipal waste waters. *Water SA*, 21(2): 117-124.

WERF. (2003). Methods for wastewater characterization in activated sludge modeling. - Water Environment Research Foundation report 99-WWF-3, ISBN13: 9781843396628. WERF (Alexandria) and IWA Publishing (London), pp. 575

Wett, B. (2007). Development and implementation of a robust deammonification process. *Water Science and Technology*, 56(7): 81-88.

Widdel, F. (2006). The genus *Desulfotomaculum*. In: The prokaryotes, Vol. 2, Dworkin, M., Falkow, S., Rosenberg, E., Schleifer, K.-H., and Stackebrandt, E. (Eds.), New York: Springer, pp: 787-794.

Wiesmann, U. (1994). Biological nitrogen removal from wastewater. In: Advances in biochemical engineering/biotechnology, Fiechter, A. (Ed.), ISBN-13: 978-1843396628, Berlin: Springer-Verlag Berlin Heidelberg. pp: 113-154.

Wik, T. and Breitholtz, C. (1996). Steady-state solution of a two-species biofilm problem. *Biotechnology and Bioengineering*, 50(6): 675-686.

Winogradsky, M.S. (1890). Reserches sur les organismes de la nitrification. *Annales de l'institut Pasteur*, 4: 213-231.

Winogradsky, M.S. (1892). Contribution à la morphologie des organismes de la nitrification. *Archives of Biological Sciences*, 1: 87-137.

Wu, D., Ekama, G.A., Wang, H.G., Wei, L., Lu, H., Chui, H.K., Liu, W.T., Brdjanovic, D., van Loosdrecht, M.C.M. and Chen, G.H. (2014). Simultaneous nitrogen and phosphorus removal in the sulphur cycle-associated Enhanced Biological Phosphorus Removal (EBPR) process. *Water Research*, 49: 251-264.

Yamamoto, R.I., Komori, T. and Matsui, S. (1991). Filamentous bulking and hindrance of phosphate removal due to sulfate reduction in activated sludge. *Water Science and Technology*, 23(4-6): 927-935.

Yang, Z., Zhou, S. and Sun, Y. (2009). Start-up of simultaneous removal of ammonium and sulfate from an anaerobic ammonium oxidation (Anammox) process in an anaerobic up-flow bioreactor. *Journal of Hazardous Materials*, 169: 113-118.

Yoshida. Y., Takahashi. K., Saito. T. and Tanaka. K. (2006). The effect of nitrite on aerobic phosphate uptake and denitrifying activity of phosphate-accumulating organisms. *Water Science and Technology*, 53(6): 21-27.

Zeng. R.J., van Loosdrecht. M.C.M., Yuan. Z. and Keller. J. (2003a). Metabolic model for glycogen-accumulating organisms in anaerobic/aerobic activated sludge systems. *Biotechnology and Bioengineering*, 81(1): 92-105.

Zeng. R.J., Saunders. A.M., Yuan. Z., Blackall. L.L. and Keller, J. (2003b). Identification and comparison of aerobic and denitrifying polyphosphate-accumulating organisms. *Biotechnology and Bioengineering*, 83(2): 140-148.

Zeng, W., Li, B., Yang, Y., Wang, X., Li, L. and Peng, Y. (2014). Impact of nitrite on aerobic phosphorus uptake by polyphosphate accumulating organisms in enhanced biological phosphorus removal sludges. *Bioprocess and Biosystems Engineering*, 37(2): 277-287.

Zhou, Y., Pijuan, M. and Yuan, Z. (2007). Free nitrous acid inhibition on anoxic phosphorus uptake and denitrification by polyphosphate accumulating organisms. *Biotechnology and Bioengineering*, 98(4): 903-912.

Zhou, Y., Pijuan, M., Zeng, R.J., Lu, H. and Yuan, Z. (2008). Could polyphosphate-accumulating organisms (PAOs) be glycogen-accumulating organisms (GAOs)? *Water Research*, 42(10): 2361-2368.

Zhou, Y., Ganda, L., Lim, M., Yuan, Z. and Ng, W.J. (2012). Response of poly-phosphate accumulating organisms to free nitrous acid inhibition under anoxic and aerobic conditions. *Bioresource Technology*, 116: 340-347.

Zietz, U. (1995). The formation of sludge bulking in the activated sludge process. *European Water Pollution Control*, 5: 21-27.

Zumft, W.G. (1997). Cell biology and molecular basis of denitrification. *Microbiology and Molecular Biology Reviews* 61: 533-616.

3

RESPIROMETRY

Authors:
Henri Spanjers
Peter A. Vanrolleghem

Reviewers:
George A. Ekama
Mathieu Spérandio

3.1 INTRODUCTION

The objective of this chapter is to provide practical guidelines for the assessment of the respiration rate of biomass. The approach will be practically oriented and method-driven. However, some biochemical background on respiration will be provided in order to understand how respiration is related to microbial substrate utilization and growth. We will explain that respiration can be assessed in terms of the uptake rate of a terminal electron acceptor, such as molecular oxygen or nitrate, or, in the case of anaerobic respiration, in terms of the production rate of methane or sulphide. The measurement of the consumption (or production) rate, i.e. respirometry, will be explained following various measuring principles, and we will provide some practical recommendations. The focus will be on laboratory tests using samples of biomass and wastewater. However, in principle most measuring principles can be automated, or have already been automated in commercial respirometers, to measure respiration rate automatically or even in-line at a wastewater treatment plant. In-line measurement of respiration rate, however, is out of the scope of this textbook. One specific method, off-gas analyses, provides an inherent way to assess the respiration rate in a pilot-scale or full-scale wastewater treatment plant. This method is described in Chapter 4.

The information that can be extracted from respirometric measurements can be divided into two types: direct and indirect (Spanjers *et al.*, 1998). Direct information, such as the aerobic respiration rate or specific methanogenic activity, provides information on the actual activity of the biomass, and can be used for example to record respirograms (time series of respiration rates) in the lab. Indirect information refers to variables that are deduced from respirometric measurements, such as microorganism concentration, substrate concentration and kinetic parameters. In this case respirometric measurements are used as input to simple arithmetic calculations or even model fitting. Chapter 5 will describe how data such as respirometric data can be used in model fitting to assess deduced variables and parameters.

Because the objective of this chapter is to provide only practical guidelines for the assessment of the respiration rate, only a basic explanation of the biochemical background will be provided, and the reader is referred to the literature on biochemistry (Alberts *et al.*, 2002; Nelson and Cox, 2008). As the practical guidelines focus on respirometric methods that can easily be carried out in most laboratories, no rigorous discussion of all the respirometric measuring principles will be given, and we

ISBN: 9781789065275 (Paperback), ISBN: 9781780404752 (eBook). Published by IWA Publishing, London, UK.

refer to Spanjers *et al.* (1998) for a more complete description of the concepts. For a discussion of the use of direct and indirect respirometric information in control of the activated sludge process the reader may consult Copp *et al.* (2002).

3.1.1 Basics of respiration

In biochemical terms, microbial respiration is the adenosine triphosphate (ATP)-generating metabolic process in which either organic or inorganic compounds serve as the electron donor and inorganic compounds serve as the terminal electron acceptor (e.g. oxygen, nitrate, sulphate). The universal energy carrier ATP is generated as electrons removed from the electron donor are transferred along the electron transport chain from one metabolic carrier to the next and, eventually, to the terminal electron acceptor. In this way, microorganisms convert the energy of intramolecular bonds in the electron donor to the high-energy phosphate bonds of ATP (catabolism). The energy is then used to synthesize the various molecular components required for cell growth (anabolism), maintenance and reproduction.

During the process of respiration the electron donor is converted to its oxidized form and the electron acceptor is converted to its reduced form. In the case of a carbonaceous donor (organic compounds), the oxidized form is carbon dioxide. If the electron acceptor is molecular oxygen then its reduced form is water. The conversion of a carbonaceous donor with oxygen as the electron acceptor is carried out by heterotrophic bacteria.

Inorganic donors that are converted to their oxidized form by aerobic microorganisms, where oxygen serves as the terminal electron acceptor, include ammonium and nitrite, ferrous (divalent iron) and sulphide, and the conversions are carried out by nitrifiers (ammonia and nitrite-oxidizing bacteria), iron-oxidizing bacteria and sulphide-oxidizing bacteria, respectively. In this case CO_2 forms the source of carbon, and the organisms are called autotrophs. Non-aerobic microorganisms use inorganic compounds other than oxygen such as nitrite, nitrate, sulphate and carbon dioxide as the terminal electron acceptor. In these cases we are talking about anoxic (nitrite, nitrate) and anaerobic processes (CO_2, sulphate). Note that in wastewater treatment various respiration processes may take place simultaneously where different microorganisms use diverse substrates and terminal e-acceptors or compete for the same substrates and terminal electron acceptors.

Figure 3.1 shows a schematic overview of some examples of metabolic conversions. Note that both the electron donor and the terminal electron acceptor may be considered as substrate, like many other components that enter the metabolic pathways. In respirometry, respiration is generally considered as the consumption of O_2, NO_2^- or NO_3^- or (in anaerobic respirometry) the production of CH_4. In general terms, the metabolic conversions involved in respiration are catabolic reactions, and several gaseous compounds consumed or produced during these reactions can be used to assess key metabolic conversions. Note, however, that in principle other substances may also be considered, such as the consumption of NH_4^+, HS^- or S^{2-}, or SO_4^{2-}, or the production of N_2. Other products that are not included in the figure but that are also linked to respiration include H^+ and heat and the associated methods are titrimetry and calorimetry, respectively. However, these are out of the scope of this chapter.

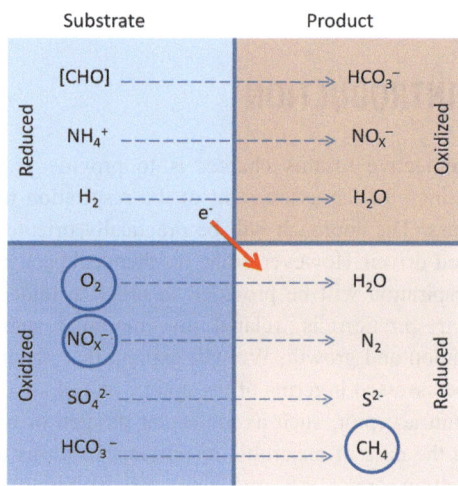

Figure 3.1 Schematic overview of some examples of metabolic conversions. e⁻ denotes the electron that is transferred from the electron donor to the terminal electron acceptor. [CHO] denotes any carbohydrate. The coloured substances are generally used as measured variables in respirometry.

Because the energy generated during the process of microbial respiration is used for cell growth and maintenance functions, such as reproduction, cell mobility, osmotic activity, etc., the respiration rate is linked to the rate of these processes. However, it is difficult to differentiate between these two processes. As an example, consider the aerobic respiration by heterotrophic microorganisms that use carbonaceous (organic) substrate as the electron donor and oxygen as the terminal electron acceptor. Only a portion (1-Y) of the consumed organic substrate is oxidised to provide energy for cell growth and maintenance. The remainder, typically half (on a weight/weight basis) of the substrate molecules (the yield Y) is reorganised into new cell mass. Hence the oxygen consumption rate is linked to the biomass growth through the yield. In anaerobic respiration of hydrogenotrophic methanogens, where H_2 substrate is used as the electron donor and CO_2 as the electron acceptor, only a small portion of Y of the substrate is rearranged into biomass, while the largest part is oxidized to produce CH_4.

In activated sludge, carbonaceous substrate removal is not the only oxygen-consuming process. In addition to the oxygen consumption by heterotrophic biomass, there are some other biological processes that may contribute to the respiration of activated sludge, such as the oxidation of inorganic compounds by nitrifiers and other bacteria, and specific microbial oxidation reactions catalysed by oxidases and mono-oxygenases. Nitrifying bacteria incorporate only a minor part of the substrate ammonia into new biomass while most of the substrate (ammonium) is oxidised for energy production. These autotrophic bacteria use dissolved carbon dioxide as a carbon source for new biomass. In comparison to heterotrophic biomass, nitrifiers need more oxygen for their growth. Nitrification occurs in two steps: the oxidation of ammonia to nitrite and the oxidation of nitrite to nitrate. Like nitrifiers, the autotrophic sulphur-oxidizing bacteria and iron-oxidizing bacteria utilise inorganic compounds instead of organic matter to obtain energy and use carbon dioxide or carbonate as a carbon source. Sulphur-oxidizing bacteria are able to oxidise hydrogen sulphide (or other reduced sulphur compounds) to sulphuric acid. Iron-oxidizing bacteria oxidise inorganic ferrous iron to the ferric form to obtain energy. In addition to bacteria, protozoa and other predating higher organisms are present in the activated sludge, and they also consume oxygen. Finally, some inorganic electron donors such as ferrous iron and sulphide can be chemically oxidised, also utilising oxygen.

All the above-mentioned oxygen-consuming processes contribute to the total respiration rate of the activated sludge. Respirometry is usually intended to measure only biological oxygen consumption and sometimes it is attempted to distinguish between different biological processes such as heterotrophic oxidation and nitrification. However, in many cases it is difficult to distinguish between specific microbial processes and to identify chemical oxygen consumption.

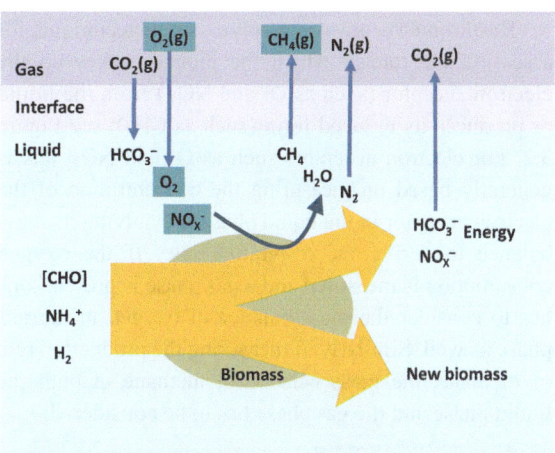

Figure 3.2 The relationship between respiration, substrate utilization and growth for three types of substrate [CHO], NH_4 and H_2 and related electron acceptor e.g. O_2, NO_3 and HCO_3 (Spanjers et al., 1998).

3.1.2 Basics of respirometry

Respirometry is generally defined as the measurement and interpretation of the rate of biological consumption of an inorganic electron acceptor under well-defined experimental conditions. In principle, all the substances depicted in the 'substrate-oxidized form' highlighted in Figure 3.1 can serve as the measured variable. An exception is anaerobic respirometry where generally the production rate of the ultimate reduced product methane is measured. This is because during anaerobic degradation many intermediates are involved and it is impracticable to measure the consumption rate of these intermediate substrates. Moreover, methanogenesis is generally not the rate-limiting step; hence, the methane production rate reflects the rate-limiting process (mostly hydrolysis in the case of a complex substrate).

Note that in principle one may also measure the consumption rate of the electron donor, such as [CHO], NH_4^+ and H_2. However, this is generally not considered to be respirometry, also because electron donor substances such as [CHO] and NH_4^+ may also be consumed by processes not related to energy generation, for example the uptake in biomass, and hence not clearly related to energy generation. Finally, measurement of CO_2 production may be considered respirometry because CO_2 production is related to energy generation (Figure 3.2). However, because CO_2 in the gas phase is associated with the carbonate system, extra measurements will be needed to measure the pH and bicarbonate concentration in the liquid phase.

Respirometry always involves some technique for assessing the rate at which the biomass takes up the electron acceptor (such as O_2 and NO_3^-) from the liquid or produces its reduced form (such as CH_4); see Figure 3.2. For electron acceptors such as O_2 and NO_3^- this is generally based on measuring the concentration of the electron acceptor in the liquid phase and solving its mass balance to derive the respiration rate. If the oxygen consumption is measured and a gas phase is present, one has to consider the mass balance of oxygen in the gas phase as well. Similarly, if measuring the production rate of methane, the mass balance of methane in both the liquid phase and the gas phase has to be considered.

For aerobic respirometry i.e. assessing the rate at which biomass takes up O_2, Spanjers et al., (1998) presented a classification of respirometric principles that was based on two simple criteria: the first one being the location of the oxygen measurement, liquid or gas phase; and the second being the state of the gas and liquid phases, both either flowing or static. It was found that the majority of the proposed respirometric devices could be put into one of the eight classes created by this structure. Moreover, for each of the classes, examples of implementations were found in the literature.

3.2 GENERAL METHODOLOGY OF RESPIROMETRY

3.2.1 Basics of respirometric methodology

The respiration rate is usually measured with a respirometer. Respirometers range from a simple, manually operated bottle equipped with a sensor to complicated instruments that operate fully automatically. In some cases the bioreactor of the treatment plant itself can serve as a respirometer. Except for the latter, a feature common to all respirometers is a reactor, separated from the bioreactor, where different components (biomass, substrate, etc.) are brought together. The operation of all respirometers involves some technique for assessing the rate at which the biomass takes up a component from the liquid or produces a component (Figure 3.2). Many techniques have been developed in the past. However, Spanjers et al. (1998) found that all measuring techniques for the respiration rate can be classified into only eight basic principles according to two criteria: (1) the phase where the concentration is measured (gas or liquid, G and L, respectively) and (2) whether or not there is input and output of liquid and gas (flowing or static, F and S, respectively). The operation of all existing respirometers

can be explained in terms of these criteria. Figure 3.3 shows a generic scheme for a respirometer. Note that the gas phase also includes bubbles dispersed in the liquid phase. In the subsequent sections, the principles will be discussed according to the above criteria. We will not discuss the usefulness of the different measuring techniques, because we believe that any technique has its merits, depending on the specific application, provided that the correct measuring conditions are satisfied.

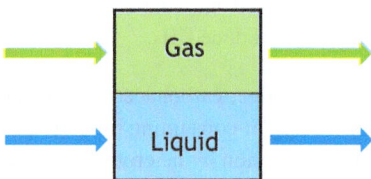

Figure 3.3 Generic scheme of a respirometer.

3.2.2 Generalized principles: beyond oxygen

3.2.2.1 Principles based on measuring in the liquid phase

The majority of the techniques based on measurement in the liquid phase use a specific electrode or sensor. A reliable respiration rate measurement is only possible if the sensor is correctly calibrated and if a number of environmental variables, such as temperature and pressure, is accounted for. Sensors also have a response time that must be accounted for in some respirometric setups.

Respirometers that are based on measuring dissolved oxygen (DO) concentration in the liquid phase use a DO mass balance over the liquid phase. Consider a system consisting of a liquid phase, containing biomass, and a gas phase both being ideally mixed and having an input and output (Figure 3.4). It is assumed that the DO concentration in the liquid phase can be measured. The DO mass balance over the liquid phase is:

$$\frac{d(V_L \cdot S_{O2})}{dt} = Q_{in} \cdot S_{O2,in} - Q_{out} \cdot S_{O2} + V_L \cdot kLa \cdot (S_{O2}^* - S_{O2}) - V_L \cdot r_{O2}$$

Eq. 3.1

Where, S_{O2} is the DO concentration in the liquid phase (mg L^{-1}), S_{O2}^* is the saturation DO concentration in the liquid phase (mg L^{-1}), $S_{O2,in}$ is the DO concentration in the liquid phase entering the system (mg L^{-1}), kLa is the oxygen mass transfer coefficient, based on liquid volume (h^{-1}), Q_{in} is the flow rate of the liquid entering the system (L h^{-1}), Q_{out} is the flow rate of the liquid leaving the system (L h^{-1}), r_{O2} is the respiration rate of the biomass in the liquid, (mg L^{-1} h^{-1}), and V_L is the volume of the liquid phase (L).

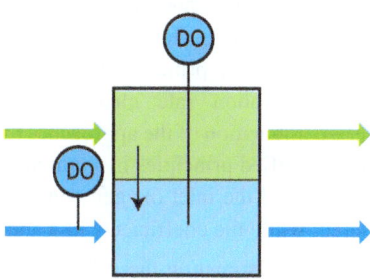

Figure 3.4 The liquid phase principle, flowing gas, flowing liquid (LFF).

Notice that, since it is a mass balance over the liquid phase, Eq. 3.1 does not contain gas flow terms. The first and second term on the right-hand side represent the advective flow of DO in the input and output liquid streams. In most systems Q_{in} and Q_{out} will be equal so that the liquid volume is constant. The third term describes the mass transfer of oxygen from the gas phase to the liquid phase. The last term contains the respiration rate to be derived from the mass balance. Therefore, S_O must be measured and all other coefficients be known or neglected (i.e. non-influential). In practice, the determination of r_{O2} can be simplified in several ways. In what follows it is assumed that the liquid volume is constant, so that the terms in Eq. 3.1 can be divided by V_L.

- **Static gas, static liquid (LSS)**

One approach is to use a method without liquid flow and oxygen mass transfer (Figure 3.5). Then the first three terms on the right-hand side of Eq. 3.1 fall away and the mass balance reduces to:

$$\frac{dS_{O2}}{dt} = -r_{O2} \qquad \text{Eq. 3.2}$$

Hence, to obtain the respiration rate only the differential term has to be determined. This can be done by measuring the decrease in DO as a function of time due to respiration, which is equivalent to approximating the differential term with a finite difference term:

$\Delta S_{O2}/\Delta t = -r_{O2}$. Typical of this principle is that the DO may become exhausted after some time so that for continued measurement of r_{O2} reaeration it is necessary to bring the DO concentration back to a higher level. DO and substrate limit the respiration when their concentrations become too low, causing a non-linear DO decrease complicating the assessment of the differential term. Note that in Figure 3.5 there is a gas phase. However, it is assumed there is no mass transfer from the gas phase into the liquid phase. In practice, in order to prevent the input of oxygen into the liquid, the gas phase may be absent. The procedure for the determination of r_O according to standard methods (APHA *et al.*, 2012) is based on this principle.

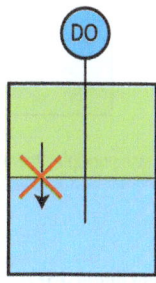

Figure 3.5 The liquid phase principle, static gas, static liquid (LSS).

- **Flowing gas, static liquid (LFS)**

The disadvantage of the need for reaerations can be eliminated by continuously aerating the biomass. Then, the oxygen mass transfer term $kLa \cdot (S_{O2}^* - S_{O2})$ must be included in the mass balance (Eq. 3.3):

$$\frac{dS_{O2}}{dt} = kLa \cdot (S_{O2}^* - S_{O2}) - r_{O2} \qquad \text{Eq. 3.3}$$

To obtain r_{O2}, both the differential term and the mass transfer term must be determined. To calculate the latter, the mass transfer coefficient (kLa) and the DO saturation concentration (S_{O2}^*) must be known. These coefficients have to be determined regularly because they depend on environmental conditions such as temperature, barometric pressure and the properties of the liquid (viscosity, salinity, etc.). The simplest approach is to determine these by using separate reaeration tests and look-up tables. Another approach is to estimate the coefficients from the dynamics of the DO concentration response by applying parameter estimation techniques. The advantage of the latter method is that the values of the aeration coefficients can be updated relatively easily. This respirometric principle allows the measurement of r_{O2} at a nearly constant DO concentration, thereby eliminating the dependency of r_{O2} on the DO concentration (provided DO >> 0 mg L^{-1}). Note that, whereas Figure 3.6 shows an input and an output in the gas phase, there is no gas flow term in Eq. 3.3. There is no need to consider gas flow terms provided S_{O2}^* is known or determined.

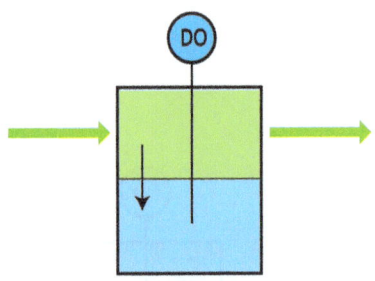

Figure 3.6 The liquid phase principle, flowing gas, static liquid (LFS).

- **Static gas, flowing liquid (LSF)**

Repetitive aeration or estimation of oxygen transfer coefficients, as with the above principles, can be avoided when liquid with a high enough input DO concentration flows continuously through a closed completely mixed cell without the gas phase (Figure 3.7). The liquid flow terms now have to be included in the mass balance (Eq. 3.4):

$$\frac{dS_{O2}}{dt} = \frac{Q_{in}}{V_L} \cdot S_{O2,in} - \frac{Q_{out}}{V_L} \cdot S_{O2} - r_{O2} \qquad \text{Eq. 3.4}$$

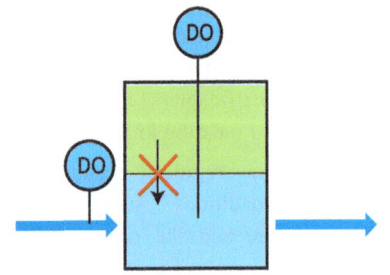

Figure 3.7 The liquid phase principle, static (no) gas, flowing liquid (LSF).

Both DO concentrations, $S_{O2,in}$ and S_{O2}, must be measured to allow calculation of r_{O2}. In a respirometer Q_{in} and V_L are instrument constants and are, therefore, assumed to be known or calibrated. This principle is in fact the continuous counterpart of the one explained in Eq. 3.2, and it is as such also sensitive to the effect of substrate and DO limitation. However, the effect of limiting substrate can be eliminated by the continuous supply of substrate (wastewater) and DO to the respiration cell.

- **Flowing gas, flowing liquid (LFF)**

Without the above simplifications the full mass balance (Eq. 3.1) holds for the principle depicted in Figure 3.4. To obtain respiration rate measurements with this principle, a combination of the approaches mentioned for the above simplified principles is required. For instance, the flow rates and the inlet oxygen concentrations must be measured, while the coefficients kLa and S_{O2}^* must be assessed, e.g. by estimating these from the dynamics of the DO concentration.

3.2.2.2 Principles based on measuring in the gas phase

Respirometric techniques based on measuring gaseous oxygen always deal with two phases: a liquid phase containing the respiring biomass and a gas phase where the oxygen measurement takes place. The main reason for measuring in the gas phase is to overcome difficulties associated with interfering contaminants common in the liquid phase (e.g. the formation of biomass film on the sensor). Gaseous oxygen is measured by physical methods such as the paramagnetic method, or gasometric methods.

Gasometric methods measure changes in the concentration of gaseous oxygen. According to the ideal gas law $P \cdot V = n \cdot R \cdot T$, these can be derived from changes in the pressure (if volume is kept constant, the manometric method) or changes in the volume (if pressure is kept constant, the volumetric method). These methods are typically applied to closed measuring systems (no input and output streams), which may provoke a need for reaerations and thus temporary interruption of the measurements. This limits the possibility for continued monitoring of the respiration rate. However, interruptions because of reaerations are not needed if the consumed oxygen is replenished at a known rate, e.g. by supplying pure oxygen from a reservoir or by using electrolysis. The rate at which oxygen is supplied is then equivalent to the biological respiration rate (assuming infinitely fast mass transfer to the liquid). Because carbon dioxide is released from the liquid phase as a result of the biological activity, this gas has to be removed from the gas phase in order to avoid interference with the oxygen measurement. In practice this is done by using alkali to chemically absorb the carbon dioxide produced.

Respirometric principles based on measuring gaseous oxygen also use oxygen mass balances to derive the respiration rate. However, in addition to the mass balance in the liquid phase (Eq. 3.1), a balance in the (ideally mixed) gas phase must be considered (Figure 3.8):

$$\frac{d}{dt}(V_G \cdot C_{O2}) = F_{in} \cdot C_{O2,in} - F_{out} \cdot C_{O2} - V_L \cdot kLa \cdot (S^*_{O2} - S_{O2}) \qquad \text{Eq. 3.5}$$

Where, C_{O2} is the O_2 concentration in the gas phase (mg L^{-1}), $C_{O2,in}$ is the O_2 concentration in the gas entering the system (mg L^{-1}), F_{in} is the flow rate of the gas entering the system (L h^{-1}), F_{out} is the flow rate of the gas leaving the system (L h^{-1}), and V_G is the volume of the gas phase (L).

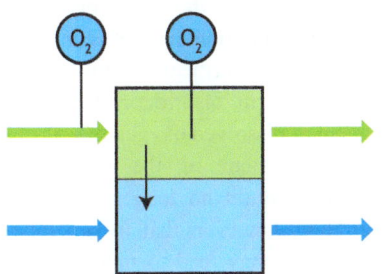

Figure 3.8 The gas phase principle, flowing gas, flowing liquid (GFF).

The term $V_L \cdot kLa \cdot (S^*_{O2} - S_{O2})$ represents the mass transfer rate of oxygen from the gas phase to the liquid phase, and it is the connection between the two phases. From mass balances (Eq. 3.1 and Eq. 3.5) it follows that, in order to calculate r_{O2}, C_{O2} must be measured (directly or using the gas law, see above) and knowledge of S_{O2} is required. However, S_{O2} is not measured in the gas phase principles. In these respirometric principles it is assumed that the oxygen concentrations in the gas and liquid phases are in equilibrium, i.e. mass transfer is sufficiently fast (kLa → ∞), so that $S_{O2} \approx S^*_{O2}$. Since, by definition, the saturation DO concentration is proportional to the O_2 concentration in the gas phase:

$$S^*_{O2} = H \cdot C_{O2} \qquad \text{Eq. 3.6}$$

it is reasonable to state that:

$$S_{O2} = H \cdot C_{O2} \qquad \text{Eq. 3.7}$$

and that:

$$\frac{dS_{O2}}{dt} = H \cdot \frac{dC_{O2}}{dt} \qquad \text{Eq. 3.8}$$

Hence, the measurement in the gas phase is a good representation of the condition in the liquid phase, provided the proportionality (Henry) constant H is known, e.g. from calibration or tables, and the mass transfer coefficient is high. The validity of this equilibrium assumption should be critically evaluated.

- **Static gas, static liquid (GSS)**

The simplest gas phase technique for measuring the respiration rate is based on a static liquid phase and a static gas phase, i.e. no input or output (Figure 3.9). In addition to the DO mass balance in the liquid phase, an oxygen mass balance in the gas phase must be considered:

$$\frac{dS_{O2}}{dt} = kLa \cdot (S^*_{O2} - S_{O2}) - r_{O2} \qquad \text{Eq. 3.9}$$

$$\frac{d(V_G \cdot C_{O2})}{dt} = -V_L \cdot kLa \cdot (S^*_{O2} - S_{O2}) \qquad \text{Eq. 3.10}$$

Hence, in order to calculate r_{O2}, the change of the oxygen concentration in the gas phase, dC_{O2}/dt, must be measured and knowledge of dS_{O2}/dt is required (Eq. 3.9). It is possible to measure dC_{O2}/dt by using an oxygen sensor. If a gasometric method is used, dC_{O2}/dt is related to the change in volume or the change in pressure (Eq. 3.10).

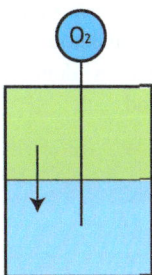

Figure 3.9 The gas phase principle, static gas, static liquid (GSS).

With this principle, the same restriction as with the simplest DO-based principle exists: when the oxygen becomes exhausted it must be replenished by, for instance, venting the gas phase in order to continue the measurement of r_{O2}.

- **Flowing gas, static liquid (GFS)**

Another technique is based on a flowing gas phase, i.e. the biomass is continuously aerated with air (or pure oxygen) so that the presence of sufficient oxygen is assured (Figure 3.10). In comparison to Eq. 3.10, two transport terms must be included in the mass balance on the gas phase:

$$\frac{dS_{O2}}{dt} = kLa \cdot (S^*_{O2} - S_{O2}) - r_{O2} \qquad \text{Eq. 3.11}$$

$$\frac{d(V_G \cdot C_{O2})}{dt} = F_{in} \cdot C_{O2,in} - F_{out} \cdot C_{O2} - V_L \cdot kLa \cdot (S^*_{O2} - S_{O2})$$

$$\text{Eq. 3.12}$$

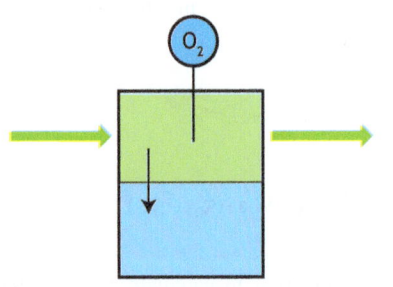

Figure 3.10 The gas phase principle, flowing gas, static liquid (GFS).

In order to allow the calculation of r_O, the gas flow rates, F_{in} and F_{out}, and the oxygen concentrations in the input and output streams, $C_{O,in}$ and C_O, must be known in addition to the variables of the previous technique. Of these, usually C_O is measured and the others are set or known. A gasometric method is not evident here, and the measurement of C_O is done for example with the paramagnetic method.

- **Static gas, flowing liquid (GSF)**

Implementations of the gas phase principle with static gas and flowing liquid (Figure 3.11) have not been found in literature or in practice so far.

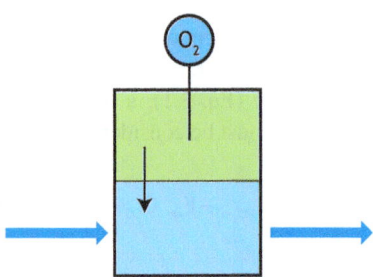

Figure 3.11 The gas phase principle, static gas, flowing liquid (GSF).

- **Flowing gas, flowing liquid (GFF)**

The gas phase principle can also be applied to a full-scale bioreactor. In this case there are liquid input and output streams for the reactor, and transport terms must be added to the mass balance in the liquid phase (Eq. 3.1). The assumption on proportionality between C_{O2} and S_{O2} (Eq. 3.7) becomes more critical because, in addition, the liquid outflow term also depends on it. Additional measurement of dissolved oxygen may then be useful for a correct assessment of the respiration rate. The technique then would no longer be a pure gas phase principle. Note, however, that in general combining L and G principles may lead to more reliable respiration rate measurements.

Table 3.1 summarises the eight measuring principles. The first column contains the names of the mass balance terms, and the second column the mathematical equivalents. The succeeding columns list the respirometric principles, the first four being the liquid-phase principles, and the others being the gas-phase principles. The mass balances for each principle are formed by multiplying the mathematical terms with the coefficients in the column of the appropriate principle and summing them up.

Table 3.1 Overview of measuring principles of respiration rates.

Respirometric principle → Process ↓	Equation ↓		Measurement in LIQUID phase				Measurement in GAS phase			
			LSS	LFS	LSF	LFF	GSS	GFS	GSF	GFF
		Figure nr. →	3.5	3.6	3.7	3.4	3.9	3.10	3.11	3.8
Respiration	$V_L \cdot r_{O2}$		-1	-1	-1	-1	-1	-1	-1	-1
Dissolved oxygen accumulation	$\frac{d}{dt}(V_L \cdot S_{O2})$		-1	-1	-1	-1	-1	-1	-1	-1
Liquid flow	$Q_{in} \cdot S_{O2,in} - Q_{out} \cdot S_{O2}$				1	1			1	1
Gas exchange	$V_L \cdot k_L a \,(S_{O2}^* - S_{O2})$			1		1	1	1	1	1
Gaseous oxygen accumulation	$\frac{d}{dt}(V_G \cdot C_{O2})$						-1	-1	-1	-1
Gas flow	$F_{in} \cdot C_{O2,in} - F_{out} \cdot C_{O2}$							1		1
Gas exchange	$V_L \cdot k_L a \,(S_{O2}^* - S_{O2})$						-1	-1	-1	-1

3.3 EQUIPMENT

3.3.1 Equipment for anaerobic respirometry

To carry out an anaerobic respirometric test there are two requirements. Firstly, a setup is needed in which anaerobic respiration takes place. This can be a small bottle or larger reactor. In this bottle or reactor a substrate, for example, primary sludge or starch, and an inoculum with the consortia required for anaerobic respiration are combined. Secondly, a system to measure the methane production is required. To quantify anaerobic respiration, the flow of electrons has to be determined. Neither the consumption of substrate nor the consumption of an electron acceptor can be measured directly and therefore the final products of anaerobic respiration, H_2 and methane, are determined in anaerobic respirometry.

3.3.1.1 Biogas composition

Methane leaves the bottle or reactor via the biogas. Besides methane, biogas also contains CO_2, H_2S and traces of other compounds. Thus, in order to quantify the methane flow, both the biogas flow and the biogas composition need to be known. To this end, the composition of biogas can be either measured or adapted. Adaptation of the biogas means removing all gas other than methane prior to flow quantification.

- **Measuring the biogas composition and correcting the measured flow**

Measuring the biogas composition can be done with gas chromatography. However, there are also cheaper and easier methods, for example, the apparatus shown in Figure 3.12.

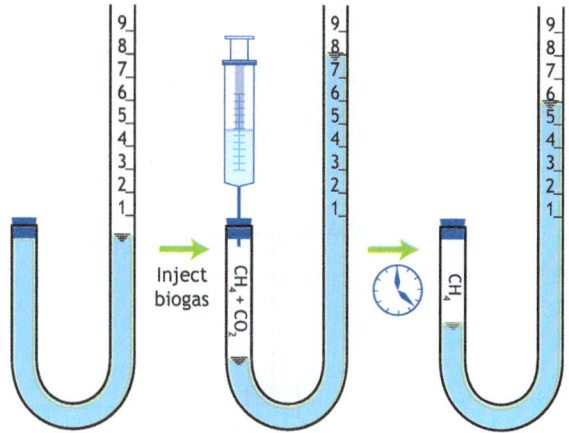

Figure 3.12 An inexpensive and simple way for determining the methane concentration in biogas.

This tube is filled with an alkaline solution (typically 3 molar of NaOH) to remove CO_2 and H_2S, which dissolve in an alkaline solution, leaving only methane in the gas phase. First, biogas is injected into the left leg (t = 0). Then CO_2 and H_2S dissolve in the alkaline solution (orange) over time until all the CO_2 and H_2S are removed (t = end). The methane content can be calculated from the differences in volume at t = 0 and t = end. In this example the total biogas volume is 10 mL, of which 4 mL is dissolved, hence, the methane content of the biogas sample is calculated as 60 %. The time required for all the CO_2 and H_2S to be absorbed into the alkaline solution is to be determined experimentally. This can be done by assessing the time required for the system to reach steady state, i.e. the gas volume does not change anymore. To calculate the CH_4 flow rate from the bottle or reactor, the volume should be corrected using the ideal gas law and the actual temperature.

- **Removing other gases from the biogas**

When CO_2 and H_2S are removed, the gas will usually contain 100% methane (some N_2 and H_2 may be present). This means that in situ removal of compounds other than methane, combined with flow measurement, yields the methane flow. In practice, this means that from the reaction vessel, the gas is led over a large surface of an alkaline solution (typically 3 molar NaOH solution) (Figure 3.13). Notice that, in contrast to what is shown in Figure 3.13, the inlet to the scrubber bottle may not be submerged. This is to prevent back flow of alkaline solution. For example, when there is an under pressure in the head space of the reaction vessel, alkaline solution would be sucked into this reaction vessel, compromising the experiment instantly. An under pressure can occur when the temperature of the head space drops, e.g. when a thermostatic water bath fails or the door of the incubator is left open for a while.

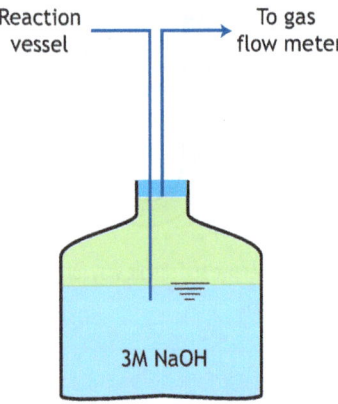

Figure 3.13 Schematic picture of the scrubber bottle.

3.3.1.2 Measuring the gas flow

There is a large variety of ways to measure gas production but in lab-scale anaerobic respirometry it is usually limited to manometric or volumetric methods.

- **Manometric methods**

Manometric methods are based upon measuring pressure increase in the head space of a reaction vessel. As biogas is produced, the pressure in the headspace increases. However, high headspace pressure can result in increased CO_2 solubility, which may significantly disturb microbial activity (Theodorou et al., 1994). Therefore, the pressure needs to be released periodically to prevent it from becoming too high (pressure release). Generally, an upper limit of 1.4 bar is applied. One must also make sure that the reaction vessel is designed to withstand the pressure. When the pressure is not released automatically, this method requires labour during the experiment. Inappropriate operation can lead to explosion of the reaction vessel. It is strongly recommended to always use safety goggles when working with the manometric method. In addition, the initial pressure measurement and the measurements after draining should also consider the temperature effect on gas pressure and water vapour pressure, i.e. an equilibrium condition shall be reached in the reaction vessels before measurement. To calculate the gas production, the pressure increases between two pressure drains are summed and with the ideal gas law the amount of produced moles of biogas is calculated from this total increase in pressure. The composition of the biogas needs to be measured as well if there is no CO_2 and H_2S scrubbing.

- **Volumetric methods**

A classical and robust volumetric method is to use Mariotte's bottle (McCarthy, 1934), where the gas is introduced into a bottle with an outlet for the liquid and it displaces the liquid (Figure 3.14). The weight or volume of the displaced liquid indicates the volume of gas that is produced. When the liquid is an alkaline solution, CO_2 or H_2S is scrubbed in situ and the weight of displaced liquid will indicate the volume of methane. A disadvantage is that the flask needs to be refilled periodically, which disrupts the pressure control. If the displaced liquid is measured with a balance connected to a computer, the gas production can be measured in real time.

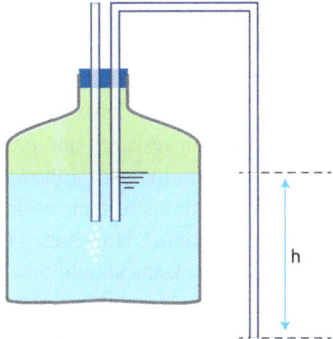

Figure 3.14 Mariotte's bottle for measuring produced gas volumes.

Another example of a volumetric measuring principle is the tilting mechanism (Figure 3.15).

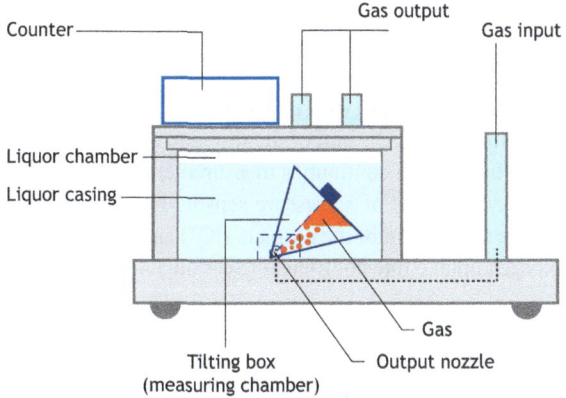

Figure 3.15 Schematic overview of the cross section of a tilting box anaerobic respirometer. The tilting box alternates between right and left as the gas is introduced. The amount of clicks is measured and registered (www.ritter.de).

The advantage is that there is no need for actively resetting the gas flow meter. It can run continuously without requiring attention as Mariotte's bottle does. Several commercial systems exist that use this principle. The measurement principle is based on a tilting box submerged in oil or water. This box is filled from the bottom with gas. This gas accumulates under the chamber and at a certain point this gas induces positive buoyancy and then the box tilts, releasing the gas and thus resetting the system. Every tilt is counted and from this a gas volume is calculated. The downside of this method is that it is rather expensive and that it has a limited flow range (up to 4 L h^{-1}).

3.3.2 Equipment for aerobic and anoxic respirometry

Similar to anaerobic respirometry, a setup is needed in which aerobic and anoxic respiration takes place. This basically consists of a stirred bottle or reactor where biomass under aerobic or anoxic conditions and wastewater, or a specific substrate, are combined. In addition, an arrangement is required to measure the uptake of the terminal electron acceptor, i.e. oxygen, nitrite or nitrate. Data handling may be manually, as is mostly the case in BOD measurements, or completely automated, for example if the measured data is to be converted to respiration rates with a high measuring frequency. In a number of sophisticated (including commercial) respirometers the operation of the equipment is so complicated that it requires an automated control system.

3.3.2.1 Reactor

The reactor is usually a vessel with a volume ranging from a few 100 mL to several litres. Depending on the application, laboratory or field, the material may be glass or plastic, and is often transparent in order to enable inspection of the content. Depending on the measuring principle (Section 3.3.1.2), the vessel is completely sealed to prevent the exchange of oxygen with the gas phase, or open to allow the transfer of oxygen from the gas phase. Open vessels may also be equipped with aeration equipment (e.g. a sparger) to enhance oxygen transfer. In some cases the vessel may be operated in both open mode (for aeration) and closed mode (for measuring oxygen uptake). In all cases, the vessel is completely mixed, with a magnetic bar, an impeller, a pump, or by aeration. In the laboratory the vessel may be thermostated, often using a double wall for cooling/heating or just a heating element if the temperature is maintained above ambient temperature. Depending on the operation principle (flowing liquid, flowing gas), the reactor may have several inlet and outlet ports, and one or more apertures to accommodate (a) sensor(s). Supplementary equipment may include valves, pumps (for biomass, wastewater, substrate, air, and gas), a mixing tank, substrate container, oxygen container, NO$_3$ supply container, sample pre-treatment unit (sieve, filter), oxygen generator, etc.

3.3.2.2 Measuring arrangement

In many cases the measuring arrangement consists of a sensor (i.e. a probe with an associated meter whether or not connected to a data acquisition system) to measure the concentration of the electron acceptor, i.e. oxygen, nitrite or nitrate. Oxygen may be measured directly in the liquid phase with a galvanic, polarographic or optical dissolved oxygen probe. In simple laboratory tests, especially a BOD test, dissolved oxygen may be measured with a titrimetric or photometric method (Section 3.4.2). The oxygen concentration in the gas phase may be directly measured with a paramagnetic oxygen analyser. However, changes in oxygen concentration may be measured by means of a pressure sensor or gas volume displacement sensor. Nitrate and nitrite concentrations in the liquid phase can be measured with an ion-selective or UV-spectrophotometric sensor (Rieger *et al.*, 2008).

Sensors may have slow response times, and it is important to ensure that the sensor is fast enough to follow the kinetics of the biochemical process. As a rule, the sensor must be 10 times faster than the measured reaction rate.

3.3.2.3 Practical implementation

Many practical implementations have been described in the literature and a number of them have been introduced onto the market. As explained in Section 3.3.1, all the measuring techniques for the respiration rate can be classified into only eight basic principles and the operation of all the existing respirometers can be explained in terms of this classification and the corresponding mass balances. However, only a limited number of respirometers have been applied in considerable numbers in research and practice, or even commercial production. In what follows we describe some respirometers in terms of their basic principle and technical implementation. However, it should be emphasized that by no means should this be understood as a recommendation for a specific method. The choice of a certain measuring principle, its technical implementation, or commercial manifestation depend on the measurement purpose, skill of the user and available budget.

- **Liquid phase, static gas, static liquid (LSS) principle**

The LSS principle can be considered as the simplest respirometric principle because the absence of flowing liquid and gas implies that no supplementary materials, such as pumps or aeration equipment, are needed. The BOD test (Section 3.4.2) is an example of the application of this principle. Figure 3.16 shows an example of a BOD bottle used in a test where DO is only measured in the beginning and at the end of the test, and an example of a BOD bottle with continuous measurement of the oxygen uptake by means of a pressure sensor. The latter allows for the assessment of the ultimate BOD and the first order oxygen uptake rate coefficient (Section 3.4.2).

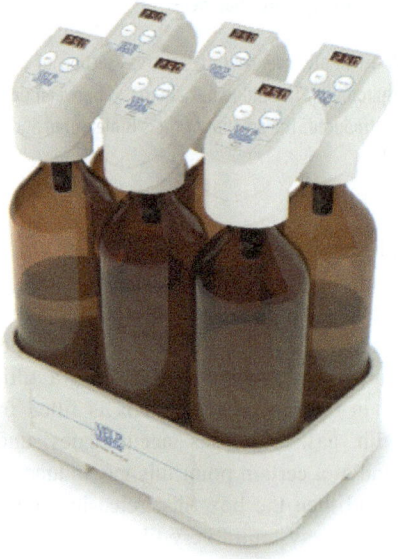

Figure 3.16 A BOD bottle for the classical BOD test (left) and bottles for the continuous measurement of oxygen uptake (right) (photos: Wheaton and VELP Scientifica).

However, the LSS principle has also been implemented in a semi-continuous version to measure the respiration rate of biomass semi-continuously, both in the lab and in the field. Because of the much higher biomass concentration than used in a typical BOD test, the DO concentration drops due to respiration within a few minutes from a near saturated concentration to a limiting concentration. The respiration rate is then calculated from the slope of the DO concentration decline. To allow repeated measurement of the respiration rate the biomass is reaerated after each measurement, which yields a typical saw-tooth DO profile (Figure 3.17). In this example the time between the on/off periods of aeration is constant. Other respirometers make the aeration switch on and off dependent on the actual DO concentration, e.g. between 4 and 6 mg O_2 L^{-1}. These upper and lower DO limits should be defined carefully; they determine the frequency of respiration rate date and their accuracy. Indeed, when the respiration rate is low, it may take a long time to lower the DO concentration from the upper to the lower limit, whereas too short declines make the calculation of the respiration rate sensitive to measurement errors since only a few DO data points are available.

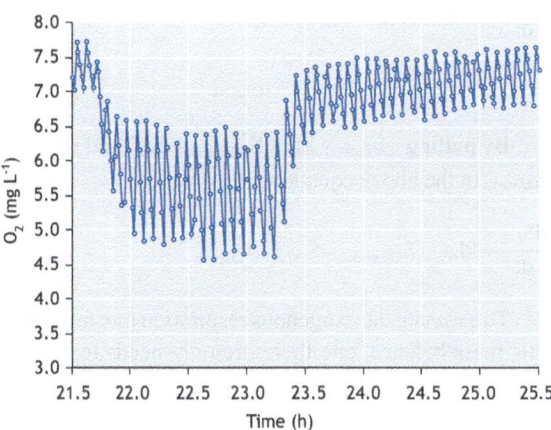

Figure 3.17 Raw DO concentration data from a LSS respirometer with re-aeration.

Potential difficulties with this technique are that during the DO decline, oxygen transfer from the gas phase to the liquid phase needs to be avoided (especially critical when the respiration rate is low) and that identifying a linear decrease of DO is not always evident. The latter is especially a challenge when the measuring technique is automated. In fact, the transition from the reaeration phase to the DO decline phase can take some time (tens of seconds) and is affected by the removal of gas bubbles from the liquid and by the transient response of the DO probe. Respirometers based on this technique allow the measurement of the respiration rate with a measuring interval ranging from typically a few minutes to several tens of minutes. They also permit, especially in the lab, the generation of respirograms by the addition of wastewater or specific substrates. Figure 3.18 is an illustration of a respirometer based on this technique. This floating ball respirometer is designed for automated sampling and discharge of activated sludge, repeated aeration and calculation of respiration rate.

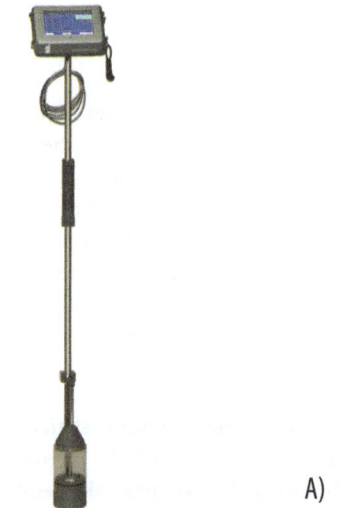

Figure 3.18 Example of a practical implementation of respirometry following the LSS principle. (A) a close up with a DO probe visible and (B) the respirometer in place, i.e. floating on the activated sludge in an aeration tank (photos: Strathkelvin Instruments Ltd.).

Figure 3.19 Example of another practical implementation following the LSS principle (photo: P.A. Vanrolleghem).

Figure 3.19 is an example of another practical implementation following the LSS principle with a closed respiration cell (the right-hand vessel) that is filled with activated sludge from the aerated tank (the left-hand vessel). The DO decline in this cell is measured until a certain minimal DO is reached (or after a given time, or a given DO variation), after which the content is exchanged for fresh, aerated activated sludge from the aerated vessel and a new cycle is started.

- **Liquid phase, flowing gas, static liquid (LFS) principle**

The disadvantage of the need for reaeration can be eliminated by continuously aerating the biomass. The continuous supply of oxygen guarantees a non-limiting DO concentration even at high respiration rates, for example at high biomass concentration and high wastewater or substrate doses. Moreover, continuous aeration allows an open vessel, which facilitates the addition of wastewater and substrate.

To obtain the respiration rate, both the mass transfer term (under process conditions) and differential term in the DO mass balance must be known (Eq. 3.3). The mass transfer term is calculated from the measured DO concentration, the mass transfer coefficient kLa and the DO saturation concentration S^*_{O2}. These two coefficients have to be determined regularly because they depend on environmental conditions such as temperature, barometric pressure and the properties of the liquid (e.g. salts and certain organics). The simplest approach is to determine these by using separate reaeration tests and look-up tables. Standard procedures for these tests under process conditions are available and are based on the disturbance of the equilibrium DO concentration by interrupting the aeration, adding hydrogen peroxide, or even by the addition of a readily biodegradable substrate. The obtained reaeration curve can then be used to estimate the kLa and DO saturation concentration S^*_{O2}. Nonlinear parameter estimation techniques as presented in Chapter 5 are recommended to obtain reliable values. The advantage of the method based on disturbance by respiration of a readily biodegradable substrate is that the values of the aeration coefficients can be updated relatively easily and frequently. However, when only a moderate DO disturbance occurs, the accuracy of the estimated kLa is low. Also, it must be assumed that the respiration rate has dropped to a constant (endogenous) rate during the reaeration part of the curve.

Estimation of S^*_{O2} is not required when one is only interested in the substrate-induced respiration, i.e. exogenous respiration $r_{O2,exo}$. Total respiration is the sum of endogenous respiration $r_{O2,endo}$ and exogenous respiration $r_{O2,exo}$. Considering $r_{O2,endo}$, kLa and S^*_{O2} to be constant over a short interval, it can be shown that the equilibrium DO concentration reached under endogenous conditions $S^*_{O2,endo}$ encapsulates the endogenous respiration (Kong et al., 1996). The mass balance for oxygen can then be rewritten as:

$$\frac{dS_{O2}}{dt} = kLa \cdot (S^*_{O2} - S_{O2}) - r_{O2} =$$

$$kLa \cdot (S^*_{O2} - S_{O2}) - r_{O2,endo} - r_{O2,exo} \quad \text{Eq. 3.13}$$

By putting $r_{O2,endo} = kLa \,(S^*_{O2,endo} - S_{O2})$ and replacing $r_{O2,endo}$ in the above equation, one obtains:

$$\frac{dS_{O2}}{dt} = kLa \cdot (S^*_{O2,endo} - S_{O2}) - r_{O2,exo}$$

Eq. 3.14

To estimate the exogenous respiration rate $r_{O2,exo}$ from this mass balance, one therefore only needs to estimate the equilibrium DO concentration $S^*_{O2,endo}$ (directly from the data, Figure 3.20) and the kLa from the reaeration part of a DO disturbance curve obtained with a readily biodegradable substrate. Another advantage of this LFS respirometric principle is that it allows the measurement of r_{O2} at a nearly constant DO concentration, thereby eliminating the dependency of the respiration on the DO concentration (provided DO $\gg$ 0 mg L^{-1}). Yet another advantage is that the time interval at which respiration rate values can be obtained is short, and is in fact only limited by the measuring frequency of the DO probe. This makes a respirometer based on the LFS principle suitable for kinetic tests and model optimization experiments.

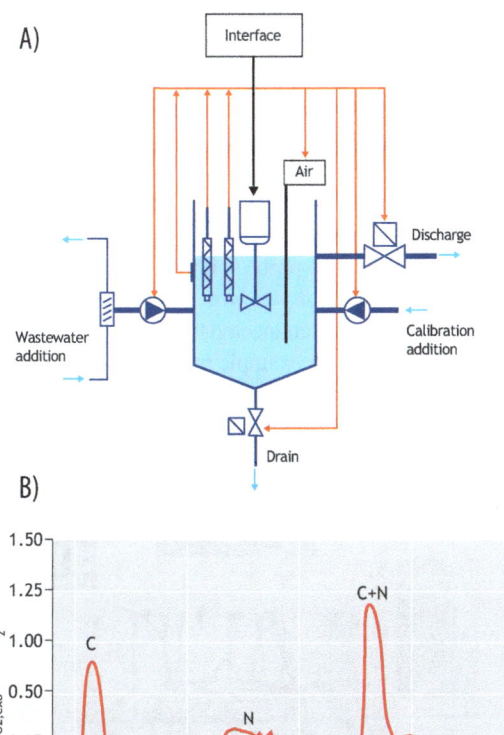

Figure 3.20 (A) Diagram showing a LFS principle-based respirometer setup (Vanrolleghem *et al.*, 1994) and (B) an example of typical raw data (Kong *et al.*, 1996).

Obviously, whereas in its basic form a respirometer following the LFS principle consists of a vessel equipped with an aerator, an impellor or recirculation pump for mixing and a DO probe, a more advanced version designed for automated (online) experiments includes supplementary equipment such as pumps for filling the vessel with biomass, wastewater and substrate addition, level switches, and valves for vessel drainage. This requires a suitable data handling and control system, besides sufficient computing capacity for the estimation procedure. Figure 3.21 shows an example of a commercial version of this measuring principle.

Figure 3.21 Example of a commercial version of a LFS principle-based respirometer using the setup depicted in Figure 3.20. In the middle is the thermostated vessel and at the bottom, left and right, pumps for wastewater and calibration addition (photo: Kelma NV).

- **Liquid phase, static gas, flowing liquid (LSF) principle**

The LSF respirometric measuring principles allow continuous sampling of a biomass stream, for example activated sludge from a lab reactor or full-scale aeration tank, while measuring the respiration rate. When the biomass flows through a closed completely mixed vessel (respirometric cell) without a gas phase, then, following Eq. 3.4, the respiration rate can be calculated if the volume of the respirometric cell and the flow through rate are known and DO concentrations at the inlet and outlet of the cell are measured. Alternatively, the DO can be measured in the source reactor (provided that the decrease in DO in the supply line is negligible) and in the respirometric cell itself. In any case, as with other principles the DO concentration in the cell must be high enough to prevent DO limitation, which requires a sufficiently high input DO concentration at a given respiration rate. A potential source of erroneous measurements is associated with the use of two DO probes to measure DO at the inlet and outlet, because when probe characteristics differ slightly, relatively large relative errors can occur in the difference between the two DO measurements needed to calculate the respiration rate. Spanjers and Olsson (1992) solved this by measuring the DO concentrations in the inlet and outlet of the cell alternately with one single probe located at one port. This was realized by periodically changing the flow direction through the vessel using four solenoid valves which were activated two by two (Figure 3.22).

If a steady state is assumed with respect to the respiration rate then the rate can be calculated using Eq. 3.4 by assuming that the derivative is zero. However, by approximating the equation by a difference equation, the respiration rate can be calculated under dynamic conditions.

Figure 3.23 shows a commercial version of the LSF principle using the one probe solution. This respirometer measures the respiration rate with an interval of typically one minute and can be connected to a lab reactor or a full-scale aeration tank, for example by using a fast loop.

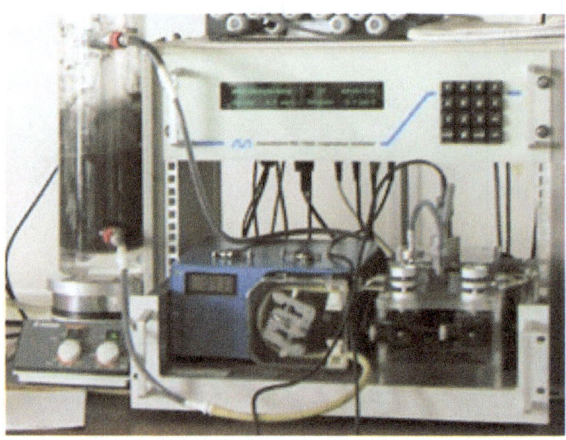

Figure 3.23 Example of a commercial version of the LSF principle using the setup depicted in Figure 3.22. On the left is a lab-scale aeration tank with activated sludge. In the equipment box on the left is the sampling pump, and on the right is the set of solenoid valves. The vessel is placed behind the valves (photo: Applitek NV).

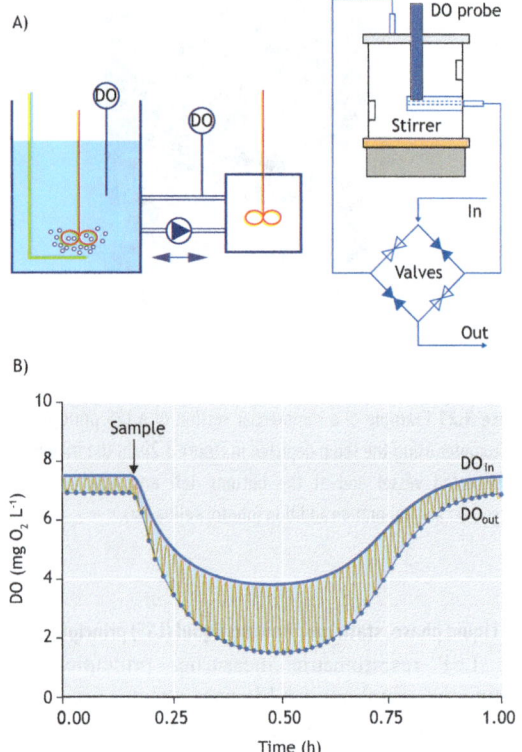

Figure 3.22 Example of a practical implementation following the LSF principle, with one DO probe to measure DO concentration at the inlet and outlet of the respiration vessel (Spanjers, 1993). (A) a schematic of the measurement arrangement and (B) a typical profile of a signal recorded by the single probe (Spanjers and Olsson, 1992). The signal represents the DO oscillating between the DO at the inlet and the DO at the outlet of the cell. This DO signal is the basis for the calculation of the respiration rate.

- **The gas phase, static gas, static liquid (GSS) principle**

Like the LSS principle which is one of the liquid phase principles, the GSS principle is the simplest gas phase principle because the absence of flowing gas and liquid implies that no supplementary materials, such as pumps and aeration equipment, are needed. However, because the calculation of the respiration rate is based on the measurement of oxygen in the gas phase and the actual respiration takes place in the liquid phase, the relation between the gas phase oxygen dynamics must be related to the respiration rate. Thus, in addition to the DO mass balance in the liquid phase, an oxygen mass balance in the gas phase must be considered and the set of equations solved for the respiration rate, assuming a transfer relation between the gas and liquid.

A typical GSS-based respirometer consists of a biomass vessel, mixing equipment and a gas measurement arrangement. Gaseous oxygen is measured by physical methods, such as the gasometric method or the paramagnetic method. Gasometric methods measure changes in the concentration of gaseous oxygen, which can be derived from changes in the pressure (if volume is kept constant, a manometric method) or changes in the volume (if pressure is kept constant, a volumetric method, e.g. Mariotte's bottle), see Section 3.3.1.2. As with the LSS principle, when the oxygen consumption is too high, these methods need replenishment of the gaseous oxygen and thus temporary interruption of the measurements. This limits the possibility for continued monitoring of the respiration rate. An important complication of the GSS principle is that, because carbon dioxide is released from the liquid phase as a result of the biological activity, this gas has to be removed from the gas phase in order to avoid interference with some of the simpler oxygen measurement principles. In practice this is done by using alkali to chemically absorb the carbon dioxide produced.

Similarly to the LSS principle, the GSS principle can be used to carry out a BOD test (Section 3.4.2), for example Oxitop (Figure 3.16).

- **The gas phase, flowing gas, static liquid (GFS) principle**

Like the LFS measuring principle, respirometers using the GFS principle are based on a flowing gas phase, i.e. the biomass is continuously aerated with air (or pure oxygen) so that the presence of sufficient oxygen in the liquid is ensured. However, the calculation of the respiration rate is based on measurement of oxygen in the gas phase, more specifically the gas leaving the liquid phase after aeration, also called off-gas. Because the off-gas may contain other components that are influenced by the metabolic processes in the liquid phase, such as carbon dioxide and nitrogen, it is obvious that these may also be measured to obtain additional information on the activity of the biomass.

A typical GFS-based respirometer consists of a biomass vessel, aeration and mixing equipment and an off-gas measurement arrangement. Gaseous oxygen is measured by physical methods, such as the gasometric method (that is: by supplying pure oxygen from a reservoir or by using electrolysis) or the paramagnetic method.

Following the gasometric method, in a closed headspace the change of pressure or change of volume is related to oxygen consumption in the liquid phase. This information can be used to activate an oxygen production system, based on an oxygen bottle or an electrolytic cell, and the oxygen flow or electrical current can be converted to the respiration rate. In fact the resulting oxygen supply serves as the aeration, i.e. flowing gas. Change of pressure can be measured by a pressure sensor. No documentation exists that describes the use of volume change to activate oxygen supply. Obviously the gasometric method is based on measurement of the oxygen transfer $[k_La \cdot (S^*_{O2}-S_{O2})]$ and not on the measurement of the oxygen concentration in the gas phase. The gas phase oxygen concentration must be assumed constant (i.e. $dC_{O2}/dt = 0$). In any case the gasometric method requires a CO_2 absorption arrangement in order to remove CO_2 from the gas. CO_2 in the gas that is produced during biodegradation and is released to the gas phase from the liquid phase, would otherwise interfere with the measurement of pressure or volume change. Typically one mole of CO_2 is produced per mole of O_2 in aerobic respiration, which by definition means that no gas pressure change will occur.

Alternatively, the oxygen concentration in the gas phase may be measured directly, which eliminates the need to capture CO_2 and allows continuous aeration of the biomass with air. However, in addition to the gaseous oxygen concentration, DO concentration in the liquid phase must be measured because besides the mass balance in the gas phase, the mass balance in the liquid phase also needs to be considered. Likewise, the flow rate of the gas needs to be measured, for example with a mass flow controller.

Oxygen is one of the few gases that show paramagnetic characteristics, so it can be measured quantitatively in a gaseous mixture by using the paramagnetic method. The method is based on the change in a magnetic field as a result of the presence of oxygen, and this change is proportional to the concentration of gaseous oxygen.

Another method to measure oxygen concentration in the gas phase is by using a mass spectrometer. Using this more expensive equipment has the advantage that other gases may also be measured, which then is usually termed off-gas analysis. Especially CO_2 may be measured in the gas phase because it is a useful indicator of biomass activity under all the redox conditions. However, if CO_2 in the off-gas is measured, and because CO_2 is associated with the carbonate system, extra equipment will be needed to measure the pH and bicarbonate concentration in the liquid phase (Pratt et al., 2003). Note that a mass spectrometer is expensive equipment that also requires special (calibration) gases, which makes this method more an advanced laboratory tool rather than a field application. An alternative may be a less expensive infrared CO_2 analyser.

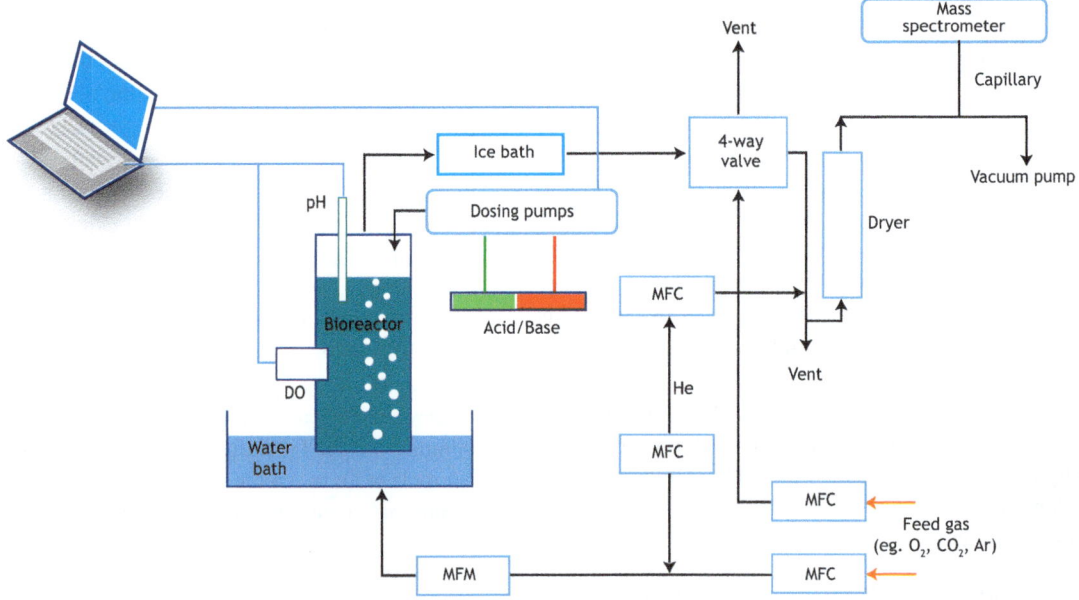

Figure 3.24 Scheme of a GFS-based respirometer setup (Pratt *et al.*, 2003).

Figure 3.24 shows an example of a practical implementation of the GFS principle. This respirometer is based on off-gas analysis using mass spectrometry and is integrated with a titration unit to account for the interaction of CO_2 production and evolution with the acid/base buffering systems in the liquid phase.

The respirometer was used to examine the two-step nitrification process, more specifically nitrite accumulation in wastewater treatment systems operated under varying environmental conditions, i.e. pH and DO concentrations (Gapes *et al.*, 2003).

3.4 WASTEWATER CHARACTERIZATION

Several methods have been developed and applied for the characterization of wastewater, both in terms of pollutant load characterization and in terms of toxicity assessment. In what follows first the different respirometric methods for evaluation of biochemical oxygen demand will be described, followed by respirometric toxicity tests and, finally, an overview of the methods for wastewater fractionation.

3.4.1 Biomethane potential (BMP)

3.4.1.1 Purpose

A biomethane potential (BMP) test is performed when the methane yield of a substrate needs to be known, e.g. when a business case for an anaerobic digester is being developed. Also, the methane production over time can be of interest to optimize the solids retention time of the digester or for the dimensioning of the biogas handling equipment.

3.4.1.2 General

The BMP test is performed to test what the methane potential of a sample is. BMP determines, to a certain extent, both the design and economic details of a biogas plant (Angelidaki *et al.*, 2009). But the BMP test can also be done to evaluate the performance of the biomass. For example, the methane production rate can be used to estimate the hydrolytic activity of the inoculum. BMP tests are often used in literature but there is a large variety in protocols. There have been some efforts to propose a standard (Angelidaki *et al.*, 2009).

Figure 3.25 depicts a typical result of a BMP test. In this case the methane production over time of coarsely filtered sewage is displayed.

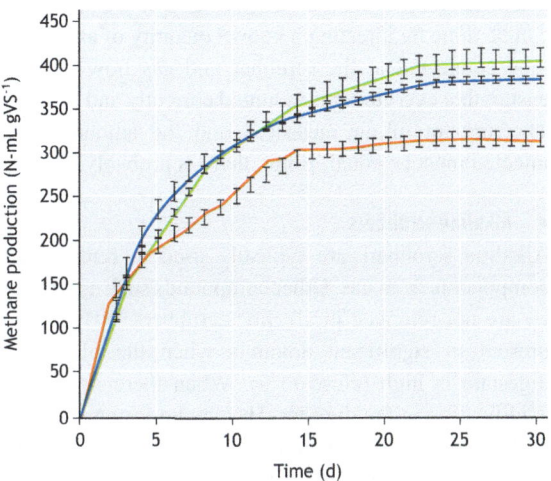

Figure 3.25 Results of the BMP test: methane production over time from coarsely-filtered sewage. Measurements were carried out in triplicate. The red line represents solids from the filter belt; green, solids from the drum filter, and blue, a mix of both solids (Kooijman, 2015; unpublished data)

3.4.1.3 Test execution

In BMP tests, the amount of inoculum (required for anaerobic digestion) is usually measured as volatile suspended solids (VSS). Volatile solids (VS) is a possible alternative. The difference between these methods is that VS considers all the volatile fractions in the inoculum, whereas VSS only considers solids larger than a certain mesh, separated by filtration. VSS is preferred over VS because viable biomass is not expected to pass through the filter in the VSS measurement and thus VSS is more closely correlated to viable biomass than VS; see also Section 3.5.1. However VS measurement can also be chosen because of its accuracy and simplicity of measurement and generally low dissolved volatile solids concentration in inoculum, compared with the volatile particulates. For VS or VSS measurement, most researchers use the standard method (APHA *et al.*, 2012).

In BMP tests, the way to measure the substrate depends on the purpose of the test and on the form of the substrate. For wastewater biomass usually VS is used. Liquid substrate such as wastewater can be quantified by COD. When the BMP in conventional anaerobic digestion is investigated, it is important that no VFAs are accumulated during the BMP tests. This means that the acidification and VFA consumption by methanogenic consortia need to be in equilibrium. The ratio between inoculum and substrate is therefore important. For biomass from a digester the following ratio can be applied, which will in most cases not lead to a net VFA production:

$$2 \geq \frac{VSS_{inoculum}}{VS_{substrate}} \qquad \text{Eq. 3.15}$$

Notice that these are masses applied in the test, expressed in grams of VS or VSS, not concentrations. To obtain the BMP of the substrate, the background production of methane from the inoculum (a blank without substrate) needs to be measured in parallel and subtracted from the produced gas of the inoculum-substrate mixtures.

To perform the BMP test, the reaction bottles need to be incubated at the desired temperature. The standard of mesophilic temperature is 35 °C. The temperature can be kept constant by using a water bath. Stirring has to be performed by impellers in the reaction bottle. When using biomass with high viscosity in BMP tests, insufficient stirring may negatively affect the digestion rates and sometimes impellers in small test bottles (200-400 mL) do not meet the stirring requirement. A solution for this can be to incubate the reaction bottles in an incubator shaker. The duration of the BMP tests depends on their purpose and the characteristics of the substrate but it is most common to use a run time of 30 days.

3.4.1.4 Data processing

In many cases anaerobic respirometric tests aim at measuring the BMP of a substrate. As shown in Figure 3.25, the BMP depends on the time of digestion, similar to the analysis of BOD (Section 3.4.2). The BMP is also usually expressed per gram of VS. For waste-activated sludge this is between 150-200 N-mL gVS^{-1}. For primary sludge this is 300-400 N-mL g VS^{-1}. Since the hydrolysis is often the rate-limiting step in anaerobic digestion (Eastman and Ferguson, 1981), the methane production rate is directly related to the hydrolysis rate and thus the slope of a respirogram as shown in Figure 3.25 is a direct measure of the hydrolysis rate at that point in time.

3.4.1.5 Recommendations

- **Pressure and temperature correction**

In order to determine the amount of methane that is produced in an experiment, the volume in combination with the pressure and temperature is required at each time instant during a test. Typically the amount of produced gas is expressed under standard conditions (usually 273.15 K, 0 °C and 1,013.25 mbar, 1 atm).

- **Methane diffusion**

Methane molecules are known to be able to diffuse through plastics such as silicon. Therefore it is crucial in the design of a BMP (or Specific Methanogenic Activity, SMA, Section 3.5.2) test that the materials that are in contact with the biogas, have poor diffusivity for methane.

- **pH indicator dye for the scrubbing solution**

It is crucial that the liquid in the scrubber bottles and in the measurement device as depicted in Figure 3.13 and Figure 3.15, respectively, has a high (> 9) pH such that CO_2 and H_2S are absorbed by the liquid. A lower pH will result in erroneous measurements. To ensure that the scrubbing liquid is not saturated, methylene blue dye can be added, which turns the liquid blue when pH > 9 making it possible to visually verify the effectiveness of the scrubber solution.

- **Inoculum activity**

The activity of methanogens is very prone to temperature differences. Especially when performing SMA tests, it is important that the activity of the methanogens is high from the beginning of the experiments. Therefore it is strongly advised to store the inoculum at 35 °C for 24 h prior to the experiment. In this way, in a BMP (or SMA) test performed at 35 °C, there will be no temperature shock for the methanogens and the BMP (or SMA) will not be affected by temperature.

- **Micro and macro nutrients**

During a BMP (or SMA) test there may be a lack of micro and macro nutrients, which may affect the conversion performance. If insufficient nutrients are present they should be added. In literature there are various suggestions for nutrient solutions for anaerobic digestion tests (Angelidaki *et al.*, 2009; Zhang *et al.*, 2014).

- **Oxygen inhibition**

When water is in equilibrium with air, the concentration of oxygen will be around 9 mg L^{-1} at room temperature and sea level. Oxygen is known to inhibit methanogens. Also oxygen will 'consume' COD in a mixture. Therefore, it can be desirable (especially in SMA tests where the rates of methanogens are measured) to remove the oxygen from the substrate solution before mixing with biomass. Flushing with N_2 gas is commonly applied.

Very gently bubbling N_2 gas for ~ 60 sec through the substrate solution will remove oxygen.

- **Gas tightness**

Prior to the BMP test, the system of gas production and gas measurement should be verified to be gas tight. This can be done by injecting a known quantity of air into the tube connected to the scrubber and gas flow meter to ensure that everything is mounted correctly and gas tight. The amount of air measured and the amount of air injected must be equal. If not, there is probably a leak.

- **Alkaline scrubbers**

Alkaline scrubbers are typically used to remove acid components from gas. Other compounds such as NH_3 and H_2 are not removed in alkaline scrubbers. NH_3 is often present in significant amounts when the pH of the digestate is high (close to 9). When there is a severe acidification in the digester, H_2 may be formed and thus be present in larger quantities in the biogas. To avoid substantial variations in the pH of the mixture during the assay, a phosphorus buffer solution may be added to the mixture.

3.4.2 Biochemical oxygen demand (BOD)

3.4.2.1 Purpose

The biochemical oxygen demand (BOD) test is performed to assess the biodegradable organic matter concentration of a water sample, e.g. to design a wastewater treatment plant (WWTP) or to evaluate its performance in terms of organic matter removal. Thanks to its sensitivity, it is also used to evaluate the organic matter concentration of the receiving waters.

3.4.2.2 General

The determination of the BOD of wastewaters, effluents, and polluted receiving waters is based on a test that measures the bacterial consumption of oxygen during a specified incubation time. The test quantifies the biochemical degradation of organic material (carbonaceous biochemical oxygen demand - CBOD) but it will also include the oxygen used to oxidize inorganic material such as sulphides and ferrous iron. Unless a nitrification inhibitor is added in the test, it may also measure the amount of oxygen used to oxidize reduced forms of nitrogen (nitrogenous biochemical oxygen demand - NBOD). To make clear what is meant, the term 'total BOD' (BOD_t) is used when no nitrification inhibitor is added, i.e. the sum of CBOD and NBOD.

Normally the incubation time is limited to five days, leading to the traditional BOD_5. However, tests can be conducted over other incubation times, e.g. seven days to facilitate lab organization, or 28, 60 up to 90 days of incubation to determine the so-called ultimate BOD (also: UBOD, BOD_∞ or BOD_U). This measures the oxygen required for the total degradation of organic material (the ultimate carbonaceous demand) and/or the oxygen to oxidize reduced nitrogen compounds (the ultimate nitrogenous demand).

Measurements that include NBOD are not generally useful for assessing the oxygen demand associated with organic material. In fact, NBOD can be estimated directly from nitrifiable nitrogen (ammonia or total Kjeldahl nitrogen) and CBOD can be estimated by subtracting the theoretical equivalent of the reduced nitrogen oxidation from the uninhibited test results. However, this method is cumbersome and is subject to considerable error. The chemical inhibition of nitrification provides a more direct and more reliable measure of CBOD.

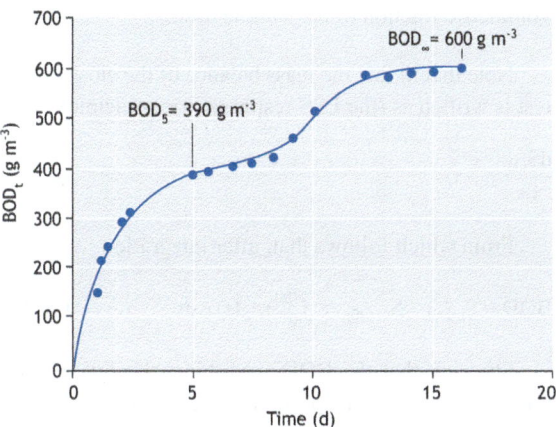

Figure 3.26 The BOD_t curve for mixed municipal and slaughterhouse wastewaters (Henze et al., 1995).

3.4.2.3 Test execution

There are basically two measuring principles for BOD. One employs a closed bottle (LSS respirometer) and the only oxygen available for oxidation of the organic matter is the oxygen dissolved in the (diluted) sample at the beginning of the test.

In other methods oxygen is supplied continuously from a gas phase present in the bottle (LFS respirometer) and the oxygen consumed is monitored.

In what follows the first test approach will be presented and subsequently the methods using oxygen supply from a gas phase are discussed.

- **BOD test with a LSS respirometer**

The method consists of completely filling an airtight bottle with a water sample and incubating it in the dark (to prevent photosynthesis) at 20 ± 0.1 °C for a specified number of days (5, 7, 28, 60, 90 days). Dissolved oxygen concentration is measured initially and after incubation, and the BOD is computed from the difference between the initial and final DO. Because the initial DO concentration is determined shortly after the dilution is made, all the oxygen uptake occurring after this measurement is included in the BOD measurement. The bottles are typically 300 mL with a ground-glass stopper and a flared mouth. Bottles should be cleaned carefully with detergent and rinsed thoroughly (to eliminate any detergent from the bottle).

Dilution

Since the only source of oxygen is that initially present, only limited oxidation can take place because DO concentration should never decrease below 2 mg O_2 L^{-1} to prevent oxygen limitation. This therefore limits sample BOD concentrations to about maximum 7 mg L^{-1}. Whereas this may be adequate for effluent and receiving water samples, wastewaters will have to be considerably diluted in this closed bottle test. Dilution comes however with problems since bacterial growth requires nutrients such as nitrogen, phosphorus, and trace metals (Mg, Ca, Fe). Without these the biodegradation of the pollutants may be limited, leading to underestimation of the BOD. Also, buffering may be needed to ensure that the pH of the incubated sample remains in a range suitable for bacterial growth. Obviously the dilution water used should not contain biodegradable matter.

Seeding

Since the test depends on bacterial activity to degrade the organic matter present in the sample, it is essential that a population of microorganisms is present that is capable of oxidizing the biodegradable matter in the sample. Domestic wastewater, non-disinfected effluents from biological WWTP and surface waters subjected to wastewater discharge contain satisfactory microbial populations. Some waters, e.g. industrial wastewaters, may require a 'seed' to initiate biodegradation. Such a seed can be obtained from the biomass or effluent of a

WWTP, but since nitrifiers may be present in such seed, it is recommended to apply a nitrification inhibitor in the test to ensure proper CBOD test results. In some cases the pollutants may require organisms other than the ones present in domestic WWTP and then it is recommended to seed with bacteria obtained from a plant that is subject to this waste or from receiving water downstream the discharge point.

Blank

Both the dilution water and seed may affect the result of the BOD test, e.g. by introducing organic matter into the bottle. In fact four situations may occur as illustrated in the scheme below:

	Seed	Dilution
1	-	-
2	x	-
3	-	x
4	x	x

Since the test quality could be affected by dilution, it should be ensured by performing a blank BOD test in which the same amount of seed as in the sample test is added to a bottle filled with water for dilution (Figure 3.27).

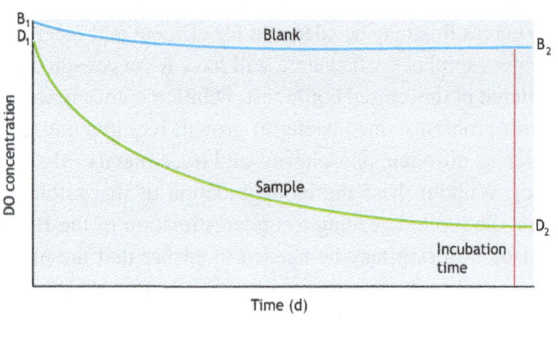

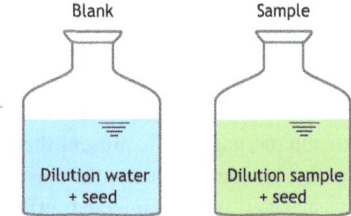

Figure 3.27 A typical BOD closed bottle test result with a sample and a blank.

DO measurement

Dissolved oxygen can be measured with the azide modification of the titrimetric iodometric method or by using a well-calibrated DO electrode. For the measurement of the ultimate BOD over extended incubation periods, only the DO electrode measurement approach is recommended because DO must be measured intermittently during the incubation (intervals of 2 to maximum 5 days, minimum 6 to 8 values).

Data processing

The calculation of the BOD of the sample is as follows (Figure 3.27):

$$BOD = \frac{(D_1 - D_2) - (B_1 - B_2)}{P} \qquad \text{Eq. 3.16}$$

Where, D_1 is the DO concentration of the diluted sample immediately after preparation (mg L^{-1}), D_2 is the DO concentration of the diluted sample at the end of the incubation period (mg L^{-1}), B_1 is the DO concentration of the blank immediately after preparation (mg L^{-1}), B_2 is the DO concentration of the blank at the end of the incubation period (mg L^{-1}), and P is the decimal volumetric fraction of the sample used.

Note that in fact the mass balance of the closed bottle test is written as (the LSS respirometric principle):

$$\frac{dS_{O2}}{dt} = -r_{O2} \qquad \text{Eq. 3.17}$$

From which follows that, after integration,

$$BOD = S_{O2,t0} - S_{O2,tfin} = \int_{t0}^{tfin} r_{O2}(t) \cdot dt \qquad \text{Eq. 3.18}$$

Showing that the BOD is nothing else but the area under a so-called respirogram, i.e. a time series of respiration rate data.

For the determination of the ultimate BOD, the following first order equation should be adjusted to the time series of DO depletion data (Figure 3.28):

$$BOD_t = BOD_U (1 - e^{-kt}) \qquad \text{Eq. 3.19}$$

Where, BOD_t is the oxygen uptake measured at time t (mg L^{-1}), BOD_U is the ultimate BOD (mg L^{-1}), and k is the first order oxygen uptake rate coefficient (d^{-1}).

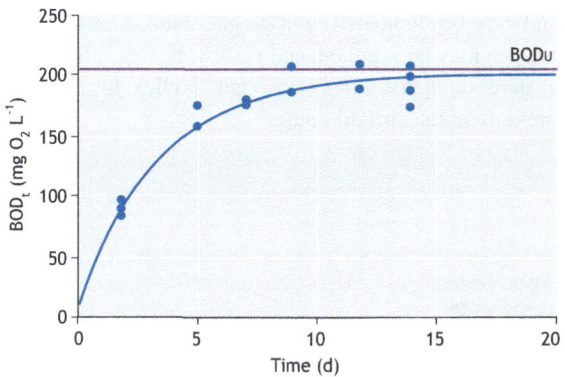

Figure 3.28 Measured BOD of the sample over a period of 14 days and fit of a first-order equation (Weijers, 2000).

Equation adjustment should preferably be performed using nonlinear regression, e.g. by using the Solver function in Excel to minimize the sum of squared errors between the measured time series of oxygen uptake and the model predictions of BOD_t for certain BOD_U and k values. Note that this approach not only yields BOD_U but also k. The latter provides information on the degradation rate of the organic matter. Note that a first-order model may not always be the best choice. Much better fits can often be obtained with alternative kinetic models, in particular consisting of a sum of two or more first-order models.

Recommendations

When the DO concentration at the end of the test is below 1 mg L^{-1} or the DO depletion is less than 2 mg L^{-1}, the test should be carried out again, but with a higher or lower dilution, respectively.

To verify whether the BOD test has been performed well, a test check can be made with a solution with known BOD. The recommended solution is a standard mixture of 150 mg L^{-1} glutamic acid and 150 mg L^{-1} glucose. A 2 % dilution of this stock solution should lead to a BOD_5 of approximately 200 ± 30 mg L^{-1}. A BOD_U of 308 mg L^{-1} can be anticipated (APHA, 2012).

Sample pre-treatment may be necessary. The temperature and pH of the sample may have to be adjusted to 20 °C and 6.5 < pH < 7.5 before dilution. If the sample has been chlorinated, it should be dechlorinated (by adding Na_2SO_3) and a seed should certainly be used. Samples supersaturated with oxygen (above 9 mg L^{-1} at 20 °C) may be encountered when the sample was cold or there has been photosynthetic activity. In such cases deoxygenation is necessary by vigorously shaking a partially filled BOD bottle or by aerating it with clean compressed air.

Nitrification can be inhibited using multiple chemicals, e.g. nitrapyrin, allylthiourea (ATU), or 2-chloro-6-(trichloromethyl)pyridine (TCMP). While recommended concentrations normally lead to adequate inhibition, adaptation to these chemicals has been reported and they can also be degraded during the test, allowing nitrification to start up at a later stage in the test. It is therefore recommended to check at the end of the test whether nitrite and nitrate have been formed.

(Field) test kits are available to measure BOD without lab equipment. Standard nutrients and seed bacteria are supplied, and DO is measured based on the photometric method.

- **A BOD test with a GFS respirometer**

To solve the problem of a limited amount of dissolved oxygen available in the closed bottle test, BOD test equipment has been developed. The oxygen is supplied to the liquid within this test equipment, thus enabling the degradation of organic matter. In this way, dilution may not be necessary or reduced significantly as the BOD measurement range can be extended significantly. The basic equipment consists of a bottle in which a gas phase is provided that can supply oxygen as it is consumed in the tested sample. Either the volume of gas contains sufficient oxygen to allow completion of the oxidation of the biodegradable matter in the sample, or the equipment is able to replenish the gas phase with fresh oxygen from an external source. In this way, besides BOD, the oxygen uptake can also be measured more or less continuously over time, which inherently provides the possibility to calculate the respiration rate over time.

There is various equipment that works according to the GFS principle. Basic principles include manometric respirometers that monitor the pressure change in the gas phase above the liquid as oxygen is consumed. Interference by CO_2 that is produced during biodegradation and released from the liquid is dealt with by capturing the CO_2 in an alkaline (KOH) solution or granules integrated in the equipment. Using the ideal gas law the recorded pressure drop can be translated into oxygen uptake. Equipment using a pressure transducer, simple calculation logic and a data logger for the BOD time series is commercially available. Volumetric respirometers record the gas volume reduction (at constant pressure) as oxygen gets depleted in the gas phase. Again, CO_2 scrubbing with alkaline solution/granules is required. Electrolytic respirometers use the principle of constant volume and gas pressure to activate the electrolytic production of oxygen and maintain the oxygen concentration in the gas phase constant (Figure 3.29). CO_2 emitted from the liquid would again interfere with the constant pressure principle and must be eliminated from the gas phase. Alternatively, oxygen may be supplied, and its flow be measured, from a pure oxygen source (e.g. gas bottle) to maintain pressure in the airtight bottle.

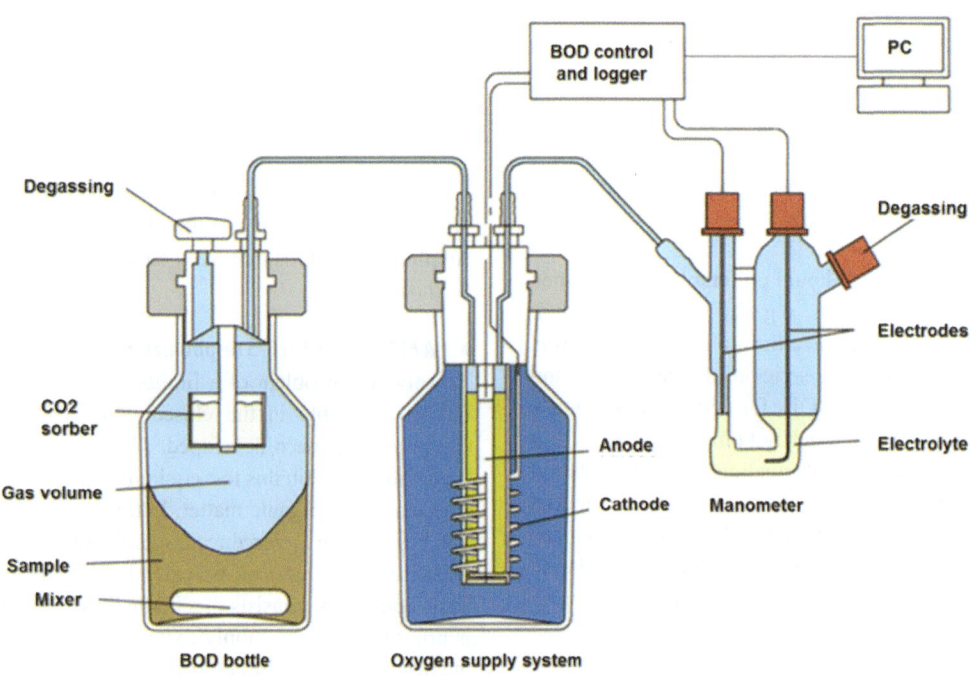

Figure 3.29 An electrolytic respirometer for BOD analysis (SELUTEC GmbH).

If no external oxygen supply is available, as in the respirometer described earlier ("BOD test with a LSS respirometer"), one should pay attention to the overall oxygen consumption that will be exerted by the sample. This should not exceed the amount of oxygen present in the gas phase above the liquid as this would lead to oxygen limitation and thus erroneous results. The volume of sample to be added in the bottle will thus depend on its BOD content and respirometer manuals will typically provide a table with volumes to be added for different BOD ranges. In any case, methods without oxygen supply suffer from decreasing DO concentration in the liquid, which may influence the oxidation rate during the test.

The instruments allow readings of the BOD as frequent as every 15 min to every 6 h. High frequency data collection may help in interpreting the results in terms of degradation kinetics or allow a better model to fit to reduce the influence of measurement noise or get a more reliable estimate of the ultimate BOD.

Recommendations
Interference by gases other than CO_2 may lead to erroneous results, but is not often reported. Temperature variations may also affect pressure and volume measurement as they affect overall pressure. In addition, atmospheric pressure changes may affect the measurement in some respirometers.

An oxygen demand as small as 0.1 mg L^{-1} can be detected, but the test precision will depend on the total amount of oxygen consumed, the precision of the pressure or volume measurement and the effect of temperature and atmospheric pressure variations.

Since oxygen must be transferred from the gas phase to the liquid phase, one must be careful that the oxygen uptake rate does not exceed the mass transfer rate by too much as this would lead to oxygen limitation of the biodegradation process and thus errors in the BOD result. The oxygen transfer is mostly dependent on the mixing conditions in these respirometers, thus limiting the oxygen uptake rates to 10 mg L^{-1} h^{-1} for low mixing devices and to 100 mg L^{-1} h^{-1} if high intensity mixing is provided. If the uptake rate exceeds the supply rate of the respirometer at hand, one can dilute the sample such that the uptake rate decreases to acceptable values. The dilution water composition must be checked as mentioned above.

Note that the BOD test is a respirometric test because it measures consumption of the terminal electron acceptor O_2, although it generally does not measure the respiration rate. In fact BOD is the cumulative oxygen consumption that may be obtained by integrating the respiration rate over a certain incubation period. As noted above, the LFS respirometry inherently provides the possibility to calculate the respiration rate over time.

Also note that BOD can be measured in similar ways by using other electron acceptors, such as nitrate.

Seeding and nutrient additions may be required for particular wastewater samples if competent biomass is not present or the wastewater is not balanced in terms of nutrients versus organic material and may thus be subject to limitations of bacterial growth and thus biased BOD results.

3.4.3 Short-term biochemical oxygen demand (BOD$_{st}$)

Temporal variation of the wastewater composition can be readily characterized by using chemical methods such as COD and TOC analysis. These methods can provide data at high measuring frequency (e.g. on an hourly basis) but they do not provide information on the bio-treatability of the pollutants. Traditional methods that rely on the monitoring of the biodegradation of the pollutants to obtain an indication of the treatability such as the aforementioned BOD$_5$ method are clearly unsuitable to provide such high-frequency information due to the large time delay between the sample introduction and measurement result. However, the principle of monitoring the oxygen uptake for assessment of the treatability and organic matter concentration of a wastewater is a very powerful one, because most wastewater treatment processes rely on aerobic degradation of the organic matter. Therefore, methods have been proposed to decrease the response time of these biologically mediated methods to such a level that their application in high-frequency monitoring becomes possible.

The short-term biochemical oxygen demand (BOD$_{st}$) is defined as the amount of oxygen consumed for biodegradation of readily biodegradable organic matter per volume of wastewater.

In the traditional BOD$_5$ test (Section 3.4.2), a small amount of biomass is added to a large wastewater sample (typically the initial substrate to biomass ratio S_0/X_0 is set to be between 10/1 and 100/1 mg BOD$_5$ mg MLVSS^{-1}). As a result, substantial growth has to occur before the available pollutants are degraded and possibly a lag phase may occur where adaptation of the sludge to the pollutants takes place. To speed up the response time (to within an hour), the techniques for BOD$_{st}$ determination are based on a low S_0/X_0 ratio (typically 1/20 to 1/200 mg BOD$_5$ mg MLVSS^{-1}). These conditions are obtained by the addition of a small aliquot of wastewater to the activated sludge present in the test vessel. In this way the degradation time can be reduced considerably (to often less than one hour) and no significant growth of biomass is to be expected given the relatively low amount of organic matter added.

Because of this short test time, it is evident that matter that biodegrades slowly in the wastewater sample will not be degraded, i.e. only the readily biodegradable fraction is measured. There is also no time for adaptation of the biomass to any new organic component that may be present in the wastewater and to which the biomass is exposed for the first time. In addition there is an important issue of not using biomass from an enhanced biological sludge removal (EBPR) plant. This biomass contains phosphorus accumulating organisms (PAO) that can store volatile fatty acids (VFA). When the VFA fraction of the readily biodegradable matter is taken up for storage it will not be oxidized and hence it will not be measured in a respirometric test. The reader is also referred to sections 3.5.3.2 and 3.5.3.3 (Figure 3.48).

However, being able to measure the readily biodegradable matter is very relevant information in particular for anticipating and optimizing the performance of denitrification and enhanced biological phosphorus removal. Note that, because a respirometric test can be used to distinguish between slowly and readily biodegradable matter, this provides a basis for the assessment of wastewater fractions in the context of a mathematical model (Section 3.4.5).

Due to the increased respiration resulting from the high degradative capacity available in the test vessel, problems may arise to meet the oxygen requirements of the sludge. Respirometric methods that are based on measurement of the decrease of the initial amount of oxygen present in the biomass are severely restricted because of the risk of oxygen limitation. As a result the dynamic concentration range of such methods is small, with maximum sample additions of 5 mg BOD_{st} L^{-1} activated sludge. Obviously, aeration of the sludge, available in many respirometric equipment, solves this problem. However, the intensity of aeration must be sufficient. Still, since in many cases the dissolved oxygen is measured in the test vessel, it can easily be checked whether the oxygen supply has been sufficient (that is: DO should always be above 2 mg L^{-1}) during the BOD_{st} test. Initial BOD_{st} concentrations in the test vessel can easily reach 100 mg BOD_{st} L^{-1} in most respirometers.

3.4.3.1 Test execution

The BOD_{st} test is performed by first bringing a volume of activated sludge in the thermostated (e.g. 20 °C) and aerated test vessel and letting it come to a stable state characterized by what is called endogenous respiration. Endogenous respiration is defined as the state the biomass gets in when there is no more external substrate available in the activated sludge. The biomass is respiring its own reserve materials or is respiring the lysis products of dead biomass. Reaching this state may take between a few hours and a day, depending on how loaded the activated sludge is with slowly biodegradable (hydrolysable) matter. Typical endogenous respiration rates are between 2 and 10 mg O_2 L^{-1} h^{-1}. Once the endogenous state has been reached a pulse of wastewater is injected, the amount of which is calculated to lead to the desired S_0/X_0-ratio of approximately 1/20 to 1/200 mg BOD_5 mg $MLVSS^{-1}$. As soon as the substrate becomes available to the biomass, biodegradation starts and the oxygen uptake rate increases quickly to a maximum rate that is determined by the activity of the biomass and the degradation rate of the substrate. If nitrifiers are present in the sludge and the wastewater contains nitrifiable nitrogen (ammonia and organic nitrogen that is ammonified rapidly) this exogenous respiration rate also includes the oxygen uptake rate for nitrification. The substrates will gradually get exhausted and the oxygen uptake rate will gradually decrease to eventually reach the endogenous rate. The substrate pulse-induced curve of oxygen uptake rates is called a respirogram, a schematic of which is given in Figure 3.30.

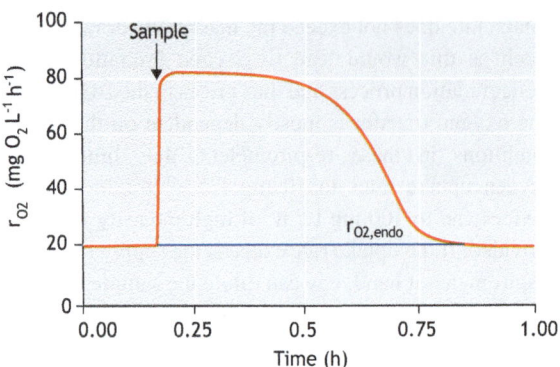

Figure 3.30 Schematic of a respirogram. After biomass has undergone endogenous respiration (no external substrate is present), a sample containing organic matter is injected. The exogenous respiration starts and continues until all the substrate has been removed, and respiration returns to the endogenous rate level.

Figure 3.31 illustrates that exogenous heterotrophic and autotrophic respiration are independent and their respirograms can be superimposed. Three respirograms are shown, one with only COD addition as acetate showing heterotrophic activity, one with ammonia addition showing nitrifier activity and one with the COD/N mixture.

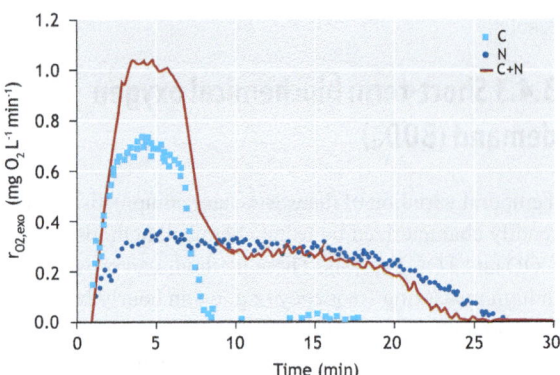

Figure 3.31 Illustration of adding together the exogenous heterotrophic and autotrophic respiration rates. The square symbols represent a respirogram with the addition of 20 mg COD L^{-1} of acetate, the circles represent the respiration rate due to nitrification after the addition of 2.5 mg N L^{-1} and the line is the respirogram of the mixture of 20 mg COD L^{-1} and 2.5 mg N L^{-1} (Kong et al., 1996)

Figure 3.32, Figure 3.33 and Figure 3.34 show some typical respirograms, some of them only showing the exogenous respiration rates that, following some respirometric principles, can be directly calculated from the DO data.

The presence of respiration due to nitrification is noteworthy in the respirogram in Figure 3.32. Indeed, when a wastewater sample supplemented with extra ammonium is injected, one can clearly observe additional exogenous respiration due to respiration for nitrification in the absence of nitrification inhibition.

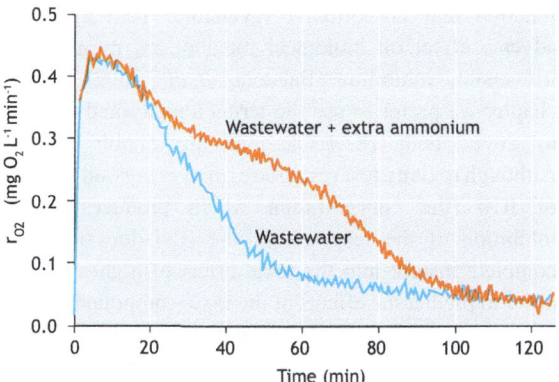

Figure 3.32 Example of respirograms obtained from BOD tests with wastewater and wastewater mixed with an additional amount of ammonia (Petersen *et al.*, 2002a).

When two experiments are performed, one in the presence of a nitrification inhibitor and one without an inhibitor, one can estimate the concentration of ammonium in the wastewater by making the difference between the BOD_{st} of both samples. This oxygen demand divided by 4.33 mg O_2 mg N^{-1} enables the ammonium concentration in the wastewater sample to be obtained (provided that full nitrification to nitrate has been achieved). Moreover, since ammonification is normally a very quick process, it is not only ammonium that can be quantified like this, but in fact the nitrifiable nitrogen in the wastewater.

Figure 3.33 shows an example of a respirogram for a typical municipal wastewater sample added to biomass without a nitrification inhibitor. In this graph only the exogenous respiration rate is depicted. The interpretation of the respirogram goes as follows (Spanjers and Vanrolleghem, 1995). Starting from the right end of the respirogram, the respiration rate increases gradually until about 50 minutes at which point a sudden decrease in the respiration rate is observed. This sudden decrease is due to the complete removal of ammonium from the activated sludge, after which only hydrolysable organic matter is being oxidized, hence the typical exponential decrease in respiration rate due to the first-order nature of hydrolysis.

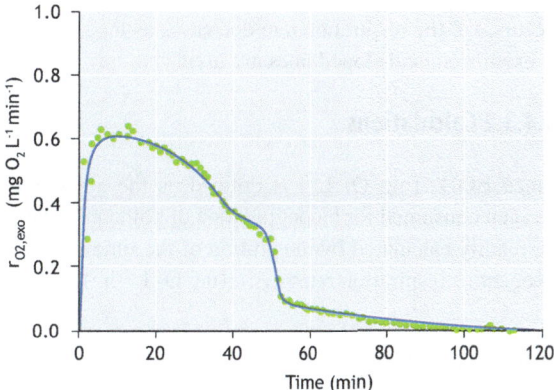

Figure 3.33 An example of respirograms obtained from BOD_{st} tests with municipal wastewater (Spanjers and Vanrolleghem, 1995).

The final example presents a respirogram of an industrial wastewater (Figure 3.34).

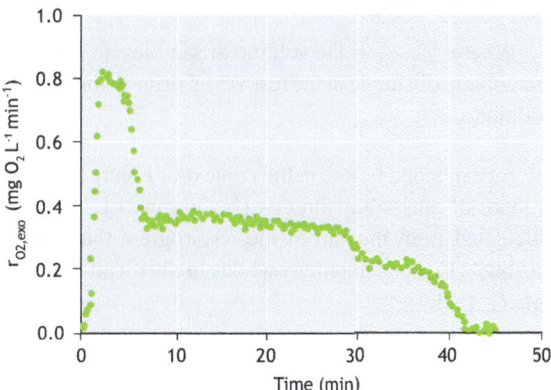

Figure 3.34 The results of the OUR test with industrial wastewater depicting three superimposed respirograms corresponding to the degradation of three different solvents (Coen *et al.*, 1998)

Again, only exogenous respiration rates are shown. The respirogram shows in fact the superposition of three respirograms. This could be explained by the presence of three major solvents in the wastewater that were degraded in parallel by the acclimated biomass. By making three horizontal lines corresponding to the plateaus, one can delineate the amounts of each of the solvents by taking the corresponding area (see the calculations below). Note that in Figure 3.34 horizontal lines can be used because the measured rates correspond with substrate saturation conditions, whereas in some other cases (for example Figure 3.39) substrate limitation occurs, i.e. the respiration rate decreases as the substrate is exhausted, and sloped lines are used.

3.4.3.2 Calculations

Since BOD_{st} (mg O_2 L^{-1}) is defined as the amount of oxygen consumed for biodegradation of pollutants, it can be readily calculated by integration of the time series of exogenous respiration rates $r_{O2,exo}$ (mg O_2 L^{-1} h^{-1}):

$$BOD_{st} = \int_{t_{pulse}}^{t_{final}} r_{O2,exo}(t) \cdot dt \qquad \text{Eq. 3.20}$$

Where, t_{pulse} is the time of pulse addition, and t_{final} is the time needed to return to the endogenous respiration rate after the sample addition.

Evidently the BOD_{st} of the injected sample is deduced from:

$$BOD_{st}^{sample} = \frac{V_{Sludge} + V_{Sample}}{V_{Sample}} \cdot BOD_{st} \qquad \text{Eq. 3.21}$$

Where, V_{Sample} is the volume of sample and V_{Sludge} is the volume of sludge in the test vessel prior to the sample addition.

As explained before in the context of Figure 3.31, the amount of nitrifiable nitrogen (N_{nit} in mg N L^{-1}) can be calculated from the part of the respirogram that can be attributed to the respiration rate due to nitrification $r_{O2,exo}^{Nit}$ (mg O_2 L^{-1} h^{-1}):

$$N_{Nit} = \frac{1}{4.57 - Y_{ANO}} \int_{t_{pulse}}^{t_{final}} r_{O2,exo}^{Nit}(t) \cdot dt \qquad \text{Eq. 3.22}$$

Where Y_{ANO} is the nitrifier's yield coefficient, typically 0.24 mg COD mg N^{-1}.

Similarly, the BOD_{st} attributed to different fractions of organic matter present in a wastewater can be calculated from the exogenous respiration rates that can be attributed to the superimposed respiration rates $r_{O2,exo}^i$ (mg O_2 L^{-1} h^{-1}), Figure 3.33:

$$BOD_{st}^i = \int_{t_{pulse}}^{t_{final}} r_{O2,exo}^i(t) \cdot dt \qquad \text{Eq. 3.23}$$

3.4.4 Toxicity and inhibition

3.4.4.1 Purpose

Respirometric techniques have been frequently preferred for the assessment of inhibitory and toxic effects of substances or wastewater on biomass (Volskay and Grady, 1990). Inhibition is the impairment of biological function and is normally reversible. Toxicity is an adverse effect on biological metabolism, normally an irreversible inhibition (Batstone et al., 2002). In this chapter we prefer to use the terms toxicity and toxicant to cover both reversible and irreversible effects. Although toxicity test results are often expressed in terms of IC_{50} (the concentration which produces 50 % inhibition of the respiration), the IC_{50} does not give complete insight into the toxic effect of a chemical. In order to predict the effects of the toxic compounds on the removal of organic matter and nutrients or to design mitigative actions in the biological treatment process, the effects of a toxicant on the biodegradation kinetics should be quantified.

3.4.4.2 Test execution

Both decreases in the endogenous and exogenous respiration rates can be used to assess toxicity. If the decrease in the endogenous respiration rate is used as an indicator of toxicity, biomass is first brought to the endogenous state, the toxicant is added and then the reduction in endogenous respiration rate measured. If the toxicant or toxic wastewater is biodegradable, exogenous respiration may occur, interfering with the evaluation of the reduction in the endogenous respiration rate. It has also been found that biomass is less sensitive to toxicants when it is in its endogenous state.

For toxicity assessment based on exogenous respiration rates, respirograms (see Section 3.4.3.1) are the basis. In an exogenous rate-based test, a reference substrate (e.g. acetate for heterotrophic toxicity tests or ammonia for nitrifier toxicity tests) is injected prior to injection of a potentially toxic sample so as to assess the reference respiration rate (i.e. biomass activity). After the toxicant has been added and has had time to affect the biomass, another pulse of reference substrate is injected and the determined activity (e.g. the maximum exogenous respiration rate) is compared to that obtained prior to the toxicant injection. It is to be noted that the time between the toxicant injection and reference substrate injection may affect the level of toxicity determined since longer-term exposure may lead to a stronger effect or may also lead to biomass adaptation (Figure 3.35). No clear indication is available on optimal exposure times. Sometimes even stimulation of respiration may occur, i.e. the maximum respiration rates for reference substrate degradation after the addition of the toxicant are higher than without the toxicant. This is due to increased energy requirements by the biomass to cope with the toxicant (e.g. energetic uncoupling in the presence of benzoic acid as the toxicant). In any case it is important to always report the applied time between the injections of the toxicant and the reference substrate.

3.4.4.3 Calculations

A toxicity severity level can then be deduced by calculating the ratio in activity levels before and after the toxicant addition.

$$\text{Toxicity (\%)} = \frac{r_{O2,exo}^{max}(\text{before}) - r_{O2,exo}^{max}(\text{after})}{r_{O2,exo}^{max}(\text{before})} \times 100$$

Eq. 3.22

Obviously the toxicity severity level will depend on the dose of toxicant applied. Dose-response curves are obtained from a series of experiments in which the toxicity response is measured at different doses, and they allow the assessment of the IC_{50} (the half maximum inhibitory concentration). Normally the testing protocol to obtain the dose-effect relationship is to perform a reference-toxicant-reference sequence at one dose, replace the biomass with fresh biomass and perform another reference-toxicant-reference sequence at a higher dose. Typically the concentration of the dose is a multiple of the previous dose. This set of reference-toxicant-reference sequences is continued until the biomass is completely inhibited.

To speed up the experimentation the step of replacing the biomass can be omitted (Kong *et al.*, 1994), but then the contact time is not under the control of the experimenter. In Figure 3.36 an example of such fast determination of the dose-effect relationship is given for copper addition to biomass in increasing doses of 0, 2.5, 5.0 and 10.0 ppm. This leads to a copper exposure concentration of 0, 2.5, 7.5 and finally 17.5 ppm, but each with different contact times. The figure clearly demonstrates that the maximum respiration rate for degradation of acetate (the reference substrate in this experiment) decreases rapidly. It is noteworthy that the area under the curve remains the same, i.e. the BOD_{st} (Section 3.4.3) remains the same since all the acetate is still degraded, albeit at a lower rate. This also means that the respirogram takes longer to complete.

From this experiment the IC_{50} can be visually deduced to be approximately 7.5 ppm since the maximum respiration rate at this copper concentration is about half the maximum respiration rate in the absence of copper (compare the first, at ~0.5 h, with the third respirogram, at ~1.7 h).

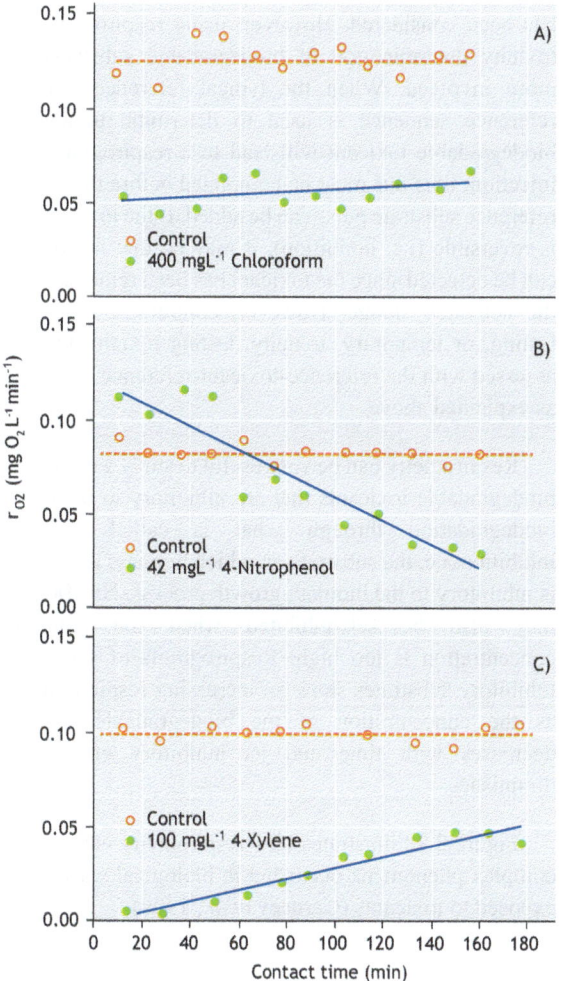

Figure 3.35 Typical data from respiration inhibition tests with different contact times (A) Chloroform, (B) 4-Nitrophenol, and (C) 4-Xylene (Volskay and Grady, 1990)

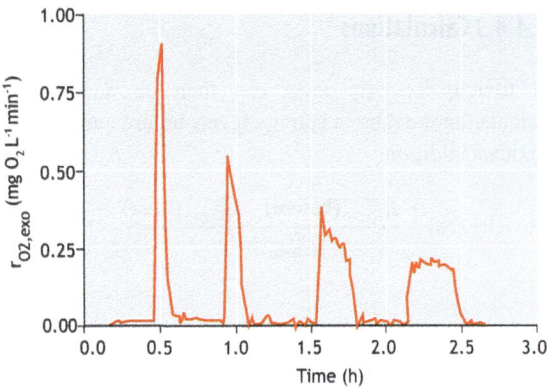

Figure 3.36 Typical respirograms obtained from an ARIKA (Automated Respiration Inhibition Kinetics Analysis) test with copper (Kong et al., 1994). The first respirogram is obtained using pure acetate and the successive respirograms are a series of results where a mixture of acetate with geometrically increasing toxicant concentration (cumulative toxicant concentration: 2.5, 7.5 and 17.5 ppm) was used.

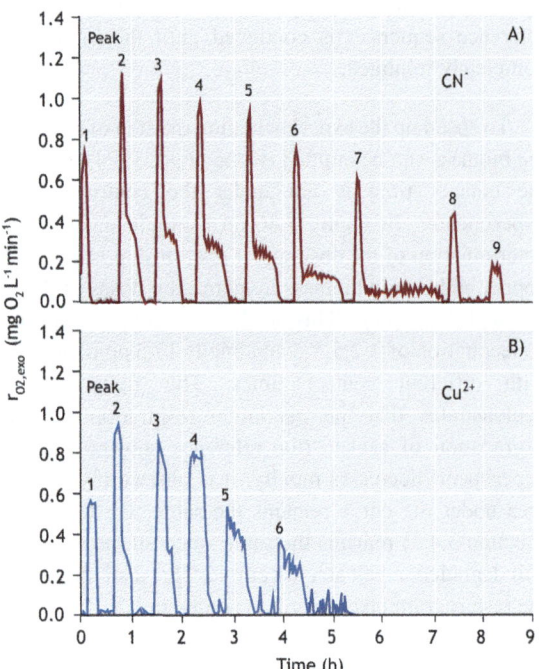

Figure 3.37 Exogenous respirograms with C and N (20 mg L^{-1} and 2 mg L^{-1}, respectively) substrate addition with Cu^{2+} (A) and CN^{-} (B) as toxicants. The first peak is the one for pure acetate, the second one is the pure mixture of C and N, followed by a series of mixtures of C and N substrate and toxicant (cumulative Cu^{2+} concentrations of 2.5, 7.5, 17.5, 37.5, 77.5 and 157.5 ppm and cumulative CN^{-} concentrations of 0.013, 0.038, 0.088, 0.188, 0.388, 0.788 and 1.588 ppm). (Kong et al., 1996).

As previously mentioned, autotrophic and heterotrophic respirograms can be superimposed and toxicity testing towards heterotrophic and autotrophic activity can thus be done in a single experiment with a mixture of COD and N as the reference substrate. In Figure 3.37 the impact of two classical toxicants, copper and cyanide, on heterotrophic and autotrophic biomass are assessed. These respirogram series with increasing toxicant concentration show that nitrifiers are more sensitive to cyanide than heterotrophs (IC$_{50}$ is approximately 0.2 ppm for nitrifiers, whereas it is about 1 ppm for heterotrophs). For copper the IC$_{50}$ for acetate oxidizers and IC$_{50}$ for nitrifiers are approximately the same as the whole respirogram decreases together.

3.4.4.4 Biodegradable toxicants

So far, only the toxicity of non-biodegradable substances has been considered. However, using respirometry for toxicity determination of biodegradable substances is more involved. When the typical reference-toxicant-reference sequence is used to determine toxicity, a biodegradable toxicant will lead to a respirogram after injection; thus this must be completed before the second reference substrate pulse can be added. If the toxic impact is reversible (i.e. inhibition), it may be that no toxicity can be detected since the toxicant has been removed from the activated sludge. However, contrary to this non-lasting, or temporary, toxicity, lasting toxicity may be assessed with the reference-toxicant-reference sequence, as explained above.

Respirometry can be very useful to study a number of biodegradable toxicants that are inhibitory to their own biodegradation through what is called substrate inhibition, i.e. the substrate on which biomass is growing is inhibitory to the biomass growth process. Nitrification may also be self-inhibited when the ammonia concentration is too high. Respirograms of such self-inhibitory substrates show an increasing respiration rate as the concentration of the biodegradable toxicant decreases with time and its inhibitory effect thus diminishes.

Figure 3.38 illustrates how respirometry can explain complex phenomena occurring in biological wastewater exposed to toxicants (Gernaey et al., 1999).

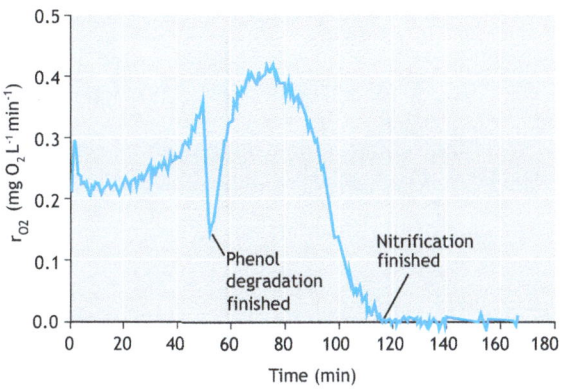

Figure 3.38 The respirometric response of a mixed heterotrophic-nitrifying biomass to the addition of a mixture of phenol (15 mg L^{-1}) and ammonium (5 mg N L^{-1}). Only the exogenous respiration rate is shown (Gernaey *et al.*, 1999).

In this example, phenol and ammonia are added to a mixed biomass consisting of nitrifying and heterotrophic bacteria. Phenol is one of the best-known examples of a toxic but biodegradable compound. In a preliminary experiment it was confirmed that phenol could be degraded by the biomass used in the experiments. What is observed in the respirometric experiment depicted in Figure 3.38 is that nitrification is inhibited by the phenol that is gradually degraded by heterotrophs. However, the phenol degradation itself is also inhibited by the toxic effect of phenol on the heterotrophs, that is: self-inhibition. In the first phase, nitrification is inhibited but the phenol is degraded, albeit slowly. Phenol degradation speeds up after about 30 minutes and comes to an end after about 50 minutes, as can be seen from the sharp decrease in the $r_{O2,exo}$ profile. As soon as the phenol degradation is completed, nitrification increases to the same rate as before the phenol addition (a separate experiment with ammonium addition may also be performed to assess the effect on nitrification alone).

Because the inhibiting concentration depends on the biomass' origin and acclimatization, the toxicity of a compound to acclimatized biomass may be interesting to study. An example of acclimatization based on a respirometric experiment, with a progressive increase in the IC$_{50}$ and inhibition coefficient, can be found in Rezouga *et al.*, (2009).

3.4.5 Wastewater fractionation

The use of dynamic models in activated sludge processes has become more and more widespread and has become a way of thinking and communicating about wastewater treatment processes. The Activated Sludge Model No.1 (ASM1) presented by the IAWQ Task Group on Mathematical Modelling for Design and Operation of Biological Wastewater Treatment Processes (Henze *et al.*, 1987) is generally accepted as state-of-the-art, and is used for simulation of wastewater treatment plants in many studies. It will form the basis of this section, which deals with the use of the respiration rate to obtain wastewater fractions in the context of ASM1.

Before establishing the relationship between the respiration rate and the fractions of the ASM1 components, it is essential to explain this model for the heterotrophic process (Table 3.2). It is assumed that the reader comprehends the Gujer matrix as depicted in Table 3.2. In the mass balance of the heterotrophic organisms X_{OHO} (component 5, in short: c. 5), the production of X_{OHO} by aerobic growth (process, or reaction 1, in short: r. 1) is counteracted by the loss of X_{OHO} by heterotrophic decay (r. 4). In this decay process component X_{OHO} (c. 5) is converted to component XC_B (c. 4). This production of XC_B is counteracted by the loss of XC_B by hydrolysis (r. 7), leading to production of component S_B (c. 2). S_S is used for heterotrophic growth (r. 1) where it is converted to component X_{OHO} (c. 5) at the expense of component oxygen S_{O2} (c. 8), i.e. respiration. A similar reasoning can be made for the processes involving the soluble and particulate nitrogen components (S_{NHx}, $S_{B,N}$ and $XC_{B,N}$) and autotrophic (nitrifying) organisms (X_{ANO}). Anoxic growth (r. 2) with nitrate S_{NOx} (c. 9) as the terminal electron acceptor is also considered.

Hence, the wastewater fractions that will be quantified using aerobic or anoxic respirometry are the biodegradable COD fractions: S_B, XC_B, the biomasses potentially present in the wastewater: X_{OHO} and X_{ANO} and the nitrogen fractions: S_{NHx}, $S_{B,N}$ and $XC_{B,N}$. Other, inert fractions can be determined as well using respirometric data, but then additional chemical analyses of total and soluble COD and total and soluble nitrogen fractions will be needed. This will not be dealt with here, but can be found in Vanrolleghem *et al.* (1999) and Petersen *et al.* (2003).

Table 3.2 Activated Sludge Model No. 1: Gujer matrix (Henze et al., 1987).

Component (i) → ↓ Process (j)	1 S_U	2 S_B	3 $X_{U,Inf}$	4 XC_B	5 X_{OHO}	6 X_{ANO}	7 $X_{U,E}$	8 S_{O2}	9 S_{NOx}	10 S_{NHx}	11 $S_{B,N}$	12 $XC_{B,N}$	13 S_{ALK}	Process rate (ρ_j)
1 Aerobic growth of heterotrophic biomass		$-\dfrac{1}{Y_{OHO}}$			1			$-\dfrac{1-Y_{OHO}}{Y_{OHO}}$		$-i_{N,XBio}$			$-\dfrac{i_{N,XBio}}{14}$	$\mu_{OHO,max} \dfrac{S_B}{K_{SB,OHO}+S_B} \dfrac{S_{O2}}{K_{O2,OHO}+S_{O2}} \cdot X_{OHO}$
2 Anoxic growth of heterotrophic biomass		$-\dfrac{1}{Y_{OHO}}$			1				$-\dfrac{1-Y_{OHO}}{2.86 \cdot Y_{OHO}}$	$-i_{N,XBio}$			$\dfrac{1-Y_{OHO}}{14 \cdot 2.86 \cdot Y_{OHO}} - \dfrac{i_{N,XBio}}{14}$	$\eta_{\mu,OHO}\,\mu_{OHO,max} \dfrac{S_B}{K_{SB,OHO}+S_B} \dfrac{K_{O2,OHO}}{K_{O2,OHO}+S_{O2}} \dfrac{S_{NOx}}{K_{NOx,OHO}+S_{NOx}} \cdot X_{OHO}$
3 Aerobic growth of autotrophic biomass						1		$-\dfrac{4.57-Y_{ANO}}{Y_{ANO}}$	$\dfrac{1}{Y_{ANO}}$	$-i_{N,XBio} - \dfrac{1}{Y_{ANO}}$			$-\dfrac{i_{N,XBio}}{14} - \dfrac{2}{14 \cdot Y_{ANO}}$	$\mu_{ANO,max} \dfrac{S_{NHx}}{K_{NHx,ANO}+S_{NHx}} \dfrac{S_{O2}}{K_{O2,ANO}+S_{O2}} \cdot X_{ANO}$
4 Decay of heterotrophic biomass				$1-f_{XU,Bio,lys}$	-1		$f_{XU,Bio,lys}$					$i_{N,XBio} - f_{XU,Bio,lys}\,i_{N,XE}$		$b_{OHO}\,X_{OHO}$
5 Decay of autotrophic biomass				$1-f_{XU,Bio,lys}$		-1	$f_{XU,Bio,lys}$					$i_{N,XBio} - f_{XU,Bio,lys}\,i_{N,XE}$		$b_{ANO}\,X_{ANO}$
6 Ammonification of soluble organic nitrogen										1	-1		$\dfrac{1}{14}$	$q_{am}\,S_{B,N}\cdot X_{OHO}$
7 Hydrolysis of slowly biodegradable substrate		1		-1										$q_{XCB\_SB,hyd} \dfrac{XC_B / X_{OHO}}{K_{XCB,hyd}+XC_B/X_{OHO}} \left[\dfrac{S_{O2}}{K_{O2,OHO}+S_{O2}} + \eta_{q,hyd,ax} \dfrac{K_{O2,OHO}}{K_{O2,OHO}+S_{O2}} \dfrac{S_{NOx}}{K_{NOx,OHO}+S_{NOx}} \right] \cdot X_{OHO}$
8 Hydrolysis of organic nitrogen											1	-1		$\rho_7 \cdot (XC_{B,N}/XC_B)$

The total respiration rate of biomass in contact with wastewater is, according to the ASM1:

$$r_{O2,tot} = \frac{1 - Y_{OHO}}{Y_{OHO}} \cdot X_{OHO} \cdot \mu_{OHO} + \frac{4.57 - Y_{ANO}}{Y_{ANO}} \cdot X_{ANO} \cdot \mu_{ANO}$$

Eq. 3.25

Where the specific growth rates μ_{OHO} and μ_{ANO} are functions of S_B and S_{NHx}, respectively (Henze *et al.*, 1987).

The concentrations of S_B and S_{NHx}, in turn, depend on the rates at which XC_B, $S_{N,B}$ and $XC_{N,B}$ are degraded (Table 3.2). It is clear that all the independent processes summarised in Table 3.2 eventually act on the mass balance of oxygen (and nitrate if the same evaluation is done for anoxic conditions).

There are two approaches for the assessment of model wastewater fractions: direct methods focus on specific fractions that can be directly evaluated from the measured respiration rates (Ekama *et al.*, 1986; Spanjers *et al.*, 1999), whereas optimisation methods use a (more or less simplified) model that is fitted to the measured data (Kappeler and Gujer, 1992; Wanner *et al.*, 1992; Spanjers and Vanrolleghem, 1995). In the latter, numerical techniques are used to find values of the unknown wastewater fractions that lead to the smallest deviation between the model predicted and the measured respiration rates. The optimisation methods for estimation, using numerical techniques, will not be discussed here, but reference is made to Chapter 5.

From the processes described in Table 3.2, it is clear that the total respiration rate is affected by the concentrations of all the biodegradable components and that the cumulative oxygen consumption (i.e. the integral of r_{O2}) is a measure of the amount of components degraded. Notice that because in direct methods integrals of respiration rates are taken, the measuring frequency and signal-to-noise ratios (measurement error compared to measurement value) are not very critical for the reliable assessment of the component concentrations. This is in contrast to kinetic characterization (Vanrolleghem *et al.*, 1999 and Petersen *et al.*, 2003), where information is obtained from changes in respiration rates (derivative of r_{O2}). This implies a much higher dependency on the parameter accuracy of the respirometric measurement quality.

Inherent with respirometric tests for wastewater characterization is the use of biomass: the assessment of wastewater components is based on the respirometric response of biomass to wastewater. Two important aspects are associated with the use of biomass. The first aspect is the amount of wastewater component with respect to biomass (S_O/X_O ratio, Sections 3.4.1.3 and 3.4.3) that is used. Second, in the death-regeneration concept adopted in ASM1, new S_B, XC_B, S_{NHx}, $S_{N,B}$ and $XC_{N,B}$ are continuously generated from decaying biomass. Within this model it is therefore difficult to distinguish between the components originating from the wastewater and from the biomass itself. In fact the transition between exogenous respiration and endogenous respiration is gradual. The respirometric test should thus be organised in such a way that these rates can be distinguished. This is one of the most challenging problems in respirometric characterization of wastewater in the context of ASM1.

Figure 3.39 shows a respirogram collected in a batch experiment where at the start wastewater is added to endogenous sludge.

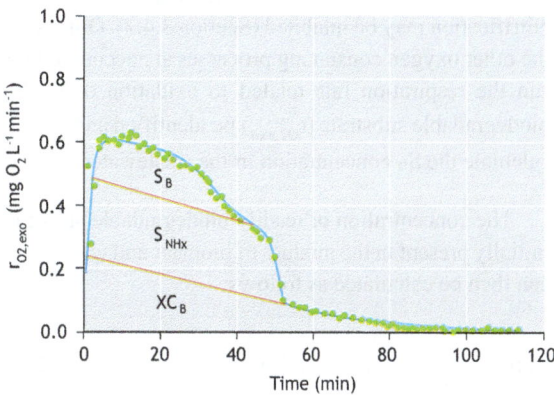

Figure 3.39 An exogenous respiration rate profile obtained after addition of 0.7 L wastewater into 1.3 L activated sludge and fractionation according to the procedure described by Spanjers and Vanrolleghem, 1995b.

In Figure 3.39 only the exogenous respiration rate is presented, i.e. the endogenous respiration is subtracted from the total respiration rate. A typical wastewater respirogram shows an initial peak brought about by the oxidation of readily biodegradable matter, followed by one or more shoulders where successively other components continue to be oxidised. The full area under the respirogram represents the total of biodegradable components $(S_B + XC_B) / (1 - Y_{OHO}) + (S_{NHx} + S_{N,B} + XC_{N,B}) / (4.57 - Y_{ANO})$, as follows from the above equation on total respiration rate. In Figure 3.39 three substrate fractions can be discerned, corresponding to S_B, XC_B and S_{NHx}.

3.4.5.1 Readily biodegradable substrate (S_B)

The readily biodegradable substrate is presumably composed of (simple/low molecular) soluble compounds, such as volatile fatty acids, alcohols, etc. (Henze *et al.*, 1992). The characteristic of these compounds is that they are degraded rapidly and hence provoke a fast respirometric response.

A typical batch test for determination of S_B (e.g. Ekama *et al.*, 1986) involves the addition of a wastewater sample to endogenous sludge, and monitoring the respiration rate until it returns to a level where it can be reasonably assumed that the readily biodegradable substrate is removed from the activated sludge. If other processes occur that are also consuming oxygen (e.g. endogenous and nitrification respiration rates), they must be identified or assumed so that their respiration rates can be subtracted from the total respiration rate (Figure 3.39). Nitrification may be inhibited (Section 3.4.2). Only when the other oxygen consuming processes are accounted for, can the respiration rate related to oxidation of readily biodegradable substrate ($r_{O2,exo}^{S_B}$) be identified and used to calculate the S_B concentration in the wastewater.

The concentration of readily biodegradable substrate initially present in the mixture of biomass and wastewater can then be calculated as follows:

$$S_B(0) = \frac{1}{1-Y_{OHO}} \left(\int_0^{t_{fin}} r_{O2,exo}^{S_B}(t)\,dt \right) \quad \text{Eq. 3.26}$$

The concentration of S_B in the wastewater is then easily calculated by taking into account the dilution. The end point t_{final} of the integration interval is the time instant where S_B is completely oxidised and where the exogenous respiration rate for S_B becomes zero. The integral can easily be obtained by determining the area under the curve, for instance by using a spreadsheet program, also known as the graphical method. An alternative to this direct method is the optimisation method, as explained above. This consists of solving the mass balance equations with a numerical integrator to predict the exogenous respiration rates for S_B in such a batch experiment. Depending on the initial value $S_B(0)$ given to the integration algorithm, the simulation will result in a different predicted respirogram. One can therefore search for the $S_B(0)$ value that gives the 'best fit' to the measured data. For this simple application the optimisation method may be a bit excessive, but for more complex estimation tasks (next paragraph), the approach becomes more straightforward than direct calculation methods.

Notice that knowledge of the heterotrophic yield coefficient Y_{OHO} is needed for the calculation of S_B from respiration rates. This stoichiometric coefficient is always involved when oxygen consumption is converted to substrate equivalents (see also next section and Sections 3.4.5.2 – 3.4.5.6). The batch test described above is also used to assess other ASM1 components and, using the optimization method, kinetic parameters. This explains its popularity in calibration procedures.

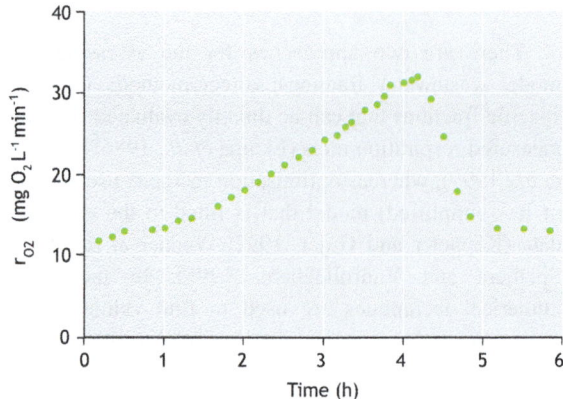

Figure 3.40 Respiration rates measured in a batch experiment (Kappeler and Gujer, 1992) for the determination of readily biodegradable substrate according to the method by Wentzel *et al.*, 1995.

Another batch test (Wentzel *et al.*, 1995) consists of monitoring the respiration rate of unsettled sewage without seed for a relatively long period (up to approximately 20 h). A respirogram similar to the one depicted in Figure 3.40 is obtained. S_B is calculated from the respiration rates observed between the start of the test up to the sudden drop (due to depletion of S_B), with correction for the increasing endogenous respiration due to the increase in biomass during the test. In addition to Y_{OHO}, knowledge of the net growth rate is required, which can be obtained from the same test.

Ekama *et al.* (1986) presented a method for determining S_B that involves respiration rate monitoring in a completely mixed reactor operated under a daily cyclic square-wave feed. It is hypothesised that the sudden drop in respiration rate to a lower level, observed upon termination of the feed (Figure 3.41), corresponds uniquely to the S_B that has entered via the influent. Hence, the concentration of readily biodegradable substrate in the wastewater can be calculated as:

$$S_B = \frac{V_{react}}{Q_{ww}} \cdot \frac{\Delta r_{O2,tot}}{1-Y_{OHO}} \qquad \text{Eq. 3.27}$$

Vanrolleghem, 1995). Alternatively, if the data of such respirometric batch tests are evaluated using the optimisation method to match the response of the model to the data, the nitrification part can easily be extracted from the respirogram (Spanjers and Vanrolleghem, 1995).

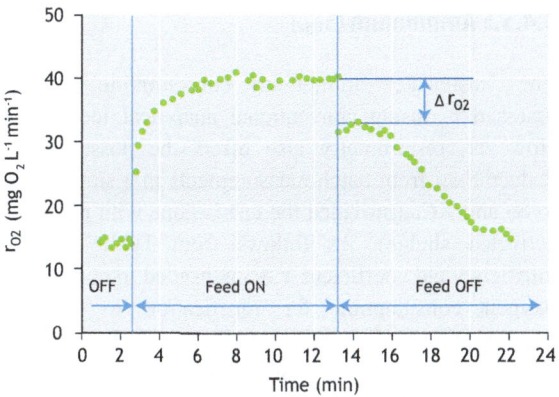

Figure 3.41 Respiration rates (r_{O2}) obtained using the experimental setup of Ekama et al., (1986) and permitting the direct assessment of readily biodegradable substrate S_B.

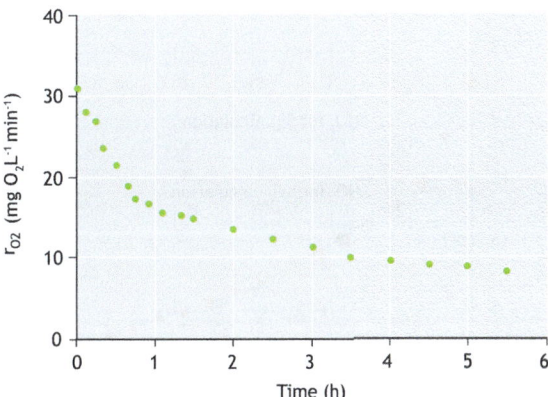

Figure 3.42 Respiration rate (r_{O2}) obtained according to Kappeler and Gujer (1992) for estimation of XC_B.

3.4.5.2 Slowly biodegradable substrate (XC_B)

It is presumed that slowly biodegradable substrate XC_B, sometimes also defined as particulate material, is composed of (high-molecular) compounds ranging from soluble to colloidal and particulate (Henze, 1992). The common feature of these compounds is that they cannot pass the cell membrane and must first undergo hydrolysis to low-molecular compounds (S_B) that can subsequently be assimilated and oxidized. Because the rate of hydrolysis is lower than the oxidation rate of S_{SB}, the respirometric response to XC_B is slower and can normally be identified quite easily as a tail to the respirogram.

In a batch test an exponentially decreasing 'tail' can frequently be observed in respirograms after the initial SS peak (Figure 3.39). In Figure 3.42, this tailing starts after approx. 0.75 hour. The wastewater concentration of XC_B can be assessed in a similar way to the above, using the appropriate $r_{O2,exo}^{XC_B}$ (Kappeler and Gujer, 1992). Simultaneously occurring oxidation processes such as nitrification might interfere with and complicate the identification of the respiration rate governed by hydrolysis. In this case a nitrification inhibitor may be used to facilitate the assessment of XC_B (Spanjers and

The example of Figure 3.43 is presented to stress that all the methods mentioned above using oxygen respiration are also applicable for anoxic respiration tests using nitrate measurements (by lab analysis or using nitrate probes). In this figure not a nitrate respirogram is presented but, in fact, its integral, i.e. the change in nitrate concentration. Backtracking through the different periods with different nitrate utilization rates (the initial period with S_B and XC_B biodegradation, followed by a period with biodegradation of XC_B only, followed by endogenous nitrate respiration) to the Y-axis allows direct assessment of the nitrate used for the different processes and thus calculation of their concomitant concentrations in the wastewater.

The calculation is to be done in a similar way as for the oxygen respiration rate, with some modifications (the respective ΔNO_X of Figure 3.43 substitute the integrals in this case):

$$S_B(0) = \frac{2.86}{1-Y_{OHO,ax}} \left(\int_0^{t_{fin}} r_{NOx,exo}^{S_B}(t)dt \right) \qquad \text{Eq. 3.28}$$

$$XC_B(0) = \frac{2.86}{1-Y_{OHO,ax}} \left(\int_0^{t_{fin}} r_{NOx,exo}^{XC_B}(t)dt \right) \qquad \text{Eq. 3.29}$$

Where the COD equivalence term of 2.86 g COD g NO_3-N^{-1} is used. It should be taken into account that the anoxic yield is somewhat lower than the aerobic yield: whereas for Y_{OHO} a value of 0.66 g $COD_{biomass}$ g $COD_{substrate}^{-1}$ is often used, the anoxic yield is suggested to be taken as 0.54 g $COD_{biomass}$ g $COD_{substrate}^{-1}$ (Muller et al., 2003).

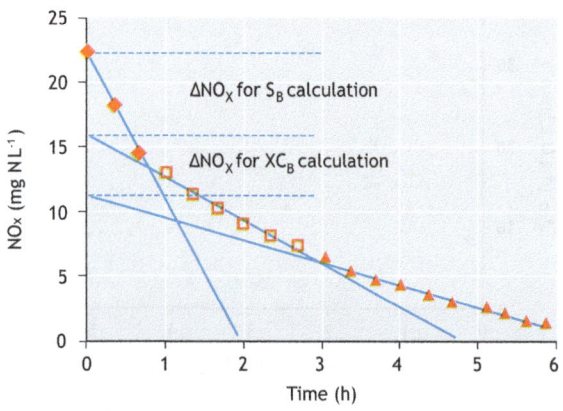

Figure 3.43 A typical nitrate concentration profile for determination of S_B and XC_B from a nitrate (i.e. anoxic) respiration experiment (Urbain et al., 1998).

3.4.5.3 Heterotrophic biomass (X_{OHO})

Some wastewaters can contain significant concentrations of heterotrophic biomass (Henze, 1992), so there is a need to quantify this component. A batch test described by Kappeler and Gujer (1992) and Wentzel et al. (1995) assessed X_{OHO} from the evolution of the respiration rate of raw wastewater without adding a biomass seed. The calculation requires Y_{OHO} and two parameters that can be assessed from the same data: the maximum specific growth rate μ_{OHO} and the decay coefficient b_{OHO}. Respirograms look like the one presented in Figure 3.40. The procedure basically backtracks the amount of heterotrophic biomass originally present in the wastewater by comparing the original respiration rate with the respiration rate after significant (hence, well quantifiable) growth of X_{OHO}.

3.4.5.4 Autotrophic (nitrifying) biomass (X_{ANO})

So far, the authors are not aware of procedures in which the autotrophic biomass concentration in wastewater is assessed. This is probably due to the fact that it is quite unlikely to find significant amounts of nitrifiers in the wastewater. However, it could be imagined that a similar procedure as the one developed for X_{OHO} is applicable, i.e. evaluate the respiration rate for nitrification $r_{O2, exo}^{Nit}$ of the autotrophs present in the wastewater and compare it to the respiration rate of a culture with known autotrophic biomass concentration X_{ANO}, e.g. after significant growth.

3.4.5.5 Ammonium (S_{NHx})

The wastewater ammonium concentration can be assessed by using conventional analytical techniques. However, respirometry also offers the possibility to deduce S_{NHx} from batch measurements in a similar way to S_B and XC_B (provided the test is done with nitrifying activated sludge). As follows from Table 3.2, the nitrifiers yield coefficient Y_{ANO} is needed to convert the oxygen consumption for nitrification to nitrogen concentration by dividing by (4.57 - Y_{ANO}). However, the value of S_{NHx} is not very sensitive to Y_{ANO} as Y_{ANO} is small compared to 4.57. Notice that ammonia is also used for assimilation (i.e. about 12 % of the biomass weight as VSS consists of nitrogen), which may require a considerable amount of the nitrogen if a large amount of COD is biodegraded ($COD^{Degraded}$). The nitrogen that can be nitrified can be approximated by:

$$N_{Nit} = S_{NHx} - i_{N,Bio} \cdot Y_{OHO} \cdot COD^{Degraded} \qquad Eq.\ 3.30$$

Where $i_{N,Bio}$ is the nitrogen content of newly formed biomass. From this equation one can easily deduce the original nitrogen concentration when $COD^{Degraded}$, and the stoichiometric parameters $i_{N,Bio}$ and Y_{OHO} are known. Fitting a model in which carbon and nitrogen oxidation are included to the respirometric data (i.e. the optimisation method) will automatically take this correction into account (Spanjers and Vanrolleghem, 1995).

3.4.5.6 Organic nitrogen fractions ($XC_{B,N}$ and $S_{B,N}$)

Probably because the hydrolysis and ammonification rates of organic nitrogen compounds are relatively fast, little attention has been devoted so far to the establishment of respirometric techniques for $XC_{B,N}$ and $S_{B,N}$ quantification (these components have even been abandoned in the subsequent Activated Sludge Models No. 2 and 3). In batch tests, these compounds are typically converted to S_{NHx} before the S_{NHx} that was originally present in the wastewater has been removed by nitrification. These fractions are thus encapsulated in the determined ammonia concentration. Therefore, $XC_{B,N}$ and $S_{B,N}$ are not directly observable in such tests.

Still, for some wastewaters the hydrolysis and ammonification steps may be considerably slower and quantification of the component concentrations may be required. In such cases, one can imagine a procedure in which the nitrification respiration rate $r_{O2,exo}^{Nit}$ is monitored and interpreted in terms of hydrolysis and ammonification, similar to the way the respiration resulting from COD degradation is interpreted in terms of the hydrolysis process. Subsequently, the amounts of nitrogen containing substrates could be assessed by taking the integral of $r_{O2,exo}^{Nit}$ for the corresponding fractions and dividing these by (4.57 − Y_{ANO}), the stoichiometric coefficient corresponding to nitrification. In case simultaneous COD removal is taking place, correction should again be made for nitrogen assimilated into the new heterotrophic biomass (Section 3.4.5.5).

3.5 BIOMASS CHARACTERIZATION

3.5.1 Volatile suspended solids

In biomass characterization tests, activity is usually normalized by expressing a conversion rate per unit of volatile suspended solids (VSS) in order to account for the varying biomass concentrations. Note, however, that VSS is not equivalent to biomass, that is: only a (usually unknown) part of the VSS consists of active biomass, and VSS is only an approximation of the concentration of biomass. In fact, the challenge of the methods described in this section is to assess the part of the VSS that represents the biomass concentration. For further information on how to assess the part of the VSS that is active biomass the reader is referred to Ekama *et al.* (1996), Still *et al.* (1996) and Lee *et al.* (2006).

3.5.2 Specific methanogenic activity (SMA)

3.5.2.1 Purpose

The specific methanogenic activity (SMA) test involves the assessment of the aceticlastic methanogenic activity of a biomass. This is in contrast to the anaerobic biomass activity tests where the overall activity of the biomass, degrading a usually complex substrate, is measured; in these tests the activity is limited by the rate of the slowest degradation step, which is usually the hydrolysis so that the hydrolytic activity of the biomass for that particulate substrate is assessed. The SMA test may be used for monitoring reactor performance or for characterizing biomass prior to its use as an inoculum for the start-up of a new reactor and thus its potential as an inoculum for that specific process (Sorensen and Ahring, 1993).

3.5.2.2 General

In SMA tests, the activity of the methanogens is quantified. This is done by supplying a substrate that can be converted directly into methane. This can be CO_2 or H_2 gas, but a more common and a more practical substrate is acetate. In literature this has become a standard test as more and more researchers are using it (Ersahin *et al.*, 2014; Jeison and van Lier, 2007). The conversion rate of acetate to methane, normalized to biomass, gives information about the activity of the methanogens in the biomass. It is usually expressed in g COD g VSS^{-1} d^{-1}.

To prevent acidification of the anaerobic digestion, VFA production and VFA consumption need to be in equilibrium (Section 3.4.1). The SMA therefore indicates the maximum acetate production during the anaerobic digestion that can be handled by the aceticlastic methanogens without the pH dropping to values that inhibit methanogenesis.

3.5.2.3 Test execution

In SMA tests (like BMP tests), the amount of inoculum (required for anaerobic digestion) is usually measured as volatile suspended solids (VSS), although volatile solids (VS) is also possible (Section 3.4.1).

In an SMA test the ratio between the inoculum and substrate is balanced such that the quantity of methane is high enough to be measured accurately; the flow is to be within the range of the gas flow meter and the concentration of acetate is below inhibition levels. What is often used is an acetate concentration of 2 g COD L^{-1} and the following substrate to inoculum ratio can be used:

$$2 = \frac{VSS_{inoculum}}{COD_{substrate}} \qquad \text{Eq. 3.31}$$

Both COD and VSS are given in masses supplied to the test, expressed in grams, not concentrations.

- **Stock solutions**
  - The stock solution needed is a phosphate buffer which contains stock solution A (0.2 M $K_2HPO_4 \cdot 3H_2O$: 45.65 g L^{-1}) and stock solution B (0.2 M $NaH_2PO_4 \cdot 2H_2O$: 31.20 g L^{-1}).

- The macronutrient solution contains per litre: 170 g NH_4Cl, 8 g $CaCl_2·2H_2O$ and 9 g $MgSO_4·7H_2O$.
- The micronutrient solution contains per litre: 2 g $FeCl_3·4H_2O$, 2 g $CoCl_2·6H_2O$, 0.5 g $MnCl_2·4H_2O$, 30 mg $CuCl_2·2H_2O$, 50 mg $ZnCl_2$, 50 mg HBO_3, 90 mg $(NH_4)_6Mo_7O_2·4H_2O$, 100 mg $Na_2SeO_3·5H_2O$, 50 mg $NiCl_2·6H_2O$, 1 g EDTA, 1 mL HCl 36%, 0.5 g resazurine, and 2 g yeast extract.

- **Inoculum**

The inoculum is typical anaerobic sludge from a full-scale biogas plant. Measure total suspended solids (TSS) and total volatile suspended solids (VSS) of the sludge.

- **Substrates**

As substrate for aceticlastic methanogenesis activity, sodium acetate-trihydrate salt ($NaC_2H_3O_2·3H_2O$) (M = 136.02 g mol^{-1}) is used. COD value: 0.4706 g COD per g $NaC_2H_3O_2·3H_2O$. Typically, a sodium acetate solution (substrate solution) with a concentration of 2.0 g COD L^{-1} is prepared, which should include the following stock solutions:
- Phosphate buffer: mix 30.5 mL of stock solution A and 19.5 ml of stock solution A (in total 50 mL) per litre substrate medium to obtain a 10 mM phosphate buffer at pH = 7.
- Macronutrients: dose 6 mL per L substrate medium.
- Micronutrients: dose 6 mL per L substrate medium.

Check the pH and measure the COD concentration of the substrate solution. For the blanks, prepare the same solution but without substrate (medium solution).

- **Preparation of the reaction vessels and test execution**

Use triplicates for both the blank and sample, i.e. three bottles as blanks (only inoculum and media solutions) and three bottles for the sample (inoculum and substrate solutions). As stated above, an inoculum to substrate ratio of 2:1 (based on VSS) is normally used in the SMA test. Choose the total volume of liquid that is suitable for the reaction vessels. To perform the SMA test, as for the BMP test, the bottles need to be incubated at a desired temperature. The standard mesophilic temperature is 35 °C. The SMA test is continued until biogas production ceases.

3.5.2.4 Data processing

The methane produced during an SMA test is measured over time. In Figure 3.44, example data of an SMA test are displayed. Notice that the test did not start at 0 N-mL. This was because the head space of the reaction bottle expanded during the warming up to 35 °C, creating a small amount of gas flow. This gas flow in the first 10 minutes does therefore not represent the actual methane flow. During the first two days a phase of low activity can be observed: the lag phase. This is explained by the fact that the methanogens at t = 0 were introduced into a new matrix where the osmotic pressure, conductivity, nutrient composition or substrate concentrations were different from the matrix where methanogens were taken from (digested sludge). This lag phase is normal for SMA tests. After t = 4 days, the methane production has almost come to a halt. This is most likely because of depletion of the substrate. Prior to the start of an SMA test, the expected methane production, based on the acetate COD, can be calculated. For 1 g COD of acetate, theoretically 350 N-mL of methane can be produced. This can be calculated as follows: 1 mole of acetate yields 1 mole of CH_4, hence 1 g Ac yields 1/59 mole of CH_4. At standard temperature and pressure (STP, that is: 273.15 K and 1013.25 mbar), 1 mole of gas is equivalent to 22.4 L, hence 1 g Ac yields 22.4/59 = 0.380 L CH_4. Since 1 g of Ac represents a COD of 1.085 we get 0.380/1.085 = 0.350 L CH_4 g COD^{-1}.

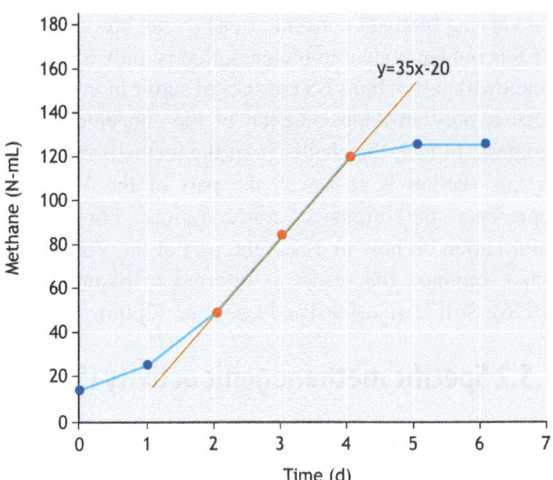

Figure 3.44 Example of an SMA test.

The data that needs to be extracted from this graph is the maximum rate of production. This maximum rate of production can be calculated from the slope of the respirogram in Figure 3.44. The slope can be calculated at any point, but the appropriate interval should be chosen. In this case the interval between t = 2 and t = 4 (in red in Figure 3.44) is chosen because the production rate is the highest here. The maximum production rate of the sludge is 35 N-mL d^{-1}. In this example 800 mg VSS of inoculum was introduced and therefore the SMA is 43 N-mL g VSS^{-1} d^{-1}. Note that the most common way of expressing the amount of methane is in COD. One g COD of methane occupies 350 N-mL of gas, at STP. Therefore the SMA of this experiment is 43/350 = 0.123 g COD g VSS^{-1} d^{-1}. This value falls within the typical range of 0.1 and 0.2 g COD g VSS^{-1} d^{-1} that is commonly found in conventional sludge digesters at WWTPs. Sludge in a UASB-type reactor usually has a higher SMA, between 0.2 and 0.4 g COD g VSS^{-1} d^{-1}.

- **Recommendations**

Especially in SMA tests it is important to pay attention to the sensitivity of methanogens to temperature and inhibition by oxygen. For further information on this and some other recommendations the reader is referred to the section on BMP tests.

The way to process the data of a blank group is different from the BMP test, that is: in the SMA test the data of the blank group is not subtracted from the data of the sample group.

3.5.3 Specific aerobic and anoxic biomass activity

Kristensen *et al.* (1992) summarized a set of laboratory procedures that have been developed to assess the activity of nitrifiers (AUR: ammonia utilization rate), denitrifying biomass (NUR: nitrate utilization rate) and heterotrophic biomass (OUR: oxygen utilization rate). The three procedures and schematic data examples are presented in Figure 3.45 and discussed below.

3.5.3.1 Maximum specific nitrification rate (AUR)

To assess the ammonia utilization rate (AUR), concentrated biomass (e.g. taken from the return or wastage line) and tap water are mixed in one litre cylinders to reach a suspended solids concentration of 3-4 g VSS L^{-1}. The activated sludge is kept in suspension by aeration through diffusors, which also provide the biomass with oxygen in a concentration range of 6-8 mg O$_2$ L^{-1}. After reaching a stable condition (endogenous respiration), ammonia is added to reach an initial NH$_4$-N concentration of 20 mg N L^{-1}. Please note that nitrification is an acidifying reaction and pH should be monitored during the test to ensure that no pH effects are occurring. The addition of alkalinity or installing a pH control system with base addition can improve the quality of the AUR determination.

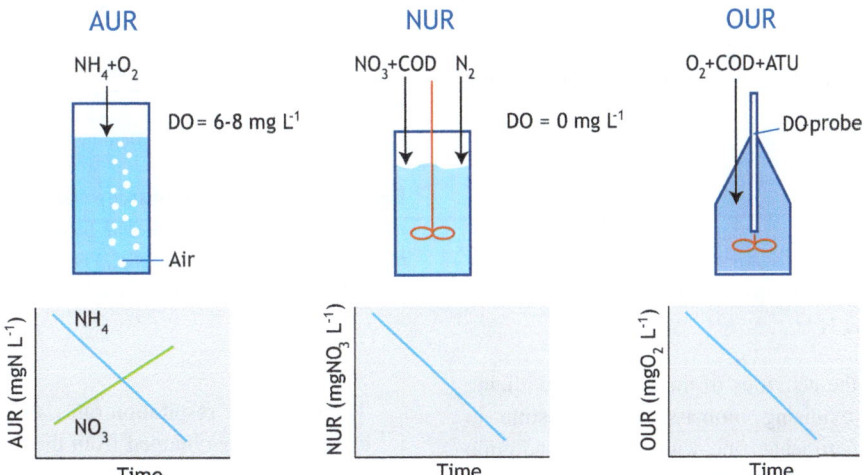

Figure 3.45 Laboratory respirometric procedures and data examples for characterization of nitrifying biomass (AUR: ammonia utilization rate), denitrifying biomass (NUR: nitrate utilization rate) and heterotrophic biomass (OUR: oxygen utilization rate) (Kristensen *et al.*, 1992).

Samples of 10 mL of activated sludge are then withdrawn at intervals of 15-30 min for 3-4 h. The samples are to be immediately filtered to stop the bio-reactions, and the filtrates can be preserved by addition of 0.1 mL of 4 M H_2SO_4 until the analysis. The samples are subsequently analysed for ammonia nitrogen, and nitrate plus nitrite nitrogen. Alternatively an ion-selective ammonia probe, or otherwise an ion-selective or UV-based nitrate and nitrite sensor, can be used directly in the aerated suspension to have more detailed time series and potentially perform biokinetic modelling studies for the nitrifiers.

The AUR (mg NH_4-N L^{-1} h^{-1}) is calculated from the slope of the resulting nitrate plus the nitrite production curve and as a control also from the ammonia utilization curve. Indeed, ammonia uptake may also be affected by endogenous heterotrophic activity due to decay and ammonia release heterotrophic growth with concomitant ammonia utilization. The produced oxidized nitrogen forms are directly due to nitrification, of course on the condition that oxygen is sufficiently high at all times to prevent denitrification. The specific AUR (SAUR, mg NH_4-N g VSS^{-1} h^{-1}) is obtained by dividing the volumetric rate by the biomass concentration (g VSS L^{-1}) set at the beginning of the experiment to be able to compare the nitrifying capacities with typical values.

Nitrification capacity can of course also be deduced from a respirometric experiment with the addition of ammonia (Section 3.4.5.5). The maximum respiration rate that can be attributed to nitrification ($r_{O,ex}^{Nit}$) in mg O_2 L^{-1} h^{-1}, i.e. after subtraction of the endogenous respiration, can be translated into an ammonia conversion rate (mg NH_4-N L^{-1} h^{-1}) using an equation similar to the one used to obtain the nitrifiable nitrogen (Section 3.4.5.5):

$$AUR = \frac{r_{ANO,O2}}{4.57 - Y_{ANO}} \qquad Eq.\ 3.32$$

Where Y_{ANO} is the nitrifier yield coefficient, typically 0.24 mg COD mg N^{-1}.

To separate the activities of the ammonia-oxidizing and the nitrite-oxidizing biomass, it is possible to perform two experiments, one with ammonia addition and one with nitrite addition. The uptake of ammonia and nitrite can be monitored in these separate experiments and translated into the respective activities. Experiments in which nitrite build-up occurs can also be used to extract both activities separately, i.e. by determining the nitrite utilization rate after all the ammonia has been oxidized, one can calculate the activity of the nitrite-oxidizing biomass. The AUR should now be obtained from the ammonia profile, and not from the nitrate profile since the latter is lagging behind the ammonia profile due to the nitrite accumulation in the activated sludge. For this type of analysis to work, it is important that the ammonia-oxidizing capacity is significantly faster than the nitrite-oxidizing capacity because sufficient nitrite must accumulate (more than 2 mg NO_2-N L^{-1} at ammonia depletion is recommended).

An alternative method for determination of the activities of the two biomass groups involved in nitrification was developed by Surmacz-Gorska et al., (1996). It is based on a respirometric experiment to which first sodium chlorate ($NaClO_3$, 20 mM, i.e. 2.13 g L^{-1}), an inhibitor for the second step in nitrification, is added when the DO has decreased by about 3 mg L^{-1}, followed by addition of ATU after DO declines with another 2 mg L^{-1} to inhibit the first step of nitrification. A typical DO concentration profile obtained in a closed bottle test is given in Figure 3.46.

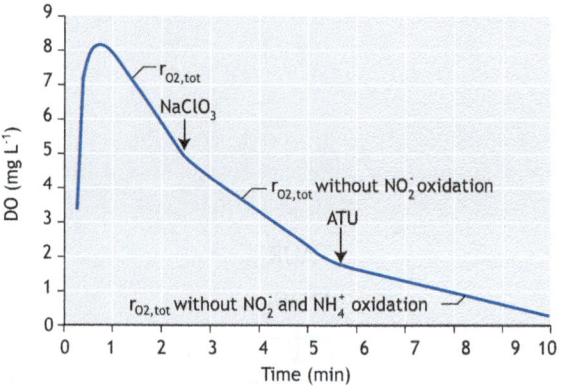

Figure 3.46 A typical DO concentration profile recorded with the two-step nitrification characterization procedure based on chlorate inhibition of nitrite oxidation (Surmacz-Gorska et al., 1996). The slopes of the DO profile are used to assess the respective r_{O2} (OURs).

The respective respiration rates, expressed in mg O_2 L^{-1} h^{-1}, are directly obtained from the three DO declines and allow oxygen consumption to be calculated for the two nitrification steps: $r_{O2,NO2,exo}$ (associated to $r_{NOO,O2}$) is calculated from the difference in the DO slope before and after chlorate addition, whereas $r_{O2,NH4,exo}$ (associated to $r_{AOO,O2}$) is obtained from the difference in the DO slope before and after ATU addition. With these respiration rates, the activities of both nitrification biomasses (in mg NH_4-N L^{-1} h^{-1} and mg NO_2-N L^{-1} h^{-1}, respectively) can be calculated as follows:

$$r_{NH4\_NO2} = \frac{r_{AOO,O2}}{3.43 - Y_{AOO}} \quad \text{Eq. 3.33}$$

$$r_{NO2\_NO3} = \frac{r_{NOO,O2}}{1.14 - Y_{NOO}} \quad \text{Eq. 3.34}$$

Where, Y_{AOO} and Y_{NOO} are the yield coefficients of the two nitrification steps, typically 0.18 mg COD mg NH_4-N^{-1} and 0.06 mg COD mg NO_2-N^{-1} respectively. Specific activities are again obtained by dividing the volumetric rates by the VSS concentration.

3.5.3.2 Maximum specific aerobic heterotrophic respiration rate (OUR)

To determine the oxygen utilization rate related to aerobic heterotrophic activity (OUR), biomass and tap water are mixed to obtain a concentration of suspended solids of 2-3 g VSS L^{-1} in a batch volume of one litre. An experiment is then conducted with COD in excess. Typically acetate is added in a concentration of typically 200 mg COD L^{-1}, i.e. an S_0/X_0 ratio of about 1/10 to 1/20. To obtain aerobic heterotrophic endogenous activity, an alternative indicator of aerobic heterotrophic activity, no COD is to be added. Nitrification must be inhibited by the addition of allylthiourea (ATU, typically 5-10 mg L^{-1}). The biomass is continuously aerated to maintain a DO concentration of 6-8 mg O_2 L^{-1}. In the procedure proposed by Kristensen *et al.* (1992), the respiration rate is measured by periodically pouring part of the batch into a 300 mL BOD flask to measure the oxygen utilization rate using an oxygen probe introduced into the flask (the LSS principle). OUR can then be calculated from the slope of the resulting DO decline. Note that here the total respiration rate in the presence of a nitrification inhibitor is used as the indicator of aerobic heterotrophic activity, i.e. $r_{O2,endo} + r_{O2,exo}$. The alternative ways of obtaining the respiration rate discussed in Sections 3.2 and 3.3 can of course be applied to obtain OUR.

The specific oxygen utilization rate (SOUR), an often used indicator of biomass activity, is calculated by dividing the OUR by the VSS concentration in the batch experiment.

An important element to consider is that the use of acetate may not be ideal in all cases because some biomasses may have adapted to a feast-or-famine regime and store COD rather than use it directly for growth. The corresponding respiration rate for storage may be quite different than the respiration rate that is preferable for heterotrophic activity assessment. If storage is detected, an alternative COD source should be used for activity assessment, e.g. the previously mentioned BOD reference substrate glutamic acid.

3.5.3.3 Maximum specific denitrification rate (NUR)

The specific nitrate utilization rate (SNUR) for denitrification can be assessed by using completely mixed, closed atmosphere, two-litre batch reactors. Concentrated biomass is collected and mixed with tap water in the reactors to obtain a suspended solids concentration of 3-4 g VSS L^{-1}. After reaching a stable condition (endogenous respiration), nitrate is added to obtain an initial concentration of 20-30 mg N L^{-1}. COD, most often as acetate, is added in excess to obtain an initial concentration of 100-150 mg COD L^{-1}. For determination of endogenous NUR, higher concentrations of biomass are to be applied (to reduce the experimentation time) and no COD is added. Since denitrification is a process that increases pH, which is amplified when acetate is the COD source (see pH control results of NUR tests with acetate and glucose in Figure 3.47; Petersen *et al.*, 2002b), the pH should be monitored and corrected for if necessary. Still, the probability of negative pH impacts is much lower than with AUR tests given the lower pH sensitivity of heterotrophic biomass and the lower pH impact of the denitrification.

Samples of 10 mL of activated sludge are withdrawn at intervals of 15-30 min for 3-4 h. Samples are best withdrawn under nitrogen gas addition in order to avoid oxygen intrusion into the reactors. The samples are to be pre-treated as mentioned above for the AUR determination and analysed for nitrate plus nitrite nitrogen. Alternatively, an ion-selective or UV-based nitrate and nitrite sensor can be used in the reactor to have more detailed time series and potentially perform biokinetic modelling studies on the denitrification process. An example of such data is given in Figure 3.47.

The NUR (mg NO_3-N L^{-1} h^{-1}) is calculated from the slope of the resulting nitrate plus nitrite utilization curve. The Specific NUR (SNUR, mg NO_3-N g VSS^{-1} h^{-1}) is obtained by dividing the volumetric rate by the biomass concentration (g VSS L^{-1}) set at the beginning of the experiment to be able to compare the denitrifying capacities with typical values.

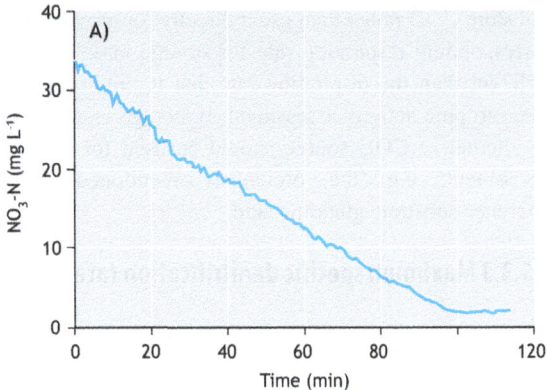

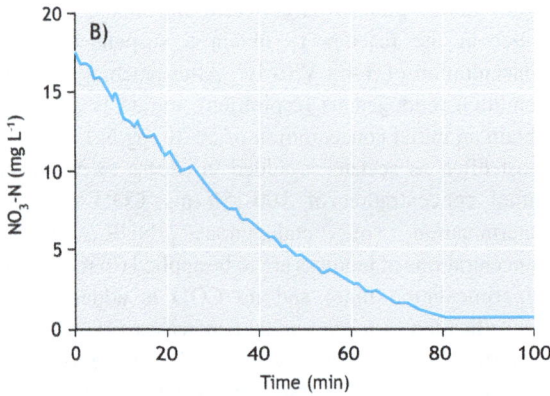

Figure 3.47 Example of a nitrate utilization rate experiment with acetate (A) and glucose (B) under pH-control with acid (A) or base addition (B) (Petersen et al., 2002b).

Anoxic and aerobic heterotrophic activities are closely related as they both reflect the capacity of heterotrophic biomass to oxidize organic matter, with either oxidized nitrogen or oxygen as the electron acceptor. Figure 3.48 shows a typical comparison of NUR and OUR respirograms obtained with the same biomass and acetate addition (Sin and Vanrolleghem, 2004). The tailing-off occurring in both experiments after return to endogenous respiration shows that the aforementioned COD storage occurs both under anoxic and aerobic conditions for this biomass sample.

When comparing respiration rates under anoxic and aerobic conditions, one should be aware that the COD conversion rates are, however, typically lower under anoxic conditions. This is explained in two ways: either the specific conversion rates are lower under anoxic conditions or only a fraction of the total biomass is capable of respiring with nitrate. An often-used biomass characteristic for this phenomenon is the so-called anoxic reduction factor η that makes the ratio between both activities on an electron-equivalent basis (hence the factor 2.86):

$$\eta = 2.86 \cdot \frac{r_{NO_3,exo}}{r_{O_2,exo}} \qquad \text{Eq. 3.35}$$

The factor η is close to one (~0.85) for readily biodegradable substrate (Ekama et al., 1996). For slowly biodegradable substrate in the activated sludge model No. 1 (ASM1), η ~0.33 (van Haandel et al., 1981) and in ASM2, it is 0.66 (Clayton et al., 1991). The reason for this difference has still not been understood.

Use of the ratio η assumes that the growth yield on nitrate or on oxygen is the same. However, the growth yield under anoxic conditions is typically lower than that under aerobic conditions (Muller et al., 2003). This means that for the same COD conversion rate more electron acceptor will be consumed under anoxic conditions (more COD must be burnt to achieve the same biomass production). Therefore this means that, in theory, the reduction factor may be above one if the COD conversion rate is the same.

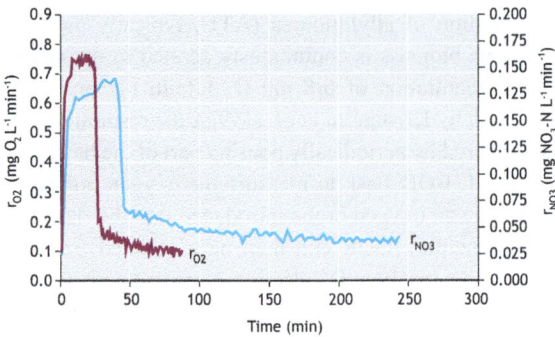

Figure 3.48 A comparison of aerobic (red) and anoxic (blue) respirograms with acetate addition to sludge from WWTP Ossemeersen: after the addition of 46.9 mg COD$_{Ac}$ L^{-1} (r_{O2}) and 38.9 mg COD$_{Ac}$ L^{-1} (r_{NO3}) (Sin and Vanrolleghem, 2004).

References

Alberts, B., Johnson A., Lewis J., Raff, M., Roberts, K. and Walter P. (2002) Molecular Biology of the Cell. 4th edition, Garland Science, 1616 pp.

Angelidaki, I., Alves, M., Bolzonella, D., Borzacconi, L., Campos, J.L., Guwy, A.J., Kalyuzhnyi, S., Jenicek, P., van Lier, J.B., 2009. Defining the biomethane potential (BMP) of solid organic wastes and energy crops : a proposed protocol for batch assays. *Water Sci Tech.* 59(5): 927-34.

American Public Health Association (APHA), American Water Works Association (AWWA), and Water Environment Federation (WEF) (2012). Standard Methods for the Examination of Water and Wastewater, 22nd Edition. New York. ISBN 9780875530130.

Batstone, D.J., Keller, J., Angelidaki, I.., Kalyuzhnyi, S., Pavlostathis, S.G., Rozzi, A., Sanders, W., Siegrist, H. and Vavilin, V. (2002) Anaerobic Digestion Model No. 1 (ADM1). IWA Publishing, London.

Clayton, J.A., Ekama, G.A, Wentzel M.C. and Marais G.v.R (1991) Denitrification kinetics in biological nitrogen and phosphorus removal systems treating municipal wastewaters. *Water Sci Tech.* 23(4/6-2): 1025-1035.

Coen, F., Petersen, B., Vanrolleghem, P.A., Vanderhaegen, B. and Henze, M. (1998) Model-based characterization of hydraulic, kinetic and influent properties of an industrial WWTP. *Water Sci Tech.* 37(12): 317-326.

Copp, J.B., Spanjers, H., Vanrolleghem, P.A. (Eds.), 2002. Respirometry in control of the activated sludge process: Benchmarking control strategies. Scientific and Technical Report No. 11, IWA Publishing, London, UK.

Eastman, J.A. and Ferguson, J.F., 1981. Solubilization of particulate organic carbon during the acid phase of anaerobic digestion. *Water Poll. Control Fed.* 53: 352-366.

Ekama, G.A., Dold, P.L. and Marais, G.v.R. (1986) Procedures for determining influent COD fractions and the maximum specific growth rate of heterotrophs in activated sludge systems. *Water Sci Tech.* 18: 91-114.

Ekama, G.A., Wentzel, M.C., Casey, T.G. and Marais, G.v.R (1996) Filamentous organism bulking in nutrient removal activated sludge systems. Paper 3: Stimulation of the selector effect under anoxic conditions. *Water SA*, 22(2):119-126.

Ersahin, M.E., Ozgun, H., Tao, Y., van Lier, J.B., 2014. Applicability of dynamic membrane technology in anaerobic membrane bioreactors. *Water Res.* 48: 420-9.

Gaps, D., Pratt, S., Yuan, Z., and Keller, J. (2003) Online titrimetric and off-gas analysis for examining nitrification processes in wastewater treatment. *Water Res.* 37: 2678-2690.

Gernaey, K., Petersen, B., Ottoy, J.P. and Vanrolleghem, P.A. (1999) Biosensing activated sludge. WQI, May/June 1999, 16-21.

Henze, M. Grady Jr., C.P.L., Gujer, W., Marais G.v.R. and Matsuo, T. (1987). Activated sludge model No. 1. Scientific and Technical Report No. 1, IAWPRC, London.

Henze, M. (1992) Characterization of wastewater for modelling of activated sludge processes. *Water Sci Tech.* 25 (6): 1-15.

Henze, M., Harremoes, P., Jansen, J. and Arvin, E. (1995). Wastewater Treatment Biological and Chemical Processes. Berlin, Springer-Verlag, pp. 383.

Jeison, D. and van Lier, J.B., (2007). Thermophilic treatment of acidified and partially acidified wastewater using an anaerobic submerged MBR: Factors affecting long-term operational flux. *Water Res.* 41: 3868-79.

Kappeler, J. and Gujer, W. (1992) Estimation of kinetic parameters of heterotrophic biomass under aerobic conditions and characterization of wastewater for activated sludge modelling. *Water Sci Tech.* 25(6): 125-139.

Kong, Z., Vanrolleghem, P.A. and Verstraete, W. (1994) Automated respiration inhibition kinetics analysis (ARIKA) with a respirographic biosensor. *Water Sci Tech.* 30(4): 275-284.

Kong, Z., Vanrolleghem, P.A., Willems, P. and Verstraete, W. (1996) Simultaneous determination of inhibition kinetics of carbon oxidation and nitrification with a respirometer. *Water Res.,* 30: 825-836.

Kristensen, G.H, Jorgensen, P.E. and Henze, M. (1992) Characterization of functional groups and substrate in activated sludge and wastewater by AUR, NUR and OUR. *Water Sci Tech.* 25(6): 43-57.

Lasaridi, K.E. and Stentiford, E.I. (1998) A simple respirometric technique for assessing compost stability. *Water Res.* 32(12): 3717-23.

Lee, B.J., Wentzel, M.C. and Ekama, G.A. (2006) Measurement and modelling of ordinary heterotrophic organism active biomass concentration in anoxic/aerobic activated sludge mixed liquor. *Water Sci Tech.* 54(1):1-10.

Muller, A., Wentzel, M.C., Ekama, G.A. and Loewenthal, R.E. (2003) Heterotrophic anoxic yield in anoxic aerobic activated sludge systems treating municipal wastewater. *Water Res.* 37: 2435-2441.

Nelson, D.L., and Cox, M.M. (2008) Lehninger Principles of Biochemistry. 5th Edition, Palgrave Macmillan, ISBN: 978-0-716-77108-1, 1100 pp.McCarthy, E.L. (1934). Mariotte's bottle. *Science.* 80, 100.

Petersen, B., Gernaey, K., Henze, M. and Vanrolleghem, P.A. (2002a) Evaluation of an ASM1 model calibration procedure on a municipal-industrial wastewater treatment plant. *J. Hydroinformatics*, 4, 15-38.

Petersen, B., Gernaey, K. and Vanrolleghem, P.A. (2002b) Anoxic activated sludge monitoring with combined nitrate and titrimetric measurements. *Water Sci Tech.* 45(4-5): 181-190.

Petersen, B., Gernaey, K., Henze, M. and Vanrolleghem, P.A. (2003) Calibration of activated sludge models: A critical review of experimental designs. In: Biotechnology for the Environment: Wastewater Treatment and Modeling, Waste Gas Handling. Eds. Agathos S.N. and Reineke W., Kluwer Academic Publishers, Dordrecht, The Netherlands. 101-186.

Pratt, S., Yuan, Z., Gapes, D., Dorigo, M., Zeng, R.J. and Keller, J. (2003) Development of a Novel Titration and Off-Gas Analysis (TOGA) Sensor for Study of Biological Processes in Wastewater Treatment Systems. *Biotechnol. Bioeng.* 81: 482-495.

Rezouga, F., Hamdi, M. and Sperandio, M. (2009) Variability of kinetic parameters due to biomass acclimation: Case of para-nitrophenol biodegradation. *Bioresource Technology.* 06/2009; 100(21): 5021-9.

Rieger, L., Langergraber, G., Kaelin, D., Siegrist H. and Vanrolleghem P.A. (2008) Long-term evaluation of a spectral sensor for nitrite and nitrate. *Water Sci Tech.* 57(10): 1563-1569.

Sin, G. and Vanrolleghem, P.A. (2004) A nitrate biosensor based methodology for monitoring anoxic activated sludge activity. *Water Sci Tech.* 50(11): 125-133.

Sorensen, A.H. and Ahring, B.K. 1993. Measurements of the specific methanogenic activity of anaerobic biomass. *Appl. Microbiol. Biotechnol.* 40: 427-431.

Spanjers, H. and Vanrolleghem, P.A. (1995a) Application of a hybrid respirometric technique to the characterization of an industrial wastewater. Proceedings 50th Purdue Industrial Waste Conference, 611-618.

Still, D.A., Ekama, G.A., Wentzel, M.C., Casey, T.G. and Marais, G.v.R (1996) Filamentous organism bulking in nutrient removal activated sludge systems. Paper 2: Stimulation of the selector effect under aerobic conditions. *Water SA,* 22(2): 97-118.

Spanjers, H. and Vanrolleghem, P.A. (1995b) Respirometry as a tool for rapid characterization of wastewater and activated sludge. *Water Sci Tech.* 31(2): 105-114.

Spanjers, H., Vanrolleghem, P.A., Olsson, G., Dold, P.L., 1998. Respirometry in control of the activated sludge process: Principles. Scientific and Technical Report No. 7, IAWQ, London, UK.

Surmacz-Gorska, J., Gernaey, K., Demuynck, C., Vanrolleghem, P.A. and Verstraete, W. (1996) Nitrification monitoring in activated sludge by oxygen uptake rate (OUR) measurements. *Water Res.,* 30:1228-1236.

Theodorou, M.K., Williams, B.A., Dhanoa, M.S., Mcallan, A.B., France, J., 1994. A simple gas production method using a pressure transducer to determine the fermentation kinetics of ruminant feeds. *Anim. Feed Sci. Technol.* 48: 185–197.

Urbain, V., Naidoo, V., Ginestet, P. and Buckley, C.A. (1998) Characterization of wastewater biodegradable organic fraction: accuracy of the nitrate utilisation rate test. In Proceedings of the Water Environmental Federation 71st Annual Conference and Exposition, October 3-7, Orlando, Florida (USA), 247–255.

van Haandel, A.C., Ekama, G.A. and Marais, G.v.R (1981) The activated sludge process Part 3 - Single sludge denitrification. *Water Res.,* 15: 1135-1152.

Vanrolleghem, P.A., Kong, Z., Rombouts, G. and Verstraete, W. (1994). An on-line respirographic biosensor for the characterization of load and toxicity of wastewaters. *J. Chem. Technol. Biotechnol.,* 59: 321-333.

Vanrolleghem, P.A., Spanjers, H., Petersen, B., Ginestet, P. and Takacs, I. (1999) Estimating (combinations of) Activated Sludge Model No.1 parameters and components by respirometry. *Water Sci Tech.* 39(1): 195-214.

Volskay, V. and Grady, C. (1990) Respiration inhibition kinetic analysis. *Water Res.,* 24, 863-874.

Weijers, S.R. (2000) Modelling, Identification and Control of Activated Sludge Plants for Nitrogen Removal. PhD thesis. Eindhoven University of Technology.

Wanner, O., Kappeler, J. and Gujer, W. (1992) Calibration of an activated sludge model based on human expertise and on a mathematical optimization technique - A comparison. *Water Sci Tech.* 25(6): 141-148.

Wentzel, M.C., Mbewe, A. and Ekama, G.A. (1995) Batch test for measurement of readily biodegradable COD and active organism concentrations in municipal waste waters. *Water SA,* 21, 117-124.

Zhang, X., Hu, J., Spanjers, H., van Lier, J.B., (2014). Performance of inorganic coagulants in treatment of backwash waters from a brackish aquaculture recirculation system and digestibility of salty sludge. *Aquac. Eng.* 61: 9–16.

4

OFF-GAS EMISSION TESTS

Authors:
Kartik Chandran
Eveline I.P. Volcke
Mark C.M. van Loosdrecht

Reviewers:
Peter A. Vanrolleghem
Sylvie Gillot

4.1 INTRODUCTION

Wastewater treatment technology and biological nutrient removal (BNR) processes are used mainly to achieve improved water quality. However, BNR processes are a potential contributor to nitrous oxide (N_2O) and methane (CH_4) emissions to the atmosphere, depending on the plant configuration and operating conditions (Kampschreur, 2009). For measuring the carbon footprint of wastewater treatment and BNR processes, this flux of greenhouse gases (GHGs) to the atmosphere should be measured. GHG emissions can be further enhanced by increasing attention on energy neutral treatment plants based on methane generation by anaerobic processes. The release of methane, and especially of nitrous oxides, is of concern because the greenhouse impact of N_2O on a mass equivalent basis is about 300 times that of carbon dioxide (CO_2) and 34 times of CH_4 (IPCC, 2013).

During the past few decades, there have been ever increasing attempts to characterize off-gas (and GHGs) emissions from full-scale wastewater treatment plants (WWTPs), and to develop a methodology to collect full-scale plant data in many parts of the world. In this chapter, the approaches applied in several such studies are described. One of the original protocols developed by researchers in the US, which was reviewed and endorsed by the United States Environmental Protection Agency (USEPA), is presented in detail, as well as methods from other studies elsewhere. It is expected and approaches presented in this chapter will enable off-gas and greenhouse gas emissions from WWTPs to be monitored and quantified using standardized experimental approaches. Because plant operation has a key role in off-gas and greenhouse gas emissions, the results of such monitoring and quantifying efforts could ultimately lead to WWTPs not just improving their aqueous discharges, but at the same time minimizing their gaseous emission. This chapter focuses on the emissions from wastewater treatment by activated sludge and similar processes. Methane gas measurements are well-established as a standard method, whereas an overview of all nitrous oxide measurement methods is given by Rapson *et al.* (2014).

ISBN: 9781789065275 (Paperback), ISBN: 9781780404752 (eBook). Published by IWA Publishing, London, UK.

Off-gas measurements of carbon dioxide and oxygen (O_2) are a good tool for evaluating plant performance. The CO_2 and O_2 content in off-gas can be measured with similar strategies described here for GHGs. A combination of off-gas measurements and titration measurements forms an excellent way to monitor biological processes (Pratt *et al.*, 2003; Gapes *et al.*, 2003). Oxygen depletion in the off-gas is a measure of the respiration rates and can be seen as a (full-plant) respirometric approach, similar to those described in Chapter 3. Carbon dioxide production is also related to the aerobic and anoxic respiration processes. However it is more difficult to interpret since the change (usually a net decrease) in alkalinity also significantly influences the carbon dioxide content in the off-gas (Hellinga *et al.*, 1996). Measuring the oxygen and carbon dioxide conversion could be of great value for better process monitoring and data handling (see Chapter 5), since they allow a better evaluation of mass balances over the wastewater treatment process (Takacs *et al.*, 2007; Puig *et al.*, 2008).

4.2 SELECTING THE SAMPLING STRATEGY

A major consideration for off-gas measurements from WWTPS is a representative sampling procedure. Given the biological nature of N_2O production and link to processes such as nitrification and denitrification, spatial and temporal variability in the measured emission fluxes of these gases is expected. A correlation with wastewater constituents such as ammonia, nitrite and nitrate could also be expected and has been observed in nearly all the measurement campaigns conducted to date. Since emissions from full-scale WWTPs of especially N_2O have proven to be highly dynamic, a measurement program has to be decided upon in a case-by-case evaluation (Daelman *et al.*, 2013). The sampling strategies employed during sampling campaigns must adequately capture such variability. Some factors that influence the sampling strategies are described below.

4.2.1 Plant performance

One of the most significant results of past studies has been the implication that N_2O emissions from WWTPs are linked to overall or local accumulation of nitrogen species such as ammonium (NH_4^+) or nitrite (NO_2^-). There is especially a need for characterization of emissions from plants that do not show a stable nitrification or denitrification performance. Such characterization is important given that many plants are not designed or operated in a BNR mode. Under certain conditions (such as high temperatures during the summer when these highly-loaded plants exhibit incomplete nitrification), these plants can emit more N_2O in comparison to well-designed and operated BNR plants. In contrast to N_2O, CH_4 emissions are likely to occur due to lack of control in an activated sludge process. For instance, poor aeration or dissolved oxygen (DO) control in the activated sludge process may lead to anaerobic zones and associated CH_4 production. Besides, CH_4 that was formed in the sewer system and enters the plant with the influent is likely to be stripped in the activated sludge tank, even though it could also be aerobically oxidized. CH_4 stripping in the activated sludge tank could be minimized while its biological conversion could be promoted through adequate process design and control (Daelman *et al.*, 2014)

4.2.2 Seasonal variations in emissions

The link between temperature and emissions of N_2O or CH_4 is not straightforward and needs to be considered in combination with the plant design. For plants operating at a high sludge retention time (SRT) and sustaining complete NH_3 removal all year round, higher emissions can be expected during the summer period, owing to overall higher microbial activity at higher temperatures. Especially for such plants, the off-gas emission may shift more downstream in plug-flow installations at lower temperatures. The slower nitrification occurs more downstream at these lower temperatures. For plants that cannot fully nitrify at low temperatures and foster non-limiting NH_3 concentrations at lower temperatures, off-gas emission can actually be higher at lower temperatures. In contrast, as these plants can achieve full nitrification during the higher temperature, their emissions could be lower. Seasonally nitrifying treatment plants could have higher emission as well. Such plants can 'wash-out' nitrifying bacteria at low temperatures and consequently have extremely low off-gas emission. As nitrification is supported during higher temperatures but often not sufficiently to achieve the required effluent NH_3 concentration, off-gas emission could be higher during the summer. For planning gas measurement locations the measurement of nitrite might be helpful since there is a general, but not unique, correlation between the presence of nitrite in the liquid phase and N_2O formation. CH_4 emissions from aeration tanks result largely from processes in sewer and sludge processing facilities, and therefore the link between temperature and measured emissions is much less obvious than for N_2O emissions.

4.2.3 Sampling objective

A range of different sampling strategies for the quantification of N_2O emissions has been reported in literature, such as 24 h online sampling, 1-week online sampling, long-term weekly grab sampling and taking a single grab sample. The difference in the sampling methods may partly explain the large variability in N_2O emissions reported, aside from the differences in the real emission, which is related to the plant performance and shows both diurnal and seasonal variations.

The optimal sampling strategy depends on the objective of the sampling campaign. Daelman *et al.* (2013) applied several N_2O sampling strategies applied in literature to the extensive dataset of a long-term online monitoring campaign. They showed that a reliable determination of the actual average N_2O emission requires long-term sampling, be it online or grab sampling, covering the entire temperature range that can possibly be encountered. Short-term sampling unavoidably ignores the long-term variation that is present. For long-term grab sampling, nighttime and weekend samples contribute significantly to a more accurate estimate. High-frequency (online) sampling is indispensable to identify correlations between the emission and the process variables that induce the emissions and in this way gain insight into the underlying N_2O formation mechanisms. A method to obtain the number of grab samples or online sampling periods required to obtain a sufficiently precise estimation of the emission was presented by Daelman *et al.*, (2013), serving as a guideline for sampling campaigns, to balance cost and precision.

4.3 PLANT ASSESSMENT AND DATA COLLECTION

4.3.1. Preparation of a sampling campaign

A sampling program (or campaign) is primarily carried out to (*i*) collect routine operating data concerning the overall WWTP performance, (*ii*) acquire information that can be used to document the performance of a given treatment operation or process, (*iii*) collect data that can be used to implement proposed new programs for sewerage and WWTP, and (*iv*) obtain records needed for reporting regulatory compliance. In general this information is:

a. Influent characterization measurements.
b. Mass balance measurements.
c. Activated sludge characterization measurements.
d. Effluent characterization.
e. All other necessary information.

A well-prepared sampling plan should not only give the type (e.g. chemical oxygen demand: COD, total Kjeldahl nitrogen: TKN, total phosphorus: TP) and location (e.g. influent, effluent or aeration tank) of the sampling point, but it should also indicate what the sampling frequency is and how the sample should be taken, handled, stored, and processed in the laboratory for analytical measurements. The goal of the campaign planning is to communicate all this information to the staff responsible for acquiring the samples (operators or plant technical personnel) and the laboratory personnel analysing the samples. Tables should be developed indicating all of this information, preferably in a single table overview. Also, a dedicated process flow diagram should be developed indicating precisely all the measurement points related to the planning.

While designing the sampling campaign one must address at least the following practical questions:
1. WHY: What is the purpose of taking the samples?
2. WHAT: Which parameters are to be sampled?
3. WHERE: Where are the sampling locations?
4. WHEN: When are the samples taken?
5. HOW OFTEN: What is the frequency for each sampled parameter?
6. HOW: How will the sampling be executed?
7. BY WHOM: Who will take the samples and who will analyse them?
8. EQUIPMENT: What will be used for collecting, storing and analysing the samples?

The six criteria for quality data are:
1. Collecting representative samples.
2. Formulating the objectives of the sampling program.
3. Proper handling and preservation of the samples.
4. Proper chain-of-custody and sample identification procedures.
5. Field quality assurance.
6. Proper analysis.

The sample, in order to serve its purpose, has to be representative, reproducible, defendable and useful. It is also important to realize that proper understanding of the sampling procedures is crucial for the design and success of the sampling program. This should not only be realized by the designer but also by all the others involved in the project. Therefore, it is advised to organize a preoperational meeting to communicate the measurement planning to all the personnel involved. During this meeting, it should be clearly stated what the goal of the project is, what will be done with the specific measurements and also the importance of acquiring accurate information. It is possible that it will be necessary to give (short practical) training in (grab) sampling techniques.

To ensure that the measurement planning is properly understood and executed, the measurement plan should relate as much as possible to the practical routines of the plant and laboratory personnel. Where possible, references should be made to standard (practical) routines. For example, measurement locations on the process flow diagram should be indicated in the same way as used by the operators. Often the designer has his own preferred way of indicating the flows (e.g. in a flow diagram using the logical numbering Q_1 to Q_n to indicate flows in the mass balance), to communicate this information effectively, and the practical names of the process units and flows should also be indicated in all the diagrams and tables.

Also, for analytical measurements the designer can have different names that do not always correspond with the names and practical methods applied in the laboratory. Therefore the measurement plan should also be checked with the laboratory personnel and the applied measurement routines (analytical methods and equipment).

For the purpose of a campaign, samples should preferably be taken in dry weather conditions unless otherwise specified. In practice it is not always clear if and when these conditions are met. Typically it is advised to plan a day of sampling after three days without (major) rain. This means that the measurement planning can never be planned on an exact date. However, it should incorporate a period of several days in which the planning can be executed, depending on the weather conditions.

Some of the samples can be done automatically with 24-h continuous (or quasi-continuous) flow proportional composite measurement devices. These automatic samplers collect samples over a 24-h period and have to be started 24 h ahead of time to provide a collected sample on the desired sampling day. It should be born in mind that operators have to be notified a day in advance to be able to collect the samples on time. All the other samples taken on the day of sampling will mostly be spot samples (also called grab, random, catch, individual, instantaneous, and snap samples). Because the sampling program contains multiple measurement points and for each measurement point, multiple parameters have to be measured, the amount of containers required to store the samples will rapidly increase; e.g. the amount of samples needed for a relatively small campaign can easily exceed 100. Also, it should be realized that for each sampling point the sampled volume should allow analytical measurements, washing of filters and vessels, possible spills and possibly duplicate or triple measurements in the laboratory. Also, the laboratory should be prepared to handle this large amount of samples in one (or more) days. Some of the samples can easily be stored to be measured at a different time. However, for other measurements it is critical that the samples are pre-treated (e.g. filtered) before they can be stored. Again, other samples cannot be stored at all (e.g. as the result of biological conversions which occur in the sample vessel). Also, this should be planned in cooperation with the laboratory personnel to avoid a possible work overload and possible inaccurate measurement results.

4.3.2 Sample identification and data sheet

As was pointed out previously, in the sample measurement plan many samples may need to be analysed for one sampling day. Clearly this requires a good administration and sample identification (sample ID). A list of the samples collected should be kept. The samples should be listed in one overview and include the information below. In addition, the individual sample containers should be labelled accordingly.

The container label requirements are:
a. Sample site location.
b. Container number if more than one container/sample.
c. Name of sample collector.
d. Facility name/location.
e. Date and time of collection.
f. Identified as a grab or composite.
g. Test parameters - list analysis to be done.
h. Preservative used.

Proper administration of the sampling program is necessary for quality assurance. The following checklist can be used for constructing custody forms supporting the measurement campaign. Often, forms of this kind will already be available and used for the regular sampling program. It can be helpful to find out if these forms and also the forms in the measurement campaign are available instead of constructing new ones.

a. Project number: assign a number to the sampling episode that can be used on the bottles to track the samples.
b. Sampler: the sampler should print and/or sign their name.
c. Project name: the project name, and if necessary, the project address.
d. Project contact name: the name of the person who is in charge of the project.
e. Project telephone number: the phone number where the project contact person can be contacted.
f. Sample date: list the dates for grab and composite samples.
g. Sample time: list the sample time for each sample.
h. Sample type: composite (collected over a 24-h period) or grab (taken as grab samples).
i. Station location: the site where the sample was taken.
j. Number of containers: list the number and type of containers for each sampling event.
k. Analysis requested: list the analysis needed that is compatible with reference to official methods.
l. Remarks: note any special requirements for the samples for the laboratory.
m. Split samples: when a sample is split into different proportions for analysis, the receiving party can accept or reject the samples and they must sign this box on the custody form.

4.3.3 Factors that can limit the validity of the results

When evaluating the results of the sampling program, there are several factors that can limit the validity of the results, such as:
a. Missing values.
b. Sampling frequencies that change over the recorded period of time.
c. Multiple observations within the sampling period.
d. Uncertainty in the sample preservation and measurement procedures.
e. Censoring the measurement signals.
f. Small sample size.
g. Improper data handling.
h. Equipment inaccuracies.

It is always useful to check the sampling procedure and the measurement results against the above criteria.

4.3.4 Practical advice for analytical measurements

Some typical re-occurring errors in the case studies in the sample procedure or analytical measurements have been documented and are listed below:

- The method used for analytical measurements of suspended solids is different for concentrated and diluted activated sludge samples; if the concentration is above a certain range, then filtration, drying and weighing the solids will no longer be possible. In this case the dry mass method is often used (evaporation of the sample). This includes the soluble salts, which especially for lower concentrated sludge samples, will not give representative results for total suspended solids. This should especially be taken into account when measuring activated sludge, thickened sludge and dirty water flows. For the more concentrated sludge flows (> 20%), dry mass analysis can be used.
- When measuring the filtrate of the sample containing suspended solids, the filter paper should not be washed other than with the sample itself and not with demineralized water, which will cause dilution of the measured filtrate.
- When handling samples containing suspended solids, it is important to take well stirred homogeneous samples when taking the samples or pouring a sample into another vessel. It should be noted that TSS measurements are essential for the model design study and at the same time are very difficult to sample accurately. It is advised to instruct all personnel on how to take reliable TSS samples.
- When carrying out the activated sludge characterization, all the samples should be taken from the same vessel as it is important to know the relative concentration of TSS (total suspended solids), VSS (volatile suspended solids), COD, TKN and TP in the sample. This is not usually a standard procedure and therefore all the personnel, especially those working in a laboratory responsible for analysing, should be well instructed in this matter.
- A common known problem is the destruction of suspended solids, which is necessary to analytically measure TP (and also COD and TKN). When the destruction procedure is not long enough, phosphorus measurements will be underestimated. Especially when there are chemical precipitates (e.g. as a result of dosing of iron-salts) including phosphorus in the sample, destruction can be difficult.

- Measurements of nitrate in samples containing activated sludge (denitrifying heterotrophic bacteria) should be done quickly before denitrification occurs.
- Measurement of volatile fatty acids (VFAs) in samples containing activated sludge (denitrifying heterotrophic bacteria) is practically impossible for the same reason; VFAs are almost immediately oxidized by living organisms. These types of samples are highly unreliable and should not be used. Measurement of volatile fatty acids in the influent is possible; however, it should be immediately filtered after sampling and stored under prepared conditions.

4.3.5 General methodology for sampling

In general, there is a standard approach developed regarding the design of a water quality monitoring system; the approach for a sampling program at a WWTP can be derived from this that consists of six distinctive steps:

Step 1: Evaluation of the existing information
- Wastewater collection and treatment.
  - Domestic wastewater.
  - Industrial effluents.
  - Combined sewer overflows and pumping stations.
  - Wastewater treatment plants.
- Lab-scale and pilot plant investigations.
  - Wastewater treatment plants.
  - Wastewater collection networks.

Step 2: Evaluation of the information expectations
- Water quality goals and objectives.
- Water quality problems and issues.
- Management goals and strategy.
- Monitoring role in management.
- Monitoring goals (as statistical hypotheses).

Step 3: Establishment of statistical design criteria
- Statistically characterise the population to be sampled.
  - Variation in quality.
  - Seasonal impacts.
  - Correlation (independence).
  - Applicable probability distributions.
  - Out of the many statistical tests select the most appropriate.

Step 4: Design the monitoring network
- Where to sample (from the monitoring role in management).
- What to measure (from the water quality goals and problems).
- When to sample (from specific circumstances).
- How frequently to sample (from the needs of the statistical tests).

Step 5: Develop the operating plans and procedures
- Sampling routes and points.
- Field sampling and analysis procedure.
- Sample preservation and transportation.
- Laboratory analysis procedures.
- Quality control procedures.
- Data management and retrieval hardware and database management systems.
- Data analysis software.

Step 6: Develop reporting procedures
- Type of format of the reports.
- Frequency of report publication.
- Distribution of reports (information).
- Evaluation of the report's ability to meet the initial information expectations.

Based on the above, a general wastewater and sludge sampling program can be carried out following the steps listed below:
- Definition of the purpose of the sampling.
- Determination of the type, scope and required accuracy of the analyses to be carried out.
- Definition of the character of the samples to be collected.
- Selection of the localities and sources to be sampled, and of the sampling points at these localities.
- Determination of the hydraulic and other parameters relating to the subject of the sampling program.
- Consideration of the occupational safety and hygiene of those collecting the samples.
- Preparation of an optimal sampling programme.
- Selection of the sampling and measuring equipment suited to the sources to be sampled, and determination of its state.
- Selection of the most suitable sampling technique in line with a fixed programme and selected equipment for the given source/site, including preparations, subsidiary measurements and observations. When implementing a sampling programme the techniques used should comply with a number of general requirements such as reliability, economy, repeatability and conservation.
- Selection of appropriate procedures, sample handling equipment and tools, transportation from site to laboratory, storage after delivery.
- Consideration of the most suitable methods of analysis on site and in the laboratory, the quickest possible interpretation of the analyses including reliability checks, and possible repetition of the

sampling, and other factors likely to influence the accuracy or representativeness of the analyses.
- Use of feedback whenever possible to modify a programme to an optimal sampling programme.
- Establishment of the conditions necessary for the immediate use of the results and for their storage as primary sources of information in the future, whether for short- or long-term use.
- Selection of an appropriate system for data management (i.e. handling, processing, transfer and manipulation).
- Selection of the method of documentation to be used throughout the programme.

Logistically, performing the measurement campaign in the design study is a complicated task and at the same time it is the most important and critical step. Therefore it is advised to carefully prepare the measurement campaign in all possible aspects. It is also advised to accompany the first rounds of the measurements together with all the personnel involved to make sure that all the instructions in the measurement plan are correct and as intended, and properly understood by all the personnel. The analytical results of a sample are only as accurate as the quality of the sample taken. If the technique for collecting samples is poor, then no matter how accurate your lab procedures are, the results will be poor. By sampling according to set procedures, one reduces the chance of error and increases the accuracy of the sample results. It should be born in mind that for a lot of samples the measurement campaign is a one-off opportunity; redoing samples at a different stage will often not lead to satisfactory results because the activated sludge plant is a dynamic system that is continuously changing over time. With this in mind, for the study the samples should be collected simultaneously in such a way that their relative information can be used for the plant assessment. Therefore, individual grab samples will often not be representative, meaning that (missing or incorrect) measurements often cannot be redone. In conclusion, data collection through sampling at WWTPs is a critical activity that needs a systematic and professional approach and the design of a sampling program is often a task for an experienced specialist.

4.3.6 Sampling in the framework of the off-gas measurements

The strong relation between N_2O emissions and operational conditions makes the development of a preliminary reconnaissance analysis crucial. The field teams need to gather process-operating data during meetings with plant operators (and process engineers) prior to the sampling campaign. The following background information is typically collected from the evaluation site:
- An overall plant description which includes general information related to the plant configuration, liquid and solids process flow diagrams, design criteria, major technological process equipment from the plant's design documentation and/or operation and maintenance manuals.
- Process units and layouts which entail information on anaerobic/aerobic/anoxic zone configuration, zone volumes, operating set points, basins in service, aeration flow and distribution, recycle streams and flow rates (if applicable), and
- Plant operating data, ideally providing a summary of a minimum of 3 months' of plant data on relevant treatment process(es) to allow for characterization of the influent and effluent, and target and actual operating set points for key operational parameters (e.g. DO, SRT). The operating data is needed to check that the plant was not in an upset condition before and during the period of investigation (sampling).

The collected data and information can be used to make analyses by conventional techniques such as the development of solids and nitrogen mass balances as well as through the use of process modeling. Details of model-based evaluation are not presented in this chapter and can be found elsewhere (Melcer, 1999; Puig *et al.*, 2008; Rieger *et al.*, 2012).

In order to collect the necessary data for the assessment it is necessary to determine the influent flow, and organic matter and nitrogen content in the influent, in preparation for the detailed liquid and air measurement campaign at the WWTP. For the initial sampling the following parameters need to be monitored regarding the activated sludge process:
- Influent flow rates (minimum of once per hour).
- Influent and effluent ammonia (up to eight times per day or continuously).
- Influent and effluent nitrite and nitrate (up to eight times per day or continuously).
- Influent and effluent COD (start with once per hour, can be reduced depending on observed variability).

Additionally, diurnal performance and in-tank profiles should be gathered at the time of the N_2O gas phase sampling. As far as feasible, all the liquid phase analyses must be according to approved methods and protocols (e.g. APHA et al., 2012). Given the important effect of nitrite on N_2O emission, it is advised to give this measurement special attention during the sampling campaign. As far as possible, the sampling team should work with the plant laboratory personnel to include data from the online analysers present at a plant, to avoid duplication of data-gathering efforts. In addition to the data presented in Table 4.1, the following diurnal performance and in-tank profiles should be gathered (Table 4.2).

Table 4.1 Overview of the data requirements for the preliminary WWTP assessment.

Sample location	Parameter and sampling frequency (number of samples per week)												
	TSS	VSS	$cBOD_{5,tot.}$[1]	$cBOD_{5,sol.}$	$COD_{tot.}$[1]	$COD_{sol.}$	$COD_{filt.floc.}$	Temp.	TKN[1]	$TKN_{sol.}$	NH_3-N	NO_3-N	NO_2-N
Settled sewage	1	1	1	1	1	1	1		1	1	1	1	1
Secondary effluent	1	1	1	1	1	1[2]	1[2]		1	1	1	1	1
Reactor MLSS	1	1						1					
RAS MLSS	1												
WAS MLSS	1[3]												
Clarifier	Sludge blanket TSS (use sludge judge 1/d and average 1/w)
Flow split and flow rate	Different measurements possible • Approximate - set a settled sewage gate and allow for natural flow regime (if there is no information on the range of flows). • Confirm the flow split by carrying out mass balance and MLSS concentration measurements. • Alternatively, take a measurement of activated sludge at each pass. • Use an optical density meter to get each pass TSS every 2-3 hours to get the moving average.
Anoxic zone mixing	Mechanical or aerator-driven.
Influent flow	Diurnal flow pattern at appropriate time intervals (15 min for periods of rapid diurnal increase, 1 h for stable periods).
RAS flow	Average weekly RAS flow, indicate the location and type of flow measurement and variability of flow.
WAS flow	Average weekly WAS flow, indicate the location and type of flow measurement, times of WAS wasting (if it is not continuous).
Dissolved oxygen	1/d (then average 1/w), indicate the location of the DO measurement along the basin length and time of measurement.
Aeration rate	Daily average, indicate the location of the airflow measurement and variability over the course of the day. SCADA output at short time intervals is preferred.
Chemicals dosage	Daily, indicate the ferric chloride equivalent (and other chemicals) strength, dosing points and dose at each point

[1] Homogenize subsample prior to the 'total' measurement. Discard the remaining sample – do not use for 'filtrate' or 'soluble' determinations.
[2] Soluble COD can be used instead of filtered flocculated COD on the secondary effluent.
[3] When RAS and WAS are from the same stream, a TSS measurement on one of these streams is sufficient.
RAS: return activated sludge; WAS: waste activated sludge.

Table 4.2 Complementary data requirements preferentially accompanying off-gas measurements.

Sample location	Parameter (number of samples per day)													
	TSS	VSS	$cBOD_{5,tot.}$[1]	$cBOD_{5,sol.}$	$COD_{tot.}$[1]	$COD_{sol.}$	$COD_{filt.floc.}$	TKN[1]	$TKN_{sol.}$	pH	Alk	NH_3-N	NO_3-N	NO_2-N
Settled sewage	8	2	8	8	8	8	8	8	8	8	8	8	8	8
Secondary effluent	8		8	8	8	8[2]	8[2]	8	8	8	8	8	8	8
RAS MLSS	8													
WAS MLSS	8[3]													
Influent flow	Diurnal flow pattern at appropriate time intervals (15 min for periods of rapid diurnal increase, 1 h for stable periods).
RAS flow	Average daily RAS flow, indicate the location and type of flow measurement and variability of flow.
WAS flow	Average daily WAS flow, indicate the location and type of flow measurement, times of WAS wasting if not continuous.
Dissolved oxygen	1/h, indicate the location of the DO measurement along the basin length and time of measurement.
Aeration rate	Daily average, indicate the location of the airflow measurement and variability over the course of the day (SCADA outputs are preferred).
In-tank profiles	TSS	VSS	pH	DO	ORP	Temp.	$COD_{filt.floc.}$	Alk.	NH_3-N	NO_3-N	NO_2-N
	8/d	2/d	8/d	8/d	8/d	8/d	8/d	8/d	8/d	8/d	8/d

The collection of conventional wastewater samples for analysis of parameters in Table 4.2 should preferably be conducted by facility personnel who usually collect operational and compliance samples. In advance of each sampling event, the team conducting the monitoring campaigns should consult with the laboratory personnel to ensure that the samples for the conventional parameters are collected during the GHG monitoring event to meet the requirements of both the research design and the WWTP's laboratory operating procedures.

The plant's sample handling and custody requirements can be utilized as far as possible for each field sampling campaign. To confirm the adequacy of the procedures, the plant's procedures for field sample handling and chain of custody should be reviewed with the project team approximately two weeks prior to the full-scale testing. At that time, if modifications are deemed necessary by the project team, they can be defined and documented in the site-specific sampling protocol.

4.3.7. Testing and measurements protocol

Table 4.3 provides examples for the sample location, the chemical parameter, sample container, preservative and holding time for wastewater samples to be collected during the measurements. This example is a starting point and can be customized and expanded as required for the specific plant where the sampling is conducted. For the full-scale field testing, the plant laboratory will follow their specific laboratory standard operating procedures for each parameter. Standard operating procedures from participating laboratories must be reviewed for adequacy and consistency prior to the measurement campaigns.

Table 4.3 Examples of sampling specifications to accompany off-gas monitoring.

Parameter	Measurement classification			Sample location	Sample volume[1] (mL)	Sample preservation	Max. holding time
	Type	Frequency	Sample equipment				
pH	C	NA	NA, *in situ*	Reactor	NA	NA	None, online
COD: Colorimetric	I, C	2/7 d	35 mL glass vial	Reactor, effluent	8	4 °C	1 d
NH$_3$-N: Potentiometric (ISE)	I, C	2/7 d	200 mL glass bottle	Effluent	80	4 °C[2]	1 d
NO$_2^-$-N: Spectrophotometric	I, C	2/7 d	200 mL glass bottle	Effluent	40	4 °C[2]	2 d
NO$_3^-$-N: Potentiometric (ISE)	I, C	2/7 d	200 mL glass bottle	Effluent	40	H$_2$SO$_4$, pH < 2	28 d
N$_2$O	C	1/7 d	Gas sampling assembly	Reactor headspace	NA	NA	NA
NOx	C	1/7 d	Gas sampling assembly	Reactor headspace	NA	NA	NA
DO	C	NA	NA, *in situ*	Reactor	NA	NA	None, online

[1]The tabulated sample volume is twice that required for routine duplicate analysis and is apportioned into two sample containers. The additional volume is collected to determine quality control measures such as accuracy (analysis of spiked samples), precision (duplicate analysis) and to account for potential sample loss while handling or analysing.
C: continuous measurement; I: intermittent measurement; Frequency of measurement applies only to continuous measurements
[2] Storage at 4 °C. Note: the biomass is removed from the sample via centrifugation at 3,500 g for 10 min. Biomass removal arrests further biochemical oxidation of NH$_4^+$-N and NO$_2^-$-N.

4.4 EMISSION MEASUREMENTS

The various methods for off-gas measurements used in the research and practice are summarized in Table 4.4. When a treatment plant is covered, the concentrations can be easily measured in the ventilation air. Combining the concentration measurements with airflow measurements by e.g. use of a pitot tube (Klopfenstein, 1998) gives a direct estimation of the emitted fluxes (Daelman *et al.*, 2015). Most treatment plants are however directly open to the air and the collection of off-gas is a problem that needs special attention for representative sampling. This can be solved by either using gas hoods for gas collection or measuring in the gas plume downwind of the treatment plant. The former is the most common approach and is therefore presented in detail in this chapter.

When off-gas measurements are difficult (e.g. in open tanks or when surface aerators prevent the placement of flux chambers), emissions can be calculated from the measurement of dissolved CH$_4$ and N$_2$O concentrations. These can be integrated with mass transfer coefficient measurements in mass balance calculations to estimate the emission fluxes (Foley *et al.*, 2010). For dissolved N$_2$O, microelectrodes can be used to measure the concentration in the water phase on line (Foley *et al.*, 2010, Section 4.7.1), but sampling and gas chromatographic analysis is also possible. For this purpose, sampling the liquid phase and salting out the soluble gasses to analyse them as a gas has been shown to be an accurate and reliable method (Daelman et al., 2012; Section 4.7.3).

Concentrations of dissolved gases such as N_2O can be continuously measured based on gas phase measurements, according to a gas-stripping method proposed by Mampaey et al. (2015) (Section 4.7.4)

Table 4.4 Comparison of different off-gas measurement techniques and approaches.

Method	Liquid phase	Gas phase	Advantages	Disadvantages
Direct measurements from off-gas in covered tanks	Sampling depending on accessibility to treatment tanks	Continuous	Well-developed and easy to apply method. Gives integral emission loads of a treatment plant, with temporal variations.	Needs covered tanks. Difficult to use for spatial variations inside the plant.
Flux chamber for open-surface process tanks	Grab sample, N_2O sensor, or using a stripping device with gas measurement	Continuous monitoring	Well-developed and widely applied protocols. Mechanisms and processes contributing to emissions can be inferred based on spatially and temporally based measurements (also zone-specific measurements).	Multiple measurements needed to address spatial and temporal variability and to obtain full emission loads of a plant.
Measuring the emission plume downwind of the treatment plant	Not sampled	Continuous	Straightforward data analysis and calculation when gases are fully mixed. Downwind plume changes can be instantaneously detected and the measurements adjusted accordingly. Flexibility to carry the equipment around either by car or small trolley. Capability to point out emissions from hotspots. Whole plant emission quantification.	A skilled operator is required. Dependence on favourable wind conditions combined with road access. Monitoring only possible with favourable wind. Very limited spatial resolution.

A third alternative for measuring the emission of a total wastewater treatment plant is the detection of the CH_4 and N_2O concentrations in the downwind plume of the plant. This needs very specialized equipment such as photoacoustic or cavity down-ring spectroscopy (Yoshida et al., 2014, Rapson et al., 2014) and is therefore only usable by specialists. The advantage is that monitoring can be done outside plant boundaries, and the full plant with all its process units is monitored. Hotspots at the plant can be identified (e.g. influent works for methane emission or ventilation air stack); however, accurate spatial resolution for e.g. the aeration tanks is not possible. Given the costs of the equipment, long-term monitoring is economically often not possible. The method is established for landfill monitoring, where emissions are estimated by sampling a few days per year with mobile equipment.

4.5 N₂O MEASUREMENT IN OPEN TANKS

The overall procedure for measuring off-gas fluxes from the headspace of open surface activated sludge tanks can involve a flux-chamber approach, which has been used by most published work to date. This works for most installations, except for surface-aerated systems. When the surface-aerated space is covered, the gas composition in the ventilation air (and its flow rate) can be measured. When the aerators are open in the air, one has to rely on the use of liquid phase balances for N_2O or CH_4 (Foley et al., 2010; Section 4.7). The principal advantage of the flux chamber approach is that it can yield some understanding of the spatial variability in emissions across different zones or the reactor of an activated sludge p rocess. On the other hand, it is operationally intensive to conduct due to the need for multiple sampling locations to get representative measurements. In the US, the off-gas measurement method is a variant of the EPA/600/8-86/008 and the South Coast Air Quality Management District (SCAQMD) tracer methods. This variant was developed to measure sources that have a relatively high surface flux rate when compared to diffusion (for instance, spilled oil containment).

Commercially available replicas of the US EPA surface emission isolation flux chamber (SEIFC, figures 4.1-4.3) are used to measure gaseous nitrogen fluxes from activated sludge reactors. The SEIFC consists of a floating enclosed space from which exhaust gas can be collected as a continuous or grab sample. Because the surface area under the SEIFC can be measured, the specific flux of the gaseous compound of interest can be indirectly determined. The SEIFC 'floats' on the activated sludge tank surface, and several replicate measurements can be taken at different locations in a single tank as well as from different tanks (nitrification, denitrification) along a treatment train.

The SEIFC is equipped with mixing (a physical mixer or via sweep gas circulation) to ensure adequate gas mixing and, ideally, an online temperature probe. The SEIFC is currently one of the few devices accepted by the US EPA for measuring gaseous fluxes (Tata et al., 2003). Continuous gas-phase analyses are conducted via infrared (N_2O/CH_4). Although not discussed in this chapter, the gas could also be measured for oxygen and carbon dioxide for measuring bio-conversions in the wastewater treatment process.

In general, sampling in open surface tanks needs to be conducted at multiple locations of the activated sludge train in each wastewater treatment facility. These locations include aerobic, anoxic, and anaerobic zones, depending upon the configuration of the given facility. In non-aerated zones a sweep gas is used to produce an effective gas flow through the flux chamber. During the course of the gas phase sampling, liquid phase samples for dissolved compounds such as ammonium, nitrate and nitrate also need to be collected adjacent to the hood location. The samples should be filtered immediately upon collection in the field and can be analysed utilizing readily available field methods (i.e. Hach kits) and standard laboratory analytical methods (APHA et al., 2012). This allows the measured gas data to be linked with the actual concentrations of relevant compounds at the point of measurement. Comparing the data with influent/effluent measurements or measurements at other spots in the tank might give biased correlation between the off-gas and liquid phase samples.

In open surface tanks, the specific locations selected are typically close to the influent or effluent end of each demarcated anoxic or aerobic zone in the WWTP. Continuous measurement at each of these specific locations is typically conducted over a minimum 24-h period or longer.

The treatment trains of selected wastewater treatment plants that accomplish nitrification and denitrification are characterized based on their liquid-phase and gas-phase nitrogen concentrations and speciation. Testing is conducted at each location during a sampling campaign during which gas-phase monitoring is conducted in real-time continuous mode and liquid-phase sampling is conducted via discrete grab sampling. Trends and variations in gaseous emissions and speciation are also ascertained. This sampling effort is intended to assist in the development of process-operating criteria that minimize both gaseous and liquid-phase nitrogen emissions from wastewater treatment facilities. The analysis of nitrogen GHG compounds and precursors in both the air and liquid phases is complemented by analysis of conventional wastewater parameters.

Monitoring of the liquid phase and the gas phase can be conducted in seasonally distinct regimes, for instance, once in warm temperature conditions (i.e. summer, early autumn), and cold temperature conditions (winter/early spring) to account for annual variability.

The overall procedure for measuring N_2O and CH_4 fluxes from the headspace of activated sludge tanks involves a variant of the EPA/600/8-86/008 and the SCAQMD tracer methods. Gas-phase analyses can be conducted via infrared (N_2O/CH_4) analysers.

In the absence of an approved method for N_2O in air or water, method modification can be necessary to measure N_2O emissions. In a recent study (Ahn et al., 2010a), in order to evaluate the performance of the measurement of N_2O fluxes using the procedure developed by the researchers, three side-by-side monitoring events were conducted along with the research procedure during the first sampling event at a step feed BNR facility. In addition to the standard research protocol, two additional side-by-side monitoring events were conducted as follows (Table 4.5).

Plant wastewater research engineers measured fluxes using the EPA isolation flux chamber and SCAQMD tracer method and with a photo acoustic analyser to directly determine N_2O. Plant consultants used the textbook EPA isolation flux chamber and SCAQMD tracer dilution method to measure the flux and the following analytical methods to measure ozone precursors and GHGs. These side-by-side tests were not designed to validate the modified analytical approach to establish an approved methodology. However, they provided an independent verification that the approach accurately measured nitrogen GHG emissions to meet the objectives of that study, for zones where concurrent side-by-side measurement was conducted. This study was part of the original Water Environment Research Foundation

(WERF) project, during which this protocol was developed.

Table 4.5 Additional comparative analytical methodologies for measuring off-gas emissions from wastewater treatment plants.

Method/Species	Technique	Application
ASTM method 1946-permanent gas analysis	GC/TCD	Relevant fixed gases: CH_4, CO, CO_2, and helium (He) as a separate analysis.
NIOSH 6600	FTIR	N_2O

Based on this side-by-side comparison, it was further recommended that subsequent studies should consider the helium (He) tracer method (based on ASTM D1946) to measure the gas flow rate from the flux chamber, the results of which are reported in (Ahn *et al.*, 2010b).

4.5.1 Protocol for measuring the surface flux of N_2O

The following protocol is intended to provide researchers and field sampling teams with a detailed description of the data collection methodology and analysis requirements to enable the calculation of gaseous nitrogen fluxes from different zones of activated sludge trains in a WWTP.

4.5.1.1 Equipment, materials and supplies

The following equipment is needed to perform the protocol; suppliers and manufacturers may vary:
1. A surface emission isolation flux chamber (commercially available from vendors or custom-built based on specifications from the US EPA (Kienbusch, 1986)).
2. Teledyne API N_2O Monitor Model 320E (Teledyne API, San Diego, CA).
3. Zero gas (containing zero ppm N_2O and CH_4), and N_2O and CH_4 gas standards (Tech Air, White Plains, NY).
4. Dwyer series 475 Mark III digital manometers to measure the flux chamber pressure from 0 to 1" (2.54 cm water column) (high sensitivity) and 0 to 100" (low sensitivity) of a water column (Dwyer Instruments Inc., Michigan City, IN).
5. A rotameter to measure the influent sweep gas flow rate, 0-30 L min^{-1} (Fisher Scientific, Fairlawn, NJ).
6. An adjustable air pump, 0-10 L min^{-1} (Fisher Scientific, Fairlawn, NJ) to provide sweep gas flow into the flux chamber.
7. A vacuum pump, 0-30 L min^{-1} (Fisher Scientific, Fairlawn, NJ) for active pumping of gas from the flux chamber, if needed.
8. 0.2 μm cartridge filters, set of 10 (Millipore, Ann Arbor, MI) to prevent fine particulates from entering the gas analysers.
9. A silica gel column for capturing moisture (Fisher Scientific, Fairlawn, NJ).
10. A glass water trap consisting of a 100 mL glass bottle placed in ice within a Styrofoam® box.
11. Teflon® tubing (approximately 0.5") and fittings.
12. A 100-300' extension cord and power strip.
13. A laptop personal computer (with at least 512 MB RAM) with data acquisition programs for N_2O and CH_4 analysers pre-installed.
14. A set of miscellaneous hand tools including adjustable wrenches, different size screwdrivers and adjustable pliers.

4.5.1.2 Experimental procedure

The overall procedure for measuring N_2O and CH_4 fluxes from the head-space of activated sludge tanks involves a variant of the EPA/600/8-86/008 and the South Coast Air Quality Management District (SCAQMD) tracer methods, which allow sampling of gaseous emissions from high surface flux rate operations.

This variant has been developed to measure those sources that have a relatively high surface flux rate when compared to diffusion; this facilitates increased sampling at composting and wastewater treatment plants.

Commercially available replicas of the US EPA surface emission isolation flux chamber (SEIFC) are used to measure gaseous N fluxes from activated sludge reactors. The US EPA SEIFC essentially consists of a floating enclosed space through which carrier gas (typically nitrogen or argon) is fed at a fixed flow rate and exhaust gas is collected in a real-time or discrete fashion. Since the surface area under the SEIFC can be calculated or measured, the specific flux of the gaseous compound of interest can thus be determined. Since the SEIFC 'floats' on the activated sludge tank surface, several replicate measurements can be taken at different locations in a single tank as well as from different tanks (nitrification, denitrification) along a treatment train. The SEIFC is also equipped for mixing (with a physical mixer or via sweep gas circulation) to ensure adequate gas and

in some cases, an online temperature probe. The SEIFC is currently one of the few devices accepted by the US EPA for measuring gaseous fluxes (Tata *et al.*, 2003). Gas-phase analyses are conducted via infrared (N_2O, CH_4) methods.

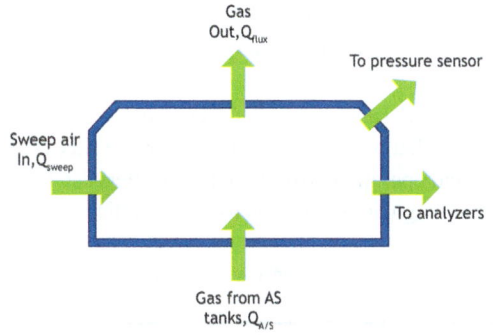

Figure 4.1 Schematic of the US EPA surface emission isolation flux chamber (modified from Tata *et al.*, 2003).

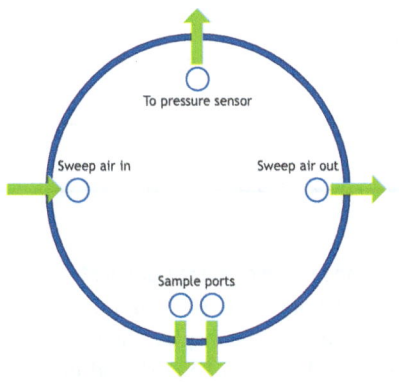

Figure 4.2 Modified schematic of the flux chamber.

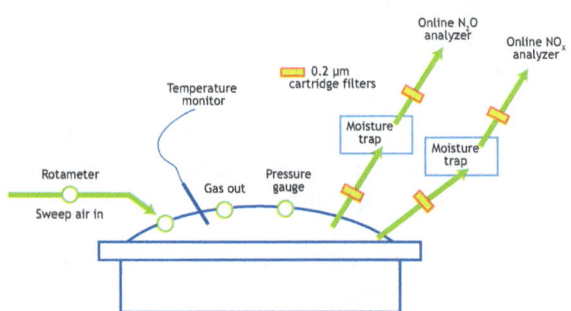

Figure 4.3 Schematic of flux-chamber set-up for gas flux measurements.

In general, sampling is conducted at multiple locations of the activated sludge train in each wastewater treatment facility. These locations are the aerobic, anoxic and anaerobic zones, depending upon the configuration of the given facility. Additionally, within each zone, multiple points (approximately three, but not less than two) must be sampled to address any variability in gas fluxes that may result from variations in mixing or flow patterns therein.

Pressure build-up can be minimized by equipping the flux chamber with multiple vents or a variable size vent and continuously monitoring the pressure drop across the hood using a sensitive pressure gauge. In all the field locations, the gas flow rate should be measured using the tracer gas technique and pressure across the flux chamber should be passively monitored if necessary. Alternatively, the aeration rate from plant records (available as an order of magnitude verification) can be used. The modified setup of the flux chamber used in this study is depicted in figures 4.1-4.3.

During the course of the gas phase sampling, liquid phase samples are collected adjacent to the hood location. The samples should be filtered immediately upon collection in the field and analysed by the host plant personnel for ammonia, nitrite and nitrate concentration, utilizing readily available field methods (i.e. a Hach Kit). As the primary purpose of these measurements is to ensure the presence of the targeted nitrogen species, without consideration for the accuracy in the concentration measurements, the simplest available field method will be used for these preliminary measurements.

4.5.1.3 Sampling methods for nitrogen GHG emissions

The gas-phase sampling method in aerobic zones
1. Seal all but one vent in the flux chamber and connect a high-sensitivity pressure gauge to the one open vent.
2. Lower the flux chamber into the aerobic zone (the bottom of the rim should be below the surface of the water by 2.5-5.0 cm minimum).
3. Wait for the N_2O analyser to equilibrate, based on the stability indicator (<0.03)
4. Pull the flux chamber up. Open two vents and connect the gas analyser. The other vents should be left open to the atmosphere.

5. Record the temperature of the gas in the flux chamber using a digital temperature gauge (Fisher Scientific number 15-077-8 or a suitable alternative).
6. Care must be taken that the flow going to the two analysers does not exceed the gas flow rate from the flux chamber. Otherwise, atmospheric air will be drawn in through the vents in the flux chamber.

Determination of the gas flow rate from the flux chamber in aerobic zones

1. Disconnect the gas analysers and connect one outlet vent to the inlet line of a field gas chromatograph equipped with a thermal conductivity detector. Close the other vent.
2. Introduce tracer gas (10 % helium, 90 % zero air) through an inlet vent into the flux chamber at a known flow rate (for instance 1 L min^{-1}).
3. Measure the concentration of helium gas exiting the flux chamber (see the protocol in Section 4.6).
4. Based on the measured helium concentrations, calculate the flow rate of the aeration tank headspace gas entering the flux chamber (Eq. 4.1, Section 4.6).

The gas-phase sampling method in anoxic zones

1. Seal all but one vent in the flux chamber and connect a high-sensitivity pressure gauge to the one open vent.
2. Lower the flux chamber into the anoxic zone with a 1-2 inch minimum submergence, into the liquid surface.
3. Wait for the N$_2$O analyser to equilibrate, based on the stability indicator (< 0.03)
4. Pull the flux chamber up. Open two vents and connect the gas analyser and the sweep gas pump (Note: sweep gas is only used during anoxic zone sampling). The other vents should be left open to the atmosphere.
5. Record the temperature of the gas in the flux chamber using a digital temperature gauge (Fisher Scientific number 15-077-8 or a suitable alternative).
6. Care must be taken never to allow the flow going to the two analysers to exceed the sweep gas rate, or dilution air will be drawn in through an opening in the chamber.

Determination of the gas flow rate from the flux chamber in the anoxic zone

1. Disconnect the gas analysers and connect one outlet vent to the inlet line of a field gas chromatograph equipped with a thermal conductivity detector. Close the other vent.
2. Introduce sweep gas to the chamber at a flow rate of 4 L min^{-1} and wait 6 min for a steady state.
3. Introduce tracer gas (10% He, 90% zero air) through an inlet vent into the flux chamber at a known flow rate (for instance 1 L min^{-1}).
4. Measure the concentration of He gas exiting the flux chamber.
5. Based on the measured He concentrations, calculate the flow rate of aeration tank headspace gas entering the flux chamber (Eq. 4.1, Section 4.6).

Table 4.6 summarizes the data recording requirement checklist that needs to be followed for the flux-chamber setup and operation. Additional parameters can be added by sampling teams based on a case-specific basis.

Table 4.6 Checklist for flux chamber set-up and operation in field.

Measurement	Sampling location 1	Sampling location 2	Sampling location 3
Pressure in flux chamber			
Gas flow rate from flux chamber			
Gas temperature in flux chamber			
Wastewater temperature			
Air-pump flow rates			

Continuous and real-time gas measurement

1. Turn on the power by pressing the on/off switch on the front panel. The display should turn on and the green (sample) status LED should blink, indicating the instrument has entered the HOLD-OFF mode. Sample mode can be entered immediately by pressing the EXIT button on the front panel. The red 'fault' light will also be on until the flows, temperatures and voltages are within operating limits. Clear the fault messages. After the warming up, review the TEST function values in the front panel display by pushing the button TEST on the left of the display.
2. Activate the DAS data acquisition software and set the sampling frequency for 1 sample per minute.
3. Start the data acquisition.
4. Connect the inlet tubing of the analyser to the outlet tubing from the SEIFC securely using a standard 1/4" compression fitting connector.
5. Acquire data for about 20 min in anoxic zones and about 10 min in aerobic zones after stable readings

are obtained, as indicated by the stability indicator on the N_2O analyser.
6. Stop the DAS software and immediately save the acquired data.
7. Repeat steps 2-5 for each sampling point and sampling locations (individual tanks).
8. Note that the measurement range is 0-1,000 ppm.
9. Before each sampling event, calibrate the instrument by using 'zero gas' and N_2O standard gas as per the manufacturer's instructions.

Principles of real-time N_2O and CH_4 measurements

Continuous N_2O and CH_4 measurements are performed via infrared (IR) gas-filter correlation, which is based on the absorption of IR radiation by N_2O and CH_4 molecules at appropriate wavelengths. As part of the measurement process, a broad wavelength IR beam is generated inside the instrument and passed through a rotating gas filter wheel, which causes the beam to alternatively pass through a gas cell filled with nitrogen (a measure cell), and a cell filled with N_2O/N_2 mixture (a reference cell) at a frequency of 30 cycles per second. N_2O concentrations are inferred based on the amount of IR absorption. Ultimately, the 'stripped' beam strikes the detector, which is a thermo-electrically cooled solid-state photo-conductor. This detector, along with its pre-amplifier converts the light signal into a modulated voltage signal.

4.5.1.4 Direct measurement of the liquid-phase N_2O content

In addition to measuring gaseous phase N_2O concentrations in the headspace of aerobic and anoxic zones, liquid phase N_2O concentrations could also be measured to discriminate between N_2O generation in the liquid phase and N_2O emission in the gas phase. Liquid-phase N_2O concentrations can be measured using a polarographic Clark-type electrode (Unisense, Aarhus, Denmark). For additional details of the liquid phase measurements summarized in this section, please refer to the protocols in Section 4.7.

1. Withdraw a sample of about 20 mL from the test reactors in 50 mL conical centrifuge tubes or alternatively similar containers (plastic or glass beakers are acceptable).
2. Take out the microsensor from the calibration chamber (containing deionized water), rinse out with deionized water, and mop dry with a tissue.
3. Immerse the microsensor into the samples. Proceed as rapidly as possible after obtaining the sample.
4. Record the numbers from the display on the picoammeter. The measurement numbers should become stable within one minute.
5. Pull out the microsensor, rinse out and place it back into the calibration chamber.
6. Repeat steps 1 to 5 for each sampling point and location.

The overall sampling effort is rather involved and thus needs to be closely coordinated. The real-time data from the analysers or probes should be automatically downloaded onto a field computer or recorded in laboratory notebooks. All the electronic data should be backed up frequently and where feasible electronic data should be stored on a temporary disk drive (in addition to the PC hard drive) during the field testing events.

4.6 MEASUREMENT OF OFF-GAS FLOW IN OPEN TANKS

This section describes a protocol for the measurement of off-gas flow rates using a helium tracer gas method (after ASTM method D1946). While measuring and reporting gaseous emission fluxes or while performing mass balances, the measurement of both headspace gas concentrations and the overall off-gas flow rates is needed. While gaseous N_2O and CH_4 concentrations from the activated sludge surface can be measured online, it is often not practical to conduct online or even discrete round-the-clock advective gas flow measurements. While blower operational data can be used as an approximation for determining advective flow, however usually only run time or electricity use are measured and conversion to actual air flow has uncertainties. Moreover, this approach suffers some serious uncertainties in the link between overall bulk blower air flow and actual distribution among different aerated zones. Direct measurement of the gas flow in aeration air pipes is an option, or for closed tanks often the off-gas flow rate can be measured. For open tanks a helium tracer method can be used, but not in a continuous mode owing to cost and plant access issues. At this point, the only way to overcome this limitation for open tank systems and reduce the inherent variability from extrapolating flow data from a few discrete measurements is to measure the advective flow rate using the He tracer method more frequently and employ a correlation analysis to translate the discrete flow

4.6.1 Protocol for aerated or aerobic zone

Advective flow of gas through the flux-chamber ($Q_{emission}$) in aerated zones is measured using a modification of the ASTM method D1946. Briefly, a tracer gas consisting of 100,000 ppmv ($C_{helium-tracer}$) helium is introduced into the flux-chamber at a known flow rate, Q_{tracer} (Eq. 4.3). Helium concentrations in the off-gas from the flux chamber ($C_{helium-FC}$) can be measured using a field gas chromatograph equipped with a thermal conductivity detector (GC-TCD). $Q_{emission}$ can then be computed using Eq. 4.1.

Protocol steps
1. Activate the field gas chromatograph approximately prior to the actual helium measurements to allow for the thermal conductivity detector (TCD) and GC column to attain the desired temperatures.
2. After measuring the gas-phase concentrations, disconnect the gas analysers and connect one outlet vent to the inlet line of the field GC. Close the other vent.
3. Introduce tracer gas (10 % helium, 90 % zero air) through an inlet vent into the flux chamber at a known flow rate (for instance 1 L min^{-1}).
4. Measure the concentration of helium gas exiting the flux chamber (as per the ASTM method D1946).
5. Based on the measured helium concentrations, calculate via linear algebra the flow rate of the aeration tank headspace gas entering the flux chamber (Eq. 4.1).

$$Q_{tracer} \cdot C_{helium-tracer} = (Q_{tracer} + Q_{emission}) \cdot C_{helium-GC}$$
$$Q_{emission} = \frac{Q_{tracer} \cdot (C_{helium-tracer} - C_{helium-GC})}{C_{helium-GC}}$$

Eq. 4.1

6. For each sampling location, repeat steps 2-5 at least three times.

4.6.2 Protocol for non-aerated zones

The only modification to the protocol to measure the emission flow rate from non-aerated zones is the introduction of sweep gas (air) or carrier gas through the flux chamber at a known flow rate (Q_{sweep}), in addition to the helium tracer gas. The corresponding $Q_{emission}$ is computed using Eq. 4.2. The addition of sweep gas is needed to promote mixing of the SEIFC contents, owing to the low advective gas flow from the anoxic-zone headspace. Sweep-air N$_2$O and CH$_4$ concentrations always need to be measured and checked to ensure that they remain below the detection limits of the N$_2$O or CH$_4$ analysers.

Protocol steps
1. The only modification to the protocol for adaptation to measuring the emission flow rate from the anoxic zone is the introduction of sweep gas.
2. Introduce sweep gas to the chamber at a flow rate of 4 L min^{-1} and wait 6 min for a steady state.
3. Follow steps 2-6 as described above for determination of the emission flow rate from aerobic zones.
4. Calculate the emission flow rate from the anoxic zone using Eq. 4.2.

$$Q_{tracer} \cdot C_{helium-tracer} = (Q_{tracer} + Q_{sweep} + Q_{emission}) \cdot C_{helium-GC}$$
$$Q_{emission} = \frac{Q_{tracer} \cdot (C_{helium-tracer} - C_{helium-GC})}{C_{helium-GC}} - Q_{sweep}$$

Eq. 4.2

Each sampling campaign consists of discrete and continuous N$_2$O measurements. During the discrete N$_2$O measurements, $Q_{emission}$ should be determined at each location in the treatment plant where N$_2$O is measured. During continuous N$_2$O measurements, $Q_{emission}$ should be determined several times a day in correspondence with liquid-phase measurements.

In the case of surface-aerated tanks measuring actual gas flow rates, the actual fluxes of greenhouse gasses is usually an even more difficult task. In these cases emissions can be best estimated from liquid-phase measurements and balances over the liquid phase.

4.7 AQUEOUS N$_2$O and CH$_4$ CONCENTRATION DETERMINATION

The aqueous concentrations of nitrous oxide can be directly measured by oxygen electrodes that are polarized differently than for oxygen measurement. For strict anaerobic or anoxic conditions (i.e. no oxygen present) these adapted electrodes can be directly used. For measuring N$_2$O concentrations in an aerated aqueous phase, a miniaturized Clark-type sensor with an internal reference and a guard cathode is available, and can be successfully applied (Foley *et al.*, 2010). The sensor is equipped with an oxygen front guard to prevent oxygen from interfering with the N$_2$O measurements. This method is described in Section 4.7.1. For methane no dissolved gas-phase measurement is available.

The concentrations of N₂O and CH₄ in a liquid sample can be determined by driving the dissolved gas into the gas phase. In a grab sample this can be done by salting out the gasses into the headspace of the sample tube, and analysing the gas phase on a GC (Section 4.7.3). With a gas-stripping device it is possible to continuously monitor the dissolved gas concentrations (Section 4.7.4).

4.7.1 Measurement protocol for dissolved N₂O measurement using polarographic electrodes

4.7.1.1 Equipment

1. A nitrous oxide microsensor N2O25 (Unisense, Aarhus, Denmark).
2. A two-channel picoammeter PA2000 (Unisense, Aarhus, Denmark).
3. A calibration chamber CAL300 (Unisense, Aarhus, Denmark).
4. Zero air and N₂O gas standard (Tech Air, White Plains, NY).
5. Teflon® tubing, silicone tubing and fittings.
6. A squeezer with deionized water.
7. Kimwipes.
8. BD Falcon 50 mL conical tubes.

4.7.1.2 Experimental procedure

The Unisense nitrous oxide microsensor is a miniaturized Clark-type sensor with an internal reference and a guard cathode. In addition, the sensor is equipped with an oxygen front guard, which prevents oxygen from interfering with the nitrous oxide measurements. The sensor is connected to a high-sensitivity picoammeter and the cathode is polarized against the internal reference. Driven by the external partial pressure, nitrous oxide from the environment will penetrate through the sensor tip membranes and be reduced at the metal cathode surface. The picoammeter converts the resulting reduction current into a signal. The internal guard cathode is also polarized and scavenges oxygen in the electrolyte, thus minimizing zero-current and pre-polarization time.

The measurement steps are as follows:

a. Turn on the power switch located on the front panel of the picoammeter.
b. Check that the 'Gain' screw for Channel 1 is turned fully counter-clockwise.
c. Turn the display switch, located on the centre of the panel, to 'Signal 1' and check that the display reads zero. If not, adjust the offset, as per the manufacturer's instructions.
d. Turn the display switch to 'Pol. 1'. Check if the polarization voltage shows -0.8 V. If not, adjust the volt and polarity switch.
e. Connect the leads of the 'pre-polarized' microsensor to the meter in the following order: (1) Signal wire (black) to 'Input' of Channel 1 on the front panel. (2) Guard wire (yellow) to 'Guard' of Channel 1.
f. Rinse out the sensor with deionized water and absorb the moisture with tissue paper.
g. Place the sensor into the calibration chamber, which contains deionized water.
h. Select the 'Normal' setting for the 'Mode' switch on the front panel, unless you need the extremely fast response.
i. Select the appropriate measuring range using the 'Range' switch on the panel. Usually 200 pA is selected, but if not suitable, select an alternative available range.
j. Withdraw about 20 mL sample from test reactors in 50 mL conical centrifuge tubes or alternatively similar containers (plastic or glass beakers are acceptable).
k. Take out the microsensor from the calibration chamber (containing deionized water), rinse out with deionized water, and mop dry with a tissue.
l. Immerse the microsensor into the samples. For (j) and (k), proceed as rapidly as possible after acquiring the sample.
m. Record the numbers from the display on the picoammeter. The measurement numbers should be stable within one minute.
n. Pull out the microsensor, rinse out and place it back in the calibration chamber.
o. Repeat steps (j) to (n) for each sampling point and location.
p. When the measurements are complete, disconnect the sensor leads in the reverse order to which they were connected.

The measurement range is adjustable, 0-0.616 ppmv-N₂O (with 500 ppm N₂O gas standard).

If the sensor is new or has not been operated for several days, then it must be polarized for at least 2 h and up to 12 h before it can be calibrated and/or used, as follows:

a. Secure the nitrous oxide sensor with its tip, immersed in nitrous oxide-free water.
b. Turn the display switch to 'Pol. 1' and adjust the polarization to -1.30 V.
c. Turn the display switch to 'Signal 1' and adjust the 'Gain' screw completely counter-clockwise. Adjust the display to zero on the 'Offset' dial, if needed.
d. Connect the signal wire (black) of the microsensor to the 'Input' terminal.
e. After 5 min, adjust the polarization to -0.8 V and then connect the guard wire (yellow) to the 'Guard' terminal.
f. Pre-polarize for up to 12 h if possible to get the maximum stability.

After the sensor has been polarized, it must be calibrated with zero air and N_2O gas standards. Typically, we have used 500 ppm N_2O gas standards for calibration. Note that N_2O gas standards are specialised items and can be purchased from vendors such as TechAir.

To be consistent in terms of units for liquid and gas phase N_2O, the results of this study are expressed in terms of N_2O. Alternatively, liquid and gas phase N_2O concentrations can also be expressed as 'N' to estimate the fraction of influent nitrogen discharged as N_2O.

4.7.2 Measurement protocol for dissolved gasses using gas chromatography

Both nitrous oxide and methane can be readily analysed by standard gas chromatographic analysis (Weiss, 1981). For liquid samples this can be performed by taking a sample and transferring it directly to a closed bottle or tube with a calibrated volume. In the bottle an equilibrium will establish for the volatile compounds between gas and liquid phase. After equilibration the gas phase concentration can be measured. With the Henry coefficient (corrected for temperature and ionic strength) the liquid concentration for methane or nitrous oxide can be calculated. Using these concentrations and the known gas and liquid volumes in the sample bottle then gives the total mass of compound in the bottle and the original concentration in the sample. This method is prone to many measurements and uncertainties for e.g. the true Henry coefficient for the actual fluid composition.

It is therefore more reliable to drive all the soluble gasses into the gas phase by e.g. a high salt concentration (the salting-out method). This has been described in detail by Daelman et al. (2012) based on the salting-out method described by Gal'chenko et al., (2004). The measurement protocol is described below.

4.7.3 Measurement protocol for dissolved gas measurement by the salting-out method

Before the start of each sampling round, serum bottles of 120 mL are filled with 20 g NaCl. At the different sampling locations at the WWTP, samples are collected with a sampling beaker. From this beaker, 50 mL of the sample is added carefully to a serum bottle filled with salt, using a syringe with a catheter tip and a 10 cm silicone tube. While emptying the syringe into the bottle, the silicon tube is held under the rising liquid surface in order to keep the liquid-gas interface as small as possible to avoid stripping. Immediately after adding the syringe content to the bottle, the bottle is sealed with a rubber stopper and an aluminium cap. The sealed bottle is shaken vigorously in order to speed up the dissolving of the salt. At 20 °C the solubility of NaCl in water is about 360 g L^{-1}, so these samples, containing 400 g NaCl L^{-1}, are oversaturated. As a consequence of the high salt concentration, the microbial activity in the sludge samples is halted, and the dissolved gases are salted out. Dissolved methane, but also the other dissolved gases such as carbon dioxide escape from the liquid phase to the headspace of the serum bottles. This results in a pressure build-up in the headspace. Before sampling the headspace of the samples for analysis with GCFID, the pressure in the headspace needs to be equilibrated with the atmosphere by allowing the gas in the headspace to expand in a submerged graduated syringe. The increase in gas volume is used to calculate the pressure build-up in the headspace. After the gas pressure in the headspace is brought to atmospheric pressure, the headspace is sampled with a gas syringe and analysed in the conventional procedure by gas chromatography. The amount of methane in the headspace before expansion is calculated from the concentration, the measured volume of the headspace of the sealed bottle and the headspace pressure after the pressure build-up that is calculated from the volume expansion of the headspace. Before the sample was saturated with salt, this amount of methane in the gas phase must have been completely dissolved in the liquid sample. By dividing this amount by the sample volume (50 mL), the original methane concentration in the liquid can be established.

4.7.3.1 Equipment

The equipment needed in this protocol is as follows:
- A sampling beaker.
- A graduated syringe with a catheter tip and tube of ca. 15 cm.
- A serum bottle of ca. 120 mL.
- A rubber stopper.
- An aluminum seal.
- A crimper for aluminium seals.
- 20 g NaCl.
- A beaker.
- A graduated syringe (or burette) with a catheter tip and without a plunger.
- A tube of ca. 30 cm.
- A hypodermic needle.

4.7.3.2 Sampling procedure

- Before sampling starts, add 20 g NaCl to the serum bottle.
- Take a sample from the liquid surface of the reactor or from the valve of a pipe using a sampling beaker.
- Suck 50 mL (V_{sample}) of the sample from the sampling beaker into the syringe with the catheter tip and tube.
- Release the content of the syringe into the serum bottle, while keeping the end of the tube under the liquid surface.
- Seal the serum bottle with the rubber stopper and the aluminium seal.
- Shake the bottle vigorously.

4.7.3.3 Measurement procedure

Measurement of the volume expansion due to pressure build-up in the bottle

a. Place the graduated syringe (or burette) in a beaker with water.
b. Connect the graduated syringe (or burette) with the catheter tip to the needle with a tube.
c. Make sure that the open end of the syringe is submerged in the water in the beaker.
d. Register the headspace volume in the syringe V_0.
e. Pierce the rubber stopper of the serum bottle with the needle connected to the syringe (Figure 4.4).
f. The headspace volume of the syringe will expand.
g. Bring the water level in the syringe to the water level in the beaker to cancel out the pressure of the water column.
h. Register the new headspace volume in the syringe V_1.
i. $V_1 - V_0 = V_s$ with V_s the volume expansion due to the pressure build-up in the serum bottle.

Figure 4.4 Measurement of the volume expansion due to the pressure build-up that is caused by the salting-out of the dissolved gases (photo: van Dongen, 2015).

Measurement of the methane or the nitrous oxide concentration in the headspace

Draw a sample from the headspace of the serum bottle with a gas syringe and measure it with a gas chromatograph equipped with a flame ionization detector, according to the appropriate method for measuring methane.

Measurement of the headspace of the serum bottle

a. Mark the level of the NaCl saturated sample in the serum bottle.
b. Mark the lower side of the rubber stopper.
c. Empty and rinse the bottle.
d. Fill the bottle with clean water up until the mark of the liquid level.
e. Register the weight of the bottle W_0.
f. Add clean water to the bottle up until the mark of the stopper.

g. Register the weight of the bottle W_1.
h. $(W_1 - W_0) \times \rho = V_{HS}$ with ρ the density of the water and V_{HS} the volume of the headspace of the serum bottle.

4.7.3.4 Calculations

Volume

$$V = V_s + V_{HS} \quad \text{Eq. 4.4}$$

Where, V is the expanded volume of the headspace (m³), V_s is the volume expansion due to the pressure build-up (m³), and V_{HS} is the headspace of the serum bottle before expansion (m³).

Amount of methane

$$n = \frac{P \cdot V}{R \cdot T} \quad \text{Eq. 4.5}$$

Where, n is the amount of methane in the expanded headspace of the serum bottle (mol), P is the atmospheric pressure (Pa), V is the expanded volume of the headspace (m³), R is the ideal gas constant: 8.314 m³ Pa mol⁻¹ K⁻¹, and T is the temperature (K).

Concentration

$$C = \frac{n}{V_{sample}} \quad \text{Eq. 4.6}$$

Where, C is the concentration (M), N is the amount of methane in the expanded headspace of the serum bottle (mol), and V_{sample} is the volume of the sample (L).

4.7.4 Measurement protocol for dissolved gas measurement by the stripping method

4.7.4.1 Operational principle

This method was first published in Mampaey *et al.* (2015). By using a gas-stripping device, it is possible to monitor dissolved gases, e.g. N_2O, on a continuous basis. The unknown liquid concentration can be calculated from the measured gas-phase concentration in the off-gas of a stripping device. The proposed method relies on a gas-stripping device, consisting of a stripping flask and a scum trap flask, as displayed in Figure 4.5.

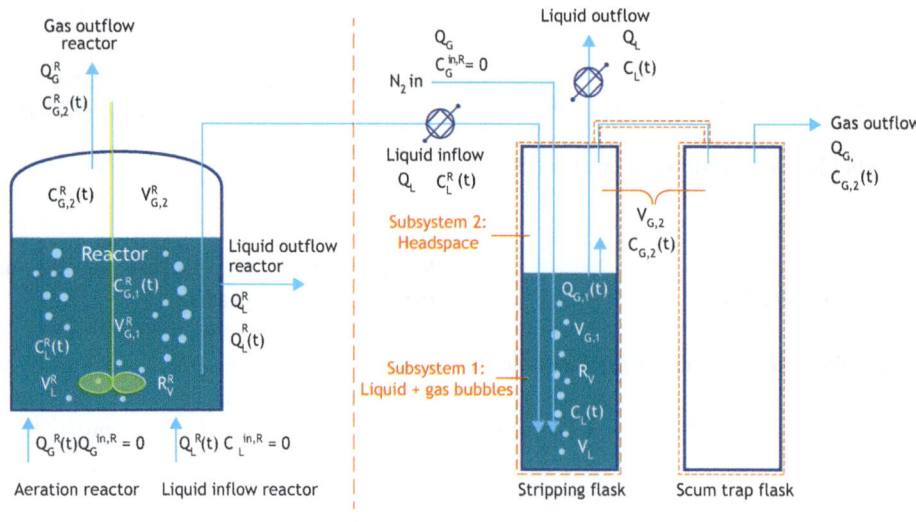

Figure 4.5 Lay-out of the reactor (left) and the gas stripping device (right) for monitoring dissolved gases (Mampaey *et al.*, 2015).

A liquid sample stream from the reactor is continuously supplied to the stripping flask at a constant flow rate Q_L, while maintaining a constant liquid volume V_L in the stripping flask. The scum trap flask is an empty bottle to collect entrained scum from the stripping flask.

Nitrogen is used as the stripping gas in the stripping flask through fine bubble aeration at a constant flow rate Q_G^{in}. The gas outflow of the gas-stripping device is analysed by an online gas-phase analyser. The dissolved concentration in the reactor, $C_L^R(t)$, is calculated from the measured gas concentration $C_{G,2}(t)$ according to Eq. 4.7.

$$C_L^R(t) = \frac{Q_G}{Q_L} \cdot \left(1 + \frac{Q_L}{a_3 \cdot V_L}\right) \cdot C_{G,2}(t) - \frac{Q_G}{Q_L} \cdot a_1 \qquad \text{Eq. 4.7}$$

The parameters a_1 and a_3 are determined from a batch-stripping test. During this test, the stripping flask is filled in batches with a liquid sample from the reactor under study from which the dissolved N_2O is subsequently stripped with N_2. The monitored gas phase profile $C_{G,2}(t)$ from the stripping device is then described by Eq. 4.8.

$$C_{G,2}(t) = a_1 + a_2 \cdot \exp(-a_3 \cdot t) - a_4 \cdot \exp(-a_5 \cdot t) \qquad \text{Eq. 4.8}$$

The gas-stripping device provides an adequate method to indirectly measure dissolved gases (N_2O or other gases) in the liquid phase, for aerated as well as non-aerated conditions/reactors, following variations both in time and in space. Its application to an intermittently aerated (on/off) partial nitritation (SHARON) reactor was demonstrated by Mampaey et al. (2015). Castro-Barros et al. (2015) applied the method to a one-stage partial nitritation-anammox reactor, subject to alternating high and low aeration. In both cases, the mass balance approach on which the liquid N_2O concentration measurement method is based also allowed the determination of the N_2O formation rate.

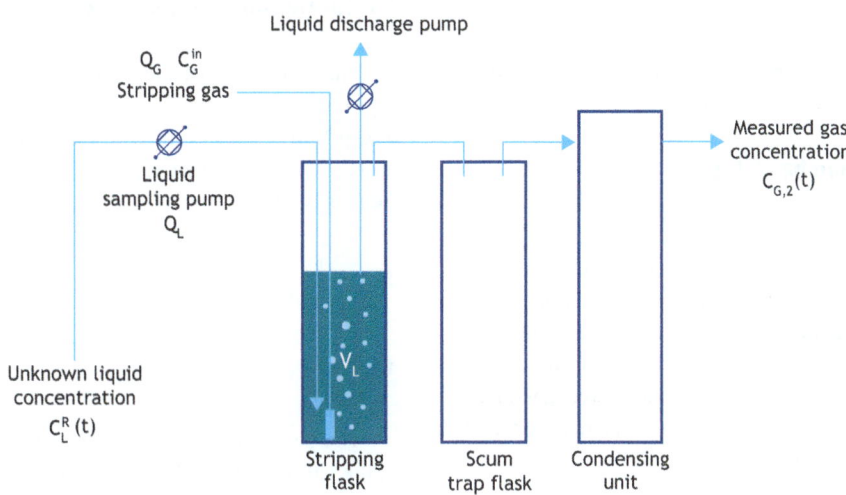

Figure 4.6 Layout of the gas-stripping device.

4.7.4.2 Equipment

The detailed layout of the gas-stripping device is shown in Figure 4.6, and consists of:

a. A stripping flask: a graded plastic cylinder of 250 mL with an (aquarium) aeration stone for the stripping gas and a rubber stopper with four connections (stripping gas in, stripping gas out, sampled liquid in and liquid out) to seal the stripping flask. The liquid is introduced at the bottom of the stripping flask, next to the aeration stone for intense mixing. The liquid is extracted 10 cm below the rubber stopper, which results in a constant liquid volume V_L of 100 mL. A larger liquid volume increases the sensitivity of the device (K) but decreases the frequency of measurements (or decreases the observability of rapid changes in concentrations).

b. A scum trap flask: a 2 L bottle to collect any entrained scum was used.

c. A liquid sampling pump: a peristaltic pump with a tube with an inside diameter of 3.1 mm (Masterflex tubing size L/S = 16), which yields a liquid flow rate of 85 mL min^{-1}. It is advised to keep the length of the liquid feed tube as short as possible. A larger liquid flow rate (Q_L) increases the sensitivity of the device.

d. A liquid discharge pump: a peristaltic pump with a tube with an inside diameter of 6.4 mm (Masterflex tubing size L/S = 17), which yields a liquid flow rate of 245 mL min^{-1}.
e. Stripping gas: a stripping gas flow rate (Q_G) of 1.2 NL min^{-1} is required for the analyser. The following two possibilities of stripping gas can be utilized: (*i*) gas cylinder N_2 is utilized, a mass flow controller can be used to control the stripping gas flow rate (at 1.2 NL min^{-1}), and (*ii*) ambient air can also be utilized as stripping gas. A diaphragm pump in combination with a mass flow controller can be used to provide the required gas flow rate (1.2 NL min^{-1}).
f. A condensing column / gas drying unit for the removal of air moisture.
g. Tubing: it is advised to use neoprene tubing to prevent diffusion of O_2.
h. A weatherproof box to protect the electronic part of the setup.

The gas phase is measured by a gas-phase measurement device as described in Section 4.5.

4.7.4.3 Calibration batch test

The parameters a_1 and a_3 are characteristics of the gas-stripping device and are determined from a batch-stripping test. During this test, the stripping flask is filled in batches with a liquid sample (containing dissolved N_2O) from the reactor under study and flushed with the stripping gas (typically N_2). The monitored gas-phase profile $C_{G,2}(t)$ from the stripping device is then described by a double exponential profile (Eq. 4.9):

$$C_{G,2}(t) = a_1 + a_2 \cdot \exp(-a_3 \cdot t) - a_4 \cdot \exp(-a_5 \cdot t) \quad \text{Eq. 4.9}$$

Eq. 4.9 is fitted to the batch test measurement to obtain the values for the coefficients. A calibration example is shown in Figure 4.7.

Parameter a_1 is related to the (N_2O) conversion rate in the stripping flask and its concentration in the stripping gas, a_3, is related to the interphase transfer rate in the stripping flask, and a_5 is related to the gas delay by the scum trap flask. The exact meanings of these parameters can be found in Mampaey *et al.* (2015).

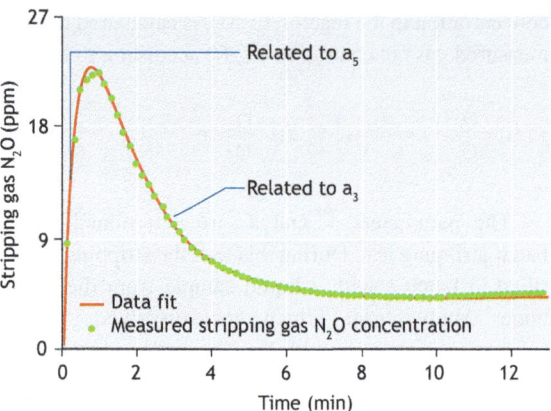

Figure 4.7 Example of a calibration batch test with data fit.

4.7.4.4 Measurement accuracy

The sensitivity of the setup can be calculated from Eq. 4.6. In the current setup, an accuracy for $C_L^R(t)$ of 0.03 g N m^{-3} is achieved. The fastest measurable changes in ($C_L^R(t)$) is estimated from Eq. 4.11, given as a frequency of sampling (f_{sample}).

$$K = \frac{a_3}{a_3 + D_L} \cdot \frac{Q_L}{Q_G} \quad \text{Eq. 4.10}$$

$$f_{sample} < a_3 + D_L \quad \text{Eq. 4.11}$$

The parameter a_3 is the decreasing exponential term in the calibration batch test (Eq. 4.8), $D_L = Q_L/V_L$ is the liquid dilution rate, Q_G is the stripping gas flow rate (1.2 NL min^{-1}), Q_L is the liquid sampling flow rate (84 mL min^{-1}), and V_L is the liquid volume in the stripping flask (100 mL).

4.7.4.5 Calculation of the N_2O formation rate in the stripping device

N_2O formation can occur in the stripping flask and is reflected in parameter a_1. The N_2O formation rate in the gas-stripping device is calculated from Eq. 4.12. C_G^{in} is the N_2O concentration in the fresh stripping gas; if no N_2O is present, $C_G^{in} = 0$.

$$R_V = \left(a_1 - C_G^{in}\right) \cdot \frac{Q_G}{V_L} \quad \text{Eq. 4.12}$$

The formation rate in the sampled water body is calculated as the sum of the gas emission and change in dissolved N₂O through a mass balance, Eq. 4.13.

$$R_V(t) = \frac{\Delta C_L^R(t)}{\Delta t} + \text{Emission} \qquad \text{Eq. 4.13}$$

4.8 DATA ANALYSIS AND PROCESSING

4.8.1 Determination of fluxes

The net flux of gaseous N₂O or CH₄ (kg m⁻² d⁻¹) can be calculated based on the gas flow rate out of the flux chamber ($Q_{emission}$, m³ d⁻¹), the gas concentration (C, kg m⁻³), and the cross-sectional area of the SEIFC (A, m²) (Eq. 4.14).

$$\text{Flux} = \frac{Q_{emission} \cdot C}{A} \qquad \text{Eq. 4.14}$$

The calculated flux should be corrected to reflect standard temperature (20 °C) and pressure (1 atm).

4.8.2 Determination of aggregated emission fractions

As previously described, the specific locations selected for measuring gaseous and aqueous concentrations of N₂O and CH₄ can be close to the influent or effluent end of each demarcated anoxic or aerobic zone in the WWTP, or, alternatively, at locations where the production of these two gases could be inferred based on initial screening of the process variable concentrations. Using these point measurements and assuming that these measured concentrations were uniform over a particular zone, the emissions from any given zone can be computed according to the following equation:

Emissions from i^{th} zone in a WWTP = Emissions from SEIFC × (Area of ith zone/Area of SEIFC).

Emission measurements must be repeated at multiple zones for any given plant. The overall emissions of the plant are then calculated by summing the emissions from each zone measured over a 24-h period (at the very minimum). The surface flux calculated from Eq. 4.14 is translated into the flux of a given zone by multiplying with the specific zone area. The N₂O emission fractions (mass/mass) for each WWTP at any given time point can be computed by normalizing the measured flux from each zone in the facility to the daily influent total Kjeldahl nitrogen (TKN) loading according to Eq. 4.15. Correspondingly, for methane, the measured flux could be normalized to the influent COD loading. Emission fractions are typically averaged over the course of the diurnal sampling period and reported as the average (avg) ± standard deviation (sd) for each individual process sampled.

During each campaign, wastewater nitrogen species and COD concentrations including influent, bioreactor, and effluent concentrations are measured simultaneously about six times per day according to APHA *et al.* (2012) to supplement the gas-phase measurements. The discrete measurements are to be averaged to generate the emission fractions described in Eq. 4.15.

$$\text{Emission fraction} = \frac{\sum_{i=1}^{n} \text{Flux}_i \times \text{Area}_i \text{ (kg N}_2\text{O-N)}}{\text{Daily influent TKN load (kg-N)}}$$

$$\text{Eq. 4.15}$$

Where, Flux_i is the N₂O emission flux calculated from the ith zone (kg N₂O-N m⁻² d⁻¹), Area_i is the surface area of the ith zone (m²), n is the number of zones in a given facility from which the N₂O fluxes are captured, and the Daily influent TKN load is the average influent load (influent flow rate × influent TKN concentrations) over the course of 24 h. It should be noted that the above calculations reflect the emission factor calculated from discrete N₂O measurements. In plants where significant diurnal variability exists, such variability will be accounted for by a combination of explicit measurements in select zones and mathematical modeling output of N₂O fluxes from the remaining zones.

Virtually identical equations can be used for methane, with the only distinction of normalizing to total COD mass loads in the influent or, as applicable, the COD mass removal rate.

On average, wastewater characterization about six times per day is recommended at each gas sampling location as well as in the tank influent and effluent. At facilities where analysis is not as frequent (for instance in the influent and effluent samples), daily composite measurements can be used. Alternatively, in some facilities, online devices (for measuring pH, DO, oxidation-reduction potential (ORP), and select N-species, including NH_4^+-N and NO_3^--N) can also be used at different locations of the activated sludge tank to facilitate the characterization of the wastewater.

4.8.3 Calculation of the emission factors

To directly compare results from off-gas monitoring campaigns with the approaches suggested by the EPA and IPCC, it becomes necessary to summarize the monitoring campaign results in terms of emission factors. Such emission factors can be computed by normalizing the total reactor N_2O or CH_4 mass flux to the unit population equivalent flow rate (100 gal PE^{-1} d^{-1}, or equivalently, 378.5 L PE^{-1} d^{-1} in the US, US EPA, 2012). The resulting emissions fluxes are expressed in units consistent with the US EPA inventory report (g N_2O or CH_4 PE^{-1} d^{-1}) (US EPA, 2012). Although the dynamic variability in WWTP off-gas emissions renders the use of point emission factors (such as the ones presented by the EPA and IPCC somewhat limited, the added step of computing emission factors could be useful for a side-by-side comparison of estimated emissions and those that are actually measured.

References

Ahn, J.-H., Kim, S., Pagilla, K., Katehis, D., and Chandran, K. (2010a). Spatial and temporal variability in N_2O generation and emission from full-scale BNR and non-BNR processes. *Water Environment Research*. 82: 2362-2372.

Ahn, J. H., Kim, S., Park, H., Rahm, B., Pagilla, K., and Chandran, K. (2010b). N_2O Emissions from Activated Sludge Processes, 2008-2009: Results of a National Monitoring Survey in the United States. *Environmental Science & Technology*. 44: 4505-4511.

APHA, AWWA and WEF, (2005). Standard methods for the examination of water and wastewater. 22nd edition. Eaton, A.D., Clesceri, L.S., and Greenberg, A.E. (eds.) Washington DC.

Castro-Barros C.M., Daelman, M.R.J., Mampaey, K.E., van Loosdrecht M.C.M., Volcke, E.I.P. (2015). Effect of aeration regime on N2O emission from partial nitritation-anammox in a full-scale granular sludge reactor, *Water Res.* 68: 793-803.

Daelman, M.R.J., van Voorthuizen E.M., van Dongen, U.G.J.M., Volcke, E.I.P., and van Loosdrecht M.C.M., (2012). Methane emission during municipal wastewater treatment. *Water Res.* 46: 3657-3670.

Daelman, M.R., de Baets, B., van Loosdrecht, M.C.M., and Volcke, E.I. (2013). Influence of sampling strategies on the estimated nitrous oxide emission from wastewater treatment plants. *Water Res.* 47(9): 3120-3130.

Daelman, M., van Eynde, T., van Loosdrecht, M.C.M., Volcke, E.I.P. (2014) Effect of Process design and operating parameters on aerobic methane oxidation in municipal WWTP. *Water Res.* 66, 308-319.

Foley, J., de Haas, D., Yuan, Z., and Lant, P. (2010). Nitrous oxide generation in full-scale biological nutrient removal wastewater treatment plants. *Water Res.* 44(3): 831-844.

Gal'chenko, V.F., Lein, A.Y., and Ivanov, M.V. (2004). Methane content in the bottom sediments and water column of the Black Sea. *Microbiology*. 73(2): 211-223.

Gapes, D., Pratt, S., Yuan, Z., and Keller, J. (2003). Online titrimetric and off-gas analysis for examining nitrification processes in wastewater treatment. *Water Res.* 37(11): 2678-2690.

Hellinga, C., Vanrolleghem, P., Van Loosdrecht, M.C.M., and Heijnen, J.J. (1996). The potential of off-gas analyses for monitoring wastewater treatment plants. *Water Sci Tech.* 33(1): 13-23.

IPCC (2013). Climate Change 2013: The Physical Science Basis. Working Group I Contribution to the Fifth Assessment Report of the Intergovernmental Panel on Climate Change. Chapter 8: Anthropogenic and Natural Radiative Forcing.

Kampschreur, M.J., Tan, N.C.G., Kleerebezem, R., Picioreanu, C., Jetten M.S.M., and van Loosdrecht M.C.M. (2008a). Effect of Dynamic Process Conditions on Nitrogen Oxides Emission from a Nitrifying Culture. *Environmental Science and Technology*. 42: 429-435.

Kampschreur, M.J., Temmink H., Kleerebezem, R., Jetten, M.S.M., and van Loosdrecht M.C.M. (2009). Nitrous oxide emission during wastewater treatment. *Water Res.* 43: 4093-4103.

Kampschreur, M. J., van der Star W.R.L., Wielders H.A., Mulder J.W., Jetten M.S.M., and van Loosdrecht M.C.M. (2008b). Dynamics of nitric oxide and nitrous oxide emission during full-scale reject water treatment. *Water Res.* 42: 812-826.

Kienbusch, M. (1986). Measurement of Gaseous Emissions Rates from Land Surfaces using an Emission Isolation Flux Chamber, User's Guide, EPA Users Guide. United States Environmental Protection Agency.

Klopfenstein Jr,R. (1998). Air velocity and flow measurement using a Pitot tube. *ISA transactions*. 37(4): 257-263.

Mampaey, K. E., van Dongen, U. G., van Loosdrecht, M.C.M., Volcke, E.I. (2015). Novel method for online monitoring of dissolved N2O concentrations through a gas stripping device. *Environmental Technology*. 36(13): 1680-1690.

Melcer, H. (1999). Methods for Wastewater Characterization in Activated Sludge Modelling. Water Environment Research Foundation, Alexandria, VA.

Puig, S., van Loosdrecht, M.C.M., Colprim, J., Meijer, S.C.F. (2008). Data evaluation of full-scale wastewater treatment plants by mass balance. *Water Res.* 42(18): 4645-4655.

Pratt, S., Yuan, Z., Gapes, D., Dorigo, M., Zeng, R.J., and Keller, J. (2003). Development of a novel titration and off-gas analysis (TOGA) sensor for study of biological processes in wastewater treatment systems. *Biotechnology and Bioengineering*. 81(4): 482-495.

Rapson, T.D., Dacres, H. (2014). Analytical techniques for measuring nitrous oxide. *TrAC Trends in Analytical Chemistry*. 54: 65-74.

Rieger, L., Gillot, S., Langergraber, G., Ohtsuki, T., Shaw, A., Tak, I., and Winkler, S. (2012). Guidelines for using activated sludge models. Water Intelligence Online, 11, 9781780401164.

Tata, P., Witherspoon J., and Lue-Hing C. (Eds). (2003). VOC Emissions from Wastewater Treatment Plants. Lewis Publishers, Boca Raton, FL.

USEPA. 2012. Inventory of U.S. Greenhouse Gas Emissions and Sinks: 1990-2006, EPA 430-R-08-005. Washington, D.C.

Weiss, R.F. (1981). Determinations of carbon dioxide and methane by dual catalyst flame ionization chromatography and nitrous oxide by electron capture chromatography. *Journal of Chromatographic Science*. 19(12): 611-616.

5

DATA HANDLING AND PARAMETER ESTIMATION

Authors:
Gürkan Sin
Krist V. Gernaey

Reviewer:
Sebastiaan C.F. Meijer
Juan A. Baeza

5.1 INTRODUCTION

Modelling is one of the key tools at the disposal of modern wastewater treatment professionals, researchers and engineers. It enables them to study and understand complex phenomena underlying the physical, chemical and biological performance of wastewater treatment plants at different temporal and spatial scales.

At full-scale wastewater treatment plants (WWTPs), mechanistic modelling using the ASM framework and concept (e.g. Henze *et al.*, 2000) has become an important part of the engineering toolbox for process engineers. It supports plant design, operation, optimization and control applications. Models have also been increasingly used to help take decisions on complex problems including the process/technology selection for retrofitting, as well as validation of control and optimization strategies (Gernaey *et al.*, 2014; Mauricio-Iglesias *et al.*, 2014; Vangsgaard *et al.*, 2014; Bozkurt *et al.*, 2015).

Models have also been used as an integral part of the comprehensive analysis and interpretation of data obtained from a range of experimental methods from the laboratory, as well as pilot-scale studies to characterise and study wastewater treatment plants. In this regard, models help to properly explain various kinetic parameters for different microbial groups and their activities in WWTPs by using parameter estimation techniques. Indeed, estimating parameters is an integral part of model development and application (Seber and Wild, 1989; Ljung, 1999; Dochain and Vanrolleghem, 2001; Omlin and Reichert, 1999; Brun *et al.*, 2002; Sin *et al.*, 2010) and can be broadly defined as follows:

Given a model and a set of data/measurements from the experimental setup in question, estimate all or some of the parameters of the model using an appropriate statistical method.

The focus of this chapter is to provide a set of tools and the techniques necessary to estimate the kinetic and stoichiometric parameters for wastewater treatment processes using data obtained from experimental batch activity tests. These methods and tools are mainly intended for practical applications, i.e. by consultants, engineers, and professionals. However, it is also expected that they will be useful both for graduate teaching as well as a stepping stone for academic researchers who wish to expand their theoretical interest in the subject. For the models selected to interpret the experimental data, this chapter uses available models from literature that are mostly based on the Activated Sludge Model (ASM) framework and their appropriate extensions (Henze et al., 2000).

The chapter presents an overview of the most commonly used methods in the estimation of parameters from experimental batch data, namely: (*i*) data handling and validation, (*ii*) parameter estimation: maximum likelihood estimation (MLE) and bootstrap methods, (*iii*) uncertainty analysis: linear error propagation and the Monte Carlo method, and (*iv*) sensitivity and identifiability analysis.

5.2 THEORY AND METHODS

5.2.1 Data handling and validation

5.2.1.1 Systematic data analysis for biological processes

Most activated sludge processes can be studied using simplified process stoichiometry models which rely on a 'black box' description of the cellular metabolism using measurement data of the concentrations of reactants (pollutants) and products e.g. CO_2, intermediate oxidised nitrogen species, etc. Likewise, the Activated Sludge Model (ASM) framework (Henze et al., 2000) relies on a black box description of aerobic and anoxic heterotrophic activities, nitrification, hydrolysis and decay processes.

A general model formulation of the process stoichiometry describing the conversion of substrates to biomass and metabolic products is formulated below (for carbon metabolism):

$$CH_2O + Y_{SO}O_2 + Y_{SN}NH_3 \rightarrow Y_{SX}X + Y_{SC}CO_2 + Y_{SP_1}P_1 + \ldots + Y_{SW}H_2O \quad \text{Eq. 5.1}$$

Equation 5.1 represents a simplification of the complex metabolic 'machinery' of cellular activity into one global relation. This simplified reaction allows the calculation of the process yields including Y_{SO} (yield of oxygen per unit substrate), Y_{SN} (yield of nitrogen per unit substrate), Y_{SX} (yield of biomass per unit substrate), Y_{SC} (yield of CO_2 per unit of substrate), Y_{SP_1} (yield of intermediate product P_1 per unit of substrate), and Y_{SW} (yield of water per unit of substrate).

The coefficients of this equation are written on the basis of 1 C-mol of carbon substrate. This includes growth yield for biomass, Y_{SX}, substrate (ammonia) consumption yields, Y_{SN}, oxygen consumption yields, Y_{SO}, yield for production of CO_2, Y_{SC}, and yield for water, Y_{SW}. The biomass, X, is also written on the basis of 1 C-mol and is assumed to have a typical composition of $CH_aO_bN_c$. The biomass composition can be measured experimentally, $CH_{1.8}O_{0.5}N_{0.2}$ being a typical value. Some of the yields are also measured experimentally from the observed rates of consumption and production of components in the process as follows:

$$Y_{ji} = \frac{r_i}{r_j} = \frac{q_i}{q_j} \quad \text{and} \quad Y_{ji} = Y_{ij}^{-1} \quad \text{Eq. 5.2}$$

Where, q_i refers to the volumetric conversion/production rate of component i, i.e. the mass of component i per unit volume of the reactor per unit time (Mass i Volume^{-1} Time^{-1}), r_i refers to the measured rate of the mass of component i per unit time per unit weight of the biomass (Mass i Time^{-1} Mass biomass^{-1}) and Y_{ji} is the yield of component i per unit of component j. In the case of biomass, x, this would refer to the specific growth rate μ:

$$\mu = r_x = \frac{q_x}{x} \quad \text{Eq. 5.3}$$

One of the advantages of using this process stoichiometry is that it allows elemental balances for C, H, N and O to be set up and to make sure that the process stoichiometry is balanced. For the process stoichiometry given in Eq 5.1, the following elemental balance for carbon will hold, assuming all the relevant yields are measured:

$$C-\text{balance}: \quad -1 + Y_{sx} + Y_{sc} + Y_{sp_1} = 0 \quad \text{Eq. 5.4}$$

Similarly to the carbon balance, the elemental balance for N, O and H can also be performed. Usually in biological process studies, the yield coefficient for water, Y_{SW}, is ignored because the production of water is negligible compared with the high flow rates typically treated in WWTPs. For this reason, H and O balances and process stoichiometry are usually not closed in wastewater applications. However, the balance for the degree of reduction is closed in wastewater treatment process stoichiometry. This is the framework on which ASM is based. The degree of reduction balance is relevant since most biological reactions involve reduction-oxidation (redox)-type chemical conversion reactions in metabolism activities.

5.2.1.2 Degree of reduction analysis

A biological process will convert a substrate i.e. the input to a metabolic pathway, into a product that is in a reduced or oxidized state relative to the substrate. In order to perform redox analysis on a biological process, a method to calculate the redox potential of substrates and products is required. In the ASM framework and other biotechnological applications (Heijnen, 1999; Villadsen *et al.*, 2011), the following methodology is used:

1) Define a standard for the redox state for the balanced elements, typically C, O, N, S and P.
2) Select H_2O, CO_2, NH_3, H_2SO_4, and H_3PO_4 as the reference redox-neutral compounds for calculating the redox state for the elements O, C, N, S, and P respectively. Moreover, a unit of redox is defined as H = 1. With these definitions, the following redox levels of the five listed elements are obtained: O = -2, C = 4, N = -3, S = 6 and P = 5.
3) Calculate the redox level of the substrate and products using the standard redox levels of the elements. Several examples are provided below:
   a) Glucose ($C_6H_{12}O_6$): $6 \cdot 4 + 12 \cdot 1 + 6 \cdot (-2) = 24$. Per 1 C-mol, the redox level of glucose becomes, $\gamma_g = 24/6 = 4$ mol e⁻ C-mol⁻¹.
   b) Acetic acid ($C_2H_4O_2$): $2 \cdot 4 + 4 \cdot 1 + 2 \cdot (-2) = 8$. Per 1 C-mol, the redox level of Hac becomes, $\gamma_a = 8/2 = 4$ mol e⁻ C-mol⁻¹
   c) Propionic acid ($C_3H_6O_2$): $3 \cdot 4 + 6 \cdot 1 + 2 \cdot (-2) = 14$. Per 1 C-mol, the redox level of HPr becomes, $\gamma_p = 14/3 = 4.67$ mol e⁻ C-mol⁻¹.
   d) Ethanol (C_2H_6O): $2 \cdot 4 + 6 \cdot 1 + 1 \cdot (-2) = 12$. Per 1 C-mol, the redox level of HAc becomes, $\gamma_e = 12/2 = 6$ mol e⁻ C-mol⁻¹.
4) Perform a degree of reduction balance over a given process stoichiometry (see Example 5.1).

Example 5.1 Elemental balance and degree of reduction analysis for aerobic glucose oxidation

General process stoichiometry for the aerobic oxidation of glucose to biomass:

$$CH_2O + Y_{SO}O_2 + Y_{SN}NH_3 \rightarrow Y_{SX}X + Y_{SC}CO_2 \qquad \text{Eq. 5.5}$$

Assuming the biomass composition X is $CH_{1.8}O_{0.5}N_{0.2}$. The degree of reduction for biomass is calculated assuming the nitrogen source is ammonia (hence the nitrogen oxidation state is -3, γ_X: $4 + 1.8 + 0.5 \cdot (-2) + 0.2 \cdot (-3) = 4.2$ mol e⁻ C-mol⁻¹.

Now C, N and the degree of reduction balances can be performed for the process stoichiometry as follows:

Carbon balance: $-1 + Y_{SX} + Y_{SC} = 0$ $\qquad$ Eq. 5.6

Nitrogen balance: $-Y_{SN} + 0.2 \cdot Y_{SX} = 0$ $\qquad$ Eq. 5.7

Redox balance:

$$-1 \cdot \gamma_g - \gamma_{O2} \cdot Y_{SO} - \gamma_{NH3} \cdot Y_{SN} + \gamma_X \cdot Y_{SX} + \gamma_{CO2} \cdot Y_{SC} = 0$$
$$-1 \cdot \gamma_g - \gamma_{O2} \cdot Y_{SO} - 0 \cdot Y_{SN} + \gamma_X \cdot Y_{SX} + 0 \cdot Y_{SC} = 0$$
$$\text{Eq. 5.8.}$$

In these balance equations, there are four unknowns (Y_{SN}, Y_{SO}, Y_{SX}, Y_{SC}). Since three equations are available, only one measurement of the yield is necessary to calculate all the others. For example, in ASM applications, biomass growth yield is usually assumed measured or known, hence the other remaining yields can be calculated as follows:

CO_2 yield: $Y_{SC} = 1 - Y_{SX}$ $\qquad$ Eq. 5.9

NH_3 yield: $Y_{SN} = 0.2 Y_{SX}$ $\qquad$ Eq. 5.10

O_2 yield: $Y_{SO} = \dfrac{\gamma_g - \gamma_X \cdot Y_{SX}}{\gamma_{O2}} = \dfrac{4 - 4.2 Y_{SX}}{4}$ $\qquad$ Eq. 5.11

With these coefficients known, the process stoichiometry model for 1 C-mol of glucose consumption becomes as follows:

$$CH_2O + \frac{4 - 4.2 Y_{SX}}{4} \cdot O_2 + 0.2 Y_{SX} \cdot NH_3 \rightarrow$$
$$Y_{SX} \cdot X + (1 - Y_{SX}) \cdot CO_2 \qquad \text{Eq. 5.12}$$

In the ASM framework, the process stoichiometry is calculated using a unit production of biomass as a reference. Hence, the coefficients of Eq. 5.12 can be re-arranged as follows:

$$\frac{1}{Y_{SX}} \cdot CH_2O + \frac{4-4.2Y_{SX}}{4Y_{SX}} \cdot O_2 + 0.2NH_3 \rightarrow$$
$$X + \left(\frac{1}{Y_{SX}} - Y_{SX}\right) \cdot CO_2 \quad \text{Eq. 5.13}$$

The unit conversion from a C-mol to a g COD basis, being the unit of ASM models, is defined using O_2 as the reference compound. Accordingly, 1 g COD is defined as -1 g O_2. From the degree of reduction of oxygen, the conversion to COD from one unit redox (mol e⁻) is calculated as follows:

$$\frac{\text{Molecular weight of } O_2}{\text{Degree of reduction of } O_2} = \frac{MW_{O2}}{\gamma_{O2}} =$$
$$\frac{32}{4} = 8 \text{ g COD L}^{-1} \text{ (mol e}^-)^{-1} \quad \text{Eq. 5.14}$$

To convert a C-mol to a g COD basis, the unit redox needs to be multiplied with the degree of reduction of the substrate as follows:

$$\left(\frac{\text{mol } \bar{e}}{C-\text{mol e}}\right) \cdot \left(\frac{g \text{ COD}}{\text{mol } \bar{e}}\right) = \gamma_g \cdot 8 \frac{g \text{ COD}}{C-\text{mol}} \quad \text{Eq. 5.15}$$

5.2.1.3 Consistency check of the experimental data

The value of performing elemental balances around data collected from experiments with biological processes is obvious: to confirm the data consistency with the first law of thermodynamics, which asserts that energy (in the form of matter, heat, etc.) is conserved. A primary and obvious requirement for performing elemental balances is that the model is checked and consistent. Experimental data needs to be checked for gross (measurement) errors that may be caused by incorrect calibration or malfunction of the instruments, equipment and/or sensors.

Inconsistency in the data can be checked from the sum of the elements that make up the substrates consumed in the reaction (e.g. glucose, ammonia, oxygen, etc.). This should equal to the sum of the elements (products) produced in the reaction (therefore also see Eq. 5.4 for the carbon balance). Deviation from this elemental balance indicates an incorrectly defined system description, a model inconsistency and/or measurement flaws.

In addition to the elemental balances, the degree of reduction balance provides information about whether the right compounds are included for a given pathway or whether a compound is missing in the process stoichiometry. Adding this check is helpful and provides consistency with the bioenergetic principles of biological processes (Roels, 1980; Heijnen, 1999; Villadsen et al., 2011).

The consistency checks and the elemental balances (in addition to the charge balances) are included in the ASM framework as a conservation matrix to verify the internal consistency of the yield coefficients (Henze et al., 2000).

The elemental composition and degree of reduction can be performed systematically using the following generic balance equation in order to test the consistency of the measured data:

$$\sum_{j=1}^{N} e_{sj} q_{sj} + e_x q_x + \sum_{j=1}^{M} e_{pj} q_{pj} = 0 \quad \text{Eq. 5.16}$$

The equation above is formulated for a biological process with N substrates and M metabolic products. In the equation, e is the elemental composition (C, H, O and N) for a component, and q the volumetric production (or consumption) rate for substrates (q_{sj}), biomass (q_x) and metabolic products (q_{pj}). Hence, the elemental balance can be formulated as follows:

$$E \cdot q = 0 \quad \text{Eq. 5.17}$$

In this equation, E is the conservation matrix and its columns refer to each conserved element and property, e.g. C, H, O, N, γ, etc. Each row of matrix E contains values of a conserved property related to substrates, products and biomass; q is a column vector including the measured volumetric rates for each compound. This is substrate as well as products and biomass.

The total number of columns in E is the number of compounds, which is the sum of substrates (N), products (M) and biomass, hence N + M + 1. The total number of constraints is 5 (C, H, O, N and γ). This means that N+M-4 is the number of degrees of freedom that needs to be measured or specified in order to calculate all the rates.

Typically, not all the rates will be measured in batch experiments. Therefore, let us assume q_m is the measured set of volumetric rates and q_u the unmeasured set of rates which need to be calculated. In this case Eq. 5.17 can be reformulated as follows:

$$E_m q_m + E_u q_u = 0 \qquad \text{Eq. 5.18}$$
$$q_u = -(E_u)^{-1} E_m q_m = 0$$

Provided that the inverse of E_u exists ($\det(E_u) \neq 0$), Eq. 5.18 provides a calculation/estimation of the unmeasured rates in a biological process. These estimated rates are valuable on their own, but can also be used for validation purposes if redundant measurements are available. This systematic method of data consistency check is highlighted in Example 5.2.

All these calculations help to verify and validate the experimental data and measurement of the process yield. The data can now be used for further kinetic analysis and parameter estimation.

5.2.2 Parameter estimation

Here we recall a state-space model formalism to describe a system of interest. Let y be a vector of outputs resulting from a dynamic model, f, employing a parameter vector, θ; input vector, u; and state variables, x:

$$\frac{dx}{dt} = f(x, \theta, u, t); \quad x(0) = x_0 \qquad \text{Eq. 5.19}$$
$$y = g(x, \theta, u, t)$$

The above equation describes a system (a batch setup or a full WWTP) in terms of a coupled ordinary differential equations (ODE) and algebraic system of equations using a state-space formalism.

The problem statement for parameter estimation reads then as follows: for a given set of measurements, *y*, with its measurement noise collected from the system of interest, and given the model structure in Eq. 5.19, estimate the unknown model parameters (θ).

The solution approaches to this problem can be broadly classified as the manual trial and error method, and formal statistical methods.

5.2.2.1 The manual trial and error method

This approach has no formal scientific basis except for a practical motivation that has to do with getting a good model fit to the data. It works as follows: the user chooses one parameter from the parameter set and then changes it incrementally (increases or decreases around its nominal value) until a reasonable model fit is obtained to the measured data. The same process may be iterated for another parameter. The fitting process is terminated when the user deems that the model fit to data is good. This is often determined by practical and/or time constraints because this procedure will never lead to an optimal fit of the model to the measured data. In addition, multiple different sets of parameter values can be obtained which may not necessarily have a physical meaning. The success of this procedure often relies on the experience of the modeller in selecting the appropriate parameters to fit certain aspects of the measured data. Although this approach is largely subjective and suboptimal, the approach is still widely used in industry as well as in the academic/research environment. Practical data quality issues do not often allow the precise determination of parameters. Also not all (commercial) modelling software platforms provide the appropriate statistical routines for parameter estimation. There are automated procedures for model calibration using algorithms such as statistical sampling techniques, optimization algorithm, etc. (Sin *et al.*, 2008). However, such procedures focus on obtaining a good fit to experimental data and not necessarily on the identifiability and/or estimation of a parameter from a data set. This is because the latter requires proper use of statistical theory.

5.2.2.2 Formal statistical methods

In this approach, a proper statistical framework is used to suggest the problem, which is then solved mathematically by using appropriate numerical solution strategies, e.g. minimization algorithms or sampling algorithms. Under this category, the following statistical frameworks are usually employed:
a. Frequentist framework (maximum likelihood, least squares, non-linear regression, etc.).
b. Bayesian framework (Metropolis-Hasting, Markov Chain Monte Carlo (MCMC), importance sampling, etc.).
c. Pragmatic/hybrid framework (employing some elements of the two schools of thought above, e.g. the bootstrap method, Monte Carlo filtering, etc.).

The above statistical methods are among the most commonly used and recommended here as well. In particular, we focus on the frequentist and bootstrap methods as they are more fit to the intended purpose of this chapter.

Frequentist method - maximum likelihood theory

In the parameter estimation problem we usually define parameter estimators, $\hat{\theta}$, to distinguish them from the true model parameters, θ. In the context of statistical estimation, model parameters are defined as unknown and statistical methods are used to infer their true value. This difference is subtle but important to understand and to interpret the results of parameter estimation, irrespective of the methods used.

Maximum likelihood is a general method for finding estimators, $\hat{\theta}$, from a given set of measurements, y. In this approach, the model parameters θ are treated as true, fixed values, but their corresponding estimators $\hat{\theta}$ are treated as random variables. The reason is that the estimators depend on the measurements, which are assumed to be a stochastic process:

$$y = f(\theta) + \varepsilon \quad \text{where} \quad \varepsilon \propto N(0,\sigma) \qquad \text{Eq. 5.20}$$

Measurement errors, ε, are defined by a probability distribution, e.g. normal distribution, N, with zero mean and standard deviation (σ). With these assumptions, the likelihood function (L) for the parameter estimation becomes as follows (Seber and Wild, 1989):

$$L(y,\theta) = \frac{1}{\sigma\sqrt{2\pi}} \exp\left(-\frac{(y-f(\theta))^2}{2\sigma^2}\right) \qquad \text{Eq. 5.21}$$

The most likely estimate of θ is found as those parameter values that maximize the likelihood function:

$$\hat{\theta}: \min_\theta L(y,\theta) \qquad \text{Eq. 5.22}$$

The solution to this problem setting (5.24) is often found by optimization algorithms such as simplex, interior point, genetic algorithms, simulated annealing, etc. The parameters obtained by calculating the maximum likelihood (Eq. 5.21) are the same as the parameters obtained by calculating the minimum cost function in Eq. 5.23.

The least squares method

This is a special case of the maximum likelihood method in which the measurements are assumed to be independent and identically distributed with white measurement errors having a known standard deviation, σ (Gaussian). The likelihood function becomes equivalent to minimizing the following cost (or objective) function, S(y,θ) (Seber and Wild, 1989):

$$S(y,\theta) = \sum \frac{(y-f(\theta))^2}{\sigma^2} \qquad \text{Eq. 5.23}$$

Where, y stands for the measurement set, $f(\theta)$ stands for the corresponding model predictions, and Σ stands for the standard deviation of the measurement errors. The solution to the objective function (Eq. 5.24) is found by minimization algorithms (e.g. Newton's method, gradient descent, interior-point, Nelder-Mead simplex, genetic, etc.).

$$\hat{\theta}: \min_\theta S(y,\theta)$$
$$\left.\frac{\partial}{\partial \theta} S(y,\theta)\right|_{\hat{\theta}} = 0 \qquad \text{Eq. 5.24}$$

The solution to the above optimization problem provides the best estimate of the parameter values. The next step is to evaluate the quality of the parameter estimators. This step requires the estimation of the confidence interval of the parameter values and the pairwise linear correlation between the parameters.

The covariance matrix of parameter estimators

As a result of stochastic measurement, estimators have a degree of uncertainty. In the frequentist framework of thought, probability is defined in terms of the frequency of the occurrence of outcomes. Hence, in this method the uncertainty of the parameter estimators is defined by a 95 % confidence interval interpreted as the range in which 95 times out of 100 the values of the parameter estimators are likely to be located. This can be explained as if one performs the same measurement 100 times, and then performs the parameter estimation on these 100 sets and observes the following: 95 occurrences of the estimator values lie in the confidence interval, while 5 occurrences are outside this interval.

In order to estimate the confidence interval, first the covariance matrix $(\text{cov}(\hat{\theta}))$, which contains complete information about the uncertainty of the parameter estimators of the estimators, needs to be estimated. One method to obtain $\text{cov}(\hat{\theta})$ is to use a linear approximation method through estimation of the Jacobian matrix (F.) of the parameter estimation problem (Seber and Wild, 1989):

$$\text{cov}(\hat{\theta}) = s^2 (F.^{'} \cdot F.)^{-1} \quad \text{where } F. = \left.\frac{\partial f(\theta)}{\partial \theta}\right|_{\theta=\hat{\theta}} \qquad \text{Eq. 5.25}$$

Where, s^2 is the unbiased estimation of σ^2 obtained from the residuals of the parameter estimation:

$$s^2 = \frac{S_{min}(y,\hat{\theta})}{n-p} \qquad \text{Eq. 5.26}$$

Here, n is the total number of measurements, p is the number of estimated parameters, n-p is the degrees of freedom, $S_{min}(y,\hat{\theta})$ is the minimum objective function value and F. is the Jacobian matrix, which corresponds to the first order derivative of the model function, f, with respect to the parameter vector θ evaluated at θ = $\hat{\theta}$.

The covariance matrix is a square matrix with (p×p) dimensions. The diagonal elements of the matrix are the variance of the parameter estimators, while the non-diagonal elements are the covariance between any pair of parameter estimators.

The 95 % confidence interval of the parameter estimators can now be approximated. Assuming a large *n*, the confidence intervals (the difference between the estimators and true parameter values), follow a student *t*-distribution, the confidence interval at 100 (1-α) % significance:

$$\hat{\theta}_{1-\alpha} = \hat{\theta} \pm t_{N-p}^{\alpha/2} \sqrt{diag\, \text{cov}(\hat{\theta})} \qquad \text{Eq. 5.27}$$

Where, $t_{N-p}^{\alpha/2}$ is the upper α/2 percentile of the t-distribution with N-p degrees of freedom, and *diag* cov($\hat{\theta}$) represents the diagonal elements of the covariance matrix of the parameters.

The pairwise linear correlation between the parameter estimators, R_{ij}, can be obtained by calculating a correlation matrix from unit standardization of the covariance matrix as follows:

$$R_{ij} = \frac{\text{cov}(\theta_i, \theta_j)}{\sigma_{\theta_i} \times \sigma_{\theta_j}} \qquad \text{Eq. 5.28}$$

This linear correlation will range from [-1 1] and indicate whether or not he parameter estimator is uniquely identifiable (if the correlation coefficient is low) or correlated (if the correlation coefficient is high).

The bootstrap method

One of the key assumptions for using the maximum likelihood estimation (MLE) method as well as its simplified version, the nonlinear least squares method, is that the underlying distribution of errors is assumed to follow a normal (Gaussian) distribution.

In many practical applications, however, this condition is rarely satisfied. Hence, theoretically the MLE method for parameter estimation cannot be applied without compromising its assumptions, which may lead to over or underestimation of the parameter estimation errors and their covariance structure.

An alternative to this approach is the bootstrap method developed by Efron (1979), which removes the assumption that the residuals follow a normal distribution. Instead, the bootstrap method works with the actual distribution of the measurement errors, which are then propagated to the parameter estimation errors by using an appropriate Monte Carlo scheme (Figure 5.1).

The bootstrap method uses the original data set D(0) with its N data points, to generate any number of synthetic data sets $D^S(1); D^S(2);...$, also with N data points. The procedure is simply to draw N data points with replacements from the set D(0). Because of the replacement, sets are obtained in which a random fraction of the original measured points, typically 1/e = 37 %, are replaced by duplicated original points. This is illustrated in Figure 5.1.

The application of the bootstrap method for parameter estimation in the field of wastewater treatment requires adjustment due to the nature of the data that is in the time series. Hence, the sampling is not performed from the original data points (which are the time series and indicate a particular trend). Instead, the sampling is performed from the residual errors and then added to the simulated model outputs (obtained by using reference parameter estimation) (Figure 5.1). This is reasonable because the measurement errors are what is assumed to be stochastic and not the main trend of the measured data points, which are caused by biological processes/mechanisms. Bearing this in mind, the theoretical background of the bootstrap method is outlined below.

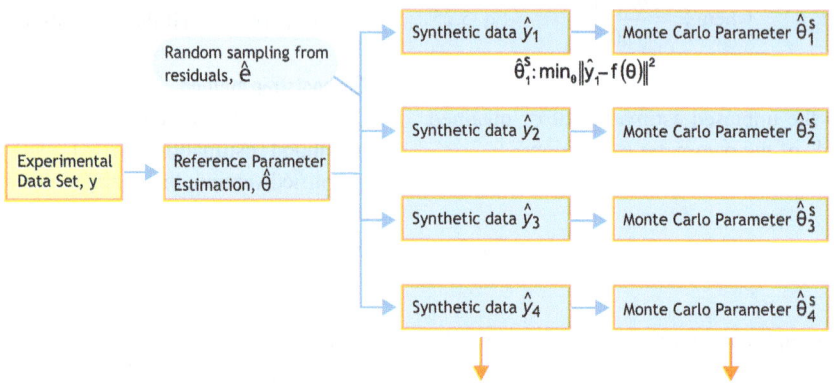

Figure 5.1 Illustration of the workflow for the bootstrap method: synthetic data sets are generated by Monte Carlo samples (random sampling with replacement) from the reference MLE. For each data set, the same estimation procedure is performed, giving M different sets of parameter estimates: $\theta^s_{(1)}$, $\theta^s_{(2)}, \ldots \theta^s_{(M)}$.

Let us define a simple nonlinear model where y_i is the ith measurement, f_i is the ith model prediction, θ is a parameter vector (of length p), and ε_i is the measurement error of y_i:

$$y_i = f_i(\theta) + \varepsilon_i \quad \text{where} \quad \varepsilon_i \infty F \qquad \text{Eq. 5.29}$$

The distribution of errors, F, is not known. This is unlike in MLE, where the distribution is assumed *a priori*. Given y, use least squares minimization, to estimate $\hat{\theta}$:

$$\hat{\theta} : \min_\theta \left\| y - f(\theta) \right\|^2 \qquad \text{Eq. 5.30}$$

The bootstrap method defines $\hat{F}$ as the sample probability distribution of $\hat{\varepsilon}$ as follows:

$$\hat{F} = \frac{1}{n} \quad \text{(density) at} \quad \varepsilon_i = (y_i - f_i(\theta)) \quad i = 1,2,\ldots n$$
$$\text{Eq. 5.31}$$

The density is the probability of the ith observation. In a uniform distribution each observation (in this case the measurement error, ε_i) has an equal probability of occurrence, where density is estimated from 1/n. The bootstrap sample, y^*, given $(\hat{\theta}, \hat{F})$, is then generated as follows:

$$y_i^* = f_i(\hat{\theta}) + \varepsilon_i^* \quad \text{where} \quad \varepsilon_i^* \propto \hat{F} \qquad \text{Eq. 5.32}$$

The realisation of measurement error in each bootstrap method, ε^*, is simulated by random sampling with replacement from the original residuals, which assigns each point with a uniform (probability) weight. By performing N random sampling with a replacement and then adding them to the model prediction (Eq. 5.31), a new synthetic data set is generated, $D^s(1) = y^*$.

By repeating the above sampling procedure M times, M data sets are generated: $D^s(1)$, $D^s(2)$, $D^s(3)$, … $D^s(M)$.

Each synthetic data set, $D^s(j)$, makes it possible to obtain a new parameter estimator $\hat{\theta}(j)$ by the same least squares minimisation method which is repeated M times:

$$\hat{\theta}_j : \min_\theta \left\| D^s(j) - f(\theta) \right\|^2 \quad \text{where} \quad j = 1,2\ldots M \quad \text{Eq. 5.33}$$

The outcome from this iteration is a matrix of parameter estimators, $\hat{\theta}(M \times p)$ (M is the number of Monte Carlo samples of synthetic data and p is the number of parameters estimated). Hence, each parameter estimator now has a column vector with values. This vector of values can be plotted as a histogram and interpreted using common frequentist parameters such as the mean, standard deviation and the 95 % percentile. The covariance and correlation matrix can be computed using $\hat{\theta}(M \times p)$ itself. This effectively provides all the needed information on the quality of the parameter estimators.

DATA HANDLING AND PARAMETER ESTIMATION

For measurement errors that follow a normal distribution, both MLE and the bootstrap method will essentially provide the same results. However, if the underlying distribution of the measurements significantly deviates from a normal distribution, the bootstrap method is expected to provide a better analysis of the confidence interval of the estimators.

5.2.3 Uncertainty analysis

5.2.3.1 Linear error propagation

In linear error propagation, the covariance matrix of the parameter estimators, $cov(\theta)$, is used to propagate measurement errors to model prediction errors and to calculate standard errors and confidence intervals of the parameter estimates. Therefore, the covariance matrix of model predictions, $cov(y)$, can be estimated using $cov(\hat{\theta})$ as follows (Seber and Wild, 1989):

$$cov(y) = (F.^{'} F.) cov(\hat{\theta}) (F.^{'} F.)^{-1} \qquad \text{Eq. 5.34}$$

In a similar fashion, the $1-\alpha$ confidence interval of the predictions, y, can be approximated as follows:

$$y_{1-\alpha} = y \pm t_{N-p}^{\alpha/2} \sqrt{diag\, cov(y)} \qquad \text{Eq. 5.35}$$

This concludes parameter estimation, confidence intervals and prediction uncertainty as viewed from the point of view of the frequentist analysis.

5.2.3.2 The Monte Carlo method

The Monte Carlo (MC) method was originally used to calculate multi-dimensional integrals and its systematic use started in the 1940s with the 'Los Alamos School' of mathematicians and physicists, namely Von Neumann, Ulam, Metropolis, Kahn, Fermi and their collaborators. The term was coined by Ulam in 1946 in honour of a relative who was keen on gambling (Metropolis and Ulam, 1949).

Within the context of uncertainty analysis, which is concerned with estimating the error propagation from a set of inputs to a set of model outputs, the integral of interest is the calculation of the mean and variance of the model outputs which are themselves indeed multidimensional integrals (the dimensionality number is determined by the length of the vector of input parameters):

$$I = \int f(x)\, dx = \int f(u_1 \ldots u_d)\, d^d x \qquad \text{Eq. 5.36}$$

Authors consider the integral of a function $f(x)$ with x as the input vector $x = (u_1, \ldots u_d)$. Hence, the integral is taken on the d variables $u_1, \ldots, u_d$ over the unit hypercube $[0, 1]^d$. In the parameter estimation, these input variables are parameters of the model that have a certain range with lower and upper bounds. We assume that f is square-integrable, which means that a real value solution exists at each integration point. As a short-hand notation we will denote a point in the unit hypercube by $x = (u_1, \ldots u_d)$ and the function evaluated at this point by $f(x) = f(u_1, \ldots u_d)$, and then the multidimensional integration operation is given by:

$$E = \frac{1}{N} \sum_1^N f(x_N) \qquad \lim_{N \to \infty} \frac{1}{N} \sum_1^N f(x_N) = I \qquad \text{Eq. 5.37}$$

The law of large numbers ensures that the MC estimate (E) converges to the true value of this integral. However, as most of the time N is finite (a sampling number from input space u with dxd dimension), there will be an error in the Monte Carlo integration of multidimensional functions. This Monte Carlo integration error is scaled like $1/\sqrt{N}$. Hence, the average Monte Carlo integration error is given by $MCerr = \sigma(f)/\sqrt{N}$, where $\sigma(f)$ is the standard deviation of the error, which can be approximated using sample variance:

$$\sigma^2(f) \approx s^2(f) = \frac{1}{N-1} \sum_1^N (f(x_N) - E)^2 \qquad \text{Eq. 5.38}$$

For notational simplicity, we consider the following simple model: $y = f(x)$, where the function f represents the model under study, $x:[x_1; \ldots x_d]$ is the vector of the model inputs, and $y:[y_1; \ldots y_n]$ is the vector of the model predictions.

The goal of an uncertainty analysis is to determine the uncertainty in the elements of y that results from uncertainty in the elements of x. Given uncertainty in the vector x characterised by the distribution functions $D=[D_1, \ldots D_d]$, where D_1 is the distribution function associated with x_1, the uncertainty in y is given by:

$$var(y) = \int ((y) - f(x))^2 dx$$
$$E(y) = \int f(x)\, dx \qquad \text{Eq. 5.39}$$

Where, var(y) and E(y) are the variance and expected value respectively of a vector of random variables, y, which are computed by the Monte Carlo sampling technique. In addition to the variance and mean values, one can also easily compute a percentile for y including the 95% upper and lower bounds.

5.2.4 Local sensitivity analysis and identifiability analysis

5.2.4.1 Local sensitivity analysis

Most of the sensitivity analysis results reported in the literature are of a local nature, and these are also called one factor at a time (OAT) methods. In OAT methods, each input variable is varied (also called perturbation) one at a time around its nominal value, and the resulting effect on the output is measured. The sensitivity analysis results from these methods are useful and valid in close proximity to the parameters analysed, hence the name local. In addition, the parameter sensitivity functions depend on the nominal values used in the analysis. Alternative methods, such as regional or global methods, expand the analysis from one point in the parameter space to cover a broader range in the entire parameter space but this is beyond the scope of this chapter (interested readers can consult literature elsewhere such as Saltelli *et al.*, 2000; Sin *et al.*, 2009).

The local sensitivity measure is commonly defined using the first order derivative of an output, y = f(x), with respect to an input parameter, x:

Absolute sensitivity: $sa = \dfrac{\partial y}{\partial x}$ Eq. 5.40

(effect on y by perturbing x around its nominal value x^0).

Relative sensitivity: $sr = \dfrac{\partial y}{\partial x} \dfrac{x^0}{y^0}$ Eq. 5.41

(relative effect of y by perturbing x with a fixed fraction of its nominal value x^0).

The relative sensitivity functions are non-dimensional with respect to units and are used to compare the effects of model inputs among each other.

These first-order derivatives can be computed analytically, for example using Maple or Matlab symbolic manipulation toolbox software. Alternatively, the derivatives can be obtained numerically by model simulations with a small positive or negative perturbation, Δx, of the model inputs around their nominal values, x^0. Depending on the direction of the perturbation, the sensitivity analysis can be approximated using the forward, backward or central difference methods:

Forward perturbation:

$$\dfrac{\partial y}{\partial x} = \dfrac{f(x^0 + \Delta x) - f(x^0)}{\Delta x}$$ Eq. 5.42

Backward perturbation:

$$\dfrac{\partial y}{\partial x} = \dfrac{f(x^0) - f(x^0 - \Delta x)}{\Delta x}$$ Eq. 5.43

Central difference:

$$\dfrac{\partial y}{\partial x} = \dfrac{f(x^0 + \Delta x) - f(x^0 - \Delta x)}{2\Delta x}$$ Eq. 5.44

When an appropriately small perturbation step, Δx, is selected (usually a perturbation factor, $\varepsilon = 10^{-3}$ is used. Hence $\Delta x = \varepsilon \cdot x$), all three methods provide exactly the same results.

Once the sensitivity functions have been calculated, they can be used to assess the parameter significance when determining the model outputs. Typically, large absolute values indicate high parameter importance, while a value close to zero implies no effect of the parameter on the model output (hence the parameter is not influential). This information is useful to assess parameter identifiability issues for the design of experiments.

5.2.4.2 Identifiability analysis using the collinearity index

The first step in parameter estimation is determining which sets of parameters can be selected for estimation. This problem is the subject of identifiability analysis, which is concerned with identifying which subsets of parameters can be identified *uniquely* from a given set of measurements. Thereby, it is assumed a model can have a number of parameters. Here the term *uniquely* is important and needs to be understood as follows: a parameter estimate is unique when its value can be estimated independently of other parameter values and with sufficiently high accuracy (i.e. a small uncertainty). This means that the correlation coefficient between any pair of parameters should be low (e.g. lower than 0.5) and the standard error of parameter estimates should be low (e.g. the relative error of the parameter estimate, σ_θ/θ, lower than e.g. 25%). As it turns out, many parameter estimation problems are ill-conditioned problems. A problem is defined ill-conditioned when the condition number of a function/matrix is very high, which is caused by multicollinearity issues. In regression problems, the condition number is used as a diagnostic tool to identify parameter identifiability issues. Such regression diagnostics are helpful in generating potential candidates of the parameter subsets for estimation which the user can select from.

There are several identifiability tests suggested in literature that are entirely based on the sensitivity functions of the parameters on the outputs. Here we are using the two-step procedure of Brun *et al.*, 2002. Accordingly, the procedure works as follows: (*i*) assessment of the parameter significance ranking, (*ii*) collinearity analysis (dependency analysis of the parameter sensitivity functions in a parameter subset):

Step 1. Rank the significance of the parameters: δ^{msqr}

$$\delta^{msqr} = \sqrt{\frac{1}{N}\sum_{i}^{N}(sr_i)} \qquad \text{Eq. 5.45}$$

Where, sr is a vector of non-dimensional sensitivity values, sr = i...N values.

Step 2. Calculate the collinearity index of a parameter subset K, γ_K.

$$\gamma_K = \frac{1}{\sqrt{\min \lambda_K}} \qquad \text{Eq. 5.46}$$

$$\lambda_K = \text{eigen}(\text{snorm}_K^T \, \text{snorm}_K) \qquad \text{Eq. 5.47}$$

$$\text{snorm} = \frac{sr}{\|sr\|} \qquad \text{Eq. 5.48}$$

Where, K indicates a parameter subset, snorm is the normalized non-dimensional sensitivity function using the Euclidian norm, and λ_K represents the eigenvalues of the normalized sensitivity matrix for parameter subset K.

In Step 1, parameters that have negligible or near-zero influence on the measured model outputs are screened out from consideration for parameter estimation. In the second step, for each parameter subset (all the combinations of the parameter subsets which include 2, 3, 4,...m parameters) the collinearity index is calculated. The collinearity index is the measure of the similarity between any two vectors of the sensitivity functions. Subsets that have highly similar sensitivity functions will tend to have a very large number ($\gamma_K \sim \inf$), while independent vectors will have a smaller value $\gamma_K \sim 1$ which is desirable. In identifiability analysis, a threshold value of 5-20 is usually used in literature (Brun *et al.*, 2001; Sin and Vanrolleghem, 2007; Sin *et al.*, 2010). It is noted that this γ_K value is to be used as guidance for selecting parameter subsets as candidates for parameter estimation. The best practice is to iterate and try a number of higher ranking subsets.

5.3 METHODOLOGY AND WORKFLOW

5.3.1 Data consistency check using an elemental balance and a degree of reduction analysis

The following workflow is involved in performing an elemental balance and a degree of reduction analysis:

Step 1. Formulate a black box process stoichiometry for the biological process.

In this step, the most relevant reactants and products consumed and produced in the biological process are identified and written down. The output is a list of reactants and products for Step 2.

Step 2. Compose the elemental composition matrices (E_m and E_u).

First establish which variables of interest are measured and then define the matrices as follows: E_m includes the elemental composition and the degree of reductions for these measured variables, while E_u includes those of unmeasured variables. To calculate the degree of reduction, use the procedure given in Section 5.2.1.2.

Step 3. Compute the unmeasured rates of the species (q_u).

Using E_m and E_u together with the vector of the measured rates (q_m), the unmeasured rates (q_u) are estimated from the solution of the linear set of equations in Eq. 5.18.

Step 4. Calculate the yield coefficients.

In this step, since all the species rates of consumption/productions are now known, the yield coefficients can be calculated using Eq. 5.2 and the process stoichiometry can be written using the yield coefficient values.

Step 5. Verify the elemental balance.

In this step, a simple check is performed to verify if the elemental balance and degree of reduction balance are closed. If not, the procedure needs to be iterated by assuming a different hypothesis concerning the formation of by-products.

5.3.2 Parameter estimation workflow for the non-linear least squares method

This workflow assumes that an appropriate and consistent mathematical model is used to describe the data. Such a model confirms the elemental balance and degree of reduction analysis (see the workflow in Section 5.3.1). Usually these models are available from literature. Most of them are modified from ASM models with appropriate simplifications and/or additions reflecting the conditions of the batch experiment.

Step 1. Initialisation.

In this step, the initial conditions for the model variables are specified as well as a nominal set of parameters for the model. The initial conditions for the model are specified according to the experimental conditions (e.g. 10 mg NH_4-N added at time 0, k_La is a certain value, oxygen saturation at a given temperature is specified, etc.). An initial guess of the model parameters is taken from literature.

Step 2. Select the experimental data and a parameter subset for the parameter estimation.

In this step, the experimental data is reviewed for the parameter estimation and which parameters need to be estimated is defined. This can be done using expert judgement or, more systematically, a sensitivity and identifiability analysis (see Section 5.3.4).

Step 3. Define and solve the parameter estimation problem.

In this step, the parameter estimation problem is defined as a minimization problem and solved using optimization algorithms (e.g. *fminsearch* in Matlab)

Step 4. Estimate the uncertainty of the parameter estimators and model outputs.

In this step, calculate the covariance matrix of the parameter estimators and compute the parameter confidence intervals as well as the parameter correlation matrix. Given the covariance matrix of the parameter estimators, estimate the covariance matrix of the model outputs by linear error propagation.

Step 5. Review and analyse the results.

In this step, review the values of the parameter values, which should be within the range of parameter values obtained from the literature. In addition, inspect the confidence intervals of the parameter estimators. Very large confidence intervals imply that the parameter in question may not be estimated reliably and should be excluded from the subset.

Further, plot and review the results from the best-fit solution. Typically, the data and model predictions should fit well.

If the results (both parameter values) and the best fit solution to the data are not satisfactory, iterate as appropriate by going back to Step 1 or Step 2.

5.3.3 Parameter estimation workflow for the bootstrap method

The workflow of the bootstrap method follows on from Step 1, Step 2 and Step 3 of the non-linear least squares method.

Step 1. Perform a reference parameter estimation using the non-linear least squares method.

This step is basically an execution of steps 1, 2 and 3 of the workflow in the non-linear least squares technique. The output is a residual vector that is passed on to the next step. The residual vector is then plotted and reviewed. If the residuals follow a systematic pattern (it should be random) or contain outliers, this is a cause for concern as it may imply the bootstrap method is not suited for this application.

Step 2. Generate synthetic data by bootstrap sampling and repeat the parameter estimation.

Synthetic data is generated using Eq. 5.29-5.32 by performing bootstrap sampling (random sampling with replacement) from the residual vector and adding it to the model prediction obtained in Step 1. For each synthetic data, the parameter estimation in Step 1 is repeated and the output (that is, the values of the parameter estimators) is recorded in a matrix.

Step 3. Review and analyse the results.

In this step, the mean, standard deviation and the correlation matrix of the parameter estimators are computed from the recorded matrix data in Step 2. Moreover, the distribution function of the parameter estimators can be estimated and plotted using the vector of the parameter values that was obtained in Step 2.

As in Step 5 of the workflow in the non-linear least squares method, the results are interpreted and evaluated using knowledge from literature and process engineering.

5.3.4 Local sensitivity and identifiability analysis workflow

The workflow of this procedure starts with the assumption that a mathematical model is available and ready to be used to describe a set of experimental data.

Step 1. Initialisation.

A framework is defined for the sensitivity analysis by defining the experimental conditions (the initial conditions for the batch experiments) as well as a set of nominal values for the model analysis. The model is solved with these initial conditions and the model outputs are plotted and reviewed before performing the sensitivity analysis.

Step 2. Compute the sensitivity functions.

Define which outputs are measured and hence should be included in the sensitivity analysis. Define the experimental data points (every 1 min versus every 5 min).

Compute the sensitivity functions of the parameters on the outputs using a numerical difference, e.g. using a forward, backward or central difference. Plot, review and analyse the results.

Step 3. Rank the parameter significance.

Calculate the delta mean-square measure, δ^{msqr}, and rank the parameters according to this measure. Exclude any parameters that have zero or negligible impact on the outputs.

Step 4. Compute the collinearity index.

For all the parameter combinations (e.g. subset size 2, 3, 4....m), the collinearity index, γ_K, is calculated. Each parameter subset is ranked according to the collinearity index value.

Step 5. Review and analyse the results.

Based on the results from Step 3 and Step 4, identify a short list of candidates (parameter subsets) that are identifiable. Exclude these parameters from any parameter subset that has near-zero or negligible sensitivity on the outputs.

5.3.5 Uncertainty analysis using the Monte Carlo method and linear error propagation

The workflow for the Monte Carlo method includes the following steps:

Step 1. Input the uncertainty definition.

Identify which inputs (parameters) have uncertainty. Define a range/distribution for each uncertainty input, e.g. normal distribution, uniform distribution, etc. The output from the parameter estimators (e.g. bootstrap) can be used as input here.

Step 2. Sampling from the input space.

Define the sampling number, N, (e.g. 50, 100, etc.) and sample from the input space using an appropriate sampling technique. The most common sampling techniques are random sampling, Latin Hypercube sampling, etc. The output from this step is a sampling matrix, X_{Nxm}, where N is the number of samples and m is the number of inputs.

Step 3. Perform the Monte Carlo simulations.

Perform N simulations with the model using the sampling matrix from Step 2. Record the outputs in an appropriate matrix form to be processed in the next step.

Step 4. Review and analyse the results.

Plot the outputs and review the results. Calculate the mean, standard deviation/variance, and percentiles (e.g. 95 %) for the outputs. Analyse the results within the context of parameter estimation quality and model prediction uncertainty. Iterate the analysis, if necessary, by going back to Step 1 or Step 2.

The workflow for linear error propagation:

The workflow is relatively straightforward as it is complementary to the covariance matrix of the parameter estimators and should be performed as part of the parameter estimation in the non-linear least squares method. It requires the covariance matrix of parameter estimators as well as the Jacobian matrix which are both obtained in Step 4 of the non-linear least squares methodology.

5.4 ADDITIONAL EXAMPLES

Example 5.2 Anaerobic fermentation of glucose

In this example, anaerobic fermentation of glucose to ethanol and glycerol as metabolic products is considered.

Step 1. Formulate the process stoichiometry.

Ammonia is assumed to be the nitrogen source for growth. The biomass composition is assumed to be $CH_{1.6}O_{0.5}N_{0.15}$. All the substrates are given on the basis of 1 C-mol, whereas nitrogen is on the basis of 1 N-mol. In this biological process, the substrates are CH_2O (glucose) and NH_3. The products are $CH_{1.61}O_{0.52}N_{0.15}$ (biomass), $CH_3O_{0.5}$ (ethanol), $CH_{8/3}O$ (glycerol) and CO_2. Water is excluded from the analysis, as its rate of production is not considered relevant to the process. This means that the H and O balances will not be considered either.

Step 2. Compose the elemental composition matrices (E_m and E_u).

As the process has six species (substrates + products) and three constraints (two elemental balances for C and N plus a degree of reduction balance), measurement of three rates is sufficient to estimate/infer the remaining rates.

To illustrate the concept, the measured rates are selected as the volumetric consumption rate of substrate ($-q_s$), the biomass production rate (q_x), and the glycerol production rate (q_g) hence the remaining rates for ammonia consumption as well as the production of ethanol and CO_2 need to be estimated using Eq. 5.18. In the measured rate vectors, a negative sign indicates the consumption of a species, while a positive sign indicates the production of a species.

Step 3 Compute the unmeasured rates of the species (q_u).

Recall Eq. 5.18, which is solved as follows:

$$E_m \cdot q_m + E_u \cdot q_u = 0$$

$$\begin{array}{c} \begin{matrix} S & X & Gly \end{matrix} \\ C \\ N \\ \gamma \end{array} \begin{bmatrix} 1 & 1 & 1 \\ 0 & 0.15 & 0 \\ 4 & 4.12 & 4.67 \end{bmatrix} \cdot \begin{pmatrix} -q_s \\ q_x \\ q_g \end{pmatrix} + \begin{array}{c}\begin{matrix} NH_3 & Eth & CO_2 \end{matrix} \\ \begin{bmatrix} 0 & 1.0 & 1.0 \\ 1 & 0 & 0 \\ 0 & 6 & 0 \end{bmatrix}\end{array} \cdot \begin{pmatrix} -q_n \\ q_e \\ q_c \end{pmatrix} = 0$$

$$q_u = -(E_u)^{-1} \cdot E_m \cdot q_m$$

$$\begin{pmatrix} -q_n \\ q_e \\ q_c \end{pmatrix} = -\left(\begin{bmatrix} 0 & 1.0 & 1.0 \\ 1 & 0 & 0 \\ 0 & 6 & 0 \end{bmatrix}\right)^{-1} \cdot \begin{bmatrix} 1 & 1 & 1 \\ 0 & 0.15 & 0 \\ 4 & 4.12 & 4.67 \end{bmatrix} \cdot \begin{pmatrix} -q_s \\ q_x \\ q_g \end{pmatrix}$$

Solving the system of linear equations above yields the following solution where the three unmeasured rates are calculated as a function of the measured rates q_s, q_g and q_x:

$$\begin{pmatrix} -q_n \\ q_e \\ q_c \end{pmatrix} = \begin{pmatrix} -0.15 q_x \\ 2q_s/3 - 467 q_g/600 - 103 q_x/150 \\ q_s/3 - 133 q_g/600 - 47 q_x/150 \end{pmatrix}$$

Step 4. Calculate the process yields.

Once the rates of all the products and substrates are estimated, one can then calculate the yield coefficients for the process by recalling Eq. 5.2 as follows:

$$Y_{sx} = \frac{q_x}{q_s} \text{ and } Y_{sg} = \frac{q_g}{q_s}$$

$$Y_{sn} = \frac{q_n}{q_s} = \frac{-0.15 q_x}{q_s} = -0.15 Y_{sx}$$

$$Y_{se} = \frac{q_e}{q_s} = \frac{(2q_s/3 - 467q_g/600 - 103q_x/150)}{q_s} =$$

$$\frac{2}{3} - 0.7783 Y_{sg} - 0.6867 Y_{sx}$$

$$Y_{sc} = \frac{q_c}{q_s} = \frac{(q_s/3 - 133q_g/600 - 47q_x/150)}{q_s} =$$

$$\frac{1}{3} - 0.2217 Y_{sg} - 0.3133 Y_{sx}$$

With these yield coefficients estimated, the simplified process stoichiometry reads as follows:

$$0 = -CH_2O - 0.15 Y_{sx} \cdot NH_3 \ldots$$
$$+ Y_{sx} \cdot CH_{1.61}O_{0.52}N_{0.15} + \left(\frac{2}{3} - 0.7783 Y_{sg} - 0.6867 Y_{sx}\right) \cdot CH_3O_{0.5} +$$
$$Y_{sg} \cdot CH_{8/3}O + \left(\frac{1}{3} - 0.2217 Y_{sg} - 0.3133 Y_{sx}\right) \cdot CO_2$$

Step 5. Verify the elemental balance.

From the process stoichiometry, it is straightforward to verify that the elemental and degree of reduction balances are closed:

$$-1 + Y_{sx} + \left(\frac{2}{3} - 0.7783 \cdot Y_{sg} - 0.6867 \cdot Y_{sx}\right) + Y_{sg} +$$
$$\left(\frac{1}{3} - 0.2217 \cdot Y_{sg} - 0.3133 \cdot Y_{sx}\right) = 0$$

The nitrogen balance:

$$-0 - 0.15 \cdot Y_{sx} + 0.15 \cdot Y_{sx} + 0 + 0 + 0 = 0$$

The degree of reduction balance:

$$-1 \cdot 4 + Y_{sx} \cdot 4.12 + \left(\frac{2}{3} - 0.7783 \cdot Y_{sg} - 0.6867 \cdot Y_{sx}\right) \cdot 6 +$$
$$Y_{sg} \cdot 4.67 = 0$$

In the above example, three measured rates were assumed available as a minimum requirement to identify the system of linear equations. In practical applications, there might be two other situations: (*i*) redundant measurements: measurements of most or perhaps all of the species rates of production/consumption are available. In this case, the additional rate measurements can be used for data quality check and validation (where some of the measured rates could be used as a validation of the estimated coefficients from the balance analysis); (*ii*) a limited set of measurements: in this case too few rates measurements are available to uniquely estimate the unmeasured rates. If data from enough variables is not available, some variables should be fixed (e.g. this could be done iteratively in a search algorithm) to reduce the degrees of freedom. If not, there are infinite solutions. Further discussion of these techniques can be found elsewhere in relevant literature (Villadsen *et al.*, 2011; Meijer *et al.*, 2002; van der Heijden *et al.*, 1994).

Example 5.3 Estimate the parameters of the ammonium and nitrite oxidation processes using data from batch tests: the non-linear least squares method

Aerobic batch tests with a sludge sample from a pre-denitrification plant are performed to measure the parameters of the nitrifying bacteria, in particular the ammonium-oxidizing organisms (AOO) and nitrite-oxidising organisms (NOO). Following the recommended experimental procedure in literature (Guisasola *et al.*, 2005), two separate batch tests were performed as follows: (*i*) batch test 1 with added ammonium of 20 mg NH$_4$-N L^{-1} and an inhibitor (sodium azide) to suppress NOO activity, (*ii*) batch test 2 with added ammonium of 20 mg NH$_4$-N L^{-1} without any inhibitor addition. In both tests, both the pH and temperature are controlled at 7.5 and 25 °C respectively. During both the batch tests, ammonium, nitrite and nitrate are measured every 5 minutes, while dissolved oxygen is measured every minute. The data collected is shown in Figure 5.2. For the sake of simplicity and to keep the focus on the demonstration of the methods and their proper interpretation, these examples use synthetic data with random (white-noise) addition.

Part 1. Estimate the parameters of the ammonium oxidation process.

Several models are suggested in literature reviewed in Sin *et al.*, 2008. We use the following mathematical model given in Table 5.1 to describe the kinetics of ammonium and nitrite oxidation. For the sake of simplicity, the following is assumed: (*i*) endogenous respiration related to heterotrophic biomass is constant (hence not modelled), (ii) the inert fraction of the biomass released during decay is negligible (hence not modelled), and (iii) the ammonium consumed for autotrophic growth of biomass is negligible. It is noted that for the sake of completeness all of the above phenomena should be described which makes the analysis more accurate. However, here the model is kept simple to focus the attention of the reader on the workflow of parameter estimation.

Table 5.1 The two-step nitrification model structure using matrix representation (adopted from Sin et al., 2008)

Variables→C_i Processes ↓j	S_{NH} mg N L^{-1}	S_O mg O$_2$ L^{-1}	S_{NO2} mg N L^{-1}	S_{NO3} mg N L^{-1}	X_{AOO} mg COD L^{-1}	X_{NOO} mg COD L^{-1}	Rates q_j
AOO growth	$\frac{-1}{Y_{AOO}}$	$1-\frac{3.43}{Y_{AOO}}$	$\frac{1}{Y_{AOO}}$		1		$\mu_{max}^{AOO} \cdot M_{NH} \cdot M_{O,AOO} \cdot X_{AOO}$
AOO decay		1			−1		$b_{AOO} \cdot X_{AOO}$
NOO growth		$1-\frac{1.14}{Y_{NOO}}$	$\frac{-1}{Y_{NOO}}$	$\frac{1}{Y_{NOO}}$		1	$\mu_{max}^{NOO} \cdot M_{NO2} \cdot M_{O,NOO} \cdot X_{NOO}$
NOO decay		1				−1	$b_{NOO} \cdot X_{NOO}$
Aeration		1					$kLa \cdot (S_O^{sat} - S_O)$

$$M_{NH}: \frac{S_{NH}}{S_{NH}+K_{s,AOO}}; M_{O,AOO}: \frac{S_O}{S_O+K_{o,AOO}}; M_{O,NOO}: \frac{S_O}{S_O+K_{o,NOO}}; M_{NO2}: \frac{S_{NO2}}{S_{NO2}+K_{s,NOO}}$$

The model has in total six ordinary differential equations (ODE), which corresponds to one mass balance for each variable of interest. Using a matrix notation, each ODE can be formulated as follows:

$$\frac{dC_i}{dt} = \sum_j v_{ij} \cdot q_j \qquad \text{Eq. 5.49}$$

The model is implemented in Matlab and solved using a standard differential equation solver (ODE45 in Matlab).

```
%% solve the ODE model:
%[time,output] = ODEsolver('Model',[starttime
simulation  endtime simulation],Initial conditions for
model variables,simulation options,model parameters);
options=odeset('RelTol',1e-7,'AbsTol',1e-8);
[t,y] = ode45(@nitmod,t,x0,options,par);
```

Step 1. Initialisation.

The model has in total 12 parameters. The nominal values as well as their range are taken from literature (Sin et al., 2008) and shown in Table 5.2

The model has six state variables, all of which need to be specified to solve the system of the ODE equations. The initial condition corresponding to batch test 1 is shown in Table 5.3.

Step 2. Select the measurements and parameter subset for parameter estimation.

We used data collected from batch test 1 which includes ammonium, nitrite and dissolved oxygen measurements. Due to the suppression of NOO activity, no nitrate production is observed. Since batch test 1 is not designed for decay rate coefficient estimation, we consider all the parameters of AOO except b_{AOO} for estimation. Hence, the following is our selection:

- Y = [NH$_4$ NO$_2$ DO]; selected measurement set, Y.
- θ = [Y_{AOO} μ_{max}^{AOO} $K_{s,AOO}$ $K_{o,AOO}$]; parameter subset for the estimation.

Step 3. Solve the parameter estimation problem.

The parameter estimation is programmed as a minimization problem using the sum of the squared errors as the cost function and solved using an unconstrained non-linear optimisation solver (*fminsearch* algorithm in Matlab) using the initial parameter guess given in Table 5.2 and initial conditions in Table 5.3. To simulate the inhibitor addition, the maximum growth rate of NOO is assumed to be zero in the model simulations. The best estimates of the parameter estimators are given in Table 5.3.

Table 5.2 Nominal values of the model parameters used as an initial guess for parameter estimation together with their upper and lower bounds.

Parameter	Symbol	Unit	Nominal value	Range
Ammonium-oxidising organisms (AOO)				
Biomass yield	Y_{AOO}	mg COD mg N^{-1}	0.15	0.11 - 0.21
Maximum growth rate	μ_{max}^{AOO}	d^{-1}	0.8	0.50 - 2.10
Substrate (NH$_4$) affinity	$K_{s,AOO}$	mg N L^{-1}	0.4	0.14 - 1.00
Oxygen affinity	$K_{o,AOO}$	mg O$_2$ L^{-1}	0.5	0.10 - 1.45
Decay rate coefficient	b_{AOO}	d^{-1}	0.1	0.07 - 0.30
Nitrite-oxidising organisms (NOO)				
Biomass yield	Y_{NOO}	mg COD mg N^{-1}	0.05	0.03 - 0.09
Maximum growth rate	μ_{max}^{NOO}	d^{-1}	0.5	0.40 - 1.05
Substrate (NO$_2$) affinity	$K_{s,NOO}$	mg N^{-1}	1.5	0.10 - 3.00
Oxygen affinity	$K_{o,NOO}$	mg O$_2$ L^{-1}	1.45	0.30 - 1.50
Decay rate coefficient	b_{NOO}	d^{-1}	0.12	0.08 - 0.20
Experimental setup				
Oxygen mass transfer	k_La	d^{-1}	360	*
Oxygen saturation	$S_{O,sat}$	mg O$_2$ L^{-1}	8	*

* Not estimated in this example but assumed known.

Table 5.3 Initial condition of the state variables for the model in batch test 1.

Variable	Symbol	Unit	Initial value	Comment
Ammonium	S_{NH}	mg N L^{-1}	20	Pulse addition
Oxygen	S_O	mg O$_2$ L^{-1}	8	Saturation
Nitrite	S_{NO2}	mg N L^{-1}	0	Post denitrified
Nitrate	S_{NO3}	mg N L^{-1}	0	Post denitrified
AOO biomass	X_{AOO}	mg COD L^{-1}	75	Ratio of AOO to NOO reflects ratio of their yields
NOO biomass	X_{NOO}	m COD L^{-1}	25	

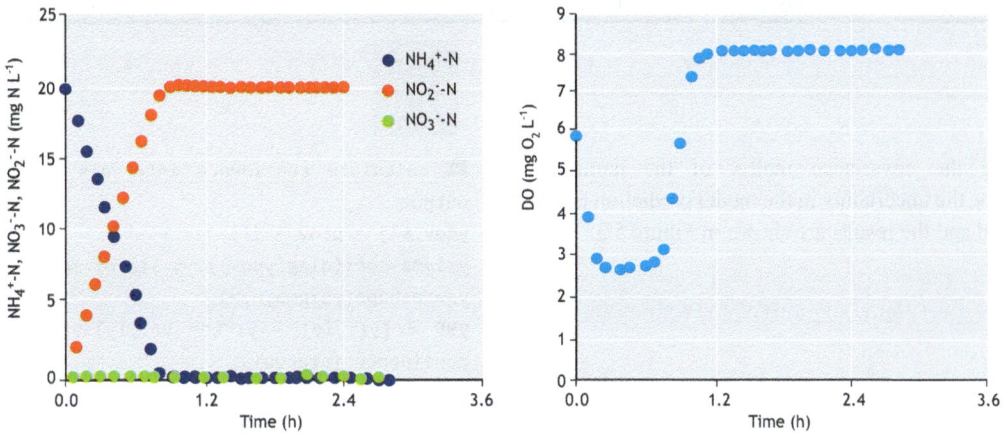

Figure 5.2 Data collected in batch test 1. NH$_4$, NO$_2$ and DO are used as the measured data set.

```matlab
%%step 3 define and solve parameter estimation
problem (as a minimization problem)
options =optimset('display',
'iter','tolfun',1.0e-06, 'tolx',1.0e-5,
'maxfunevals', 1000);
[pmin,sse]=fminsearch(@costf,pinit,options,td,yd
,idx,iy);
```

Step 4. Estimate the uncertainty of the parameter estimators and the model prediction uncertainty.

In this step, the covariance matrix of the parameter estimators is computed. From the covariance matrix, the standard deviation, 95 % confidence interval as well as the correlation matrix are obtained. The results are shown in Table 5.4.

Table 5.4 Optimal values of the parameter estimators after the solution of the parameter estimation problem.

Parameter	Initial guess, $\theta°$	Optimal values, $\hat{\theta}$
Y_{AOO}	0.1	0.15
μ_{max}^{AOO}	0.8	1.45
$K_{s,AOO}$	0.4	0.50
$K_{o,AOO}$	0.5	0.69

```matlab
%% get the Jacobian matrix. use built-in
"lsqnonlin.m" but with no iteration.
options =optimset('display',
'iter','tolfun',1.0e-06, 'tolx',1.0e-5,
'maxfunevals', 0);
[~,~,residual,~,~,~,jacobian]=lsqnonlin(@costl,p
min,[],[],options,td,yd,idx,iy);
j(:,:)=jacobian; e=residual;
s=e'*e/dof; %variance of errors
%% calculate the covariance of parameter
estimators
pcov = s*inv(j'*j) ; %covariance of parameters
psigma=sqrt(diag(pcov))'; % standard deviation
parameters
pcor = pcov ./ [psigma'*psigma]; % correlation
matrix
alfa=0.025; % significance level
tcr=tinv((1-alfa),dof); % critical t-dist value
at alfa
p95 =[pmin-psigma*tcr; pmin+psigma*tcr]; %+-95%
confidence intervals
```

Table 5.5 Parameter estimation quality for the ammonium oxidation process: standard deviation, 95% confidence intervals and correlation matrix.

Parameter	Optimal value, $\hat{\theta}$	Standard deviation, σ_θ	95 % confidence interval (CI)		Correlation matrix			
					Y_{AOO}	μ_{max}^{AOO}	$K_{s,AOO}$	$K_{o,AOO}$
Y_{AOO}	0.15	0.0076	0.130	0.160	1	0.96	0.0520	0.17
μ_{max}^{AOO}	1.45	0.0810	1.290	1.610		1	0.0083	0.42
$K_{s,AOO}$	0.50	0.0180	0.470	0.540			1	-0.26
$K_{o,AOO}$	0.69	0.0590	0.570	0.800				1

Using the covariance matrix of the parameter estimators, the uncertainty in the model prediction is also calculated and the results are shown in Figure 5.3.

```matlab
%% calculate confidence intervals on the model
output
ycov = j * pcov * j';
ysigma=sqrt(diag(ycov)); % std of model outputs
ys=reshape(ysigma,n,m);
y95 = [y(:,iy) - ys*tcr y(:,iy)+ys*tcr]; % 95%
confidence intervals
```

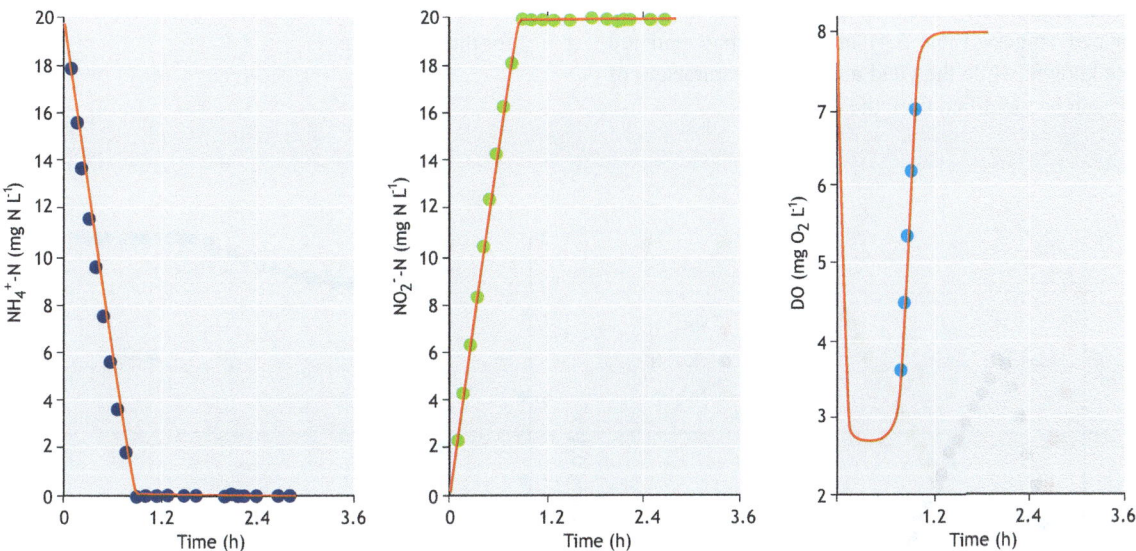

Figure 5.3 Model outputs including 95 % confidence intervals calculated using linear error propagation (red lines). The results are compared with the experimental data set.

Step 5. Review and analyse the results.

The estimated parameter values (Table 5.5) are found to be within the range reported in literature. This is an indication that the parameter values are credible. The uncertainty of these parameter estimators is found to be quite low. For example, the relative error (e.g. standard deviation/mean value of parameter values) is less than 10 %, which is also reflected in the small confidence interval. This indicates that the parameter estimation quality is good. It is usually noted that relative error higher than 50 % is indicative of bad estimation quality, while relative error below 10 % is good.

Regarding the correlation matrix, typically from estimating parameters from batch data for Monod-like models, the growth yield is significantly correlated with the maximum growth rate (the linear correlation coefficient is 0.96). Also notable is the correlation between the maximum growth rate and the oxygen affinity constant. This means that a unique estimation of the yield and maximum growth rate is not possible. Further investigation of the correlation requires a sensitivity analysis, which is demonstrated in Example 5.5.

Since the parameter estimation uncertainty is low, the uncertainty in the model predictions is also observed to be small. In Figure 5.3, the mean (or average) model prediction and the 95 % upper and lower bounds are quite close to each other. This means that the model prediction uncertainty due to parameter estimation uncertainty is negligible. It is noted that a comprehensive uncertainty analysis of the model predictions will require analysis of all the other sources of uncertainty including other model parameters as well as the initial conditions. However, this is outside the scope of this example and can be seen elsewhere (Sin *et al.*, 2010). Measurement error uncertainty is considered in Example 5.6.

This concludes the analysis of parameter estimation using the non-linear least squares method for the AOO parameters.

Part 2. Estimate the parameters for the NOO step.

Steps 1 and 2. Initial conditions and selection of data and parameter subsets for the parameter estimation.

The same initial condition for batch test 1 is used in batch test 2 but without any inhibitor addition, meaning that in this example the nitration is active. The data collected from batch test 2 is shown in Figure 5.3, which includes ammonium, nitrite, nitrate and DO measurements.

- $Y_2 = [NH_4\ NO_2\ NO_3\ DO]$; selected measurement set, Y.

The parameter values of AOO were set to the estimated values (Table 5.4) in the first part and are hence known, while the yield and kinetic parameters of NOO can be identified from the data:

- $\theta_2 = [Y_{NOO}\ \mu_{max}^{NOO}\ K_{s,NOO}\ K_{o,NOO}]$; parameter subset for the estimation.

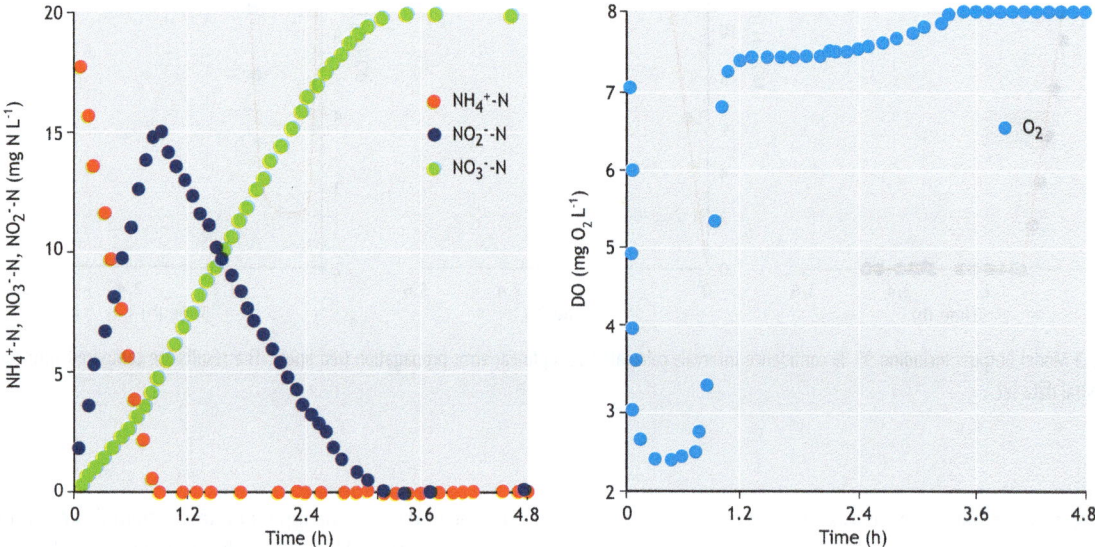

Figure 5.4 Measured data from batch test 2.

Steps 3 and 4. Solve the parameter estimation problem and calculate the parameter estimation uncertainties.

The AOO parameters are previously estimated in Part 1. The results of the solution of the parameter estimation problem as well as the parameter uncertainties for NOO are shown in Table 5.6.

Table 5.6 Optimal values of the parameter estimators after solution of the parameter estimation problem.

Parameter	Optimal values, $\hat{\theta}$	Standard deviation, σ_θ	95 % confidence interval (CI)		Correlation matrix			
			Lower bound	Upper bound	Y_{NOO}	μ_{max}^{NOO}	$K_{s,NOO}$	$K_{o,NOO}$
Y_{NOO}	0.04	0.01	0.01	0.07	1.00	1.00	0.54	-0.86
μ_{max}^{NOO}	0.41	0.13	0.15	0.66		1.00	0.55	-0.86
$K_{s,NOO}$	1.48	0.03	1.42	1.55			1.00	-0.37
$K_{o,NOO}$	1.50	0.05	1.39	1.60				1.00

The linear propagation of the parameter estimation error (covariance matrix) to the model prediction uncertainty is shown in Figure 5.5.

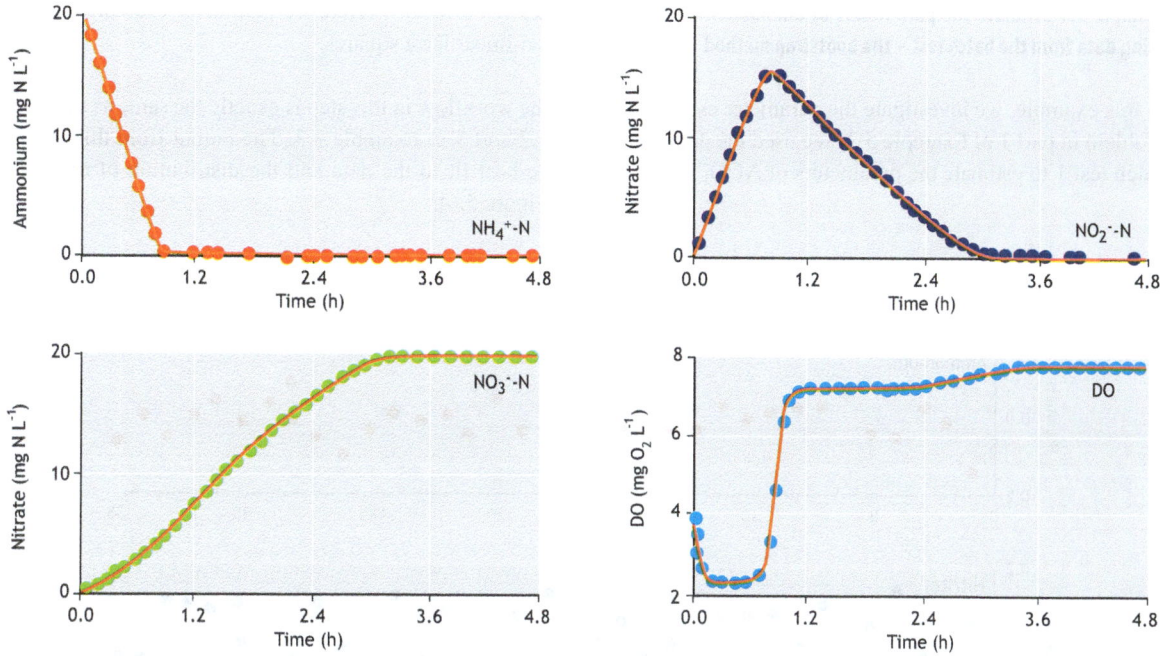

Figure 5.5 Model outputs including 95 % confidence intervals compared with the experimental data set.

Step 5. Review and analyse the results

The estimated parameter values are within the range reported for the NOO parameters in literature, which makes them credible. However, this time the parameter estimation error is noticeably higher, e.g. the relative error (the ratio of standard deviation to the optimal parameter value) is more than 30%, especially for the yield and maximum growth rate. This is not surprising since the estimation of both the yield and maximum growth rate is fully correlated (the pairwise linear correlation coefficient is 1). These statistics mean that a unique parameter estimation for the yield, maximum growth rate and oxygen half-saturation coefficient of NOO (the pairwise linear correlation coefficient is 0.86) is not possible with this batch experiment. Hence, this parameter subset should be considered as a subset that provides a good fit to the experimental data, while individually each parameter value may not have sensible/physical meaning.

The propagation of the parameter covariance matrix to the model prediction uncertainty indicates low uncertainty on the model outputs. This means that although parameters themselves are not uniquely identifiable, they can still be used to perform model predictions, e.g. to describe batch test data. While performing simulations with the model, however, one needs to report the 95% confidence intervals of the simulated values as well. The latter reflects how the covariance of the parameter estimates (implying the parameter estimation quality) affects the model prediction quality. For example, if the 95% confidence interval of the model predictions is low, then the effect of the parameter estimation error is negligible.

Part 1 and Part 2 conclude the parameter estimation for the two-step nitrification step. The results show that the quality of the parameter estimation for AOO is relatively higher than that of NOO using batch data for these experiments. This poor identifiability will be investigated later on, using sensitivity analysis to improve the identifiability of individual parameters of the model.

Regarding the model prediction errors, the 95% confidence interval of the model outputs is quite low. This means that the effects of the parameter estimation errors on the model outputs are low.

Example 5.4 Estimate the parameters of ammonium oxidation using data from the batch test – the bootstrap method

In this example, we investigate the parameter estimation problem in part 1 of Example 5.3. We used the data from batch test 1 to estimate the parameters of AOO.

Step 1. Perform a reference parameter estimation using non-linear least squares.

The workflow in this step is exactly the same as the steps 1, 2 and 3 in Example 5.3. The output from this step is the best fit to the data and the distribution of residuals (Figure 5.6).

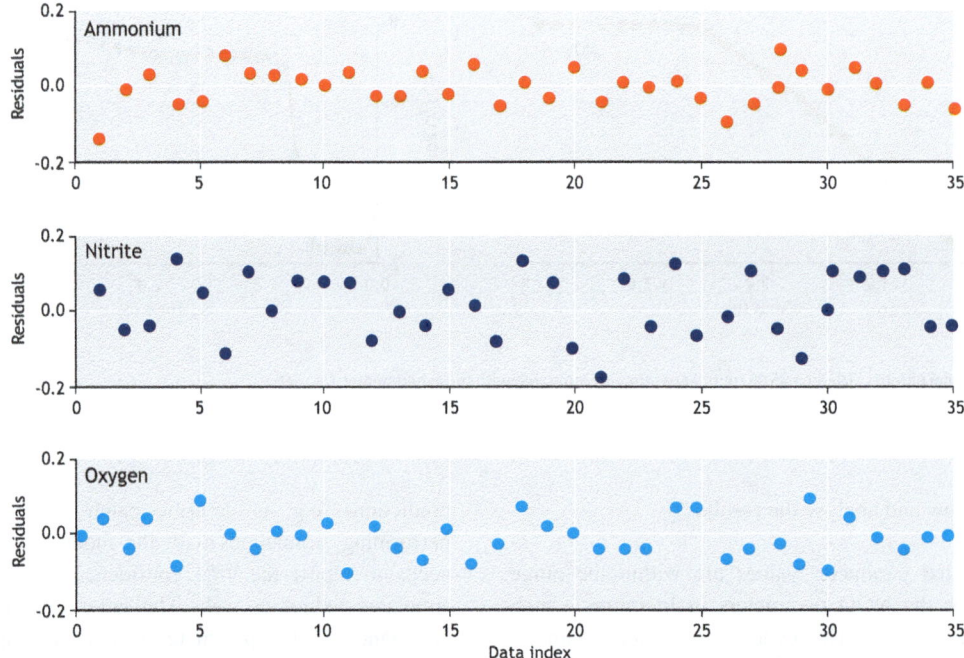

Figure 5.6 Residuals from the reference parameter estimation.

Step 2. Generate synthetic data by bootstrap sampling and repeat the parameter estimation.

In this step, bootstrap sampling from residuals is performed.

```
nboot=50; % bootstrap samples
for i=1:nboot
  disp(['the iteration number is : ',num2str(i)])
  onesam =ceil(n*rand(n,m)); % random sampling with replacement
  rsam =res(onesam); % measurement errors for each variable
  ybt = y(:,iy) + rsam ; % synthetic data: error + model (ref PE)
  options =optimset('display','iter','tolfun',1.0e-06,'tolx',1.0e-5,'maxfunevals',1000);
  [pmin(i,:),sse(i,:)]=lsqnonlin(@cost1,pmin1,plo,phi,options,td,ybt,idx,iy);
  bootsam(:,:,i)=ybt; % record samples
end
```

Fifty bootstrap samples from residuals (random sampling with replacement) are performed and added to the model, thereby yielding the 50 synthetic measurement data sets shown in Figure 5.7.

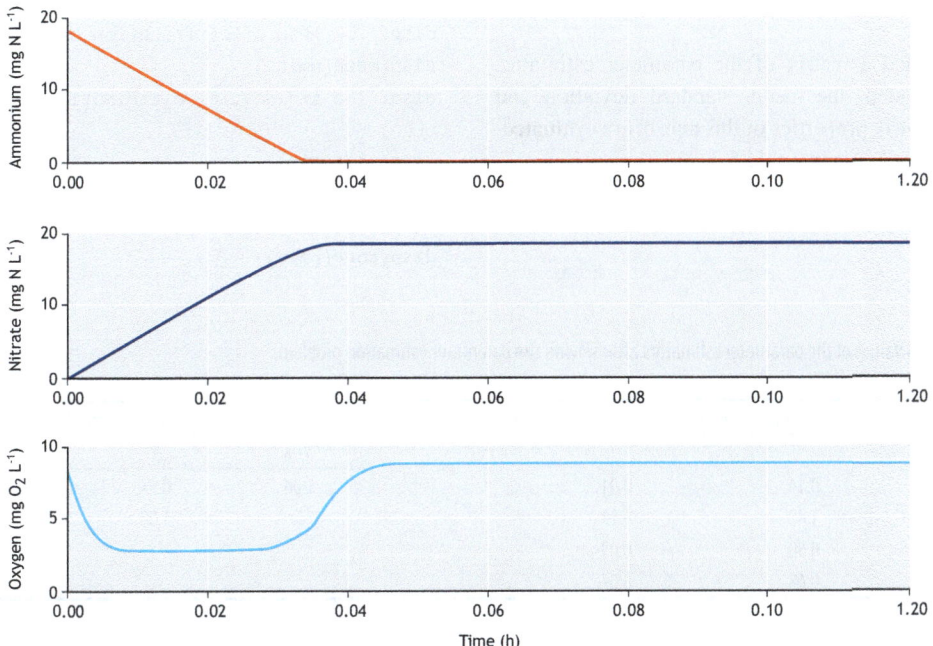

Figure 5.7 Generation of synthetic data using bootstrap sampling from the residuals (50 samples in total).

For each of this synthetic data (a bootstrap sample), a parameter estimation is performed and the results are recorded for analysis. Because 50 synthetic data sets are generated, this means that 50 different estimates of parameters are obtained. The results are shown as a histogram for each parameter estimate in Figure 5.8.

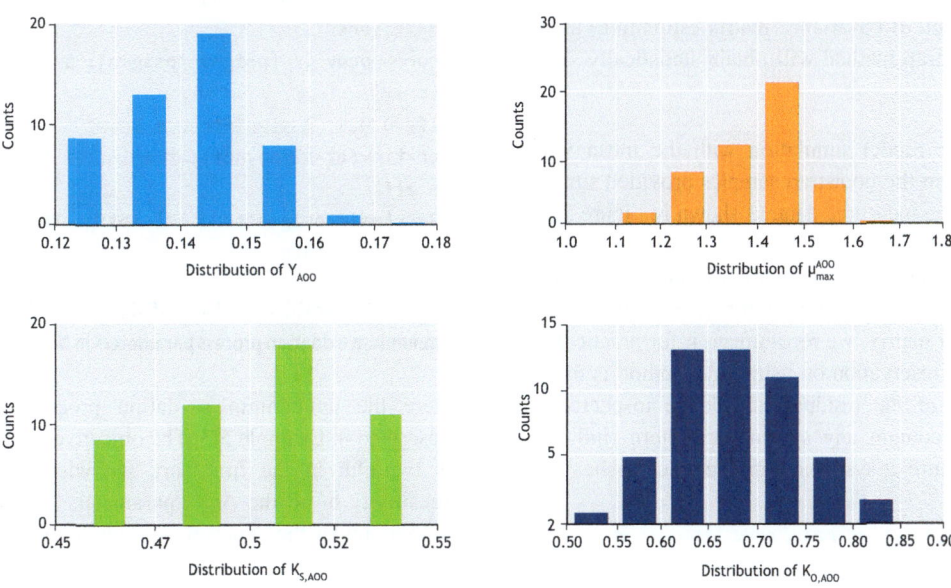

Figure 5.8 Distribution of the parameter estimates obtained using the bootstrap method (each distribution contains 50 estimated values for each parameter).

Step 3. Review and analyse the results.

Step 2 provided a matrix of the parameter estimates, $\theta_{50\times 4}$. In this step, the mean, standard deviation and correlation matrix properties of this matrix are evaluated. The results are shown in Table 5.7.

```
.%step 3 Evaluate/interpret distribution of theta
disp('The mean of distribution of theta are')
disp(mean(pmin))
disp('The std.dev. of distribution of theta are')
disp(std(pmin))
disp('')
disp('The correlation of parameters')
disp(corr(pmin))
```

Table 5.7 Optimal values of the parameter estimators after solving the parameter estimation problem.

Parameter	Optimal value, $\hat{\theta}$	Standard deviation, σ_θ	Correlation matrix			
			Y_{AOO}	μ_{max}^{AOO}	$K_{s,AOO}$	$K_{o,AOO}$
Y_{AOO}	0.14	0.01	1.00	0.97	-0.03	0.20
μ_{max}^{AOO}	1.40	0.11		1.00	-0.07	0.41
$K_{s,AOO}$	0.50	0.02			1.00	-0.28
$K_{o,AOO}$	0.68	0.07				1.00

All the results, including the mean parameter estimates, their standard deviation and the correlation matrix are in good agreement with the parameter estimates obtained from the non-linear least squares method (compare with Table 5.6.). This is expected, since the distribution of residuals is found to be quite similar to a normal distribution (Figure 5.6). In this case, both the non-linear least squares (and the linear approximation of covariance matrix estimation) as well as the bootstrap method will obtain statistically similar results.

Also the model simulation with the mean values obtained from the bootstrap samples provided similarly good fit to the measured data, as shown in Figure 5.3.

Because the bootstrap method is intuitively simple and straightforward and does not require a calculation of the Jacobian matrix, we recommend it for practical use. However, a reservation on using this method is that the distribution of the residuals should be inspected and should not contain any systematic pattern (indicating model structure or systematic measurement issues).

```
%% get the Jacobian matrix. use built-in
"lsqnonlin.m" but with no iteration.
options =optimset('display',
'iter','tolfun',1.0e-06, 'tolx',1.0e-5,
'maxfunevals', 0);
[~,~,residual,~,~,~,jacobian]=lsqnonlin(@cost1,p
min,[],[],options,td,yd,idx,iy);
j(:,:)=jacobian; e=residual;
s=e'*e/dof; %variance of errors
%% calculate the covariance of parameter
estimators
pcov = s*inv(j'*j) ; %covariance of parameters
psigma=sqrt(diag(pcov))'; % standard deviation
parameters
pcor = pcov ./ [psigma'*psigma]; % correlation
matrix
alfa=0.025; % significance level
tcr=tinv((1-alfa),dof); % critical t-dist value
at alfa
p95 =[pmin-psigma*tcr; pmin+psigma*tcr]; %+-95%
confidence intervals
```

Example 5.5 Sensitivity and identifiability analysis of the ammonium oxidation process parameters in batch tests

Here the ammonium oxidation process is used as described in Example 5.3. The objective of this example is twofold: in the first part, we wish to assess the sensitivity of all the AOO parameters to all the model outputs under the experimental conditions of batch test 1. In the second part we wish to examine, given the measured data set, which parameter subsets are potentially identifiable and compare them with the parameter subset already used in the parameter estimation in Example 5.3.

Step 1. Initialisation. We use the initial conditions of batch test 1 as described in Table 5.3 as well as the nominal values of AOO model parameters as given in Table 5.2.

The model outputs of interest are:

- $y = [NH_4\ NO_2\ NO_3\ DO\ AOO\ NOO]$

The parameter set of interest is:

- $\theta = [Y_{AOO}\ \mu_{max}^{AOO}\ K_{s,AOO}\ K_{o,AOO}\ b_{AOO}]$

Step 2. Compute and analyse the sensitivity functions.

In this step, the absolute sensitivity functions are computed using numerical differentiation and the results are recorded for analysis.

```
for i=1:m; %for each parameter
  dp(i) = pert(i) * abs(ps(i)); % parameter
perturbation
  p(i)  = ps(i) + dp(i);    % forward
perturbation
  [t1,y1] = ode45(@nitmod,td,x0,options,p);
  p(i) = ps(i) - dp(i); %backward perturbation
  [t2,y2] = ode45(@nitmod,td,x0,options,p);
  dydpc(:,:,i) = (y1-y2) ./ (2 * dp(i));
%central difference
  dydpf(:,:,i) = (y1-y) ./ dp(i); %forward
difference
  dydpb(:,:,i) = (y-y2) ./ dp(i); %backward
difference
  p(i)=ps(i); % reset parameter to its reference
value
end
```

The output sensitivity functions (absolute) are plotted in Figure 5.9 for one parameter, namely the yield of AOO growth for the purpose of detailed examination. The interpretation of a sensitivity function is as follows: (*i*) higher magnitude (positive or negative alike) means higher influence, while lower or near zero magnitude means negligible/zero influence of the parameter on the output, (*ii*) negative sensitivity means that an increase in a parameter value would decrease the model output, and (*iii*) positive sensitivity means that an increase in a parameter value would increase the model output. With this in mind, it is noted that the yield of AOO has a positive effect on ammonium and an equally negative impact on nitrite. This is expected from the model structure where there is an inverse relationship between the yield and ammonium (substrate) consumption. A higher yield means less ammonium is consumed per unit growth of biomass, and hence it would also mean more ammonium present in the batch test. Since less ammonium is consumed, less nitrite would be produced (hence the negative correlation).

On the other hand, it is also noted that the sensitivity of the yield parameter increases gradually during the linear growth phase and starts to decrease as we are nearer to the depletion of ammonium. Once the ammonium is depleted, the sensitivity becomes nil as expected. As predicted, the yield has a positive impact on AOO growth since a higher yield means higher biomass production. Regarding oxygen, the yield first has a positive impact that becomes negative towards the completion of ammonium. This means there is a rather non-linear relationship between the oxygen profile and the yield parameter. As expected, the yield of AOO has no impact on the nitrate and NOO outputs in batch test 1, because of the addition of the inhibitor that effectively suppressed the second step of nitrification.

In the sensitivity analysis, what is informative is to compare the sensitivity functions among each other. This is done in Figure 5.10 using non-dimensional sensitivity functions, which are obtained by scaling the absolute sensitivity function with their respective nominal values of parameters and outputs (Eq. 5.41). Figure 5.10 plots the sensitivity of all the model parameters with respect to the six model outputs. Each subplot in the figure presents the sensitivity functions of all the parameters with respect to one model output shown in the legend. The *y*-axis indicates the non-dimensional sensitivity measure, while the *x*-axis indicates the time during the batch activity. For example, we observe that the sensitivity of parameters to nitrate and NOO is zero. This is logical since NOO activity is assumed to be zero in this simulation.

For the model outputs for ammonium, nitrite and oxygen, the sensitivity functions of the yield and maximum growth rate for AOO follow an inversely proportional trend/pattern. This inversely proportional relation is the reason why the parameter estimation problem is an ill-conditioned problem. This means that if the search algorithm increases the yield and yet at the same time decreases the maximum growth rate with a certain fraction, the effect on the model output could be cancelled out. The result is that many combinations of parameter values for the yield and maximum growth rate can have a similar effect on the model output. This is the reason why a high correlation coefficient is obtained after the parameter estimation has been performed. This means that for a parameter to be uniquely identifiable, their sensitivity functions should be unique and not correlated with the sensitivity function of the other parameters.

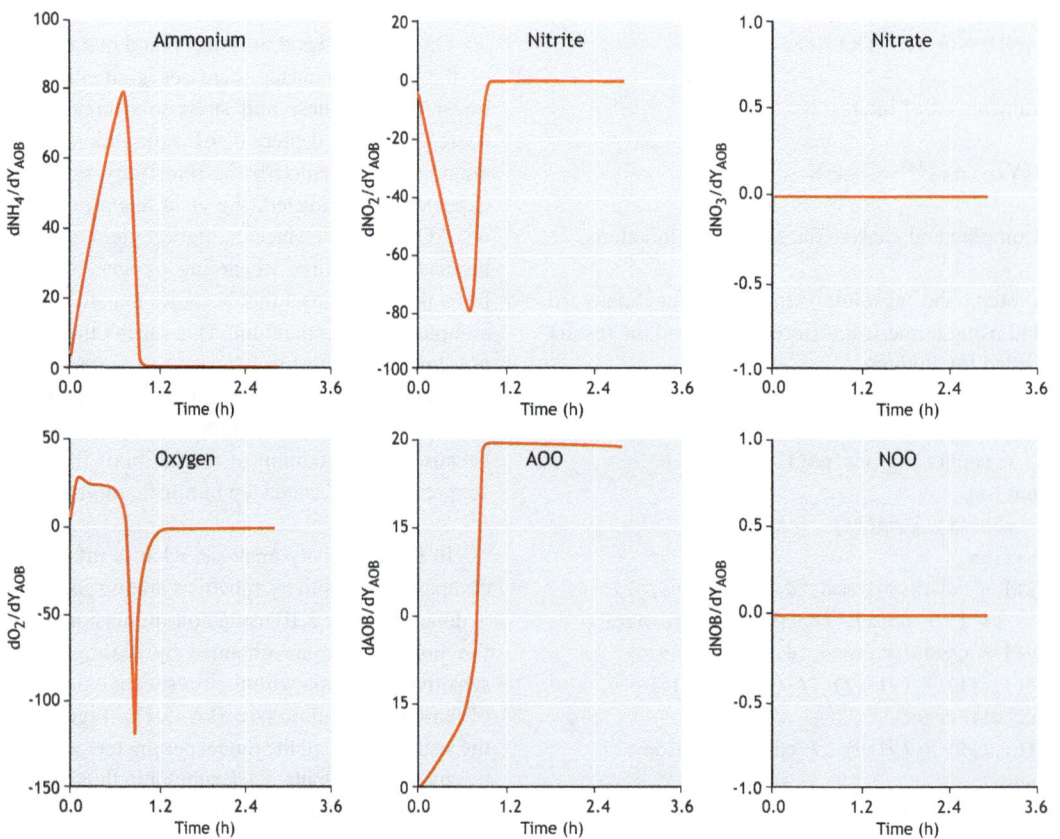

Figure 5.9 Absolute sensitivity of the AOO yield on all the model outputs.

Another point of interest regarding these plots is that the relative effect (that is, the magnitude of values on the y-axis) of the parameters on ammonium, oxygen and nitrite is quite similar. This means that all three of these variables are equally relevant and important for estimating these parameters.

Step 3. Parameter-significance ranking.

In this step the significance of parameters is ranked by summarizing the non-dimensional sensitivity functions of the parameters to model outputs using the δ^{msqr} measure. The results are shown in Figure 5.11.

The results show that the decay rate of AOO has almost zero effect on all three of the measured variables (ammonium, nitrite and oxygen) and therefore cannot be estimated. This is known from process engineering and for this reason, short-term batch tests are not used to determine decay constants. This result therefore is a confirmation of the correctness of the sensitivity analysis. With regards to the maximum growth rate and yield, these parameters are equally important followed by the affinity constant for oxygen and ammonium. This indicates that at least four parameters can potentially be estimated from the data set.

DATA HANDLING AND PARAMETER ESTIMATION

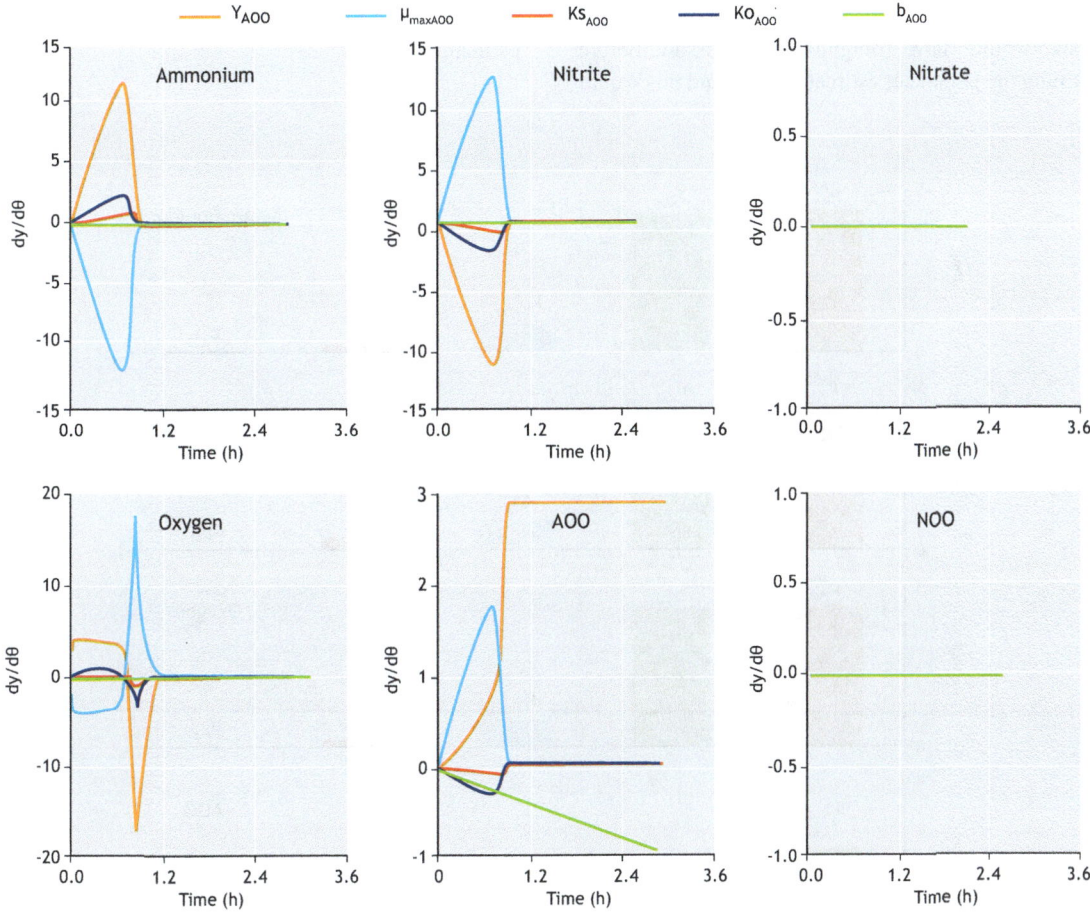

Figure 5.10 Relative sensitivity functions of the AOO parameters on the model outputs.

Step 5. Identifiability analysis.

In this step, normalized sensitivity functions are used to assess which parameter subsets have a small collinearity index. The collinearity index is a measure of how two sensitivity functions are aligned together, therefore implying linear dependency.

```
for i = 2:subset
    combos = combnk(set,i); % all possible
parameter combinations of different subset size
(2,3,4...)
    for j=1:n
        tempn    = snormy(:,combos(j,:)) ;
        tempa    = say(:,combos(j,:)) ;
        nsm      = tempn'*tempn; % normalized
sensitivity matrix
        asm      = tempa'*tempa;   % absolute
sensitivity matrix, fim
        dtm      = sqrt(det(asm))^(1/(i*2));
%determinant index
        col      = 1/sqrt(min(eig(nsm))); %
collinearity index
        subs(j,:)  = [k i col dtm] ;
    end
end
```

The identifiability analysis indicates that there are 26 different combinations of the parameter subsets that can potentially be used for parameter estimation using the ammonium, nitrite and oxygen measurements (Table 5.8). The collinearity index value was changed from 1.2 to 53 and in general tends to increase for larger parameter subset sizes. The parameter subset K#21 is the one used in the parameter estimation above (see examples 5.3 and 5.4). This subset has a collinearity index of 45, which is far higher than typically considered threshold values of 5-15 for a subset to be considered practically identifiable

(Brun *et al.*, 2002; Sin *et al.*, 2010). As shown here, the analysis would have diagnosed the issue before performing the parameter estimation (PE) and this would have indicated that this subset was not suitable for the estimation.

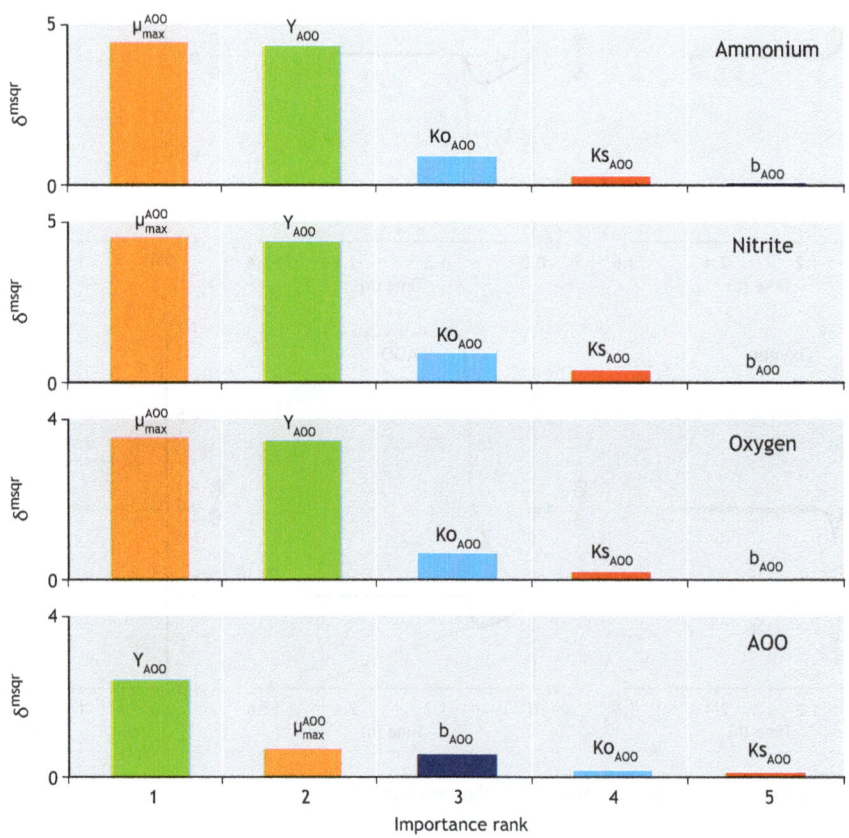

Figure 5.2 Significance ranking of the AOO parameters with respect to the model outputs.

However, given that the sensitivity of b_{AOO} was not influential on the outputs (see Step 3), any subset containing this parameter would not be recommended for parameter estimation. Nevertheless there remain many subsets that meet a threshold of 5-15 for γ_K that can be considered for the parameter estimation problem. The parameter subsets shaded in Table 5.8 meet these identifiability criteria, and therefore can be used for parameter estimation. The best practice is to start with the parameter subset with the largest size (of parameters) and lowest γ_K. Taking these considerations of the sensitivity and collinearity index of the parameter subsets into account helps to avoid the ill-conditioned parameter estimation problem and to improve the quality of the parameter estimates.

Example 5.6 Estimate the model prediction uncertainty of the nitrification model – the Monte Carlo method

In this example, we wish to propagate the parameter uncertainties resulting from parameter estimation (e.g. Example 5.3 and Example 5.4) to model output uncertainty using the Monte Carlo method.

For the uncertainty analysis, the problem is defined as follows: (*i*) only the uncertainty in the estimated AOO parameters is considered, (*ii*) the experimental conditions of batch test 1 are taken in account (Table 5.3), and (*iii*) the model in Table 5.1 is used to describe the system and nominal parameter values in Table 5.2.

Table 5.8 The collinearity index calculation for all the parameter combinations.

Subset K	Subset size	Parameter combination					γ_K
1	2	$K_{o,AOO}$	b_{AOO}				1.32
2	2	$K_{s,AOO}$	b_{AOO}				1.26
3	2	$K_{s,AOO}$	$K_{o,AOO}$				2.09
4	2	μ_{max}^{AOO}	b_{AOO}				1.30
5	2	μ_{max}^{AOO}	$K_{o,AOO}$				13.92
6	2	μ_{max}^{AOO}	$K_{s,AOO}$				2.03
7	2	Y_{AOO}	b_{AOO}				1.28
8	2	Y_{AOO}	$K_{o,AOO}$				12.55
9	2	Y_{AOO}	$K_{s,AOO}$				2.02
10	2	Y_{AOO}	μ_{max}^{AOO}				42.93
11	3	$K_{s,AOO}$	$K_{o,AOO}$	b_{AOO}			2.10
12	3	μ_{max}^{AOO}	$K_{o,AOO}$	b_{AOO}			14.05
13	3	μ_{max}^{AOO}	$K_{s,AOO}$	b_{AOO}			2.03
14	3	μ_{max}^{AOO}	$K_{s,AOO}$	$K_{o,AOO}$			14.23
15	3	Y_{AOO}	$K_{o,AOO}$	b_{AOO}			13.09
16	3	Y_{AOO}	$K_{s,AOO}$	b_{AOO}			2.02
17	3	Y_{AOO}	$K_{s,AOO}$	$K_{o,AOO}$			12.89
18	3	Y_{AOO}	μ_{max}^{AOO}	b_{AOO}			51.25
19	3	Y_{AOO}	μ_{max}^{AOO}	$K_{o,AOO}$			45.87
20	3	Y_{AOO}	μ_{max}^{AOO}	$K_{s,AOO}$			43.37
21	4	Y_{AOO}	μ_{max}^{AOO}	$K_{s,AOO}$	$K_{o,AOO}$		45.91
22	4	Y_{AOO}	μ_{max}^{AOO}	$K_{s,AOO}$	b_{AOO}		51.25
23	4	Y_{AOO}	μ_{max}^{AOO}	$K_{o,AOO}$	b_{AOO}		53.01
24	4	Y_{AOO}	$K_{s,AOO}$	$K_{o,AOO}$	b_{AOO}		13.30
25	4	μ_{max}^{AOO}	$K_{s,AOO}$	$K_{o,AOO}$	b_{AOO}		14.30
26	5	Y_{AOO}	μ_{max}^{AOO}	$K_{s,AOO}$	$K_{o,AOO}$	b_{AOO}	53.07

Step 1. Input uncertainty definition.

As defined in the above problem definition, only the uncertainties in the estimated AOO parameters are taken into account:

- $\theta_{input} = [Y_{AOO}\ \mu_{max}^{AOO}\ K_{s,AOO}\ K_{o,AOO}]$.

Mean and standard deviation estimates are taken as obtained from the bootstrap method together with their correlation matrix (Table 5.7). Further it is assumed that these parameters follow a normal distribution or multivariate normal distribution since they have a covariance matrix and are correlated. This assumption can be verified by calculating the empirical density function for each parameter using the parameter estimates matrix ($\theta_{50 \times 4}$) and shown in Figure 5.12.

```
figure
labels=['\theta_1';'\theta_2';'\theta_3';'\theta
_4']; %or better the name of parameter
for i=1:4
    subplot(2,2,i)
    [f xi]=ksdensity(pmin(:,i));
    plot(xi,f)
xlabel(labels(i,:),'FontSize',fs,'FontWeight','b
old')
end
```

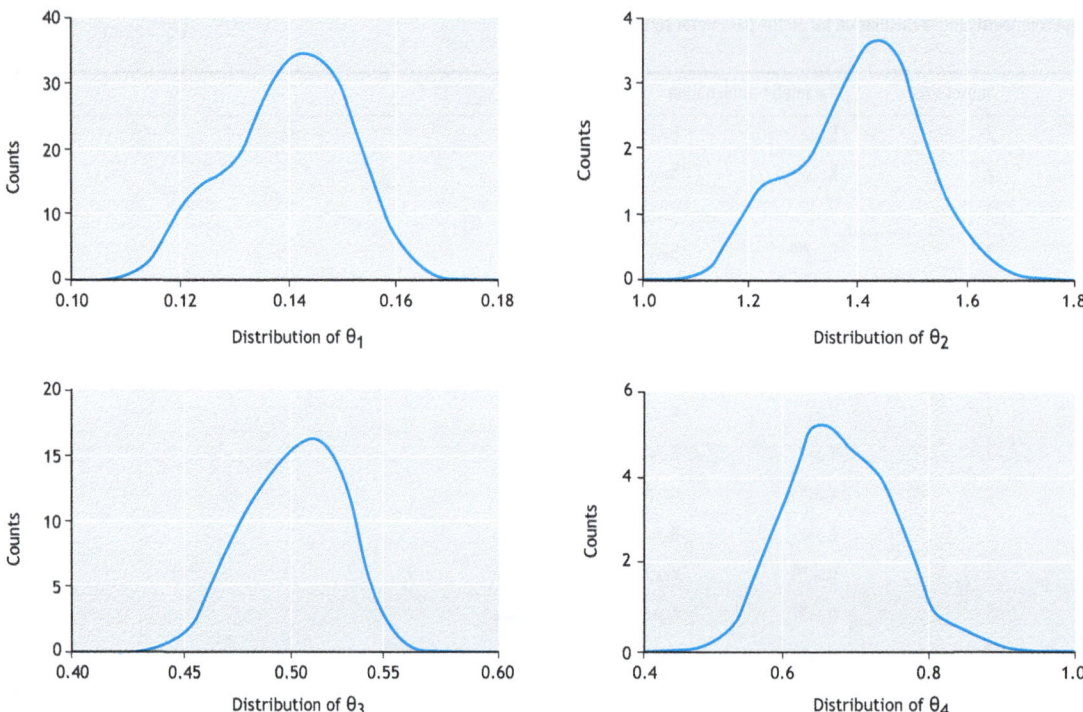

Figure 5.12 Empirical probability density estimates for the AOO parameters as obtained by the bootstrap method.

Step 2. Sampling from the input space

Since the input parameters have a known covariance matrix, any sampling technique must take this into account. In this example, since the parameters are defined to follow a normal distribution, the input uncertainty space is represented by a multivariate normal distribution. A random sampling technique is used to sample from this space:

```
%% do random sampling
N= 100; %% sampling number
mu=mean(pmin); %% mean values of parameters
sigma=cov(pmin); %% covariance matrix (includes
stand dev and correlation information)
X = mvnrnd(mu,sigma,N); % sample parameter space
using multivariate random sampling
```

The output from this step is a sampling matrix, $X_{N \times m}$, where N is the sampling number and m is the number of inputs. The sampled values can be viewed using a matrix plot as in Figure 5.13. In this figure, which is a matrix plot, the diagonal subplots are the histogram of the parameter values while the non-diagonal subplots show the sampled values of the two pairs of parameters. In this case the most important observations are that (*i*) the parameter input space is sampled randomly and (*ii*) the parameter correlation structure is preserved in the sampled values.

Step 3. Perform the Monte Carlo simulations.

In this step, N model simulations are performed using the sampling matrix from Step 2 ($X_{N \times m}$) and the model outputs are recorded in a matrix form to be processed in the next step.

```
%%step 2 perform monte carlo simulations for
each parameter value
% Solution of the model
initcond;options=odeset('RelTol',1e-
7,'AbsTol',1e-8);
for i=1:nboot
    disp(['the iteration number is :
',num2str(i)])
    par(idx) = X(i,:) ; %read a sample from
sampling matrix
    [t,y1] = ode45(@nitmod,td,x0,options,par); ;
%solve the model
    y(:,:,i)=y1; %record the outputs
end
```

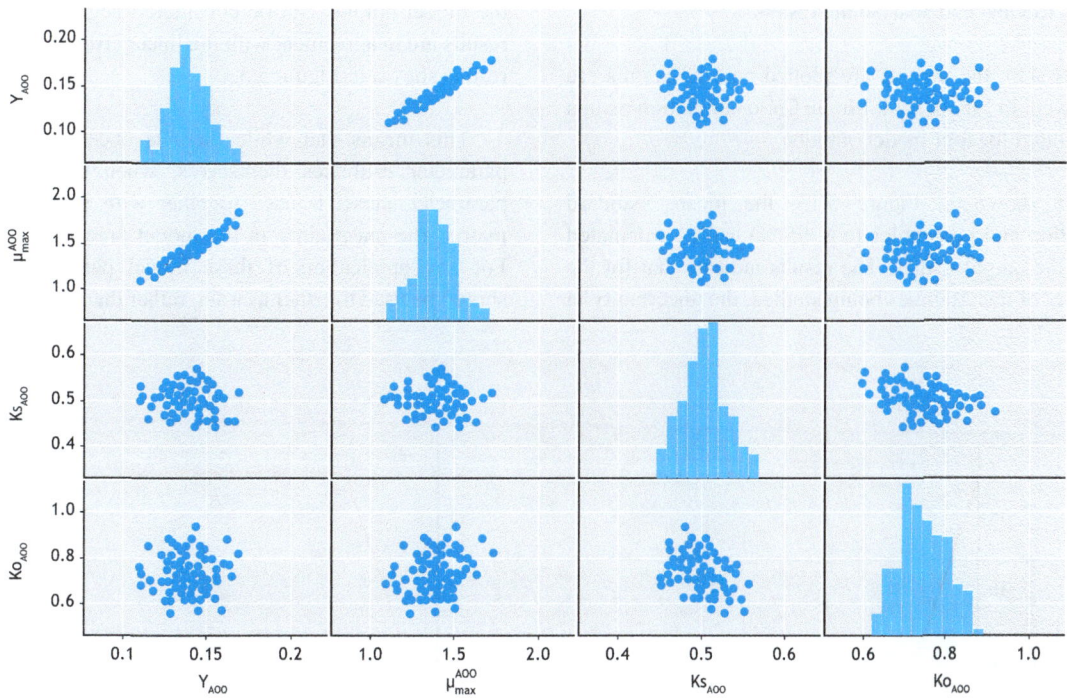

Figure 5.13 Plotting of the sampling matrix of the input space, $X_{N \times m}$ – the multivariate random sampling technique with a known covariance matrix.

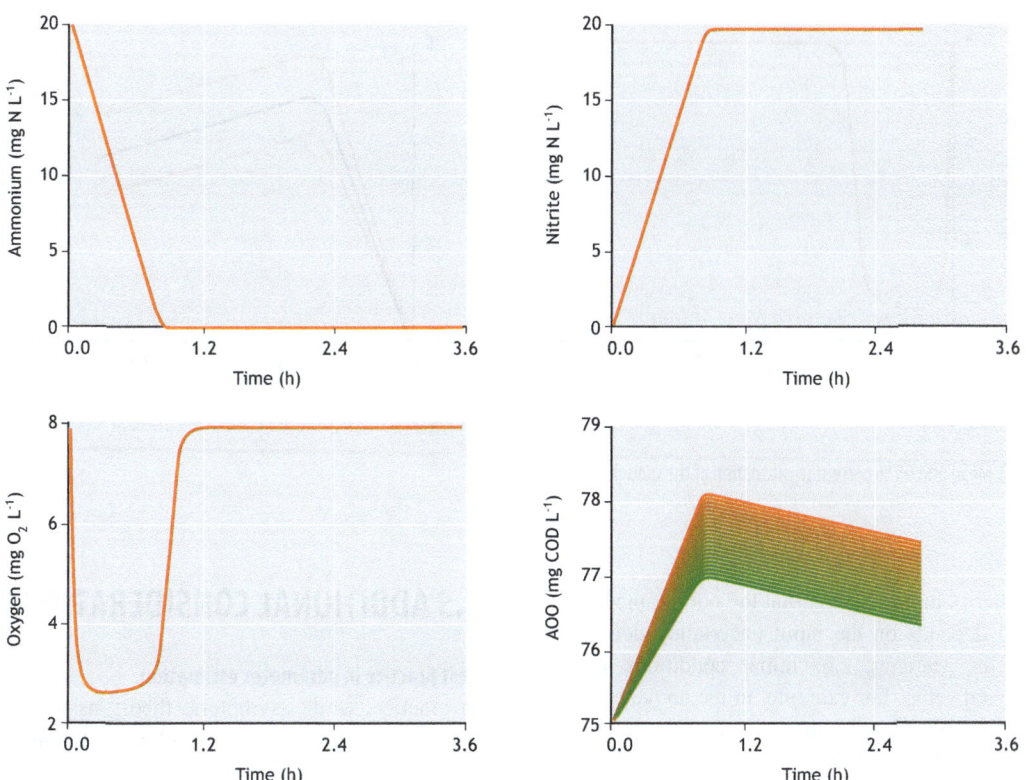

Figure 5.14 Monte Carlo simulations (N = 100) of the model outputs.

Step 4. Review and analyse the results.

In this step, the outputs are plotted and the results are reviewed. In Figure 5.14, Monte Carlo simulation results are plotted for four model outputs.

As shown in Figure 5.15, the mean, standard deviation and percentiles (e.g. 95 %) can be calculated from the output matrix. The results indicate that for the sources of uncertainties being studied, the uncertainty in the model outputs can be considered negligible. These results are in agreement with the linear error propagation results shown in Figure 5.5.

This means that while there is uncertainty in the parameter estimates themselves, when the estimated parameter subset is used together with its covariance matrix, the uncertainty in the model prediction is low. For any application of these model parameters they should be used together as a set, rather than individually.

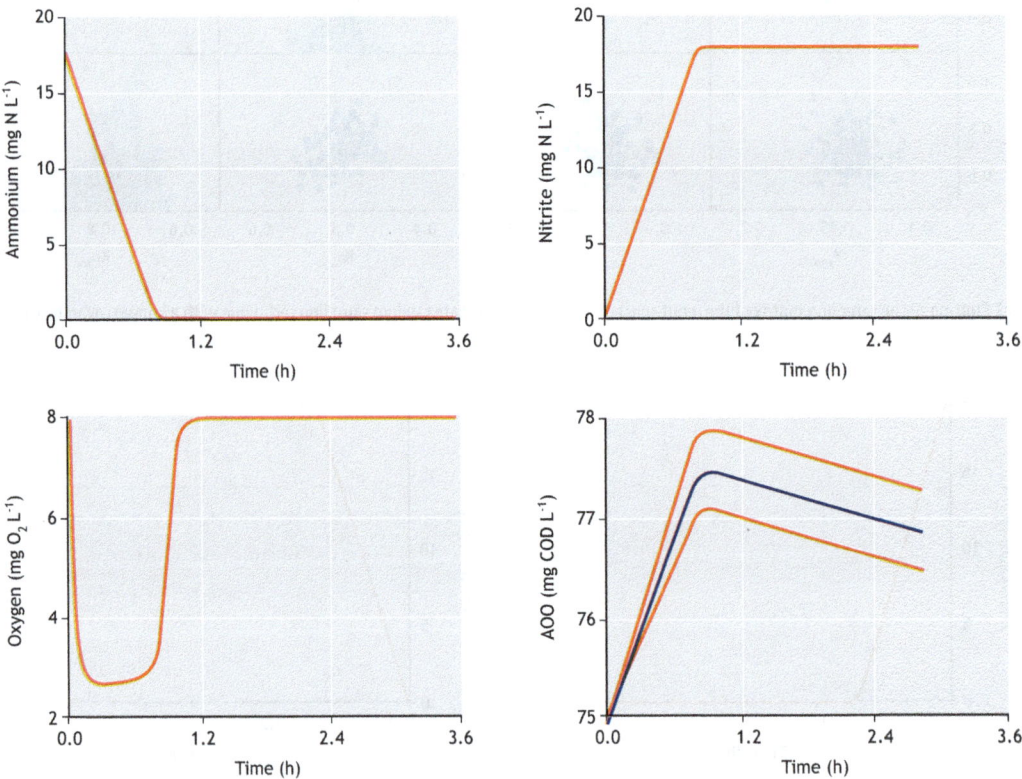

Figure 5.15 Mean and 95 % percentile calculation of the model output uncertainty.

Another point to make is that the output uncertainty evaluated depends on the input uncertainty defined as well as the framing, e.g. initial conditions of the experimental setup. For example, in the above example what was not considered is the measurement uncertainty or uncertainty due to other fixed parameters (decay) and initial conditions (the initial concentration of autotrophic bacteria). Therefore these results need to be interpreted within the context where they are generated.

5.5 ADDITIONAL CONSIDERATIONS

Best practice in parameter estimation

In practice, while asymptotic theory assumption gives reasonable results, there are often deviations from the assumptions. In particular:

- The measurement errors are often auto-correlated, meaning that too many observations are redundant and not independent (non-independently and identically distributed (iid) random variables). This tends to cause an underestimation of asymptotic confidence intervals due to smaller sample variance, σ^2. A practical solution to this problem is to check the autocorrelation function of the residuals and filter them or perform subsampling such that autocorrelation is decreased in the data set. The parameter estimation can then be redone using the subsample data set.
- Parameter estimation algorithms may stop at local minima, resulting in an incorrect linearization result (the point at which the non-linear least squares are linearized). To alleviate this issue, parameter estimation needs to be performed several times with either different initial guesses, different search algorithms and/or an identifiability analysis.

Afterwards it is important to verify that the minimum solution is consistent with different minimization algorithms.

Identifiability or ill-conditioning problem: Not all the parameters can be estimated accurately. This can be caused by a too large confidence interval compared to the mean or optimized value of the parameter estimators. The solution is to perform an identifiability analysis or re-parameterisation of the model, so that a lower number of parameters needs to be estimated.

While we have robust and extensive statistical theories and methods relevant for estimation of model parameters as demonstrated above, the definition of the parameter estimation problem itself, which is concerned with stating what is the data available, what is the candidate model structure, and what is the starting point for the parameter values, is taken for granted. Hence a proper analysis and definition of the parameter estimation problem will always require a good engineering judgment. For robust parameter estimation in practice, due to the empirical/experiental nature of parameter definition, the statistical methods (including MLE estimates) should be treated within the context/definition of the problem of interest.

With regards to bootstrap sampling, the most important issue is whether or not the residuals are representative of typical measurement error. For a more detailed discussion of this issue, refer to Efron (1979).

Best practice in uncertainty analysis

When performing uncertainty analysis, the most important issue is the framing and the corresponding definition of the input uncertainty sources. Hence, the outcome from an uncertainty analysis should not be treated as absolute but dependent on the framing of the analysis. A detailed discussion of these issues can be found elsewhere (e.g. Sin *et al.*, 2009; Sin *et al.*, 2010).

Another important issue is the covariance matrix of the parameters (or correlation matrix), which should be obtained from a parameter estimation technique. Assuming the correlation matrix is negligible may lead to over or under estimation of the model output uncertainty. Hence, in a sampling step the appropriate correlation matrix should be defined for inputs (*e.g.* parameters) considered for the analysis.

Regarding the sampling number, one needs to iterate several times to see if the results differ from one iteration to another. Since the models used for the parameter estimation are relatively simple to solve numerically, it is recommended to use a sufficiently high number of iterations e.g. 250 or 500.

References

Brun, R., Kühni, M., Siegrist, H., Gujer, W., Reichert, P. (2002). Practical identifiability of ASM2d parameters - systematic selection and tuning of parameter subsets. *Water Res.* 36(16): 4113-4127.

Brun, R., Reichert, P., and Künsch, H. R. (2001). Practical identifiability analysis of large environmental simulation models. *Water Resources Research*, 37(4):1015-1030.

Bozkurt, H., Quaglia, A., Gernaey, K.V., Sin, G. (2015). A mathematical programming framework for early stage design of wastewater treatment plants. *Environmental Modelling & Software*, 64: 164-176.

Dochain, D., Vanrolleghem, P.A. (2001). Dynamical Modelling and Estimation in Wastewater Treatment Processes. London UK: IWA Publishing.

Efron, B. (1979). Bootstrap methods: another look at the jackknife. *The Annals of Statistics*, 7(1):1-26.

Gernaey, K.V., Jeppsson, U., Vanrolleghem, P.A., Copp, J.B. (Eds.). (2014). *Benchmarking of control strategies for wastewater treatment plants*. IWA Publishing.

Guisasola, A., Jubany, I., Baeza, J.A., Carrera, J., Lafuente, J. (2005). Respirometric estimation of the oxygen affinity constants for biological ammonium and nitrite oxidation. *Journal of Chemical Technology and Biotechnology*, 80(4): 388-396.

Heijnen, J.J. (1999). Bioenergetics of microbial growth. *Encyclopaedia of Bioprocess Technology*.

Henze, M., Gujer, W., Mino, T., van Loosdrecht, M.C.M., (2000). ASM2, ASM2d and ASM3. *IWA Scientific and Technical Report*, 9. London UK.

Ljung L. (1999). System identification - Theory for the user. 2nd edition. Prentice-Hall.

Mauricio-Iglesias, M., Vangsgaard, A.K., Gernaey, K.V., Smets, B.F., Sin, G. (2015). A novel control strategy for single-stage autotrophic nitrogen removal in SBR. *Chemical Engineering Journal*, 260: 64-73.

Meijer, S.C.F., Van Der Spoel, H., Susanti, S., Heijnen, J.J., van Loosdrecht, M.C.M. (2002). Error diagnostics and data reconciliation for activated sludge modelling using mass balances. *Water Sci Tech*. 45(6): 145-156.

Metropolis, N., Ulam, S. (1949). The Monte Carlo method. *Journal of the American Statistical Association*, 44(247): 335-341.

Omlin, M. and Reichert, P. (1999). A comparison of techniques for the estimation of model prediction uncertainty. *Ecol. Model.*, 115: 45-59.

Roels, J.A. (1980). Application of macroscopic principles to microbial metabolism. *Biotechnology and Bioengineering*, 22(12): 2457-2514.

Saltelli, A., Tarantola, S., and Campolongo, F. (2000). Sensitivity analysis as an ingredient of modeling. *Statistical Science*, 15(4):377-395.

Seber G. and Wild C. (1989) Non-linear regression. Wiley, New York.

Sin, G., Gernaey, K.V., Neumann, M.B., van Loosdrecht, M.C.M., Gujer, W. (2009). Uncertainty analysis in WWTP model applications: a critical discussion using an example from design. *Water Res.* 43(11): 2894-2906.

Sin, G., Gernaey, K.V., Neumann, M.B., van Loosdrecht, M.C.M., Gujer, W. (2011). Global sensitivity analysis in wastewater treatment plant model applications: prioritizing sources of uncertainty. *Water Research,* 45(2): 639-651.

Sin, G., de Pauw, D.J.W., Weijers, S., Vanrolleghem, P.A. (2008). An efficient approach to automate the manual trial and error calibration of activated sludge models. *Biotechnology and Bioengineering.* 100(3): 516-528.

Sin, G., Meyer, A.S., Gernaey, K.V. (2010). Assessing reliability of cellulose hydrolysis models to support biofuel process design-identifiability and uncertainty analysis. *Computers & Chemical Engineering*, 34(9): 1385-1392.

Sin, G., Vanrolleghem, P.A. (2007). Extensions to modeling aerobic carbon degradation using combined respirometric–titrimetric measurements in view of activated sludge model calibration. *Water Res.* 41(15): 3345-3358.

Vangsgaard, A.K., Mauricio-Iglesias, M., Gernaey, K.V., Sin, G. (2014). Development of novel control strategies for single-stage autotrophic nitrogen removal: A process oriented approach. *Computers & Chemical Engineering*, 66: 71-81.

van der Heijden, R.T.J.M., Romein, B., Heijnen, J.J., Hellinga, C., Luyben, K. (1994). Linear constraint relations in biochemical reaction systems: II. Diagnosis and estimation of gross errors. *Biotechnology and bioengineering*, 43(1): 11-20.

Villadsen, J., Nielsen, J., Lidén, G. (2011). Elemental and Redox Balances. In *Bioreaction Engineering Principles* (pp. 63-118). Springer, US.

6

SETTLING TESTS

Authors:
Elena Torfs
Ingmar Nopens
Mari K.H. Winkler
Peter A. Vanrolleghem
Sophie Balemans
Ilse Y. Smets

Reviewers:
Glen T. Daigger
Imre Takács

6.1 INTRODUCTION

Settling is an important process in several of the unit operations in wastewater treatment plants (WWTPs). The most commonly known of these unit processes are primary settling tanks (PSTs), which are a treatment units before the biological reactor, and secondary settling tanks (SSTs), which are a clarification step prior to discharge into a receiving water. Moreover, settling also plays an important role in new technologies that are being developed such as granular sludge reactors. Due to the different nature of the settleable components (raw wastewater, activated sludge, and granular sludge) and the concentration at which these compounds occur, these unit processes are characterised by distinctly different settling behaviours. This section presents an overview of the different settling regimes that a particle-liquid suspension can undergo and relates these regimes to the specific settling behaviour observed in SSTs, PSTs and granular sludge reactors.

The settling behaviour of a suspension (e.g. secondary sludge, raw wastewater or granular sludge) is governed by its concentration and flocculation tendency and can be classified into four regimes (Figure 6.1): discrete non-flocculent settling (Class I), discrete flocculent settling (Class II), zone settling or hindered settling (Class III) and compressive settling (Class IV).

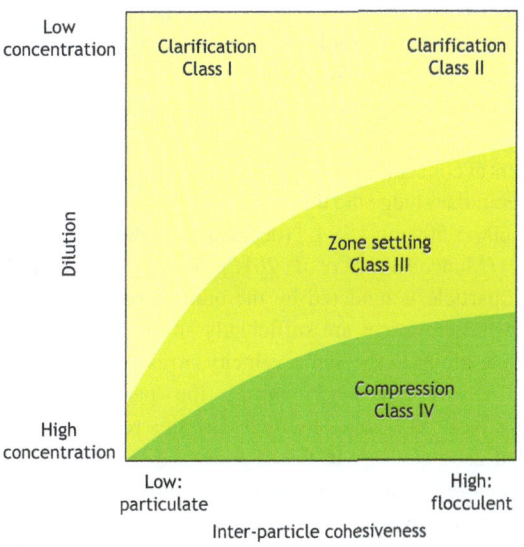

Figure 6.1 Settling regimes (Ekama *et al.*, 1997).

At low concentrations (classes I and II), the particles are completely dispersed, there is no physical contact between them and the concentration is typically too diluted for particles to influence each other's settling behaviour. Each particle settles at its own characteristic terminal velocity, which depends on individual particle properties such as shape, size, porosity and density. If these dilute particles show no tendency to flocculate (for example, granular sludge), this regime is called discrete non-flocculent settling (Class I). However, certain suspensions (among which raw wastewater solids and activated sludge flocs) have a natural tendency to flocculate even at low concentrations (Ekama *et al.*, 1997). Through subsequent processes of collision and cohesion, larger flocs are formed causing their settling velocity to change over time. This regime is called discrete flocculent settling (Class II). It is important to note that during discrete flocculent settling, the formed flocs will still settle at their own characteristic terminal velocity. Hence, both discrete non-flocculent and discrete flocculent settling undergo basically the same settling dynamics. The difference lies in the fact that, for discrete flocculent settling, an additional flocculation process is occurring simultaneously with the settling process, which alters the particles' individual properties and consequently their terminal settling velocity.

The transition from discrete settling to hindered settling (Class III) occurs if the solid concentration in the tank exceeds a threshold concentration where the particles no longer settle independently of one another. As can be seen from Figure 6.1, this threshold concentration depends on the flocculation state of the sludge. For secondary sludge the transition typically occurs at concentrations of 600 - 700 mg TSS L^{-1} whereas for granular sludge the threshold concentration can go up to 1,600-5,500 mg TSS L^{-1} (depending on the granulation state) (Mancell-Egala *et al.*, 2016). Above this threshold, each particle is hindered by the other particles and the inter-particle forces are sufficiently strong to drag each particle along at the same velocity, irrespective of size and density. In other words, the particles settle collectively as a zone, and therefore this regime is also called zone settling. In this regime, a distinct interface between the clear supernatant and the subsiding particles is formed. When the solids concentration further increases above a critical concentration (5-10 g L^{-1}), the settling behaviour changes to compressive settling (Class IV). The exact transition concentration depends once more on the flocculation state of the particles (De Clercq *et al.*, 2008). At these elevated concentrations, the solids come into physical contact with one another and are subjected to compaction due to the weight of overlying particles. The settling velocity will be much lower than in the hindered settling regime.

Activated sludge with a proper biological make-up shows a natural tendency to flocculate and, depending on its concentration, can cover the entire right side of Figure 6.1. Hence, in an SST, different settling regimes occur simultaneously at different locations throughout the tank. Low concentrations in the upper regions of the SST favour discrete (flocculent) settling whereas the concentrations of the incoming sludge are typically in the range for hindered settling, and sludge thickening inside the sludge blanket is governed by compressive settling. In contrast to this, a specific characteristic of the granular sludge technology is the granules' low tendency to coagulate under reduced hydrodynamic shear (de Kreuk and van Loosdrecht, 2004), thus positioning them on the left side of Figure 6.1. This feature causes granular sludge to undergo discrete (non-flocculent) settling at concentrations where conventional activated sludge undergoes hindered or compression settling. Finally, PSTs are fed by incoming wastewater containing a relatively low concentration of suspended solids (i.e. the upper part Figure 6.1). Hence, the dominant settling regime in these tanks is discrete settling (classes I and II).

As each unit process is characterised by its own settling behaviour, different experimental methods are required to assess their performance. This chapter provides an overview of experimental methods to analyse the settling and flocculation behaviour in secondary settling tanks (Section 6.2 and 6.3), granular sludge reactors (Section 6.4) and primary settling tanks (Section 6.5).

6.2 MEASURING SLUDGE SETTLEABILITY IN SSTs

To evaluate the performance of an SST, it is essential to quantify the settling behaviour of the activated sludge in the system. Batch settling experiments are an interesting information source in this respect since they eliminate the hydraulic influences of in- and outgoing flows on the settling behaviour. Therefore, several methods aim to determine the settling characteristics of the activated sludge by measuring certain properties during the settling of activated sludge in a batch reservoir.

Several types of measurements can be performed by means of a batch settling test. These measurements range from very simple experiments providing a rough indication of the sample's general settleability (Section 6.2.1) to more labour-intensive experiments that measure the specific settling velocity (Section 6.2.2) or even determine a relation between the settling velocity and the sludge concentration (Section 6.2.3). Moreover, some useful recommendations to consider when performing batch settling experiments are provided in Section 6.2.4 and an overview of recent developments with respect to these type of experiments can be found in Section 6.2.5.

6.2.1 Sludge settleability parameters

6.2.1.1 Goal and application

A number of parameters have been developed to obtain a quantitative measure of the settleability of an activated sludge sample. These Sludge Settleability Parameters (SSPs) are based on the volume that sludge occupies after a fixed period of settling. Among these, the Sludge Volume Index (SVI) (Mohlman, 1934) is the most known. A number of issues have been reported with the SVI as a measure of sludge settleability (Dick and Vesilind, 1969; Ekama et al., 1997) of which the most important one is its dependency on the sludge concentration. Particularly at higher concentrations, measured SVI values can deviate significantly between sludge concentrations (Dick and Vesilind, 1969). Moreover, SVI measurements have been found to be influenced by the dimensions of the settling cylinder. These problems can be significantly reduced by conducting the test under certain prescribed conditions. Hence, a number of modifications have been proposed to the standard SVI test in order to yield more consistent information (Stobbe, 1964; White, 1976, 1975). Stobbe (1964) proposed conducting the SVI test with diluted sludge and called it the Diluted SVI (DSVI). White (1975, 1976) proposed the Stirred Specific Volume Index (SSVI$_{3.5}$) where the sludge sample is stirred during settlement. Although each of these SSPs are described in more detail below, it is important to note that the SSVI$_{3.5}$ is known to provide the most consistent results.

6.2.1.2 Equipment

a. A graduated (minimum resolution 50 mL) cylindrical reservoir with a volume of 1 litre (for SVI and DSVI) or with dimensions specified by White (1975) for SSVI.
b. A digital timer displaying accuracy in seconds.
c. A sludge sample from either the recycle flow of the SST or the feed flow into the SST. The latter can be collected from the bioreactor or the splitter structure.
d. Effluent from the same WWTP (in case dilution is needed).
e. Equipment for the Total Suspended Solids (TSS) test (according to method 2540 D in APHA et al., 2012).
f. A stirrer for the SSVI test.

6.2.1.3 The Sludge Volume Index (SVI)

The Sludge Volume Index (SVI) (Mohlman, 1934) is defined as the volume (in mL) occupied by 1 g of sludge after 30 min settling in a 1 L unstirred cylinder.

- **Protocol**
1. Measure the concentration of the sludge sample with a TSS test according to method 2540 D of Standard Methods (APHA et al., 2012).
2. Fill a 1 L graduated cylinder with the sludge sample and allow the sample to settle.
3. After 30 min of settling, read the volume occupied by the sludge from the graduated cylinder (SV$_{30}$ in mL L^{-1}).
4. Calculate the SVI from Eq. 6.1, with X$_{TSS}$ being the measured concentration of the sample in g L^{-1}:

$$\text{SVI} = \frac{\text{SV}_{30}}{\text{X}_{TSS}} \qquad \text{Eq. 6.1}$$

- **Example**

In this example an SVI test is performed with a sludge sample from the bioreactor in the WWTP at Destelbergen. The concentration of the sample is measured at 2.93 g L^{-1}.

A graduated cylinder is filled with the sludge sample and the sludge is allowed to settle. After 30 min of settling, the sludge occupies a volume of 290 mL (SV$_{30}$).

Hence the sample has an SVI of:

$$\text{SVI} = \frac{290 \text{ mL L}^{-1}}{2.93 \text{ g L}^{-1}} = 99 \text{ mL g}^{-1}$$

This result indicates a sludge with good settling properties. Typical SVI values for AS can be found between 50-400 mL g^{-1} where 50 mL g^{-1} indicates a sample with very good settleability and 400 mL g^{-1} a sample with poor settling properties.

6.2.1.4 The Diluted Sludge Volume Index (DSVI)

The Diluted Sludge Volume Index: DSVI (Stobbe, 1964) differs from the standard SVI by performing an additional dilution step prior to settling. The sludge is hereby diluted with effluent until the settled volume after 30 min is between 150 ml L^{-1} and 250 ml L^{-1}. Note that all the dilutions must be made with effluent (before chemical disinfection) from the plant where the sludge is obtained to reduce the possibility of foreign substances affecting the settling behaviour.

- **Protocol**
1. Dilute the sludge sample with effluent until the settled volume after 30 min is between 150 mL L^{-1} and 250 mL L^{-1}.
2. Perform steps 1-3 from the standard SVI test.
3. Calculate the DSVI from Eq. 6.1, with X_{TSS} being the concentration of the diluted sample in g L^{-1}.

The advantage of the DSVI lies in its insensitivity to the sludge concentration, allowing for consistent comparison of sludge settleability between different activated sludge plants.

6.2.1.5 The Stirred Specific Volume Index (SSVI$_{3.5}$)

The Stirred Specific Volume Index (SSVI$_{3.5}$) was presented by White (1975, 1976) who found that stirring the sample during settling reduces wall effects, short circuiting and bridge formation effects, thereby creating conditions more closely related to those prevailing in the sludge blanket in SSTs.

The SSVI$_{3.5}$ is determined by performing an SVI test at a specific concentration of 3.5 g L^{-1} while the sludge is gently stirred at a speed of about 1 rpm. To determine the SSVI$_{3.5}$, the sludge concentration is measured with a TSS test and subsequently diluted with effluent to a concentration of 3.5 g L^{-1}. In some cases further concentration of the sample may be necessary if the plant is operating at MLSS values below 3.5 g L^{-1} and sampling from the return activated sludge flow is not possible.

Compared to the DSVI, the SSVI$_{3.5}$ has the additional advantage that it not only overcomes the concentration dependency but is shown to be relatively insensitive to the dimensions of the settling column, provided that it is not smaller than the dimensions specified by White (1976), i.e. a depth to diameter ratio of between 5:1 and 6:1, and a volume of more than 4 L. As the measurement is performed in a larger reservoir, the SSVI$_{3.5}$ cannot be directly calculated from Eq. 6.1 but the volume of the particular column has to be reduced to an equivalent 1 L column. This is done by expressing the settled sludge volume at 30 min as a fraction of the column volume (f_{sv}) and multiplying this fraction by 1,000 mL to obtain the equivalent 1 L stirred settled volume SSV$_{30}$ (Eq. 6.2).

$$SSV_{30} = f_{sv} \cdot 1,000 \qquad \text{Eq. 6.2}$$

This value can then be used in Eq. 6.1 to calculate the SSVI$_{3.5}$ with X_{TSS} = 3.5 g L^{-1}.

Although the SSVI$_{3.5}$ is not as easily executed as the DSVI due to the specified stirring equipment required, it does provide the most consistent results (Ekama et al., 1997; Lee et al., 1983).

- **Protocol**
1. Measure the concentration of the sludge sample with a TSS test according to method 2540 D of Standard Methods (APHA et al., 2012).
2. Dilute the sample with effluent to a concentration of 3.5 g L^{-1}.
3. Fill a graduated cylinder with the minimum dimensions specified by White (1976).
4. After 30 min of settling during which the sample is stirred at a speed of 1 rpm, read off the volume occupied by the sludge from the graduated cylinder and calculate the SSV$_{30}$ from Eq. 6.2.
5. Calculate the SSVI$_{3.5}$ from Eq. 6.1 with SV$_{30}$ = SSV$_{30}$ and X_{TSS} = 3.5 g L^{-1}.

6.2.2 The batch settling curve and hindered settling velocity

6.2.2.1 Goal and application

The sludge settleability parameters presented above provide a low level measurement of the general settleability. However, it should be stressed that they represent only a momentary recording of the settling behaviour. In reality, the volume of a sludge sample after 30 min of settling will depend on both its hindered settling and compression behaviour, which are both influenced by a number of factors such as the composition of the activated sludge (for example, the population of filamentous organisms), floc size distributions, surface properties, rheology, etc. Consequently, two sludge samples with different settling behaviour can result in similar values for the sludge settleability parameters.

More detailed information on the settling behaviour of a sludge sample can be obtained from a batch settling curve which makes it possible to investigate the settling behaviour of sludge at different settling times.

Batch settling curves can serve different purposes. They can be used either qualitatively to determine operational or seasonal trends in the settling behaviour or they can be used quantitatively to determine the SST's capacity limit. In the former, a simple graduated cylinder can be used (for example, the cylindrical reservoir in Figure 6.2). In the latter, the selection of an appropriate settling reservoir to avoid wall effects during the test is imperative. More information on the optimal shape and size of batch settling reservoirs is provided in Section 6.2.4.1.

6.2.2.2 Equipment

For this test the following equipment is needed:
a. A graduated (minimum resolution 50 mL) cylindrical reservoir.
b. A digital timer displaying accuracy in seconds.
c. A sludge sample from either the recycle flow of the SST or the feed flow into the SST. The latter can be collected from the bioreactor or the splitter structure.
d. Equipment for a TSS test (APHA *et al.*, 2012).
e. Stirring equipment if the results are to be used for quantitative analysis of the SST's capacity.

6.2.2.3 Experimental procedure

To measure a batch settling curve, a reservoir is filled with a sludge sample and a timer is started to keep track of the duration of the experiment. The sludge is allowed to settle and the position of the suspension-liquid interface is measured at different time intervals. This methodology is illustrated in Figure 6.2 where the position of the suspension-liquid interface is indicated by the red arrow. Recording the height of the suspension-liquid interface at several time intervals results in a curve with the evolution of the sludge blanket height over time. Standard measurement times for a batch settling curve are 0, 0.5, 1, 2, 3, 4, 5, 10, 15, 20, 30 and 45 min but these can be adapted depending on the settling dynamics of a specific sludge sample (for more information see Section 6.2.4). At the start of the test, the suspension-liquid interface is typically measured more frequently, as the sludge is settling at a relatively fast pace. Later in the test, the frequency of the measurements is decreased, because the interface is moving more slowly.

- **Protocol**
1. Homogenize the sludge sample. Do not shake the sample vigorously as this will disturb the sludge and alter its settling properties.
2. Fill the cylindrical reservoir with the sample. Pour gently and in a steady flow so as not to disturb the sludge too much nor to allow it to settle again in the container.
3. Start the timer immediately after filling the column.
4. Measure the sludge water interface at the following time intervals: 0, 0.5, 1, 2, 3, 4, 5, 10, 15, 20, 30 and 45 min.

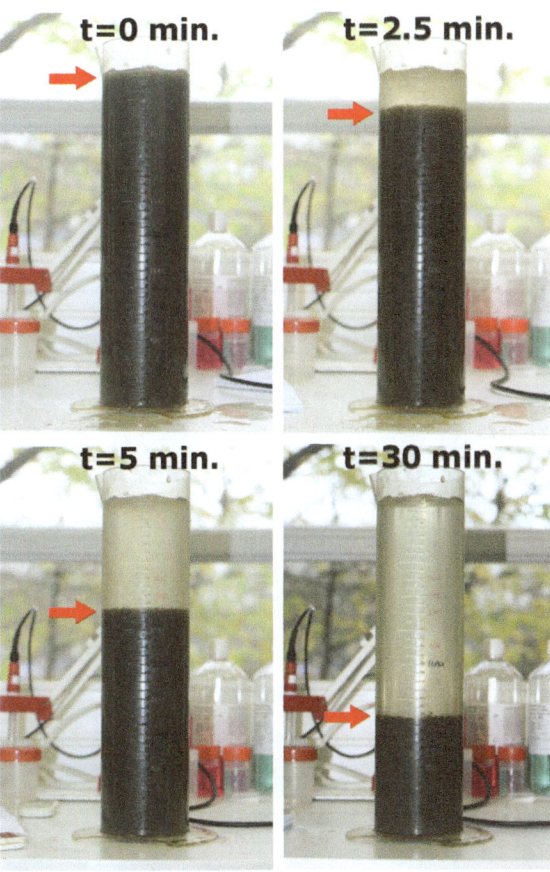

Figure 6.2 Photograph of the batch settling column at different settling times, indicating the suspension-liquid interface (photo: E. Torfs).

- **Example**

A batch settling curve is measured with a sludge sample from the bioreactor in the WWTP at Destelbergen. The concentration of the sample is measured as 2.93 g L^{-1}. The measured heights of the suspension-liquid interface (i.e. the Sludge Blanket Height: SBH) at different settling times are provided in Table 6.2 and the batch curve is shown in Figure 6.3.

Table 6.2 Measured sludge blanket height during a single batch settling test.

Time (min)	SBH (m)
0	0.250
0.5	0.247
1	0.244
2	0.238
3	0.206
4	0.184
5	0.169
10	0.122
15	0.106
20	0.098
30	0.088
45	0.081

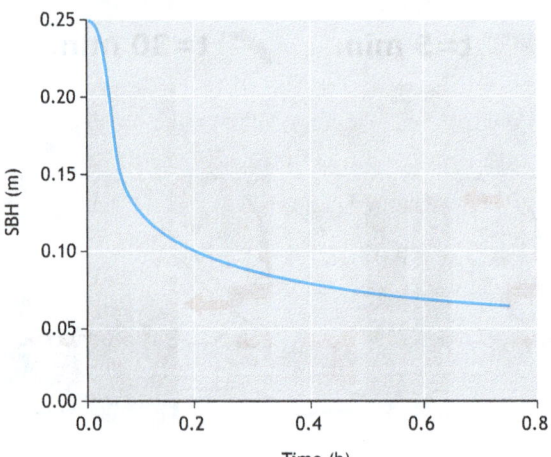

Figure 6.3 Measured batch settling curve.

6.2.2.4 Interpreting a batch settling curve

Typically, four different phases can be observed in a batch settling curve. Each phase marks a change in the settling behaviour at the suspension-liquid interface. Figure 6.4 shows the evolution of the sludge blanket height over time during a batch settling test, indicating the four phases. It is important to note that a batch settling curve only provides information on the settling behaviour at the sludge-water interface. At any specific time, the settling behaviour at different depths throughout the column may differ from the settling behaviour at the interface depending on the local concentrations.

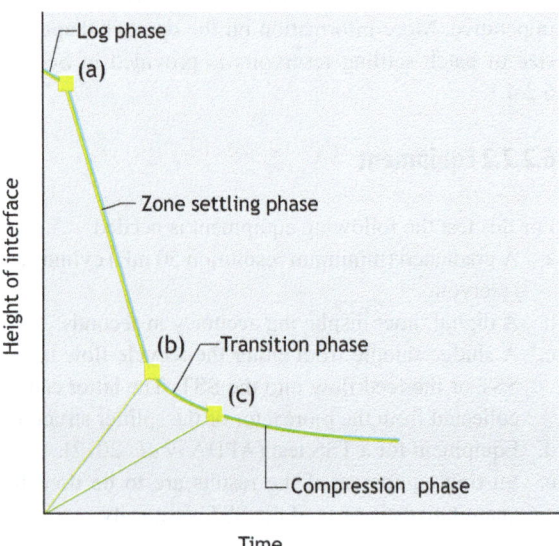

Figure 6.4 Evolution of the sludge blanket height over time, indicating the four phases (Rushton et al., 2000).

Figure 6.5 represents the distribution of settling regions over the depth of the column at different times during a batch settling experiment.

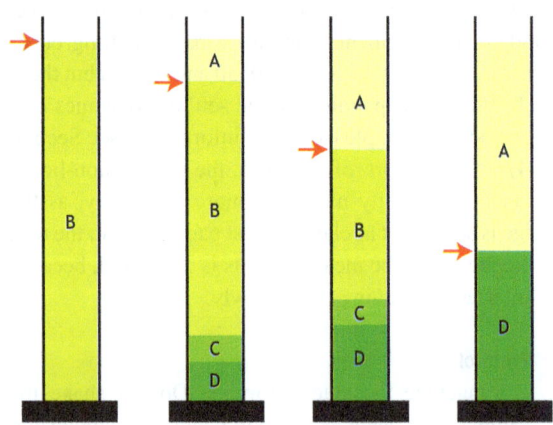

Figure 6.5 Chronological process of a batch settling test (Ekama et al., 1997).

Almost immediately after the start-up of the experiment, four regions are formed at increasing depth. The top region (region A) consists of supernatant. Below region A, regions B, C and D are formed where respectively zone settling, transition settling and compression settling take place (Ekama *et al.,* 1997). The position of the sludge/water interface is indicated with a red arrow. Hence, the phases that are recorded in a batch settling curve occur as the suspension-liquid interface passes through these different settling regions.

From the beginning of the test up to point (a) (Figure 6.4), the suspension-liquid interface is in the lag phase. In this phase, the activated sludge needs to recover from disturbances due to turbulence caused by the filling of the batch column.

During the hindered settling phase or zone settling phase which starts at point (a) and ends at point (b) (Figure 6.4), the interface is located in the hindered settling region (region B). This phase is characterised by a distinct linear decline in the batch curve. An equilibrium between the gravitational forces causing the particles to settle and the hydraulic friction forces resisting this motion results in the same settling velocity for all the particles in the region. If the column is not stirred, then the velocity at which the interface moves downward is called the hindered settling velocity v_{hs} at the inlet solids concentration. If the dimensions and conditions for the batch test were set so as to avoid wall effects (Section 6.2.4.1), the measured settling velocity corresponds with the zone settling velocity in an actual SST.

The transition phase starts (point (b)) when the sludge blanket reaches the transition layer (region C). The transition layer is a layer of constant thickness and is formed by particles coming from the decreasing hindered settling layer and particles coming from the increasing compression layer. Although during this phase the same characteristics exist as in the zone settling regime, the settling velocity decreases because the concentration gradient increases with depth. The transition phase ends when the sludge blanket reaches the compression layer.

The last phase starts at point (c) and is called the compression phase. The time at which the compression phase starts, called the compression point, is difficult to identify. During the compression phase the particles undergo compaction, thus creating an increasing concentration gradient as well as a decreasing settling velocity.

6.2.2.5 Measuring the hindered settling velocity

At moderate sludge concentrations (between approx. 1 g L^{-1} and 6 g L^{-1}), sludge will initially settle according to the zone or hindered settling regime. The slope of the linear part of a batch settling curve corresponds to the hindered settling velocity v_{hs}.

- **Example**

The hindered settling velocity for the data in Figure 6.3 is computed by determining the steepest slope between three consecutive data points (which can be performed in any software). The procedure is illustrated in Figure 6.6 and results in a settling velocity of 1.374 m h^{-1}.

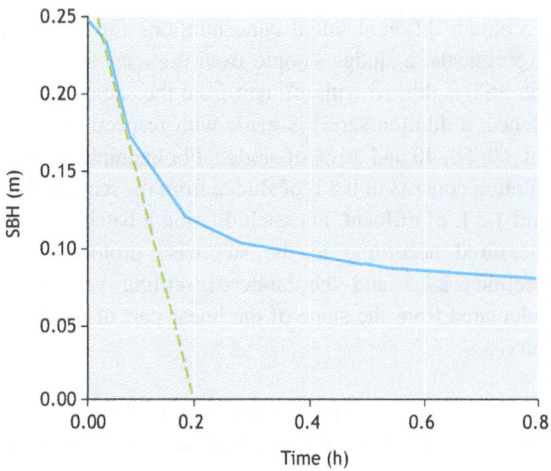

Figure 6.6 Calculation of the steepest slopes from a batch settling curve at a concentration of 2.93 g L^{-1}.

6.2.3 v_{hs}-X relation

6.2.3.1 Goal and application

The concentration in the hindered settling region is uniform and equal to the initial solids concentration of the batch. By calculating the slopes of the linear part of the batch curves for different initial concentrations, the hindered settling velocity can be determined as a function of the solids concentration. The relation between hindered settling velocity and concentration is of particular importance for the design of SSTs as it governs the determination of the limiting flux and thus the SST's surface area. As stated previously, in order to use the v_{hs}-X relation for quantitative calculations such as the determination of the SST's surface area, the dimensions and conditions of the batch settling test need to be set according to the specifications given in Section 6.2.4.1.

6.2.3.2 Equipment

For this test the following equipment is needed:
a. A graduated (minimum resolution 50 mL) cylindrical reservoir.
b. A digital timer displaying accuracy in seconds.
c. A sludge sample from the recycle flow of the SST.
d. Effluent from the same WWTP (for dilution).
e. Equipment for the TSS test (APHA *et al.*, 2012).
f. Stirring equipment if the results are to be used for quantitative analysis of the SST's capacity.

6.2.3.3 Experimental procedure

To obtain different initial concentrations for the batch experiments, a sludge sample from the recycle flow of the SST is diluted with effluent from the same WWTP. Hence, a dilution series is made with respectively 100, 80, 60, 50, 40 and 20 % of sludge. For example, a 40 % dilution consists of 0.8 L of sludge from the recycle flow and 1.2 L of effluent. For each dilution a batch curve is measured according to the step-wise protocol from Section 6.2.2.3 and the hindered settling velocity is calculated from the slope of the linear part of the batch curve.

In order to obtain reliable results for the v_{hs}-X relation, it is important to have an accurate measure of the initial concentration in each experiment. The concentrations in the dilution series are often determined by measuring the concentration in the recycle flow and then calculating the concentration of the dilution assuming the effluent concentration is negligible. However, this procedure is prone to errors if the recycle flow sample is not fully mixed at any time during the filling of the batch. A more reliable approach is to measure the TSS of each dilution experiment separately. This can be done by mixing up the content of the batch reservoir at the end of each experiment and subsequently taking a sample for the TSS measurement. This approach requires some additional work as more TSS tests need to be performed but it ensures a reliable measurement of the diluted concentration.

- **Protocol**
1. Perform Step 1 from the protocol to measure the batch curves with a sludge sample from the recycle flow.
2. Combine a certain volume of the sludge sample with the effluent until the required dilution is obtained.
3. Perform steps 2 to 4 from the protocol to measure the batch curves.
4. After 45 min of settling, homogenise the sample in the cylindrical reservoir again and take a sample to determine the sludge concentration with a TSS test.

- **Example**

Samples were collected from the recycle flow and the effluent at the WWTP in Destelbergen (Belgium). The sludge/water interface during settling was measured for different initial concentrations (Table 6.3). The resulting settling curves are shown in Figure 6.7.

Table 6.3 Measured sludge blanket height (in m) during batch settling tests at different initial concentrations.

Time	1.37 g L^{-1}	2.37 g L^{-1}	3.42 g L^{-1}	4.10 g L^{-1}	5.46 g L^{-1}	6.83 g L^{-1}
0	0.248	0.248	0.248	0.248	0.248	0.248
0.5	0.243	0.244	0.246	0.248	0.247	0.248
1	0.215	0.236	0.241	0.247	0.246	0.248
2	0.107	0.198	0.214	0.244	0.245	0.248
3	0.074	0.163	0.186	0.242	0.243	0.247
4	0.064	0.144	0.165	0.239	0.243	0.246
5	0.059	0.130	0.149	0.234	0.241	0.245
10	0.046	0.102	0.115	0.195	0.234	0.241
15	0.041	0.091	0.102	0.172	0.227	0.239
20	0.038	0.083	0.092	0.156	0.219	0.236
30	0.033	0.073	0.083	0.132	0.205	0.231
45	0.031	0.064	0.074	0.114	0.182	0.223

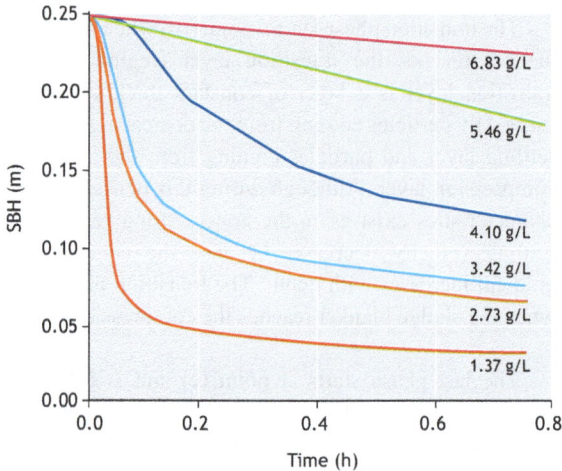

Figure 6.7 Batch settling curves at different initial concentrations.

The hindered settling velocities for the data in Figure 6.7 are computed by determining the steepest slope between three consecutive data points (Figure 6.8A). The resulting velocities are presented in Figure 6.8 B and Table 6.4.

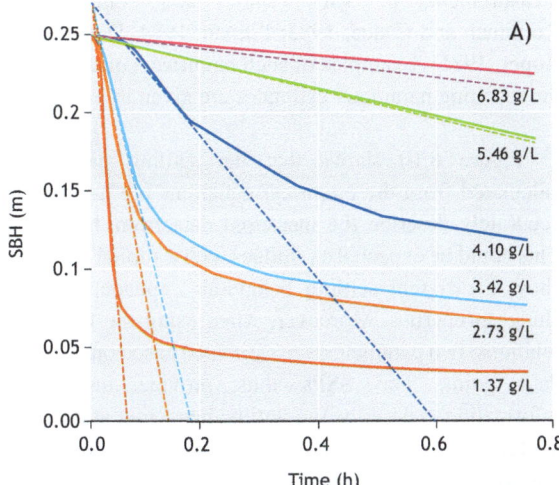

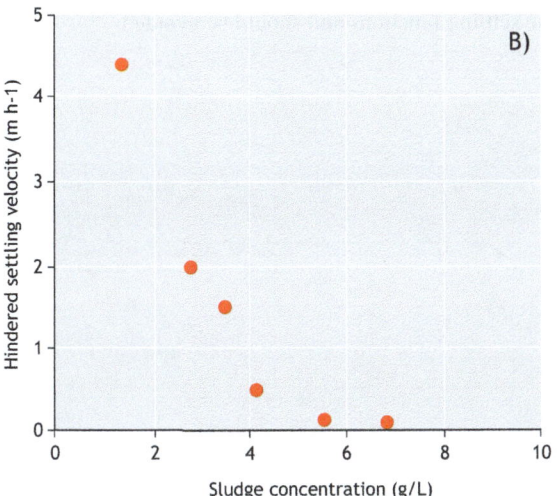

Figure 6.8 (A) Batch settling experiments at different initial solids concentration indicating the maximum slope for each curve. (B) The maximal slope represents a measurement of the hindered settling velocity.

The hindered settling velocity slows down at higher concentrations because the settling particles will be increasingly hindered by surrounding particles. Note that for the concentrations 5.46 g L^{-1} and 6.83 g L^{-1}, it becomes increasingly difficult to determine the steepest slope and the validity of these curves to measure hindered settling may be questioned. More information can be found in Section 6.2.4.3.

Table 6.4 Measured hindered settling velocities at different initial concentrations.

Concentration (g L^{-1})	v_{hs} (m h^{-1})
1.37	4.39
2.73	2.01
3.42	1.53
4.10	0.46
5.46	0.09
6.83	0.05

6.2.3.4 Determination of the zone settling parameters

Mathematically, the relation between the sludge concentration and the zone settling velocity can be described by an exponential decaying function (Eq. 6.3) (Vesilind, 1968). In this equation v_{hs} represents the hindered settling velocity of the sludge, V_0 the maximum settling velocity, X_{TSS} the solids concentration and r_V a model parameter. The parameters V_0 and r_V in this function provide information on the sludge settleability and are frequently used in SST design procedures. More information on design procedures can be found in Ekama et al. (1997).

$$v_{hs}(X) = V_0 \cdot e^{-r_V \cdot X_{TSS}} \qquad \text{Eq. 6.3}$$

The parameters V_0 and r_V can be estimated from the experimental data by minimising the Sum of Squared Errors (SSE) in Eq. 6.4 In this equation N is the number of data points, $v_{hs,i}$ the measured hindered settling velocity at concentration i, and $\tilde{v}_{hs,i}$ is the corresponding prediction by the function of Vesilind (1968) for a particular parameter set $[V_0\, r_V]$.

$$\text{SSE} = \sum_{i=1}^{N} \left(v_{hs,i} - \tilde{v}_{hs,i}(V_0, r_V)\right)^2 \qquad \text{Eq. 6.4}$$

An estimation of the zone settling parameters and calculation of the confidence intervals for these estimates can be performed as explained in Chapter 5. Minimising the SSE from Eq. 6.4 will give more weight to the fit of high settling velocities (i.e. at low concentrations). In order to give equal weight to all the measured settling velocities, a logarithmic fit can be performed.

- **Example**

Table 6.5 provides the initial parameter estimates, the optimal parameters after optimisation and the 95 % confidence intervals for the estimated parameters for the data in Table 6.4. The simulation results of calibrated functions vs. the experimental data points are shown in Figure 6.9.

Table 6.5 Initial values, optimal values and confidence intervals of the estimated parameters for the settling function.

Parameter	Initial value	Optimal value	Confidence interval
V_0 (m h^{-1})	9.647	10.608	± 1.265
r_V (L g^{-1})	0.488	0.634	± 0.038

6.2.3.5 Calibration by empirical relations based on SSPs

As the measurement of batch settling curves is much more time-consuming than the measurements of simple SSPs, several empirical equations have been developed that relate the settling parameters V_0 and r_V to simple measurements of SSPs (Härtel and Pöpel, 1992; Koopman and Cadee, 1983; Pitman, 1984, Daigger and Roper, 1985). Examples of such empirical equations and the resulting parameter estimates are given in Table 6.6.

Figure 6.10 shows that the settling parameters calculated from the empirical equations are not able to accurately describe the measured data from Table 6.4. This could be expected as sludges with a similar SVI may show a different settling behaviour dependent on the sludge properties. Moreover, when using the empirical relations, two parameters are estimated based on only one data point. The SSPs thus provide insufficient information to describe the settling behaviour at different sludge concentrations. For these reasons, the use of empirical relations based on SSPs is not an accurate method to estimate the hindered settling parameters of the settling functions and should be avoided.

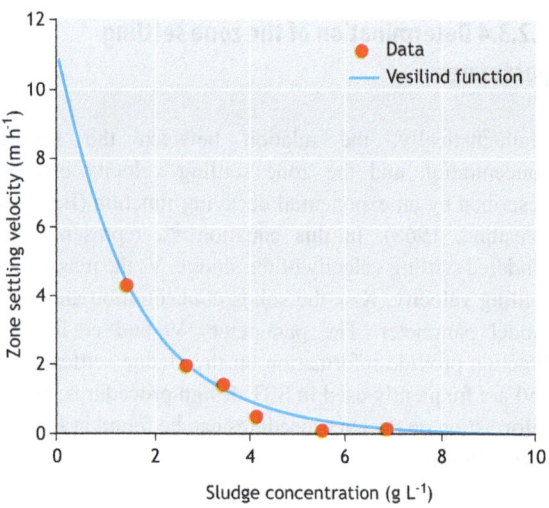

Figure 6.9 Settling velocity as a function of the solids concentration. The circles represent measured settling velocities and the line the calculated settling velocities after calibration of the function by Vesilind (1968).

Table 6.6 Estimated values for V_0 (m h^{-1}) and r_V (L g^{-1}) by empirical relations based on SSPs.

Reference	Equations	Parameter values	Equation nr.
Härtel and Pöpel (1992)	$V_0 = 17.4\, e^{-0.0113\, SVI}$	$V_0 = 9.647$	Eq. 6.5
	$r_V = -0.9834\, e^{-0.00581\, SVI} + 1.043$	$r_V = 0.488$	Eq. 6.6
Koopman and Cadee (1983)	$\ln(V_0) = 2.605 - 0.00365\, DSVI$	$V_0 = 9.993$	Eq. 6.7
	$r_V = 0.249 + 0.002191\, DSVI$	$r_V = 0.431$	Eq. 6.8
Pitman (1984)	$\dfrac{V_0}{r_V} = 67.9\, e^{-0.016\, SSVI_{3.5}}$	$V_0 = 5.669$	Eq. 6.9
	$r_V = 0.88 - 0.393\, \log\left(\dfrac{V_0}{r_V}\right)$	$r_V = 0.446$	Eq. 6.10

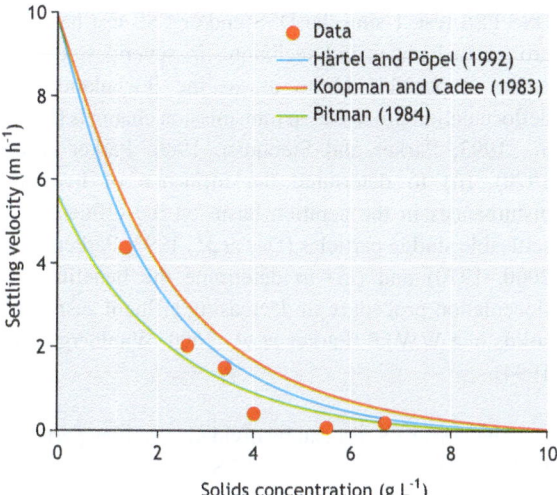

Figure 6.10 Settling velocity as a function of the solids concentration. The circles represent the measured settling velocities from the batch settling tests and the lines represent the settling velocities calculated by the function of Vesilind (1986) with parameter values based on empirical equations.

6.2.4 Recommendations for performing batch settling tests

6.2.4.1 Shape and size of the batch reservoir

It is recommended to use a cylindrical reservoir for batch settling tests. In conical reservoirs, such as Imhoff cones, no zone settling region can be measured because the reducing width of the cross section will inevitably cause a concentration gradient. When the goal of the measurements is to use the zone settling parameters for a qualitative analysis of the limiting flux and the SST's surface area, care should be taken to avoid any influence of the settling reservoir on the settling behaviour (so-called wall effects). This can be achieved by performing the batch test in a column of at least 100 mm in diameter and 1 m deep, and by gently stirring (1 rpm) the sample during settling.

6.2.4.2 Sample handling and transport

Long-distance transport and prolonged storage of the sample should be avoided. Agitation of the sample during transport and biological activity during storage may severely influence the settling behaviour. If possible, perform the measurements on-site at the WWTP immediately after sampling.

6.2.4.3 Concentration range

Hindered settling occurs typically between concentrations of 1 g L^{-1} and 6 g L^{-1}. However, these limits are dependent on the flocculation state of the sludge and may be different between WWTPs. Therefore, there should always be a visual check on whether the recorded batch settling curves are within the hindered settling region. At low initial concentrations, it becomes increasingly difficult to track the solid-liquid interface as the sludge enters the discrete settling regime. If no distinct interface can be observed, this concentration should not be considered in the analysis. On the other hand, at high concentrations, sludge starts to undergo compression. If the initial concentration is high enough for compression to occur at the very beginning of the experiment, the batch curve will no longer show a clear linear descent. If no linear decrease in the batch curve can be seen, the high concentration should also not be considered in the analysis.

6.2.4.4 Measurement frequency

The measurement times and dilution series described in sections 6.2.2 and 6.2.3 can be considered as a minimum set of measurements for a batch curve. However, more frequent measurement times or additional dilutions can be added depending on the case-specific conditions or requirements. For example, if the solid-liquid interface shows a very rapid initial increase then additional measurements can be recorded during the first 2 min of sampling. If the appropriate specialised equipment and experience is available then the solid-liquid interface may even be tracked automatically (Vanderhasselt and Vanrolleghem, 2000). Depending on the concentration of the recycle flow sample, the standard dilution series (100, 80, 60, 50, 40 and 20 %) may not provide sufficient coverage of the concentration range for hindered settling. Additional dilutions such as 55 %, 30 % etc. can be added.

6.2.5 Recent advances in batch settling tests

By measuring a batch curve, the velocity at which the suspension-liquid interface passes through different settling regions can be investigated. However, the drawback of this method is that it only provides information on the suspension-liquid interface. No information on the settling behaviour inside the sludge blanket or the actual build-up of the sludge blanket is recorded. Nor does it allow the settling velocity in the compression stage to be calculated. More advanced measurement techniques aiming to provide more detailed information on the sludge settling behaviour over time and depth have been presented in dedicated literature. Examples include detailed spatio-temporal solids concentration measurements by means of a radioactive tracer (De Clercq et al., 2005) and velocity measurements throughout the depth of the sludge blanket with an ultrasonic transducer (Locatelli et al., 2015). However, these techniques require specialised equipment and cannot be routinely performed.

6.3 MEASURING FLOCCULATION STATE OF ACTIVATED SLUDGE

As can be seen from Figure 6.1, the settling behaviour of a sludge sample is not only influenced by its concentration but also by its flocculation state. Hence, the ability of an SST to act successfully as a clarifier is highly dependent on the potential of the microorganisms to form a flocculent biomass which settles and compacts well, producing a clear effluent (Das et al., 1993). The aim of the flocculation process is to combine individual flocs into large and dense flocs that settle rapidly and to incorporate discrete particles that normally would not settle alone. If the flocculation process or breakup of flocs fails during the activated sludge process, a fraction of the particles is not incorporated into flocs. They remain in the supernatant of the SST due to lack of sufficient mass and are carried over the effluent weir, reducing the effluent quality. Failure of the settling process can have multiple causes: (*i*) denitrifying sludge, (*ii*) excessively high sludge blankets, (*iii*) poor flocculation, or (*iv*) poor hydrodynamics. In order to take appropriate remedial actions, it is important to be able to pinpoint the cause of the failure. Denitrifying sludge and high sludge blankets can be easily recognised and corrected (Parker et al., 2000). Distinguishing between flocculation problems and poor hydrodynamics is more challenging but can be accomplished by means of a Dispersed Suspended Solids/Flocculated Suspended Solids (DSS/FSS) test (Wahlberg et al., 1995)

6.3.1 DSS/FSS test

6.3.1.1 Goal and application

Wahlberg et al. (1995) proposed a procedure that makes it possible to distinguish between hydraulic and flocculation problems in a given SST, the so-called DSS/FSS test. Using the DSS and/or FSS test has been proven to be a useful technique in several studies: it makes it possible (*i*) to assess the flocculation and deflocculation processes in transmission channels (Das et al., 1993; Parker and Stenquist, 1986; Parker et al., 1970), (*ii*) to determine the influence of hydraulic disturbances in the aeration basin on the effluent non-settleable sludge particles (Das et al., 1993; Parker et al., 2000, 1970) and (*iii*) to determine the benefits of a flocculation procedure in decreasing effluent suspended solids in a WWTP (Parker et al., 2000; Wahlberg et al., 1994).

The DSS/FSS test can be divided into three parts: the Effluent Suspended Solids (ESS) test, the Dispersed Suspended Solids (DSS) test, and the Flocculated Suspended Solids (FSS) test. The ESS test consists of a simple TSS test to determine the effluent concentration. The procedures for the DSS and FSS test are provided in sections 6.3.1.3 and 6.3.1.4.

6.3.1.2 Equipment

The following equipment is needed for the execution of ESS, DSS and FSS tests:

General
a. Equipment for the TSS test (APHA et al., 2012).
b. A stopwatch.

ESS test
a. An effluent sample (minimum 0.5 L).

DSS test
b. A Kemmerer sampler.
c. A siphon for supernatant sampling.

FSS test
a. A square flocculation jar of at least 2.0 L.
b. Six paddle stirrers.
c. An activated sludge sample (minimum 1.5 L).

6.3.1.3 DSS test

Dispersed Suspended Solids (DSS) are defined as the concentration of SS remaining in the supernatant after 30 min of settling (Parker et al., 1970). The DSS test thus quantifies an activated sludge's state of flocculation at the moment and location the sample is taken. This is accomplished by the use of a single container (i.e. a Kemmerer sampler) for sampling and settling in order to protect the biological flocs in the sample from any secondary flocculation or breakup effects caused by an intermediate transfer step.

A Kemmerer sampler is a clear 4.2 L container, 105 mm in diameter and 600 mm tall with upper and lower closures (Figure 6.11). The sample is collected in the Kemmerer sampler and allowed to settle for 30 min, after which time the supernatant is sampled using a siphon and analysed for SS concentration (Wahlberg et al., 1995).

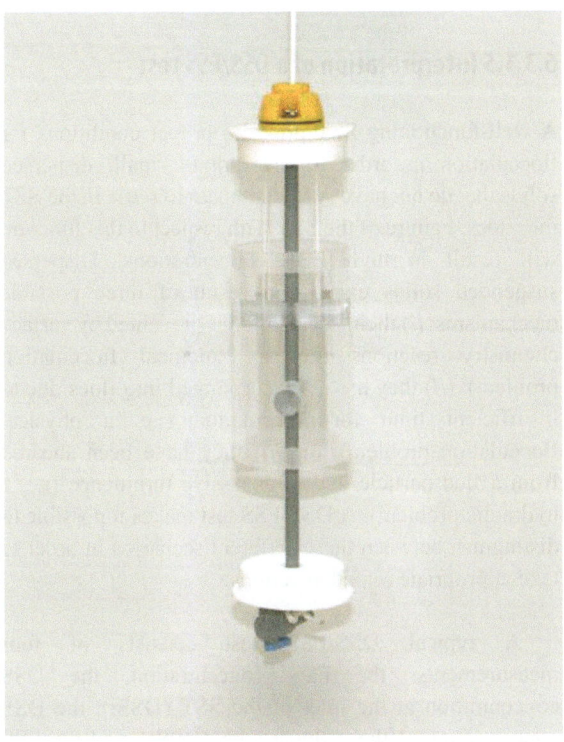

Figure 6.11 A Kemmerer sampler with open upper and lower closures (photo: Royal Eijkelkamp).

The large, settleable flocs settle in the 30 min period, whereas the dispersed, primary particles not incorporated in the settling sludge remain in the supernatant. DSS concentrations have been shown to closely approximate the ESS concentration for a well-designed and operated SST (Parker and Stenquist, 1986). Hence, large deviations between ESS and DSS indicate clarification problems. DSS tests can be performed with samples at several locations in the SST (for example: at the SST inlet, at the outlet of the flocculation well, or near the effluent weir) to analyse where potential problems (and for instance flocculation or breakup) are occurring.

- **Protocol**
1. Immerse the Kemmerer sampler at the desired location in the SST to grab a sample.
2. Allow this sample to settle for 30 min.
3. After 30 min sample 500 mL of the supernatant (be careful not to disturb the settled sludge).
4. Analyse the sampled supernatant for SS concentration.

- **Example**

Parker et al. (2000) illustrated the use of a DSS test for SST failure troubleshooting in the case of the Central Marin Sanitation Agency plant in California. A DSS test was performed at the plant as high ESS values were being observed during peak flows. The results of the DSS test are provided in Table 6.7.

Table 6.7 Measured DSS values at the Central Marin Sanitation Agency plant in California (Parker et al., 2000).

DSS (mg L^{-1}) Inlet centre well	DSS (mg L^{-1}) Outlet centre well	DSS (mg L^{-1}) Effluent weir	ESS (mg L^{-1}) Effluent weir
10.4	11.0	3.6	8.5

From these results, it became clear that no flocculation was occurring in the centre well as the DSS values at the inlet and outlet of the centre well are similar. However, the significant reduction in DSS values between the outlet of the centre well and the effluent weir showed that flocculation was occurring in the sedimentation tank. This indicated that the sludge had a good flocculation tendency but merely lacked proper conditions for flocculation in the existing centre well, which was contributing to its small diameter.

Moreover, the DSS results uncovered a significant hydraulic problem in the clarifiers. The high ESS concentration compared to the DSS at the effluent weir signifies the wash out of settleable solids over the effluent weir of the clarifier.

6.3.1.4 FSS test

Wahlberg et al. (1995) developed a complimentary test to the DSS test called the Flocculated Suspended Solids (FSS) test. Whereas the DSS test assesses the state of flocculation of a sample at a specific location in the SST at a moment in time, the FSS test quantifies the flocculation potential of an activated sludge sample by flocculating the sample under ideal conditions prior to settling.

For this test, an activated sludge sample is collected in a square flocculation jar (minimum jar volume 2 L). The sample volume should be at least 1.5 L. The sample is gently stirred for 30 min at 50 rpm (Figure 6.12) before it is allowed to settle for 30 min and the concentration in the supernatant is measured. Flocculation is maximized by stirring, and settling is performed in an ideal device (without hydraulic disturbances). Hence, the measured FSS is considered to be the minimal possible ESS.

Because the FSS concentration is measured under conditions of maximum flocculation and ideal settling, it will not change between the aeration basin and the SST. Therefore the activated sludge sample for the FSS test can be collected anywhere between the aeration basin and the SST.

- **Protocol**
1. Collect an activated sludge sample of 1.5 L from the WWTP.
2. Pour this sample into a square flocculation jar with a volume of 2 L.
3. Stir the sample for 30 min. at 50 rpm.
4. Stop stirring and allow the sample to settle for 30 min.
5. After 30 min, sample 500 mL of the supernatant (being careful not to disturb the settled sludge).
6. Analyse the sampled supernatant for suspended solids concentration.

6.3.1.5 Interpretation of a DSS/FSS test

A well-functioning SST provides proper conditions for flocculation in order to incorporate small, dispersed solids that do not have sufficient mass to settle in the SST into flocs. Failure of the SST with respect to this function will result in high ESS concentrations. Dispersed suspended solids exist as a result of three possible mechanisms; (*i*) their flocculation is prevented by surface chemistry reactions (i.e. a biological flocculation problem), (*ii*) they are not incorporated into flocs due to insufficient time for flocculation (i.e. a physical flocculation problem), or (*iii*) they have been sheared from a floc particle due to excessive turbulence (i.e. a hydraulic problem). A DSS/FSS test makes it possible to distinguish between these different scenarios in order to take appropriate remedial actions.

A typical DSS/FSS test consists of four measurements: the ESS concentration, the DSS concentration at the inlet of the SST (DSSi), the DSS concentration at the effluent weir (DSSo) and the FSS concentration. If a flocculation well is present, then DSSi should be measured after this structure. Assuming that the system under study is struggling with high ESS concentrations, because DSS/FSS tests are typically performed in the case of clarification failure, four scenarios can be defined. A DSS/FSS troubleshooting matrix which shows the cause of the poor performance under the various testing scenarios is provided in Table 6.8 (Kinnear, 2000).

Figure 6.12 Experimental setup for the FSS test. An activated sludge sample is stirred for 30 min in a square flocculation jar (photo: E. Torfs).

Table 6.8 DSS/FSS troubleshooting matrix (Kinnear, 2000).

ESS high and:		FSS	
		High	Low
DSS	High	Biological flocculation	Physical flocculation
	Low	Not possible	Hydraulics

The different testing scenarios can be interpreted as follows.

High DSSi - low FSS

This scenario indicates poor flocculation in the SST even though the activated sludge has good flocculating properties. Either the activated sludge is not receiving adequate time to flocculate or significant floc breakup is occurring prior to settlement (for example, by excessive shear in a conveyance structure). The clarification failure can be contributing to a flocculation problem of a physical nature that can be solved by either removing the cause of breakup or by incorporating an additional flocculation step prior to settling.

High DSSi - high FSS

As in the high DSSi - low FSS case, these results indicate poor flocculation in the SST. However, even under the ideal flocculation circumstances provided by the FSS test, the clarification cannot be improved. Hence, additional flocculation will not improve the clarification and the problem is most likely of a biological nature resulting in a sludge with poor flocculation properties. Modifications to the SST will not solve this problem; attention must be directed upstream of the SST.

Low DSSi - low FSS

The low DSSi suggests that the incoming activated sludge is in a well-flocculated state. As the DSSi is already in the same range as the FSS concentration, further flocculation will not improve the SST performance and the problem is most likely a hydraulic one (for example, due to short-circuiting). Comparing the DSSo and ESS concentration can provide further confirmation; if the DSSo concentration is significantly lower than the ESS concentration, then hydraulic scouring of settleable solids from the sludge blanket is indicated. To improve the clarification in this case, the tank's hydrodynamics need to be investigated by means of dye tests and/or 2-3D CFD (computational flow dynamics) modelling.

Low DSSi - high FSS

This outcome is theoretically not possible. Should it occur it is recommended to repeat the test.

- **Example**

A DSS/FSS test was used to assess the performance of the existing clarifiers prior to plant expansion at the Greeley Water Pollution Control Facility in Colorado (Brischke *et al.*, 1997; Parker *et al.*, 2000). The DSS test was performed at two locations i.e. at the inlet to the SST and at the effluent weir. The measured DSS, FSS and ESS values are shown in Table 6.9.

Table 6.9 Measured DSS, FSS and ESS values at the Greeley Water Pollution Control Facility in Colorado (Brischke *et al.*, 1997).

DSSi (mg L^{-1})	DSSo (mg L^{-1})	FSS (mg L^{-1})	ESS (mg L^{-1})
29.2	22.0	8.2	25.5

High DSS values both at the inlet and near the effluent weir indicate that no flocculation is occurring in the tank. The much lower FSS value signifies that the sludge has a high potential to flocculate but lacks appropriate conditions for flocculation in the tank. The high ESS concentrations can thus be attributed to a physical flocculation problem that can be solved by physically modifying the tank in order to provide suitable flocculation conditions. In this specific case this was accomplished through modifications of the centre well.

6.3.2 Recommendations

6.3.2.1 Flocculation conditions

Proper execution of an FSS test requires ideal flocculation conditions. Therefore, it is important to use a square flocculation jar in order to avoid the formation of a vortex during mixing. Moreover, make sure that the sample is completely mixed (i.e. no dead zones at either the top or bottom of the flocculation jar).

6.3.2.2 Temperature influence

The sample volume for the FSS test is relatively small in comparison to the volume of the Kemmerer sampler. Hence, some precautions should be taken to ensure that the samples do not change drastically in temperature during the 1 h it takes to conduct the test. For example: do not perform the test in direct sunlight.

6.3.2.3 Supernatant sampling

Regardless of the specific supernatant sampling technique, care should be taken not to pull any floating debris or settled solids into the supernatant sample as this can severely alter the results. Moreover, as the concentration in effluent and supernatant is generally very low, the sampled volume for the TSS test should be sufficiently large ($\pm$ 500 mL).

6.3.3 Advances in the measurement of the flocculation state

From the above it becomes clear that 'good bioflocculation' of the activated sludge is a prerequisite for 'good sedimentation' and a good effluent quality. Hence, what is the definition of a well-flocculated activated sludge floc? An activated sludge floc is composed of: (*i*) a backbone of filamentous organisms, onto which, (*ii*) microcolonies (i.e. clusters of micro-organisms) can attach and this aggregation of micro-organisms is then embedded in a matrix of (*iii*) extracellular polymer substances (EPS) (Figure 6.13). Whenever one of these components is not in balance with the rest then problems might be encountered.

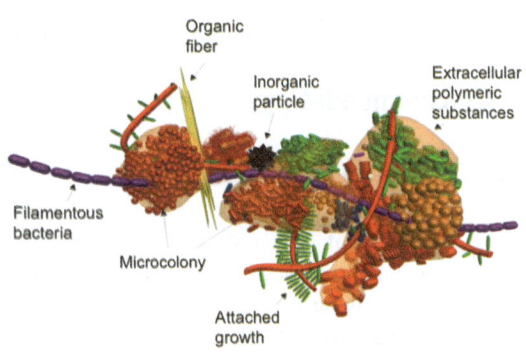

Figure 6.13 The structural makeup of an activated sludge floc: microcolonies attach to filamentous bacteria which form the backbone of the floc, while extracellular polymeric substances constitute the embedding matrix (Nielsen *et al.*, 2012).

One of the most common settling problems is that of filamentous bulking where there is a dominance of filamentous organisms that will make the floc structure very open and will retain a lot of water. Such filaments might even entangle with other protruding filaments so that a network is formed that prevents the sludge from settling. In contrast, when there are not enough filaments to form the backbone, so-called pinpoint flocs are observed. These are small clusters of micro-colonies that do not settle well. Hence, a good floc that settles well should be dense (not open) and sufficiently large.

Such characteristics of activated sludge can be quantified by microscopic image analysis. Changes in the average floc size, filament length, floc roundness, floc fractal dimension etc. can reveal a lot of information on the settling behaviour of sludge. While some research groups have been developing specific image analysis software to infer this information (Amaral and Ferreira, 2005; Da Motta *et al.*, 2001a, 2001b; Jenneé *et al.*, 2004; van Dierdonck *et al.*, 2013), also freeware software is available (such as ImageJ or FIJI (http://fiji.sc/Fiji)) to perform a basic analysis. A comprehensive overview of what is currently available in this image analysis domain is available in dedicated literature (Mesquita *et al.*, 2013). For microscopic monitoring that is more focused on revealing the specific microbial communities present, FISH analysis can be interesting; the reader is referred to the FISH handbook for biological wastewater treatment (Nielsen *et al.*, 2009).

As well as image analysis, three additional bioflocculation-related monitoring tools can be mentioned that are more focused on the forces that hold the floc components together. On the one hand, the global floc strength measurement (Mikkelsen and Keiding, 2002) compares the turbidity of the supernatant before and after shearing of a sludge sample. Lower turbidity after shearing indicates better bioflocculation. On the other hand, relative hydrophobicity and surface charge can be measured. The relative hydrophobicity is related to hydrophobic interactions that prove to be important in keeping the floc aggregated. The value that results from such an analysis (e.g. the MATH test, Chang and Lee, 1998) which assesses the microbial adhesion tendency to hydrocarbons) should not be taken as an absolute value but can be interesting in revealing changes over a certain operational period. The surface charge is related to electrostatic forces; with a more neutral activated sludge surface, the aggregates experience less repulsion resulting in improved coagulation and flocculation. The measurement of surface charge is based on a colloid titration technique (Kawamura and Tanaka, 1966; Kawamura *et al.*, 1967; Morgan *et al.*, 1990).

6.4 MEASURING THE SETTLING BEHAVIOUR OF GRANULAR SLUDGE

6.4.1 Goal and application

In recent years new technologies have been developed to improve the separation of sludge from the treated effluent. One of these technologies is the use of aerobic granular sludge. Aerobic granules are spherical biofilms with a typical shape factor of 0.7 - 0.8 (Beun *et al.*, 2002). Whereas conventional activated sludge is characterized by settling velocities below 5 m h^{-1} (Vanderhasselt and Vanrolleghem, 2000), granular sludge settles significantly faster with settling velocities in the range of 10 up to 100 m h^{-1} (Bassin et al., 2012; Etterer and Wilderer, 2001; Winkler et al., 2012, 2011a). Moreover, granules show a low flocculating tendency positioning their settling behaviour at the far left side of Figure 6.1. The granules will thus settle independent even at higher concentrations (with almost no hindered and compression regime present) and directly form a compact sludge bed. For this reason the SVI after 5 minutes will be approximately the same as the SVI measured after 30 minutes. Typical SVI values for granules are less than 30 ml/g (de Kreuk and van Loosdrecht, 2004; Liu et al., 2005; Liu and Tay, 2007; Tay et al., 2004).

Granular sludge is developed in Sequencing Batch Reactors (SBR) as these systems fulfil a number of specific requirements for the formation of granules: a feast - famine regime for the selection of appropriate microorganisms (Beun et al., 1999), short settling times to ensure retention of granular biomass and wash-out of flocculent biomass (Qin et al., 2004) and sufficient shear force to ensure an optimal physical granule integrity (Tay et al., 2001).

One of the most important parameters to select for granular sludge is the settling velocity. By applying short settling times in an SBR, only large biomass aggregates that settle well are selected, while flocculent sludge is washed out (Beun et al., 2000). The parameters determining the settling velocity of particles and in turn biomass washout are of crucial importance to granular sludge technology. The balances of forces for the sedimentation of a spherical particle depend on the buoyancy, gravity and drag force (Giancoli, 1995). From this relation, the settling velocity is influenced by the water viscosity, particle size and shape, and the difference between the density of the water and the particles. Hence, the settling velocity of granules is influenced by the density and size of the particles where an increase in diameter affects the settling velocity more severely (Winkler et al., 2012, 2011a). Therefore, this section presents a method to measure the density and size of granules as well as a procedure to calculate the theoretical settling velocity of granules under different temperature conditions.

6.4.2 Equipment

For granular sludge tests the following equipment is needed:

a. A pycnometer.
b. A microbalance.
c. Sieves or a microscope with an image analyser.

6.4.3 Density measurements

Granule densities between 1,036 - 1,048 kg m^{-3} have been reported (Etterer and Wilderer, 2001), which are comparable to densities of conventional activated sludge (1,020-1,060 kg m^{-3}) (Andreadakis, 1993; Dammel and Schroeder, 1991). However, precipitates may form within the granule core (Lee and Chen, 2015; Mañas et al., 2012) and significantly increase the granule density up to 1,300 kg m^{-3} (Juang et al., 2010; Winkler et al., 2013).

The specific biomass density can be measured with a pycnometer (Figure 6.14). A pycnometer is a simple and inexpensive glass flask, with an exactly calibrated volume. The pycnometer flask is closed with a glass stopper that acts as a valve; it contains a small groove, through which excess water is forced out when closed. The pycnometer has a known volume V.

Figure 6.14 Picture of a pycnometer (photo: E. Torfs).

- **Protocol**
1. Measure the weight of the pycnometer (closed with the glass stopper) in a completely dry state (m_0).
2. Fill the pycnometer with water and measure its weight again (m_T). It is very important to dry the pycnometer carefully before the weight is determined in order to remove all excess water from the outside of the pycnometer.

3. Calculate the mass of water in the pycnometer (m_{H2O}) as:

$$m_{H2O} = m_T - m_0 \qquad \text{Eq. 6.11}$$

4. Measure the weight of the granule sample for which you want to determine the density (m_s).
5. Place the granule sample inside the pycnometer and determine the weight of the pycnometer together with the inserted sample ($m_0 + m_s$).
6. Fill the pycnometer (containing the solids sample) further with water and weigh its mass again (m_{TS}). As in Step two, make sure that the pycnometer is dried carefully before the weight is determined.
7. Calculate the weight of the added water (m'_{H2O}) as:

$$m'_{H_2O} = m_{TS} - m_0 - m_s \qquad \text{Eq. 6.12}$$

8. Determine the volume of added water (V'_{H2O}) according to:

$$V'_{H_2O} = \frac{m'_{H_2O}}{\rho_{H_2O}} \qquad \text{Eq. 6.13}$$

9. Calculate the volume of measured solids V_s from the difference between the volume of water that fills the empty pycnometer (V) and the previously determined volume of water (V'_{H2O}).

$$V_s = V - V'_{H_2O} = \frac{m_{H_2O} - m'_{H_2O}}{\rho_{H_2O}} \qquad \text{Eq. 6.14}$$

10. Finally, calculate the density of the granules ρ_s as:

$$\rho_s = \frac{m_s}{V_s} \qquad \text{Eq. 6.15}$$

- **Example**

A pycnometer with a volume (V) of 0.1 L is used to determine the density of a granular sludge sample. All the measurements and calculations are provided in Table 6.10 (the density of water, ρ_{H_2O}, at 20 °C is 998 g L^{-1}).

Table 6.10 Density calculation for a granular sludge sample.

Variable	Symbol	Procedure	Value
Mass empty pycnometer (g)	m_0	Measured	54.51
Mass pycnometer and water (g)	m_T	Measured	153.70
Mass water (g)	m_{H2O}	Calculated	99.19
Mass solids (g)	m_s	Measured	19.56
Mass pycnometer, solids and water (g)	m_{TS}	Measured	154.55
Mass added water (g)	m'_{H2O}	Calculated	80.48
Volume added water (L)	V'_{H2O}	Calculated	0.08
Volume solids (L)	V_s	Calculated	0.02
Density solids (g L^{-1})	ρ_s	Calculated	1,010.40

6.4.4 Granular biomass size determination

Although there is no common consensus on the minimum diameter (Bathe et al., 2005), sieves with a diameter of 0.2 mm have been used to determine the minimum size of granular biomass (Bin et al., 2011; de Kreuk, 2006; Li et al., 2009). The largest diameter reported is 16 mm (Zheng et al., 2006), but typically diameters range between 0.5-3 mm (de Kreuk and van Loosdrecht, 2004; Shi et al., 2009; Winkler et al., 2011b). The size distribution can be measured either by simple sieving tests or by means of an image analyser.

6.4.4.1 Sieving

Granule size can be determined by means of sieves with different mesh sizes. The screening can be performed with sieves with mesh openings of, i.e. 2.0, 1.0, 0.5 and 0.3 mm, making it possible to cover the most common granule size range.

Figure 6.15 Stacked sieves with different mesh openings (photo: Fieldmaster).

- **Protocol**
1. Measure the total wet weight of a sample.
2. Mount the sieves vertically one on top of the other in increasing order of mesh opening (from bottom to top) so that the coarsest mesh is at the top (Figure 6.15).
3. Pour the granule sample onto the sieves.
4. Wash each sieve successively to allow the granules to move from one sieve to the next.
5. Filter the liquid (which has trickled through all the sieves) in order to collect particles smaller than 0.3 mm.
6. Backwash each sieve to retain each granule fraction in a separate beaker.
7. Determine the wet weight, and if needed the dry weight (TSS), the ash content and the VSS of each fraction, which will result in the theoretical percentage of each class size (Laguna et al., 1999).

6.4.4.2 Image analyser

Alternatively, the granule size can be measured by means of an image analyser using the averaged projected surface area of the granules. For this method, a sample is transferred into a petri dish and placed under a stereo microscope with a fixed magnification (e.g. 7.5 × magnification). Each image analysed is recorded by the image analyser. The Petri dish needs to be turned multiple times in order to measure different granules. Different image analysers are available on the market and each requires different handling. An example of granule size distribution data, created during the measuring procedure, is presented in Figure 6.16.

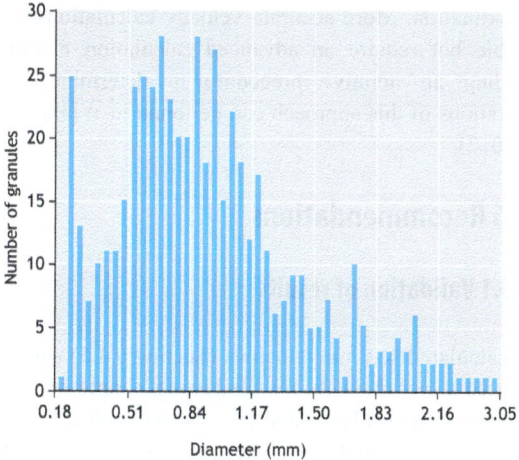

Figure 6.16 Particle size distribution of a granular sludge reactor.

6.4.5 Calculating the settling velocity of granules

- **Protocol**

The measured average density and diameter of granules can be used to calculate the theoretical settling velocity. For particle Reynolds numbers smaller than or equal to 1, Stokes' law can be used to calculate the settling velocity of a particle.

$$v_s = \frac{g}{18} \cdot \frac{\rho_p - \rho_w}{\rho_w} \cdot \frac{d_p^2}{v_w}.$$ Eq. 6.16

The Reynolds numbers can be calculated with the following equation:

$$Re = d_p \cdot \frac{v_s}{v_w}$$ Eq. 6.17

Where, v_s is the sedimentation velocity of a single particle (m s^{-1}), d_p is the particle diameter (m), ρ_p is the density of a particle (kg m^{-3}), ρ_w is the density of the fluid (kg m^{-3}), g is the gravitational constant (9.81 m s^{-2}), v_w is the kinematic viscosity of water (m^2 s^{-1}), and Re$_p$ is the particle Reynolds number.

The density and viscosity of the medium depend on the temperature and the solutes present in the water. With increasing temperature, the viscosity and density of the water decrease. At high temperature, water molecules are more mobile than at low temperature, resulting in a decrease of viscosity by a factor of two between 10 and 40 °C (Podolsky, 2000). A table with density and viscosity values of water at different temperatures can be found elsewhere.

- **Example**

An example of a calculation for the settling velocity of a particle with a diameter of 0.4 mm and 1,010 kg m^{-3} at different temperatures is given in Table 6.11 and plotted in Figure 6.17. Note that the condition of Reynolds numbers smaller than or equal to 1 is not met for every temperature. The implications of this will be discussed further on in this section.

Table 6.11 Settling velocity of a granule with a particle diameter of 0.4 mm and a particle density of 1,010 kg m^{-3} at different temperatures.

T	°C	5	10	15	20	25	30	35	40
v_w	m^2 s^{-1}	1.5e-06	1.3e-06	1.1e-06	1.0e-06	9.4e-07	8.2e-07	7.4e-07	6.6e-07
ρ_w	kg m^{-3}	1,000	1,000	999	998	997	996	994	992
Re	-	0.2	0.2	0.2	0.4	0.5	0.7	1.0	1.5
v_s	m h^{-1}	2.1	2.5	3.0	3.6	4.2	5.5	6.8	8.7

Table 6.12 Settling velocity of a granule with a particle diameter of 0.4 mm and a particle density of 1,050 kg m^{-3} at different temperatures.

T	°C	5	10	15	20	25	30	35	40
v_w	m^2 s^{-1}	1.5e-06	1.3e-06	1.1e-06	1.0e-06	9.4e-07	8.2e-07	7.4e-07	6.6e-07
ρ_w	kg m^{-3}	1,000	1,000	999	998	997	996	994	992
Re	-	0.8	1.0	1.4	1.7	2.1	2.8	3.6	4.7
v_s	m h^{-1}	10.7	12.1	14.0	15.8	17.7	20.9	23.9	27.9

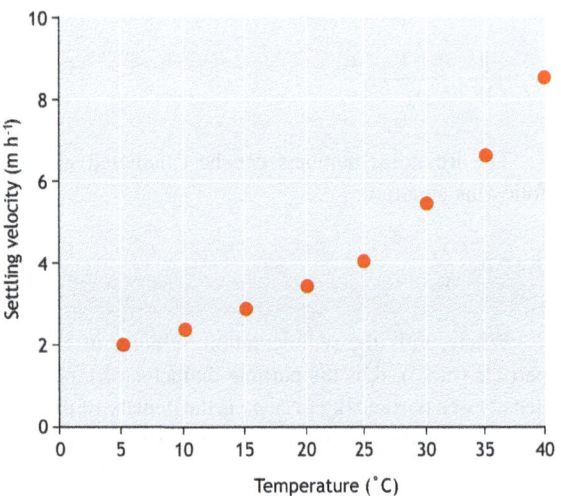

Figure 6.17 Calculated settling velocities at different temperatures for granules with a diameter of 0.4 mm and a density of 1,010 kg m^{-3}.

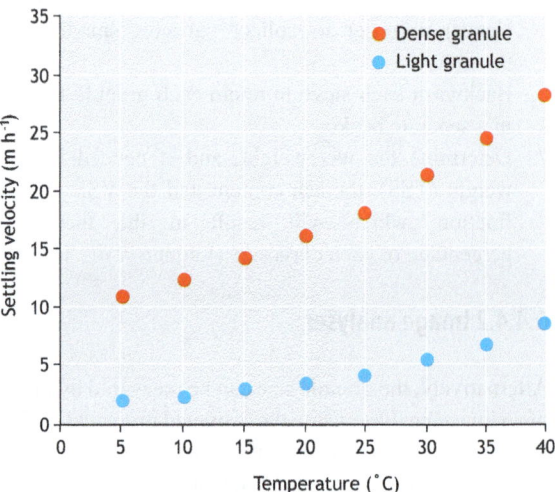

Figure 6.18 Calculated settling velocities at different temperatures for a light (1,010 kg m^{-3}) and dense (1,050 kg m^{-3}) granular sludge particle at a diameter of 0.4 mm.

The example in Figure 6.17 shows that the settling velocity of a small granule with a diameter of 400 μm and a density of 1,010 kg m^{-3} varies between 2 m h^{-1} and 9 m h^{-1} for temperatures ranging between 5 – 40 °C (Winkler et al., 2012). At low temperatures the separation of small and light granules from flocs (with typical settling velocities below ± 5 m h^{-1}) can therefore become troublesome. Earlier research has experimentally proven that a start-up process at cold temperatures is difficult. In addition all microbial processes run slower at low temperatures (Brdjanovic et al., 1997; Kettunen and Rintala, 1997; Lettinga et al., 2001), hence limiting granulation at a lower temperature even further. An increase in density or diameter of the granules will significantly increase the settling velocity and thus facilitate the separation.

- **Example**

Calculated settling velocities for a particle with the same diameter (0.4 mm) but a higher density (1,050 kg m^{-3}) at different temperatures are shown in Table 6.12 and Figure 6.18.

Larger and denser particles result in particle Reynolds numbers larger than 1. For these particles the theoretical settling velocities calculated according to Stokes' law will deviate from the true settling velocities. For most applications, these errors (< 10 %) are acceptable and Stokes' law remains a good approximation. More accurate velocity calculations are possible but require an advanced calculation method including an iterative procedure to determine Re. Illustrations of this approach can be found in Winkler et al. (2012).

6.4.6 Recommendations

6.4.6.1 Validation of results

The calculated settling velocities (Section 6.4.5) can be validated experimentally by conducting simple settling tests with granules harvested from sieves with different mesh sizes (Section 6.4.4). The different size fractions can be poured into the reactor column itself or into a cylinder and the time is measured until the granule reaches the bottom of the reactor in order to express its settling velocity in meters per hour.

6.4.6.2 Application for flocculent sludge

The experimental methods described in sections 6.4.3 and 6.4.4 can also be applied to activated sludge in SSTs (particularly in the top region of SSTs where discrete settling is known to occur). However, this type of application is currently not very common because the high flocculation potential of sludge causes attention to be mainly directed towards hindered settling (as this is the dominant regime at the concentration where activated sludge enters the SST and has been mainly used to determine SSTs' capacity and design). For granular sludge on the other hand, the dominant settling regime is discrete thus causing the focus to shift to size and density measurements.

6.5 MEASURING SETTLING VELOCITY DISTRIBUTION IN PSTs

6.5.1 Introduction

Primary settling tanks (PSTs) are used as a pre-treatment in WWTPs. Measurement campaigns conducted since the early 1970s on urban discharges have clearly shown that many pollutants occur in particulate form. Moreover, particles transported in suspension have also emerged as highly settleable despite their relatively small particle size (30 to 40 µm). Hence, gravitational separation in PSTs can serve as a valuable tool to separate coarse, settleable particles in the raw wastewater prior to further treatment in the biological reactors.

As concentrations of particulate matter in raw wastewater are relatively low compared to concentrations in SSTs, the settling regime in PSTs has a discrete (non-flocculent and flocculent) nature and the settling velocity depends on the individual properties of particles. Given the variety of densities, shapes and sizes of suspended particles in wastewaters (have in mind Stokes' law), it is challenging and time-consuming to calculate the settling velocities of different particles from size and density measurements. Therefore, this section presents a method to directly measure the distribution of settling velocities in representative wastewater samples. This measurement can be performed using the ViCAs protocol, developed by Chebbo and Gromaire (2009). The ViCAs ('Vitesse de Chute en Assainissement', which is French for 'settling velocity in sanitation') is a test to measure the settling velocity of particles in a column under static conditions. The test provides insight into the behaviour of particles present in a wastewater sample in order to obtain an idea about its composition. As such it can serve as important information in different application domains. In primary settling tanks, results from ViCAs experiments can be used as input to primary clarifier models (Bachis *et al.*, 2015) or to study the effect of chemical dosage on the settling velocity distribution in order to improve chemically enhanced primary treatment (CEPT) performance. Furthermore, ViCAs experiments can be applied in the design of combined sewer retention tanks where knowledge of the settling velocity distribution can be used to determine an optimal HRT and corresponding load reduction to the treatment plant.

6.5.2 General principle

A ViCAs test is performed by settling a sample in a ViCAs column (Figure 6.19).

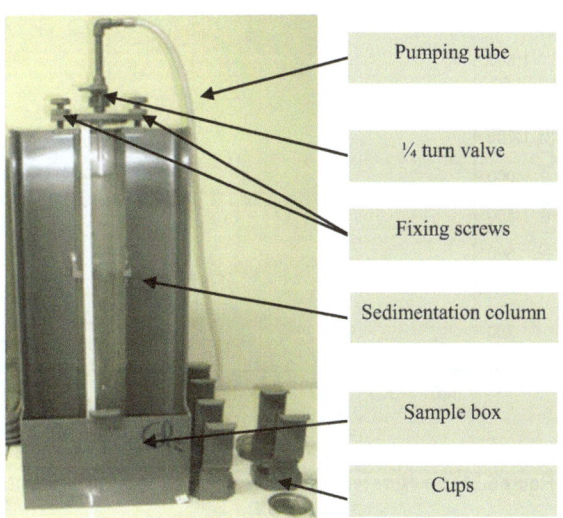

Figure 6.19 The ViCAs test equipment (photo: Chebbo and Gromaire, 2009).

The ViCAs protocol is based on the principle of homogeneous suspensions (Figure 6.20). At the beginning of the measurement, the solids are uniformly distributed over the whole sedimentation height. Then the particles are assumed to settle independently of each other, without forming aggregates and without diffusion. The solids, after having settled for a predetermined period of time, are recovered at the bottom of the sedimentation column in cups.

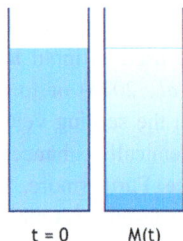

Figure 6.20 Principle of the homogeneous suspension.

Their mass is thus recovered from the cups, which allows the evolution of the cumulative mass M(t) of settled material as a function of time t to be determined (Figure 6.21).

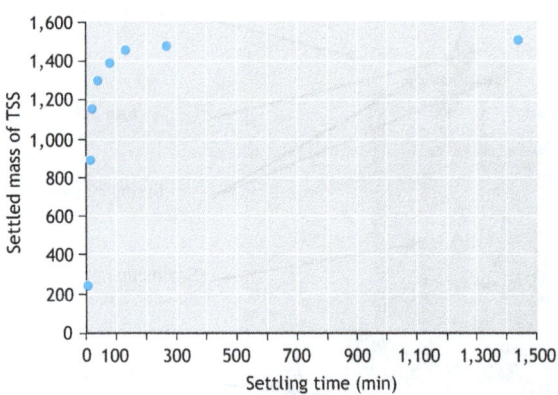

Figure 6.21 Cumulative evolution of the settled mass as a function of time.

In practice, the cumulative curve of settled mass consists of n points (7 < n < 12), corresponding to n samples taken after different settling times.

The measurements of the settled mass as a function of time make it possible to calculate the settling velocity distribution $f(v_s)$. Figure 6.22 shows an example of a settling velocity distribution curve for a typical wastewater sample.

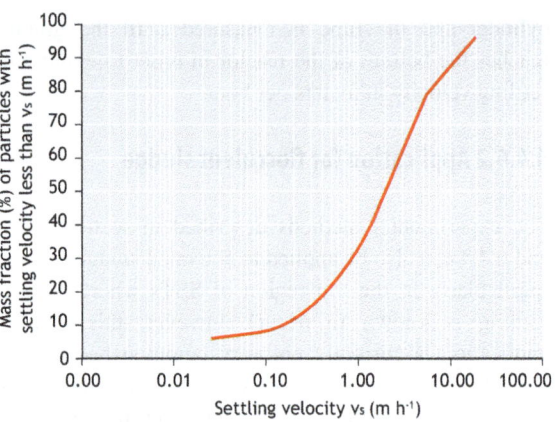

Figure 6.22 Settling velocity distribution curve $f(v_s)$ for a typical wastewater sample.

6.5.3 Sampling and sample preservation

The test can be carried out using a composite sample or by mixing several samples with similar composition (e.g. harvest 5 bottles of 1 L over an interval of 10 min). During sampling, suspended solids that may disturb the test are removed and, if necessary, the sample can be directly filtered with a coarse filter, which does not affect the sample composition. The analyses must be made within a maximum of 24 hours after collection, to avoid flocculation. If the initial concentration exceeds 1,000 mg L^{-1} then it is necessary to dilute the sample. To perform the dilution, the sample is split into two, so as to have 5 L on the one hand, and 15 L on the other. From the 15 L sample, 5 L of supernatant are withdrawn after 24 h of settling, and used to perform the dilution.

6.5.4 Equipment

For the execution of ViCAs protocol following equipment is needed:
1. A ViCAs column with its support and two associated cups.
2. A rubber band to hold the column in place.
3. A timer (to measure the time steps).
4. A beaker of 5 L.
5. A spatula (for homogenisation).
6. A vacuum pump.
7. A plastic connection tube.
8. A filtration Erlenmeyer of 1,000 mL to create a vacuum buffer and protect the pump.

SETTLING TESTS

Figure 6.23 shows the ViCAs equipment and the filling system ready for use.

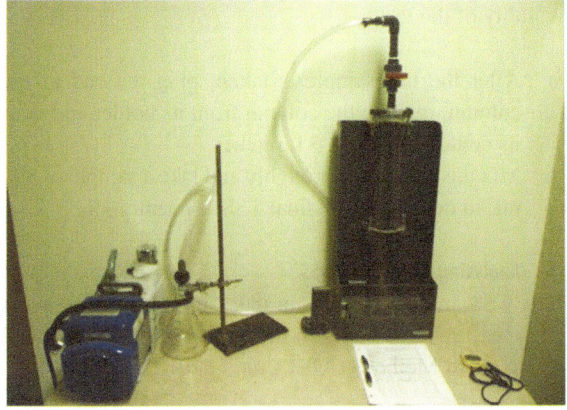

Figure 6.23 ViCAs apparatus with filling equipment (photo: Gromaire M.C. and Chebbo G., ViCAs manual).

6.5.5 Analytical protocol

The sample to be analysed is poured into a sample box (Figure 6.19) and is rapidly aspirated by vacuum within a column. It is then maintained under vacuum for the whole duration of the test, i.e. the sample is hanging in the column. The particles settled during certain periods of time Δt are collected in cups placed under the column; both the cups and column have the same diameter. The cups are first filled with tap water and then each in turn immersed in the sample box and moved under the column. At the end of each particular period Δt, the content of the cups is filtered and the TSS and VSS of the recovered solids are measured.

- **Sample preparation**
a. Homogenize the sample using the spatula and pour 5 L into an appropriate beaker.
b. Stir again and take 1 sample of 500 mL that will be used to determine the initial TSS concentration in the column.

- **Filling the column**
a. Mix the sample of 4.5 L before it is poured quickly into the sample box (Figure 6.24A).
b. Suck the liquid into the column (in 2 to 5 seconds) and then close the valve by a ¼ turn (Figure 6.24B)
c. Stop the vacuum pump.

It is important to note that:
a. The filling phase requires some training because it has to be done very quickly.
b. For a more successful test, it is better to have two operators.
c. An insufficient volume of the sample or closing a valve too late may lead to air leakage into the column. The filling will then have to be repeated.
d. The use of a protective device such as a Woulff bottle is indispensable.

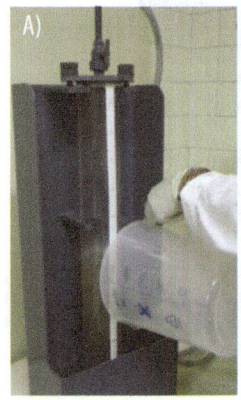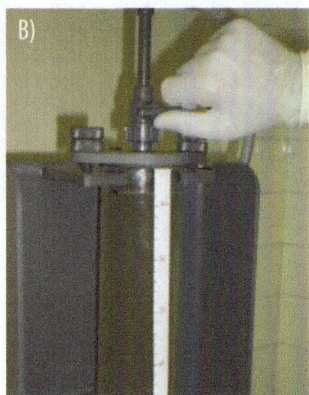

Figure 6.24 Filling the box with a sample (A) and closing the valve towards the vacuum (B) (photo: Gromaire M.C. and Chebbo G., ViCAs manual).

- **Start of the settling test**
a. Immediately after closing the valve with a ¼ turn, slide the first cup under the column: gently place the cup in the sample box, and slide it under the base of the column.
b. Start the timer and disconnect the pumping equipment.
c. Place a piece of adhesive tape to measure the height of water in the column at the end of the test (there may be a variation in water height due to the exchange of cups).

- **Changing the cups**
a. Ten seconds before the time of change, gently introduce a water-filled cup into the groove (Figure 6.25).
b. Slide the two cups gently to the new position and remove the old cup (Figure 6.26).

It is important to note that:
a. For a full analysis the change of cup is carried out at 2 min, 6 min, 14 min, 30 min, 1 h, 2 h, 4 h and over 22 h for a total of 8 cups.
b. This step is tricky because one must not lose the content of the cup and one must minimize any turbulence caused by moving the cups below the column.

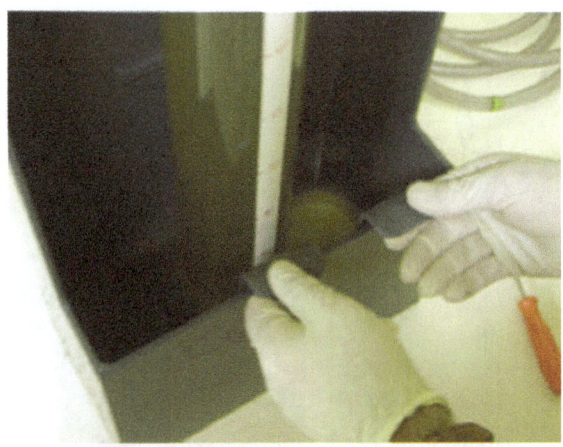

Figure 6.25 Introducing a new water-filled cup (photo: Gromaire M.C. and Chebbo G., ViCAs manual).

Figure 6.26 Moving the cups to their new positions (photo: Gromaire M.C. and Chebbo G., ViCAs manual).

- **Determination of the final concentration**

The final concentration in the column is determined by collecting the total volume of the column and analysing its content. This step makes it possible to perform a mass balance check, which is essential to determine the validity of the test.

a. After the last sample is taken, plug the end of the column, remove the column from its holder and pour its contents into the 5 L pitcher.
b. Mix the contents thoroughly and take a sample of 500 mL to determine the final TSS concentration.

- **Analysing the TSS and VSS**

The TSS and VSS of the initial, final and cup contents are measured according to methods 2540 D and 2540 E of Standard Methods (APHA *et al.*, 2012).

6.5.6 Calculations and result presentation

6.5.6.1 Mass balance check

A mass balance calculation is performed to estimate losses (or gains) of solids during the experiment and thus to assess the quality of the measurement.

The percentage mass balance error E (%) can be calculated as follows:

$$E = \frac{M_{ini} - (M_{set} + M_{fin})}{M_{ini}} \quad \text{Eq. 6.18}$$

Where, M_{ini} is the initial mass in the column (mg), M_{fin} is the final mass in the column (mg), and M_{set} is the sum of the masses recovered in the cups (mg).

6.5.6.2 Calculation of the settling velocity distribution

A theoretical analysis (Chebbo and Bachoc, 1992; Chancellor *et al.*, 1998) shows that the cumulative curve M(t) can be written as:

$$M(t) = S(t) + t\frac{dM(t)}{dt} \quad \text{Eq. 6.19}$$

Where, M(t) is the cumulated mass of particles settled to the bottom of the column between t = 0 and t, S(t) is the mass of particles settled between t = 0 and t that have a settling velocity above H/t, with H the water height in the column, $t\frac{dM(t)}{dt}$ is the mass of particles settled at time t that have a settling velocity below H/t (and thus initially located at a height in the water column less than H).

In order to obtain the settling velocity distribution of the sample, it is necessary to determine the curve S(t) that can subsequently be transformed into the cumulative settling velocity distribution $f(v_s)$ (Figure 6.22).

Practically, a continuous function M(t) is numerically fitted to the measured values $M(t_i)$, and then used to analytically solve Eq. 6.20.

The following expression can be used for M(t):

$$M(t) = \frac{b}{1+\left(\frac{c}{t}\right)^d} \qquad \text{Eq. 6.20}$$

Where b, c and d are three numerical parameters that can be determined by the least squares method. The following constraints must be respected: $0 < b \leq M_{init}$, $c > 0$ and $0 < d < 1$. An example of the result of fitting the curve is given in Figure 6.27.

From which follows:

$$f(v_s) = 100 \cdot \left(1 - \frac{S(t)}{M_{set} + M_{fin}}\right) \text{ with } v_s = \frac{H}{t} \qquad \text{Eq. 6.22}$$

6.5.6.3 Recommendations

- **Manipulations**

Careful manipulations will allow mass balance errors below 10% to be achieved. An error exceeding 15 % should lead to an invalidation of the ViCAs analysis.

- **Sample frequency**

The sample times are those generally used to make the TSS fractionation of the sample. They may be subject to change depending on the project and the type of sample. Also, the user is free to change the Δt used, as long as a minimum interval of 7 and a maximum of 15 intervals is respected.

- **Reproducibility**

Reproducibility tests can be performed in order to confirm the results for a single ViCAs test. The success of the test is dependent on the meticulousness of the person in following the protocol and handling the equipment. Especially important is the changing of the cups below the column and the filtration of the recovered masses. Figure 6.28 shows the results of a reproducibility test on primary effluent from the Québec-Est wastewater treatment plant.

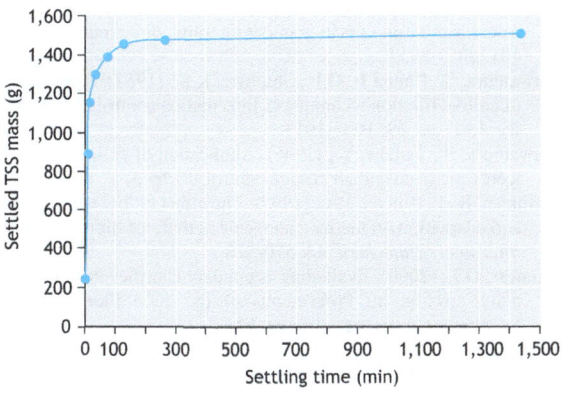

Figure 6.27 Example of a curve fitted to a cumulative series of settled masses.

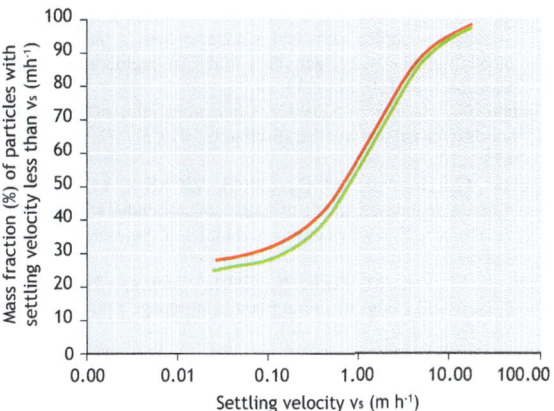

Figure 6.28 ViCAs reproducibility test.

The fitted curve M(t) can then be used to calculate S(t):

$$S(t) = M(t) - t \cdot \frac{dM(t)}{dt} = \frac{b \cdot \left(1+(1-d)\cdot\left(\frac{c}{t}\right)^d\right)}{\left(1+\left(\frac{c}{t}\right)^d\right)^2} \qquad \text{Eq. 6.21}$$

References

Amaral, A., Ferreira, E., (2005). Activated sludge monitoring of a wastewater treatment plant using image analysis and partial least squares regression. *Anal. Chim. Acta.* 544: 246-253.

Andreadakis, A.D., (1993). Density of activated sludge solids. *Water Res.* 27: 1707-1714.

Bachis, G., Maruéjould, T., Tik, S., Amerlinck, Y., Melcer, H., Nopens, I., Lessard, P., Vanrolleghem, P.A., 2015. Modelling and characterisation of primary settlers in view of whole plant and resource recovery modelling. *Wat. Sci. Tech.* 72: 2251-2261.

Bassin, J.P., Winkler, M.-K.H., Kleerebezem, R., Dezotti, M., van Loosdrecht, M.C.M., (2012). Improved phosphate removal by selective sludge discharge in aerobic granular sludge reactors. *Biotechnol. Bioeng.* 109: 1919–1928.

Bathe, S., de Kreuk, M.K., McSwain, B.S., Schwarzenbeck, N., (2005). Discussion Outcomes, In Aerobic Granular Sludge. IWA Publishing. ISBN 978-1843395096.

Beun, J.J., Hendriks, A., van Loosdrecht, M.C.M., Morgenroth, E., Wilderer, P.A., Heijnen, J.J., (1999). Aerobic granulation in a sequencing batch reactor. *Water Res.* 33: 2283-2290.

Beun, J.J., van Loosdrecht, M. c. M., Heijnen, J.J., (2002). Aerobic granulation in a sequencing batch airlift reactor. *Water Res.* 36: 702-712.

Beun, J.J., van Loosdrecht, M.C.M., Heijnen, J.J., (2000). Aerobic granulation. *Water Sci. Technol.* 41: 41–48.

Bin, Z., Zhe, C., Zhigang, Q., Min, J., Zhiqiang, C., Zhaoli, C., Junwen, L., Xuan, W., Jingfeng, W., (2011). Dynamic and distribution of ammonia-oxidizing bacteria communities during sludge granulation in an anaerobic-aerobic sequencing batch reactor. *Water Res.* 45: 6207-6216.

Brdjanovic, D., van Loosdrecht, M.C.M., Hooijmans, C.M., Alaerts, G.J., Heijnen, J.J., (1997). Temperature effects on physiology of biological phosphorus removal. *J. Environ. Eng.* 123: 144–153.

Brischke, K., Wahlberg, W., Dingeman, T., Schump, D., (1997). Performance quantified: the impact of final clarifier performance on effluent quality, in: Procs. 70th Annual WEF Conference and Exposition. 237-244.

Chancelier, J., Chebbo, G., Lucas-Aiguier, E., (1998). Estimation of settling velocities. *Water Res.* 32: 3461-3471.

Chang, I., Lee, C., (1998). Membrane filtration characteristics in membrane-coupled activated sludge system - the effect of physiological states of activated sludge on membrane fouling. *Desalination.* 120: 221–233.

Chebbo, G., Bachoc, A., (1992). Characterization of suspended solids in urban wet weather discharges. *Wat. Sci. Tech.* 25: 171-179.

Chebbo, G., Gromaire, M.-C., (2009). VICAS - An Operating protocol to measure the distributions of suspended solid settling velocities within urban drainage samples. *J. Environ. Eng.* 135: 768-775.

da Motta, M., Pons, M., Roche, N., (2001a). Automated monitoring of activated sludge in a pilot plant using image analysis. *Water Sci. Technol.* 43: 91-96.

da Motta, M., Pons, M., Roche, N., Vivier, H., (2001b). Characterization of activated sludge by automated image analysis. *Biochem. Eng. J.* 9: 165-173.

Daiger, G.T., Roper, E.T., (1985). The relationship between SVI and activated-sludge settling characteristics. *J. Water Poll. Control Fed.* 57: 859-866.

Dammel, E.E., Schroeder, E.D., (1991). Density of activated sludge solids. *Scanning.* 25: 841–846.

Das, D., Keinath, T.M., Parker, D., Wahlberg, E.J., (1993). Floc breakup in activated sludge plants. *Water Environ. Res.* 65, 138-145.

De Clercq, J., Jacobs, F., Kinnear, D.J., Nopens, I., Dierckx, R. a, Defrancq, J., Vanrolleghem, P.A., (2005). Detailed spatio-temporal solids concentration profiling during batch settling of activated sludge using a radiotracer. *Water Res.* 39: 2125-2135.

De Clercq, J., Nopens, I., Defrancq, J., Vanrolleghem, P.A., (2008). Extending and calibrating a mechanistic hindered and compression settling model for activated sludge using in-depth batch experiments. *Water Res.* 42: 781-791.

de Kreuk, M.K., (2006). Aerobic Granular Sludge: Scaling up a new technology. T.U. Delft.

de Kreuk, M.K., van Loosdrecht, M.C.M., (2004). Selection of slow growing organisms as a means for improving aerobic granular sludge stability. *Water Sci. Technol.* 49.

Dick, R.I., Vesilind, P.A., (1969). The Sludge Volume Index - What is it? *J. WPCF.* 41:1285–1291.

Eaton, A.D., Clesceri, L.S., Greenberg, A.E., (1995). Standard Methods for Examination of Water & Wastewater, 19th edition. American Public Health Association, Washington, DC.

Ekama, G.A., Barnard, J.L., Gunthert, F.W., Krebs, P., McCorquodale, J.A., Parker, D.S., (1997). Secondary Settling Tanks: Theory, Modelling, Design and Operation. International Association on Water Quality.

Etterer, T., Wilderer, P.A., (2001). Generation and properties of aerobic granular sludge. *Water Sci. Technol.* 43: 19-26.

Giancoli, D.C., (1995). Physics. Principles with applications. New Jersey: Prentice Hall. ISBN 9780130606204.

Härtel, L., Pöpel, H.J., (1992). A dynamic secondary clarifier model including processes of sludge thickening. *Water Sci. Technol.* 25: 267-284.

Jenneé, R., Banadda, E., Smets, I., Van Impe, J., (2004). Monitoring activated sludge settling properties using image analysis. *Water Sci. Technol.* 50: 285.

Juang, Y.-C., Sunil, S.A., Lee, D.J., Tay, J.H., (2010). Stable aerobic granules for continuous-flow reactors: Precipitating calcium and iron salts in granular interiors. *Bioresour. Technol.* 101: 8051-8057.

Kawamura, S., Hanna Jr, G.P., Shumate, K.S., (1967). Application of colloid titration technique to flocculation control. *Am. Water Work. Assoc.* 59: 1003-1013.

Kawamura, S., Tanaka, Y., (1966). Application of colloid titration technique to coagulant dosage control. *Water Sew. Work.*

Kettunen, R.H., Rintala, J.A., (1997). The effect of low temperature and adaptation on the methanogenic activity of biomass. *Appl. Microbiol. Biotechnol.* 48: 570-576.

Kinnear, D.J., (2000). Evaluating Secondary Clarifier Performance and Capacity, in: Proceedings of the 2000 Florida Water Resources Conference, Tampa, FL.

Koopman, B., Cadee, K., (1983). Prediction of thickening capacity using diluted sludge volume index. *Water Environ. Res.* 17: 1427-1431.

Lee, D.-J., Chen, Y.-Y., (2015). Magnesium carbonate precipitate strengthened aerobic granules. *Bioresour. Technol.* 183: 136-140.

Lee, S.-E., Koopman, B., Bode, H., Jenkins, D., (1983). Evaluation of alternative sludge settleability indices. *Water Res.* 17: 1421-1426.

Lettinga, G., Rebac, S., Zeeman, G., (2001). Challenge of psychrophilic anaerobic wastewater treatment. *Trends Biotechnol.* 19: 363-370.

Li, X.-M., Liu, Q.Q., Yang, Q., Guo, L., Zeng, G.M., Hu, J.M., Zheng, W., (2009). {Enhanced aerobic sludge granulation in sequencing batch reactor by Mg^{2+} augmentation. *Bioresour. Technol.* 100: 64-67.

Liu, Q.S., Liu, Y., Tay, S., Show, K., Ivanov, V., Benjamin, M., Tay, J.H., (2005). Startup of Pilot-Scale Aerobic Granular

Sludge Reactor by Stored Granules. *Environ. Technol.* 26: 1363-1370.

Liu, Y.-Q., Tay, J.-H., (2007). Cultivation of aerobic granules in a bubble column and an airlift reactor with divided draft tubes at low aeration rate. *Biochem. Eng. J.* 34: 1-7.

Locatelli, F., François, P., Laurent, J., Lawniczak, F., Dufresne, M., Vazquez, J., Bekkour, K., (2015). Detailed velocity and concentration profiles measurement during activated sludge batch settling using an ultrasonic transducer. *Separ. Sci. Technol.* 50: 1059-1065.

Mañas, A., Pocquet, M., Biscans, B., Sperandio, M., (2012). Parameters influencing calcium phosphate precipitation in granular sludge sequencing batch reactor. *Chem. Eng. Sci.* 77: 165-175.

Mancell-Egala, W., Kinnear, D., Jones, K., De Clippeleir, H., Takács, I., Murthy, S., (2016). Limit of stokesian settling concentration characterizes sludge settling velocity. *Water Res.* 90: 100-110.

Mesquita, D.P., Amaral, A.L., Ferreira, E.C., (2013). Activated sludge characterization through microscopy: a review on quantitative image analysis and chemometric techniques. *Anal. Chim. Acta* 802: 14-28.

Mikkelsen, L.H., Keiding, K., (2002). The shear sensitivity of activated sludge: an evaluation of the possibility for a standardised floc strength test. *Water Res.* 36: 2931-2940.

Mohlman, F.M., (1934). The Sludge Index. *Sewage Work. J.* 6: 119-122.

Morgan, J.W., Forster, C.F., Evison, L., (1990). A comparative study of the nature of biopolymers extracted from anaerobic and activated sludges. *Water Res.* 24: 743-750.

Nielsen, P.H., Daims, H., Lemmer, H., (2009). FISH Handbook for Biological Wastewater Treatment: Identification and quantification of microorganisms in activated sludge and biofilms by FISH. IWA Publishing. ISBN 9781843392316.

Nielsen, P.H., Saunders, A.M., Hansen, A.A., Larsen, P., Nielsen, J.L., (2012). Microbial communities involved in enhanced biological phosphorus removal from wastewater - a model system in environmental biotechnology. *Curr. Opin. Biotechnol.* 23: 452-459.

Parker, D.S., Kaufman, W.J., Jenkins, D., (1970). Characteristics of biological flocs in turbulent regimes. *SERL Report No. 70-5.* University of California, Berkeley, CA.

Parker, D.S., Stenquist, R.J., (1986). Flocculator-Clarifier performance. *J. Water Pollut. Control Fed.* 58: 214-219.

Parker, D.S., Wahlberg, E.J., Gerges, H.Z., (2000). Improving secondary clarifier performance and capacity using a structured diagnostics approach. *Water Sci. Technol.* 41: 201-208.

Pitman, A.R., (1984). Settling of nutrient removal activated sludges. *Water Sci. Technol.* 17; 493-504.

Podolsky, R.D., (2000). Temperature and water viscosity: Physiological versus mechanical effects on suspension feeding. *Science.* 265: 100-103.

Qin, L., Liu, Y., Tay, J.-H., (2004). Effect of settling time on aerobic granulation in sequencing batch reactor. *Biochem. Eng. J.* 21: 47-52.

Shi, X.-Y., Yu, H.-Q., Sun, Y.-J., Huang, X., (2009). Characteristics of aerobic granules rich in autotrophic ammonium-oxidizing bacteria in a sequencing batch reactor. *Chem. Eng. J.* 147: 102-109.

Stobbe, C.T., (1964). Über das Verhalten des belebten Schlammes in aufsteigender Wasserbewegung. Technischen Hochschule Hannover.

Tay, J., Pan, S., He, Y., Tay, S., (2004). Effect of Organic Loading Rate on Aerobic Granulation. I: Reactor Performance. *J. Environ. Eng.-ASCE* 130: 1094-1101.

Tay, J.-H., Liu, Q.-S., Liu, Y., (2001). The effects of shear force on the formation, structure and metabolism of aerobic granules. *Appl. Microbiol. Biotechnol.* 57: 227-233.

Van Dierdonck, J., den Broeck, R., Vansant, A., Van Impe, J., Smets, I., (2013). Microscopic image analysis versus sludge volume index to monitor activated sludge bioflocculation - a case study. *Sep. Sci. Technol.* 48: 1433–1441.

Vanderhasselt, A., Vanrolleghem, P.A., (2000). Estimation of sludge sedimentation parameters from single batch settling curves. *Water Res.* 34: 395-406.

Vesilind, P.A., (1968). Design of prototype thickeners from batch settling tests. *Water Sew. Work.* 115: 302-307.

Wahlberg, E.J., Keinath, T.M., Parker, D.S., (1994). Influence of activated sludge flocculation time on secondary clarification. *Water Environ. Res.* 66: 779-786.

Wahlberg, E.J., Merrill, D.T., Parker, D.S., (1995). Troubleshooting activated sludge secondary clarifier performance using simple diagnostic tests, in: Procs. 68*th* Annual WEF Conference and Exposition. 435–444.

White, M.J.D., (1976). Design and control of secondary settling tanks. *Water Pollut. Control.* 74: 459-467.

White, M.J.D., (1975). Settling of activated sludge. Technical Report TR11. Water Research Centre, England.

Winkler, M.K., Kleerebezem, R., Strous, M., Chandran, K., van Loosdrecht, M.C.M., (2013). Factors influencing the density of aerobic granular sludge. *Appl. Microbiol. Biotechnol.* 97: 7459-7468.

Winkler, M.K.H., Bassin, J.P., Kleerebezem, R., de Bruin, L.M.M., van den Brand, T.P.H., van Loosdrecht, M.C.M., (2011a). Selective sludge removal in a segregated aerobic granular biomass system as a strategy to control PAO-GAO competition at high temperatures. *Water Res.* 45: 3291-3299.

Winkler, M.K.H., Kleerebezem, R., Kuenen, J.G., Yang, J.J., van Loosdrecht, M.C.M., (2011b). Segregation of Biomass in Cyclic Anaerobic/Aerobic Granular Sludge Allows the Enrichment of Anaerobic Ammonium Oxidizing Bacteria at Low Temperatures. *Environ. Sci. Technol.* 45: 7330-7337.

Winkler, M.K.H., Bassin, J.P., Kleerebezem, R., van der Lans, R.G.J.M., van Loosdrecht, M.C.M., (2012). Temperature and salt effects on settling velocity in granular sludge technology. *Water Res.* 46: 3897-3902.

Zheng, Y.-M., Yu, H.Q., Liu, S.H., Liu, X.Z., (2006). Formation and instability of aerobic granules under high organic loading conditions. *Chemosphere.* 63: 1791-1800.

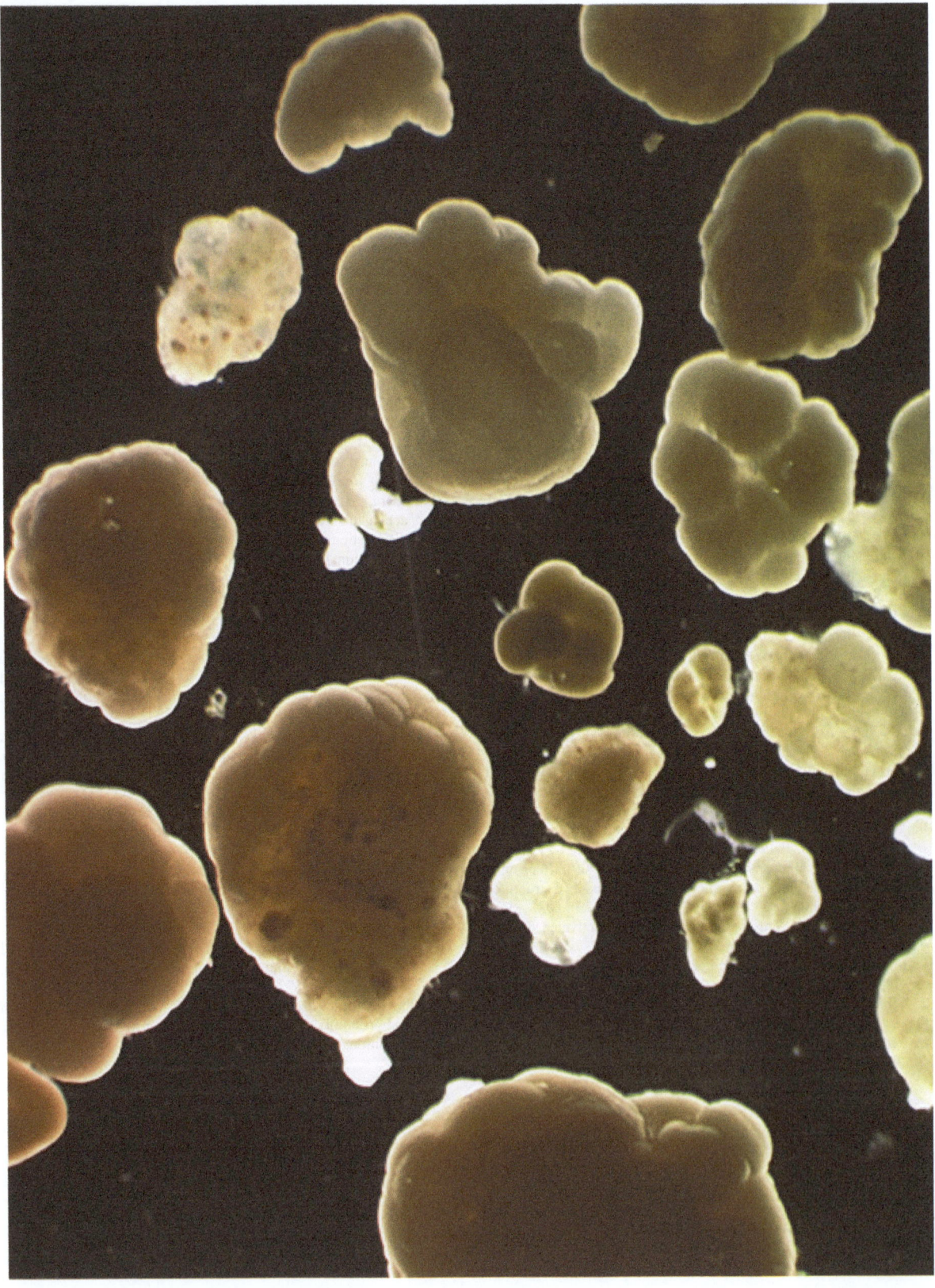

Figure 6.29 Granular sludge (photo: Beun *et al.*, 1999).

7

MICROSCOPY

Authors:
Jeppe L. Nielsen
Robert J. Seviour
Per H. Nielsen

Reviewer:
Jiři Wanner

7.1 INTRODUCTION

One widely applied method for studying the microbiology of activated sludge is microscopy. It provides an insight into the hidden world of microbes, which cannot be seen otherwise by the naked eye. In this chapter we describe protocols for the microscopic examination of activated sludge samples. These include staining techniques to give an understanding the huge taxonomic and functional diversity found among the microbes in activated sludge. Direct observations and staining can enable the differences between bacterial, fungal and protozoan populations to be distinguished. Studying these microorganisms effectively requires the correct use of the microscope to reveal differences in their shape and size and to diagnose cellular structures. Standard light microscopes with appropriate performance for routine purposes are available from several manufacturers, but using more sophisticated microscopic techniques can enhance the level of information generated.

In this chapter the basic principles of the light and fluorescence microscope are explained and methodologies for relevant staining techniques and data interpretation are provided. The aim of the experimental protocols outlined here is to serve as a user-friendly guide for cell morphological examination characterization, staining techniques (e.g. DAPI: 4',6-diamidino-2-phenylindole dihydrochloride, Neisser, Gram, Nile Blue) for detecting cell viability and intracellular accumulation of storage compounds including poly-hydroxy-alkanoates (PHA) and poly-phosphate granules (poly-P) and *in situ* identification of targeted microbial populations using Fluorescence *in situ* Hybridization (FISH). Combinations of these techniques provide powerful tools for elucidating metabolic features of cells at a single cell level. Combining FISH with staining techniques is often problematic, and so careful planning is required to ensure that the interpretation of this information is unequivocal. This chapter provides the basis for carrying out standard protocols.

7.2 THE LIGHT MICROSCOPE

The purpose of the microscope is to provide sufficient magnification to distinguish between the objects examined. The most commonly used microscope is a bright field microscope that projects a focussed beam of light onto the image on a glass slide. Nowadays, almost all these microscopes are (*i*) binocular (Figure 7.1), meaning that both eyes are used to view the object, making persistent use less tiring, and (*ii*) compound, where more than one lens system is used to achieve the required sample resolution.

ISBN: 9781789065275 (Paperback), ISBN: 9781780404752 (eBook). Published by IWA Publishing, London, UK.

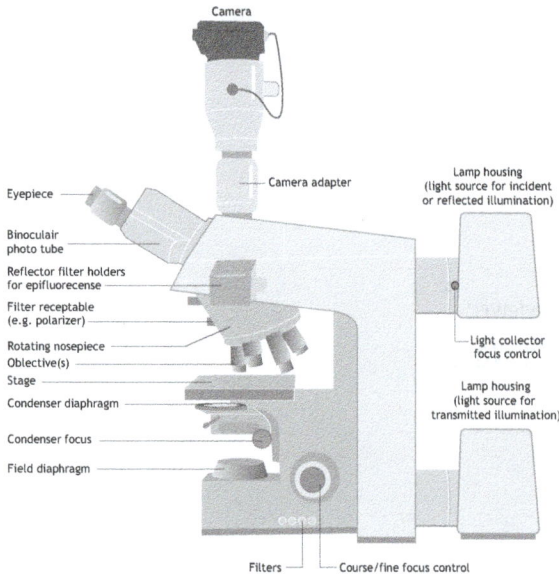

Figure 7.1 The compound microscope.

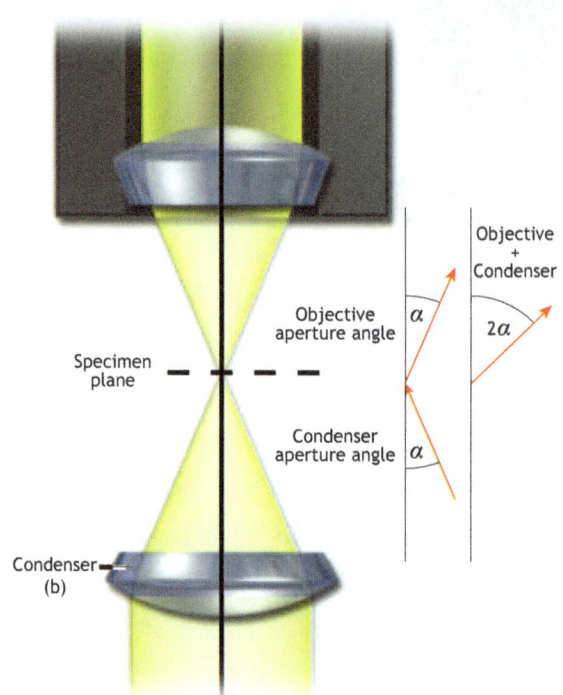

Figure 7.2 The light path in microscopy illustrating the concept of the numerical aperture and condensers.

The eyepiece (ocular) lenses and the objective lenses provide the resolving and magnifying power of the microscope, while the condenser lens system, by focussing the light source onto the specimen (Figure 7.2), maximises the resolution of the systems by increasing the numerical aperture or light-capturing ability of the objective lenses. Its position is critical for optimization of the performance, and therefore it should supply a cone of light capable of filling the objective lens aperture. The resolution reflects the ability of the microscope to discriminate between two closely positioned entities. If they are closer than the resolution distance then they appear blurred in the microscope image. Actual size, image size and magnification are related as follows:

Image size = Actual size · Magnification   Eq. 7.1

The distance between two distinct objects is referred to as the resolution d and its relationship to numerical aperture (NA) and wavelength (λ):

$$d = \frac{1.22 \cdot \lambda}{N \cdot \sin \alpha} \quad \text{or} \quad d = \frac{1.22 \cdot \lambda}{NA_{objective} + NA_{condenser}} \quad \text{Eq. 7.2}$$

Where,
λ is the light wavelength;
α equals one-half of the objective's opening angle and;
N is the refractive index of the immersion medium used below the objective lens.

At the highest magnifications used (100×), because it has a low light-capturing ability (high NA), it is necessary to increase the refractive index (RI) of the medium between the specimen and the objective lens. For light microscopes, the highest resolution is obtained when the condenser aperture angle matches the objective. Lenses in air as a medium cannot have NA that exceed 0.65, whereas in water and certain immersions such as oils the objective lens NA can be increased theoretically to ca. 1.515, and the resolution of the microscope is then approx. 0.2 µm, suitable for viewing/studying most bacteria, but not viruses.

Reflections and loss of illumination intensity can be eliminated by using immersion oil with a RI that matches the RI of the lens glass. The NA of an objective is also partially dependent upon the amount of correction for any optical aberration. Highly corrected objectives have much larger NA for the respective magnification (Table 7.1).

Table 7.1 Examples of objectives and their numerical aperture (NA) and optical correction.

Magnification	Plan Achromat (NA)	Plan Fluorite (NA)	Plan Apochromat (NA)
2×	0.06	0.08	0.10
4×	0.10	0.13	0.20
10×	0.25	0.30	0.45
20×	0.40	0.50	0.75
40×	0.65	0.75	0.95
40× (oil)	n/a	1.30	1.40
63×	0.75	0.85	0.95
63× (oil)	n/a	1.30	1.40
100× (oil)	1.25	1.30	1.40

7.2.1 Standard applications of light microscopy

The microscope must be set up and used properly to generate high quality images. The routine for setting up the microscope as described below is recommended. More experienced users may omit some of the steps in the procedure as they may have gained familiarity with the microscope and the samples under examination already. Microscopes are precision instruments, sensitive and expensive to repair or replace. Thus, they require careful handling and maintenance. The following advice should be observed:
- Avoid large temperature variations, store the equipment in a cool and dark place, and protect it from dust. Always use a dust cover when it is not in use.
- Objective lenses must be cleaned using a lens tissue. Do not dry wipe any lens as this may lead to scratching. Begin by blowing off dust or any loose material with a pressurized optical duster.
- Be careful not to spread residual immersion oil onto other oculars or clean objectives. Use a commercial cleaner or solvent (e.g. 70 % ethyl alcohol - EtOH) to remove any oil or grease. Flooding or placing solvents directly onto the lens is discouraged, and a lens tissue must always be used.
- Keep the microscope clean at all times.
- Objectives immersed in oil must be cleaned immediately after use by wiping with a lens tissue. If not, the oil may dry and thus need removal with a suitable lens cleaning solution. Do not use any solvents (e.g. alcohol) as this might lead to lens damage.

7.2.2 Low power objective

At the first setup of the microscope for optimal performance, one should always start using a low magnification, e.g. 10× objective or lower. The eyepieces should be adjusted to a position suitable for viewing with both eyes. Then the next steps should be followed:
- Lower the stage and insert the prepared slide so that the specimen is centred in the cone of light after passing through the condenser.
- Lower the lens until it is located a few millimetres from the slide surface. Look down the ocular while carefully turning the coarse focus knob until the image becomes visible, and then by adjustment with the fine focus knob bring the specimen image into sharp focus.
- Adjust the iris diaphragm to optimise the contrast to where the image is sharp. With the eyepiece taken out, check that approx. $2/3$ to $3/4$ of the back lens is light-filled. If not, adjust the condenser position until the iris diaphragm is sharply focussed and, if necessary, readjust the iris diaphragm until it superimposes on the circle of light. The best resolution is when the condenser is raised to near its maximum height.
- Reinsert the eyepiece and examine the preparation using the fine focus adjustment and adjustable mechanical stage.

7.2.3 High power objective

In a good quality microscope the objectives should be parfocal, meaning that all the objectives have very similar focus settings.
- Rotate the nosepiece to the high power objective and then the fine adjustment to sharpen the image.
- Adjust the iris diaphragm to increase the illumination and optimise contrast and then re-examine the specimen.

7.2.4 Immersion objective

Check the required medium designated for the high magnification objective (usually written on the objective or indicated by a black (oil) or blue (water) ring). Add a small drop of immersion oil or water (< 2 cm in diameter) on the slide while rotating the objective lens into place. The space between the slide and the lens should be filled with the immersion medium.

Some microscopes are only equipped with oil or water objectives. These should be used by adding a small drop of the required medium onto the slide and slowly lowering the lens until it touches the medium, shown by a clear change in the light cone. Then perform fine adjustment while viewing through the ocular lenses until the image becomes sharp. If this fails, repeat the whole procedure. Avoid using any media other than those designated. In other words, do not shift from an oil immersion objective to a water or air objective or any other combinations. Clean and dry the slide before changing objectives.

7.2.5 Important considerations

When carrying out microscopic investigations one should consider:
- Empty magnification occurs when the image is enlarged beyond the physical resolving power of the microscope. Therefore optimal magnifications are usually achieved using 500-1,000 times the NA of the objective. For this reason avoid using eyepieces of high additional magnification, if the objective does not supply sufficient resolution. For example:
63× (objective) · 12.5× (eyepiece) = 787.5.
- Once the condenser has been focused properly, it should not be necessary to further adjust it.
- The iris diaphragm needs to be changed, so the cone of the light matches the NA of the objective and should therefore be readjusted each time the objective is changed.
- Less experienced users have a tendency to only use one eye while looking through the oculars. Keep using both eyes open. This might require some practice to get used to it.
- Air bubbles can markedly impair the image quality when using immersion lenses. These can be detected by removing the eyepiece and examining the objective rear focal plane through the microscope observation tubes. If observed, gently squeeze the cover slip with a nail. Alternatively, clean the objective lenses and specimen slide and then carefully reapply the oil.
- Dirt or dried oil on the front lens of the objective (or on the slide from previous inspections) might cause blurred images from unwanted scattered light. Clean the lens with a dedicated lens tissue.
- Uneven focus in the field of view or difficulties in obtaining sharp images might be because of how the slide is arranged on the stage. Check that the slide is positioned correctly.
- Problems with focusing might be because the cover slip is adhering to the objective lens. Affix the cover slip with clamps, glue or tape.
- Use the correct thickness for the cover slip. The objectives used on most biological microscopes are designed for use with No. 1.5 cover slips (0.17 mm thickness). Check this with the supplier of the cover slips.
- Use the correct immersion oil and avoid mixing oils with different RI. Usually immersion oil with a RI of 1.515 can be used for all objectives having NA greater than 0.95. Check with the supplier if in doubt.
- Air bubbles can produce shadows or unclear zones in the field of view, but can usually be overcome by elevating the objective and re-focussing, or by checking to see whether sufficient immersion medium is present between the objective and the cover glass. Remove and clean the slide and the objective if the nuisance persists.
- Insufficient illumination can be due to incorrect adjustment of the condenser or an inadequately opening the sub-stage diaphragm.

7.2.6 Bright-field and dark-field illumination

The principle of the light microscope is based on a bending and scattering of the light passing from the light source through the sample into the eyepiece lenses. Variations in the RI of the components in the specimen allow the generation of images based on contrasts, transmissions and reflections.

An efficient imaging with high resolution and proper control of contrast and depth is obtained by aligning the microscope to optimize the brightness and illumination of the image (see the section on Köhler illumination).

The most commonly used light microscope technique is with Bright-field (BF) illumination. This technique is suited for objects with high natural absorption such as plant cells or pigmented cells. However, most bacterial cells are small and appear as transparent objects in BF thus lacking sufficient contrasts. Staining can compensate for this low contrast, and most of the common microbiological staining protocols (Gram, Neisser etc.) are used together with BF.

Phase contrast (Ph) microscopy is an optical-microscopy technique that converts phase shifts in light passing through a transparent specimen to brightness changes in the image. It allows samples with low contrasts to be studied and therefore is suited for examining unstained fixed specimens and live cells. These samples types have very low absorbance differences compared to their surrounding medium, and such differences in refractive index cannot be detected by BF microscopy. In Ph microscopy an annulus in the condenser aperture generates a hollow cone of zero-order illumination that is projected onto the back focal plane of the objective. A matching phase ring in the objective lens absorbs non-diffracted light and shifts the wavelength slightly, thereby enhancing the contrast. This approach works best with relatively thin specimens (< 5 μm). Ph requires the use of designated objectives in conjunction with aligned phase rings.

7.2.7 Fluorescence microscopy

Some molecules contain fluorochromes that can absorb photons from light with specific energies and thereby become 'excited'. This state lasts for approximately 10^{-7} seconds and it ends by emitting a photon with a slightly lower energy, and hence longer wavelength. These fluorochromes exist as components of naturally occurring compounds frequently found in nature (e.g. pigments such as chlorophylls, vitamins etc.). However, new chemically synthesised compounds with high extinction coefficients are nowadays widely applied to label specific biomolecules and to stain tissues and cells for their identification. The diversity of these fluorescent markers, stains and labels targeting biomolecules, and dynamic dyes for measuring chemical conditions (pH, Ca^{2+}, oxygen etc.) is growing rapidly (Table 7.2).

Table 7.2 Dyes commonly used in microscopy of complex microbial systems.

Stain	Synonym or chemical name	Target	Chemical structure	Excitation/Emission
DAPI	4',6-diamidino-2-phenylindole dihydrochloride	A cell-permeable DNA-binding dye preferential to adenine and thymine-rich DNA	$C_{16}H_{15}N_5 \cdot 2HCl$	$\lambda ex \sim 359$ nm, $\lambda em \sim 461$ nm; λex 340 nm, λem 488 nm (only DAPI), λex 364 nm, λem 454 nm (DAPI-DNA-Komplex:100 mM NaCl, 10 mM EDTA, 10 mM Tris, pH 7)
Propidium Iodide	2,7-diamino-9-phenyl-10 (diethylaminopropyl)- phenanthridium iodide methiodide	A membrane-impermeable fluorescent nucleic acid stain	$C_{27}H_{34}N_4 \cdot 2I$	Excitable at 536 nm and emits at 617 nm (red)
Calcofluor	Calcofluor White	Ca^{2+} fluorescent probe; nonspecific fluorochrome binding to alginate, cellulose and chitin in cell walls	$C_{40}H_{44}N_{12}O_{10}S_2$	λex 355 nm, λem 433 nm
Nile Red	9-(diethylamino)-5H-benzo[a]phenoxazin-5-one	Lipophilic stain	$C_{20}H_{18}N_2O_2$	λex 552 nm, λem 636 nm
Nile Blue	Basic Blue 12, Nile Blue sulphate	Lipophilic stain targeting neutral fats	$C_{20}H_{20}N_3O \cdot 4SO_4$	λex 630 nm, λem 665 nm
DCFDA	2',7'-dichlorofluorescein diacetate	A cell-permeable fluorogenic stress probe targets reactive oxygen species (ROS) and nitric oxide (NO)	$C_{24}H_{14}Cl_2O_7$	λex 504 nm, λem 524 nm
Hoechst 33342	Bisbenzimide	A lipophilic fluorescent stain for DNA labeling; an A/T-specific DNA minor groove ligand	$C_{27}H_{28}N_6O \cdot 3HCl$	$\lambda ex < 380$ nm, λem 450-495 nm
Congo Red	C.I. Direct Red; Cosmos Red; Cotton Red; Direct Red	An amyloidophylic dye that specifically stains stacked β sheet aggregates	$C_{32}H_{22}N_6O_6S_2 \cdot 2Na$	N/A

Name	Synonym	Description	Formula	Fluorescence
Crystal Violet	Gentian Violet; Hexamethylpararosaniline chloride	Component of Gram staining that allows one to recognize the difference between gram-positive and gram-negative bacteria	$C_{25}H_{30}N_3 \cdot Cl$	N/A
Sudan Black B	Solvent Black 3; Fat Black HB	A fat-soluble diazo dye used for staining of neutral triglycerides and lipids	$C_{29}H_{24}N_6$	λex ~596-605 nm, stains blue-black
Neutral Red	Basic Red 5; Toluylene Red	Vital dye used as an indicator and biological stain	$C_{15}H_{17}N_4 \cdot Cl$	N/A
BCECF-AM	Spiro(isobenzofuran-1(3H), 9'-(9H)xanthene)-2',7'-dipropanoic acid	A membrane-permeable fluorescent indicator for measurement of cytoplasmic pH	Mixture of $C_{30}H_{20}O_{11}$, $C_{35}H_{28}O_{15}$ and $C_{40}H_{36}O_{19}$	λex 505 nm, λem 520 nm
Acridine Orange	3,6-Bis(dimethylamino)acridine	A fluorescent nucleic acid binding dye which interacts with both DNA and RNA	$C_{17}H_{19}N_3$	When bound to DNA, it emits green fluorescence (Em = 525 nm) and when bound to RNA, it emits red fluorescence (Em = ~650 nm).
Phalloidin-TRITC	Phalloidin-tetramethylrhodamine B isothiocyanate	A fluorescent stain used to identify filamentous actin	$C_{62}H_{72}N_{12}O_{12}S_4$	λex 540-545 nm, λem 570-573 nm
Bisbenzimide	Hoechst 33342 trihydrochloride	A/T-specific DNA minor groove ligand widely used fluorochrome for visualizing cellular DNA	$C_{27}H_{28}N_6O \cdot 3HCl$	λex 350 nm, λem 461 nm
Fluorescein isothiocyanate	FITC; Fluorescein 5-isothiocyanate	Amine-reactive reagent for the FITC labelling of nucleotides and proteins	$C_{21}H_{11}NO_5S$	λex 492 nm, λem 518 nm
Rhodamine B isothiocyanate	RBITC	A fluorescent probe for labelling of nucleotides and proteins	$C_{29}H_{30}ClN_3O_3S$	λex 543 nm, λem 580 nm
Sudan IV	Scarlet Red, Fat Ponceau R	Staining triglycerides, lipids and lipoproteins	$C_{24}H_{20}N_4O$	N/A
Safranin	Basic Red 2	Stain Gram negative bacteria	$C_{20}H_{19}ClN_4$	N/A
1,3,6,8-Pyrenetetrasulfonic acid tetrasodium salt	Tetrasodium 1,3,6,8-pyrenetetrasulfonate; HPTS	A fluorescent probe and pH indicator	$C_{16}H_6O_{12}S_4 \cdot 4Na$	λex 454 nm, λem 511 nm
5-Cyano-2,3-di-(p-tolyl) tetrazolium chloride	CTC	Vital stain	$C_{16}H_{14}ClN_5$	λex 490 nm, λem 630 nm

Fluorescent molecules have very specific absorption and emission wavelengths (high specificity) and follow the Lambert-Beers law, making them suitable for quantitative determination under a large range of physiological and physical conditions. Application of appropriate light source and filters for discriminating shorter-wavelength excitation and longer-wavelength emission light provides highly selective conditions for visualizing an individual fluorescent molecule with low interferences. Most fluorescence microscopes are equipped with epi-illumination/excitation and incorporate specific dichroic mirrors or chromatic beam splitters to allow a separation of the desired fluorescence emission light from any unabsorbed reflected excitation light and transmission of light not deriving from the applied fluorochrome. Fluorescence microscopy can be readily combined with other microscopic techniques including Ph microscopy.

Alignment of the microscope (Köhler illumination) is the most critical step for obtaining optimal resolution for all light microscopes. The following steps are advised:

- Centre and focus the light source. This can be carried out by removing the light diffuser (if applicable) and then, without the condenser, project the light onto a piece of paper placed in the condenser carrier. Fill the circular area with the image of the light source.
- Adjust the iris diaphragm located nearest to the light source (field iris) until you see sharp edges.
- Focus by adjusting the condenser-focusing knob. Both the specimen and the iris diaphragm should be in focus.
- Centre the image of the field iris using the condenser-centring knobs (usually located on the condenser).
- Open the field iris so that the edges lie just beyond the field of view.
- Optimize the image contrast, depending on the specimen, by adjusting the condenser iris diaphragm. Do not use this aperture to control light intensity.
- Adjust light intensity with the illuminator rheostat or by inserting neutral density filters (for colour photography).

Specimen contrast with a correct Köhler illuminated microscope is obtained by adjusting the condenser diaphragm. Illumination intensity is varied by adjusting the voltage to the light source or by neutral density filters in front of the illuminator.

7.2.8 Confocal laser scanning microscopy

Confocal laser scanning microscopy (CLSM) has become widely used among microbial ecologists due to its superior image quality. This increase in image quality comes from exchanging the light source used in epifluorescent microscopy with lasers that have much high intensities per area. This allows so-called pinholes, which prevent light that derives from out-of-focus information from reaching the photomultipliers (Figure 7.3), to be applied. This provides the basis for less blurry images and furthermore allows optical sectioning. The equipment used for CLSM is considered specialized equipment that requires skilled operators, and is outside the scope of this chapter.

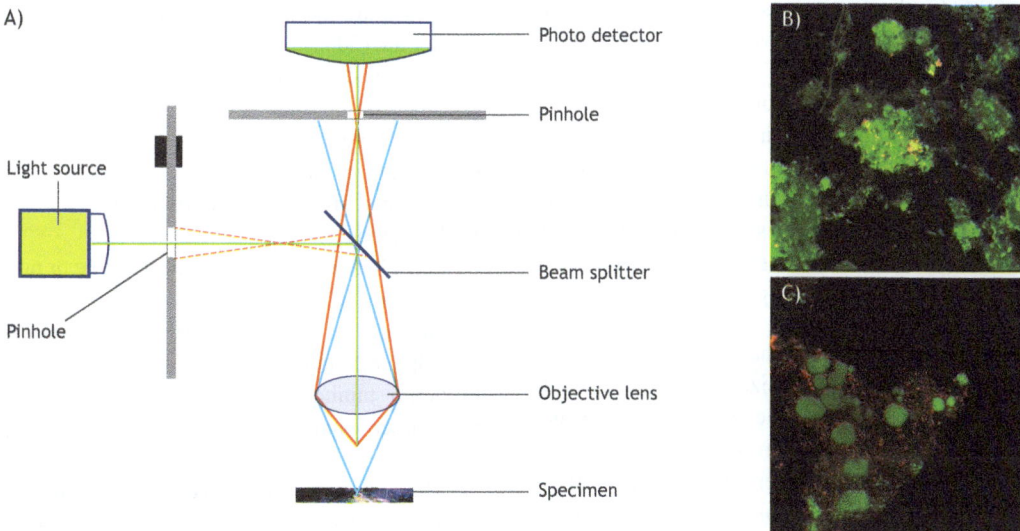

Figure 7.3 (A) Principle of the confocal laser scanning microscope. Images (B) and (C) show the same sample after fluorescence *in situ* hybridization (FISH) visualized by epifluorescence microscopy, and by a confocal laser scanning microscope, respectively (images: J.L. Nielsen).

7.3 MORPHOLOGICAL INVESTIGATIONS

In activated sludge systems most bacteria responsible for the removal of pollutants (e.g. C, N and P) grow in flocs and facilitate the settling and separation of the treated effluent. This floc-formation process is a result of the aggregation of wastewater components and an active excretion of extracellular polymeric substances (EPS) by 'floc-forming bacteria'. The flocs consist of bacteria often clustered into micro-colonies, organic and inorganic particles, and of the filamentous bacteria that provide the scaffolding or matrix backbone of the floc around which microorganisms aggregate.

Settling and the compaction properties of the activated sludge are directly related to the floc size and structure, and reflect its chemical, physical and biological components. The presence and abundance of filamentous bacteria are especially important for good sludge and settling properties. Imbalance between floc-forming bacteria and filamentous bacteria can lead to sludge bulking, where filaments extend from the floc surface and form inter-floc bridges, or disperse growth (pin point flocs, which results from too few floc-forming bacteria). Filamentous bulking can also result from the generation of large, irregular and open flocs. Excessive amounts of (certain) floc-forming bacteria occasionally lead to an overproduction of hydrophobic EPS, which may result in the formation of weak and buoyant flocs (so-called zoogloeal bulking) (Eikelboom, 2000; Jenkins *et al.*, 2004).

Morphological examinations alone or in combination with chemical analyses (EPS, ions etc.) and physical measurements (settling properties, compactness etc.) of the sludge can therefore reveal important plant operational and maintenance information. It is recommended to combine the information obtained from microscopic examination with direct plant observation, e.g. colour of the mixed liquor (activated sludge), percentage of the aeration basin surface covered by foam, depth of the sludge blanket in the secondary clarifier, the type of floating material on the surface of the clarifier (effervescent with bubbles *versus* greasy and sticky, etc.).

The general microscopical characterisation of flocs includes descriptions of their size, form and overall structure, the presence of organic fibres and inorganic particles, or single cells, and the recognition of different morphotypes of filamentous bacteria and their quantities. More details about the microscopical characterisation of activated sludge can be found in the manuals by Eikelboom (2000), Jenkins *et al.* (2004) and Seviour and Nielsen (2010).

7.3.1 Microscopic identification of filamentous microorganisms

For an unequivocal and precise identification of these filamentous bacteria it is considered important to find suitable control measures for bulking (Nielsen *et al.*, 2009; Seviour and Nielsen, 2010). This precise identification requires the application of molecular tools including FISH and 16S rRNA amplicon sequencing (see Chapter 8). A few filamentous morphotypes in activated sludge can be characterized and to some extent tentatively identified using the manuals by Eikelboom (2000) and Jenkins *et al.* (2004). However, nowadays it is clear that most of filamentous microorganisms cannot be characterised adequately in this way. Therefore, the recommendation here is to use microscopy in association with molecular methods to allow for the identification *in situ* of these and other bacteria, and to restrict the morphological studies to preliminary characterizations and, where appropriate, to confirm the molecular data. Since there has been a long tradition of microscopic morphological characterisation and 'identification' of filamentous bacteria in activated sludge, especially by the industry, and since this approach is still widely used, these methods will be outlined here.

Characterization of filamentous bacteria is based on the analysis of their morphological features, reactions to the Gram and Neisser staining, and physiological traits, such as the accumulation of intracellular elemental sulphur granules. Eleven features were assessed for distinguishing the filamentous species in the manual by Eikelboom (2000):

- Shape and length of the filaments.
- Shape of the cells.
- Filament diameter.
- Motility-gliding.
- Attached growth of other bacteria.
- Branching.
- Septa between adjoining cells (visible or not visible).
- Presence of a sheath.
- Granules of stored internal compounds (especially sulphur granules).
- Gram staining.
- Neisser staining.

Microscopy has been used to estimate the abundance of filamentous bacteria in activated sludge and their immediate influence on settling properties. For indexing into levels, the so-called Filament Index (FI) is an efficient semi-quantitative measure. The scale ranges from 0 (filaments absent) to 5 (excessive numbers of filaments) (Figure 7.4). This evaluation is simple to perform at low magnification (10×) phase contrast, and is considered reliable. It is recommended by Eikelboom (2000) to carry out such quantification regularly (every 1-2 weeks) or when changes in sludge settling properties are observed.

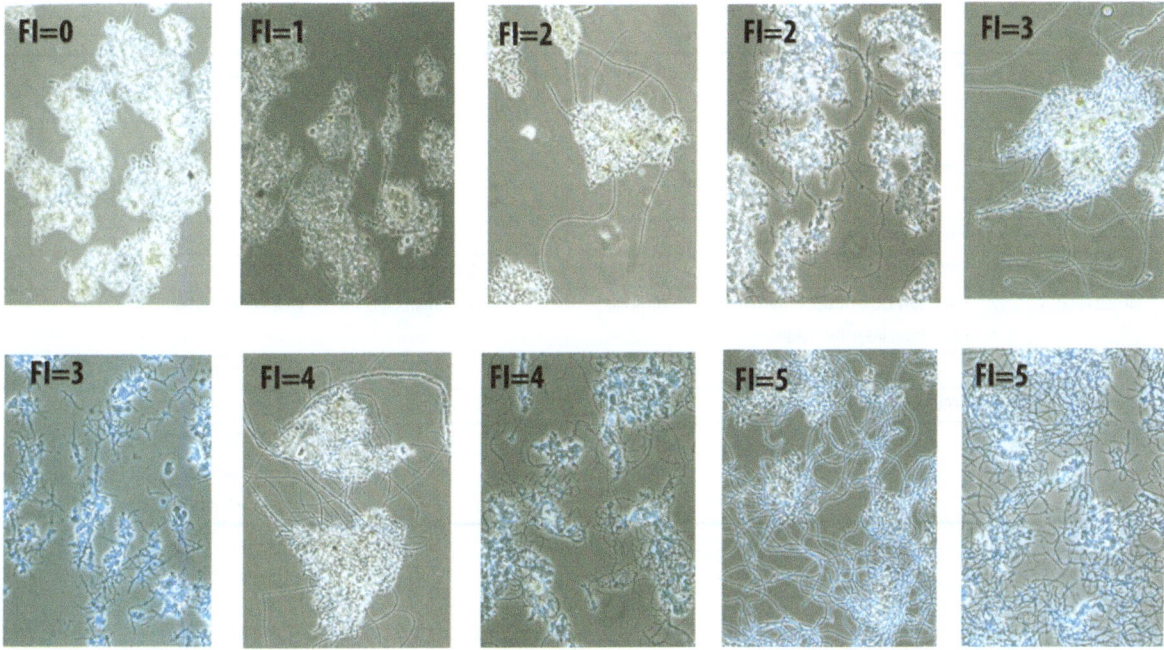

Figure 7.4 Filament indices (photos: Eikelboom, 2000 by courtesy of IWA Publishing).

7.3.2 Identification of protozoa and metazoa

The microbial communities in activated sludge consist, besides prokaryotes, of bacteriovorous and carnivorous protozoa and metazoa. These are important for shaping the composition of the sludge community and improving the quality of the final effluent, and are widely used to monitor plant performance. Their ecological importance comes from feeding on floc-associated and freely suspended bacteria, and they therefore reduce the turbidity of the final discharged liquid phase. The protozoa can be divided into flagellates, amoebae, free-swimming and crawling ciliates, attached ciliates, while the metazoa include rotifers, nematodes and oligochaete worms. The description of the protozoal and metazoan fauna may provide clues for reasons for the poor plant operation and often correlates with its physicochemical parameters. Table 7.3 shows some generalizations commonly used to predict plant performance and effluent quality. These generalizations should be used with caution (Seviour and Nielsen, 2010). Frequent examinations of the presence of protozoa can be used to obtain information about under- or overloading of the plant, whether the plant is sufficiently aerated, has suitable sludge retention time, etc.

More specific bio-indicators for plant performance might be found when applying higher taxonomical resolution. However, it is a time-consuming task that requires a high degree of expertise and experience. Furthermore it is still unclear how important some of these protozoa are as indicators of the wastewater treatment plant performance. Recent studies have applied FISH to identify protozoa (Xia et al., 2014), but because of practical difficulties, only a small number of FISH probes for these have been published so far. Protozoan species such as *Vorticella picta* have been used as indicators for low organic sludge loading, while *Vorticella microstoma* and *Opercularia coarctata* are indicators for high organic loading (Madoni, 1994). Other studies have proposed that *Acineta tuberosa*, *Zoothamnium* and *Euplotes sp.* are good indicators for high-quality effluent and thus well-functioning plants (Salvado et al., 1995).

Table 7.3 Generalizations of the presence and abundance of protozoa and metazoa for the evaluation of the performance of an activated sludge plant.

Microorganism	Characteristics	Examples	Overabundance indications	Manifestation in wastewater treatment plant
Flagellates	High numbers found during recovery from toxic discharge or low dissolved oxygen (DO) concentration	-	High organic loading	Poor performance
Amoebae	High during start-up and recovery	-	High organic loading	Unstable operation
Free-swimming ciliates	Occur under good floc formation	Euplotes, Aspidisca	High organic loading	Well-operating plant
Attached ciliates	Under low DO or toxicity, stalked ciliates will leave their stalks	Vorticella	Low organic loading	Well-operating plant
Crawling ciliates	-	-	-	Good performance
Rotifers	Require high DO	Euchlanis	Low organic loading	Well-functioning and stable plant
Higher invertebrates	-	Nematodes, Targrades, Annelids	Low organic loading	High ammonia loads as Tardigrades and Annelids are susceptible to ammonia toxicity

7.4 EXAMINING ACTIVATED SLUDGE SAMPLES MICROSCOPICALLY

7.4.1 Mounting the activated sludge sample

Two approaches are used for light microscopic examination: (*i*) wet mounts suitable for examining sludge characteristics such as floc structure, filament morphology, protozoa and metazoa identification, and (*ii*) fixed smears for stained preparations.

A wet mount is prepared by adding a drop (ca. 1 cm diameter) of fresh activated sludge onto a glass slide and then gently placing a cover slip on top of the drop. If the drop is too large, then it might be more difficult to focus on the sample or the cover slip will float and the specimen will then leak around it and risk contaminating the microscope. If too small, the specimen will contain interfering air bubbles and will prematurely dry out. Sealing the edges along the cover slip with nail polish or vaseline will reduce these problems. The slide is then ready for examination.

A fixed smear immobilises the biomass onto the slide, but may reduce specimen contrast. The sample is thinly smeared onto the slide and then can be easily spread using a pipette. The smear is allowed to air dry at room temperature. Heat fixing by passing the slide through a flame is usually not necessary for activated sludge samples and runs the risk of harming the sample.

Hydrophobic slide surfaces from remnant grease used in the cutting process frequently create problems as they can cause the smear to dry out irregularly. Thus, slides should be degreased before use. Acid washing of slides and coating the glass slide with gelatine or poly-L-lysine are recommended. Acid-washed slides are prepared by leaving them in a preheated (60 °C) 1 M HCl solution overnight and then rinsing them first in distilled water (dH$_2$O) followed by 95 % ethanol solution before drying. For improved sample adhesion, dip the slides in 0.5 % (*w/v*: weight/volume percent) gelatine solution for 5 min at 70 °C. Alternatively, acid-washed slides can be dipped in 0.01 % poly L-lysine solution for 5 min at room temperature. Slides are then air-dried in a vertical position in a dust-free environment. Fixing the sample at a higher temperature (40-60 °C) to the gelatine or poly-L-lysine coated slides improves its adherence to the coated slides.

Immediate deposition of the activated sludge sample onto the slide will usually maintain the spatial structure of the flocs. However, a gentle homogenization of flocs prior to immobilization of the sample often helps staining protocols to visualize cell structures of interest, and this can be achieved by gently rubbing two glass slides with a 20 µL sample against each other, or more efficiently with a tissue grinder.

The cryosectioning procedure can reveal high specimen resolution while keeping its spatial organisation intact. This procedure involves embedding the sample in paraffin or a polymerizing resin (e.g. Tissue-Tek O.C.T.). Paraffin-embedding usually ensures undamaged tissue slices but requires sample heating and de-waxing with xylene, while cold polymerizing resin application is performed at ambient temperature and requires no further chemical treatment. A simple cryosectioning procedure involves mixing the activated sludge (or biofilm) sample and embedding material (Tissue-Tek O.C.T.) in the lid of an Eppendorf tube. After allowing the embedding material to migrate into the sample overnight at 4 °C, it is transferred to liquid nitrogen. Sectioning into 5-20 µm thin slices is carried out on a cryotome at -20 °C. The slices should be immediately placed on a slide (at room temperature), where they melt and are allowed to dry on the bench for 3 hours. Afterwards, further staining of the sample can now be done.

7.4.2 Gram staining

The Gram stain is a differential stain to distinguish between two major bacterial groups, Gram-positive (stained purple) and Gram-negative (stained red) cells (Figure 7.5).

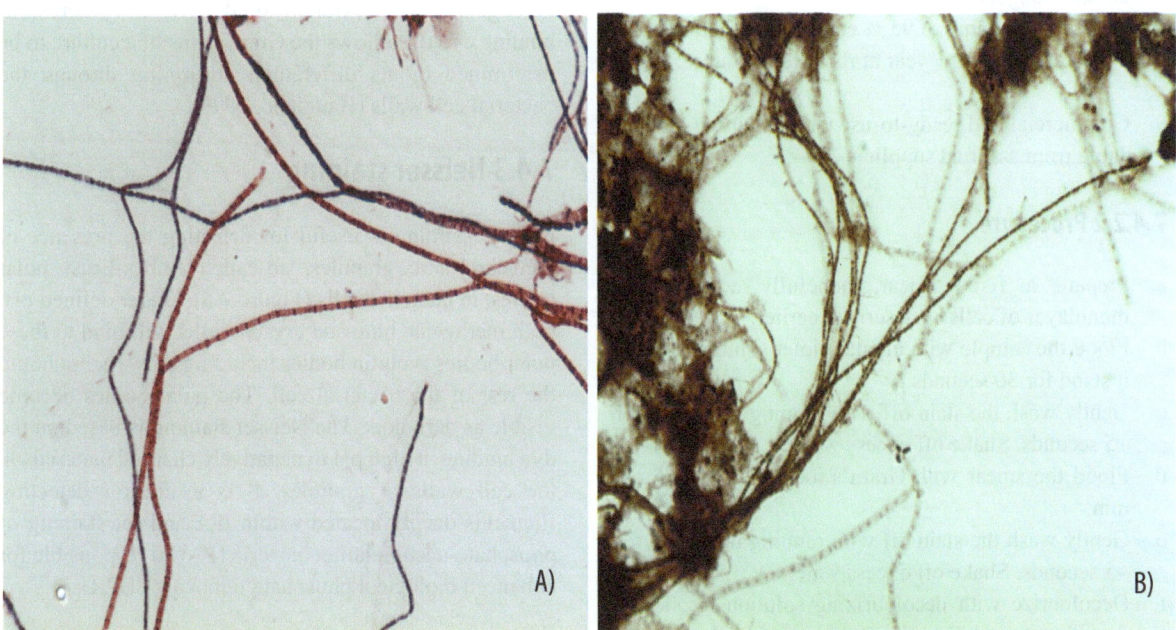

Figure 7.5 Examples of typical Gram (A) and Neisser (B) staining of an activated sludge sample (photo: B. McIlroy).

This staining technique reflects basic differences in their cell wall composition and organisation. Gram-positive cell walls contain high levels of peptidoglycan, which contracts in the presence of the decolouriser (70 % ethanol), hence retaining the crystal violet/iodine complex. In Gram-negative bacteria, layers of peptidoglycan are present, but in much lower quantities. Furthermore Gram-negative cells have a lipopolysaccharide membrane lying outside the peptidoglycan. In the presence of the decolouriser, this membrane is disrupted, and the cells therefore become leaky, allowing the crystal violet/iodine complex to be washed out. A suitable counterstain (safranin) is then applied to the cells, which are therefore stained red.

7.4.2.1 Reagents and solutions for Gram staining

- **Crystal violet solution**
  Crystal violet stock (solution A): Dissolve 20 g Crystal violet (85 % dye) in 100 mL of 95 % ethanol. Filter the sample.

Oxalate stock (solution B): Dissolve 1 g ammonium oxalate in 100 mL water.

Working solution: Dilute solution A (1:10) in solution B. The working solution can be stored at room temperature up to one year in the dark.

- **Gram's iodine solution**
  Dissolve 1 g iodine crystals and 2 g potassium iodide in 5 mL water. Then add 240 mL distilled water and 60 mL of 5 % (w/v) sodium bicarbonate solution. Mix thoroughly. The solution can be stored at room temperature for up to one year in the dark.

- **Counterstain**
  Mix 2.5 g safranin O with 100 mL of 95 % ethanol. Can be stored for up to 1 year at room temperature.

- **Decolourizing solution**
  Mix equal volumes of 95 % ethanol and acetone. Can be stored for up to 1 year at room temperature.

Commercial and ready-to-use Gram staining kits are available from assorted suppliers.

7.4.2.2 Procedure

a. Prepare a fixed smear, hopefully achieving a monolayer of cells as described earlier.
b. Flood the sample with crystal violet solution and let it stand for 30 seconds.
c. Gently wash the stain off with running tap water for ~5 seconds. Shake off excess water.
d. Flood the smear with Gram's iodine solution for 1 min.
e. Gently wash the stain off with running tap water for ~5 seconds. Shake off excess water.
f. Decolourize with decolourizing solution by slowly dropping the solution above the smear on the slightly tilted slide. This should allow the decolourizing solution to slowly flow down across the smear. Continue until no more purple colour is eluted from the smear (excessive decolourization will decolourize Gram-positive cells, and too moderate decolourization may give false Gram-positive cells; false Gram stain reactions signals can also be seen with stored and starved samples, where Gram-positive cells stain Gram negatively).
g. Gently rinse off the stain with running water for ~5 seconds. Shake off excess water.
h. Cover the slide with counterstain (safranin) for 30 seconds.
i. Gently rinse off the stain with running water for ~5 seconds. Shake off excess water.
j. Allow the slide to air-dry, or carefully blot with filter paper.

A well-prepared smear should be barely visible to the naked eye. Examine the slide by adding a drop of immersion oil in the centre, and examine using a 100× Bright field objective. Do not use a cover slip. A Gram-stained slide can be stored indefinitely at room temperature in the dark. Gram-positive cells appear purple and Gram-negative cells appear red. The colour can vary from blue to almost black. A blue filter will enhance the contrast. Some cells can appear as intermediates or unevenly stained, and are referred to as Gram-variable.

Problems with large flocs and too dense smears can be overcome by sample dilution or gentle sample homogenisation. Hexidium iodide (HI) is a fluorescent binding dye that allows the Gram status of a culture to be determined by its differential absorption through the bacterial cell walls (Haugland, 1999).

7.4.3 Neisser staining

Neisser staining is useful for detecting the presence of metachromatic granules, so-called Babes-Ernst polar bodies, in bacterial cells (Figure 7.5). Under defined pH, both methylene blue and crystal violet will bind to these polar bodies (volutin bodies including poly-P), but not to the rest of the bacterial cell. The polar bodies become visible as dark dots. The Neisser staining is based on the dye binding at high pH to negatively charged materials in the cell walls or granules. It is useful for detecting filaments deeply located within floc and for staining of phosphate-accumulating bacteria (PAOs) responsible for enhanced biological phosphate removal (EBPR).

7.4.3.1 Reagents and solutions for Neisser staining

- **Methylene blue solution**
  Dissolve 0.1 g methylene blue in 100 mL distilled water. Add 5 mL 96 % ethanol and 5 mL glacial acetic acid. Filter the sample.

- **Crystal violet solution**
  Dissolve 0.33 g crystal violet in 100 mL distilled water and 3.3 mL 96 % ethanol. Filter the sample.

- **Counter-staining solution**
  Mix 33.3 mL 1 % chrysoidine solution with 100 mL distilled water. Alternatively use 0.2 % Bismark brown.

- **Working solution**
  Prepare a fresh mixture containing two parts methylene blue solution and one part crystal violet solution.

Commercial and ready-to-use Neisser staining kits are available from several suppliers.

7.4.3.2 Procedure

a. Prepare a fixed smear as described earlier.
b. Flood the sample with the freshly prepared working solution and leave to stand for 15 s.
c. Gently rinse off the stain with running water and ensure both sides of the slide are rinsed.
d. Flood the smear with the counter-staining solution and leave for 1 minute.
e. Gently rinse off the stain with running water and ensure both sides of the slide are rinsed.
f. Let the slide air dry or use filter paper.
g. Examine under the microscope with oil immersion using a high magnification Bright field objective, direct illumination and no cover slip.

A well-prepared smear should be barely visible to the naked eye. A Neisser stained slide can be stored at room temperature in the dark indefinitely. Neisser-positive cells appear blue/grey or purple while Neisser-negative cells appear yellow/brownish. Frequently, exocellular sheaths and polysaccharide capsular material also stain. Neisser-negative cells have little contrast and may be difficult to see.

7.4.4 DAPI staining

The DAPI nucleic acid stain preferentially stains double-stranded (ds) DNA which then emits a blue-fluorescent light. Its affinity appears to be associated with clusters rich in adenine and thymine (AT), and DAPI binds especially to the minor groove of the DNA double helix, where the natural fluorescence is enhanced because of the displacement of water molecules from both DAPI and the minor DNA groove (Trotta and Paci, 1998). DAPI also binds to RNA, although in a different binding mode that might involve AU-selective intercalation (Tanious et al., 1992). Other molecules have been reported to adsorb DAPI, including poly-P, where the complex emits light of a wavelength in the yellow range. The specificity of the DAPI stain depends on the protocol and on the concentration used. Cells containing poly-P in concentrations higher than 400 μmol g^{-1} dry weight can be visualized when DAPI is applied at a concentration of at least 5 to 50 μg mL^{-1}. At higher concentrations DAPI also binds to, and fluoresces with, other cellular constituents such as lipids. At lower concentrations, the resulting blue fluorescence is related primarily to its binding to DNA. Two forms of the dye are frequently used, DAPI dihydrochloride and the more water soluble DAPI dilactate.

7.4.4.1 Reagents and solutions for DAPI staining

- **DAPI stock solution**
  Prepare a 5 mg mL^{-1} DAPI stock solution by dissolving it in distilled water (dH$_2$O) or dimethylformamide (DMF). The DAPI dihydrochloride may take some time to completely dissolve and may require mild sonication. Other solvents (e.g. PBS: Phosphate-buffered saline) are problematic because of solubility difficulties. For long-term storage, the DAPI stock solution can be aliquoted and stored at -20 °C. For short-term storage it can be kept at 4 °C, if protected from direct light exposure.

7.4.4.2 Procedure

Counter-staining by DAPI can be performed directly with liquid samples or fixed smear samples as described earlier.

a. For samples in suspension: mix DAPI stock solution to a final concentration of 1 μg mL^{-1} directly into a small volume sub-sample. Allow the dye to migrate into the sample and bind at room temperature for 5 to 30 min (depending on the nature of the sample). Protect it from direct exposure to light. Remove excess dye by centrifugation and rinse the sample with sterile water until any background fluorescence is minimized.
b. For immobilized samples: Embed the smeared sample with a 1 μg mL^{-1} DAPI solution. Allow it to migrate into the sample and react at room temperature for 5 to 30 min, depending on the nature of the sample. Keep it protected from direct exposure to light. Remove any excess DAPI by centrifugation and rinse with plenty of sterile water until any background fluorescence is minimal. Immobilize the suspended sample as described for the smeared sample. Let it air dry. A DAPI-stained slide can be stored at -20 °C indefinitely. Apply a drop of antifading agent (Citifluor, Vectashield or a mixture hereof) directly onto the slide with the sample. Cover it with a cover slip and examine it under epifluorescence microscopy. The excitation

maximum for DAPI bound to dsDNA is 358 nm, and the emission maximum is 461 nm.

7.4.5 CTC staining

The 5-cyano-2,3-ditolyl tetrazolium chloride (CTC) is a redox dye that produces fluorescent formazan (CTF) crystals in reduced environments such as in active respiring cells with an active electron transport. The formazan is deposited intracellularly as large crystals and can therefore be used as a cellular redox indicator. Although it has been described as an non-specific indicator it is still valuable as a fast indicator of filamentous activity in activated sludge samples. An advantage of the CTC activity stain is that it does not usually require any washing step, as the non-reduced CTC does not absorb light above 400 nm.

7.4.5.1 Reagents and solutions for CTC staining

- 1 % (v/v) aqueous solution of CTC (5-cyano-2,3-ditolyl tetrazolium chloride).

7.4.5.2 Procedure

a. 200 µL 50 mM are added to a test tube containing 2 mL of the sludge sample. Let the sample incubate for 1-4 h at room temperature with moderate agitation.
b. The stained cells can be visualized by filtering the sample onto a 0.2 µm black polycarbonate filter or simply add a small drop on a microscope slide and let it air-dry.
c. Filters are air-dried and mounted with immersion oil on glass slides. Examine through the microscope using an excitation of 450 nm and an emission of around 630 nm.
d. Formazan-containing cells are relatively photo-stable and can be stored at 4 °C for several days if required.

7.5 FLUORESCENCE in situ HYBRIDIZATION

Hybridization with fluorescently-labelled DNA oligonucleotides has allowed visualization of individual microbial cells in complex environments and their in situ identification. This technique can provide important information for quantitative enumeration of targeted microbial groups of interest and their spatial organization within the sample. It does not discriminate between cells that can or cannot yet be cultured.

The FISH technique, described in details elsewhere (Nielsen, 2009), is based on the principle of hybridizing fluorescently-labelled DNA probes to target sites of ribosomal rRNA in permeabilized whole cells of interest (Figure 7.6).

The probes are small DNA fragments designed to hybridize specifically to their complementary target sequences in the rRNA structures in metabolically active target cells. One powerful aspect of FISH is that it is possible to design a probe to target a narrow phylogenetic group (down to the species level) or one which can target members of a whole bacterial phylum or any other higher phylogenetic hierarchal group. Only cells with the proper target sequence in their 16S rRNA molecule will hybridize with the DNA probe used, and because an individual cell will contain multiple ribosomes, a sufficient fluorescence signal is generated to allow its detection microscopically.

Despite the fact that the method can be applied widely in microbiology, some limitations of FISH have been recognized. Although activated sludge is usually characterized by a high level of metabolic activity resulting in high cellular ribosomal counts and high fluorescence signals, some cells might emit detectable levels of fluorescence, for several reasons. Such low emission signals might be compensated for partially by using fluorochromes with higher extinction coefficients, enzymes that amplify signal intensities (Card-FISH, see detailed protocols in Pernthaler and Pernthaler, 2007) or by using multiple labelled probes (DOPE-FISH, Stoecker et al., 2010). The FISH technique is sometimes limited by penetration of the probe through the wall/membrane and into the targeted cell, and so a false negative is the outcome, as often seen with the foam-stabilising Mycolata (Seviour and Nielsen, 2010). Thus, some cell pre-permeabilisation is essential. Nor will they fluoresce if the probe is designed to target an inaccessible region of the rRNA, and helper probes will need to be designed and applied to overcome this problem (Fuchs et al., 2000). Sample autofluorescence is also often a serious problem with activated sludge samples. Methodological problems and pitfalls of the FISH technique have been reviewed elsewhere (Moter and Göbel, 2000). Important factors that influence the sensitivity and quality of the FISH technique are described by Bouvier and Del Giorgio (2003). These can be summarized as follows:
- Auto-fluorescence from naturally fluorescing compounds in the sample or cells.
- Probe sequence (binding energy, self-complementarity, etc.).

- Probe concentration.
- Type of probe.
- Microorganism (the nature of the membrane might influence permeability and undergo cell lysis).
- Target site on rRNA (accessibility might be hindered).
- Fixation procedure (might chemically modify the sample/cells).
- Presence of deoxyribonucleases (DNases), or peroxidases (for Card-FISH).
- Hybridization time.
- Temperature.
- Salt concentration.
- Formamide concentration.
- Washing time.
- Adsorption phenomena.

Information on probe sequence and stringency can be found in literature and in probeBase (Loy *et al.,* 2003) and can be coupled to multiple fluorochromes. In general, fluorochromes such as Cy3, Cy5, FLUOS, TAMRA, and various Alexa variants, which provide low fading and high intensity signals without compromising permeabilization, work well. Custom synthesis of oligonucleotide probes with or without fluorochromes can be purchased from plenty of internet-based suppliers.

The FISH procedure for each probe/sample should be optimized based on an empirical approach.

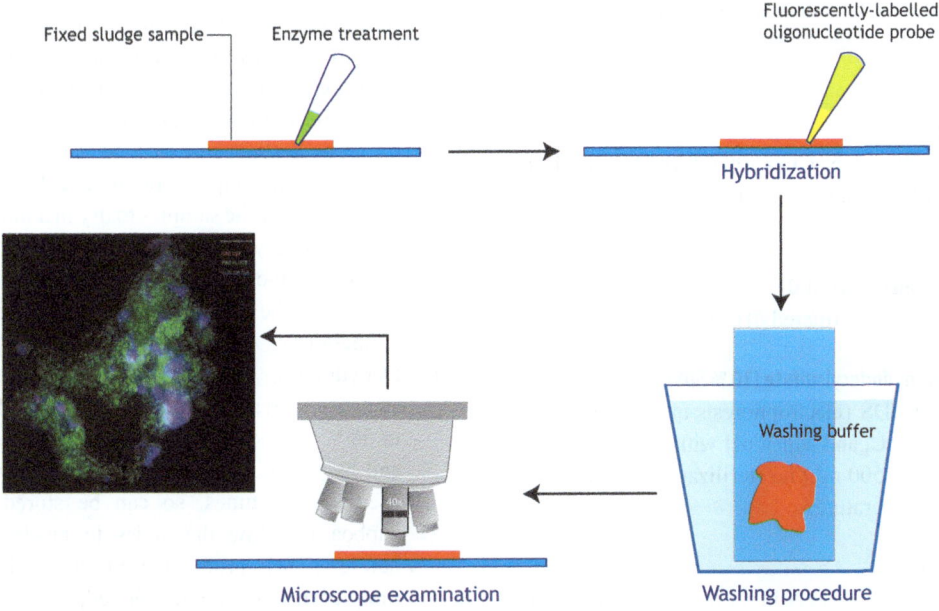

Figure 7.6 Schematic presentation of the FISH procedure.

7.5.1 Reagents and solutions for FISH

- **Fixative (8 % Paraformaldehyde, PFA) for Gram-negative cells**
  Mix 4 g PFA in 30 mL dH$_2$O at 60 °C, add a drop of 2 N NaOH to facilitate it dissolving (work in the fume hood); add 16.6 mL 3 × PBS (see below). Adjust to 50 mL with dH$_2$O. Filter the solution through a 0.22 µm polycarbonate filter to remove any particulate material. The solution should be used fresh, but it can be stored for a few days in the fridge or in aliquots at -20 °C.

- **Lysozyme for cell permeabilisation**
  Dissolve lysozyme to a final concentration of 10 mg mL^{-1} (~360,000 U mL^{-1}) in 0.05 M EDTA 0.1 M^{-1} Tris-HCl (pH 8.0). Prepare fresh when needed, and keep it on ice; it can be stored in aliquots at -20 °C.

- **Proteinase K for cell permeabilisation**
  Dissolve proteinase K (from *Tritirachium album*) to 20,000 U mL^{-1} in Tris-EDTA buffer (0.01 M EDTA, 0.1 M^{-1} Tris-HCl, pH 8.0). Prepare fresh when needed, and keep it on ice; it can be stored in aliquots at -20 °C.

- **3 × phosphate-buffered saline (3 × PBS)**
  Mix 0.1 M NaH$_2$PO$_4$ with 0.1 M Na$_2$HPO$_4$ until pH 7.4 is reached. Mix 22.8 g NaCl into 300 mL of this phosphate solution and add dH$_2$O to 1,000 mL. Autoclave, and store at room temperature.

- **Tris-EDTA buffer (TE buffer)**
  0.01 M EDTA, 0.1 M Tris-HCl, pH 8.0. Sterilize by filtration and store at 4 °C.

- **1 M Tris-HCl, pH 8.0**
  Dissolve 121.1 g Tris in 800 mL dH$_2$O, add 42 mL concentrated HCl, allow it to cool, adjust pH and fill to 1 L with dH$_2$O. Autoclave, and store at room temperature.

- **5 M NaCl**
  Dissolve 292.2 g NaCl in 800 mL dH$_2$O, and fill to 1 L with dH$_2$O. Sterile filtration, and store at room temperature.

- **Sterile distilled H$_2$O (dH$_2$O)**
  Autoclave sterile filtered dH$_2$O.

- **10 % Sodium dodecylsulfate (10 % SDS)**
  Heat 50 g SDS (electrophoresis-quality) in 400 mL dH$_2$O to 70 °C, and adjust pH with concentrated HCl to 7.2, fill to 500 mL; no sterilization required. Store at room temperature.

- **0.5 M EDTA**
  Dissolve 18.6 g EDTA in 80 mL dH$_2$O by adjusting the pH to 8.0 (ca. 2 g NaOH pellets required), and fill to 100 mL with dH$_2$O. Sterilize by filtration and store at 4 °C.

7.5.2 Procedure

a. Sample collection. Fresh samples are collected and fixed immediately as follows (can be kept at 4 °C for 2-3 days without impact on the cells).
b. Fixation. The sample is usually fixed for both Gram-negative and Gram-positive cells, unless the target population is known.
   (i) Gram-positive cells: Mix equal volumes of activated sludge (mixed liquor) and 96 % ethanol. Keep in the freezer (-20 °C).
   (ii) Gram-negative fixation: Centrifuge 5 mL sample for 8 min at approximately 3,400 × g. Remove the supernatant and replace it with cold 4 % PFA/PBS. The cell suspension is fixed for 3 h at ~4 °C. Centrifuge for 8 min at approximately 3,400 × g. Remove the supernatant (PFA) and discard appropriately. Add 5 mL of cold tap water and mix before centrifugation for 8 min at approximately 3,400×g. Remove the supernatant. Repeat this step once more. Add 5 mL of cold sterile filtered tap water and mix. The sample is now ready for the FISH procedure.

  The fixed sample can be kept at -20 °C for several months. Prior to usage it must be treated as follows: Centrifuge for 8 min at approximately 3,400×g. Remove the supernatant and replace by a cold solution of 1:1 PBS/EtOH. If the sample has been kept in 1:1 PBS/EtOH, it must be washed once and re-suspended in tap water prior to immobilization.

c. Immobilisation on slides. Spread 15 µL of the sample on a cover slip or in each well of a Teflon-coated slide. Use the pipette tip to evenly distribute the sample. Allow the samples to dry in a fume hood until completely dry (15-30 min). Drying at 46 °C appears to improve the binding of more dilute samples to the surface of the slide compared with drying at room temperature.

d. Dehydration. Dehydrate the slides in EtOH by stepwise increasing concentrations as follows: 3 min in 50 % ethanol followed by 3 min in 80 % ethanol and by 3 min in 96 % ethanol. The ethanol can be used several times, so can be stored in a fume cupboard. Allow the slides to air-dry completely before FISH probe hybridisation. DAPI-stained samples must not be dehydrated, as they fade rapidly in the presence of ethanol.

e. Permeabilization. Some cells (e.g. Mycolata) require pretreatment to bring about permeabilization of their cell wall/envelope. This can best be achieved enzymatically with exposure to Lysozyme and proteinase K, or chemically (mild acid treatment). Which enzymes are used, their concentrations and incubation conditions etc. depend on the nature of the sample and cells, and should be optimized for each sample. However, many activated sludge samples can be FISH-probed without any enzymatic or chemical permeabilization or by applying mild treatment. Where necessary the methods detailed in the following protocols should be used:

(i) Permeabilization with lysozyme
- Apply 10-15 μL cold lysozyme (36,000-360,000 U mL^{-1}) to the sample on a slide or slide well. Place the slide in a horizontal position, in a 50 mL polyethylene tube, containing a tissue paper soaked with 2 mL of dH$_2$O.
- Incubate the slide for 10-60 min, depending on the nature of the cells, at 37 °C.
- Wash the slide 3 times in dH$_2$O, and once in absolute ethanol, and allow it to air-dry.

The slide can now be stored at -20 °C for several months.

(ii) Permeabilization with proteinase K
- Apply 10-15 μL cold proteinase K (2,000-20,000 U mL^{-1}) per slide or slide well and transfer to a 50 mL polyethylene tube lined with a moisturized tissue paper. Incubate for 20-60 min at 37 °C.
- Wash the slide 3 times in dH$_2$O, then once in absolute ethanol, and let the slide air-dry.

At this point the slide can be stored at -20 °C for several months.

(iii) Permeabilization with mild acid hydrolysis
- Submerge the slides with dehydrated cells in hydrochloric acid (1 M HCl) at 37 °C for 30 min.
- Wash the slide with dH$_2$O, then once in absolute ethanol, and let the slide air-dry.

f. Hybridisation. For optimal hybridization of fluorescent oligonucleotide probes all the factors affecting the success and outcome of the approach need to be taken into account every time a new system is examined.

(i) Prepare hybridisation buffers (Table 7.4) containing appropriate hybridization stringency conditions for each oligonucleotide probe that can be found in probeBase (Loy et al., 2003). If the probe has not been properly tested, these conditions will need to be determined for each new probe in an empirical fashion.

(ii) Transfer 8 μL of hybridisation buffer onto the slide within an area of 1-2 cm^2 on a glass slide or into each well of a Teflon-coated slide with one or more wells.

(iii) Add 1 μL of each gene probe (probe concentration 50 ng μL^{-1}) and mix carefully (avoid contact with the sample) with the hybridization buffer (sterile pipettes must be used for all work with gene probes). If more gene probes are added to the same well the order is unimportant. Equimolar concentrations of each probe should be used. Place the slide horizontally into a sterile Greiner tube (50 mL) with a piece of cotton paper wetted with 1-2 mL of the same hybridisation buffer. Place the tube in the hybridisation oven (46 °C) for 1½ hour.

Table 7.4 Hybridization buffer (46 °C) for the FISH procedure.

FA %	FA μL	dH$_2$O μL	5 M NaCl μL	1 M Tris/HCl μL	10 % SDS μL
0	0	1,600	360	40	2
5	100	1,500	360	40	2
10	200	1,400	360	40	2
15	300	1,300	360	40	2
20	400	1,200	360	40	2
25	500	1,100	360	40	2
30	600	1,000	360	40	2
35	700	900	360	40	2
40	800	800	360	40	2
45	900	700	360	40	2
50	1,000	600	360	40	2
60	1,200	400	360	40	2
65	1,300	300	360	40	2
70	1,400	200	360	40	2

g. Washing.
(i) Prepare 50 mL washing buffer (Table 7.5) and preheat in a water bath to 48 °C before washing.
(ii) After hybridisation, carefully remove the slide from the Greiner tube using a pair of tweezers.
(iii) Pour a few millilitres of the preheated washing buffer on the top of the slide (Caution! Not directly on the sample) to remove excess probe solution from the wells.
(iv) Put the slide into the preheated (48 °C) 50 mL Greiner tube containing the washing buffer for 15 min in a 48 °C water bath to wash out the unbound probes.
(v) Again remove the slide carefully with a pair of tweezers. Rinse with cold dH$_2$O by dipping the slide in a glass beaker to remove any crystallised NaCl.
(vi) Allow the slides to air-dry.

Table 7.5 Washing buffer (48 °C) for the FISH procedure.

FA %	1 M Tris/HCl (pH 8,0) µL	10 % SDS µL	5 M NaCl µL	0.5 M EDTA µL
0	1,000	50	9000	0
5	1,000	50	6,300	0
10	1,000	50	4,500	0
15	1,000	50	3,180	0
20	1,000	50	2,150	500
25	1,000	50	1,490	500
30	1,000	50	1,020	500
35	1,000	50	700	500
40	1,000	50	460	500
45	1,000	50	300	500
50	1,000	50	180	500
55	1,000	50	100	500

h. Microscopy (EPI-fluorescence and CLSM). The dried slides can now be examined microscopically.
  (i) Place a small drop of Citifluor (or Vectashield, or a mixture hereof) mounting fluid on the slides and put a cover slip on top. Make sure that the Citifluor is distributed to all the wells before microscopy.
  (ii) If the slides are not evaluated on the same day of hybridisation they can be stored at -20 °C for some weeks, where the intensity of the fluorescence signal remains virtually unchanged.

7.6 COMBINED STAINING TECHNIQUES

Many different staining techniques can be combined to provide a better understanding of the relationships between bacterial identification and their function. This approach is very powerful and can provide ecophysiological information at the single cell level in complex communities. Thus they allow detection of intracellular storage compounds such as poly-P and PHA in populations of interest, and help explain which organisms are responsible for the chemical transformations in processes such as enhanced biological phosphate removal. They can also identify cells that are able to synthesise specific extracellular enzymes or cells whose surfaces are hydrophobic, which is an important cell feature linked to foaming (Table 7.6).

Most of the bacteria in complex mixed natural communities such as those in activated sludge cannot be isolated and grown in pure culture, and even establishing enrichment cultures in the laboratory is likely to modify the selective natural biotic and abiotic parameters. So it is necessary to study their functions directly in the natural habitats. Many such studies have been carried out through use of combined methodologies as those described here, with the hope that other investigators will adopt them in their studies (e.g. Nielsen et al., 2010a,b).

Table 7.6 Staining techniques that can be combined to link identification with function.

Identification	FISH, Gram	References
Substrate uptake	Microautoradiography	Nierychlo et al., 2015
Surface properties	Microsphere adhesion to cell	Nielsen et al., 2001
Surface components	Lectins, antibodies	Böckelmann et al., 2002
Exoenzymatic activity	Enzyme-linked fluorescence	Nielsen et al., 2002, 2010a Kragelund et al., 2007 van Ommen Kloecke and Geesey, 1999
Internal storage products	PHA, poly-P, Sulphur	Nielsen et al., 2010b

An example is given below on how to apply the FISH technique in combination with (*i*) DAPI staining and (*ii*) PHA staining to detect and identify the bacteria capable of taking up poly-P (PAOs) and bacteria accumulating large quantities of bioplastics in the form of PHA in activated sludge systems designed to remove phosphorus microbiologically. Such data will contribute toward a better understanding of the complex microbiology of these processes (e.g. how PAOs may outcompete other organisms) to pave the way for an eventual knowledge-based optimization of the wastewater treatment plants with EBPR. It is remarkable that most PAOs have never been isolated in pure culture, but are studied either as enrichment cultures or in full-scale systems.

DAPI staining can detect all bacteria from its non-specific binding to DNA. However, DAPI has also been known to stain intracellular poly-phosphate granules in bacterial cells when applied at elevated concentrations. The fluorescence spectra of DAPI-DNA complexes exhibit a maximum fluorescence emission at around 450 nm (blue), whereas DAPI-poly-P complexes produce a so-called bathochromic shift with emission around 525-550 nm (green-yellow). While DNA is usually stained by working solutions containing 1 µg mL^{-1} DAPI, poly-P detection is carried out by incubating cells with 50 µg mL^{-1} DAPI.

7.6.1 FISH – DAPI staining

7.6.1.1 Reagents and solutions for DAPI staining

- **DAPI stock solution and storage** (see Section 7.4.4.1)

- **Reagents used for FISH** (see Section 7.5.1)

7.6.1.2 Procedure

a. Perform FISH as described above using FISH probes designed to target the PAO, and the appropriate conditions for fixation, immobilisation and probe hybridisation.
b. After incubation in the dark for 30 min at room temperature (longer staining might be required for some samples), remove any excess DAPI by centrifugation and rinse with plenty of sterile water until any background fluorescence is minimal.
c. Allow to air-dry. A FISH and DAPI stained slide can be stored at -20 °C indefinitely.
d. Apply a drop of anti-fading agent (Citifluor, Vectashield or a mixture) directly onto the sample. Cover it with a cover slip and examine under the epifluorescence microscope. The excitation maximum for DAPI bound to dsDNA is 358 nm and 525-550 nm for poly-P. The emission maximum is 461 nm.

A filter set with a bandpass excitation maximum at 350 nm and a 500 nm long pass emission filter, is combined with the appropriate dichroic mirror (see the manufacturers' instructions).

The best results for the combined use of FISH with DAPI-poly-P staining are obtained with probes tagged with red-shifted fluorescence dyes such as Cy3 for FISH.

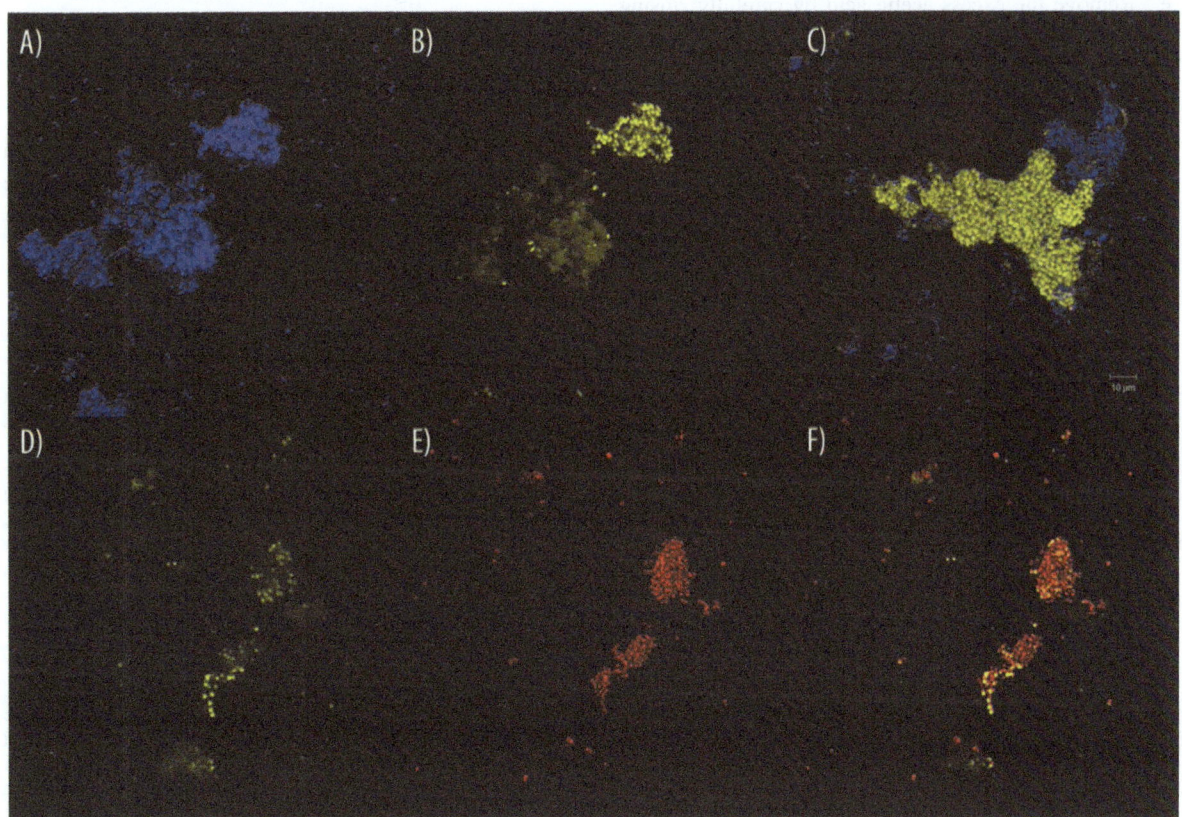

Figure 7.7 Combined use of DAPI staining (blue) for poly-P staining (yellow) and FISH (red) for identification of the cells. Image (A) and image (B), and (D), (E) and (F) show the same microscopic field, respectively.

7.6.2 FISH – PHA staining

7.6.2.1 Reagents and solutions for PHA staining

- **Reagents used for PHA staining**
  (i) 1 % (v/v) aqueous solution of Nile Blue
  (ii) 8 % (v/v) acetic acid

- **Reagents used for FISH** (see Section 7.5.1)

7.6.2.2 Procedure

a. Dilute freshly collected activated sludge with water to a final concentration of approximately 1 g MLSS L^{-1}, and transfer 20 µL to a gelatine-coated microscope slide. Allow to dry. Homogenization can be helpful to increase resolution.
b. Stain the sample by dipping the slide into a suspension of 1 % aqueous solution of Nile Blue (heated to 55 °C) for 10 min.
c. Remove the excess stain by carefully rinsing with dH_2O at room temperature.
d. Wash the stained cells for 1 min in 8 % acetic acid.
e. Remove the excess acetic acid by carefully rinsing with dH_2O.
f. Allow the slide to air-dry.
g. Add a drop of dH_2O to the slide, and examine with an epifluorescence microscope (excitation wavelength 630 nm). A positive PHA staining response will be shown by the presence of fluorescent PHA granules inside the cell, and be readily distinguishable from negatively-stained cells.
h. Record any field of interest with a charge coupled device (CCD) camera or a laser-scanning microscope and store the coordinates for that field on the microscopic micrometre stage. Alternatively mark the position of the light cone with a pencil on the side of the slide.
i. Remove the slide from the stage, and let the slide air-dry.
j. Fix the sample directly on the slide with PFA or EtOH followed by a permeabilization step (if required) as described under the FISH procedure. Allow the slide to air-dry.
k. FISH is carried out as described in the Section 7.5. For clear differentiation of the fluorescence signals from PHA staining and FISH hybridization, ensure oligonucleotide probes labelled with a strong fluorochrome clearly distinguishable from the PHA emission spectra (such as FLUOS, Cy5) are used.
l. Relocate the co-ordinates where the PHA images were acquired on the microscope stage, and evaluate the FISH signal. Obtain the images before and after applying the FISH and evaluate the identity of those cells with positive PHA staining.

PHA staining can be carried out on ethanol or PFA fixed samples, but fixation directly on the slide between the PHA staining and the FISH procedure should be avoided.

References

Bouvier, T., Del Giorgio, P.A. (2003). Factors influencing the detection of bacterial cells using fluorescence in situ hybridization (FISH): A quantitative review of published reports. *FEMS Microbiol Ecol.* 44(1): 3-15.

Böckelmann, U., Manz, W., Neu, T.R., and Szewzyk, U. (2002). Investigation of lotic microbial aggregates by a combined technique of fluorescent in situ hybridization and lectin-binding-analysis. *Journal of Microbiological Methods,* 49:75-87.

Eikelboom, D.H. (2000) Process control of activated sludge plants by microscopic investigation. IWA Publishing, London, UK. ISBN 1900222302.

Eikelboom, D.H. (2006) Identification and control of filamentous microorganisms in industrial wastewater treatment plants. IWA Publishing, London, UK. ISBN 1843390965.

Fuchs, B.M., Glöckner, F.O., Wulf, J., and Amann, R. (2000). Unlabelled helper oligonucleotides increase the in situ accessibility to 16S rRNA of fluorescently labelled oligonucleotide probes. *Applied and Environmental Microbiology,* 66(8): 3603-3607.

Haugland, R.P. (1999). Molecular probes. Handbook of fluorescent probes and research chemicals, 7th ed. Molecular Probes, Eugene, Oregon, US. ISBN 0965224007.

Jenkins, D., Richard, M.G., and Daigger, G.T. (2004) Manual on the causes and control of activated sludge bulking, foaming, and other solids separation problems. 3rd ed. IWA Publishing, London, UK. ISBN 9781566706476.

Kragelund, C., Remesova, Z., Nielsen, J.L., Thomsen, T.R., Eales, K.L., Seviour, R.J., Wanner, J., and Nielsen, P.H. (2007). Ecophysiology of mycolic acid containing Actinobacteria (Mycolata) in activated sludge foams. *FEMS Microbiol. Ecol.* 61:174-184.

Lopez-Vazquez, C.M., Hooijmans, C.M., Brdjanovic, D., Gijzen, H.J., van Loosdrecht, M.C.M. (2008). Factors affecting the microbial populations at full-scale Enhanced Biological Phosphorus Removal (EBPR) wastewater treatment plants in The Netherlands. *Water Res.,* 42(10-11): 2349-60

Loy, A., Horn, M., and Wagner, M. (2003). probeBase: an online resource for rRNA-targeted oligonucleotide probes. *Nucleic Acids Research,* 31(1): 514-516.

Moter, A., and Göbel, U.B. (2000). Fluorescence in situ hybridization (FISH) for direct visualization of microorganisms. *Journal of Microbiological Methods*, 41: 85-112.

Nierychlo, M., Nielsen, J.L., and Nielsen, P.H. (2015). Studies of the ecophysiology of single cells in microbial communities by (quantitative) microautoradiography and fluorescence in situ hybridization (MAR-FISH). In Hydrocarbon and lipid microbiology protocols, McGenity *et al.* (eds.), Springer Protocols Handbooks, Springer-Verlag, Berlin Heidelberg, Germany. ISBN 9783662451793.

Nielsen, P.H., Kragelund, C., Seviour, R.J., and Nielsen, J.L. (2009). Identity and ecophysiology of filamentous bacteria in activated sludge. *FEMS Mic. Review.* 33(6): 969-998.

Nielsen J.L. (2009). Protocol for Fluorescence in situ Hybridization (FISH) with rRNA-targeted oligonucleotides. In: FISH Handbook for biological wastewater treatment, Identification and quantification of microorganisms in activated sludge and biofilms by FISH. Nielsen, P.H., Daims, H., and Lemmer, H. (Eds). ISBN 978-1843392316.

Nielsen, J. L., Mikkelse, L. H., and Nielsen, P. H. (2001) *In situ* detection of cell surface hydrophobicity of probe-defined bacteria in activated sludge. *Water Sci. Tech.* 43: 97-103.

Nielsen, J.L., Kragelund, C., and Nielsen, P.H. (2010a). Ecophysiological analysis of microorganisms in complex microbial systems by combination of fluorescence in situ hybridization with extracellular staining techniques. In: Methods in Molecular Biology. Bioremediation. Humana Press Inc. Cummings, S. (Ed.). 599: 117-128. ISBN 9780444010827.

Nielsen, J.L., Kragelund, C., and Nielsen, P.H. (2010b). Combination of fluorescence in situ hybridization with staining techniques for cell viability and accumulation of PHA and poly-P in microorganisms in complex microbial systems. In: Methods in Molecular Biology. Bioremediation. Humana Press Inc. Cummings, S. (Ed.). 599: 103-116. ISBN 9780444010827.

Nielsen, P. H., Roslev, P., Dueholm, T.E., and Nielsen, J.L. (2002). *Microthrix parvicella*, a specialized lipid consumer in anaerobic-aerobic activated sludge plants. *Water Sci. Technol.* 46: 73-80.

Pernthaler, A., and Pernthaler, J. (2007). Fluorescence *in situ* hybridization for the identification of environmental microbes. Methods in Molecular Biology, vol. 353: Protocols for nucleic acid analysis by nonradioactive probes. 2nd ed. Hilario E. and Mackay J. (Eds). Humana Press Inc., Totowa, NJ. ISBN 9780444010827.

Salvadó, H., Gracia, M.P., Amigó, J.M. (1995), Capability of ciliated protozoa as indicators of effluent quality in activated sludge plants, *Wat. Res.,* 29: 1041-1050.

Seviour, R.J. and Nielsen, P.H. (2010). Microbial ecology of activated sludge. Seviour, R.J. and Nielsen, P.H. (Eds). IWA publishing, London, UK. ISBN 9781843390329.

Stoecker, K., Dorninger, C., Daims, H., and Wagner, M. (2010). Double labelling of oligonucleotide probes for fluorescence in situ hybridization (DOPE-FISH) improves signal intensity and increases rRNA accessibility. *Applied and Environmental Microbiology.* 76(3): 922-926.

Trotta, E., and Paci, M. (1998). Solution structure of DAPI selectively bound in the minor groove of a DNA T.T mismatch-containing site: NMR and molecular dynamics studies. *Nucleic Acids Res.* 26(20): 4706-4713.

van Ommen Kloecke, F., and Geesey, G.G. (1999) Localization and identification of populations of phosphatase-active bacterial cells associated with activated sludge flocs. *Microb. Ecol.* 38: 201-214.

Xia, Y., Kong, Y.H., Seviour, R., Forster, R.J., Kisidayova, S., and McAllister, T.A. (2014). Fluorescence in situ hybridization probing of protozoal Entodinium spp. and their methanogenic colonizers in the rumen of cattle fed alfalfa hay or triticale straw. *Journal of Applied Microbiology.* 116: 14-22.

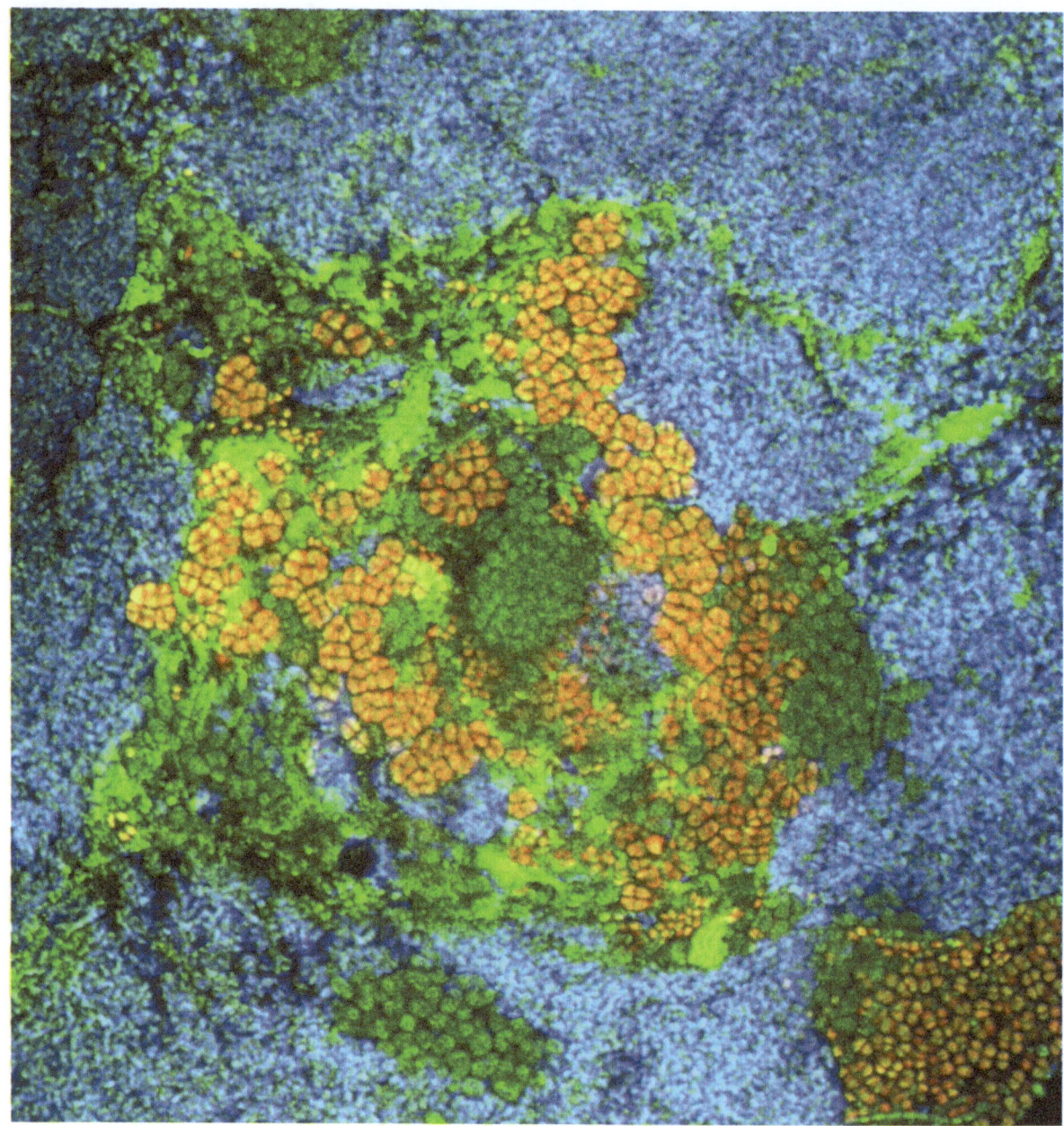

Figure 7.8 Result of the application of FISH procedure on the sludge sample from an EBPR removal reactor. Blue: Accumulibacter (PAOs targeted by probe PAOmix – Cy5); red: Defluviicoccus (GAOs targeted probe DF2mix – Cy3); green: Eubacteria (targeted by probe EUB338 mix), (image: McIlroy, 2016).

8

MOLECULAR METHODS

Authors:
Søren M. Karst
Mads Albertsen
Rasmus H. Kirkegaard
Morten S. Dueholm
Per H. Nielsen

Reviewer:
Holger Daims

8.1 INTRODUCTION

Molecular methods can be applied in the wastewater field for a fast, reliable and cheap identification of relevant microorganisms. In some cases, it is also possible to link the identification to a function, but surprises are frequently encountered. A recent example is that certain nitrifiers, which were believed to have a simple and well described physiology, are now known to be much more diverse than hitherto known (Daims et al., 2015). The function concerns both their physiology (heterotrophs, nitrifiers, fermenters etc.) as well as their morphology (filamentous or single cells), which is important for their overall effect in the wastewater systems. Through knowledge of the identity and function of the microorganisms, it may be possible to manipulate their presence to optimize the plant performance, e.g. ensure the presence of nitrifiers or removal of foam-forming filamentous species.

The molecular identification of microorganisms is usually based on the 16S rRNA gene. However, in some cases the identification is done using sequencing of functional genes instead, such as those encoding the ammonium monooxygenase enzyme (AMO) for ammonia oxidizers (Rotthauwe et al., 1997; Okano et al., 2004), as these provide higher phylogenetic resolution, which might be useful for fine-scale studies.

The most common methods applied in the wastewater field for identification have been real-time quantitative PCR (qPCR), clone library generation and fingerprinting techniques such as denaturing gradient gel electrophoresis (DGGE; Muyzer, 1999) or terminal restriction fragment length polymorphism (T-RFLP) (Marsh, 1999; Marzorati et al., 2008). However, these fingerprinting methods are hardly used anymore as they are often more difficult to use and provide less information compared to their high-throughput sequencing-based counterparts and their continued use cannot be recommended.

High-throughput sequencing can be applied for metagenomics or metatranscriptomics, where all the DNA or expressed genes (mRNA) from a certain community is sequenced. We do not, however, regard these methods as relevant for most readers of this book, as they require significant skills in molecular biology and bioinformatics.

Instead, high-throughput amplicon sequencing is recommended for routine analyses of microbial communities and will be described in greater detail. The method provides a list of microbes and an estimate of their relative abundance. One of the first sequencing platforms to be used for high-throughput amplicon sequencing was the Roche 454 (often termed 'pyrosequencing'). It is, however, now outdated (ultimo 2016) and the Illumina platform is presently dominating the amplicon sequencing market. By using the Illumina platform it is possible to analyse hundreds of samples in a fast, easy and cheap way compared to prior techniques.

The identification of microorganisms is usually done by comparison of the unknown sequences to a known reference set with a defined taxonomy. In this chapter we recommend the MiDAS database (midasfieldguide.org), which is a curated database that specifically targets microorganisms in the wastewater treatment field. Canonical or putative names for most common genus-level taxa are included and can be used as a common vocabulary for all researchers in the field to refer to the same organisms. The MiDAS database also provides all the available functional information about the 150 most abundant microorganisms encountered in Danish wastewater treatment plants (WWTP) and probably also worldwide (McIlroy et al., 2015).

In this chapter we will focus on the methods of choice today and the next few years in wastewater microbiology: DNA extraction, qPCR and amplicon sequencing.

8.2 EXTRACTION OF DNA

8.2.1 General considerations

An optimized and standardized protocol for DNA extraction is essential to any analysis of microbial composition using DNA sequencing. This is due to the simple fact that microbes differ enormously in their resistance to different lysing methods (Thomas et al., 2012; Guillén-Navarro et al., 2015). Hence, microbes with cell walls that are difficult to lyse will effectively seem less abundant if sub-optimal extraction protocols are used (Bollet et al., 1991; Filippidou et al., 2015). Furthermore, activated sludge samples contain various chemicals that render some techniques unsuccessful due to inhibition (Guo and Zhang, 2013). Thus, the method for DNA extraction needs to be robust in order to cope with the challenges presented by activated sludge. However, despite much research effort into different DNA extraction protocols, it seems unlikely that there will ever be a perfect protocol. The biases introduced in DNA extraction can only be minimised not circumvented (Guo and Zhang, 2013; Albertsen et al., 2015). The aim of this section is to give a brief overview of the steps involved in DNA extraction and to make general recommendations when working with activated sludge. In addition, a protocol optimized for the use in activated sludge is presented, based on the protocol developed by Albertsen et al. (2015).

8.2.2 Sampling

It is of key importance that the sample is representative of the activated sludge in the process tank of the plant or in a lab-scale reactor. For large-scale systems it is recommended to sample a larger volume (1 L) from a well-mixed tank, then perform homogenisation and finally sub-sample 3×2 mL aliquots that can be readily frozen and stored for years at -20 °C until analysed. Ideally, biological replicates are stored to ensure that sample variance can be analysed and that extra biomass is available in case something goes wrong. It is important to minimise the time from sampling to freezing as the changed conditions outside the original environment might favour the growth of some species over others, rendering the sample unsuitable for comparative analysis (Guo and Zhang, 2013). Sampling should preferably take place quite often, e.g. every week and the samples stored frozen in a 'bio-bank' for later use. As the number of samples grows fast, it is important to label each of them clearly and keep a log with sample IDs, reactor ID, dates, related chemical measurements, and any additional information that might be relevant for a later microbial analysis.

8.2.3 DNA extraction

DNA extraction involves a few general steps that are modified and combined in different ways in a range of commercial kits depending on the target organisms, type of environment, and the purpose for the extracted DNA. The common steps are disruption and cell lysis, protein removal, chemical removal and DNA elution (Figure 8.1).

8.2.3.1 Cell lysis

Various methods have been developed to lyse the cells in order to release their DNA. Some methods use chemicals to burst the cells, some use enzymatic degradation of cell structures, and others use physical stress such as freeze-thaw cycles, or mechanical stress such as ultra sound or bead beating (Bollet et al., 1991; Tsai and Olson, 1991;

Zhou *et al.*, 1996). Off-the-shelf kits have been developed that use combinations of these strategies optimised for different cell and sample types. The difficult nature of activated sludge presents a challenge for several of these approaches due to different kinds of inhibition (Tullis and Rubin, 1980). However, mechanical lysis has proven very robust and does not suffer from inhibition effects (Salonen *et al.*, 2010; Guo and Zhang, 2013; Albertsen *et al.*, 2015). The lysis of cells is often carried out in solutions with detergents and surfactants that support the disruption and removal of cell membrane constituents such as lipids by a subsequent centrifugation step.

If any parameters are modified in the cell lysis step, it is important to test the effects on yield and integrity of the extracted DNA. Increasing the intensity or duration of some steps will likely increase the yield until some level of saturation. However, the increased duration and intensity might shear DNA apart and fewer samples can be handled in a reasonable time (Bollet *et al.*, 1991; Bürgmann *et al.*, 2001).

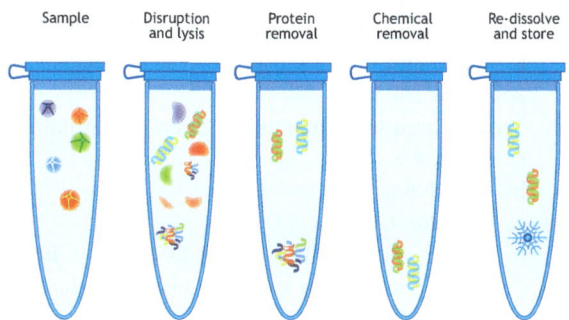

Figure 8.1 The main steps in DNA extraction.

8.2.3.2 Nuclease activity inhibition and protein removal

Microbial cells possess numerous enzymes that are specialised in degrading DNA (nucleases), and it is therefore essential that these are removed or inhibited as soon as possible after cell lysis. A common method to remove nuclease activity is by the addition of proteases, which are specialised in the degradation of proteins, including the nucleases. Afterwards the proteins are removed by increasing the salt concentration, which will cause them to precipitate. The precipitate can subsequently be removed by centrifugation, leaving the DNA in solution, and the proteins in the pellet (Miller *et al.*, 1999).

8.2.3.3 Purification

Along with the extracted DNA comes another type of nucleic acid, which is also found in the cell, the RNA. RNA molecules are often removed by the addition of the enzyme RNase that cleaves RNA specifically and leaves the DNA intact (Miller *et al.*, 1999). Following these steps, it is important to remove unwanted salts, detergents, proteins and other reagents used in the cell lysis process. This purification is usually carried out by precipitating the DNA with ethanol, as DNA is insoluble in this and can be pelleted by a centrifugation step (Bollet *et al.*, 1991). The DNA can be washed by replacing the supernatant with new ethanol. Alternatively, the DNA can be adsorbed to a matrix on a filter or a silica that allows washing steps and subsequent release by altering the salt concentration.

8.2.3.4 Elution and storage

After DNA has been isolated from the other cell constituents, and chemicals used in the extraction, the ethanol can be removed by evaporation and the DNA can be dissolved in DNase-free water or in a protective buffer solution such as Tris-EDTA (TE) buffer (Miller *et al.*, 1999). The purified DNA can be frozen and stored for years, or kept in the fridge for a few weeks.

8.2.4 Quantification and integrity

Depending on the downstream processing, it is important to quantify the DNA and check that it is not overly fragmented. The best quantification is provided using a fluorescence-based method that can distinguish between DNA and RNA, such as is implemented in the Qubit dsDNA kits. These kits can quantify DNA accurately at the very low concentrations needed for most sequencing-based methods (Singer *et al.*, 1997). However, the popular spectrophotometry-based nanodrop system can provide decent estimates for higher concentrations of DNA (> 20 ng μL^{-1}) when the DNA is very pure. The spectrum recorded by the nanodrop system also provides additional information about the purity of the DNA, as the ratio between the intensity at given wavelengths indicates the type of potential contaminants. This kind of information can be used to evaluate whether further purification steps are needed (Wilfinger *et al.*, 1997).

The size distribution of the DNA can be determined using classical gel electrophoresis (McMaster and Carmichael, 1977) or newer, more sensitive and typically faster techniques implemented in dedicated and fully automated systems such as the 'bioanalyzer' and 'tapestation' instruments (Panaro et al., 2000; Padmanaban et al., 2013). Size determination relies on the fact that longer DNA fragments will move slower through a gel matrix when exposed to an electric current than smaller fragments. Comparing the travelled distance of your extracted DNA with the travelled distance of fragments of known sizes, it is possible to estimate the length distribution of your DNA fragments (McMaster and Carmichael, 1977). Accurate estimates of the length and concentration of DNA molecules are crucial for some types of molecular analysis.

8.2.5 Optimised DNA extraction from wastewater activated sludge

This protocol explains DNA extraction from activated sludge from WWTP. The protocol is based on the FastDNA spin kit for soil protocol (MP Biomedicals) with some modifications, mainly streamlining and longer bead beating, as published by Albertsen et al. (2015).

The key in DNA extraction is consistency and hence this protocol should be followed to the letter. If you choose to deviate from the protocol do it consistently, for all samples, throughout your experiment.

8.2.5.1 Materials

Materials needed for DNA extraction are:

- A FastDNA spin kit for soil (MP Biomedicals).
- A FastPrep-24 (MP Biomedicals).
- A microcentrifuge (preferably with a cooler).
- Spintubes (DNAse-free), 1.5 mL.
- Falcon tubes, 15 mL.
- Ice.
- Ethanol.
- Pipettes (range 1 µL to 1000 µL).
- DNAse-free tips (10 µL, 300 µL and 1000 µL).
- Nuclease-free H$_2$O (Qiagen).
- A permanent marker (freeze-resistant).
- A label printer (optional).
- PPE: lab coat, safety glasses, gloves.

8.2.5.2 DNA Extraction

The total time needed for DNA extraction from a sample is approximately 4 h.

1. Sample input
   a. Target volume: 500 µL.
   b. Target Total Solids (TS): 2 mg.
   ▲ Critical step Never spin the sample down to increase concentration!
2. Prepare and mark tubes for the whole workflow (per sample):
   a. 1 × Lysing Matrix E tube (from the kit).
   b. 1 × SPIN™ filter (from the kit).
   c. 1 × catch tube (from the kit).
   d. 3 × 1.5 mL DNAse-free tubes.
   e. 1 × 15 mL Falcon tube.
3. Thaw the sample aliquot at room temperature and store on ice until used.
4. Add 480 µL Sodium Phosphate Buffer: PBS (pH 8.0) and 120 µL MT Buffer to each of the Lysing Matrix E tubes.
5. Add 250 µL PPS (Protein Precipitation Solution) to one 1.5 mL spintube for each sample.
6. Re-suspend the binding matrix and add 1.0 mL to each of the 15 mL Falcon tubes.

- **Bead-beating**
1. Mix the sample before use e.g. by vortexing.
2. Transfer a sample volume equal to 2 mg TS[1] to a Lysing Matrix E tube and add PBS so the total added volume is 500 µL[2].
3. Perform bead-beating in the FastPrep-24 instrument
   a. Time: 4 × 40 s.
   b. Speed: 6 m/s.
   c. Adaptor: Custom.
   d. ▲ Critical step Remember to load the bead-beating tubes in a balanced loading pattern. A balance tube may be required.
   e. ▲ Critical step Between each 40-seconds interval, the samples should be kept on ice for 2 min to cool down.

- **Protein precipitation and binding of DNA to matrix**
1. Spin down the samples at > 10,000 × g for 10 min, preferably at 4 °C.
2. After centrifugation, transfer the supernatants to 1.5 mL spintubes with PPS and then shake the tubes 10 times by hand.
3. ▲ Critical step Keep the tubes on ice until all the samples have been processed.
4. Centrifuge the tubes at 14,000 × g for 5 min to pellet the precipitate.

---

[1] 1-4 mg of TS is usually acceptable.

[2] Use a pipette tip with a wide orifice so that large granules are also selected.

5. Transfer the supernatant to the 15 mL tube with the binding matrix suspension.
6. Invert by hand for 2 min to allow binding of DNA to the matrix.
7. Place the tube in a rack for 3-5 min (or until the liquid appears clear) to allow settling of the silica matrix.
8. Remove and discard up to 2 × 750 µL of supernatant, being careful to avoid the settled binding matrix.
9. Re-suspend the binding matrix in the remaining amount of supernatant.

- **DNA washing and elution**
1. Transfer approximately 750 µL of the mixture to a SPIN™ filter and then centrifuge at 14,000 × g for 1 min.[3]
2. Empty the catch tube.
3. ▲ Critical step Ensure that ethanol has been added to the concentrated SEWS-M.
4. Add 500 µL prepared SEWS-M and gently re-suspend the pellet using the force of the liquid from the pipette tip - or by stirring with a pipette tip.
5. Centrifuge at 14,000 × g for 1 min.
6. Empty the catch tube and use it again.
7. Centrifuge at 14,000 × g for 2 min to 'dry' the matrix of residual wash solution.
8. Discard the catch tube and replace it with a new tube.[4]
9. Allow the SPIN™ filter to dry for 5 min at room temperature with an open lid.
10. Gently re-suspend the Binding Matrix (above the SPIN filter) in 60 µL of DES. Use a pipette tip to stir the matrix until it becomes liquid. Make sure not to disrupt the filter.
11. Centrifuge at 14,000 × g for 1 min to bring the eluted DNA into the clean catch tube. Discard the SPIN filter.
12. Label the tube appropriately either with a printed label or with a freeze-resistant marker.
13. ■ Pause Point Store DNA at -20 °C for short-term storage and -80 °C for long-term storage.

8.3 REAL-TIME QUANTITATIVE PCR (qPCR)

8.3.1 General considerations

Although high-throughput techniques such as metagenome and amplicon sequencing (see Section 8.4) have revolutionized the way we interrogate microbial communities, real-time quantitative PCR (qPCR) still remains the most sensitive technique for quantification of specific DNA species. Also, under optimal conditions it allows the detection of a single target sequence within the analysed sample, although such conditions are rarely achieved for environmental samples due to PCR inhibitory substances. Furthermore, qPCR may be used to convert the relative abundance data obtained from amplicon sequencing into absolute quantities, although this is rarely required.

In wastewater treatment systems, qPCR can be used to estimate the overall bacterial abundance (Horz et al., 2005) or to quantify bacteria belonging to specific taxonomic groups (Matsuda et al., 2007) using primers that target conserved or variable regions of the rRNA (16S or 23S) genes, respectively. It may also be used to estimate the abundance of bacteria belonging to specific functional groups, such as nitrifiers or polyphosphate-accumulating bacteria, using primers that target key functional genes (Ge et al., 2015). qPCR is also a useful tool for determining the fate of individual bioaugmentation strains. This can be done using primers that target unique genomic regions (Dueholm et al., 2015). qPCR may furthermore be used to track the spread of antibiotic resistance genes (Volkmann et al., 2004) and infectious viruses (Kitajima et al., 2014). By combining qPCR with reverse transcription (RT-qPCR), the activity (transcription) of specific genes can be quantified (Nolan et al., 2006), but this will not be covered here. qPCR is a refinement of the classical PCR (see Section 8.4.3) (Saiki et al., 1985) in which the PCR products are detected after each PCR cycle (Figure 8.2). The technique relies on the fact that in a typical PCR, the target sequence is amplified approximately two-fold for each PCR cycle until one or more reagents become limiting (Kubista et al., 2006). The PCR cycle, for which a detectable product appears, is known as the quantification cycle (C_q), and it relates to the abundance of the target sequence in the original sample (Brzoska and Hassan, 2014). The most widely used qPCR technologies are based on fluorescence reporters that allow the PCR products to be detected in real-time on thermal cyclers equipped with fluorescence detectors (Brzoska and Hassan, 2014). Two different chemistries are widely used for the detection, each with its benefits and drawbacks (Table 8.1). The first relies on the nucleic acid stain SYBR Green I, which becomes highly fluorescent upon intercalation with double-stranded DNA (dsDNA) (Figure 8.3A) (Zipper et al., 2004). The second relies on single-stranded DNA (ssDNA) hydrolysis probes that contain a fluorophore

---

[3] If you have more sample than 750 µL, you should repeat this step.

[4] The new tube is the tube that the sample is to be stored in so be sure to label it properly.

with an intrinsically strong fluorescence at the 5' end and a quencher molecule at the 3' end (Figure 8.3B) (Holland *et al.*, 1991). The intact probes do not fluoresce as the close proximity of the fluorophore and the quencher results in energy transfer from the fluorophore to the quencher by fluorescence resonance energy transfer (FRET) (Holland *et al.*, 1991). During the annealing step the probes bind to a DNA segment between the sequencing primers and are subsequently hydrolysed during the elongation by the 5'-3' exonuclease activity of DNA polymerase. This liberates the fluorophore from the quencher, leading to a fluorescence signal (Holland *et al.*, 1991). Other less common techniques have also been reviewed (Kubista *et al.*, 2006).

Table 8.1 Comparison of SYBR Green I and hydrolysis probe-based detection.

	SYBR Green I	Hydrolysis probe
Specificity	Detects all amplified double-stranded DNA, including non-specific reaction products.	Detects specific amplification products only.
Sensitivity	Depends on template quality and primer design and optimization.	1-10 copies.
Advantages	• Can detect the amplification of any double-stranded DNA sequence. • No probe is required, which can reduce the assay setup and running costs. • Multiple dyes can bind to a single amplified molecule, which increases the sensitivity for detecting amplification products. • Relative low cost of primers.	• Specific hybridization between the probe and target is required to generate a fluorescent signal, significantly reducing background and false positives. • Probes can be labelled with different, distinguishable reporter dyes. This makes it possible to have multiplex qPCR in one reaction tube. • Post-PCR processing is eliminated, which reduces the assay labour and material costs.
Disadvantages	• Because SYBR Green I dye binds to any double-stranded DNA (including non-specific double-stranded DNA sequences) it may generate false positive signals. • Cannot be used for multiplex qPCR (more then one primer set is used simultaneously).	• A different probe has to be synthesized for each unique target sequence. • Relative high cost of labelled probe.

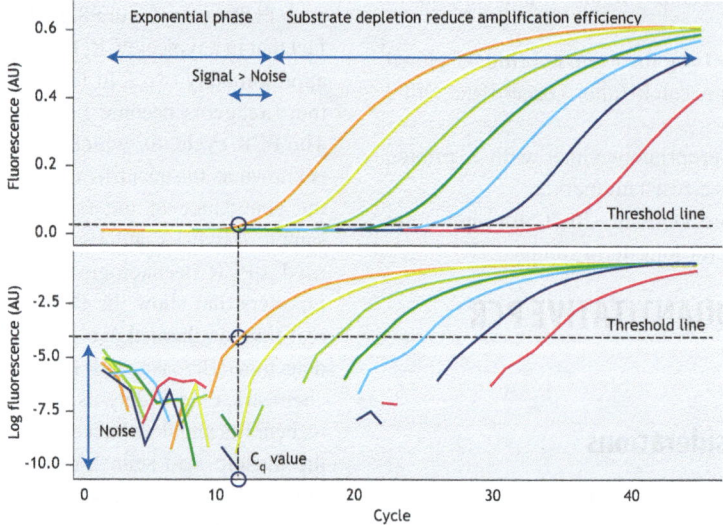

Figure 8.2 qPCR amplification curves. The upper panel shows traditional amplification plots and the lower shows the log-scaled data. The C_q value is determined from the qPCR cycle where the amplification curve intersects a threshold line placed where the fluorescence signal is significantly above the noise level. The threshold line can be determined manually from the log-scaled amplification plot (see the lower plot), but more often it is calculated by the qPCR software. An example is shown for the red amplification curve.

Before starting working with qPCR, it is recommended to read the minimal information for publication of quantitative real-time PCR experiment (MIQE) guidelines (Bustin *et al.*, 2009). The aim of this section is to introduce the reader to the basic theory behind qPCR and provide details on how to adapt qPCR assays from the literature for wastewater research. Special emphasis will be on the required controls and the pitfalls in respect to data interpretation.

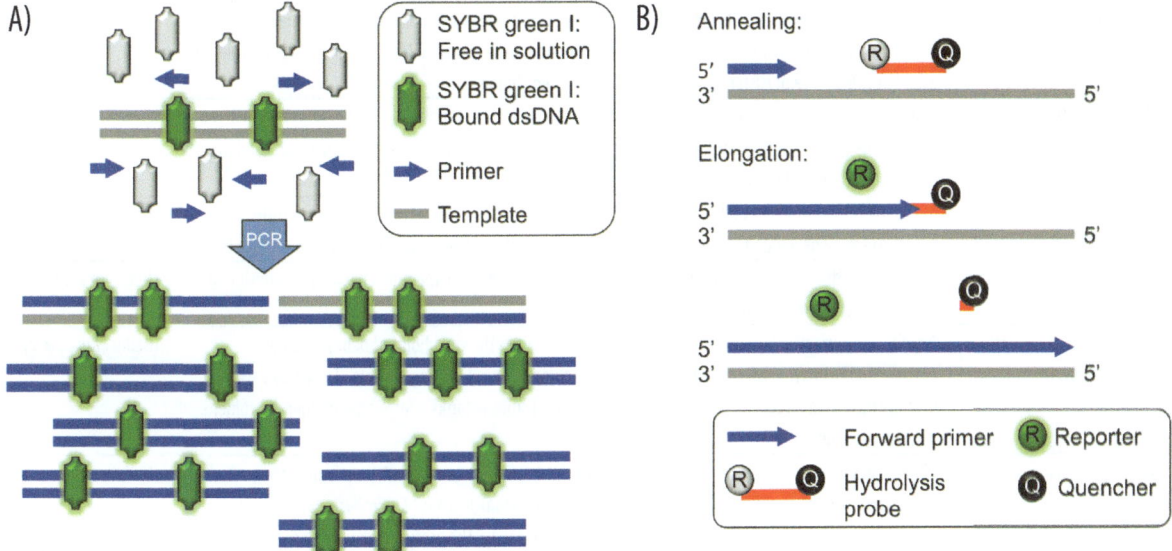

Figure 8.3 Mechanisms of common reporters used in qPCR. (A) SYBR Green I assay. The fluorescent reporter SYBR Green I is an asymmetric cyanine with two aromatic systems. When free in solution, SYBR Green I has virtually no fluorescence due to vibrations engaging both aromatic systems, which convert electronic excitation energy into heat that dissipates to the surrounding solvent. However, when SYBR Green I interacts with dsDNA, the vibrations are restricted and the asymmetric cyanine becomes highly fluorescent (Nygren *et al.*, 1998). SYBR Green I fluorescence consequently reflects the total abundance of dsDNA. (B) Hydrolysis probe-based assay. The hydrolysis probe contains a reporter fluorescent dye in the 5' end and a quencher dye in the 3' end that greatly reduces the fluorescence of the reporter of the intact probe due to fluorescence resonance energy transfer (FRET). The probe anneals to the coding strand between the forward and reverse primer-binding sites. During each elongation step, the DNA polymerase cleaves the reporter dye from the probe. Once separated from the quencher, the reporter dye emits its characteristic fluorescence (Holland *et al.*, 1991). The reporter fluorescence consequently reflects the total number of target amplifications.

8.3.2 Materials

- **Primers**

A description of some generally applicable qPCR assays relevant for wastewater treatment is provided in Table 8.2. Primers and probes should be ordered at a concentration of 100 μM as desalted and HPLC-purified stocks, respectively. Probes designed with the combination of a 5' 6-FAM reporter and a 3' TAMRA quencher are suitable for most applications. However, higher sensitivity can be obtained by replacing TAMRA with a non-fluorescent quencher, such as the black hole quencher-1 (BHQ-1) (Biosearch Technologies, USA). The design of new qPCR assays requires advanced bioinformatics skills and will not be covered here. For instructions on how to custom-design qPCR assays, consult the following literature (Basu, 2015; Brzoska and Hassan, 2014).

Table 8.2 Examples of generally applicable qPCR assays for wastewater treatment.

Target and chemistry[1]	Description	Reference
Most bacteria (HP)	A universal qPCR assay for the amplification of the 16S rRNA gene from the domain bacteria was designed and evaluated. The assay allows quantification of the total bacterial abundance within a sample.	(Nadkarni et al., 2002)
Most archaea (HP)	The coverage of multiple primers for the 16S rRNA gene from the domain Archaea and bacteria was evaluated. Specific primers allow the relative abundance of Archaea and bacteria to be determined in microbial communities from various habitats.	(Wang and Qian, 2009)
Most fungi (HP)	A universal qPCR assay for the amplification of the fungal 18S rRNA gene was designed and evaluated *in silico* and *in vitro*. The assay allows quantification of the total fungal abundance within a sample.	(Liu et al., 2012)
Individual strains (HP)	A general technique for the development of strain-specific qPCR assays was presented and used to design a qPCR assay for the bioaugmentation strain *Pseudomonas monteilii* SB3074. The assay was subsequently used to evaluate the persistence of the strain in activated sludge.	(Dueholm et al., 2015)
Nitrification (HP)	A qPCR assay that targets part of the ammonia-monooxygenase sub-unit alpha gene (*amoA*), which is a key enzyme in ammonia oxidation by ammonia-oxidizing bacteria (AOB), was developed and used to estimate the population size of AOB in soil samples.	(Okano et al., 2004)
Nitrate reduction (SG)	qPCR assays that targets membrane-bound (*narG*) and periplasmic (*napA*) protobacterial nitrate reductases were developed and used to determine their relative abundance in various environments.	(Bru et al., 2007)
Anaerobic ammonia oxidation (anammox) (SG)	A qPCR assay that specifically targets the 16S rRNA gene of all known anammox bacteria was designed and used to determine their abundance in wetland soils.	(Humbert et al., 2012)
Degradation of aromatic hydrocarbons (HP)	A qPCR assay was developed that targets *bssA*, which encodes the α-subunit of benzylsuccinate synthase, a key enzyme associated with anaerobic toluene and xylene degradation. The assay was used to study how gasohol releases from leaking underground storage tanks affected the indigenous toluene-degrading bacteria.	(Beller et al., 2002)
Antibiotic resistance (HP)	qPCR assays were designed that targeted the antibiotic-resistance genes *vanA*, *ampC*, and *mecA*, which are related to vancomycin-resistant enterococci (VRE), β-lactam-resistant *Enterobactreiaceae*, and methicillin-resistant *Staphylococcus aureus* (MRSA), respectively. The assays were used to detect the resistance genes in municipal and clinical wastewater.	(Volkmann et al., 2004)
Waterborne pathogenic viruses (HP)	qPCR assays were used to determine the relative abundance of 11 different viruses in the influent and effluent of two wastewater treatment plants.	(Kitajima et al., 2014)

[1] SG: SYBR Green I; HP: Hydrolysis probe.

- **Real-time thermal cycler**

Real-time thermal cyclers can be obtained from a large range of suppliers, including Agilent Technologies (USA), Applied Biosystems Inc. (USA), Bio-Rad (USA); Eppendorf International (Germany), and Roche Applied Science (Switzerland). Good experience was reported using the Agilent Mx3005P qPCR system from Agilent Technologies.

- **qPCR reagents**

There is a wealth of commercial qPCR kits available. We have good experience with the Brilliant III Ultra-Fast SYBR Green qPCR Master Mix (Agilent Technologies) for SYBR Green I-based assays and the EXPRESS qPCR Supermix (Life Technologies, USA) for hydrolysis probe-based assays.

- **Equipment for measuring DNA concentration**

Authors recommend the use of a Qubit fluorometer (Life Technologies, USA) or a similar probe-based technique for the determination of DNA concentrations. A NanoDrop spectrophotometer (Thermo Scientific, USA) may also be used, but this technique is more sensitive to sample impurities (see Section 8.2.4).

8.3.3 Methods

- **Preparation of qPCR standards**

The absolute concentration of the target DNA sequence is determined by comparing the C_q value of the sample to a standard dilution series with known concentrations of the target sequence (amplicon) (Figure 8.4).

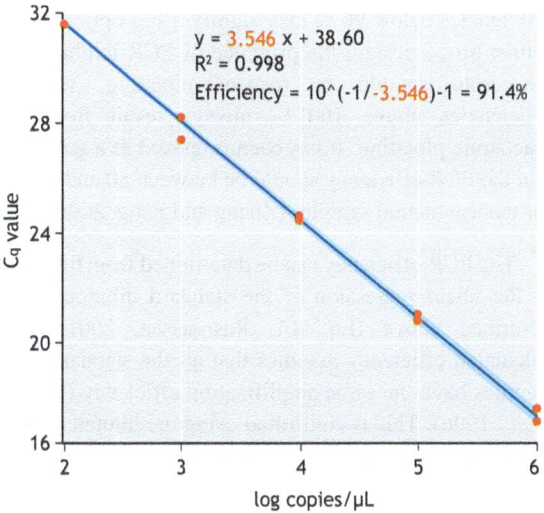

Figure 8.4 Evaluation of amplification efficiency from the slope of the linear regression of a standard dilution series.

The standards can be made from genomic DNA, plasmids, or PCR products. Although PCR products of the target sequence are easily obtained by PCR with the qPCR primers, these products often lead to a poor standard dilution series, as the small size makes it difficult to create reproducible dilutions. For routine qPCR assays, authors therefore recommend the use of linearized plasmids containing the amplicon. Linearization is important as the supercoiled circular confirmation of plasmid DNA may suppress PCR amplification (Hou et al., 2010). Linearized plasmid qPCR standards are easily made as described below. Kits and enzymes should be used according to manufacturers' recommendations unless otherwise stated.

1. Amplify the target sequence using the qPCR primers and a standard Taq-polymerase.
2. Clone the PCR product into the pCR4-TOPO plasmid using the TOPO TA cloning kit for sequencing and *E. coli* One Shot TOP10 cells (Life Technologies, USA).
3. Inoculate 10 mL LB medium containing 50 μg mL^{-1} kanamycin with a positive clone and grow the culture overnight (37 °C, 200 rpm).
4. Purify plasmids from the culture using the QIAprep Spin Miniprep Kit (Qiagen, USA).
5. Linearize the plasmids using FastDigest ScaI or FastDigest SspI (Thermo Scientific, USA).
6. Purify the linearized plasmids using the QIAEX II Suspension kit (Qiagen, USA).
7. Determine the DNA concentration using a Qubit fluorometer or a NanoDrop spectrophotometer.
8. Calculate the molecular weight of the linearized plasmid with an insert using the equation below.

   MW = final plasmid size in bp × 607.4 g mol^{-1}

9. Calculate the target sequence abundance in the sample using the equation below.

   Target sequence copies per μL = concentration in ng μL^{-1} × 10^{-9} × 6.022 × 10^{23} / MW

10. Dilute the amplicon stock to 10^9 copies per μL with 10 mM tris buffer, pH 8.5.
11. Create a 10-fold dilution series ranging from 10^8 to 10^1 copies per μL with 10 mM tris buffer, pH 8.5. Use a new pipette tip and vortex the sample after each dilution.
12. Transfer 100 μL of the standards to 200 μL PCR 8-tubes strips and store at -18 °C until used.

- **Sample preparation**
1. Purify DNA from the samples as described in 'DNA extraction' in Section 8.2.5.
2. Determine the DNA concentration using a Qubit fluorometer or a NanoDrop spectrophotometer.

- **qPCR reaction setup**
1. Prepare the qPCR master mix according to the protocol supplied with the qPCR kit. The master mix is usually prepared so that it is suited for 5 μL samples.
2. Load the master mix into a qPCR assay plate.
3. Centrifuge the assay plate at 2,200 × g for 5 min.
4. Add duplicates of the standard dilution series into the first two columns of the qPCR plate.
5. Add duplicates of the samples to the qPCR plate.
6. Add the appropriate controls described below to the qPCR plate.
   a. NTC (No template control): NTC is prepared with DNA-free water instead of a DNA template. It serves as a general control for extraneous nucleic acid contamination. When using SYBR Green chemistry, it also serves as an important control for primer dimer formation.
   b. NAC (No amplification control): NAC is prepared without the DNA polymerase. It functions as a control for background fluorescence that is not a function of the PCR. Such fluorescence is typically caused by the use of partly degraded hydrolysis probes.

Consequently, the NAC is unnecessary in SYBR Green assays.

c. Diluted sample controls: These are used to determine whether the sample contains PCR inhibitors. This is the case if the diluted sample yields a significantly higher copy number than the sample after correction for the dilution factor.

d. Amplicon-spiked samples: Selected samples are spiked with a known high concentration of the amplicon and serve as controls for the presence of PCR inhibitors.

7. Centrifuge the assay plate at 2,200 × g for 5 min.
8. Run the qPCR according to the protocol supplied with the qPCR kit. Adjust the annealing temperature and elongation time according to the assay description.
9. If using a SYBR Green assay, end the qPCR run with a melting curve analysis. This may identify primer dimer formation and the production of unspecific products. Both can be seen as additional peaks in plots of the first derivatives of melting curves. Primer dimers have considerable lower melting temperatures than the target amplicon.
10. When applying a new qPCR assay for the first time, it is always a good idea to validate the assay. To do this, purify the produced PCR product and send a small aliquot and either the forward or reverse primer to a company that performs Sanger sequencing. From the sequencing data, confirm that the amplified product is indeed the target sequence. The purified product may also be analysed on an agarose gel. A single band at the predicted length should be observed.

8.3.4 Data handling

- **Determination of sample copy numbers**

Most real-time thermal cyclers are able to carry out the data handling automatically and provide the copy number for each sample. However, the copy number may also be calculated by comparing the C_q values of the sample to those of the standard dilution series manually. To do this, plot the C_q values of the standard dilution series against the logarithm ($\log_{10}$) of the target sequence abundance and then perform linear regression. The obtained equation may subsequently be used to determine the target abundance from the samples C_q values (Figure 8.4).

- **Evaluation of PCR efficiency**

The PCR efficiency describes how the amplification deviates from the ideal situation, where the amplicon concentration doubles after each PCR cycle. PCR efficiencies below 90 % may signify a sub-optimal PCR primer/probe design, the presence of PCR inhibitors or inaccurate sample or reagent pipetting, whereas efficiencies above 100 % always result from the inaccurate pipetting. It has been proposed as a guideline that the PCR efficiency should be between 80 and 115 % for environmental samples (Zhang and Fang, 2006).

The PCR efficiency can be determined from the slope of the linear regression of the standard dilution series described above (Eq. 8.1, Rasmussen, 2001). The calculated efficiency assumes that all the standards and samples have the same amplification efficiency (Souazé et al., 1996). This is confirmed using the diluted sample or amplicon-spiked sample controls described earlier.

$$\text{Efficiency} = 10^{\frac{-1}{\text{slope}}} - 1 \qquad \text{Eq. 8.1}$$

8.3.5. Data output and interpretation

The final output of a qPCR analysis is a list of target sequence abundances for each sample. However, there are some important considerations to bear in mind when analysing the data that will significantly affect the final conclusions (Kim et al., 2013).

- **Extraction of nucleic acids is biased**

Samples from wastewater treatment systems contain a large diversity of microorganisms, which display considerable variation in their cell wall architecture (Saunders et al., 2015). Some of these are easy to lyse, whereas others are more difficult. There is consequently a significant bias introduced by the choice of nucleic acid extraction procedure (Albertsen et al., 2015). Accordingly, it may be very difficult to compare absolute quantification across studies. The DNA extraction protocol described in Section 8.2.5.2 provides results that are comparable to those of quantitative FISH analysis for wastewater treatment samples.

- **Quality of the template DNA**

Environmental samples, such as those from wastewater treatment systems, often contain compounds that have adverse effects on the PCR amplification (Bessetti, 2007). These may be humic acids, heavy metals, polysaccharides, phenolic compounds, or urea. Such inhibitors can be removed by sample polishing using adsorbent compounds, chemical washing or gel purification (Schriewer et al., 2011). However, it is important to always evaluate the removal of inhibitory compounds empirically as described in the Section 8.3.3 (Stults et al., 2001).

- **Specificity of broad-range qPCR assays**

The qPCR assays used in wastewater treatment often target microbial groups rather than individual species or strains. The assays consequently use generic primers and probes that are designed based on the known degeneracy of the target sequence. However, the known degeneracy might not always reflect that what is seen in nature, resulting in over or under estimation of the target sequence. The use of highly degenerate primers and probes also poses another problem. If the microbial community is heavily enriched in specific organisms, the perfect matching primers for these organisms will be quickly depleted, whereas primers for low-abundant organisms will be present for longer. The amplification will consequently be biased toward the low-abundant organism, resulting in an underestimation of the target sequence abundance. Finally, there may be differences in the amplification efficiency for each organism due to the variation in their GC content (Kim *et al.*, 2013).

- **Amplification of extracellular DNA (eDNA)**

Biological processes such as wastewater treatment rely on the active population of microorganisms. However, qPCR is unable to distinguish between DNA originating from active bacteria and extracellular DNA (eDNA) originating from dead and lysed cells. As wastewater treatment samples contain considerable amounts of eDNA, this may bias the data (Dominiak *et al.*, 2011). Care should therefore be taken when making conclusions about activity based on qPCR results.

- **Variation in the gene copy number**

Microbial genomes show a large variation in the copy number of metabolically important genes such as the 16S rRNA gene (Větrovský and Baldrian, 2013). This can bias quantification of the specific bacterial number unless the copy number is known. In addition, the number of whole genomes per cell may vary depending on the growth state of the bacteria (Ludwig and Schleifer, 2000). If the relative abundance of a specific bacterial species needs to be investigated, it is recommended to apply 16S rRNA amplicon data (see Section 8.4).

8.3.6 Troubleshooting

- **The sample contains PCR inhibitors**

There are three ways to circumvent PCR-inhibitor effects. The simplest option is dilution of the sample. PCR inhibitors are only effective above a certain concentration. However, dilution of the sample will also reduce the signal, leading to a less sensitive assay. The second option is to polish the DNA by applying an additional purification. This requires a concentrated sample, as material is always lost during purification. Large-scale sample polishing can be carried out in 96-well PCR plates using magnetic bead-based purification kits. The third and final solution is to purify the DNA from the original sample using another purification kit that is optimised for the given inhibitor.

- **The primer or probe design is not optimal**

A qPCR assay based on poor primers and probes should never be used. Instead, design and evaluate new primers and a probe set. Guidelines are given by several authors (Basu, 2015; Brzoska and Hassan, 2014).

- **Inaccurate sample and reagent pipetting**

Inaccurate calibration of pipettes is detrimental to qPCR. It is therefore recommended to keep a dedicated set of pipettes for qPCR that are regularly checked. The use of a multi-dispersal pipette is also recommended as it simplifies sample handling. Finally, it may be a good idea to review your pipetting techniques before carrying out qPCR.

8.3.7 Example

Enhanced degradation of specific pollutants can be achieved by the addition of catabolically relevant bacterial strains to the activated sludge in WWTPs (El Fantroussi and Agathos, 2005). This is known as bioaugmentation. Successful bioaugmentation requires that the introduced strains are able to thrive in the new environment (Thompson *et al.*, 2005). qPCR may be used to evaluate the persistence of bioaugmentation strains *in situ*. A strain-specific qPCR assay can be developed based on unique genomic sequences in the bioaugmentation strains. The abundance of the strains can subsequently be determined using DNA extracted from activated sludge at various time points after the addition of the bioaugmentation strain (Dueholm *et al.*, 2015). Here we show an example of how qPCR has been used to evaluate the persistence of the bioaugmentation strains *Pseudomonas monteilii* SB3078 and SB3101, which are used for the degradation of aromatic hydrocarbons (Dueholm *et al.*, 2014; 2015).

8.3.7.1 Samples

The bioaugmentation strain *P. monteilii* SB3078 or SB3101 was introduced into 100 mL fresh activated sludge obtained from Aalborg East WWTP in an abundance of 1 % based on the cell number. It was roughly estimated that an suspended solids (SS) = 1 g L^{-1} (activated sludge) and an OD$_{600\ nm}$ = 1 (*Pseudomonas* pure culture) both correspond to approximately 10^9 cells mL^{-1} as previously shown (Frølund *et al.*, 1996). Benzene was added to 10 μg mL^{-1}. The culturing flasks were crimp-sealed using butyl rubber stoppers and incubated at 25 °C, 150 rpm, for 4 days. The flasks were opened every 12 h for 30 min to allow evaporation of the remaining trace levels of benzene and settling of the sludge particles. 50 mL of effluent water was then removed and replaced by 50 mL of primarily settled wastewater, simulating a hydraulic retention time of 24 h. Samples for DNA extraction were collected and benzene reintroduced to 10 μg mL^{-1}. The flasks were then sealed and the incubations continued (Dueholm *et al.*, 2015). DNA was essentially extracted as described in Section 8.2.

8.3.7.2 qPCR reaction setup

1. Prepare the qPCR master mix as described below. The primers and probe target both SB3078 and SB3101.

Item	Final concentration	Per reaction (20μL)	100 × reactions
EXPRESS qPCR Supermix	1 ×	10 μL	1,000 μL
ROX (25 μM)	50 nM	0.04 μL	4 μL
Forward primer (100 μM)	500 nM	0.10 μL	10 μL
Reverse primer (100 μM)	500 nM	0.10 μL	10 μL
Hydrolysis probe (100 μM)	200 nM	0.04 μL	4 μL
DEPC water	-	4.72 μL	472 μL
Aliquot	-	15 μL	-

2. Load the master mixes into a qPCR assay plate and add 5 μL of the samples and controls (see above).
3. Centrifuge the assay plate at 2,200 × g for 5 min.
4. Run the qPCR according to the following program:
   - 50 °C for 2 min (UDG incubation).
   - 95 °C for 2 min.
   - 45 cycles of:
     - 95 °C for 15 s.
     - 60 °C for 1 min.

8.3.7.3 Results

We started by evaluating the amplification efficiency. The C_q values of the standard dilution series were plotted against the logarithm of the copy number and linear regression was then performed. This yielded the following fitting equation:

$y = -3.349 \cdot x + 39.80; R^2 = 1$

The efficiency was then calculated from the slope as:

Efficiency = $10^{(-1/slope)} - 1 = 10^{(-1/-3.349)} - 1 = 98.9\%$

This was well within the acceptable regime for environmental samples of 80-115 % (Zhang and Fang, 2006).

Next, we evaluated the controls. The NTC and NAC control did not amplify within the 45 cycles. This confirmed that there were no amplifiable contaminants present in the reagents and that the probes were stable, respectively. A sample containing DNA extracted from untreated wastewater was also analysed. This control did not amplify either, confirming the specificity of the qPCR assay. Finally, we investigated the amplification of a few diluted samples. These produced similar results to the undiluted samples, confirming no significant effect of inhibitors. And we had a look at the experimental data (Figure 8.5).

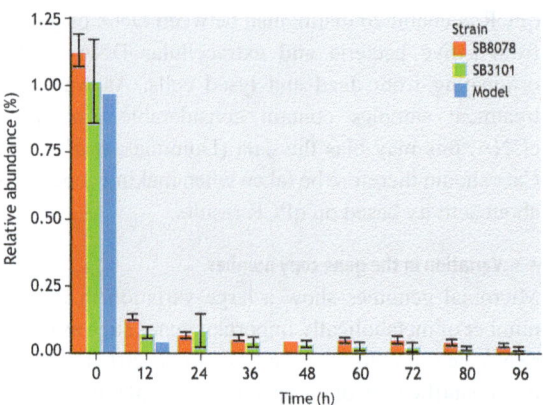

Figure 8.5 Persistence of *P. monteilii* SB3078 and SB3101 in activated sludge treated as a sequential batch reactor. Relative abundance of the bioaugmentation strains was determined using the strain-specific qPCR assay. The modelled data represents the theoretical decrease in cells that would be observed if there were no net growth and all bioaugmentation cells were planktonic. The volume occupied by solid material was calculated based on the diluted sludge volume index and this information was used together with the hydraulic retention time to calculate the rate by which planktonic cells were washed away.

The qPCR assay showed that approximately 90 % of the added bioaugmentation strains were lost within the first 24 h. This was probably due to the removal of planktonic cells with the effluent water, as it is replaced by fresh wastewater every 12 h. The remaining bioaugmentation cells were stabilised within the sludge and were able to survive throughout the experiment (4 d). The data furthermore showed that SB3078 was more persistent than SB3101 in the activated sludge settings.

8.4 AMPLICON SEQUENCING

8.4.1 General considerations

The first step of trying to understand how the bacteria in activated sludge impact the performance of a wastewater treatment plant is to get an overview of the bacterial community. This involves identifying the bacteria, their abundance and knowledge about what they are doing. Advances in DNA sequencing have made it possible to identify bacteria with high resolution and throughput by reading the 16S ribosomal RNA (rRNA) genes of the bacteria and using them as 'fingerprints'.

The approach is called 16S rRNA amplicon sequencing and consists of a number of steps, which are depicted in Figure 8.6.

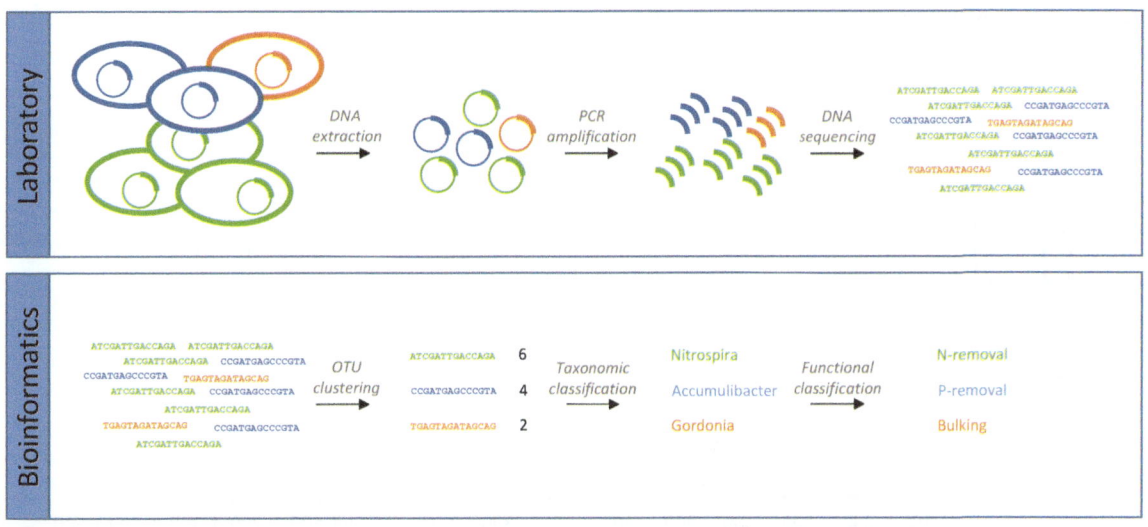

Figure 8.6 Overview of the basic steps in the analysis of microbial communities using 16S rRNA amplicon sequencing.

The first step is '*DNA extraction*', whereby genomic DNA from all the bacteria in a sample is extracted and purified (see Section 8.2). Afterwards, the 16S rRNA genes are amplified and prepared for sequencing by 'Polymerase chain reaction (PCR)'. The 16S rRNA gene amplicons are then read, or 'DNA sequenced', on a next-generation sequencing instrument. The output is the sequences of all the 16S rRNA genes in the sample. These sequences are then clustered into groups that represent a species, and the relative number of 16S rRNA genes belonging to each group is counted. Each group of species is identified through 'Taxonomic classification' by comparing the representative 16S rRNA gene sequence of each group to a database of known bacteria. The result is a table containing the name of each species in the sample and their relative abundance. This table is the basis for visualizing and analysing the bacterial community. The name of the species can also be used to link to functional information found in the literature or public databases such as MiDAS, midasfieldguide.org (McIlroy et al., 2015), e.g. if some of the identified species are known foam formers or nitrifiers.

In the Sections 8.4.2 to 8.4.7, each step of the 16S rRNA amplicon sequencing approach will be described in detail. The descriptions are based on 16S rRNA amplicon sequencing using the Illumina sequencing platform. However, the basic idea is the same for all sequencing platforms.

8.4.2 The 16S rRNA gene as a phylogenetic marker gene

A phylogenetic marker gene encodes an essential function, which is shared by all the organisms that are to be targeted and which have not been subjected to lateral gene transfer. In addition, the marker gene also has to have both evolutionary highly conserved positions as well as highly variable positions in its nucleotide sequence. The conserved parts make it possible to target the gene with a PCR and are needed for correct phylogenetic analysis, while the variable parts enable us to distinguish between different organisms and to investigate their relatedness (phylogeny).

Ribosomal genes have been the choice for phylogenetic analysis since Woese and Fox used them to show the division of life into three separate kingdoms in 1977 (Woese and Fox, 1977; Pace et al., 2012). Today the 16S rRNA gene is by far the most applied phylogenetic marker gene in environmental studies of bacterial diversity.

The 16S rRNA gene encode for a piece of RNA that makes up a functional part of the bacterial ribosome. Ribosomes are the protein factories of all cellular life forms and developed early in evolution. Conserved regions are crucial for correct ribosome structure and function, which means that most mutations in these regions are strongly selected against. Variable regions have more freedom for change, and mutations happen much more frequently (Madigan and Martinko, 2006). Therefore, the 16S rRNA gene contains several conserved islands with variable regions in between, called variable regions 1 to 9 (V1 to V9) (Ashelford et al., 2005); see Figure 8.7.

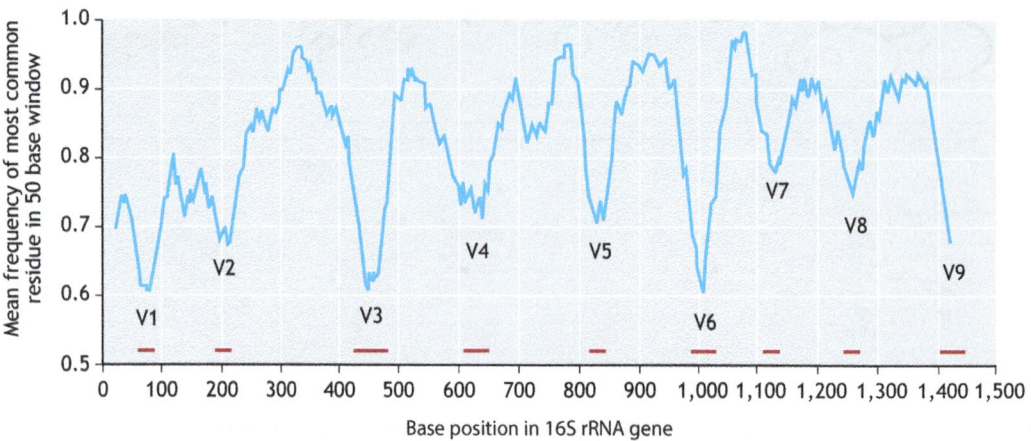

Figure 8.7 Variability in sequence composition across the 16S rRNA gene (adopted from Ashelford et al., 2005).

This makes the 16S rRNA gene an excellent marker gene as it can be targeted in its entirety or in fragments, which provides some technical flexibility. The 16S rRNA gene has been used as a marker gene for many years, and the accumulated knowledge is compiled in extensive databases that are used for comparing 16S rRNA gene data and determining the phylogenetic affiliation of new 16S rRNA genes and assigning them to a group within the bacterial taxonomy. Usually it is possible to determine the taxonomy of bacteria down to the species level by its 16S rRNA gene sequences. The ubiquity and the resolution of the 16S rRNA gene, together with the database resources, make it the preferred marker gene for bacterial community analysis.

Other marker genes are used for bacterial community analysis as well but to a lesser extent, and often to obtain higher phylogenetic resolution e.g. to strain level. It is common for these marker genes to be more variable in their sequence composition, which provides higher phylogenetic resolution. However, this also means that they only target specific subgroups of bacteria. Examples of such marker genes are the *amoA* to target ammonium-oxidizing organisms (AOO) and the *mcrA* in methanogens; see also Section 8.3 about qPCR. The principles for analysing these other marker genes are similar to the 16S rRNA gene, although protein-coding genes such as *amoA* can also be analysed at the level of amino acid sequence. For strain resolution, nucleotide sequences are needed, but they must be aligned codon-wise based on an amino-acid alignment (Juretschko et al., 2000).

8.4.3 PCR amplification

The first step in 16S rRNA amplicon sequencing is amplification of the 16S rRNA genes by PCR.

8.4.3.1 PCR reaction

The PCR is used to selectively amplify the 16S RNA gene from the background genomic DNA, so there is enough material that it is practical to analyse with the DNA sequencer (Figure 8.8).

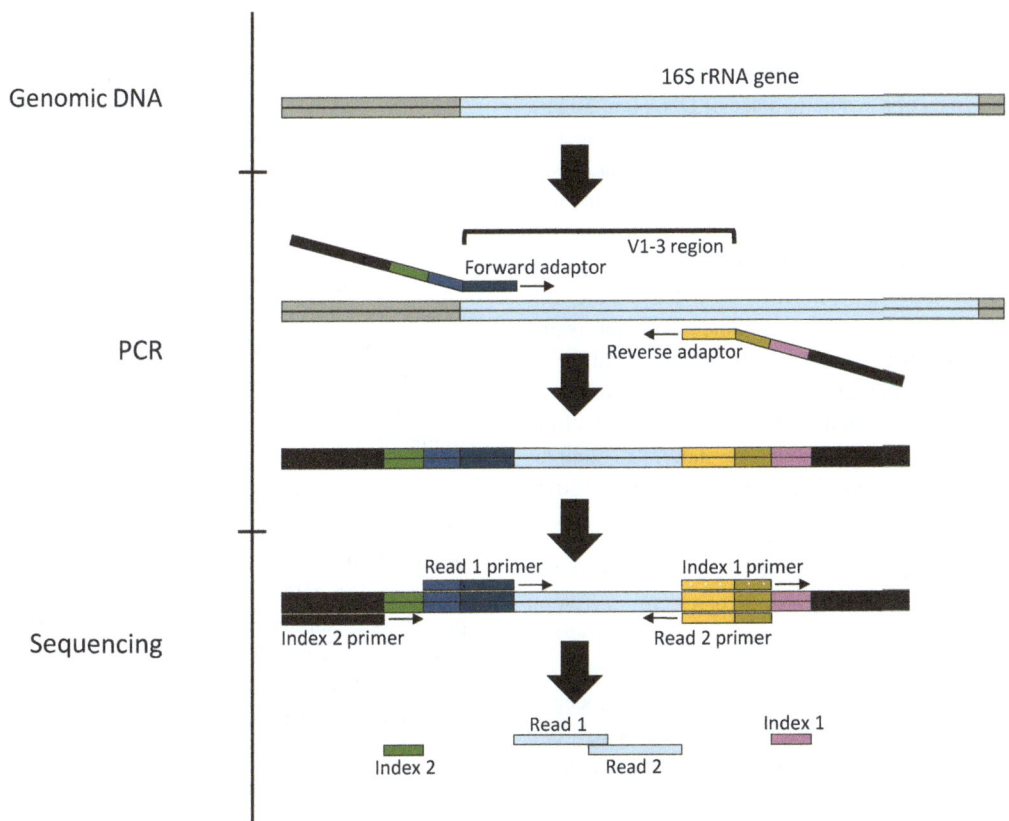

Figure 8.8 The basic steps in PCR amplification and sequencing of the 16S rRNA gene. Grey represents genomic DNA and light blue represents the 16S rRNA gene. The coloured parts of the adaptors and primers represent different functional sequences, and these are described in Section 8.4.11. The arrows signify the elongation/sequencing direction.

PCR uses a thermostable polymerase enzyme to copy DNA. The polymerase copies the DNA using small pieces of synthetic DNA (15-25 base pair: bp), called primers, as the starting point for the amplification. When analysing the 16S rRNA genes, the primers are designed to selectively match the conserved parts of the 16S rRNA gene, which enables a selective amplification of the region in between. During the PCR, cycles of heating and cooling are used to enable the amplification. First, the dsDNA is denatured by heating, which splits the two strands from each other. Secondly, the reaction is cooled to enable the primers to anneal to their target sites on the DNA. Thirdly, the temperature is increased again to provide an optimal temperature for the polymerase activity, which starts the copying of the 16S rRNA gene by extending the primer. One cycle creates two copies of the 16S rRNA gene from one original copy, and the copies can themselves function as templates. The cycle is repeated 25-35 times during the PCR, which results in an exponential amplification of the 16S rRNA gene. The copied gene product from PCR amplification is called an amplicon, hence the name 16S rRNA amplicon sequencing. Other than the DNA, primers and polymerase, the reaction contains the nucleotides that are incorporated into the new DNA along with buffers and other additives (Mg^{2+} salts etc.) that provide the optimal conditions for primer annealing and polymerase activity. For a detailed discussion of the principles behind PCR see Green and Sambrook (2012).

Another important role of the PCR is to enable sequencing of the 16S rRNA amplicons. This is achieved by having a sequencing adaptor attached to the end of the primers used in the PCR. Hereby, the adaptor is attached to all the 16S rRNA amplicons that are produced. The adaptor consists of a synthetic piece of DNA (ca. 50 bp), which has different 'active' components (see Section 8.4.11). These components enable the sequencing machine to catch the 16S rRNA amplicons, to start reading them and to recognize which sample each 16S rRNA amplicon originated from, which is referred to as barcoding or indexing. The idea is to tag all the 16S rRNA amplicons from one sample with the same barcode. This enables a large number of different samples to be mixed together (multiplexing) and to be read at the same time on the DNA sequencer, and still to be able to separate the data from the individual samples afterwards (de-multiplexing) (Illumina Inc., 2015; Caporaso et al., 2010).

The final product after PCR is called a 16S rRNA amplicon sequencing library. There are different strategies that can be used when preparing 16S rRNA amplicon sequencing libraries, but the overall idea is the same. The strategy described above uses a single step to amplify and attach adaptors; other strategies use two separate PCRs. These strategies have pros and cons in respect to cost, time and sequencing requirements. Currently, it is only feasible to sequence long 16S rRNA gene fragments with the one-step PCR strategy described. This includes the V1-3 fragment commonly used in activated sludge (Albertsen et al., 2015).

8.4.3.2 PCR biases

Different types of biases can be introduced in the PCR step, which will affect the final observed community structure. Primer bias is one of the most significant ones, and this will be addressed in the following paragraph (Albertsen et al., 2015). PCR drift is a bias introduced by stochastic events in the first cycles of the PCR, where relatively few molecules are involved in the replication or simply by reagent/sample handling variations (pipetting, position in thermocycler etc.). PCR selection bias is due to varying amplification efficiencies caused by physical properties of the 16S rRNA gene nucleotide sequence (Polz et al., 1998; Kennedy et al., 2014). To reduce the impact of PCR drift and selection, replicate PCR reactions are performed, the number of PCR cycles is kept to a minimum and the amount of template DNA should be around 10 ng. Most standard amplicon sequencing protocols have optimized these parameters.

8.4.3.3 Primer choice

As mentioned above, the primer set matches conserved parts of the 16S rRNA gene. However, it is unavoidable to have some variability in the 'conserved' part of the 16S rRNA gene. Therefore, primers will match some bacteria better than others and for some they will not match at all (Klindworth et al., 2013). This introduces a significant primer bias to the whole analysis, which is important to be aware of.

When analysing the 16S rRNA gene it would be ideal to sequence the entire gene (approximately 1,600 bp) as this provides maximum phylogenetic resolution. However, due to limitations in the Illumina sequencing technology, it is currently only possible to sequence fragments up to 550 bp of the 16S rRNA gene.

As a consequence of the primer bias and the limitation in reading length, many primer sets have been designed that target different variable regions of the bacterial 16S rRNA gene. The most commonly used primer sets target the V1-3, V4 and V3-4 regions (Albertsen et al., 2015). The primer sets have different biases, and when selecting a primer set several things should be considered.

a. The primer set should have least possible bias against the bacteria in your samples that you are most interested in. While it is possible to get an idea of the primer bias via *in silico* analysis (Klindworth et al., 2013), it is always recommended to test the primers by sequencing.
b. The primer set should be the same as the studies you want to compare to. The MiDAS database, which attempts to summarize all the current knowledge about important bacteria in activated sludge, is based on primers targeting the V1-3 region. This is due to a good resolution and broad coverage of bacteria that are responsible for the processes of interest in the activated sludge community (Albertsen et al., 2015).

For samples from specialized activated sludge systems it can be a good idea to test other primer sets. For example, the most common anammox bacteria are not targeted very well by the V1-3 primer set, and it is better to use V4 primer sets instead (Laureni et al., 2015; Gilbert et al., 2014). It is also possible to design new primers, but usually this is not recommended, since it requires expert knowledge of microbial ecosystems and phylogeny as well as extensive laboratory time for optimization and validation.

8.4.4 DNA sequencing

8.4.4.1 Sequencing platform

After the 16S rRNA amplicon libraries have been prepared, there are a number of different options for DNA sequencing. However, each method employs markedly different strategies for sequencing, which means the resulting data is suited for different purposes and needs. In respect to 16S rRNA amplicon sequencing, the most important criteria are sequencing length (> 200 bp), sequencing quality (< 1 % errors), data yield (> 10,000 reads per sample), turnover time, cost and how easy library preparation is. Currently, early 2016, the Illumina MiSeq platform is the method of choice, when compromising between these criteria. The Illumina MiSeq enables analysis of up to 400 16S rRNA amplicon libraries (50,000 reads per sample) in a single sequencing run (56 h). The MiSeq is currently able to sequence 301 bp from each end of the 16S rRNA amplicons. This is also termed paired-end (PE) sequencing and each of the two 301 bp sequences are termed 'reads'. During data processing the two reads are merged together at the overlapping ends to obtain a maximum length of approximately 550 bp. For more specialized use, which requires longer read length or shorter turn-around time, other platforms are better suited e.g. Pacbio RS II (Pacific Biosciences) or the Ion Proton System (Thermo Fisher Scientific Inc), and soon the MinION (Oxford Nanopore Technologies). Be aware that different library preparation protocols have to be used for the different sequencing platforms.

8.4.4.2 Sequencing depth

When performing amplicon sequencing, it is important to have a rough estimate of the sequencing depth needed (number of reads per sample) in order to answer the questions posed through the experimental design. For general bacterial community analysis targeting the V1-3 of 16S rRNA gene in activated sludge, 50,000 raw PE reads per sample are routinely used. This is done from the rationale that it is often rather similar communities that are compared and it is usually important to get robust estimations of the abundance of the individual community members. A rule of thumb is not to have less than 100 reads from the bacteria of interest. Below 100 reads the final results become very uncertain, due to biological and technical variation (Albertsen et al., 2015). If higher resolution is needed, the best option is to include more biological replicates. However, if this is not an option, deeper sequencing can also be used. Activated sludge contains thousands of different bacteria, and the most important 100 of these account for more than 70 % of the total community abundance (Saunders et al., 2015). On average each of these 100 species make up > 0.5 % of the total community. To obtain > 100 reads from each of these species at least 20,000 bioinformatic processed reads per sample are needed or > 30,000 raw PE reads, depending on the sequencing quality. It should be noted that the sequencing cost is usually not the most expensive part of the analysis and hence it is often preferred to make sure that enough reads are sequenced.

8.4.5 Bioinformatic processing

8.4.5.1 Available software

A rigorous standard procedure for the handling of 16S rRNA sequencing data has still not been defined, and probably will not be anytime soon due to a rapidly developing field. However, the general idea remains the same and is depicted in Figure 8.9.

Many research groups make custom workflows and code performing processing of the data, and some have made comprehensive software bundles of these that can perform almost completely automatically. The most popular are QIIME (Caporaso et al., 2010), Mothur (Schloss et al., 2009) and UPARSE (Edgar, 2013). Their settings and underlying assumptions differ, also between versions, which will produce somewhat different results even from the same sequence data. Therefore, results from different software packages and versions should not be compared. Generally, it is advised to analyse all the data from an experiment with one software package in one session and to re-run all the analysis if some data is added or if settings are changed. In the following paragraphs, some general observations are made regarding the bioinformatic processing.

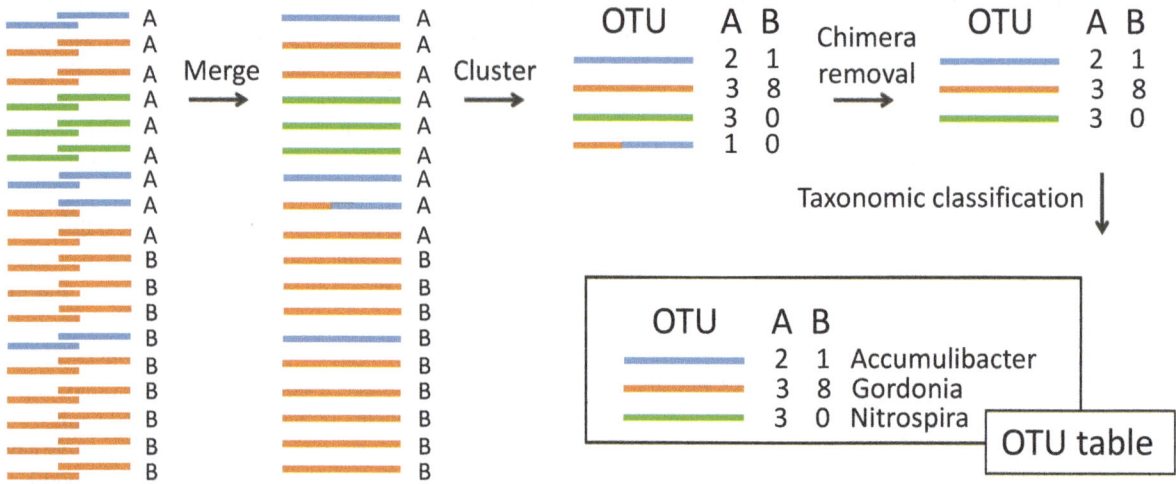

Figure 8.9 The basic steps in bioinformatic processing of 16S rRNA data from the Illumina MiSeq platform. The colours signify sequences that are > 97 % identical. The letters A a dn B signify what sample a sequence is from.

8.4.5.2 Raw data

The output from the Illumina MiSeq sequencer is a pair of files for each sample. Each file pair contains all Read1 of the paired reads in the first file and all Read2 in the second file. The files are fastq files (.fastq or .fq), which are simple text files with specific rules for listing sequence information (Figure 8.10).

Each read in a fastq file takes up four lines: *i*) The ID of the specific read starting with '@', *ii*) The raw sequence letters of the specific read, *iii*) '+' which works as a description field, but most often blank, and *iv*) Phred quality scores denoted by ASCII letters for each of the bases in the read (Cock *et al.*, 2010).

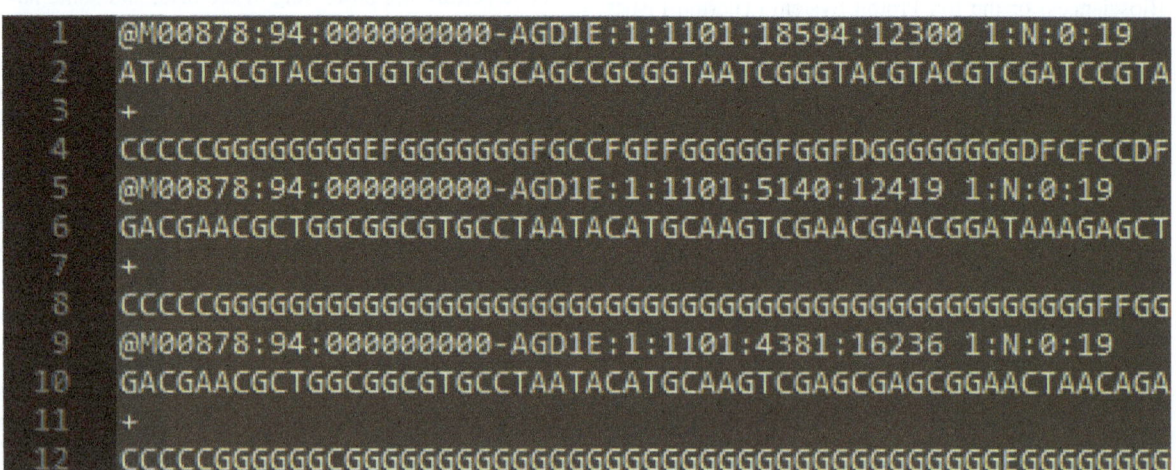

Figure 8.10 Example of a fastq file with three reads. The reads are truncated for readability.

8.4.5.3 Quality scores and filtering

The Phred quality score is a measure of the probability that a specific base contains an error (Q10 = 10 %, Q20 = 1 % and Q30 = 0.1 %) (Cock et al., 2010). For 16S rRNA amplicon sequencing, Q20 and above is preferred. The quality can be assessed by plotting the Phred quality scores, and the error rate of an internal standard. For 2 × 301 bp paired-end sequencing on a MiSeq of high-quality amplicon libraries, the average quality is usually above Q30 for the first 250 bp of Read1 and 200 bp of Read2, and after that the quality of the reads usually deteriorates (Figure 8.11). If the majority of the read quality is below Q20, something might have gone wrong during the sequencing. The bad quality data is removed through trimming and filtering. Often, if long stretches of a read have below Q20, the stretches are trimmed off from the 3' end. When sequencing V1-3 16S rRNA amplicons, reads that < 275 bp after trimming are discarded, since they are not suited for merging.

8.4.5.4 Merging paired-end reads

After quality filtering, the paired reads in the remaining high quality data are merged to create one continuous 16S rRNA gene sequence of a length of 250-550 bp depending on the targeted variable regions. After the filtering and merging there are usually around 70 % reads left, depending on the quality of the specific sequencing run. However, it is critical to be careful in the merging step as large biases can be introduced here. The length of the variable regions differ between species, e.g. the V1-3 region of the 16S rRNA gene can be between 425 and 525 bp. If the read quality is poor, 2 × 300 bp paired reads might be trimmed to 2 × 245 bp (490 bp), which is too short for merging of read pairs originating from species with large V1-3 fragments. Read pairs that doesn't merge are always discarded and thereby a bias is introduced. This can be prevented by discarding all reads < 275 bp before merging as mentioned before.

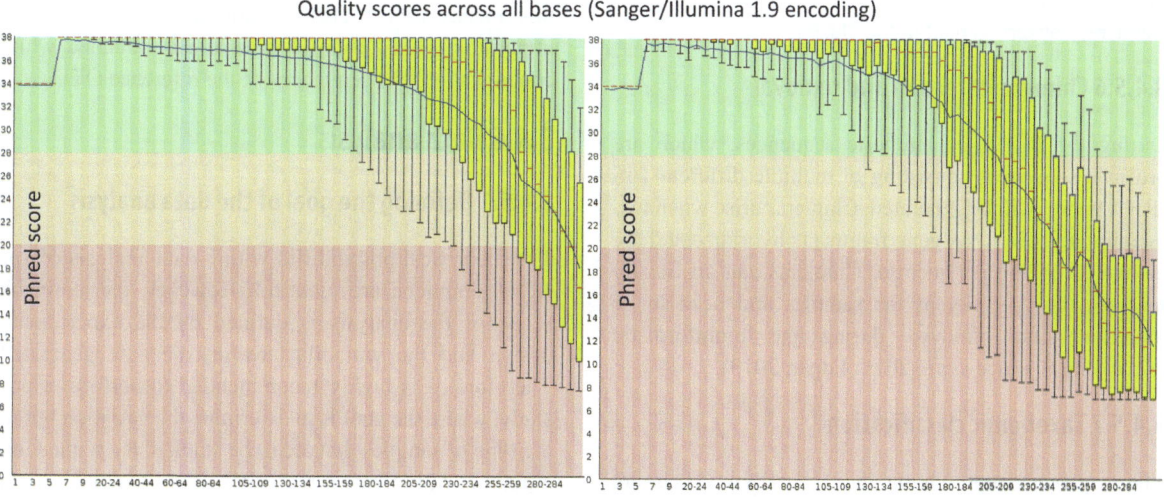

Figure 8.11 Example of Phred quality scores for 2 × 301 bp paired end sequencing of V1-3 16S amplicon libraries on an Illumina MiSeq. To the left is a typical plot for Read1 and to the right is a typical plot for Read2. Read position is plotted versus Phred score. The blue line represents the mean score and the boxplots visualize the score spread. The green area indicates good quality, yellow is reasonable and red is poor. The quality drops towards the end of the read. The quality overview was generated using the FastQC software v0.11.3.

8.4.5.5 OTU clustering

The pre-processed reads are then grouped by sequence identity. These groups are called operational taxonomic units (OTU) and the grouping process is called OTU clustering. The sequence variability in each group may arise from the presence of very closely related strains or because some sequencing errors have been introduced (Huse et al., 2010). Even with a very low error rate of 0.1 % per base, the chance of a perfect 500 bp read is only 61 % (0.999^{500}). As sequencing errors are randomly distributed, sequencing 1,000 reads from a single 16S rRNA gene would result in $(1 - 0.61) \times 1{,}000 = 390$

different reads. To circumvent this inflated diversity the reads are clustered by similarity into OTUs. When analysing 16S rRNA gene fragments from bacteria, the clustering criterion is a sequence identity of 97 %, which very roughly translates to species level depending on which variable region is used. It is important to understand the often-cited 97 % sequence identity threshold. The 97 % criterion is used for full-length 16S rRNA sequences to state that sequences with less than 97 % similarity belong to different species. The criterion cannot be used in the opposite direction, e.g. sequences that are 97 % similar or more do not necessarily belong to the same species (Janda et al., 2007). The choice of clustering algorithm will alter the resulting OTUs (Edgar, 2013; Flynn et al., 2015). Although it is still most popular to use a single threshold for defining OTUs, several algorithms are emerging that take advantage of the error profile in order to use variable clustering thresholds so that maximum resolution can be obtained (Mahé et al., 2014). After clustering, the number of reads belonging to each OTU is counted and a representative sequence from the OTU cluster is chosen. This is usually the most frequently observed read in the cluster.

8.4.5.6 Chimera detection and removal

During the PCR step chimeric sequences, which are artificial sequences consisting of multiple different 16S rRNA genes, will be generated. Chimeras arise when two incomplete 16S rRNA gene fragments hybridize and are elongated by the polymerase. Chimeras can artificially increase the diversity of the samples and need to be identified and removed during the bioinformatic processing (Quince et al., 2011; Edgar, 2013).

8.4.5.7 Taxonomic classification

The sequences representing the 16S rRNA OTUs are taxonomically classified by comparing them with a database of existing sequences. The classification is very dependent on the algorithm and the database. There are three large universal databases that are commonly used: SILVA (Quast et al., 2013), RDP (Cole et al., 2014) and Greengenes (McDonald et al., 2012) that all attempt to cover all presently known microbes. However, given their broad scope they are not curated for specific ecosystems. The MiDAS database is an expert-curated version of the SILVA database, which provides a genus-level name for most of the abundant species in the activated sludge ecosystem (McIlroy et al., 2015). Names are important as they provide a link to other studies and to the literature where functional information may be found.

The algorithm used to compare the obtained OTUs with the database can use different strategies for classification. Some of the common algorithms use different versions of the lowest common ancestor (LCA) approach (Pruesse et al., 2012). This approach takes into account that an OTU sequence can be similar to more than one sequence in the database. If so, the algorithm assigns the lowest shared taxonomy between these database sequences to the OTU.

8.4.5.8 The OTU table

The final result of the bioinformatic processing is an OTU table. The table rows represent the different OTUs and the columns represent each sample in the analysis. The table cells show the count values for the respective OTUs in the respective samples. Each OTU also has a taxonomic classification. The classification is often a delimited (separated by comma or similar) text string containing the classification at each taxonomic rank (kingdom, phylum, class, order, family, genus and species). If a classification at a certain rank is missing, this means robust classification at those levels was impossible. In addition, a fasta file is obtained which contains the DNA sequences of the reference OTUs.

8.4.6 Data analysis

8.4.6.1 Defining the goal of the data analysis

The theoretical possibilities within data analysis of 16S rRNA amplicon are plentiful. However, the attainable scope of the analysis is defined by the experimental design and the variability within the data generated. Therefore, it is highly recommended to perform a pre-study, where the goal is to determine the variation within the type of samples that are to be studied. By sequencing a number of biological replicates, it is possible to make informed decisions regarding the experimental design and the replicates needed to answer the questions posed.

In the following sections, short examples of different types of data analysis will be presented with links to exemplary studies that can serve as further inspiration. For more specific and data-driven examples it is recommended to consult the online documentation of QIIME (Carporaso et al., 2010), Mothur (Schloss et al., 2009), PhyloSeq (McMurdie and Holmes, 2013), vegan (Oksanen et al., 2015) and ampvis (Albertsen et al., 2015).

8.4.6.2 Data validation and sanity check

Before starting the main analysis, it is recommended to make a small exploratory analysis to validate the data and detect possible processing errors. This is especially important if the sequencing and bioinformatic processing has been outsourced. The amplicon-based workflow has numerous steps where errors can be introduced. These are typically sample mix-up, cross-contamination, wrong bioinformatic processing or poor sequencing quality. They can often be detected easily by getting an objective overview of the data with general statistics and simple overview plots.

For all the samples the number of raw reads before and after bioinformatic processing should be examined. If a sample generally has few reads this might indicate that there was something wrong with the library preparation. If many reads are lost during bioinformatic processing it might indicate that the data quality is poor. Sample mix-up can be detected by making principal component analysis plots (PCA) based on OTU counts of all the samples. In PCA plots samples with similar microbial communities will cluster together. Hence, use common sense and a visual inspection of whether the samples group as expected. For example, do the replicates group together?

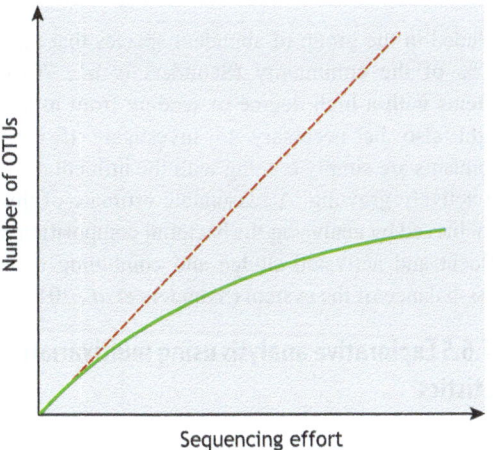

Figure 8.12 Evaluation of the sequencing effort using rarefaction curves. The red line indicates that every new read results in identification of a new OTU. The green line indicates that the discovery of new OTUs decreases with sequencing effort. Hence, the green sample is sufficiently sampled.

When the data has been sanity-checked it is advisable to check if the number of sequenced reads were enough to cover the observed diversity in the samples. This is done by a rarefaction analysis, where a curve is generated that depicts the number of identified OTUs at different sequencing depths through subsampling. The curve should flatten as a function of increasing sequencing depth (Figure 8.12), which indicates that the majority of diversity has been captured (Schloss and Handelsman, 2005).

8.4.6.3 Communities or individual species?

The unit, or perspective, for analysis can roughly be divided into two categories: communities and individual species. In the community-based analysis the parameter for comparison is whether there is an overall difference in community structure or diversity between the samples, while the individual species perspective tries to understand the role and effect of the individual species in the system.

- **The community perspective**

Community-based analysis can be divided into analysis of either alpha-diversity (within samples) or beta-diversity (between samples).

Alpha-diversity analyses are often used to investigate if a particular treatment has an effect on the number of different species observed (also called richness) or on the evenness in abundance of the species in the sample. To compare samples, a single metric is calculated for each sample, which can then be compared across all the samples (Magurran, 2004; Lozupone and Knight, 2008).

Beta-diversity analyses are used to compare the shared diversity between samples, either as the number of shared species or as the amount of shared phylogenetic diversity (Lozupone and Knight, 2008).

- **The species perspective**

While many highly complicated statistical analyses can be performed on 16S rRNA amplicon data, the vast majority of analyses are simply to identify the most abundant bacteria and link them with functional information, e.g. to work out the abundance of nitrifiers or filamentous bacteria in the sample (Figure 8.13). If the MiDAS taxonomy (McIlroy *et al.*, 2015) is used for taxonomic assignment, the MiDAS Field Guide (midasfieldguide.org) can be used to manually look up functional information of the particular species in relation to activated sludge. Alternatively, the ampvis R package (Albertsen *et al.*, 2015) can be used to link the 16S rRNA amplicon data directly with the functional information in the MiDAS Field Guide.

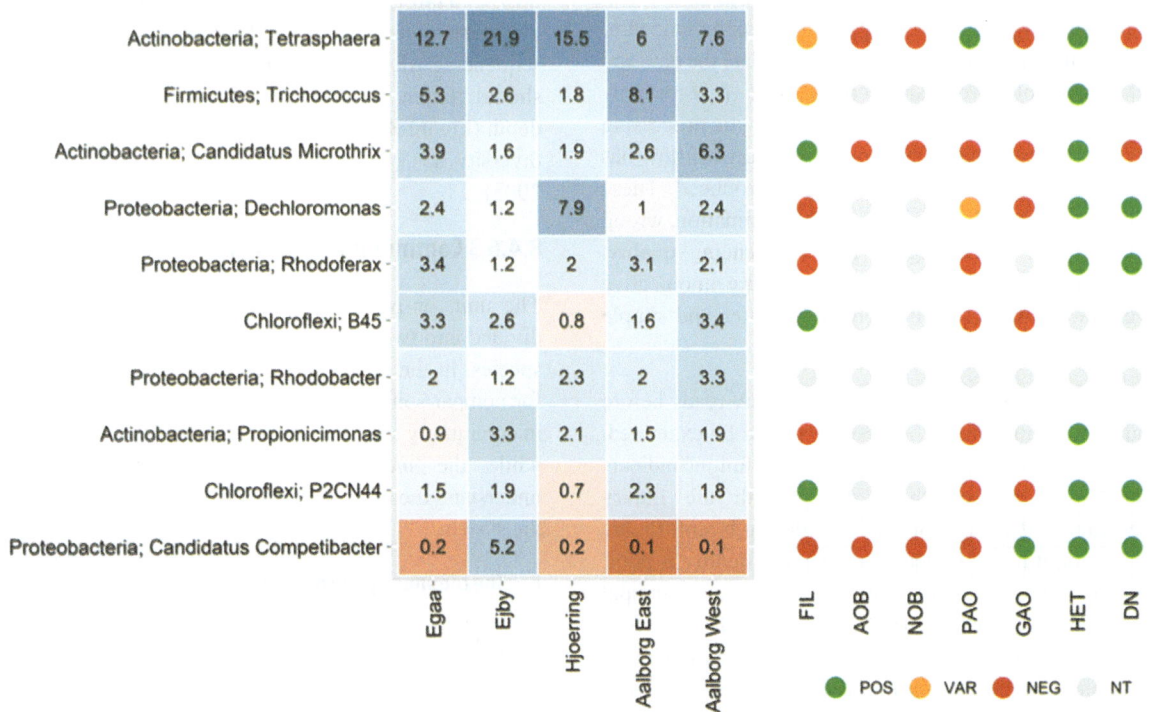

Figure 8.13 Combining species abundance with functional information is a starting point for most analysis. The figure was made through the ampvis R package (Albertsen *et al.*, 2015), which directly links genus names with functional information in the MiDAS Field Guide (www.midasfieldguide.org).

8.4.6.4 Identifying core and transient species

When investigating a specific system it is often of interest to identify the species that are important for the processes in the system and those that might cause problems. A starting point for this analysis is to identify core and transient species (Grime, 1998; Gibson *et al.*, 1999). This analysis needs multiple samples and could either be carried out in a single wastewater treatment plant (time series) or across a number of different wastewater treatment plants.

The traditional definition of core species is those species that are present in all the samples investigated (Saunders *et al.*, 2015). However, given the high sensitivity of the 16S rRNA amplicon sequencing approach, this also includes a high number of very low-abundant species as core species. These species are present in all the samples, but presumably do not contribute significantly to the main ecosystem process. Hence, often a more practical definition of the abundant core community is used, which determines a bacterium to be abundant if it, in eight out of ten samples, is included in the group of abundant species that make up 80 % of the community (Saunders *et al.*, 2015). In systems with a high degree of seeding from influent, it might also be necessary to investigate if the core organisms are simply coming with the influent or if they are actively growing. A reasonable estimate of this can be achieved by analysing the bacterial composition in the influent and activated sludge and combining it with a mass-balance of the system (Saunders *et al.*, 2015).

8.4.6.5 Explorative analysis using multivariate statistics.

The datasets generated by 16S rRNA amplicon sequencing can be extremely large and even with ten samples (each with 1,000s of species) it becomes difficult to get an overview of the data and to identify interesting patterns. Hence, often different varieties of ordination methods are applied, e.g. PCA. These methods can be used to visually group samples according to how similar their microbial communities are and to identify which species are responsible for the observed sample grouping (Figure 8.14).

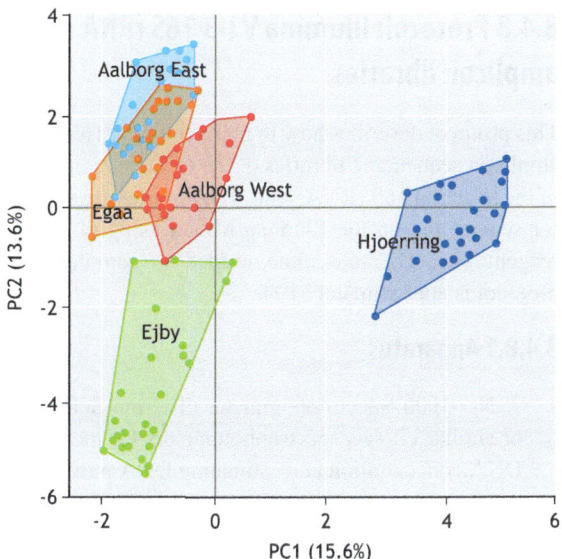

Figure 8.14 Explorative analysis using PCA. Samples are seen to group by five wastewater treatment plants in Denmark. The plot was made using the ampvis R package (Albertsen et al., 2015).

Explorative plots, where the samples are coloured by different environmental variables, are often a good place to start before formal statistical hypothesis testing is applied. However, the specific choice of the ordination method and data transformations is highly dependent on the question posed, the number of samples and the distribution of the abundance. Hence, it is recommended to consult specific literature to understand the applications of different methods (Legendre and Gallagher, 2001; Ramette, 2007; Zuur et al., 2007).

8.4.6.6 Correlation analysis

Another possibility with 16S rRNA amplicon data is to investigate correlations between different species or between species and environmental factors. Are some species co-occurring, are there keystone species that are needed for eco-system functioning or is the abundance of specific species correlated with environmental factors such as temperature?. Due to the number of possible correlations, the results are often displayed in network graphs that depict positive or negative correlations between species and environmental variables (Faust et al., 2012; 2015). However, there are many pitfalls and choices to be made in correlation analysis of 16S rRNA amplicon data. For example, the correlation analysis is often made on time-series data and correlations might be time-lagged. A sudden increase in ammonium concentration might result in an increase of nitrifiers over a period of several weeks.

8.4.6.7 Effect of treatments on individual species

To statistically test the impact of specific treatments on individual species, a number of different statistical methods have been developed that take into account the nature of the count-based amplicon data. Most methods have been modified or directly imported from statistical frameworks developed for analysis of gene expression (mRNAseq) data (Robinson et al., 2010; Love et al., 2014).

8.4.7 General observations

8.4.7.1 A relative analysis

It is important to notice that microbial community analysis using 16S rRNA amplicon sequencing is a relative analysis. This means that the abundance of the reads representing the individual species is given as a percentage of the total. Hence, there is no link from the percentage to the absolute count of cells. If needed, such a link may be established by qPCR (see Section 8.3).

8.4.7.2 Copy number bias

While the 16S rRNA gene is universal among bacteria, different bacteria can have from one to over fifteen copies of the 16S rRNA gene in their genome (Farrelly et al., 1995; Angly et al., 2014). Hence, a bacterial species with ten copies of the 16S rRNA gene would appear ten times more abundantly than a species with a single 16S rRNA gene, if they were present in equal cell counts. In addition, some bacteria have more than one copy of their genome, which will further bias the abundance estimation (Mendell et al., 2008; Pecoraro et al., 2011). Hence, instead of referring to the percentages obtained from 16S rRNA amplicon data as 'abundances' they should be referred to as 'read abundances' to imply the inherent biases.

8.4.7.3 Primer bias

As touched upon earlier, the choice of primer and PCR conditions also has a large effect on the observed microbial community. The 'universal' primers are designed by comparing the sequence of conserved regions of 16S rRNA genes from thousands of different bacteria, and then making a consensus sequence that targets most bacteria. However, it is not possible to design primers that target all bacteria equally well. In addition, biases are often taxon-specific, which means that whole taxonomic groups are under-represented or even overlooked completely. Many of the new candidate phyla that are being discovered using primer-independent methods, such as metagenomics, show large deviations in the conserved primer sites or even large sequence insertions in the 16S rRNA gene itself, which is the main reason for them having escaped our attention for decades (Brown et al., 2015).

8.4.7.4 Standardization

Many attempts have been made to eliminate the biases described above, but with a complex system such as activated sludge, this is probably never going to succeed. Furthermore, it is very difficult to validate whether the biases have been removed, as it is often difficult to make proper controls. At first sight this seems to undermine the analysis method. However, it just sets some limitations to what questions can be answered, and how the experiments should be designed.

The core message is that if the biases are the same for all the samples analysed, 16S rRNA amplicon sequencing is a very powerful tool for relative comparisons and observations of the presence or absence of specific bacteria. To ensure all biases are the same, the samples in one experiment have to be treated the exact same way throughout processing. It is preferable that they are sampled and stored, DNA is extracted, sequencing libraries are prepared and the data is processed in the exact same way.

8.4.7.5 Impact of the method

Despite some limitations, 16S rRNA amplicon sequencing is one of the most essential tools in microbial ecology today. The main reasons are the resolution and the throughput it enables, at which the preceding techniques were far inferior. In 2010, clone libraries and DGGE were still the standard techniques supplemented with FISH for *in situ* identification and quantification. A large study back then would include 100-1,000 16S rRNA sequences divided into 10 samples. Today, for the same cost, hundreds of samples can be analysed with thousands of reads per sample, in a single week. However, this also sets new requirements for the skill of the microbial ecologists in order to handle highly complex experiments and enormous amounts of data.

8.4.8 Protocol: Illumina V1-3 16S rRNA amplicon libraries

This protocol describes how to make Illumina 16S rRNA amplicon sequencing libraries of the variable regions 1-3 of the bacterial 16S rRNA gene. The libraries are suitable for sequencing on the Illumina MiSeq using 600-cycle reagent kits. The total time needed to complete this protocol is approximately 10 h.

8.4.8.1 Apparatus

- A ND-1000 Spectrophotometer (Thermo Scientific) or similar UV-vis spectrophotometers for measuring DNA concentration and estimating DNA purity.
- A Nanodrop user manual (NanoDrop Technologies Inc. 2007).
- A Qubit 2.0 Fluorometer (Life Technologies), Infinite M1000 PRO (Tecan) or similar fluorometer for precise measurement of DNA concentration by use of DNA binding dyes.
- Qubit assay user guides (Thermo Scientific 2015a; 2015b).
- A standard PCR thermocycler with heated lid.
- A magnetic stand for 96-well plates used for DNA purification e.g. MagneSphere Technology Magnetic Separation Stand (Promega) or Magnetic Stand-96 (AM10027, Ambion).
- A tapestation 2200 (Agilent) gel electrophoresis setup for checking sequencing library quality. Alternatively use a conventional gel electrophoresis setup (Green and Sambrook, 2012).
- Tapestation 2200 manuals (Agilent Technologies, 2012; 2013; 2015).
- Pipettes (range 1 µL to 1,000 µL).

8.4.8.2 Materials

PCR Plate Spinner
- DNAse-free tips (10 µL, 300 µL and 1,000 µL).
- DNAse-free tubes (1.5 mL).
- DNAse-free thin-wall, clear, PCR tubes (500 µL).
- 96-well PCR plates (#82006-664, VWR).
- PCR strip caps.
- An OptiPlate-96 Black microplates (Perkin Elmer).
- Nuclease-free H_2O (Qiagen).
- Fluorescent DNA-binding dyes e.g. a Qubit dsDNA HS assay kit (Life Technologies), a Quant-iT dsDNA assay kit, broad range (Life Technologies), a Quant-iT dsDNA assay kit, high sensitivity (Life Technologies).
- A Platinum Taq DNA Polymerase High Fidelity kit (Life Technologies).

- dNTP mix.
- Barcoded V1-3 16S rRNA gene adaptor mixes (5 μM of each forward and reverse adaptor), see Section 8.4.11.
- Agencourt AMPure XP (Beckman Coulter).
- Ethanol, 99 %.
- D1000 Screentape (Agilent) and Genomic DNA Screentapes (Agilent). Alternatively use reagents for conventional gel electrophoresis setup (Green and Sambrook, 2012).

8.4.8.3. Protocol

- **Sample DNA quality control and dilution (2.5 h)**

In this section, the quality of the extracted genomic DNA is checked and the DNA is diluted to a concentration suitable for PCR. See Section 8.4.9 Interpretation and troubleshoot for further information.

1. Fluorescence DNA concentration measurement
   a. Use a Qubit dsDNA BR assay or a Quant-iT dsDNA broad range assay following the protocol recommended by the manufacturer.
   b. Use 2 μL of each sample per reaction.
   c. Perform one measurement per sample.
2. UV-vis quality check (optional)
   a. Use a Nanodrop1000 following the protocol recommended by the manufacturer (Nanodrop Technologies Inc., 2007).
   b. For large sample batches consider measuring a subset of samples (e.g. 8 out of 96).
   c. Initialize and blank the instrument with the same buffer that the samples are eluted in.
   d. Use 2 μL sample per measurement.
   e. Perform one measurement per sample.
3. Gel electrophoresis (optional)
   a. Use a Tapestation 2200 following the protocol recommended by the manufacturer.
   b. Use Genomic Screentapes with reference DNA ladder.
   c. Perform one measurement per sample.
   d. For large sample batches consider measuring a subset of samples (e.g. 7 samples out of 96 + 1 ladder).
4. Sample dilution
   a. Based on fluorescence DNA concentration measurements, calculate how much nuclease-free water is required for each of the samples to dilute them to 5 ng μL^{-1}. The formula used is:

$$\frac{V_{sample} \cdot C_{sample}}{C_{final}} - V_{sample} =$$

$$V_{H_2O} \rightarrow \frac{5\mu L \cdot C_{sample}}{5 \frac{ng}{\mu L}} - 5\mu L = V_{H_2O} \quad \text{Eq. 8.2}$$

   b. Transfer a 5 μL sample to an empty well in a 96-well PCR plate.
   c. Dilute the sample with the calculated amount of nuclease-free water.
   ▲ Critical step If the sample concentration is 5 ng μL^{-1} or less, use the sample undiluted or discard the sample. If the dilution requires > 150 μL nuclease water, a pre-dilution might be necessary.
   d. Repeat for all the samples.
   e. Seal the plate with PCR strip caps.
   ■ Pause Point The diluted samples can be stored at -20 °C for at least a month.

- **Library PCR (2.0 h)**

This section concerns the preparation for sequencing libraries by PCR amplification. The template input is genomic DNA (5 ng μL^{-1}) and the output is V1-3 16S rRNA amplicons with a size of approximately 614 bp.

1. Preparation
   a. Library PCR reaction is run in duplicate for each sample.
   b. Remember to include a negative control (nuclease-free H$_2$O) and positive control (microbial community DNA known to amplify with 16S rRNA PCR).
   c. Note which V1-3 adaptors with unique barcodes are assigned to which samples.
2. Mix the PCR reaction
   a. Prepare the mastermix for (samples + controls) × 2 + 3. Any spare mastermix will make up for loss during pipetting.
   b. Add the reagents in the given order to produce a mastermix.
   c. Transfer 13 μL of the mastermix to the wells of a new 96-well PCR plate.
   d. Add 10 μL of assigned barcoded V1-3 16S rRNA adaptor mix (1 μM) to each well and pipette up and down 10 times to mix. The final adapter concentration is 400 nM.
   ▲ Critical step High risk of mixing up samples and adaptors. Stay alert.
   e. Add 2 μL of template DNA (total of 10 ng of DNA) and mix by pipetting up and down. The final volume is 25 μL.
   ▲ Critical step High risk of mixing up samples.
   f. Seal the 96-well PCR plate with PCR strip caps.
   g. Spin the 96-well PCR plate to settle the PCR reaction mix at the bottom of the plate.

Reagents	Final conc. in 25 μL reaction	Volume (μL) for 1 r×n
Nuclease-free water	-	7.65
×10 buffer Platinum High Fidelity	×1	2.5
dNTP (5 mM)	400 uM	2
MgSO$_4$ (50 mM)	1.5 mM	0.75
Platinum Taq DNA Polymerase High Fidelity (5 U μL^{-1})	0.02 U μL^{-1}	0.1
Total volume		13

3. Run PCR incubation
   a. Program the thermocycler with the following program:

Step	Temperature	Time
Denaturation	95 °C	2 min
30 cycles		
Denaturation	95 °C	20 s
Annealing	56 °C	30 s
Amplification	72 °C	60 s
Amplification	72 °C	5 min
Storage	4 °C	Forever

   b. After the PCR reaction spin down the PCR reaction mix again.
   c. Remove the PCR strip caps and pool the duplicate PCR reactions for each individual sample. The final volume is 50 μL.
   d. After the PCR the samples are now referred to as 'sequencing libraries' or 'short libraries'.
      ■ Pause Point The libraries can be stored at -20 °C for at least a month.

- **Library cleanup (2.0 h)**

This section involves the cleanup of the PCR reactions. The aim here is to remove leftover reagents and possible short (< 200 bp) unspecific PCR products. The output is clean sequencing libraries that only consist of the 16S rRNA amplicons (ca. 614 bp).

1. Preparation
   a. Gently shake the Agencourt AMPure XP bottle to re-suspend the beads, remove the required volume 40 μL beads × [n(samples) + 3] and let it equilibrate to room temperature.
   b. Prepare a fresh solution of 80 % ethanol by transferring 20 mL of ethanol (99 %) to a greiner (or falcon) tube (50 mL) and add 5 mL Nuclease-free water. Mix by inverting the tube.
2. Bind the libraries to the beads
   a. Transfer 40 μL of the bead solution per well in a new 96-well PCR plate corresponding to the number of samples.
   b. Add 50 μL of library to each well with beads and mix by pipetting up and down 10 times.
      ▲ Critical step It is important that the bead to sample ratio is 4:5. If for some reason there is less/more than the 50-μL sample, adjust the bead volume used accordingly.
   c. Incubate for 5 min at room temperature.
3. Wash the bound libraries
   a. Place the 96-well PCR plate on the magnetic stand and wait until the liquid is clear (2-4 min). All the subsequent steps are performed with the plate on the magnetic stand.
   b. Remove and discard as much liquid as possible with a pipette.
      ▲ Critical step Take care not to transfer the beads (brown pellets).
   c. Wash the bead-pellets with 200 μL ethanol (80 %) by gently dispensing it over the beads with a pipette. Let it rest for 30 s and then remove the liquid.
   d. Repeat the above step (3c).
   e. Ensure no excess liquid is left after the washes. If there is, remove it with a 10 μL pipette.
   f. Let the pellets dry for 7 min.
      ▲ Critical step Avoid excessive drying by heating or long drying times as this will make the DNA elution difficult.
4. Elute the library
   a. Remove the 96-well PCR plate with the dried bead pellets from the magnetic stand.
   b. Add 33 μL of nuclease-free water and mix by pipetting up and down 10 times to re-suspend the beads.
   c. Incubate for 2 min at room temperature.
   d. Place the 96-well PCR plate back on the magnetic stand and wait until the liquid clears (1-2 min).
   e. Transfer 30 μL of the liquid to an empty well in a new 96-well plate.
      ▲ Critical step Take care not to transfer any of the beads.
      ■ Pause Point The purified libraries can be stored at -20 °C for at least 6 months (see the paragraph on Storage and transport below).

MOLECULAR METHODS

- **Library quality control (1.5 h)**

This section concerns the DNA concentration measurement of the cleaned libraries and subsequent quality check using gel electrophoresis. It is necessary to confirm that only the target 16S rRNA amplicon is present in the libraries.

1. Fluorescence DNA concentration measurement
   a. Use a Qubit dsDNA High sensitivity assay or a Quant-iT dsDNA High sensitivity assay following the protocol recommended by the manufacturer.
   b. Use 2 µL of each sample per reaction.
   c. Perform one measurement per sample.
2. Gel electrophoresis
   a. Use Tapestation 2200 following the protocol recommended by the manufacturer.
   b. Use D1000 Screentapes with the reference DNA ladder.
   c. Perform one measurement per sample.
   d. For large sample batches consider measuring a subset of samples (e.g. 15 out of 96). Always include the negative and the positive control in the analysis.

- **Library pooling (2.0 h)**

This section concerns the pooling of all the libraries. The aim is to obtain a final single sample containing equimolar concentrations (the same number of molecules) of each library. The volumes are calculated from the DNA concentrations of the libraries obtained above.

1. Calculate the required volume of each sample
   a. Libraries with a concentration of less than 1 ng µL^{-1} should be excluded (either leave out or re-run PCR).
   b. Detect the sample with the lowest concentration and multiply this concentration with 15 µL (e.g. 1 ng µL^{-1} × 15 µL = 15 ng). This is the amount of library wanted from each library.
   c. Calculate the volumes required to obtain the same amount of library for each of the other libraries.
   d. If volumes less than 1 µL are required for some libraries, consider diluting the libraries and re-calculate the required volumes.
2. Pool libraries
   a. Use a new tube (1.5 mL).
   b. Transfer the calculated volume of each sample to the tube.
   c. Mix well after all the samples have been added.

3. Fluorescence DNA concentration measurement
   a. Use a Qubit dsDNA High sensitivity assay following the protocol recommended by the manufacturer.
   b. Use 2 µL of the library pool per reaction.
   c. Perform the measurement in triplicate.
   d. Calculate the average concentration.
   e. Based on the concentration, calculate the nanomolar concentration with the following formula.

$$c_{nM} = \frac{c_{ng/\mu L} \cdot 1{,}000{,}000\ \frac{\mu L}{L}}{650\ \frac{g/mol}{bp} \cdot 614\ bp} \qquad \text{Eq. 8.3}$$

   f. If the concentration is less than 4 nM, concentrate the sample with ampure bead purification (see the section on library cleanup earlier in this chapter). Be sure to use a bead to sample ratio of 4:5 (e.g. if you have 100 µL sample pool you should use 80 µL ampure bead solution). Make sure to calculate how much nuclease-free water is required for elution to get a concentration of 4 nM or above. Also anticipate the loss of up to 50 % of the product. So, if a 2× concentrate is necessary then a 100 µL input library pool should be eluted in 25 µL nuclease-free water). Repeat the DNA concentration measurement of the concentrated pool.

- **Storage and transport**

This section concerns the storage of libraries and the library pool.

1. If storage of the library pool is planned do not dilute it. DNA withstands storage better if kept in its concentrated form (> 5 ng µL^{-1}).
2. For short-term storage or transport (< 14 d), purified library DNA can be kept at ambient temperature.
3. For medium-term storage (< 12 months), the libraries can be kept at -20 °C.
4. For long-term storage (> 12 months), the libraries should be kept at -80 °C.

8.4.9 Interpretation and troubleshooting

8.4.9.1 Sample DNA quality control and dilution

There are three things to consider regarding the input DNA: (*i*) the amount, (*ii*) the quality and (*iii*) potential contaminants. These characteristics are investigated using a fluorescence DNA concentration assay, UV-vis spectrophotometry and gel electrophoresis.

The recommended amount of microbial community DNA for a PCR is 1-100 ng total DNA, where 10 ng (approximately 2×10^6 cells) is often used. If more DNA is used then the risk of amplifying random pieces of DNA is increased and inhibition of the PCR reaction may occur. Random amplicons are problematic, since they result in decreased sequencing data yield and poor quality. Low DNA concentration increases the risk of PCR failure and introduces variance for low abundant members of the community (Kennedy et al., 2014). UV-vis spectrophotometry can be used to measure DNA concentration, but the estimate can be uncertain as it is influenced by reagent contamination, the presence of nucleotides and RNA. Fluorescence-based methods are always better, and UV-vis spectrophotometry should only be used as backup (Li et al., 2014).

The quality of the DNA is important since it influences the effective amount of DNA available for the PCR reaction. If the DNA is heavily degraded (most DNA < 5,000 bp), the risk of the marker genes being broken increases, which makes them unavailable for PCR (Beers et al., 2006; Wilson et al., 1997).

Non-DNA contaminants refer to chemicals or organic molecules introduced from the sample (i.e. humic acids and complex sugars) or during the DNA extraction (i.e. SDS, alcohols, chaotropic salts), and they can inhibit the PCR reaction, reducing efficiency (Wilson et al., 1997). A UV-vis spectrum of clean DNA has a very distinct curve. Anomalies in this curve are indications of contamination and they are often screened for by looking at the ratio of the absorbance at 260 nm and 280 nm (A260/280) and the ratio at 260 nm and 230 nm (A260/230). A260/280 of clean DNA is usually around 1.8, but if it is very different (e.g. ± 0.4) then this might indicate protein contamination or residual reagents such as alcohols from the extraction kit. A260/230 of clean DNA is usually between 2.0-2.2, and if it is very different this might indicate residual carbohydrates or DNA extraction reagents. UV-vis spectrophotometry is a very sensitive method, and the type of buffer and pH can cloud the results (Thermo Scientific, 2015).

DNA contaminants can be DNA from life forms not of interest when targeting bacteria i.e. eukaryotic DNA from fungi or plants. The presence of contaminating DNA reduces the effective amount of target DNA in the sample and reduces PCR efficiency (Tebbe et al., 1993). This cannot be measured beforehand, but clues might be obtained by visually inspecting the biomass before DNA extraction (e.g. is there visible plant material?).

If there is a suspicion of contamination, it is a good idea to perform a test PCR with a few samples to see if it is a problem. Often PCR still works despite contamination, however if the PCR fails, sometimes extra purification steps can be the solution. Remember to use a clean-up method that can handle high molecular genomic DNA (> 10,000 bp) e.g. Agencourt AMPure XP beads. Many column-based purification kits are designed for DNA < 10,000 bp and will therefore remove genomic DNA.

8.4.9.2 Library PCR

When making a PCR the composition of the mastermix and the incubation settings will influence the observed community composition in the final data. In a research setting, it is common to optimize the PCR conditions to increase the amount of target amplicon produced and to reduce the amount of unwanted, random PCR products. However, for 16S rRNA amplicon sequencing, PCR settings should be kept the same for all samples that are to be compared. To ensure consistency it is recommended to use the standard settings from tested protocols. For completeness a short overview of what is usually changed during PCR optimization is given below.

The Mg^{2+} concentration and the annealing temperature will influence how specific the primers are when finding their targets. If they are too stringent a negative bias against certain species will be observed. If the stringency is too low, there will be an increased risk of amplifying random pieces of DNA. The number of PCR cycles determines how much product is produced, and many times apparently failing PCRs can succeed simply by increasing the number of cycles. However, an increased number of PCR cycles is also reported to introduce PCR selection bias (Polz et al., 1998; Kennedy et al., 2014). Also, if too many cycles are run and the primers and other reagents are depleted there is an increased risk of chimeric products (Qiu et al., 2001).

8.4.9.3 Library cleanup

The library cleanup steps removes leftover reagents (primers, nucleotides, polymerase etc.) as well as small (< 200 bp), random DNA pieces. All these contaminants can interfere with the DNA sequencing, either reducing the quality of the data or making the sequencing fail completely.

The purification method applied in the protocol is based on precipitation of the DNA onto small, magnetic plastic balls called SPRI beads (Solid Phase Reversible Immobilization). The principle is not well described in the literature (DeAngelis *et al.*, 1995), but different blogs present some hypotheses (Hadfield, 2012). Due to its chemical composition, DNA has an overall negative charge and is easily dissolved in water where it electrostatically interacts with the polar water molecules. To precipitate the DNA from the solution NaCl salt is added. Na^+ ions are formed, which shield the negative charges of the DNA and makes it less soluble thereby promoting precipitation. The crowding agent, PEG, increases the effective concentration of Na^+ and drastically increases the shielding effect. When the DNA precipitates it prefers the surface of the SPRI beads. This preference is counter intuitive since the surface of the beads is negatively charged and therefore should repel DNA in negatively charged or shielded state. The topic is much debated and no clear explanation exist, however some propose the DNA/SPRI bead interaction is mediated by a layer of water or Na+ ions coating the SPRI beads. Hence, by mixing dissolved DNA and beads in a solution containing crowding agents the result will be that the DNA precipitates on the beads. The crowding solution can be removed and the beads with the DNA can be washed using 80 % ethanol. The composition of the washing solution is optimized to dissolve small contaminants such as salt, nucleotides etc. but ensures that the DNA is not re-dissolved. After washing, the DNA is re-dissolved in nuclease-free water or buffer and is ready to be used.

Small DNA molecules are more likely to stay in solution compared to larger DNA molecules, when adding a certain amount of crowding agent and salt. This means that small pieces of DNA can be removed by adding the correct amount of crowding solution. In this protocol, 40 µL Agencourt AMPure XP bead solution (beads + crowding solution) is used for 50 µL sample. This bead to sample ratio of 4:5 favours the binding of DNA > 200 bp to the beads and smaller pieces remain in the solution and are subsequently removed. Care has to be taken when dispensing the sample or bead solution. If the bead/sample ratio is < 0.5, the 16S rRNA amplicon is also removed. If the ratio is > 1.0 then the contamination will not be removed properly.

When drying the residual ethanol from the beads after washing, take care not to overdo it. Do not use heat or dry them for too long. If the bead pellet becomes too dry (many cracks in the pellet), re-dissolving the DNA will be difficult and the product will be lost.

Other purification methods can be used instead of beads, such as column-based methods e.g. QIAquick PCR Purification Kit (Qiagen). The overall purification principles are similar.

8.4.9.4 Library quality control

The DNA concentration is measured to provide the basis for library pooling. The concentration measurement does not need to be very precise (± 20 % is appropriate), hence only one measurement is made.

Gel electrophoresis is run to ensure that all the contaminating DNA (primers and randomly amplified DNA) is removed and only the target 16S rRNA amplicon is present (614 bp). Generally, only a subset of all the samples is run since gel electrophoresis is expensive and time consuming, and a random subset should give an indication of the purification success. When picking the samples for the subset, always include the negative and the positive samples. Also, based on the concentration measurement, include the samples that have low concentration, to see if there is V1-3 16S rRNA amplicon product present. If no V1-3 16S rRNA amplicon product is present then re-run the PCR on these samples.

The V1-3 16S rRNA amplicon product has an average size of 614 bp, however the size can vary as much as ± 100 bp between the different species. For bacterial communities with many different species this usually manifests itself as a broad peak with a peak maximum around 614 bp when using Tapestation electrophoresis. If a community with only a few members is analyzed, multiple peaks between 500 and 700 bp might show up on the Tapestation electropherogram.

There should be no DNA pieces left below 300 bp. If there are, consider revising the purification step and redo the purification. If the contaminant make up < 1 % of the total DNA in the sample, sequencing should still produce a useful result and redoing the purification can be omitted. Rarely, unknown contamination above 700 bp can be seen. If this is present, try to redo the PCR. If it remains, try sequencing the library anyway. It will be possible to filter out the contaminant using bioinformatics.

The positive sample should show a clear product around 614 bp. The negative sample should show no product, or a faint product that is < 1 ‰ of the total DNA of the other samples. It is often difficult to completely avoid contamination, but as long as the level of contamination is relatively low it is acceptable. If the negative control has a relatively high level of contamination, it is likely that all the samples are contaminated. Revise your PCR setup and redo the PCR reactions for all the samples, possibly with new batches of reagent.

8.4.9.5 Library pooling

Equimolar library pooling is performed to ensure that each sample gets equal amounts of data during sequencing. Errors in pooling will have a direct effect on the amount of data obtained. If mistakes are made in pooling, start again if possible.

8.4.9.6 Pool quality control and dilution

The concentration measurement of the library pool is very important since this is used to determine how much of the library is to be loaded on the sequencer. If the concentration is measured as higher than it really is, less data will be obtained from the sequencing run. If the measured concentration is lower than it actually is, the sequencer can be overloaded, which results in poor data quality or the sequencer crashing completely.

Therefore, the library pool should be measured in triplicate and the concentrations averaged. Be sure to calibrate the fluorometer before use with reference samples that are known not to be contaminated or degraded.

8.4.9.7 Storage

The DNA should be stored in ultra-pure water or TE buffer (pH 7-8). There have been reports of degradation of DNA after long-term storage in ultrapure water, which is probably due to a fragile pH balance, as no buffer is present. Pure, dissolved DNA is very stable even at room temperature and can be safely stored or transported at ambient temperature for a short time. For medium or long-term storage, freezing the DNA at -20 or -80 °C is recommended (Smith and Monn, 2005). Avoid repeated freeze-thaw cycles of the DNA as this degrades the DNA. If a sample is to be used multiple times, prepare aliquots.

8.4.10 Protocol: Illumina V1-3 16S amplicon sequencing

This protocol describes important protocol additions when preparing sequencing of Illumina V1-3 16S rRNA Amplicon Libraries on the Illumina MiSeq. For details about the steps following the standard procedure, refer to the MiSeq System User Guide (Illumina Inc. 2014b). The time needed for the execution of the protocol is approximately 3.5 h, however sequencing requires 56 h.

8.4.10.1 Apparatus

For the execution of the protocol one needs the following apparatus:
- A MiSeq (Illumina).
- Micropipettes (range 1 uL to 1,000 uL).
- A MiSeq System User Guide (Illumina Inc. 2014b).
- An Illumina Experimental Manager v1.9 (illumina.com).

8.4.10.2 Reagents

For the execution of the protocol one needs the following reagents:
- Ice.
- A MiSeq Reagent kit v3, 600 cycles (Illumina). Includes a reagent cartridge and HT1 buffer.
- 2 M NaOH, molecular grade.
- Sequencing primers (Read1, Read2, and Index), see Section 8.4.11.
- DNAse-free tips (10 µL, 300 µL and 1,000 µL).
- DNAse-free tubes (1.5 mL).
- Nuclease-free water (Qiagen).
- PhiX control library v3, 10 nM (Illumina).
- Ethanol, 70 % (molecular grade).
- Laboratory wipes, lint-free.
- Microscope lens cleaning wipes.

8.4.10.3 Protocol

- **Prepare MiSeq (2.0 h)**

This section describes the thawing of reagents, and preparation of the MiSeq instrument and the Sample sheet file.

1. Instrument washing
   a. According to the recommendations in the User Guide.
2. Reboot the MiSeq to reset the memory
   a. Under *Manage Instrument* in the MiSeq control software, press *Reboot* and wait (this might take 10 min).
   b. The MiSeq control software will start initializing the instrument after reboot. When done the *Control interface* will appear.
3. Thaw the reagents
   a. Place the reagent cartridge and HT1 buffer in the water bath at room temperature for 1 h. After thawing store at 4 °C until use.
     - ■ Pause Point The thawed MiSeq reagent cartridge can be stored for a week at 4 °C.

b. Reagent cartridge inspection: invert the cartridge ten times, inspect for precipitates, and then tap the cartridge on the table to remove any bubbles.
c. Thaw the sequencing primers (Read1, Read2 and index) and 2 M NaOH at room temperature. Place the primers on ice after thawing, and leave 2 M NaOH at room temperature until use.
4. Prepare the Sample Sheet
   a. Open a MiSeq 'SampleSheet.csv' template in Notepad++ or another simple text editor.
      ▲ Critical step The 'SampleSheet.csv' is a comma-separated value text file (.csv) but should not be opened in Microsoft Excel! Excel might corrupt the formatting of the file.
   b. Change the project and sample-specific info: [Header] Investigator, Project name, Experiment Name, Date; [Data] Sample_ID, Sample_Name, index and index2.
      ▲ Critical step Important information is: [Header] Chemistry, [Reads] and [Data] index and index2 columns. This information has significant impact on how the run is performed. All the other information can be changed later if it is wrong.
   c. After filling out the sample sheet, check the integrity of the 'SampleSheet.csv' by loading it into the Illumina 'Experimental Manager'. If the sheet can be loaded then it should be compatible with the MiSeq.
   d. Transfer 'SampleSheet.csv' to MiSeq with a USB-stick.

- **Prepare sequencing libraries (1.0 h)**

This section describes the denaturation and dilution of the sequencing libraries.

1. Sample overview
   a. Control library: PhiX control library v3, 10 nM.
   b. Library pool (up to 400 samples), > 4 nM.
   c. Follow the following steps for both the PhiX control library and the library pool.
2. Thaw the libraries and store on ice.
3. Dilute the sequencing libraries to 4 nM with nuclease-free water.
4. Prepare 0.1 M NaOH solution.
   a. 475 µL DNA H₂O and 25 µL 2 M NaOH
5. Denature the sequencing libraries
   a. Mix 5 µL library + 5 µL 0.1 M NaOH. The final library concentration is 2 nM.
   b. Pipette up and down 10 times to mix.
   c. Incubate for 5 min at room temperature.
6. Dilute the denatured libraries (2 nM) to 20 pM.
   a. Mix 10 µL denatured library with 990 pre-chilled HT1. The concentration is 20 pM.
7. Mix the PhiX library (20 pM) with the library pool (20 pM) so they make up 20 % and 80 %, respectively, of the final mix.
   a. Mix 120 µL PhiX library with 480 µL of the library pool.
   b. Place the PhiX/pool mix on ice until use.
   ■ Pause Point Diluted libraries can be stored at -20 °C for up to a month. Longer storing times might result in reduced concentration, and consequently reduced sequencing output.

- **Load the sample and primers on the reagent cartridge**
1. Adding custom-sequencing primers to the reagent cartridge.
   a. Primer destinations:
   Read1 = well 12.
   Index = well 13.
   Read2 = well 14.
      ▲ Critical step Well numbering on the reagent cartridge can be confusing. Take the time to make sure the correct wells are used.
   b. For each primer: Pierce the tinfoil covering the target well with the tip of the 1,000 µL pipette, and then transfer 100 µL of the well content to a spin tube. Add 3.4 µL of the respective primer and mix well. Transfer the solution back to the well it originated from and mix. Repeat for all the primers.
2. Adding the sample to the reagent cartridge
   a. Pierce well no. 17 with a tip and add 600 µL PhiX/pool mix.

- **Sequencing (0.5 + 56 h)**

This section describes the startup of the sequencing run.

1. Press *Sequence* on the MiSeq Control software interface and follow the instructions in the MiSeq System User Guide for preparing/loading the flow cell, loading the reagent cartridge, referring to the sample sheet and starting the sequencing run.
2. During sequencing the progress can be monitored by opening the run folder in the Sequence analysis Viewer software and looking at the run.

8.4.10.4. Interpretation and troubleshooting

- **Prepare MiSeq and metadata**

Making sure a wash has been performed on the MiSeq prior to sequencing is important. Cross-contamination between runs can be a problem. Studies have shown that there is a bleed-over of samples from run-to-run. Usually the bleed-over will not impact 16S amplicon analysis from activated sludge, and therefore the default *Post Run Wash* is acceptable between washes. However, for delicate samples consider performing one or two *Maintenance Washes* in between runs and/or a dedicated wash of the sample line. These washes are more thorough and leftover contamination is diluted. Otherwise follow the wash instructions in the MiSeq System user guide.

```
[Header]
IEMFileVersion,4
Investigator Name,SMK
Project Name, DNASense-NDJ-RHK
Experiment Name,J214
Date,08/11/15
Workflow,GenerateFASTQ
Application,FASTQ Only
Assay,TruSeq HT
Description
Chemistry,Amplicon

[Reads]
301
301

[Settings]

[Data]
Sample_ID,Sample_Name,index   ,index2  ,Description
LIB-CP034,16SAMP-6380,ACGTGTAC,GAGCTCTC,bV13fr-337
```

Figure 8.15 SampleSheet example. The SampleSheet.csv file has been opened in Notepad++. The section titles are marked by [ ].

A MiSeq reboot helps reset the instrument computer, and makes it less likely to crash during sequencing.

Take care when thawing the MiSeq reagents. The quality of the reagents impacts the sequencing quality, especially in the read 3' ends. Experience shows that sequencing can still succeed with reagents that have been stored for 24 h at room temperature but no guarantee is given.

The 'SampleSheet.csv' file tells the MiSeq instrument how the sequencing run should be performed, how to demultiplex the samples and some basic metadata related to the samples. The metadata sheet cannot be prepared by the Illumina Experiment Manager when using adaptors and barcodes not purchased from Illumina. The 'SampleSheet.csv' file has to be prepared in a text editor such as Notepad++. Find a template online or create your own from the Illumina Experiment Manager. For detailed information about the SampleSheet.csv see Illumina Inc. (2013).

Explanation of critical information in the 'SampleSheet.csv' file:

Chemistry, Amplicon
The amplicon setting allows use of two indexes (index1 and index2)
[Reads]
301
301
Orders the MiSeq to perform paired end sequencing, where each read is 300 bp long.

Sample_ID, Sample_Name, index, index2, Description
Sample_ID column: The ID of your samples.
Sample_Name column: The name of the samples. The data output will be called by these names.
Index column: The first read barcode of your samples. Type in the sequence of the barcode. The presence and length of a barcode lets the sequencer know you want to sequence the barcode.
Index2 column: Similar to above.
Description column: Notes about samples.

- **Prepare sequencing libraries**
In this step the sequencing libraries (the PhiX control library and the library pool) are denatured with high pH (NaOH) to obtain the library amplicons in single-stranded form. The library amplicons need to be single-stranded in order for them to be captured by the MiSeq.

The pH is very important, and both too high and too low pH will prevent the library amplicons from being captured. Preferably use 2 M molecular grade NaOH.

After denaturation the libraries are diluted. The concentration is extremely important as it directly determines the amount of data produced. 20 pM produces approximately 18-25 million reads. If the concentration is lower or higher the output will be proportionally lower or higher. Outside the range of 2-25 pM there is a great risk of the sequencing run crashing completely.

The PhiX control library is used to estimate the error rate and to assist in calibration of the instrument during sequencing. This is especially important for library pools that have low complexity. Low complexity means the sequences analysed have similar sequence composition. As the 16S rRNA gene has conserved regions this is the case for 16S amplicon libraries.

- **Load the sample and primers on the reagent cartridge**
Adding sequencing primers to the MiSeq reagent cartridge is needed when sequencing libraries prepared

with adaptors not bought from Illumina. The primers initiate the reading of Read1, Read2 and Index1. The run will fail if they are not present.

- **Sequencing**

After the sequencing has started the first stats will appear after approximately 4 h. Yield per sample can be obtained after 32 h and the run will complete in 56 h.

The cluster density reveals the estimated output of the run. An average run prepared by the above protocols will produce between 700-1,000 k mm^{-2}, which produces 17-25 million PE reads.

Cluster PF reveals how much of the data meets basic internal quality requirements. With sequencing of V1-3 16S rRNA Amplicon libraries, a value of > 90 % is standard. This might change a little during sequencing.

% ≥ Q30 reveals the number of bases in the whole run expected to have a quality score above Q30. This changes a lot during sequencing. For read1 the average is usually > 70 % and for Read2 the average is usually > 60 %.

The error rate explains the actual measured error rate in the sequencing of the PhiX control library. During sequencing the MiSeq recognizes library amplicons from PhiX and compares them to a reference PhiX genome, to detect and measure sequencing mistakes. In simple terms, the quality score is the theoretical quality where the error rate is the empirical quality.

The aligned statistics explains how much the PhiX control library makes up of the whole run. It should be close to 20 % if the above protocol was followed.

8.4.11 Design of Illumina 16S amplicon sequencing adaptors

The section describes the adaptor/primer designs and the functions of the different parts.

For preparing 16S rRNA amplicon libraries, so-called adaptors are used. They come in pairs consisting of a forward and a reverse adaptor. Each adaptor consists of an adaptor part and a primer part. During the library PCR, the primer parts of the forward and reverse adaptors are used to specifically amplify the variable region 1 to 3 (V1-3) of the 16S rRNA gene (Figure 8.7). The final library amplicons contain the V1-3 16S rRNA sequences and the adaptor parts. The primers are adopted from the Human microbiome project (HMP, 2010) and are called 27F and 534R. The adaptor parts are adopted from Caporaso *et al.* (2011; 2012) and from Illumina Inc. (2014a).

During sequencing, sequencing primers attach to the adaptors and initiate sequencing (Figure 8.16).

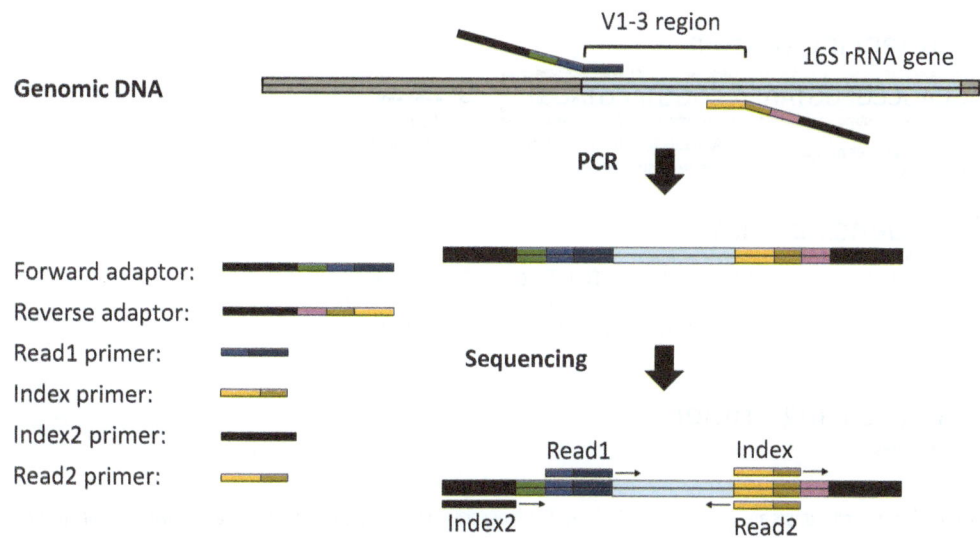

Figure 8.16 Conceptual overview of the oligos used in 16S amplicon sequencing. The adaptors are introduced during PCR and the rest of the oligos are used for sequencing. The coloured parts of the adaptors/primers represent different functional sequences.

In total, four sequences are read: Read1 and Read2, which span the V1-3 16S rRNA gene part and make up the sequences of the paired end reads, and index and index2, which make up the barcoding part which is used to identify what sample each library amplicon originates from. The barcoding part is processed directly on the MiSeq and is not a part of the data output.

The adaptors and primers are synthesized DNA oligos, which can be ordered from any large reagent company. The sequences of the respective DNA oligos can be found in Figure 8.17. The different parts of the oligos have names and a designated role in the library preparation and/or sequencing. The index part of the adaptors (NNNNNNNN) is different for each sample that is to be sequenced. For example, if 96 samples are to be sequenced then there will usually be 8 forward adaptors and 12 reverse adaptors, all of which have a unique index. Hence, a total of 96 unique combinations of the forward and reverse adaptors can be obtained.

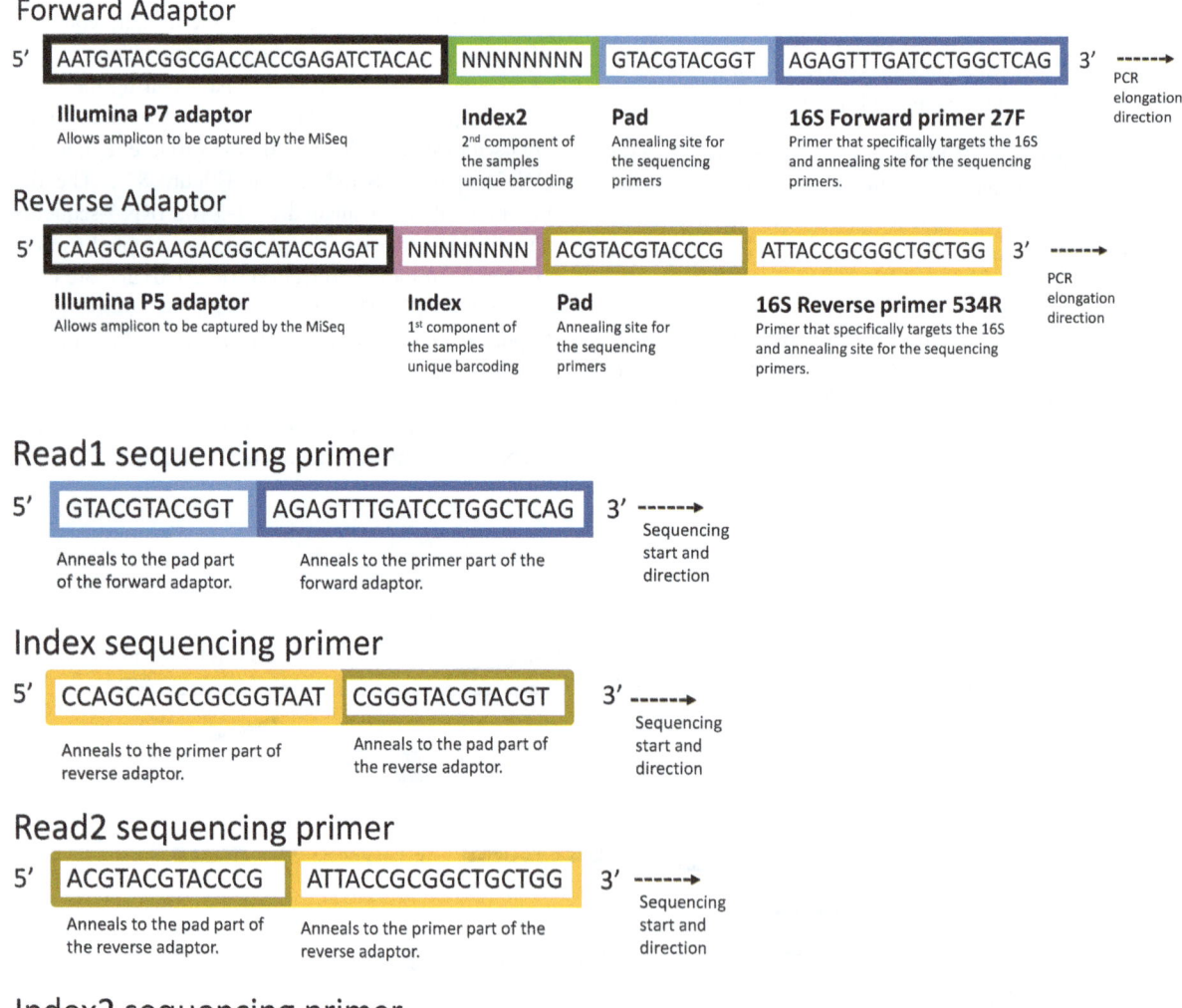

Figure 8.17 The adaptors and primers are synthesized DNA oligos. The sequences of the respective DNA oligos are shown. The different parts of the oligos have names and a designated role in the library preparation and/or sequencing. The index part of the adaptors (NNNNNNNN) is different for each sample to be sequenced.

8.5 OTHER METHODS

Fluorescence *in situ* hybridization (FISH) is an independent method to visualize microorganisms by fluorescently labelled 16S rRNA-targeted oligonucleotide probes. FISH is in itself a very powerful method for microbial community analysis, but it is also an excellent supporting technique for validating results from amplicon sequencing. FISH is described in detail in Chapter 7.

Advanced sequencing-based techniques, such as metagenomics, metatranscriptomics and metaproteomics, exist that enable the culture-independent analysis of functions in microbial communities. These techniques are starting to be applied to activated-sludge systems by the research community. However, these techniques are complicated and will not be mature enough to be valuable for widespread use in the foreseeable future. Therefore, the description of these techniques will be limited to a basic overview.

Metagenomics, or environmental genomics (Wooley *et al.*, 2010), is the study of all the community DNA recovered directly from environmental samples. Figure 8.18 shows the workflow.

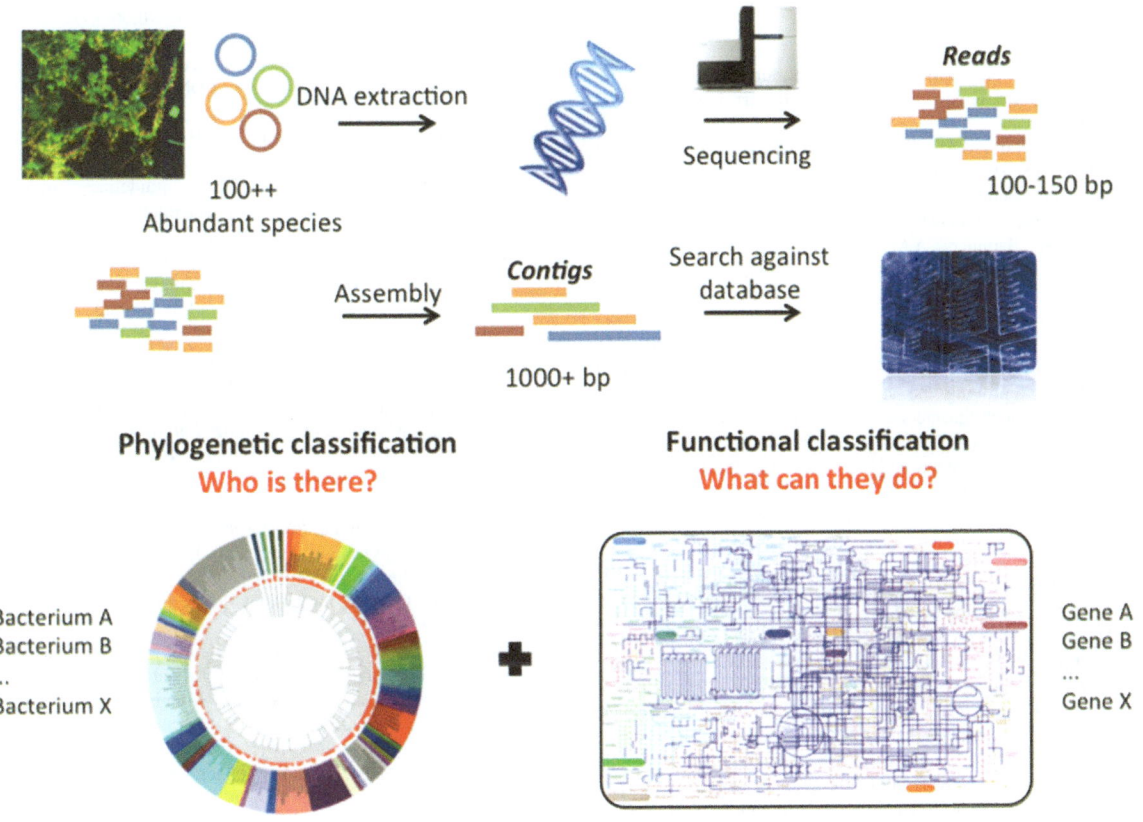

Figure 8.18 Overview of the workflow in metagenomics and the potential outcome.

Community DNA is extracted and purified before it is sequenced, here by the Illumina platform, providing short reads. The reads are assembled into progressively longer contiguous sequences (contigs). These can be applied to obtain information about the taxonomy and the function of the community members by comparing the sequence information to reference databases. However, the resulting information is unfortunately very biased. Proper taxonomic classification of DNA sequences requires closely related, annotated genomes in the reference databases. Only relatively few genomes are available in the reference databases today, so often the resulting classification is unreliable. Furthermore, a detailed functional characterization of the genes is also difficult, again due to incomplete reference databases (Albertsen et al., 2013).

Metatranscriptomics and metaproteomics describe the complete set of expressed genes and proteins in the microorganisms in an environmental sample. These techniques are rarely applied in systems related to wastewater treatment for several reasons. The main reason is the lack of good reference genomes that are a prerequisite for a reliable study of the expressed genes and proteins. Furthermore, for metaproteomics, it is difficult to obtain sufficient protein extraction efficiency (Seifert et al., 2013; Jensen et al., 2014).

Microarray technologies were regarded as very promising for taxonomic and functional analyses of communities, but the fast development in sequencing technologies has largely outpaced these techniques.

References

Agilent Technologies (2012). Agilent 2200 TapeStation User Manual edition 8. Manual Part number G2966-90001.

Agilent Technologies (2013). Agilent D1000 ScreenTape System Quick Guide edition 10. Manual Part number G2964-90032 Rev. B.

Agilent Technologies (2015). Agilent Genomic DNA ScreenTape System Quick Guide edition 6. Manual Part number G2964-90040 Rev. D.

Albertsen, M, Karst, S.M., Ziegler, A.S., Kirkegaard R.H., Nielsen P.H. (2015). Back to Basics – The influence of DNA extraction and primer choice on phylogenetic analysis of activated sludge communities. *PLoS ONE* 10:e0132783.

Albertsen M., Saunders, A.M., Nielsen K.L. and Nielsen P.H. (2013): Metagenomes obtained by "deep sequencing" - what do they tell about the EBPR communities? *Wat. Sci. Tech.*, 68: 1959-1968.

Angly, F.E., Dennis, P.G., Skarshewski, A., Vanwonterghem, I., Hugenholtz, P., and Tyson, G.W. (2014). CopyRighter: a rapid tool for improving the accuracy of microbial community profiles through lineage-specific gene copy number correction. *Microbiome*, 2, 11.

Ashelford, K.E., Chuzhanova, N.A., Fry, J.C., Jones, A. J., and Weightman, A.J. (2005). At least 1 in 20 16S rRNA sequence records currently held in public repositories is estimated to contain substantial anomalies. *Appl. Environ. Microbiol.* 71: 7724-7736.

Basu, C. (2015). PCR primer design (New York: Humana Pr).

Beers, E.H. Van, Joosse, S.A., Ligtenberg, M.J., Fles, R., Hogervorst, F.B.L., Verhoef, S., and Nederlof, P.M. (2006). A multiplex PCR predictor for aCGH success of FFPE samples. *British J. Cancer* 94: 333-337.

Beller, H.R., Kane, S.R., Legler, T.C., and Alvarez, P.J.J. (2002). A real-time polymerase chain reaction method for monitoring anaerobic, hydrocarbon-degrading bacteria based on a catabolic gene. *Environ. Sci. & Technol.* 36: 3977–3984.

Bessetti, J. (2007). An introduction to PCR inhibitors. *J. Microbiol. Meth.* 28: 159-167.

Bollet C, Gevaudan M.J., de Lamballerie X., Zandotti C., de Micco P. 1991. A simple method for the isolation of chromosomal DNA from Gram positive or acid-fast bacteria. *Nucleic Acids Research* 19:1955.

Brown, C.T., Hug, L.A, Thomas, B.C., Sharon, I., Castelle, C.J., Singh, A., ... Banfield, J. F. (2015). Unusual biology across a group comprising more than 15 % of domain Bacteria. *Nature*, 523: 208-211.

Bru, D., Sarr, A., and Philippot, L. (2007). Relative abundances of proteobacterial membrane-bound and periplasmic nitrate reductases in selected environments. *Appl. Environ. Microbiol.* 73: 5971-5974.

Brzoska, A.J., and Hassan, K.A. (2014). Quantitative PCR for detection of mRNA and gDNA in environmental isolates. In Environmental Microbiology, I.T. Paulsen, and A.J. Holmes, eds. (Totowa, NJ: Humana Press), pp. 25-42.

Bürgmann H, Pesaro M, Widmer F, Zeyer J. (2001). A strategy for optimizing quality and quantity of DNA extracted from soil. *J Microbiol. Methods* 45: 7-20.

Bustin, S.A., Benes, V., Garson, J.A., Hellemans, J., Huggett, J., Kubista, M., Mueller, R., Nolan, T., Pfaffl, M.W., Shipley, G.L., et al. (2009). The MIQE guidelines: minimum information for publication of quantitative real-time PCR experiments. *Clin. Chem.* 55: 611-622.

Caporaso, J.G., Kuczynski, J., Stombaugh, J., Bittinger, K., Bushman, F.D., Costello, E.K., ... Knight, R. (2010). QIIME allows analysis of high-throughput community sequencing data. *Nature Methods*. 7: 335-6.

Caporaso, J.G., Lauber, C.L., Walters, W.A, Berg-Lyons, D., Huntley, J., Fierer, N., ... Knight, R. (2012). Ultra-high-throughput microbial community analysis on the Illumina HiSeq and MiSeq platforms. *ISME Journal* 6: 1621-1624.

Caporaso, J.G., Lauber, C.L., Walters, W.A, Berg-Lyons, D., Lozupone, C.A., Turnbaugh, P.J., ... Knight, R. (2011). Global patterns of 16S rRNA diversity at a depth of millions of sequences per sample. *PNAS* 108. Suppl 1, 4516-4522.

Cole, J.R., Wang, Q., Fish, J. a., Chai, B., McGarrell, D.M., Sun, Y., ... Tiedje, J.M. (2014). Ribosomal Database Project: Data and tools for high throughput rRNA analysis. *Nucleic Acids Research*, 42: 633–642.

Cock, P.J.A., Fields, C.J., Goto, N., Heuer, M.L., and Rice, P.M. (2010). The Sanger FASTQ file format for sequences with quality scores, and the Solexa / Illumina FASTQ variants, *Nucleic Acids Research* 38: 1767-1771.

Daims, H., Lebedeva, E.V., Pjevac, P., Han, P., Herbold, C., Albertsen, M., ... and Wagner, M. (2015). Complete nitrification by *Nitrospira* bacteria. *Nature* Doi: 10.1038/nature16461.

DeAngelis M.M., Wang D.G., Hawkins T.L. (1995). Solid-phase reversible immobilization for the isolation of PCR products. *Nucleic Acids Research* 23: 4742-4743.

Dominiak, D.M., Nielsen, J.L., and Nielsen, P.H. (2011). Extracellular DNA is abundant and important for microcolony strength in mixed microbial biofilms. *Environ. Microbiol.* 13: 710-721.

Dueholm, M.S., Albertsen, M., D'Imperio, S., Tale, V.P., Lewis, D., Nielsen, P.H., and Nielsen, J.L. (2014). Complete genome sequences of *Pseudomonas monteilii* SB3078 and SB3101, two benzene-, toluene-, and ethylbenzene-degrading bacteria used for bioaugmentation. *Genome Announc.* 2: 524-14.

Dueholm, M.S., Marques, I.G., Karst, S.M., D'Imperio, S., Tale, V.P., Lewis, D., Nielsen, P.H., and Nielsen, J.L. (2015). Survival and activity of individual bioaugmentation strains. *Biores. Technol.* 186: 192-199.

Edgar, R. C. (2013). UPARSE: highly accurate OTU sequences from microbial amplicon reads. *Nature Methods,* 10: 996-8.

El Fantroussi, S., and Agathos, S.N. (2005). Is bioaugmentation a feasible strategy for pollutant removal and site remediation? *Curr Opin Microbiol* 8: 268-275.

Farrelly, V., Rainey, F.A, and Stackebrandt, E. (1995). Effect of genome size and rrn gene copy number on PCR amplification of 16S rRNA genes from a mixture of bacterial species. *Appl. Environ. Microbiol.* 67: 2798-2801.

Faust, K., and Raes, J. (2012). Microbial interactions: from networks to models. *Nature Reviews. Microbiology*, 10, 538–50.

Faust, K., Lahti, L., Gonze, D., de Vos, W. M., and Raes, J. (2015). Metagenomics meets time series analysis: unraveling microbial community dynamics. *Current Opin Microbiol.*, 25: 56-66.

Filippidou, S., Junier, T., Wunderlin, T., Lo, C-C., Li, P-E., Chain, P.S., Junier, P. (2015). Under-detection of endospore-forming Firmicutes in metagenomic data. *Comput. and Structural Biotechnol. J.* 13: 299-306.

Fisher, S., Barry, A., Abreu, J. (2011). A scalable, fully automated process for construction of sequence-ready human exome targeted capture libraries. *Genome Biology*, 12, R1 doi:10.1186/gb-2011-12-1-r1

Flynn, J. M., Brown, E. a., Chain, F. J. J., MacIsaac, H. J., and Cristescu, M. E. (2015). Toward accurate molecular identification of species in complex environmental samples: testing the performance of sequence filtering and clustering methods. *Ecol. Evol.*, 5: 2252-2266.

Frølund, B., Palmgren, R., Keiding, K., and Nielsen, P.H. (1996). Extraction of extracellular polymers from activated sludge using a cation exchange resin. *Wat. Res.* 30: 1749-1758.

Ge, S., Wang, S., Yang, X., Qiu, S., Li, B., and Peng, Y. (2015). Detection of nitrifiers and evaluation of partial nitrification for wastewater treatment: A review. *Chemosphere,* 140: 85-98.

Gibson, D.J., Ely, J.S., Collins, S.L. (1999). The core-satellite species hypothesis provides a theoretical basis for Grime's classification of dominant, subordinate, and transient species. *J. Ecol.* 87: 1064-1067.

Gilbert, E.M. ,S. Agrawal, S.M. Karst, H. Horn, P.H. Nielsen, S. Lackner (2014). Low temperature partial nitritation/anammox in a moving bed biofilm reactor treating low strength wastewater. *Environ. Sci. Technol.,* 48: 8784-8792.

Green, M.R., Sambrook, J. (2012). Molecular Cloning: A Laboratory Manual (Fourth Edition). Cold Spring Harbor Laboratory Press. Cold Spring Harbor, New York.

Grime, J. (1998). Benefits of plant diversity to ecosystems: immediate, filter and founder effects. *J. Ecol.* 86: 902-910.

Guillén-Navarro, K., Herrera-López, D., López-Chávez, M.Y., Cancino-Gómez, M., Reyes-Reyes, A.L. (2015). Assessment of methods to recover DNA from bacteria, fungi and archaea in complex environmental samples. *Folia Microbiol.* 60: 551-558.

Guo, F., Zhang, T. (2013). Biases during DNA extraction of activated sludge samples revealed by high throughput sequencing. *Appl. Microbiol. Biotechnol.* 97: 4607-4616.

Holland, P.M., Abramson, R.D., Watson, R., and Gelfand, D.H. (1991). Detection of specific polymerase chain reaction product by utilizing the 5'----3' exonuclease activity of *Thermus aquaticus* DNA polymerase. *PNAS* 88: 7276-7280.

Horz, H.P., Vianna, M.E., Gomes, B.P.F.A., and Conrads, G. (2005). Evaluation of universal probes and primer sets for assessing total bacterial load in clinical samples: General implications and practical use in endodontic antimicrobial therapy. *J. Clin. Microbiol.* 43: 5332-5337.

Hou, Y., Zhang, H., Miranda, L., and Lin, S. (2010). Serious overestimation in quantitative pcr by circular (supercoiled) plasmid standard: Microalgal pcna as the model gene. *PLoS ONE* 5, e9545.

Humbert, S., Zopfi, J., and Tarnawski, S.-E. (2012). Abundance of anammox bacteria in different wetland soils. *Environ. Microbiol. Rep.* 4: 484-490.

Huse, S. M., Welch, D. M., Morrison, H. G., and Sogin, M. L. (2010). Ironing out the wrinkles in the rare biosphere through improved OTU clustering. *Environ. Microbiol.* 12: 1889-1898.

Illumina Inc. (2013) MiSeq Sample Sheet, Quick Reference Guide, Part # 15028392 Rev. J

Illumina Inc. (2014a). Illumina Customer Sequence Letter. Oligonucleotide sequences © 2007-2013 Illumina, Inc. All rights reserved. Derivative works created by Illumina customers are authorized for use with Illumina instruments and products only. All other uses are strictly prohibited.

Illumina Inc. (2014b) MiSeq System User Guide, Part # 15027617 Rev. O

Illumina, Inc. (2015). An Introduction to Next-Generation Sequencing Technology, www.illumina.com

Janda, J. M., and Abbott, S. L. (2007). Minireview: 16S rRNA gene sequencing for bacterial identification in the diagnostic laboratory: Pluses, perils, and pitfalls. *J. Clin. Microbiol.* 45: 2761-2764.

Jensen S.H., A. Stensballe, P.H. Nielsen, F.-A. Herbst (2014). Metaproteomics: Evaluation of protein extraction from activated sludge. *Proteomics* 14(21-22), 2535-2539.

Hadfield, J. (2012). How do SPRI beads work? http://core-genomics.blogspot.dk/2012/04/how-do-spri-beads-work.html

Human Microbiome Project (HMP) (2010). Jumpstart Consortium Human Microbiome Project Data Generation Working Group (2010). 16S 454 Sequencing Protocol HMP Consortium. Version 4.2.2.

Juretschko, S., Purkhold, U., Pommerening-ro, A., Schmid, M. C., Koops, H., and Wagner, M. (2000). Phylogeny of all recognized species of ammonia oxidizers based on comparative 16S rRNA and amoA sequence analysis: Implications for molecular diversity surveys. *Appl. Environ. Microbiol.* 66, 5368–5382.

Kim, J., Lim, J., and Lee, C. (2013). Quantitative real-time PCR approaches for microbial community studies in wastewater treatment systems: Applications and considerations. *Biotechnology Advances* 31: 1358-1373.

Kitajima, M., Iker, B.C., Pepper, I.L., and Gerba, C.P. (2014). Relative abundance and treatment reduction of viruses during wastewater treatment processes identification of potential viral indicators. *Sci. Total Environ.* 488-489: 290-296.

Klindworth, A., Pruesse, E., Schweer, T., Peplies, J., Quast, C., Horn, M., and Glöckner, F. O. (2013). Evaluation of general 16S ribosomal RNA gene PCR primers for classical and next-generation sequencing-based diversity studies. *Nucleic Acids Research*, 41, e1.

Kubista, M., Andrade, J.M., Bengtsson, M., Forootan, A., Jonák, J., Lind, K., Sindelka, R., Sjöback, R., Sjögreen, B., Strömbom, L., *et al.* (2006). The real-time polymerase chain reaction. *Mol. Aspects Medicine* 27: 95-125.

Laureni, M., Weissbrodt, D.G., Szivák, I., Robin, O., Nielsen, J.L., Morgenroth, E., Joss, A. (2015). Activity and growth of anammox biomass on aerobically pre-treated municipal wastewater. *Wat. Res.,* 80: 325-336.

Legendre, P., and Gallagher, E. (2001). Ecologically meaningful transformations for ordination of species data. *Oecologia*, 129: 271-280.

Li, X., Wu, Y., Zhang, L., Cao, Y., Li, Y., Li, J., ... Wu, G. (2014). Comparison of three common DNA concentration measurement methods. *Analytical Biochemistry*, 451: 18-24.

Liu, C.M., Kachur, S., Dwan, M.G., Abraham, A.G., Aziz, M., Hsueh, P.-R., Huang, Y.-T., Busch, J.D., Lamit, L.J., Gehring, C.A., et al. (2012). FungiQuant: A broad-coverage fungal quantitative real-time PCR assay. *BMC Microbiol.* 12: 255.

Love, M.I., Huber, W., and Anders, S. (2014). Moderated estimation of fold change and dispersion for RNA-seq data with DESeq2. *Genome Biology,* 15: 550.

Lozupone, C.A., and Knight, R. (2008). Species divergence and the measurement of microbial diversity. *FEMS Microbiol. Rev.,* 32: 557-578.

Ludwig, W., and Schleifer, K.-H. (2000). How quantitative is quantitative PCR with respect to cell counts? *Syst. Appl. Microbiol.* 23: 556–562.

Madigan, M.T., Martinko, J.M. (2006). Brock Biology of Microorganisms, 11th ed. Pearson Education International. ISBN 0131968939.

Magurran, A. E. (2004). Measuring biological diversity. Oxford, UK: Blackwell Publishing.

Mahé, F., Rognes, T., Quince, C., de Vargas, C., and Dunthorn, M. (2014). Swarm: robust and fast clustering method for amplicon-based studies. *PeerJ,* 1-12.

Marsh, T.L., 1999. Terminal restriction fragment length polymorphism (T-RFLP): an emerging method for characterizing diversity among homologous populations of amplification products. *Curr. Opin. Microbiol.* 2: 323-327.

Marzorati, M., Wittebolle, T., Boon, N., Daffonchio, D., Verstraete., W. (2008). How to get more out of molecular fingerprints: Practical tools for microbial ecology. *Environ. Microbiol.* 10: 1571-1581.

Matsuda, K., Tsuji, H., Asahara, T., Kado, Y., and Nomoto, K. (2007). Sensitive quantitative detection of commensal bacteria by rRNA-targeted reverse transcription-PCR. *Appl. Environ. Microbiol.* 73: 32-39.

McDonald, D., Price, M. N., Goodrich, J., Nawrocki, E. P., DeSantis, T. Z., Probst, A., ... Hugenholtz, P. (2012). An improved Greengenes taxonomy with explicit ranks for ecological and evolutionary analyses of bacteria and archaea. *ISME Journal,* 6: 610-618.

McIlroy, S.J., Saunders, A.M., Albertsen, M., Nierychlo, M., McIlroy, B., Hansen A.A., Karst, S.M., Nielsen, J.L., Nielsen, P.H. (2015). MiDAS: the field guide to the microbes of activated sludge. *Database* bav062. doi: 10.1093/database/bav062.

McMaster, G.K., Carmichael, G.G. (1977). Analysis of single- and double-stranded nucleic acids on polyacrylamide and agarose gels by using glyoxal and acridine orange. *PNAS* 74: 4835-4838.

McMurdie, P.J., and Holmes, S. (2013). Phyloseq: an R package for reproducible interactive analysis and graphics of microbiome census data. *PloS One,* 8, e61217.

Mendell, J.E., Clements, K.D., Choat, J.H., and Angert, E.R. (2008). Extreme polyploidy in a large bacterium. *PNAS* 105: 6730-6734.

Miller, D.N., Bryant, J.E., Madsen, E.L., Ghiorse, W.C. (1999). Evaluation and optimization of DNA extraction and purification procedures for soil and sediment samples. *Appl. Environ. Microbiol.* 65: 4715-4724.

Muyzer, G. (1999).DGGE/TGGE a method for identifying genes from natural ecosystems. *Curr. Opin. Microbiol.* 2: 317-322.

Nadkarni, M.A., Martin, F.E., Jacques, N.A., and Hunter, N. (2002). Determination of bacterial load by real-time PCR using a broad-range (universal) probe and primers set. *Microbiology* 148: 257-266.

Nanodrop Technologies, Inc. (2007). ND-1000 Spectrophotometer V3.3 User's manual, rev.3.

Nolan, T., Hands, R.E., and Bustin, S.A. (2006). Quantification of mRNA using real-time RT-PCR. *Nature Protocols* 1: 1559-1582.

Nygren, J., Svanvik, N., and Kubista, M. (1998). The interactions between the fluorescent dye thiazole orange and DNA. *Biopolymers* 46: 39-51.

Okano, Y., Hristova, K.R., Leutenegger, C.M., Jackson, L.E., Denison, R.F., Gebreyesus, B., Lebauer, D., and Scow, K.M. (2004). Application of real-time PCR to study effects of ammonium on population size of ammonia-oxidizing bacteria in soil. *Appl. Environ. Microbiol.* 70: 1008-1016.

Oksanen, J., Blanchet, G. F., Kindt, R., Legendre, P., Minchin, P. R., O'Hara, R. B., ... Wagner, H. (2015). Vegan: Community Ecology Package. http://r-forge.r-project.org/projects/vegan/.

Pace, N. R., Sapp, J., and Goldenfeld, N. (2012). Phylogeny and beyond: Scientific, historical, and conceptual significance of the first tree of life. *PNAS,* 109: 1011-1018

Padmanaban, A., Inche, A., Gassmann, M., Salowsky, R.. 2013. High-Throughput DNA Sample QC Using the Agilent 2200 Tapestation System. *J. Biomol. Techn. JBT* 24:S41.

Panaro, N.J., Yuen, P.K., Sakazume, T., Fortina, P., Kricka, L.J., Wilding, P. 2000. Evaluation of DNA fragment sizing and quantification by the Agilent 2100 bioanalyzer. *Clinical Chemistry* 46: 1851-1853.

Pecoraro, V., Zerulla, K., Lange, C., and Soppa, J. (2011). Quantification of ploidy in proteobacteria revealed the existence of monoploid, (mero-)oligoploid and polyploid species. *PLoS ONE,* 6. e16392. doi:10.1371/journal.pone.0016392.

Polz, M.F., and Cavanaugh, C.M. (1998). Bias in template-to-product ratios in multitemplate PCR, *Appl Environ Microbiol.* 64(10): 3724-3730.

Pruesse, E., Peplies, J., and Glöckner, F. O. (2012). SINA: Accurate high-throughput multiple sequence alignment of ribosomal RNA genes. *Bioinformatics* 28(14): 1823-1829.

Qiu, X., Wu, L., Huang, H., Donel, P. E. M. C., Palumbo, A. V, Tiedje, J. M., and Zhou, J. (2001). Evaluation of PCR-generated chimeras, mutations, and heteroduplexes with 16S rRNA gene-based cloning, *Appl Environ Microbiol.* 67(2): 880-887.

Quast, C., Pruesse, E., Yilmaz, P., Gerken, J., Schweer, T., Yarza, P., ... Glöckner, F. O. (2013). The SILVA ribosomal RNA gene database project: Improved data processing and web-based tools. *Nucleic Acids Research,* 41. doi:10.1093/nar/gks1219

Quince, C., Lanzen, A., Davenport, R. J., and Turnbaugh, P. J. (2011). Removing noise from pyrosequenced amplicons. *BMC Bioinformatics,* 12: 38.

Pruesse, E., Peplies, J., and Glöckner, F. O. (2012). SINA: Accurate high-throughput multiple sequence alignment of ribosomal RNA genes, *Bioinformatics* 28(14): 1823-1829.

Ramette, A. (2007). Multivariate analyses in microbial ecology. *FEMS Microbiol. Ecol.,* 62: 142-60.

Rasmussen, R. (2001). Quantification on the LightCycler. In Rapid Cycle Real-Time PCR, P.D. med S. Meuer, P.D.C. Wittwer, and D.K.-I. Nakagawara, eds. (Springer Berlin Heidelberg), pp. 21-34.

Robinson, M.D., McCarthy, D.J., and Smyth, G.K. (2010). EdgeR: a Bioconductor package for differential expression analysis of digital gene expression data. *Bioinformatics,* 26: 139-40.

Rotthauwe, J.H., Witzel, K.P., Liesack, W., (1997). The ammonia monooxygenase structural gene amoA as a functional marker: Molecular fine-scale analysis of natural ammonia-oxidizing populations. *Appl Environ Microbiol.* 63: 4704-4712.

Saiki, R.K., Scharf, S., Faloona, F., Mullis, K.B., Horn, G.T., Erlich, H.A., and Arnheim, N. (1985). Enzymatic amplification of beta-globin genomic sequences and restriction site analysis for diagnosis of sickle cell anemia. *Science* 230: 1350-1354.

Salonen, A., Nikkilä, J., Jalanka-Tuovinen, J., Immonen, O., Rajilić-Stojanović, M., Kekkonen, R. a., Palva, A., de Vos, W.M. (2010). Comparative analysis of fecal DNA extraction methods with phylogenetic microarray: effective recovery of bacterial and archaeal DNA using mechanical cell lysis. *J. Microbiological Methods* 81: 127-34.

Saunders, A.M., Albertsen, M., Vollertsen, J., and Nielsen, P.H. (2015). The activated sludge ecosystem contains a core community of abundant organisms. *ISME J.*,doi:10.1038/ismej.2015.117

Schloss, P.D. and Handelsman, J. (2005). Metagenomics for studying unculturable microorganisms: cutting the Gordian knot. *Genome Biol.* 6: 229.

Schloss, P.D., Westcott, S.L., Ryabin, T., Hall, J.R., Hartmann, M., Hollister, E.B., ... Weber, C.F. (2009). Introducing mothur: open-source, platform-independent, community-supported software for describing and comparing microbial communities. *Appl. Environ. Microbiol.*, 75: 7537-41.

Schriewer, A., Wehlmann, A., and Wuertz, S. (2011). Improving qPCR efficiency in environmental samples by selective removal of humic acids with DAX-8. *J. Microbiol. Methods* 85: 16-21.

Seifert, J.F.-A. Herbst, P.H. Nielsen, F.J. Planes, M. Ferrer, M. von Bergen (2013) Bioinformatic progress and applications in metaproteogenomics for bridging the gap between genomic sequences and metabolic functions in microbial communities. *Proteomics* 13: 2786-2804. doi: 10.1002/pmic.201200566.

Singer, V.L., Jones, L.J., Yue, S.T., Haugland, R.P. (1997). Characterization of PicoGreen reagent and development of a fluorescence-based solution assay for double-stranded DNA quantitation. *Analytical Biochemistry* 249: 228-238.

Smith, S., Morin, P.A. (2005). Optimal storage conditions for highly dilute dna samples: a role for trehalose as a preserving agent. *J. Forensic Sci.*, 50: 1101-1108.

Souazé, F., Ntodou-Thomé, A., Tran, C.Y., Rostène, W., and Forgez, P. (1996). Quantitative RT-PCR: limits and accuracy. *BioTechniques* 21: 280-285.

Stults, J.R., Snoeyenbos-West, O., Methe, B., Lovley, D.R., and Chandler, D.P. (2001). Application of the 5 fluorogenic exonuclease assay (taqman) for quantitative ribosomal DNA and rRNA analysis in sediments. *Appl. Environ. Microbiol.* 67: 2781-2789.

Tebbe, C.C., and Vahjen, W. (1993). Interference of humic acids and DNA extracted directly from soil in detection and transformation of recombinant DNA from bacteria and a yeast. *Appl. Environ. Microbiol* 59(8): 2657-2665.

Thermo Scientific (2015). Assessment of Nucleic Acid Purity, T042-TECHNICAL BULLETIN, T042 Rev 1/11.

Thermo Scientific (2015a), Qubit dsDNA HS Assay Kits – User Guide, MAN0002326, P32851, Revision: B.0

Thermo Scientific (2015b), Qubit dsDNA BR Assay Kits – User Guide, MAN0002325, P32850, Revision: A.0

Thomas, T., Gilbert, J., Meyer, F. (2012). Metagenomics - a guide from sampling to data analysis. *Microb. Inform. Exp.* 2:3. 10.1186/2042-5783-2-3.

Thompson, I.P., Van Der Gast, C.J., Ciric, L., and Singer, A.C. (2005). Bioaugmentation for bioremediation: the challenge of strain selection. *Environ. Microbiol* .7: 909-915.

Tsai, Y., Olson, B. (1991). Rapid method for direct extraction of DNA from soil and sediments. *Appl. Environ. Microbiol.* 57: 1070-1074.

Tullis, R.H., Rubin, H. (1980). Calcium protects DNase I from proteinase K: a new method for the removal of contaminating RNase from DNase I. *Analyt. Biochem.*107: 260-264.

Větrovský, T., and Baldrian, P. (2013). The variability of the 16S rRNAS gene in bacterial genomes and its consequences for bacterial community analyses. *PLoS ONE 8.* e57923. doi:10.1371/journal.pone.0057923.

Volkmann, H., Schwartz, T., Bischoff, P., Kirchen, S., and Obst, U. (2004). Detection of clinically relevant antibiotic-resistance genes in municipal wastewater using real-time PCR (TaqMan). *J. Microbiol. Methods* 56: 277-286.

Wang, Y., and Qian, P.-Y. (2009). Conservative fragments in bacterial 16S rRNA genes and primer design for 16s ribosomal DNA amplicons in metagenomic studies. *PLoS ONE 4*, e7401. doi:10.1371/journal.pone.0007401.

Wilfinger, W.W., Mackey, K., Chomczynski, P. (1997). Effect of pH and ionic strength on the spectrophotometric assessment of nucleic acid purity. *BioTechniques* 22: 474-476.

Wilson, I.G. (1997). Inhibition and facilitation of nucleic acid amplification. *Appl. Environ. Microbiol.*, 63(10): 3741-3751.

Woese, C.R., and Fox, G.E. (1977). Phylogenetic structure of the prokaryotic domain: the primary kingdoms. *PNAS*, 74: 5088-90.

Wooley, J.C., Godzik, A., Friedberg, I. (2010). A primer on metagenomics. *PLoS Comput. Biol.* 6(2): e1000667. doi:10.1371/journal.

Zhang, T., and Fang, H.H.P. (2006). Applications of real-time polymerase chain reaction for quantification of microorganisms in environmental samples. *Appl.Microbiol. Biotechnol.* 70: 281-289.

Zhou, J., Bruns, M.A., Tiedje, J.M. (1996). DNA recovery from soils of diverse composition. *Appl. Environ. Microbiol.* 62: 316-322.

Zipper, H., Brunner, H., Bernhagen, J., and Vitzthum, F. (2004). Investigations on DNA intercalation and surface binding by SYBR Green I, its structure determination and methodological implications. *Nucl. Acids Res.* 32: e103–e103.

Zuur, A.F., Ieno, E.N. and Smith, G.M. (2007) Analysing Ecological Data. Springer, New York.

SYMBOLS AND ABBREVIATIONS

1 Symbols — INTRODUCTION

N/A

1 Abbreviations — INTRODUCTION

AMX	Anammox organisms
AOO	Ammonium oxidizing organisms
CH_4	Methane
DNA	Deoxyribonucleic acid
DPAO	Denitrifying phosphorus- or polyphosphate-accumulating organisms
FISH	Fluorescence *in situ* hybridization
GAO	Glycogen-accumulating organisms
GHG	Greenhouse gas emissions
N_2O	Nitrous oxide
NO_2^-	Nitrite
NO_3^-	Nitrate
NOO	Nitrite oxidizing organisms
PAO	Phosphorus- or polyphosphate-accumulating organism
PCR	Polymerase chain reaction
RBCOD	Readily biodegradable COD also known as readily biodegradable organics
SBCOD	Slowly biodegradable COD also known as slowly biodegradable organics
SRB	Sulphate reducing bacteria or SRO - Sulphate reducing organisms

2 Symbols — ACTIVATED SLUDGE ACTIVITY TESTS

Ac	Concentrations of acetate and acetic acid, mg Ac L^{-1}
Ac_{cons}	Ac concentration consumed in a batch activity test, C-mol L^{-1} or mg C L^{-1}
Ac_{final}	Final concentration of Ac in the bulk liquid at the end of the batch activity test, C-mol L^{-1} or mg C L^{-1}
Ac_{ini}	Initial concentration of Ac in the bulk liquid at the start of the batch activity test, C-mol L^{-1} or mg C L^{-1}
b_{AOO}	Decay or endogenous respiration rate of ammonia oxidizing organisms, d^{-1}
b_{NOO}	Decay or endogenous respiration rate of ammonia oxidizing organisms, d^{-1}
BOD	Biochemical oxygen demand, mg O_2 L^{-1}
BOD_5	Biochemical oxygen demand after 5 d, mg O_2 L^{-1}
C	COD content of an organic compound or substrate, g COD mol^{-1} organic substrate or g COD g^{-1} organic substrate
C_i	Concentration of a compound or element in the gas phase or headspace of a reactor or system, ppmv
C_{CH4}	Concentration of methane in the gas phase, ppmv
C_{CO2}	Concentration of carbon dioxide in the gas phase, ppmv
C_{H2S}	Concentration of sulphide in the gas phase, ppmv
C_{N2}	Concentration of nitrogen in the gas phase, ppmv
C_{O2}	Concentration of oxygen in the gas phase, ppmv
CO_2	Carbon dioxide

COD	Chemical oxygen demand, mg COD L^{-1}
COD$_{B,cons}$	Concentration of biodegradable substrate consumed expressed as COD, mg COD L^{-1}
COD$_{Bio,prod}$	Concentration of biomass produced expressed as COD, mg COD L^{-1}
COD$_{GLY,cons}$	Concentration of glycogen consumed expressed as COD, mg COD L^{-1}
COD$_{GLY,cons}$	Concentration of glycogen consumed expressed as COD in a batch test, mg COD L^{-1}
COD$_{GLY,prod}$	Concentration of glycogen produced expressed as COD in a batch test, mg COD L^{-1}
COD$_{input}$	Total COD concentration that gets into a system or reactor, mg COD L^{-1}
COD$_{organics}$	COD concentration of organic compounds, mg COD L^{-1}
COD$_{organics,in}$	COD concentration of organic compounds in the inlet or influent of a reactor or system, mg COD L^{-1}
COD$_{organics,out}$	COD concentration of organic compounds flowing out of a reactor or system, mg COD L^{-1}
COD$_{output}$	Total COD concentration that exits a reactor or system, mg COD L^{-1}
COD$_{PHA,cons}$	Concentration of PHA consumed expressed as COD, mg COD L^{-1}
COD$_{PHA,prod}$	Concentration of PHA produced expressed as COD, mg COD L^{-1}
COD$_{sulphide}$	Concentration of sulphide expressed as COD, mg COD L^{-1}
COD$_{total}$	Total COD concentration, mg COD L^{-1}
DP$_{SB}$	Denitrification potential of the RBCOD in wastewater, mg N L^{-1}
DP$_{XCB}$	Denitrification potential of the SBCOD in wastewater, mg N L^{-1}
E$_a$	Activation energy of a bioprocess, kJ mol^{-1} K^{-1}
E$_{a,S}$	Activation energy of a bioprocess when consuming a generic substrate S, kJ mol^{-1} K^{-1}
f$_{CV}$	Chemical oxygen demand to volatile suspended solids ratio of an organic compound, mg COD mg VSS^{-1}
F	Flux of a gas or a compound, Mass Time^{-1}
F$_{in}$	Flux of a gas or a compound that enters into a reactor or system, mol h^{-1} or mg h^{-1}
F$_{out}$	Flux of a gas or a compound that flows out or leaves a reactor or system, mol h^{-1} or mg h^{-1}
F$_{AMX,NHx\_N2}$	Maximum N$_2$ production flux rate in an anammox test, mmol N$_2$ min^{-1}
F$_{N\_NHx}$	Ammonification flux rate, mg N min^{-1}
F$_{NHx}$	Ammonium oxidation flux rate, mg N min^{-1}
F$_{NHx,H2O2}$	Ammonium oxidation flux rate determined based on the H$_2$O$_2$ titration flux rate, mg N min^{-1}
F$_{NHx,NaOH}$	Ammonium oxidation flux rate determined based on the NaOH titration flux rate, mg N min^{-1}
F$_{NO2}$	Nitrite oxidation flux rate in a titrimetric nitrification test, mg N min^{-1}
H$^+$	Proton
H$_2$O	Water
H$_2$S	Sulphide
H$_2$S$_{in}$	Concentration of sulphide that enters into a reactor or system, mg S L^{-1} or mg COD L^{-1}
H$_2$S$_{out}$	Concentration of sulphide that leaves a reactor or system, mg S L^{-1} or mg COD L^{-1}
HA	Organic acid
HRT	Hydraulic retention time, d
HS^{-1}	Dissociated sulphide
i$_{N,ANO}$	Nitrogen content of autotrophic nitrifying organisms, g N g VSS^{-1} or g N g COD^{-1}
i$_{N,Bio}$	Nitrogen content of a biomass or bacterial culture, g N g VSS^{-1} or g N g COD^{-1}
i$_{N,OHO}$	Nitrogen content in ordinary heterotrophic organisms, g N g VSS^{-1} or g N g COD^{-1}
k	Sulphate-reduction rate, nmol cm^{-3} d^{-1}
ln	Natural logarithm
m	Maximum specific maintenance or endogenous rate, Mass Active Biomass^{-1} Time^{-1}
m$_{ATP,An}$	Anaerobic ATP maintenance coefficient, mol ATP C-mol^{-1} h^{-1} or mg ATP mg active biomass^{-1} h^{-1}
m$_{ATP,Ax}$	Anoxic ATP maintenance coefficient, mol ATP C-mol^{-1} h^{-1} or mg ATP mg active biomass^{-1} h^{-1}
m$_{ATP,Ox}$	Aerobic ATP maintenance coefficient, mol ATP C-mol^{-1} h^{-1} or mg ATP mg active biomass^{-1} h^{-1}
m$_{Ax}$	Anoxic maintenance rate of a biomass, C-mol C-mol^{-1} h^{-1} or mg C mg active biomass^{-1} h^{-1}
m$_{Ox}$	Aerobic maintenance rate of a biomass, C-mol C-mol^{-1} h^{-1} or mg C mg active biomass^{-1} h^{-1}
m$_{O2}$	Aerobic endogenous respiration rate of a culture, mol O$_2$ C-mol^{-1} h^{-1} or mg O$_2$ mg active biomass^{-1} h^{-1}
m$_{NOx}$	Anoxic endogenous respiration rate on NO$_X$ of a biomass, N-mol C-mol^{-1} h^{-1} or mg NO$_X$ mg VSS^{-1} h^{-1}
m$_{NO3}$	Anoxic endogenous respiration rate on NO$_3$ of a biomass, N-mol C-mol^{-1} h^{-1} or mg NO$_3$ mg VSS^{-1} h^{-1}
m$_{NOx\_N2}$	Biomass specific endogenous denitrification rate, mg N g VSS^{-1} h^{-1}
m$_{GAO,Ax}$	Anoxic maintenance rate of GAO, C-mol C-mol^{-1} h^{-1} or mg C mg active biomass^{-1} h^{-1}
m$_{GAO,Ox}$	Aerobic maintenance rate of GAO, C-mol C-mol^{-1} h^{-1} or mg C mg active biomass^{-1} h^{-1}
m$_{GAO,NO3}$	Anoxic endogenous respiration rate of GAO on NO$_3$, N-mol C-mol^{-1} h^{-1} or mg NO$_3$ mg active biomass^{-1} h^{-1}
m$_{GAO,NOx}$	Anoxic endogenous respiration rate of GAO on NO$_X$, N-mol C-mol^{-1} h^{-1} or mg NO$_X$ mg active biomass^{-1} h^{-1}

SYMBOLS AND ABBREVIATIONS

$m_{GAO,O2}$	Aerobic endogenous respiration rate of GAO, mol O_2 C-mol^{-1} h^{-1} or mg O_2 mg active biomass^{-1} h^{-1}
$m_{GAO,ATP,Ax}$	Anoxic ATP maintenance coefficient of GAO, mol ATP C-mol^{-1} h^{-1} or mg ATP mg active biomass^{-1} h^{-1}
$m_{GAO,ATP,An}$	Anaerobic ATP maintenance coefficient of GAO, mol ATP C-mol^{-1} h^{-1} or mg ATP mg active biomass^{-1} h^{-1}
$m_{GAO,ATP,Ox}$	Aerobic ATP maintenance coefficient of GAO, mol ATP C-mol^{-1} h^{-1} or mg ATP mg active biomass^{-1} h^{-1}
$m_{PAO,Ax}$	Anoxic maintenance rate of PAO, C-mol Stor C-mol^{-1} h^{-1} or mg Stor mg active biomass^{-1} h^{-1}
$m_{PAO,Ox}$	Aerobic maintenance rate of PAO, C-mol C-mol^{-1} h^{-1} or mg C mg active biomass^{-1} h^{-1}
$m_{PAO,NO3}$	Anoxic endogenous respiration rate of PAO on NO_3, N-mol C-mol^{-1} h^{-1} or mg NO_3 mg active biomass^{-1} h^{-1}
$m_{PAO,NOx}$	Anoxic endogenous respiration rate of PAO on NO_X, N-mol C-mol^{-1} h^{-1} or mg NO_X mg active biomass^{-1} h^{-1}
$m_{PAO,O2}$	Aerobic endogenous respiration rate of PAO, mol O_2 C-mol^{-1} h^{-1} or mg O_2 mg active biomass^{-1} h^{-1}
$m_{PAO,ATP,Ax}$	Anoxic ATP maintenance coefficient of PAO, mol ATP C-mol^{-1} h^{-1} or mg ATP mg active biomass^{-1} h^{-1}
$m_{PAO,ATP,Ox}$	Aerobic ATP maintenance coefficient of PAO, mol ATP C-mol^{-1} h^{-1} or mg ATP mg active biomass^{-1} h^{-1}
$m_{PAO,ATP,An}$	Anaerobic ATP maintenance coefficient of PAO, mol ATP C-mol^{-1} h^{-1} or mg ATP mg active biomass^{-1} h^{-1}
$m_{PAO,PP\_PO4,An}$	Anaerobic endogenous orthophoshate release rate of PAO, P-mol C-mol^{-1} h^{-1} or mg P mg active biomass^{-1} h^{-1}
$m_{PAO,PP\_PO4,Sec,An}$	Anaerobic secondary orthophoshate release rate of PAO, P-mol C-mol^{-1} h^{-1} or mg P mg active biomass^{-1} h^{-1}
$m_{PAO,PP\_PO4,T,An}$	Anaerobic orthophoshate endogenous release rate of PAO at a temperature T, P-mol C-mol^{-1} h^{-1} or mg P mg active biomass^{-1} h^{-1}
$m_{PP\_PO4,An}$	Anaerobic endogenous orthophoshate release rate of a biomass, P-mol C-mol^{-1} h^{-1} or mg P mg active biomass^{-1} h^{-1}
$m_{PP\_PO4,Sec,An}$	Anaerobic secondary orthophoshate release rate of a biomass, P-mol C-mol^{-1} h^{-1} or mg P mg active biomass^{-1} h^{-1}
$m_{PP\_PO4,T,An}$	Anaerobic orthophoshate endogenous release rate of a biomass at a temperature T, P-mol C-mol^{-1} h^{-1} or mg P mg active biomass^{-1} h^{-1}
m, n, p	pH-dependent factors in the computation of the nitrogen to proton ratio in titration tests
M_i	Mass of a component
M_{N2}	Mass of nitrogen gas generated by denitrification in a manometric tests, N-mol or mg N
M_{NO2_N2}	Mass of nitrite converted during an anammox batch test, N-mol or mg N
$M_{NOx,ini}$	Mass of nitrate and nitrite added at the beginning of a denitrification titrimetric test, N-mol or mg N
MW	Molecular weight, g mol^{-1}
n(t)	N_2 moles present in the volume of the headspace (VHS) at time t, mol N_2
N	Nitrogen
N_2	Nitrogen gas
N_2H_4	Hydrazine
Net $P_{released}$	Net concentration of orthophosphate released into the bulk liquid after excluding the anaerobic maintenance release, P-mol L^{-1} or mg P L^{-1}
N_{req}	Concentration of nitrogen required for biomass growth or synthesis, mg N mg VSS^{-1}
N_S	Nitrogen requirements for biomass growth or synthesis, mg N mg VSS^{-1}
N_T	Concentration of the titrant solution in a denitrification tests, meq mL^{-1}
O_2	Molecular oxygen
P	Pressure, Pa or torr
P(t)	Pressure measured in the headspace at time t, atm
P_{atm}	Atmospheric pressure, atm
P_{max}	Maximum pressure in a manometric test, atm
pK_a	Dissociation constant
pK_1	Dissociation constant for carbonic acid
pK_2	Dissociation constant for bicarbonate
pK_{NH4}	Dissociation constant for ammonium
PO_4-P	Orthophosphate
Pr	Concentration of propionate and propionic acid, C-mol L^{-1}, mg Pr L^{-1}, mg COD L^{-1}
$P_{released}$	Concentration of orthophosphate released into the bulk liquid, P-mol L^{-1} or mg P L^{-1}
P_{req}	Concentration of phosphorus required for biomass growth or synthesis, mg P mg VSS^{-1}
P_S	Phosphorus requirements for biomass growth or synthesis, P-mol C-mol or mg P mg VSS^{-1}
Q	Flowrate or generic titration flowrate, mL h^{-1} or mL h^{-1}
q	Maximum specific formation or degradation rate of a compound or component, Mass Biomass^{-1} Time^{-1} or Mass Active Biomass^{-1} Time^{-1}
$q_{Ac,An}$	Maximum specific anaerobic acetate uptake rate of a biomass, C-mol C-mol^{-1} h^{-1} or mg Ac mg VSS^{-1} h^{-1}
$q_{Ac,T,An}$	Maximum specific anaerobic acetate uptake rate at a temperature T of a biomass, C-mol C-mol^{-1} h^{-1} or mg Ac mg active biomass^{-1} h^{-1}
$q_{AMX,NH4\_N2}$	Maximum biomass specific activity of anammox bacteria, g N-N_2 g VSS^{-1} d^{-1} or g N-NH_4^+ g VSS^{-1} d^{-1}

$q_{AMX,NH4\_NO3}$	Maximum biomass specific nitrate production rate in an anammox test, mg N g VSS^{-1} h^{-1}
$q_{AMX,NO2\_N2}$	Maximum biomass specific nitrite removal rate in an anammox test, mg N g VSS^{-1} h^{-1}
$q_{AOO,NH4}$	Maximum biomass specific ammonium oxidation rate of AOO, g N g VSS^{-1} d^{-1} or in mg N g VSS^{-1} h^{-1}
$q_{Bio,Ax}$	Maximum specific anoxic biomass growth rate, C-mol C-mol^{-1} h^{-1} or mg C mg active biomass^{-1} h^{-1}
$q_{Bio,Ox}$	Maximum specific aerobic biomass growth rate, C-mol C-mol^{-1} h^{-1} or mg C mg active biomass^{-1} h^{-1}
$q_{Bio,Ox}$	Maximum specific aerobic biomass growth rate of a culture, C-mol C-mol^{-1} h^{-1} or mg C mg active biomass^{-1} h^{-1}
$q_{GAO,Ac,An}$	Maximum specific anaerobic acetate uptake rate of GAO, C-mol C-mol^{-1} h^{-1} or mg Ac mg active biomass^{-1} h^{-1}
$q_{GAO,Ac,T,An}$	Maximum specific anaerobic acetate uptake rate of GAO at a temperature T, C-mol C-mol^{-1} h^{-1} or mg Ac mg active biomass^{-1} h^{-1}
$q_{GAO,NH4,Ox}$	Maximum specific aerobic ammonia consumption rate of GAO for biomass growth, N-mol C-mol^{-1} h^{-1} or mg N mg active biomass^{-1} h^{-1}
$q_{GAO,Ox}$	Maximum specific aerobic GAO biomass growth rate, C-mol C-mol^{-1} h^{-1} or mg C mg active biomass^{-1} h^{-1}
$q_{GAO,PHA,Ax}$	Maximum specific anoxic PHA degradation rate of GAO, C-mol C-mol^{-1} h^{-1} or mg C mg active biomass^{-1} h^{-1}
$q_{GAO,PHA,Ox}$	Maximum specific aerobic PHA degradation rate of GAO, C-mol C-mol^{-1} h^{-1} or mg C mg active biomass^{-1} h^{-1}
$q_{GAO,PHA\_Gly,Ax}$	Maximum specific anoxic glycogen formation rate of GAO, C-mol C-mol^{-1} h^{-1} or mg C mg active biomass^{-1} h^{-1}
$q_{GAO,PHA\_Gly,Ox}$	Maximum specific aerobic glycogen formation rate of GAO, C-mol C-mol^{-1} h^{-1} or mg C mg active biomass^{-1} h^{-1}
$q_{GAO,VFA,An}$	Maximum specific anaerobic volatile fatty acids uptake rate of GAO, C-mol C-mol^{-1} h^{-1} or mg VFA mg active biomass^{-1} h^{-1}
q_{N_NHx}	Biomass specific ammonification rate, mg N g VSS^{-1} h^{-1}
$q_{NH4\_Bio,Ox}$	Maximum specific aerobic ammonia consumption rate for biomass growth of a biomass, N-mol C-mol^{-1} h^{-1} or mg N mg VSS^{-1} h^{-1}
$q_{NOO,NO2\_NO3}$	Maximum biomass specific nitrite oxidation rate by NOO, g N g VSS^{-1} d^{-1} or in mg N g VSS^{-1} h^{-1}
$q_{NOO,NO2\_NO3,T}$	Maximum biomass specific consumption rate of a generic substrate (S) evaluated at a certain operative temperature (T), g S g VSS^{-1} d^{-1}
q_{NOx_N2}	Maximum biomass specific denitrification rate, g N g VSS^{-1} d^{-1} or mg N g VSS^{-1} h^{-1}
$q_{NOx\_N2,SB}$	Maximum biomass specific denitrification rate on RBCOD, mg N g VSS^{-1} h^{-1}
$q_{NOx\_N2,XCB}$	Maximum biomass specific denitrification rate on SBCOD, mg N g VSS^{-1} h^{-1}
$q_{OHO,COD,Ox}$	Maximum specific aerobic organic matter removal rate expressed in COD by ordinary heterotrophic organisms, mg COD mg VSS^{-1} h^{-1}
$q_{OHO,NH4,Ox}$	Maximum specific aerobic ammonia consumption rate of OHO for biomass growth, N-mol C-mol^{-1} h^{-1} or mg N mg VSS^{-1} h^{-1}
$q_{PAO,Ac,An}$	Maximum specific anaerobic acetate uptake rate of PAO, C-mol C-mol^{-1} h^{-1} or mg Ac mg active biomass^{-1} h^{-1}
$q_{PAO,Ac,T,An}$	Maximum specific anaerobic acetate uptake rate of PAO at a temperature T, C-mol C-mol^{-1} h^{-1} or mg Ac mg active biomass^{-1} h^{-1}
$q_{PAO,Ax}$	Maximum specific anoxic PAO biomass growth rate, C-mol C-mol^{-1} h^{-1} or mg C mg active biomass^{-1} h^{-1}
$q_{PAO,NH4,Ox}$	Maximum specific aerobic ammonia consumption rate of PAO for biomass growth, N-mol C-mol^{-1} h^{-1} or mg N mg active biomass^{-1} h^{-1}
$q_{PAO,Ox}$	Maximum specific aerobic PAO biomass growth rate, C-mol C-mol^{-1} h^{-1} or mg C mg active biomass^{-1} h^{-1}
$q_{PAO,PHA,Ax}$	Maximum specific anoxic PHA degradation rate of PAO, C-mol C-mol^{-1} h^{-1} or mg C mg active biomass^{-1} h^{-1}
$q_{PAO,PHA,Ox}$	Maximum specific aerobic PHA degradation rate of PAO, C-mol C-mol^{-1} h^{-1} or mg C mg active biomass^{-1} h^{-1}
$q_{PAO,PHA\_Gly,Ax}$	Maximum specific anoxic glycogen formation rate of PAO, C-mol C-mol^{-1} h^{-1} or mg C mg active biomass^{-1} h^{-1}
$q_{PAO,PHA\_Gly,Ox}$	Maximum specific aerobic glycogen formation rate of PAO, C-mol C-mol^{-1} h^{-1} or mg C mg active biomass^{-1} h^{-1}
$q_{PAO,PO4\_PP,Ax}$	Maximum specific anoxic poly-P formation rate of PAO, P-mol C-mol^{-1} h^{-1} or mg P mg active biomass^{-1} h^{-1}
$q_{PAO,PO4\_PP,Ox}$	Maximum specific aerobic orthophosphate uptake or poly-P formation rate of PAO, P-mol C-mol^{-1} h^{-1} or mg P mg active biomass^{-1} h^{-1}
$q_{PAO,PP\_PO4,An}$	Maximum specific anaerobic orthophosphate release rate of PAO, P-mol C-mol^{-1} h^{-1} or mg P mg active biomass^{-1} h^{-1}
$q_{PAO,Pr,An}$	Maximum specific anaerobic propionate uptake rate of PAO, C-mol C-mol^{-1} h^{-1} or mg Pr mg active biomass^{-1} h^{-1}
$q_{PAO,VFA,An}$	Maximum specific anaerobic volatile fatty acids uptake rate of PAO, C-mol C-mol^{-1} h^{-1} or mg VFA mg active biomass^{-1} h^{-1}
$q_{PHA,Ax}$	Maximum specific anoxic PHA degradation rate of a biomass, C-mol C-mol^{-1} h^{-1} or mg C mg VSS^{-1} h^{-1}
$q_{PHA,Ox}$	Maximum specific aerobic PHA degradation rate of a biomass, C-mol C-mol^{-1} h^{-1} or mg C mg active biomass^{-1} h^{-1}
$q_{PHA\_Gly,Ax}$	Maximum specific anoxic glycogen formation rate of a culture, C-mol C-mol^{-1} h^{-1} or mg C mg VSS^{-1} h^{-1}
$q_{PHA\_Gly,Ox}$	Maximum specific aerobic glycogen formation rate of a biomass, C-mol C-mol^{-1} h^{-1} or mg C mg VSS^{-1} h^{-1}
$q_{PO4\_PP,Ax}$	Maximum specific anoxic poly-P formation rate of a biomass, P-mol C-mol^{-1} h^{-1} or mg P mg VSS^{-1} h^{-1}

SYMBOLS AND ABBREVIATIONS

$q_{PO4\_PP,Ox}$	Maximum specific aerobic orthophosphate uptake or poly-P formation rate of a biomass, P-mol C-mol^{-1} h^{-1} or mg P mg VSS^{-1} h^{-1}
$q_{PP\_PO4,An}$	Maximum specific anaerobic orthophosphate release rate of a biomass, P-mol C-mol^{-1} h^{-1} or mg P mg VSS^{-1} h^{-1}
$q_{Pr,An}$	Maximum specific anaerobic propionate uptake rate of a biomass, C-mol C-mol^{-1} h^{-1} or mg Pr mg VSS^{-1} h^{-1}
$q_{SRB,SO4,An}$	Maximum specific anaerobic sulphate reduction rate by sulphate reducing organisms, S-mol C-mol^{-1} h^{-1} or mg S mg VSS^{-1} h^{-1}
$q_{SRB,VFA,An}$	Maximum specific anaerobic volatile fatty acids consumption rate by sulphate reducing organisms, C-mol C-mol^{-1} h^{-1} or mg VFA mg VSS^{-1} h^{-1}
$q_{VFA,An}$	Maximum specific anaerobic volatile fatty acids uptake rate, C-mol C-mol^{-1} h^{-1} or mg VFA mg active biomass^{-1} h^{-1}
$q_{VFA,An}$	Maximum specific anaerobic volatile fatty acids uptake rate of a biomass, C-mol C-mol^{-1} h^{-1} or mg VFA mg VSS^{-1} h^{-1}
$q_{VFA\_PHA,An}$	Maximum specific anaerobic PHA production rate, C-mol C-mol^{-1} h^{-1} or mg PHA mg active biomass^{-1} h^{-1}
Q_{in}	Influent flowrate that enters into a reactor or system, L d^{-1}, m^3 d^{-1}, mL min^{-1}
Q_{H2O2}	Titration flowrate of the oxygenated titrant, mL min^{-1}
$Q_{H2O2,final}$	Final background titration flowrate of the oxygenated titrant, mL min^{-1}
$Q_{H2O2,ini}$	Initial or background titration flowrate of the oxygenated titrant, mL min^{-1}
$Q_{H2O2,NH4}$	Titration flowrate of the oxygenated titrant during ammonium oxidation, mL min^{-1}
$Q_{H2O2,NO2}$	Titration flowrate of the oxygenated titrant during nitrite oxidation, mL min^{-1}
Q_{NaOH}	Titration rate of the NaOH titrant, mL min^{-1}
$Q_{NaOH,final}$	Final background titration rate of the NaOH titrant, mL min^{-1}
$Q_{NaOH,ini}$	Initial or background titration rate of the NaOH titrant, mL min^{-1}
$Q_{NaOH,NH4}$	Titration rate of the NaOH titrant during ammonium oxidation, mL min^{-1}
Q_{tit}	Acid titration flow rate during a denitrification test, mL min^{-1}
r	Maximum volumetric production or consumption rate of a compound or element, Mass Volume^{-1} Time^{-1}
$r_{Ac,An}$	Maximum volumetric anaerobic acetate uptake rate of a biomass, C-mol L^{-1} h^{-1} or mg Ac L^{-1} h^{-1}
$r_{AMX,NH4}$	Maximum volumetric ammonium consumption rate in an anammox test, mg N L^{-1} h^{-1}
$r_{AMX,NH4\_NO3}$	Maximum volumetric nitrate production rate in an anammox test, mg N L^{-1} h^{-1}
$r_{AMX,NO2}$	Maximum volumetric nitrite consumption rate in an anammox test, mg N L^{-1} h^{-1}
$r_{ANO,O2}$	Maximum volumetric oxygen uptake rate by autotrophic nitrifying organisms, mg O$_2$ L^{-1} h^{-1}
$r_{AOO,O2}$	Maximum volumetric oxygen uptake rate by ammonium oxidazing organisms, mg O$_2$ L^{-1} h^{-1}
$r_{B,Ox}$	Maximum volumetric aerobic organic matter removal rate expressed as COD of a biomass, mg COD L^{-1} h^{-1}
r_{COD}	Maximum volumetric COD consumption rate, mg COD L^{-1} min^{-1}
$r_{GAO,Ac,An}$	Maximum volumetric anaerobic acetate uptake rate of a GAO culture, C-mol L^{-1} h^{-1} or mg Ac L^{-1} h^{-1}
$r_{NH4\_Bio,Ox}$	Maximum volumetric aerobic ammonia consumption rate for biomass growth, N-mol L^{-1} h^{-1} or mg N L^{-1} h^{-1}
r_{NO3}	Maximum volumetric nitrate removal or uptake rate, mg N L^{-1} h^{-1}
r_{NO3_N2}	Maximum volumetric nitrate removal or uptake rate for denitrification to N$_2$, mg N L^{-1} h^{-1}
$r_{NOO,O2}$	Maximum volumetric oxygen uptake rate by nitrite oxidazing organisms, mg O$_2$ L^{-1} h^{-1}
$r_{NOx\_N2,endo}$	Maximum volumetric endogenous denitrification rate, mg N L^{-1} min^{-1}
$r_{NOx\_N2,exo}$	Maximum volumetric exogenous denitrification rate, mg N L^{-1} min^{-1}
$r_{NOx\_N2,SB}$	Maximum volumetric denitrification rate on readily biodegradable organics (RBCOD), mg N L^{-1} min^{-1}
$r_{NOx\_N2,XCB}$	Maximum volumetric denitrification rate on slowly biodegradable organics (SBCOD), mg N L^{-1} min^{-1}
$r_{O2,endo}$	Maximum volumetric endogenous oxygen uptake rate, mg O$_2$ L^{-1} min^{-1}
$r_{O2,exo}$	Maximum volumetric exogenous oxygen uptake rate, mg O$_2$ L^{-1} min^{-1}
$r_{OHO,COD,Ox}$	Maximum volumetric aerobic organic matter removal rate of ordinary heterotrophic organisms, mg COD L^{-1} h^{-1}
r_P	Pressure increase rate during a manometric test, atm min^{-1} or in atm h^{-1}
$r_{PAO,Ac,An}$	Maximum volumetric anaerobic acetate uptake rate of a PAO culture, C-mol L^{-1} h^{-1} or mg Ac mg L^{-1} h^{-1}
$r_{PO4\_PP,Ax}$	Maximum volumetric anoxic orthophosphate uptake (or poly-P formation) rate, P-mol L^{-1} h^{-1} or mg P L^{-1} h^{-1}
$r_{PO4\_PP,Ox}$	Maximum volumetric aerobic orthophosphate uptake (or poly-P formation) rate, P-mol L^{-1} h^{-1} or mg P L^{-1} h^{-1}
$r_{PP\_PO4,An}$	Maximum volumetric anaerobic orthophosphate release rate, P-mol L^{-1} h^{-1} or mg P L^{-1} h^{-1}
$r_{PP\_PO4,Sec,An}$	Anaerobic volumetric secondary orthophosphate release rate, C-mol L^{-1} h^{-1} or mg C L^{-1} h^{-1}
$r_{SO4,An}$	Maximum volumetric anaerobic sulphate reduction rate, S-mol L^{-1} h^{-1} or mg S L^{-1} h^{-1}
$r_{VFA,An}$	Maximum volumetric anaerobic volatile fatty acids uptake rate, C-mol L^{-1} h^{-1} or mg VFA L^{-1} h^{-1}
R	Ideal (or universal) gas constant, 8.314 J K^{-1} mol^{-1}
RBCOD$_{removed}$	Concentration of readily biodegradable organic matter removed in a batch activity test, C-mol L^{-1} or mg COD L^{-1}
S_i	Concentration of a component in a liquid or in the water phase, Mass Volume^{-1}

$S_{Ac,COD}$	Acetate concentration in COD units, mg COD L^{-1}
S_{ALK}	Alkalinity concentration, mmol L^{-1}
S_B	Concentration of readily biodegradable organics (as COD), mg COD L^{-1}
S_{CO2}	Concentration of carbon dioxide in the bulk liquid, C-mol L^{-1}
S_{H2O2}	Concentration of the H_2O_2 titrant, mmol O_2 mL^{-1}
S_{H2S}	Sulphide concentration in the liquid phase, mg S L^{-1} or mg COD L^{-1}
$S_{H2S,in}$	Concentration of sulphide in the inlet or that enters into a reactor or system, S-mol L^{-1} or mg S L^{-1}
$S_{H2S,out}$	Concentration of sulphide that flows out or leaves a reactor or system, S-mol L^{-1} or mg S L^{-1}
S_{IC}	Inorganic carbon, mol L^{-1}
S_{N2}	Dissolved nitrogen gas concentration, mg N L^{-1}
S_{NaOH}	Concentration of the NaOH titrant, meq mL^{-1}
S_{NH3}	Ammonia concentration, N-mol L^{-1} or mg N L^{-1}
S_{NH4}	Ammonium concentration, N-mol L^{-1} or mg N L^{-1}
S_{NHx}	Ammonium and ammonia concentration, N-mol L^{-1} or mg N L^{-1}
S_{NO2}	Nitrite concentration, N-mol L^{-1} or mg N L^{-1}
$S_{NO2,Ax}$	Nitrite concentration in a denitrification test, N-mol L^{-1} or mg N L^{-1}
$S_{NO2,ini}$	Initial nitrite concentration, N-mol L^{-1} or mg N L^{-1}
S_{NO3}	Nitrate concentration, N-mol L^{-1} or mg N L^{-1}
$S_{NO3,Ax}$	Nitrate concentration in a denitrification test, N-mol L^{-1} or mg N L^{-1}
$S_{NO3,Eq}$	Oxidized equivalents of nitrogen in a denitrification test, N-mol L^{-1} or mg N L^{-1}
$S_{NO3/SB,eq}$	Amount on nitrate equivalents that are consumed on RBCOD, N-mol L^{-1} or mg N L^{-1}
$S_{NO3/XCB,eq}$	Amount on nitrate equivalents that are consumed on SBCOD, N-mol L^{-1} or mg N L^{-1}
$S_{NO3\_N2,Ax}$	Concentration of nitrate converted into nitrogen gas by denitrification in a manometric test, N-mol L^{-1} or mg N L^{-1}
S_{NOx}	Nitrate or nitrite concentration, N-mol L^{-1} or mg N L^{-1}
S_{O2}	Dissolved oxygen (DO) concentration, mg O_2 L^{-1}
$S_{O2,in}$	Dissolved oxygen concentration in the influent, mgO_2 L^{-1}
$S_{O2,ini}$	Initial concentration of dissolved oxygen in the bulk liquid, mgO_2 L^{-1}
SO_4	Sulphate
$SO_{4,final}$	Final concentration of sulphate in the bulk liquid at the end of the batch activity test, S-mol L^{-1} or mg S L^{-1}
$SO_{4,ini}$	Initial concentration of sulphate in the bulk liquid at the beginning of the batch activity test, S-mol L^{-1} or mg S L^{-1}
$S_{SO4,in}$	Concentration of sulphate in the inlet or that enters into a reactor or system, S-mol L^{-1} or mg S L^{-1}
$S_{SO4,out}$	Concentration of sulphate that flows out or leaves a reactor or system, S-mol L^{-1} or mg S L^{-1}
$S_{PO4,final}$	Final orthophosphate concentration in the bulk liquid at the end of the batch activity test, P-mol L^{-1} or mg P L^{-1}
$S_{PO4,ini}$	Initial orthophosphate concentration in the bulk liquid at the start of the batch activity test, P-mol L^{-1} or mg P L^{-1}
SRT	Solids retention time, d
Stor	Intracellular storage compound, C-mol or P-mol, mg C or mg P
T	Temperature, °C or K
T_C	Operative temperature in Celsius degrees, °C
T_K	Operative temperature in Kelvin degrees, K
T_{ref}	Reference absolute temperature, K
V	Volume of a system, reactor or closed system, L or mL
V_G	Gas volume, L
$V_G(t)$	Gas volume at time t, L
V_{H2O2}	Volume of the oxygenated titrant, mL
V_{HS}	Headspace volume in a manometric test, L or mL
V_L	Volume of liquid in a reactor or system, L or mL
V_{ML}	Volume of a mixed liquor sample, L
V_{NaOH}	Volume of NaOH titrant, mL
V_T	Total volume of acid titrant added during a denitrification test, mL
V_{tit}	Volume of acid titrant during a denitrification test, mL
V_{WW}	Volume of wastewater, L
x	Number of carbon moles per mole of organic substrate
X_{ANO}	Concentration of autotrophic nitrifying organisms, mg VSS L^{-1} or mg COD L^{-1}
X_{AOA}	Concentration of ammonia-oxidizing archaea, mg VSS L^{-1} or mg COD L^{-1}
X_{AOO}	Concentration of ammonia oxidizing organisms, mg VSS L^{-1} or mg COD L^{-1}
X_{AOO}	Concentration of ammonia oxidizing organisms, mg VSS L^{-1} or mg COD L^{-1}

SYMBOLS AND ABBREVIATIONS

X_{Bio}	Biomass concentration, mg VSS L^{-1} or mg COD L^{-1}
$X_{Bio,COD}$	Biomass concentration in COD units, mg COD L^{-1}
XC_B	Concentration of slowly biodegradable organics, mg COD L^{-1}
X_{final}	Final concentration of a particulate compound, mg L^{-1}
X_{ini}	Initial concentration of a particulate compound, mg L^{-1}
X_{NOO}	Concentration of nitrite oxidizing organisms, mg VSS L^{-1} or mg COD L^{-1}
X_{NOO}	Concentration of nitrite oxidizing organisms, mg VSS L^{-1} or mg COD L^{-1}
X_{OHO}	Concentration of ordinary heterotrophic organisms, mg VSS L^{-1} or mg COD L^{-1}
X_{TSS}	Mixed liquor suspended solids concentration, g SS L^{-1}
X_{VSS}	Mixed liquor volatile suspended solids concentration, g VSS L^{-1}
Y	Stoichiometric yield ratio, Mass Mass^{-1}
Y_A	Growth yield of autotrophic nitrifying organisms, g COD g N^{-1}
$Y_{Ac\_PH2MV,An}$	Anaerobic PH2MV formation to acetate uptake ratio, C-mol C-mol^{-1} or mg C mg Ac^{-1}
$Y_{Ac\_PHA,An}$	Anaerobic PHA formation to acetate uptake ratio, C-mol C-mol^{-1} or mg C mg Ac^{-1}
$Y_{Ac\_PHB,An}$	Anaerobic PHB formation to acetate uptake ratio, C-mol C-mol^{-1} or mg C mg Ac^{-1}
$Y_{Ac\_PHV,An}$	Anaerobic PHV formation to acetate uptake ratio, C-mol C-mol^{-1} or mg C mg Ac^{-1}
$Y_{Ac\_PO4,An}$	Anaerobic orthophosphate released to acetate uptake ratio, P-mol C-mol^{-1} or mg P mg Ac^{-1}
$Y_{AMX,NH4}$	Growth yield of anammox bacteria on ammonium consumption, C-mol N-mol^{-1}
Y_{ANO}	Growth yield of autotrophic nitrifying organisms, g COD g N^{-1}
Y_{AOO}	Growth yield of ammonia oxidizing organisms, g COD g N^{-1}
$Y_{C\_PO4,An}$	Anaerobic orthophosphate released to carbon uptake ratio, P-mol C-mol^{-1} or mg P mg Ac^{-1}
Y_{CO2}	Yield of CO$_2$ per substrate consumed, C-mol C-mol^{-1}
Y_{Gly}	Aerobic glycogen formation to oxygen consumption ratio, C-mol mol O$_2^{-1}$ or mg C mg O$_2^{-1}$
$Y_{Gly,GAO}$	Aerobic glycogen formation to oxygen consumption ratio of GAO, C-mol mol O$_2^{-1}$ or mg C mg O$_2^{-1}$
$Y_{Gly,PAO}$	Aerobic glycogen formation to oxygen consumption ratio of PAO, C-mol mol O$_2^{-1}$ or mg C mg O$_2^{-1}$
$Y_{Gly/Ac,An}$	Anaerobic glycogen utilization to acetate uptake ratio, C-mol C-mol^{-1} or mg C mg Ac^{-1}
$Y_{Gly/Pr,An}$	Anaerobic glycogen utilization to propionate uptake ratio, C-mol C-mol^{-1} or mg C mg Pr^{-1}
$Y_{Gly/VFA,An}$	Anaerobic glycogen utilization to volatile fatty acids uptake ratio, C-mol C-mol^{-1} or mg C mg VFA^{-1}
$Y_{NH4/O2\_NO2}$	Ratio between ammonium oxidation and oxygen consumption for ammonium oxidation to nitrite, g N g O$_2^{-1}$
$Y_{NH4/O2\_NO3}$	Ratio between ammonium oxidation and oxygen consumption for ammonium oxidation to nitrate, g N g O$_2^{-1}$
Y_{NH4_H+}	Ratio between ammonium oxidation and proton production, mol Protons g N^{-1}
$Y_{NH4\_NO3,AMX}$	Ratio between nitrate production and ammonium consumption in anammox metabolism, g N g N^{-1} or N-mol N-mol^{-1}
$Y_{NO2/NH4,AMX}$	Ratio between nitrite and ammonium consumption in anammox metabolism, g N g N^{-1} or N-mol N-mol^{-1}
$Y_{NO2/O2\_NO3}$	Ratio between nitrite oxidation and oxygen consumption for nitrite oxidation to nitrate, g N g O$_2^{-1}$
Y_{NO2_H+}	Ratio between nitrite consumption and proton removed, mol Protons g N^{-1}
$Y_{NO3\_Bio,Ax}$	Anoxic biomass growth to NO$_3$ consumption ratio, C-mol N-mol^{-1} or mg C mg N^{-1}
$Y_{NO3\_Gly,Ax}$	Anoxic glycogen formation to NO$_3$ consumption ratio, C-mol N-mol^{-1} or mg C mg N^{-1}
Y_{NO3_H+}	Ratio between nitrate consumption and proton removed, mol Protons^{-1} g N
$Y_{NO3\_H+,Ax}$	Ratio between nitrate consumption and proton removed in denitrification, mol Protons^{-1} g N
$Y_{NO3\_PAO,Ax}$	Anoxic PAO biomass growth to NO$_3$ consumption ratio, C-mol N-mol^{-1} or mg C mg N^{-1}
$Y_{NO3\_PHA,Ax}$	Anoxic PHA degradation to NO$_3$ consumption ratio, C-mol N-mol^{-1} or mg C mg N^{-1}
$Y_{NO3\_PP,Ax}$	Anoxic poly-P formation to NO$_3$ consumption ratio, P-mol N-mol^{-1} or mg P mg N^{-1}
Y_{NOO}	Growth yield of nitrite oxidizing organisms, g COD g N^{-1}
$Y_{NOx\_Bio,Ax}$	Anoxic biomass growth to NO$_X$ consumption ratio, C-mol N-mol^{-1} or mg C mg N^{-1}
$Y_{NOx\_Gly,Ax}$	Anoxic glycogen formation to NO$_X$ consumption ratio, C-mol N-mol^{-1} or mg C mg N^{-1}
$Y_{NOx\_PAO,Ax}$	Anoxic PAO biomass growth to NO$_X$ consumption ratio, C-mol N-mol^{-1} or mg C mg N^{-1}
$Y_{NOx\_PHA,Ax}$	Anoxic PHA degradation to NO$_X$ consumption ratio, C-mol N-mol^{-1} or mg C mg N^{-1}
$Y_{NOx\_PP,Ax}$	Anoxic poly-P formation to NO$_X$ consumption ratio, P-mol N-mol^{-1} or mg P mg N^{-1}
Y_{OHO}	Growth yield of heterotrophic microorganisms under aerobic conditions, mg VSS COD^{-1} or g COD g COD^{-1}
$Y_{OHO,Ax}$	Growth yield of heterotrophic microorganisms under anoxic conditions, mg VSS COD^{-1} or g COD g COD^{-1}
Y_{PAO}	Aerobic PAO biomass growth to oxygen consumption ratio, C-mol mol O$_2^{-1}$ or mg C mg O$_2^{-1}$
Y_{PHA}	Aerobic PHA degradation to oxygen consumption ratio, C-mol mol O$_2^{-1}$ or mg C mg O$_2^{-1}$
$Y_{PHA,GAO}$	Aerobic PHA degradation to oxygen consumption ratio of GAO, C-mol mol O$_2^{-1}$ or mg C mg O$_2^{-1}$
$Y_{PHA,PAO}$	Aerobic PHA degradation to oxygen consumption ratio of PAO, C-mol mol O$_2^{-1}$ or mg C mg O$_2^{-1}$
$Y_{PHA\_Bio,Ax}$	Anoxic biomass growth to PHA consumption ratio, C-mol C-mol^{-1} or mg C^{-1}

$Y_{PHA\_Bio,Ox}$	Aerobic biomass growth to PHA consumption ratio, C-mol C-mol^{-1} or mg C mg C^{-1}
$Y_{PHA\_Gly,Ax}$	Anoxic glycogen formation to PHA consumption ratio, C-mol C-mol^{-1} or mg C mg C^{-1}
$Y_{PHA\_Gly,Ox}$	Aerobic glycogen formation to PHA consumption ratio, C-mol C-mol^{-1} or mg C mg C^{-1}
$Y_{PHA\_PAO,Ax}$	Anoxic PAO biomass growth to PHA consumption ratio, C-mol C-mol^{-1} or mg C^{-1}
$Y_{PHA\_PAO,Ox}$	Aerobic PAO biomass growth to PHA consumption ratio, C-mol C-mol^{-1} or mg C mg C^{-1}
$Y_{PHA\_PP,Ax}$	Anoxic poly-P formation to PHA consumption ratio, P-mol C-mol^{-1} or mg P mg C^{-1}
$Y_{PHA\_PP,Ox}$	Aerobic poly-P formation to PHA consumption ratio, P-mol C-mol^{-1} or mg P mg C^{-1}
$Y_{PHV/PHB,An}$	Anaerobic PHV formation to PHB formation ratio, C-mol C-mol^{-1} or mg C mg C^{-1}
$Y_{PO4\_PP,Ax}$	Anoxic poly-P formation to orthophoshate uptake ratio, P-mol P-mol^{-1} or mg P mg P^{-1}
$Y_{PO4\_PP,Ox}$	Aerobic poly-P formation to orthophoshate uptake ratio, P-mol P-mol^{-1} or mg P mg P^{-1}
Y_{PP}	Aerobic poly-P formation to oxygen consumption ratio, P-mol mol O$_2^{-1}$ or mg P mg O$_2^{-1}$
$Y_{Pr\_PH2MV,An}$	Anaerobic PH$_2$MV formation to propionate uptake ratio, C-mol C-mol^{-1} or mg C mg Pr^{-1}
$Y_{Pr\_PHA,An}$	Anaerobic PHA formation to propionate uptake ratio, C-mol C-mol^{-1} or mg C mg Pr^{-1}
$Y_{Pr\_PHB,An}$	Anaerobic PHB formation to propionate uptake ratio, C-mol C-mol^{-1} or mg C mg Pr^{-1}
$Y_{Pr\_PHV,An}$	Anaerobic PHV formation to propionate uptake ratio, C-mol C-mol^{-1} or mg C mg Pr^{-1}
$Y_{Pr\_PO4,An}$	Anaerobic orthophosphate released to propionate uptake ratio, P-mol C-mol^{-1} or mg P mg Pr^{-1}
$Y_{SO4/VFA,An}$	Anaerobic sulphate reduction to VFA consumption ratio, S-mol C-mol^{-1} or mg S mg Ac^{-1}
$Y_{VFA\_H2S,An}$	Anaerobic reduction of sulphate to H$_2$S to volatile fatty acids consumption ratio, S-mol C-mol^{-1} or mg S mg VFA^{-1}
$Y_{VFA\_PH2MV,An}$	Anaerobic PH$_2$MV formation to volatile fatty acids uptake ratio, C-mol C-mol^{-1} or mg C mg VFA^{-1}
$Y_{VFA\_PHA,An}$	Anaerobic PHA formation to volatile fatty acids uptake ratio, C-mol C-mol^{-1} or mg C mg VFA^{-1}
$Y_{VFA\_PHB,An}$	Anaerobic PHB formation to volatile fatty acids uptake ratio, C-mol C-mol^{-1} or mg C mg VFA^{-1}
$Y_{VFA\_PHV,An}$	Anaerobic PHV formation to volatile fatty acids uptake ratio, C-mol C-mol^{-1} or mg C mg VFA^{-1}
$Y_{VFA\_PO4,An}$	Anaerobic orthophosphate released to volatile fatty acids uptake ratio, P-mol C-mol^{-1} or mg P mg VFA^{-1}
Y_{XBio}	Biomass growth yield, C-mol C-mol^{-1} or mg VSS g COD^{-1}
Net $P_{released}$	Net concentration of orthophosphate released into the bulk liquid only due to Ac uptake, mg P L^{-1}
α	Dimensionless distribution coefficient for H$_2$S liquid-gas phases equilibrium
β	Oxygen equivalent of oxydized nitrogen
δ-ratio	ATP produced per O$_2$ consumed under aerobic conditions, mol ATP mol O$_2^{-1}$
η	Oxygen equivalents of nitrate, mg O$_2$ mg N^{-1} or mg COD mg N^{-1}
μ	Specific growth rate of biomass, Mass Time^{-1} Volume^{-1}
μ_{OHO}	Maximum specific biomass growth rate of ordinary heterotrophic organisms under aerobic conditions, h^{-1} or d^{-1}
$\mu_{OHO,Ax}$	Maximum specific biomass growth rate of ordinary heterotrophic organisms under anoxic conditions, h^{-1} or d^{-1}
μ_{AOO}	Maximum specific biomass growth rate of ammonia oxidizing organisms, d^{-1}
μ_{NOO}	Maximum specific biomass growth rate of nitrite oxidizing organisms, d^{-1}
ρ	Density, g L^{-1} or g mL^{-1}
$\Delta COD(\%)$	COD balance, %
ΔCOD_{cons}	Total concentration of COD consumed in a reactor or system, mg COD L^{-1}
$\Delta G°´$	Gibb's free energy, kJ mol^{-1}
$\Delta O_{2,cons}$	Total concentration of oxygen consumed, mg COD L^{-1}
$\Delta t_{SB,Ax}$	Duration of the anoxic denitrification phase that uses RBCOD as electron donor, min
$\Delta t_{XCB,Ax}$	Duration of the anoxic denitrification phase that uses SBCOD as electron donor, min

2 Abbreviations

ACTIVATED SLUDGE ACTIVITY TESTS

A/O	Anaerobic-oxic (aerobic) system
A^2	Anaerobic-anoxic system
A^{2}O	Anaerobic-anoxic-aerobic system
AB	Active biomass
AC	Acetogens or acetogenic bacteria
AMO	Ammonia monooxygenase
AMP	Adenosine monophosphate

SYMBOLS AND ABBREVIATIONS

ANAMMOX	Anaerobic ammonium oxidation
ANO	Autotrophic nitrifying organisms
ANS	Anaerobic Sludge
ANS	Anaerobic sludge system
AOA	Ammonium oxidizing archaea
AOO	Ammonia oxidizing organisms
APS	Adenosine phosphosulphate
ASM	Activated sludge model
ATP	Adenosin triphosphate
BIODENIPHO	Biological denitrification and phosphorus removal system
BNR	Biological nutrient removal
CAS	Conventional activated sludge
CSTR	Continuous stirred tank reactor
DGGE	Denaturing gradient gel electrophoresis
DPAO	Denitrifying phosphorus- or polyphosphate-accumulating organisms
EBPR	Enhanced biological phosphorus removal
FISH	Fluorescence *in situ* hybridization
FNA	Free nitrous acid
GAO	Glycogen-accumulating organisms
HAO	Hydroxylamine oxidoreductase
HDH	Hydrazine dehydrogenase
HPLC	High-performance liquid chromatography
HZS	Hydrazine synthase enzyme
MBR	Membrane bioreactor
MET	Methanogenic bacteria
MLSS	Mixed liquor suspended solids
MLVSS	Mixed liquor volatile suspended solids
Modified UCT	Modified University of Cape Town system
NADH	Nicotinamide adenine dinucleotide
NAR	Nitrate reductase
Nir	Nitrite oxidoreductase
NO	Nitrous oxide
NOO	Nitrite oxidizing organisms
NOR (or NXR)	Nitrite oxido reductase
NOS	Nitrous oxide reductase
OHO	Ordinary heterotrophic organisms
OUR	Oxygen uptake rate
PAO	Phosphorus- or polyphosphate-accumulating organisms
PH_2MV	Poly-β-hydroxy-2-methyl-valerate
PHA	Poly-β-hydroxy-alkanoates
PHB	Poly-β-hydroxy-butyrate
PHV	Poly-β-hydroxy-valerate
Phoredox	Phosphorus reduction oxidation system
PhoStrip	Phosphorus stripping system
PN	Partial nitrition process
PNA	Partial nitrition-anammox process
Poly-P	Polyphosphate
PPi	Pyrophosphate
RBC	Rotating biological contactor
RBCOD	Readily biodegradable organic matter measured as COD
rDNA	Ribosomal DNA
SAA	Specific anammox activity
SANI®	Sulphate-reduction, autotrophic denitrification and nitrification integrated process
SBCOD	Slowly biodegradable organic matter measured as COD
SBR	Sequencing batch reactor
SRB	Sulphate reducing bacteria

TOC	Total organic carbon	
TUDelft	Delft University of Technology	
UASB	Upflow anaerobic sludge blanket	
UCT	University of Cape Town	
VSS	Volatile suspended solids	
WWTP	Wastewater treatment plant	

Symbols

RESPIROMETRY

[CHO]	Any carbohydrate
μ_{OHO}	Maximum specific biomass growth rate of ordinary heterotrophic organisms under aerobic conditions, h^{-1} or d^{-1}
μ_{ANO}	Maximum specific biomass growth rate of autotrophic nitrifying organisms, d^{-1}
Ac	Concentrations of acetate and acetic acid, mg Ac L^{-1}
ASM1	Activated sludge model No. 1
AUR	Ammonia utilization or uptake rate, mg N L^{-1} h^{-1}
BOD	Biochemical oxygen demand, mg O_2 L^{-1}
BOD_5	Biochemical oxygen demand after 5 d, mg O_2 L^{-1}
$BOD\infty$	Ultimate biochemical oxygen demand, mg O_2 L^{-1}
BOD^i_{st}	Short-term biochemical oxygen demand attributed to a specific organic matter present in wastewater, mg O_2 L^{-1}
BOD_{st}	Short-term biochemical oxygen demand, mg O_2 L^{-1}
BOD^{st}_{sample}	Short-term biochemical oxygen demand of a sample, mg O_2 L^{-1}
BOD_t	Oxygen uptake measured at time t, mg O_2 L^{-1}
BOD_U	Ultimate biochemical oxygen demand, mg O_2 L^{-1}
CBOD	Carbonaceous biochemical oxygen demand, mg O_2 L^{-1}
CH_4	Methane
C_i	Concentration of a compound or element in the gas phase or headspace of a reactor or system, ppmv
CN^{-1}	Cyanide, mg L^{-1}
CO_2	Carbon dioxide
C_{O2}	Concentration of oxygen in the gas phase, ppmv
$C_{O2,in}$	Concentration of oxygen in the flux of a gas that enters into a reactor or system, ppmv
COD	Chemical oxygen demand, mg COD L^{-1}
COD_{Ac}	Concentration of acetate and acetic acid expressed as COD, mg COD L^{-1}
$COD^{Degraded}$	Concentration of degraded biodegradable substrate, mg COD L^{-1}
$COD_{substrate}$	Concentration of substrate expressed as COD, mg COD L^{-1}
DO	Dissolved oxygen
F_{in}	Flux of a gas or a compound that enters into a reactor or system, mol h^{-1} or mg h^{-1}
F_{out}	Flux of a gas or a compound that flows out or leaves a reactor or system, mol h^{-1} or mg h^{-1}
H	Henry's proportionality constant for the solubility of a gas, mol $m3^{-1}$ Pa^{-1}
H_2	Hydrogen
H_2S	Sulphide
HCO_3	Bicarbonate or alkalinity
IC_{50}	Concentration that produces 50% inhibition of the respiration process, mg L^{-1}
$i_{N,Bio}$	Nitrogen content of a biomass or bacterial culture, g N g VSS^{-1} or g N g COD^{-1}
k	First order oxygen uptake rate coefficient for the ultimate biochemical oxygen demand determination, d^{-1}
kLa	Volumetric mass transfer coefficient, d^{-1}
MLSS	Mixed liquor suspended solids, mg SS L^{-1}
MLVSS	Mixed liquor volatile suspended solids, mg VSS L^{-1}
n	Moles of a gas present in the volume of a headspace, mol
N_2	Nitrogen gas
NBOD	Nitrogenous biochemical oxygen demand, mg O_2 L^{-1}
NH_4	Ammonium
N_{Nit}	Concentration of nitrogen available for nitrification, mg N L^{-1}
NO_3	Nitrate

SYMBOLS AND ABBREVIATIONS

NO_x^-	Nitrate and nitrite
NUR	Nitrate uptake rate (r_{NO3}), N-mol L^{-1} h^{-1} or mg N L^{-1} h^{-1}
NUR	Nitrate utilization or uptake rate, mg N L^{-1} h^{-1}
O_2	Oxygen
OUR	Oxygen uptake rate (r_{O2}), mol O_2 L^{-1} h^{-1} or mg O_2 L^{-1} h^{-1}
OUR	Oxygen utilization or uptake rate, mg O_2 L^{-1} h^{-1}
P	Pressure, Pa or torr
Q_{in}	Influent flowrate that enters into a reactor or system, L d^{-1}, m^3 d^{-1}, mL min^{-1}
Q_{out}	Flowrate that leaves a reactor or system, L d^{-1}, m^3 d^{-1}, mL min^{-1}
Q_{ww}	Wastewater flowrate that enters into a reactor or system, L d^{-1}, m^3 d^{-1}, mL min^{-1}
r_{NO2_NO3}	Aerobic oxidation rate of nitrite to nitrate, mg N L^{-1} h^{-1}
R	Ideal (or universal) gas constant, 8.314 J K^{-1} mol^{-1}
$r_{ANO,O2}$	Exogenous respiration rate of autotrophic nitrifying organisms, mg O_2 L^{-1} h^{-1}
$r_{AOO,O2}$	Respiration rate of ammonia oxidation organisms, mg O_2 L^{-1} h^{-1}
$r^i_{O2,exo}(t)$	Time series of exogenous respiration rates associated to the oxidation of a specific component present in wastewater, mg O_2 L^{-1} h^{-1}
$r^{max}_{O2,exo}$(after)	Maximum volumetric exogenous oxygen uptake rate after the addition of a toxic compound, mg O_2 L^{-1} min^{-1}
$r^{max}_{O2,exo}$(before)	Maximum volumetric exogenous oxygen uptake rate before the addition of a toxic compound, mg O_2 L^{-1} min^{-1}
r_{NH4_NO2}	Aerobic oxidation rate of ammonia to nitrite, mg N L^{-1} h^{-1}
$r^{Nit}_{O2,exo}$	Exogenous respiration rate due to nitrification, mg O_2 L^{-1} h^{-1}
$r^{Nit}_{O2,exo}(t)$	Time series of exogenous respiration rates due to nitrification ($r^{Nit}_{O2,exo}$), mg O_2 L^{-1} h^{-1}
r_{NO3}	Volumetric nitrate uptake rate, mg N L^{-1} min^{-1}
$r_{NO3,exo}$	Volumetric exogenous nitrate uptake rate, mg N L^{-1} min^{-1}
$r_{NOO,O2}$	Respiration rate of nitrite oxidation organisms, mg O_2 L^{-1} h^{-1}
r_{O2}	Maximum volumetric oxygen uptake rate, mg O_2 L^{-1} min^{-1}
r_{O2}	Oxygen uptake rate, mg O_2 L^{-1} h^{-1}
$r_{O2,endo}$	Volumetric endogenous oxygen uptake rate, mg O_2 L^{-1} min^{-1}
$r_{O2,exo}$	Volumetric exogenous oxygen uptake rate, mg O_2 L^{-1} min^{-1}
$r_{O2,exo}(t)$	Time series of exogenous respiration rates $r_{O2,exo}$, mg O_2 L^{-1} h^{-1}
$r_{O2,NH4,exo}$	Exogenous respiration rate associated to the oxidation of ammonia, mg O_2 L^{-1} h^{-1}
$r_{O2,NO2,exo}$	Exogenous respiration rate associated to the oxidation of nitrite, mg O_2 L^{-1} h^{-1}
$r_{O2,tot}$	Total oxygen uptake rate of biomass, mg O_2 L^{-1} min^{-1}
$r^{SB}_{NOx,exo}$	Exogenous nitrate uptake rate associated to denitrification using readily biodegradable organics, mg N L^{-1} h^{-1}
$r^{SB}_{NOx,exo}(t)$	Time series of exogenous nitrate uptake rate associated to denitrification using readily biodegradable organics, mg N L^{-1} h^{-1}
$r^{SB}_{O2,exo}$	Exogenous respiration rate associated to the oxidation of readily biodegradable organics, mg O_2 L^{-1} h^{-1}
$r^{SB}_{O2,exo}(t)$	Time series of exogenous respiration rates associated to the oxidation of readily biodegradable organics, mg O_2 L^{-1} h^{-1}
$r^{XCB}_{NOx,exo}$	Exogenous nitrate uptake rate associated to denitrification using slowly biodegradable organics, mg N L^{-1} h^{-1}
$r^{XCB}_{NOx,exo}(t)$	Time series of exogenous nitrate uptake rate associated to denitrification using slowly biodegradable organics, mg N L^{-1} h^{-1}
$r^{XCB}_{O2,exo}$	Exogenous respiration rate associated to the oxidation of slowly biodegradable organics, mg O_2 L^{-1} h^{-1}
S^*_{O2}	Saturation concentration of dissolved oxygen in the bulk liquid at local conditions, mg O_2 L^{-1}
$S^*_{O2,endo}$	Saturation concentration of dissolved oxygen in the bulk liquid under endogenous conditions, mg O_2 L^{-1}
SAUR	Specific ammonia utilization or uptake rate, mg N g VSS^{-1} h^{-1}
S_B	Concentration of readily biodegradable organics (as COD), mg COD L^{-1}
$S_B(0)$	Initial concentration of readily biodegradable organics (as COD), mg COD L^{-1}
$S_{B,N}$	Concentration of nitrogen associated to the soluble biodegradable organics, N-mol L^{-1} or mg N L^{-1}
S_{NHx}	Ammonium and ammonia concentration, N-mol L^{-1} or mg N L^{-1}
S_{NOx}	Nitrate or nitrite concentration, N-mol L^{-1} or mg N L^{-1}
SNUR	Specific nitrate utilization or uptake rate, mg N g VSS h^{-1}
S_O	Initial substrate concentration, mg L^{-1}
S_{O2}	Dissolved oxygen (DO) concentration, mg O_2 L^{-1}
S_{O2}	Dissolved oxygen (DO) concentration, mg O_2 L^{-1}
$S_{O2,in}$	Dissolved oxygen concentration in the influent, mgO_2 L^{-1}
SOUR	Specific oxygen utilization or uptake rate, mg O_2 g VSS h^{-1}

STP	Standard temperature and pressure, 273.15 K and 1013.25 bar
T	Temperature, °C or K
t_{final}	Time required to return to the endogenous respiration rate after sample addition, min or h
TOC	Total organic carbon, C-mol L^{-1} or mg C L^{-1}
t_{pulse}	Time of pulse addition of the sample, min or h
TSS	Concentration of total suspended solids, mg TSS L^{-1}
UBOD	Ultimate biochemical oxygen demand, mg O$_2$ L^{-1}
V	Volume of a system, reactor or closed system, L or mL
V_G	Gas volume, L
V_L	Volume of liquid in a reactor or system, L or mL
V_{react}	Volume of a system, reactor or closed system, L or mL
VS	Volatile solids, mg VS L-1
V_{sample}	Volume of the sample added to the test vessel, L
V_{sludge}	Volume of the sludge in the test vessel prior to the sample addition, L
VSS	Concentration of volatile suspended solids, mg VSS L^{-1}
VSS	Volatile suspended solids, mg VSS L-1
$VSS_{inoculum}$	Concentration of volatile suspended solids present in the inoculum, mg VSS L^{-1}
X_{ANO}	Concentration of autotrophic nitrifying organisms, mg VSS L^{-1} or mg COD L^{-1}
XC_B	Concentration of slowly biodegradable organics, mg COD L^{-1}
$XC_{B,N}$	Concentration of nitrogen associated to the slowly biodegradable organics, N-mol L^{-1} or mg N L^{-1}
X_O	Initial biomass concentration, mg L^{-1}
X_{OHO}	Concentration of ordinary heterotrophic organisms, mg VSS L^{-1} or mg COD L^{-1}
Y	Stoichiometric growth yield ratio, Mass Mass^{-1}
Y_{ANO}	Growth yield of autotrophic nitrifying organisms, g COD g N^{-1}
Y_{AOO}	Growth yield of ammonia oxidizing organisms, g COD g N^{-1} or g VSS N^{-1}
Y_{NOO}	Growth yield of nitrite oxidizing organisms, g COD g N^{-1} or g VSS N^{-1}
Y_{OHO}	Growth yield of heterotrophic microorganisms under aerobic conditions, mg VSS COD^{-1} or g COD g COD^{-1}
$Y_{OHO,Ax}$	Growth yield of heterotrophic microorganisms under anoxic conditions, mg VSS COD^{-1} or g COD g COD^{-1}
η	Oxygen equivalents of nitrate, mg O$_2$ mg N^{-1} or mg COD mg N^{-1}
ΔNO_X	Difference in nitrate uptake rates associated to the denitrification rates on readily or slowly biodegradable organics, mg N L^{-1} min^{-1}
$\Delta r_{O2,tot}$	Difference in oxygen uptake rates before and after the continuous addition of wastewater, mg O$_2$ L^{-1} min^{-1}

3 RESPIROMETRY

Abbreviations

Ar	Argon
ARIKA	Automated respiration inhibition kinetics analysis
ATP	Adenosin triphosphate
ATU	Allylthiourea
BMP	Biomethane potential
EBPR	Enhanced biological phosphorus removal
G	Gas
GFF	Flowing gas, flowing liquid
GFS	Flowing gas, static liquid
GSF	Static gas, flowing liquid
GSS	Static gas, static liquid
IAWQ	International Association on Water Quality
L	Liquid
LFF	Flowing gas, flowing liquid
LFS	Flowing gas, static liquid
LSF	Static gas, flowing liquid
LSS	Liquid phase, static gas, static liquid

SYMBOLS AND ABBREVIATIONS

MFC	Mass flow controller
NaOH	Sodium hydroxide
PAO	Polyphosphate-accumulating organisms
SMA	Specific methanogenic activity
TCMP	2-chloro-6-(trichloromethyl)pyridine
UV	Ultraviolet light
VFA	Volatile fatty acids

OFF-GAS EMISSION TESTS

Symbols

A	Cross-sectional area of the surface emission isolation flux chamber, m^2
a1, a2, a3, a4, a5	Gas stripping parameters determined through batch tests and parameter estimation or linear regression for the description of the gas concentrations in a stripping method
Alk	Alkalinity, mg eq L-1 or mg CaCO$_3$ L^{-1}
BOD$_5$	Biochemical oxygen demand determined after 5 days, mg O$_2$ L^{-1}
C	Gas concentration, M, mol L^{-1}, mg L^{-1}, g L^{-1} or kg m^{-3} $^{-1}$
CH$_4$	Methane, ppmv, % or mg COD L^{-1}
CO$_2$	Carbon dioxide, ppmv, %, C-mol L^{-1} or mg C L^{-1}
cBOD$_{5,filtered}$	Carbonaceous biochemical oxygen demand determined after 5 days in a sample subject to filtration, mg O$_2$ L^{-1}
cBOD$_{5,total}$	Carbonaceous biochemical oxygen demand determined after 5 days in a raw non-filtered sample, mg O$_2$ L^{-1}
COD	Chemical oxygen demand, mg COD L^{-1}
COD$_{filt,floc}$	Chemical oxygen demand determined in a sample that has been subject to coagulation-flocculation and filtration, mg COD L^{-1}
COD$_{soluble}$	Chemical oxygen demand determined in a sample subject to filtration, mg COD L^{-1}
C$_{helium-FC}$	Helium concentration in the off-gas from the flux chamber, ppmv or %
C$_{helium-GC}$	Helium concentration measured in the gas chromatograph, ppmv or %
C$_{helium-tracer}$	Helium concentration in the tracer gas, ppmv or %
C$_{G,2}(t)$	Concentration of gas in the headspace of subsystem 2 in the stripping method as a function of time, ppmv or %
C$_G^{in}$	Gas concentration in the gas flow supplied to stripping device, ppmv or %
C$_G^{in,R}$	Gas concentration entering into the stripping flask, ppmv or %
C$_L(t)$	Concentration of gas in subsystem 1 in the stripping method as a function of time, ppmv, % or mg L^{-1}
C$_L^{in,R}$	Concentration of gas in the inflow to the reactor, mg L^{-1}
C$^R_{G,1}(t)$	Concentration of gas in the reactor as a function of time, ppmv or %
C$^R_{G,2}(t)$	Concentration of gas in the gas outflow as a function of time, ppmv or %
C$^R_L(t)$	Concentration of gas present in the liquid phase in the reactor as a function ot time, mg L^{-1}, ppmv or %
D$_L$	Liquid dilution rate, L L^{-1} h^{-1} or m^3 m3 $^{-1}$ d^{-1}
DO	Dissolved oxygen, mol O$_2$ L^{-1} or mg O$_2$ L^{-1}
f$_{sample}$	Frequency of sampling
H$_2$SO$_4$	Sulphuric acid, mol L^{-1} or %
He	Helium, ppmv or %
K	Sensitivity of a stripping device
MLSS	Concentration of mixed liquor suspended solids, mg SS L^{-1}
MLVSS	Concentration of mixed liquor volatile suspended solids, mg VSS L^{-1}
n	Amount of methane in the expanded headspace of the serum bottle, mol
N$_2$	Dinitrogen gas, ppmv, %, N-mol L^{-1} or mg N L^{-1}
N$_2$in	Nitrogen gas supplied into a gas stripping device, ppmv
N$_2$O	Nitrous oxide, ppmv, %, N-mol L^{-1} or mg N L^{-1}
NaCl	Sodium chloride or common salt, mg, % or mg L^{-1}
NH$_3$	Ammonia, N-mol L^{-1} or mg N L^{-1}
NH$_3$-N	Concentration of ammonia and ammonium as nitrogen, N-mol L^{-1} or mg N L^{-1}
NH$_4^+$	Ammonium, N-mol L^{-1} or mg N L^{-1}
NO$_2^-$	Nitrite, N-mol L^{-1} or mg N L^{-1}

NO_2-N	Nitrite concentration as nitrogen, N-mol L^{-1} or mg N L^{-1}
NO_3^-	Nitrate, N-mol L^{-1} or mg N L^{-1}
NO_3-N	Nitrate concentration as nitrogen, N-mol L^{-1} or mg N L^{-1}
NO_X	Concentration of nitrate and nitrite, N-mol L^{-1} or mg N L^{-1}
O_2	Oxygen, ppmv, %, O-mol L^{-1} or mg O_2 L^{-1}
ORP	Oxidation-reduction or redox potential, mV
P	Atmospheric pressure, Pa
Q	Flowrate, mL min^{-1}, L h^{-1} or m^3 d^{-1}
Q_1	Florate at the point of reference 1, mL min^{-1}, L h^{-1} or m^3 d^{-1}
$Q_{A/S}$	Flowrate supplied to the activated sludge mixed liquor system, m^3 d^{-1}
$Q_{emission}$	Advective gas flowrate through the flux-chamber, m^3 d^{-1}
Q_{flux}	Flowrate of gas leaving the surface emission isolation flux chamber, m^3 d^{-1}
Q_G	Stripping gas flowrate, mL min^{-1}, L h^{-1} or m^3 d^{-1}
$Q_{G,1}(t)$	Gas flowrate stripped out of subsystem 1 as a function of time, mL min^{-1}, L h^{-1} or m^3 d^{-1}
$Q_G^{in,R}$	Gas flowrate supplied to the reactor, mL min^{-1}, L h^{-1} or m^3 d^{-1}
$Q_R^G(t)$	Gas inflow into the reactor as a function of time, mL min^{-1}, L h^{-1} or m^3 d^{-1}
Q_L	Liquid inflow into a stripping flask, mL min^{-1}, L h^{-1} or m^3 d^{-1}
Q_L	Constant flow rate of a liquid sample from the reactor to a stripping flask, mL min^{-1}, L h^{-1} or m^3 d^{-1}
$Q^R_L(t)$	Liquid influent flowrate into the reactor as a function of time, mL min^{-1}, L h^{-1} or m^3 d^{-1}
Q_n	Flowrate at the sampling point or point of reference n, mL min^{-1}, L h^{-1} or m^3 d^{-1}
Q^R_G	Gas outflow, mL min^{-1}, L h^{-1} or m^3 d^{-1}
Q^R_L	Liquid outflow, mL min-1, L h^{-1} or m^3 d^{-1}
$Q^R_L(t)$	Liquid outflow as a function of time, mL min^{-1}, L h^{-1} or m^3 d^{-1}
Q_{sweep}	Flowrate of sweep or carrier gas entering into the surface emission isolation flux chamber, m^3 d^{-1}
Q_{tracer}	Tracer gas flowrate introduced into the flux-chamber, m^3 d^{-1}
R	Ideal gas constant, 8.314 m^3 Pa mol^{-1} K^{-1}
RAS	Return of activated sludge, m^3 d^{-1}
R^R_V	Volume of the reactor, L or m^3
R_V	Volume of liquid in subsystem 1 in the stripping method, L or m^3
$R_V(t)$	Volume of liquid in subsystem 1 in the stripping method as a function of time, L or m^3
SRT	Solids retention time, d^{-1}
t	Time, h or d
T	Temperature, °C or K
TKN	Total Kjeldahl nitrogen, N-mol L^{-1} or mg N L^{-1}
$TKN_{soluble}$	Total Kjeldahl nitrogen determined in a sample subject to filtration, N-mol L^{-1} or mg N L^{-1}
TP	Total phosphorus, P-mol L^{-1} or mg P L^{-1}
TSS	Total suspended solids, mg TSS L^{-1}
V	Expanded volume of the headspace in the end of the test, L or m^3
V_1	Headspace volume in the syringe at the end of the test, mL or L
VFA	Volatile suspended solids, C-mol L^{-1}, mg COD L^{-1} or mg VFA L^{-1}
$V_{G,1}$	Volume of gas in subsystem 1 in the stripping method, L or m^3
$V_{G,2}$	Volume of headspace in subsystem 2 in the stripping method, L or m^3
V_{HS}	Headspace of the serum bottle before expansion, L or m^3
V_L	Constant liquid volume in the stripping flask, L or m^3
V_O	Initial volume of the headspace of the sampling syringe, mL or L
$V^R_{G,1}$	Volume of the reactor, L or m^3
$V^R_{G,2}$	Headspace volume, L or m^3
V^R_L	Volume of liquid in the reactor, L or m^3
V_S	Volume expansion in the sampling syringe due to the pressure build-up in the serum bottle, mL or L
V_{sample}	Volume of the sample, L
V_{sample}	Volume of the sample, mL or L
VSS	Volatile suspended solids, mg VSS L^{-1}
WAS	Waste activated sludge, m^3 d^{-1}
W_1	Weight of the bottle after filling up the bottle with clean water up to the mark of the stopper, mL or L
W_O	Weight of the bottle after the addition of the initial water volume, mL or L
ρ	Density, g L^{-1}

SYMBOLS AND ABBREVIATIONS

4 Abbreviations — OFF-GAS EMISSION TESTS

ASTM	American Society for Testing and Materials
BNR	Biological nutrient removal
C	Continuously collected sample
DAS	Data acquisition software
EPA	Environmental Protection Agency
FTIR	Fourier transform infrared spectroscopy
GC	Gas chromatograph
GCFID	Gas chromatograph equipped with flux injector detector
GC-TCD	Gas chromatograph equipped with a thermal conductivity detector
GHG	Greenhouse gas
I	Intermittent collected sample
IPCC	Intergovernmental Panel on Climate Change
IR	Infrared light
ISE	Ion selective electrode
NA	Not applicable
PE	Person equivalent
SCADA	Supervisory control and data acquisition
SCAQMD	South Coast Air Quality Management District
SEIFC	Surface emission isolation flux chamber
SEIFC	Surface emission isolation flux chamber
SHARON	Single reactor high activity ammonia removal over nitrite
TCD	Thermal conductivity detector
US	United States
USEPA	United States Environmental Protection Agency
WWTP	Wastewater treatment plant

5 Symbols — DATA HANDLING AND PARAMETER ESTIMATION

$\text{cov}(\hat{\theta})$	Covariance matrix of estimators
$\hat{F}$	Probability distribution of residuals, $\hat{\varepsilon}$
$t_{N-p}^{\alpha/2}$	Upper $\alpha/2$ percentile of the t-distribution with N-p degrees of freedom
$\hat{\theta}$	Parameter estimators
σ	Standard deviation (of a normal distribution function)
θ	Parameter vector of a dynamic model
σ_+	Standard deviation of parameter estimates
$\theta^{\Sigma}_{(1)}$	Parameter vector estimated using data set $D^S(i)$
μ	Specific growth rate of biomass, Mass Time^{-1} Volume^{-1}
μ_{max}^{AOO}	Maximum growth rate of AOO, d^{-1}
μ_{max}^{NOO}	Maximum growth rate of NOO, d^{-1}
b_{AOO}	Decay rate of AOO biomass, d^{-1}
b_{NOO}	Decay rate of NOO biomass, d^{-1}
CH_2O	Reduced carbon source as substrate, C-mol
C_i	Component i, Mass Volume^{-1}
CO_2	Carbon dioxide, C-mol
$\text{cov}(y)$	Covariance matrix of model predictions
$D(0)$	Original data set with N data points
diag	Diagonal elements of a matrix

$D^S(i)$	i^{th} synthetic data set
E	Conservation matrix
E()	Expected value of a vector of random variable, y
F.	Jacobian matrix
H_2O	Water
iid	Independent and identically distributed
kLa	Volumetric mass transfer coefficient, d^{-1}
$K_{o,AOO}$	Oxygen affinity of AOO, mg O_2 L^{-1}
$K_{o,NOO}$	Oxygen affinity of NOO, mg O_2 L^{-1}
$K_{s,AOO}$	Substrate (NH_4) affinity of AOO, mg N L^{-1}
$K_{s,NOO}$	Substrate (NO_2) affinity of NOO, mg N L^{-1}
M_i	Monod term for component i
NH_3	Ammonia as nitrogen source for growth, N-mol
O_2	Molecular oxygen, O-mol
P_1	Product, C-mol
q_i	Volumetric conversion/production rate of component i, Mass i $Volume^{-1}$ $Time^{-1}$
q_m	Measured set of volumetric rates
q_u	Unmeasured set of volumetric rates
r_i	Rate of mass of component i per unit time per unit weight of biomass, Mass i $Time^{-1}$ Mass $biomass^{-1}$
R_{ij}	Pairwise linear correlation between parameter estimators
$S(y,\theta)$	Cost (or objective) function
s^2	Unbiased estimation of variance of residuals
Sa	Vector of absolute sensitivity function
S_{NH}	Concentration of ammonium nitrogen, mg N L^{-1}
S_{NO2}	Concentration of nitrite nitrogen, mg N L^{-1}
S_{NO3}	Concentration of nitrate nitrogen, mg N L^{-1}
S_O	Oxygen concentration, mg O_2 L^{-1}
S_O^{sat}	Oxygen saturation concentration, mg O_2 L^{-1}
Sr	Vector of relative sensitivity function
u	Input vector of a dynamic model
var()	Variance of a vector of random variable, y
v_{ij}	Stoichiometric coefficient of component i in process j
X	Biomass, C-mol
x	State variables in a dynamic model
X_{AOO}	Biomass concentration of AOO, mg COD L^{-1}
X_{NOO}	Biomass concentration of NOO, mg COD L^{-1}
y	Vector of outputs of a dynamic model
y*	The bootstrap sample
Y_{AOO}	Biomass (AOO) yield over substrate (NH_4), mg COD mg N^{-1}
Y_{ji}	Yield of component i per component
Y_{NOO}	Biomass (NOO) yield over substrate (NO_2), mg COD mg N^{-1}
Y_{SC}	Yield of CO_2 per unit substrate, C-mol $C-mol^{-1}$
Y_{SN}	Yield of nitrogen per unit substrate, N-mol $C-mol^{-1}$
Y_{SO}	Yield of oxygen per unit substrate, O-mol $C-mol^{-1}$
Y_{SP1}	Yield of intermediate product P_1 per substrate, C-mol $C-mol^{-1}$
Y_{SW}	Yield of water per unit of substrate, H-mol $C-mol^{-1}$
Y_{SX}	Yield of biomass per unit substrate, C-mol $C-mol^{-1}$
α	Confidence level
γ_g	Degree of reduction of glucose, mol e- $C-mol^{-1}$
γ_i	Degree of reduction of component i, mol e- mol^{1-}
γ_K	Collinearity index of a parameter subset K
γ_{O2}	Degree of reduction of oxygen, mol e- $O-mol^{-1}$
γ_x	Degree of reduction of biomass, mol e- $C-mol^{-1}$
δ^{msqr}	Delta mean square based sensitivity measure
Δx	Perturbation of the model inputs around their nominal values, x^0

SYMBOLS AND ABBREVIATIONS

ε	Measurement errors
λ$_K$	Eigen values of normalized sensitivity matrix for parameter subset K
σ(f)	Standard deviation of the Monte Carlo integration error

5 Abbreviations — DATA HANDLING AND PARAMETER ESTIMATION

AOO	Ammonium oxidizing organisms
ASM	Activated sludge model
COD	Chemical oxygen demand
MCMC	Markov-Chain Monte-Carlo
MLE	Maximum likelihood estimation
MW	Molecular weight
NOO	Nitrite oxidising organisms
OAT	One factor at a time
ODE	Ordinary differential equations
WWTP	Wastewater treatment plants

6 Symbols — SETTLING TESTS

m_{H_2O}	Mass of water in completely filled pyknometer, g
m'_{H_2O}	Mass of water added to pycnometer with solids sample, g
μ_w	Dynamic viscosity water, kg m^{-1} s^{-1}
ν_w	Kinematic viscosity water, m^2 s^{-1}
C_d	Continuum, intermediate
d_p	Particle diameter, m
DSS	Dispersed suspended solids concentration, mg L^{-1}
DSS$_i$	Dispersed suspended solids concentration at the inlet of the clarifier, mg L^{-1}
DSS$_o$	Dispersed suspended solids concentration at the effluent weir of the clarifier, mg L^{-1}
DSVI	Diluted sludge volume index, mL g^{-1}
E	Percentage of mass balance error, %
ESS	Effluent suspended solids concentration, mg L^{-1}
f(v_s)	Mass fraction of particles with a settling velocity smaller than v_s, %
FSS	Flocculated suspended solids concentration, mg L^{-1}
f_{sv}	Fraction of the settling column occupied by the settled sludge after 30 minutes of settling
g	Gravitational constant, m s^{-2}
H	Height of the ViCAs column, m
K	Particle-liquid constant
M(t)	ICumulated mass of particles settled to the bottom of the ViCAs column between t=0 and t, mg
m_0	Mass of empty pycnometer, g
M_{fin}	Final mass in the ViCAs column, mg
M_{ini}	Initial mass in the ViCAs column, mg
m_s	Mass of solid sample, g
M_{set}	Sum of the settled mass recovered in the cups at the bottom of the ViCAs column, mg
m_T	Mass of pycnometer filled with water, g
m_{TS}	Mass of pyknometer filled with solids sample and water, g
Re$_p$	Particle Reynolds number
r_V	Settling parameter, L g^{-1}
S(t)	Mass of particles settled in the ViCAs column between t=0 and t that have a settling velocity above H/t, mg
SSVI	Stirred specific volume index, mL g^{-1}
SV$_{30}$	Volume of settling column occupied by sludge after 30 min. of settling, mL L^{-1}

SVI	Sludge volume index, mL g^{-1}	
V'$_{H_2O}$	Volume of water added to pycnometer with solids sample, L	
V$_0$	Maximum settling velocity, m h^{-1}	
v$_{hs}$	Hindered settling velocity, m h^{-1}	
v$_s$	Sedimentation velocity of a single particle, m s^{-1}	
V$_s$	Volume of solid sample, L	
v$_{zs}$	Zone settling velocity, m h^{-1}	
X$_{TSS}$	Total suspended solids concentration, g L^{-1}	
ρ$_p$	Density of particle, kg m^{-3}	
ρ$_s$	Density of solids sample, g L^{-1}	
ρ$_w$	Density of fluid, kg m^{-3}	

6 Abbreviations — SETTLING TESTS

ViCAs   Vitesse de chute en assainissement (settling velocity in sanitation, in French)

7 Symbols — MICROSCOPY

d	Resolution of a microscope
N	Refractive index of the immersion medium used below the objective lens
α	One-half of the objective's opening angle, degree
λ	Light wavelength, m
λem	Emission light wavelength, m
λex	Excitation light wavelength, m

7 Abbreviations — MICROSCOPY

BF	Bright-field
Card-FISH	Catalyzed reporter deposition for fluorescence *in situ* hybridization
CCD	Charge coupled device
CLSM	Confocal laser scanning microscopy
CTC	5-cyano-2,3-ditolyl tetrazolium chloride
CTF	Fluorescent formazan
DAPI	4',6-diamidino-2-phenylindole dihydrochloride/dilactate
dH$_2$O	Distilled water
DMF	Dimethylformamide
DO	Dissolved oxygen
DOPE-FISH	Double labeling of oligonucleotide probes for fluorescence *in situ* hybridization
dsDNA	Doublestranded DNA
EBPR	Enhanced biological phosphate removal
EDTA	Ethylenediaminetetraacetic acid
EPS	Extracellular polymeric substances
EtOH	Ethanol
FA	Formamide
FI	Filament Index
FISH	Fluorescence *in situ* hybridization
GAO	Glycogen-accumulating organism

SYMBOLS AND ABBREVIATIONS

HI	Hexidium iodide
MLSS	Mixed liquor suspended solids
NA	Numerical aperture
PAO	Polyphosphate-accumulating organism
PBS	Phosphate-buffered saline
PFA	Paraformaldehyde
PHA	Poly-β-hydroxy-alkanoates
Ph	Phase contrast
poly-P	Poly-phosphate
RI	Refractive index (RI)
SDS	Sodium dodecylsulfate
TE	Tris-EDTA
WWTP	Wastewater treatment plant

8 MOLECULAR METHODS

Symbols

C_q	Quantification cycle in qPCR experiments
C or c	Concentration
g	g-force, G
V	Volume

Abbreviations

A260/230	Light absorbance ratio at 260 nm and 280 nm
A260/280	Light absorbance ratio at 260 nm and 280 nm
AOB	Ammonia-oxidizing bacteria
BHQ-1	Black hole quencher-1
bp	Base pairs
Cluster PF	Sequencing clusters passing filter
DGGE	Denaturing gradient gel electrophoresis
DN	Denitrifying bacteria
dsDNA	Double-stranded DNA
eDNA	Extracellular DNA
FAM	6-carboxyfluorescein
FIL	Filamentous bacteria
FISH	Fluorescence *in situ* hybridization
FRET	Fluorescence resonance energy transfer
GAO	Glycogen-accumulating organisms
HET	Heterotrophic bacteria
HPLC	High performance liquid chromatography
LCA	Least common ancestor
MIQE	Minimal information for publication of quantitative real-time PCR experiments
MRSA	Methicillin-resistant *Staphylococcus aureus*
MW	Molecular weight
NAC	No amplification control
NOB	Nitrite-oxidizing bacteria
NTC	No template control
OTU	Operational taxonomic unit
PAO	Polyphosphate-accumulating organisms

PCA	Principle component analysis
PCR	Polymerase chain reaction
PE	Paired-end
PEG	Polyethylene glycol
PPE	Personal protection equipment
Q10, Q20, Q30	Sequencing quality scores
qPCR	Real-time quantitative polymerase chain reaction
ROX	Reference dye used for qPCR
rpm	Revolutions (or rotations) per minute
rRNA	Ribosomal RNA
RT-qPCR	Reverse transcription real-time quantitative polymerase chain reaction
SDS	Sodium dodecyl sulfate
SPRI	Solid phase reversible immobilization
SS	Suspended solids
ssDNA	Single-stranded DNA
TAMRA	Tetramethylrhodamine
TE	Tris EDTA
TS	Total solids
UDG	Uracil-DNA glycosylase
UV-vis	Ultraviolet–visible
V1-V9	rRNA variable region 1 to 9
VRE	Vancomycin-resistant enterococci
WWTP	Wastewater treatment plant

IWA Publishing's authorised EU representative for General Product Safety Regulations is Diane D'Arras, 15 rue Duret, 75116 Paris, France, e-mail: safety@iwap.co.uk.

Printed and bound by CPI Group (UK) Ltd, Croydon, CR0 4YY
27/03/2026
02079978-0001